PIPELINE INTEGRITY ASSURANCE—A PRACTICAL APPROACH

ASME PRESS

ASME Press books are available at special quantity discounts to use as premiums or for use in corporate training programs. For more information, contact Special Sales at infocentral@asme.org

Library of Congress Cataloging-in-Publication Data

Mohitpour, Mo.
Pipeline integrity assurance: a practical approach/Mo Mohitpour, Alan Murray, Michael McManus; with contributions by Iain Colquhoun.
p. cm.
Includes bibliographical references.
ISBN 978-0-7918-5956-8
1. Pipelines—Protection. 2. Pipelines—Reliability. 3. Pipeline failures—Prevention. 4. Pipelines—Design and construction—Quality control. I. Murray, A. (Matthew Alan) II. McManus, Michael, 1964- III. Title.
TJ930.M5723 2010
665.5'44—dc22

2010009930

Cover Graphics: New MagnaScan MFL ILI tool
Courtesy of GE Oil & Gas – PII Pipeline Integrity Solutions

DEDICATION

This book is dedicated to the giants of our industry:

John F. Kiefner
Carl Jaske
Andrew Palmer
Winston Revie
Willard A. Maxey
Pat. H. Vieth
Bob Coote
Alan Glover
Brian Rothwell
Phil Hopkins
Michael J. Rosenfeld and many others

Who, over the years, have relentlessly through their dedication and significant contributions assured integrity, safety, and reliability for our pipelines. They lead the industry for programs to assure failures will be remote within the life cycle of the pipeline system through devising responsible pipeline integrity management systems, research, investigations, and effective maintenance and monitoring methods.

It is to them that the authors owe their inspiration for compiling this document. In comparison, it is a meager contribution. But by combining the entire spectrum of pipeline integrity assurance techniques, it is hoped that it would inform the new entrant to the industry of issues and as well captures the foresight of those responsible for our pipeline networks for ensuring the continued safety, efficiency, reliability, and economic viability of our pipeline network with least impact on the public and the environment.

It is a salute to the leaders and developers of the Pipeline Integrity Technology.

It is dedicated to our pipeline energy world and to the continued prosperity of our industry.

M. Mohitpour
A. Murray
M. McManus
I. Colquhoun

TABLE OF CONTENTS

PREFACE

Pipeline Integrity Assurance — A Practical Approach complements the series of books initiated by ASME press and written by the authors since the year 2000.

Pipelines are known to be the safest and most economic means of transporting hydrocarbon products. Keeping the product in the line ensures that public safety and the environment are protected. This is the sole purpose of pipeline integrity assurance. It is a simple notion but quite complex task. The book is designed to cover the spectrum of integrity management for the pipeline life cycle from design, through threat identification, monitoring, risk, and defect assessment to maintenance repairs and dealing with emergency situations. It follows the now familiar ASME B31.8S (gas pipelines) and API 1160 (liquid pipelines) approach to pipeline integrity management in examining and mitigating the major threats to which pipeline systems are the subject.

The intended readership is both seasoned pipeline professionals, as well as new entrants to the pipeline industry. Since we cast our net so wide, the book contains a mixture of introductory material as well as the more advanced procedures currently in use by the industry. Our aims have been to provide, in one volume, reference material for day-to-day use, as well as practical knowledge and thinking processes when undertaking various pipeline integrity maintenance tasks. Some may find the text useful as a training tool.

The content is a compilation of the knowledge and experience gained by the authors over the past 30 years to 40 years in the pipeline industry augmented by the technical literature and other sources. Chapter 1 introduces the topic and provides some historical information on pipeline incidents. This is followed in Chapter 2 by a description of Integrity Management Systems and how such a system can be developed. Since corrosion is a major threat to pipeline integrity, it is addressed in Chapter 3 in the context of the pipeline's first-line of defense, namely, coatings, while its secondary defense, cathodic protection, is the subject of Chapter 4. There are a number of effective means for assessing the current condition of a pipe, and these are described in Chapter 5. Determining the fitness for service of a pipeline containing defects is the subject of Chapter 6. Having located and assessed the impact of defects on the serviceability of the pipe, it may be necessary to make a repair or to replace a pipe section, and a description of the various ways of performing this task is provided in Chapter 7. It is often the case in many pipeline operating companies to have an internal group responsible for developing and executing the various integrity related tasks; their functions are set out in Chapter 8. Even with the best-laid plans, incidents can and do occur, and responding to them in an effective manner is critical to minimizing the extent of any possible damage. Chapter 9 examines this topic from the perspectives of incident management and emergency response. Lastly, the book concludes with a chapter on pipeline risk assessment from our colleague Dr. Iain Colquhoun.

Every care has been exercised by the authors to ensure correctness of the content, permissions, and referencing documents/Web site and names. It is not intended that specific techniques, examples, or applications be applied or copied for turnkey use. Readers are encouraged to check and assess all details prior to use. Notification of corrections, omissions, and attributions are welcomed by the authors and ASME Press.

M. Mohitpour White Rock, BC, Canada
Alan Murray, Mike McManus & Iain Colquhoun Calgary AB, Canada
2010

FOREWORDS

Foreword By Enbridge International Inc.

Pipeline Integrity Assurance — A Practical Approach is the natural follow-up and progression to the author's series of books published by the ASME Press, New York. The first book *Pipeline Design and Construction* was published in 2000 followed by *Pipeline Operation and Maintenance*, *Pipeline Pumping and Compression Systems*, and *Energy Supply and Pipeline Transportation*. These first books instructed us on designing and operating pipelines and facilities.

This book *Pipeline Integrity Assurance – A Practical Approach* instructs us on ensuring that the assets that we have designed, constructed, and are now operating are maintained to ensure the safety and life-long financial effectiveness of the capital invested. Successful pipeline systems are conceived and devised by considering all aspects of relevant influencing elements from design through to construction and operational life. The latter must include a comprehensive integrity program comprised of activities required to anticipate and prevent pipeline failures. In most modern pipeline companies, these activities are integrated in the general Operations and Maintenance processes and are related by an identifiable framework that describes the Integrity Management Program.

Pipeline Integrity Management Programs, while not an influencing factor for design, is a method of ensuring safe and reliable operation throughout pipeline life. It is insurance for the expenditure invested. *Pipeline Integrity Assurance — A Practical Approach* provides a valuable tool for understanding and developing systems to ensure the safe and responsible operation of pipeline systems.

This book reflects over 100 years of combined author and co-author experience in design, construction, operations, maintenance, regulatory compliance, and integrity assurance of pipelines. I congratulate Dr. Mohitpour, Dr. Murray, Mr. McManus, and Dr. Colquhoun for pulling together a comprehensive practical reference book that can be used by engineering and operations personnel alike. To my knowledge, this accomplishment is "practically" unprecedented.

Bill Trefanenko, P.Eng
Director, International Operations
Enbridge International Inc.

ENBRIDGE

Foreword By Transcanada Pipelines

The advancement of pipeline technology has been well described through the previous American Society of Mechanical Engineers (ASME) Press books including;

- *Pipeline Design and Construction — A Practical Approach*; Mohitpour, Golshan, and Murray
- *Pipeline Operation and Maintenance — A Practical Approach*; Mohitpour, Szabo, and Van Hardeveld
- *Energy Supply and Pipeline Transportation — Challenges and Opportunities*; Mohitpour
- *Pipeline Pumping and Compression Systems — A Practical Approach*: Mohitpour, Botros, and Van Hardeveld

The books collectively provide a summary of the fundamental principals of design, construction, operation, and energy supply for pipeline systems, and the content of all these books is a compilation of the considerable knowledge and experience gained by the authors over many years in the pipeline industry.

Pipelines are the lifeline of our modern day society. They provide the fundamental structure by which hydrocarbon energy is transported safely, reliably, and economically. However, no matter how well pipelines are designed, constructed, operated, and maintained, they are subject to degradation caused by their operating environment and third-party impacts. Thus, a successful pipeline system must not only be properly conceived, studied, designed, built, and operated, but its operation must be complemented by a robust asset management system that comprises all of the activities required to anticipate and prevent a pipeline failure.

Pipeline Integrity Assurance — A Practical Approach completes the series. It complements the full life cycle of a pipeline by considering the entire spectrum of pipeline integrity management, ranging from data gathering (tools and methods) and assessment techniques (condition and risks verification) to repairs and emergency response including incident management.

This book will provide additional useful information to those in the industry concerned with, or responsible for, the safety and integrity of pipeline systems.

We are very pleased to continue our support for this series of ASME Press books and efforts made in capturing the knowledge important to our industry.

Andrew Jenkins
Vice President, TransCanada Pipelines Limited.

TransCanada

Foreword By GE Oil & Gas — PII Pipeline Solutions

The importance of maintaining the integrity of the world's pipeline systems and infrastructure has been underscored by well-publicized tragic accidents involving loss of life and damage to property and to the environment. These accidents have also caused lost production, delivery interruptions, and legal proceedings against operating companies. The environment in which pipelines operate constantly exposes them to time-dependent and time-independent threats that can lead to deterioration of their structural integrity. It is the purpose of pipeline integrity management to ensure that pipelines do not cause harm to people or the environment, while at the same time providing reliable and secure service to pipeline operators and customers through the entire life cycle of the pipeline facilities. When properly maintained, the life of pipelines can be extended indefinitely.

Pipelines are without question the safest and most reliable and most efficient way to transport the many hydrocarbon products that society relies so heavily upon today. Nonetheless, this reliability and efficiency requires constant vigilance on the part of the operators to maintain the integrity, functionality, and security of their assets.

Pipeline Integrity Assurance — A Practical Approach provides a valuable handbook to the complex procedures and assessments necessary to ensure safe and responsible operation of pipeline systems. GE Oil & Gas–PII Pipeline Solutions is dedicated to the worldwide effort to maintain safe and reliable pipeline assets through activities aligned with those described and illustrated by Dr. Mohitpour, Dr. Murray, Michael McManus and Dr. Colquhoun in this book.

Geoff Foreman B. Eng
Global Growth and Strategy Leader
GE Oil & Gas–PII Pipeline Solutions

GE Oil & Gas
PII Pipeline Solutions

ACKNOWLEDGMENTS

Cooperation between ASME Press, the authors, reviewers, and the encouragement from industry leaders and colleagues is the backbone of initiation, preparation, and completion of this book. Because of such cooperation, writing the book was a distinct pleasure for the authors.

First, we would very much like to acknowledge and thank ASME Press publications (specifically Mr. Phillip Di Vietro, Ms. Mary Grace Stefanchik, and Ms. Tara Smith and staff) as well as the ASME marketing group (Craig Backhus, Annette Missouri, and staff) for their support, encouragement, and patience that have made this book a reality.

As a consequence of their efforts to develop the material for the book and assemble and review the various manuscripts required, the authors re-acquainted themselves with many colleagues in the pipeline industry and benefited greatly from their encouragement and knowledge and experience. The authors wish to express sincere thanks to and to acknowledge the valuable contribution of all of these colleagues. In particular, the authors would like to thank the following people:

Chad Bunch, National Energy Board (NEB)
Ms. Sarah Jane Dawson, GE Oil & Gas–PII Pipeline Solutions
John Kazakoff, Silver Fox Engineering Consultants Ltd.
Gordana Mesar, National Energy Board (NEB)
Carl Mikkola, Enbridge Energy Company, Inc.
Rafael Mora, National Energy Board (NEB)
Hal Oliver, P. Eng
Joe Paviglianiti, National Energy Board (NEB)
Edmund Seiders PE., Willbros Tulsa Oklahoma
Bill Trefanenko, Enbridge International Inc.

Mo. Mohitpour, Ph.D., P.E., P.Eng., F.I.Mech. Eng., FASME, FEIC
mo.mohitpour@shaw.ca
Alan Mathew Murray, Ph.D., P. Eng.
ma-murray@shaw.ca
Michael McManus, P.Eng.
m.mcmanus@shaw.ca
Iain Colquhoun, Ph.D., P.Eng.
iain.colquhoun@ge.com

PERMISSION ACKNOWLEDGMENTS

The authors would like to thank all the organizations who have granted their permission to use and reproduce figures, data, and other material in this book. These have been obtained from their Web sites and published technical materials and are referenced accordingly. We apologize in advance for any reference or attributions that we may have overlooked and will be pleased to remedy those brought to our attention in any subsequent editions of the book.

The authors gratefully acknowledge the following organizations for permission granted to reproduce various items that appear in the text:

American Petroleum Institute (API)
American Society of Mechanical Engineers (ASME)
BJ Pipeline Inspection Services
BJ Process and Pipeline Services
CFER Technologies
Corrpro
CRC Press
Enbridge Pipelines Inc.
GE Oil & Gas–PII Pipeline Solutions
Hyundai Heavy Industries
JFE Steel Corporation
Kent Muhlbauer NCF Industries Inc. CA
NDT Systems & Services AG)
PensPen Integrity
Pergam Suisse
Petersen®
PetroSleeve®
Pigging Products Services Association (PPSA), UK
Pipeline Products and Services Inc., Edmonton, AB.
Prentice Hall Inc., Englewood Cliffs, NJ
ROSEN Group
RTD Inc.
Strategic Maintenance Solutions Inc.
T.D. Williamson
TMS Sumas
TransCanada PipeLines Limited
Via+
Western Instruments Inc.

GLOSSARY OF TERMS

The following table provides the glossary of terms as commonly used by the industry. Most of the terms are used throughout the chapters of the book. However, Chapter 9 "Pipeline Emergency Response and Incident Management" has its own glossary of terms defining emergency response situations. A few terms may appear in both tables but only where there are correlations between the emergency response and other subjects.

Term	Definition
Abrasion Resistance	Resistance of a pipe coating material to abrasion.
Aerobic	Oxygen containing.
Alternating Current Voltage Gradient (ACVG)	A method of measuring the change in leakage current in the soil along and around a pipeline to locate coating holidays and characterize corrosion activity.
Anaerobic	Oxygen deficient.
Anchor Pattern	Roughness of the pipe surface to facilitate coating adhesion.
Anode	The electrode of an electrochemical cell at which oxidation occurs. Electrons flow away from the anode in the external circuit. Corrosion usually occurs and metal ions enter the solution at the anode.
Anomaly	An indication usually detected by a non-destructive examination, of any deviation from nominal conditions in sound weld material, the external wall of a pipe, its coating, or the electromagnetic conditions around the pipe which may or may not be due to an actual flaw.
Arc burn	Localized areas of surface melting caused by an inadvertent striking of the pipe surface by an electric arc.
Aspect Ratio	Ratio of crack length to crack depth.
Asphalt Coating	Asphalt based anti-corrosion coating.
B31 G	A semi-empirical analytic method (from the ASME B31 standard) used to estimate acceptability of the pressure-carrying capacity of a corroded pipe.

Bellhole	An excavation that minimizes surface damage yet provides sufficient room for inspection or repair of buried facilities.
Brittle Fracture	A rapid fracture mechanism that occurs without significant plastic deformation.
Buckle	Partial or complete collapse of the pipe wall due to excessive compressive load.
Cathode	The electrode of an electrochemical cell at which reduction is the principal reaction. Electrons flow toward the cathode in the external circuit.
Cathodic Disbondment	The destruction of adhesion between a coating and the coated surface caused by products of a cathodic reaction.
Cathodic Protection (CP)	A technique to reduce the corrosion of a metal surface by making that surface the cathode of an electrochemical cell by which underground metallic pipe is protected against external corrosion deterioration (pitting and/or general corrosion).
Charpy Test	A mechanical test to measure the fracture energy of a material under impact loading.
Class location	A geographical area classified according to its density of buildings intended for human occupancy and other characteristics that are considered when designing and pressure testing a pipeline. It extends 200 m on either side of the centerline of any continuous length of pipe (generally 1.6 km [1 mile] long). Class location units are categorized as Class 1 through 4. Class 1 locations are more rural, and Class 4 is urban with a prevalence of high buildings.
CIS (Close Interval Survey)	A series of above-ground pipe-to-soil potential measurements taken at meter intervals or so, along the pipeline and used to provide information on the effectiveness of the cathodic protection system.
Cluster or Colony	Two or more adjacent metal loss anomalies in the pipe body or weld zone that that are deemed to interact per adopted interaction rules, thereby further weaken the pipe.
Collinear	Lying along the same line (coaxial). a term used to describe spatial relationship of adjacent cracks.
Corrosion Activity	A state in which corrosion is active and ongoing at a rate that is sufficient to reduce the pressure-carrying capacity of a pipe during the pipeline design life.

Composite repair sleeve	A permanent repair method using composite sleeve material, which is applied with an adhesive. Most composites are composed of fiberglass materials but may also comprise other types (such as carbon fibers).
Crack	A 2-dimensional, planar feature with separation of the fracture surfaces.
Crack Coalescence	Joining of cracks that are in close proximity to form one large crack.
Critical Flaw Size	The dimensions (length and depth) of a flaw that would fail at a given stress level.
Current Attenuation Survey	A method of measuring the overall condition of the coating on a pipeline based on the application of electromagnetic field propagation theory. Concurrent data collected may include burial depth, coating resistance and conductance, anomaly location, and anomaly type.
DCVG (Direct Current Voltage Gradient)	An above-ground coating integrity assessment method that will identify areas of coating defects, which could indicate potential locations with external corrosion or mechanical damage.
Defect	An anomaly in the pipe wall that reduces the pressure-carrying capacity of the pipe. An imperfection of a type or a magnitude exceeding acceptable criteria (definition based on API 570).
Dent	Localized deformation of the pipe wall caused by mechanical means that produces a visible change in the curvature of the wall of the pipe and a change in the pipe diameter but not necessarily wall thinning.
Design pressure	The pressure selected by the designer equal to or greater than the total intended maximum operating pressure of the pipeline segment. The pipeline in general may not be continuously operated at pressures above the design pressure.
Direct Assessment	A formal process used to assess the integrity of a pipe segment based on pre assessment/data integration, indirect inspections, direct examinations, and post-assessment analysis (see, for example, NACE RP0502-02).
Direct Examination	Inspections and measurements made on the pipe surface at excavations as part of Direct Assessment.
Disbonded Coating	Any loss of adhesion between the protective coating and a pipe surface as a result of adhesive failure, chemical attack, mechanical damage, hydrogen concentrations, soil stress, etc. Disbonded coating may or may not be associated with a coating holiday.

DSAW Pipe (Double Submerged-Arc Welded Pipe)	A method of welding the long seam of a pipe in which the seam is submerged under a solid flux while being welded from both the internal and externals surfaces of the pipe.
Ductility	A measure of the capacity of a material to be deformed plastically before fracturing.
Elastic Limit	The maximum stress or strain that a material will withstand without permanent deformation. Also referred to as the proportional limit since it denotes the end of the linear portion of the stress strain curve.
Elastic-Plastic Fracture Mechanics	The consideration of both elastic and plastic behavior to predict the deformation and fracture behavior of a material.
Electrolyte	A chemical substance containing ions that migrate in an electric field. For the purposes of this book, an electrolyte refers to the soil or liquid adjacent to and in contact with a buried or submerged metallic pipeline system.
Electromagnetic Inspection Technique	An above-ground survey technique used to locate coating defects on buried pipelines by measuring changes in the magnetic field that are caused by the defects.
ERP (Emergency Response Plan)	Written procedures developed to minimize the hazard resulting from a pipeline emergency including receiving, identifying, and classifying notices of events, establishing internal and external communications, prompt and effective response to a notice, actions taken to protect people and property, and restoration of service.
ERW Pipe (Electric Resistance Welded Pipe)	A method of welding the long seam of a pipe in which the two sides of the seam are first heated by the application of an electrical current and forced together to form a bond (note that ERW pipe can be manufactured by both high-frequency and low-frequency methods — each of which produces distinct levels of reliability that have to be accounted for in assessing the integrity of ERW pipelines).
External Corrosion Direct Assessment (ECDA)	See Direct Assessment definition above.
Failure	A general term used to imply that a part in service has become (1) completely inoperable, (2) is still operable but is incapable of satisfactorily performing its intended function, or (3) has deteriorated seriously, to the point that is has become unreliable or unsafe for continued use.
Fatigue	Cracking of a material as a result of cyclic stressing.
Fault	Any anomaly in the pipeline coating, including disbonded areas and holidays.

Feature	An indication of an anomaly generated by non-destructive testing of the pipeline.
Flow Stress	An arbitrarily defined stress between yield and ultimate, which is used to predict plastic collapse.
Fracture Mechanics	A field of materials science in which fracture behavior of materials is predicted based on material properties and crack geometry.
Fracture Toughness	A measure of a materials resistance to static or dynamic crack extension of a crack.
GIS (Geographical Information System)	A system comprising computer software, hardware, data, and personnel to help manipulate, analyze, and present information that is tied to a geographic location.
GPS (Global Positioning System)	A system used to identify the latitude and longitudinal coordinates of locations using GPS satellites.
Gouge	A surface imperfection with localized metal loss generally caused by mechanical damage.
Heat-Affected Zone	Metal adjacent to a weld whose microstructure and mechanical properties have been altered by the heat of welding.
High-pH SCC	A form of SCC on underground pipelines that is intergranular and typically branched and is associated with an alkaline electrolyte (pH about 9.3). Also referred to as classical SCC.
Holiday	A breach in the pipe protective coating that exposes the unprotected surface to the environment.
Holiday Detector	Device used to detect areas of the pipe which are not being protected by a coating.
Hydrogen Embrittlement	A reduction in the ductility of a material as a result of the absorption of hydrogen. See also hydrogen-induced damage.
Hydrostatic Test	A pass/fail test used to validate the integrity of a pipeline and its associated equipment by assessing its strength and leak tightness. The component is filled with liquid, usually water, sealed, and pressurized until the nominal hoop stress reaches a specified value that is then held for a defined period of time.

Hydrogen-Induced Damage	A form of degradation of metals caused by exposure to environments (liquid or gas) that causes absorption of hydrogen into the material. Examples of hydrogen-induced damage are: (1) formation of internal cracks, blisters, or voids in steels, (2) embrittlement (i.e., loss of ductility, (3) high-temperature hydrogen attack (i.e., surface decarbonation and chemical reaction with hydrogen).
HIC (Hydrogen-Induced Cracking)	A form of hydrogen-induced damage that includes cracking of the metal.
ILI (In-line Inspection)	The inspection of a pipeline from the interior of the pipe using non destructive examination techniques. The tools are used to conduct an in-line inspection (ILI) are generally known as pigs, smart pigs, or intelligent pigs and can detect metal loss and deformation anomalies.
Immediate Indication	An indication that requires remediation or repair in a relatively short time span.
Imperfection	A flaw or other discontinuity noted during inspection that may be subject to acceptance criteria during an engineering and inspection analysis (definition based on API 570).
Indication	A finding of a nondestructive testing or inspection technique showing a deviation from the norm (see also Anomaly).
Indirect Inspection	Equipment and practices used to take measurements at the ground surface above or near a pipeline to locate or characterize corrosion activity, coating holidays, or other anomalies.
Intelligent Pig	A device capable of transiting internally a section of pipeline while performing a non-destructive examination.
Intergranular Cracking	Cracking in which the crack path is between the grains in a metal. Typically associated with high-pH SCC.
Investigative Dig	An inspection of a section of a pipeline in which the section is physically exposed to allow for a detailed examination of the pipe surface.
Lamination	A layered separation within the wall of the pipe usually introduced during manufacture.
Magnetic Particle Inspection (MPI)	A non-destructive inspection technique for locating surface cracks in a steel using fine magnetic particles and a magnetic field.

Term	Definition
MAOP (Maximum allowable operating pressure)	The maximum internal pressure at which the pipeline may be continuously operated per the appropriate code or standard. (Note that emergency shutdown valves may be set up to 10% above this pressure.) The MAOP is generally based on some specified fraction of the hydrostatic commissioning test pressure. It is common practice to select a hydrostatic test pressure such that the MAOP equals the Design Pressure (see also Design Pressure). In some codes and standards, the MAOP is referred to as the Maximum Operating Pressure.
Mechanical Damage	A type of anomaly in a pipe caused by the application of an external force. Mechanical damage can include denting, coating removal, metal removal, metal movement, cold working of the underlying metal, and residual stresses, any one of which can be detrimental.
MFL (Magnetic Flux Leakage)	A type of in-line inspection process that induces a magnetic field in a pipe wall between 2 poles of a magnet. Sensors record changes in the magnetic flux (flow) that can be used to measure metal loss.
Microbiologically Influenced Corrosion (MIC)	Corrosion or deterioration of metals resulting from the metabolic activity of microorganisms. Such corrosion may be initiated or accelerated by microbial activity.
Mitigation	The limitation or reduction of the probability of occurrence or expected consequence for a particular event or measures taken to reduce the impact of the event should it occur.
NDE (Nondestructive Examination)	An inspection method that does not damage the item being examined. This technique includes visual, radiography, ultrasonic, electromagnetic, and dye penetrant methods.
Near-Neutral pH SCC	A form of SCC on underground pipelines that is transgranular and is associated with a near neutral pH electrolyte. Typically, this form of cracking is non-branched and is associated with some corrosion of the pipe surface. Also referred to as low-pH or non-classical SCC.
Performance-Based Integrity Management Program	A non-prescriptive integrity management process in which the pipeline operator combines knowledge of the system with risk management principles to determine prevention, detection, and mitigation actions and their timing.
pH	Measure of the acidity or alkalinity of an aqueous solution.
Pig	A device that is used to transit a pipeline while performing various internal activities such as cleaning or inspection.
Pigging	Running of a pig or an in line inspection tool.

Piggability	The ability of a pipeline or segment to be inspected by an ILI tool or "smart pig".
Pig Trap	An appurtenance consisting of an enclosure and valves that is used to introduce or remove a pig from a pipeline.
Pipe Grade	Part of the material specification per an applicable pipe manufacturing standard. Usually indicates the specified minimum yield strength of the pipe steel.
Pipeline	All the parts of the physical facilities through which the fluid product moves in transportation, including pipe, valves, fittings, flanges, regulators, pressure vessels, pulsation dampeners, and other appurtenances attached to pipe.
Pipeline System	Pipelines, stations, and other facilities required for the measurement, processing, storage, and transportation of oil and gas industry fluids. Note that Pipeline and Pipeline System are often used interchangeably.
Pipe-to-Electrolyte Potential	The potential difference between the pipe metallic surface and the electrolyte that is measured with reference to an electrode in contact with the electrolyte. This measurement is commonly termed pipe-to-soil.
Plastic Collapse	A failure mechanism in which there is unstable plastic deformation.
Prescriptive Integrity Management Program	An integrity management process that follows preset conditions that result in fixed inspection and mitigation activities and timelines.
Probability of Detection	The likelihood, expressed as a percentage, that a method/tool will correctly predicts anomalies.
Probability of Exceedance	The probability that a given value will be met or exceeded (such as might be measured during a specific in-line inspection for example).
Probability of Identification	The likelihood in % that an anomaly or will be identified by a tool or analysis.
Residual Stress	Stress present in a component in the absence of any external loading, which results from the previous manufacturing process, heat treatment, or mechanical working of the material.
Risk	A measure of potential loss in terms of both the incident likelihood of occurrence and the magnitude of the consequences.
Risk Assessment	A systematic process in which potential hazards to the pipeline system are identified, and the likelihood and consequences of potential adverse events are estimated.

Risk Management	An overall program consisting of: identifying potential threats to an area or equipment; assessing the risk associated with those threats in terms of incident likelihood and consequences; mitigating risk by reducing the likelihood, the consequences, or both; and measuring the risk reduction results achieved.
Root Cause Analysis	A family of processes implemented to determine the primary cause of an event. These processes all seek to examine cause-and effect relationship through the organization and analysis of data. Such processes are often used in failure analyses.
R Ratio	A measure of the magnitude of a cyclic pressure or stress fluctuation. The ratio of the minimum to maximum stress or pressure.
RSTRENG	A computer program designed to calculate the pressure-carrying capacity of corroded pipe.
Rupture	A large-scale failure of a pipeline that results in crack propagation and an uncontrolled release of the product.
SCC (Stress Corrosion Cracking)	Brittle cracking of a metal due to the result of the combined effects from localized corrosion and tensile stress.
Shielding	Preventing or diverting the cathodic protection current from its intended path to the structure to be protected.
Significant SCC	An SCC cluster is assessed to be "significant" by the Canadian Energy Pipeline Association (CEPA) if the deepest crack, in a series of interacting cracks, is greater than 10% of the wall thickness and the total interacting length of the cracks is equal to or greater than 75% of the critical length of a 50% through wall flaw that would failure at a stress level of 110% of SMYS. CEPA also defines the interaction criteria. The presence of extensive and "significant" SCC typically will trigger an SCC mitigation program.
SMYS (Specified Minimum Yield Strength)	The minimum yield strength of a material prescribed by the specification or standard to which the material is manufactured.
Soils Model	A model or tool that predicts the SCC susceptibility of a section of pipeline based on soil characteristics such as topography, drainage, and soil type.
Stress Concentrator	A discontinuity in a structure or change in curvature that causes a local increase in stress.

Stress-Intensity Factor	The stress intensity factor relates the nominal stress to the fracture stress intensity at the crack tip. It is generally given the symbol *K*. Fracture stress intensity is a function of the applied stress, the crack dimensions, and the deformation mode.
Subcritical Crack	A crack that is not large enough to cause spontaneous failure at a given pressure or stress.
Tenting	A tent-shaped void formed along the seam weld or girth weld of a pipeline when the external coating bridges from the top of the weld to the pipe.
Third-Party Damage	Damage to a gas pipeline facility by an outside party other than those performing work for the operator. Third-party damage is a particular case of mechanical damage, which can be caused by first, second, or third parties.
Transgranular Cracking	Cracking in which the crack path is through the grains in a metal. Typically associated with near-neutral-pH SCC.
Ultrasonic Inspection	The use of high-frequency sound to determine wall thickness and to detect the presence of flaws within a weld and pipe wall.
Ultraviolet Resistance	The resistance of a pipeline coating material to degradation by ultraviolet light, such as that found in sunlight.
Weld Anomaly	Anomaly in the body or heat affected zone of a weld.
Wet Fluorescent MPI	An MPI technique that uses a suspension of fluorescent magnetic particles.
Wrap-Around Sleeve	A family of heat shrinkable products that are shrink wrapped-around pipe and held in place with a separate closure seal. The sleeves can be provided in long length rolls or pre-cut to fit a specific size pipe.
Wrinkle Bend	A pipe bend or fitting produced by a specialized field bending machine or controlled process resulting in abrupt contour discontinuities on the inner radius (note: does not refer to rippling of pipe caused by normal field cold bending methods).
Yield Strength	The stress level of a material at a specified elongation. The yield strength of line pipe steel is typically based on a total elongation (elastic and plastic) of 0.5%.

UNIT OF MEASUREMENT

The pipeline industry uses a mix of imperial and SI units interchangeably (*Système international d'unités* — often referred to erroneously as metric units). This book also utilizes a mix of such units to assure understanding of subject matters, however, where possible corresponding conversions are provided. The following table provides an SI conversion of some common units. (Note that for convenience, the table includes some metric units such as centimeters and bar as customary units.)

TO CONVERT FROM CUSTOMARY UNIT	TO DEFINE UNIT 1	SYMBOL	MULTIPLY BY
barrel per hour	liters per second	L/s	0.044 163
barrel per day	cubic meters per day	m^3/d	0.158 987
MMBOD	cubic meters per day	m^3/d	$0.158\,987 \times 10^6$
Btu/second	kilowatt	kW	1.055 056
Btu/hour	watt	W	0.293 071
Btu/lb_m	kilojoule per kilogram	kJ/kg	2.326
Btu/lb_m-°F	kilojoule per kilogram-Kelvin	kJ (kg·K)	4.1868
Btu/lb_m-mole-°R	joule per mole-Kelvin	J/(mol·K)	4.1868
Btu/°R	kilojoule per Kelvin	kJ/K	1.8991
Btu/ft^2-hr	joule per sq. meter-second	$J / (m^2 \cdot s)$	3.154 591
Btu/ft-hr-°F	joule per meter-second-kelvin	J / (m/s·K)	1.730 735
$Btu/ft\text{-}hr^2$-°F	joule per square meter-second kelvin	$J / (m^2 \cdot s \cdot k)$	5.678 263
foot-pound force (ft. lb_f)	joule	J	1.355 818
$foot^2$	square meter	m^2	0.092 903
$foot^3$	cubic meter	m^3	0.028 316 85
$foot^3$/minute	liter per second	L/s	0.471 947
$foot^3$/hour	cubic meter per day	m^3/d	0.679 604
MMSCFD	cubic meter per second	m^3/s	0.327 774
gallon/minute (GPM)	liter per second	L/s	0.063 090
$inch^2$	square centimeter	cm^2	6.451 600
$inch^3$	cubic centimeter	cm^3	16.387 064
kilowatt-hour (kWh)	megajoule	MJ	3.6
mile per hour	kilometer per hour	km/h	1.609 344
pound	kilogram	kg	0.453 592 37
pound force/$foot^2$ (psf)	Pascal	Pa	47.880 258
pound mass/$foot^3$ (lb_m/ft^3)	kilogram per cubic meter	kg/m^3	16.018 463
pound mass/gallon	kilogram per liter	kg/L	0.119 826
pound mass/hour	kilogram per hour	kg/h	0.453 592
psi	kilopascal	kPa	6.894 757
bar	kilopascal	kPa	100

TO CONVERT FROM CUSTOMARY UNIT	TO DEFINE UNIT 1	SYMBOL	MULTIPLY BY
psi/foot	kilopascal per meter	kPa/m	22.620 59
psi/mile	Pascal per meter	Pa/m	4.284 203
Watt-hour	kilojoule	kJ	3.6
yard^2	square meter	m^2	0.836 127
yard^3	cubic meter	m^3	0.764 555
acre	square meter	m^2	4046.856
atmosphere (std)	kilopascal	kPa	101.325
barrel (42 gal)	cubic meter	m^3	0.158 987
Btu (International Table)	kilojoule	kJ	1.055 056
calorie (Thermochemical)	joule	J	4.184
Fahrenheit degree (F°)	Celsius degree (range)	C°	5/9
Degree Fahrenheit (°F)	Degree Celsius (temperature)	°C	(9/5)*C°+32
degree R	degree Kelvin	K	5/9
foot	meter	m	0.3048
gallon (US liquid)	liter	L	3.785 412
horsepower (US)	kilowatt	kW	0.7457
inch (US)	millimeter	mm	25.4
inch of mercury (60°F)	kilopascal	kPa	3.376 85
inch of water (60°F)	kilopascal	kPa	0.248 843
mil	micrometer	μm	25.4
mile (US Statute)	kilometer	km	1.609 344
ounce (US fluid)	milliliter	mL	29.573 53
poise	pascal-second	Pa·s	0.1
stokes	square centimeter per second	cm^2/s	1
ton, long (2240 lb_m)	ton (1000 kg)	t	1.016 047
ton, short (2000 lb_m)	ton (1000 kg)	t	0.907 184 74
ton of refrigeration	kilowatt	kW	3.516 853
yard (US)	meter	m	0.9144

NOTE: Multiply factors for compounds units. For example:

1. To convert lb/ft^3 to kg/m^3, multiply

$$\frac{1\text{lb}}{\text{ft}^3} \times \frac{045536\,\text{kg}}{\text{lb}} \times \frac{\text{ft}^3}{(0.3048)^3\,\text{m}^3}$$

2. To convert a viscosity at 25°C of 0.548 centistoke to viscosity in centipoise, obtain

$$\frac{0.548\text{ centistoke}}{1} \times \frac{1\text{cm}^2\text{s}}{1\text{ centistoke}} \times \frac{\text{mm}^2/\text{s}}{100\text{ cm}^2/\text{s}} = 0.00548\text{ mm}^2/\text{s}$$

Now multiply by the flow density $\rho\,\frac{\text{kg}}{\text{mm}^3}$ to determine the viscosity in centipoise:

$$\frac{0.00548\,\text{mm}^2}{\text{s}} \times \rho\,\frac{\text{kg}}{\text{mm}^3} = 0.000548\,\rho\text{ centipoise}$$

Chapter 1

GENERAL AND HISTORICAL INFORMATION

INTRODUCTION

Oil and gas, in general, have been used since early human history to keep fires ablaze and also for warfare. The importance of oil in the world economy evolved slowly (Figure 1-1) (McKnight 1998). Wood and coal were generally used for heating and cooking. The Industrial Revolution subsequently generated an increasing need for energy, which was fuelled mainly by coal. During this period, it was discovered that kerosene could be extracted from crude oil and used as a light and heating fuel. Petroleum was in great demand by the end of the 1800s, and its use thus spread (Yergin 1992).

The 19th century saw the development of the most technologically efficient means of transportation and the 20th century the battle over control of transportation capital. In the present century, one of the challenges of the energy industry, in general, is the security of supply and, specifically for the pipeline industry, is the integrity and security of pipeline systems that constitute the most effective, environmentally friendly, and safe means of oil and gas transportation systems.

Pipeline networks are valuable assets, generally owned, operated, and maintained by pipeline companies. In any country, pipelines are an irreplaceable core of a hydrocarbon transportation system and means by which hydrocarbon fluids are delivered from the source of supply to the market area. It is of no coincidence that wherever there is the largest pipeline network, there is also the highest standard of living and technological progress.

There are about 2 million km of transmission pipelines worldwide (CIA 2009). This is comprised of natural gas (including NGL), oil (including heavy and extra heavy crudes), petroleum products (including "condensate" + "condensate/gas"), refined products, LPG (including LNG) pipelines, and others (including water/oily water, etc.). This statistic does not include any distribution systems.

World statistical information with respect to hydrocarbon pipelines is summarized in Table 1-1. Figure 1-2 indicates the percentage breakdown of these pipelines per major product transported. It may be noted that Table 1-1 and Figure 1-2 do not include gathering and special transmission lines. Special transmission pipelines cover carbon dioxide (CO_2), hydrogen, slurry, LNG, and similar specialty products. However, liquid transmission lines reported cover single product as well as batched product pipelines. Major market shares of world pipelines (exceeding 30,000 km in length) are indicated in Figure 1-3.

Figure 1-4 (A and B) indicates where vast majorities of gas and oil pipelines are located regionally (CIA 2009). As illustrated in Table 1-1 and Figure 1-4 (A and B), USA and Russia rank the highest for natural gas transmission, and Russia only ranks the highest for oil (namely, crude oil) transmission pipelines (Figure 1-5). However, it may be noted from Table 1-1 that Canadian natural gas network is lumped into LPG/LNG pipelines and, hence, the reason such length does not appear in Figure 1-4.

It is also significant to note that oil pipeline transmission developments are mostly for short hauls, while gas pipelines are for long haul and cross-border intercontinental

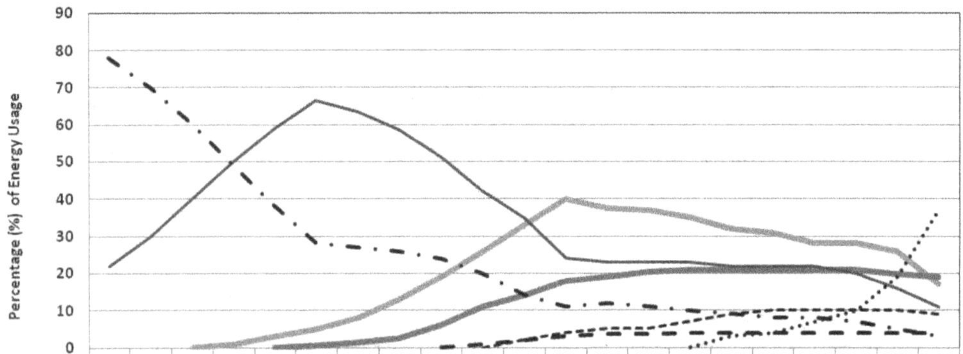

YEAR	1860	1870	1880	1890	1900	1910	1920	1930	1940	1950	1960	1970	1980	1990	2000	2010	2020	2030	2040	2050	2060
Oil			0	1	3	5	8	13	19	26	33	40	37.5	37	35	32	31	28	28	26	17
Gas					0	0.5	1.5	2.5	6	11	14	18	19	20.5	21	21	21	21	21	20	19
Coal	22	30	40	50	59	66.5	63.5	58.5	51	42	35	24	23	23	23	22	22	22	20	16	11
Nuclear									0	1	2	3	3.5	3.5	4	4	4	4	4	4	4
Hydroelectric										0	2	4	5	5	7	9	10	10	10	10	9
Biomass	78	70	60	49	38	28	27	26	24	20	14	11	12	11	10	9	8	8	7	5	3
New Commercial Sources															0	3	4	7	10	19	37

Figure 1-1. Evolvement of energy usage (after McKnight 1998)

transportation (Figure 1-5). The significance of this can be attributed to five main factors:

1. Environmental pollution, in the case of a pipeline rupture, oil is much more significant than gas
2. Gas (unless liquefied such as LNG) has to be transported by pipeline
3. Acceptability of natural gas as clean burning fuel versus oil
4. Ease of pipeline system operations for natural gas
5. Natural gas is a cheaper source of fuel

More economical form of transportation than rail or truck, pipelines are economically preferable and usually fairly secure within a nation or between friendly neighbors. Figure 1-6 illustrates major global cross-border natural gas movements. This is further illustrated in Figure 1-7. It may be noted that major cross-border gas movements is by pipeline only.

Global trade of natural gas by pipeline in 2008 amounted to 587.21 billion cubic meters (BCM). Global LNG trade movement in 2008 had amounted to 226.51 BCM. The breakdown of magnitude of LNG imports and exports by country is indicated in Figure 1-8.

The breakdown of oil including product imports and exports is provided in Table 1-2. Figure 1-9 indicates the corresponding world major oil trade movements.

From Table 1-2, it can be inferred that a net movement of about 2000 billion barrels of oil is realized annually on a global basis.

TRENDS IN PIPELINE SYSTEM DEVELOPMENT

Types of Pipelines

The basic types and categories of pipeline systems include:

- Crude oil pipelines
  - o Gathering lines
  - o Trunk lines

TABLE 1-1. World onshore operating pipelines as of 2009 (CIA 2009)

Country	Gas	Oil	Condensate-gas	LPG/LNG	Refine Product	Water/Oily Water/others
Afghanistan	468					
Albania	339	207				
Algeria	114,648	7,579	1,937	2,933		
Angola	2	87				
Argentina	28,138	5,939		41	3,629	13
Armenia	2,233					
Australia	27,105	3,258		240		1
Austria	2,721	663			157	
Azerbaijan	3,361	1,412	1			
Bahrain	20	32				
Bangladesh	2,597					
Belarus	5,250	1,528			1,730	
Belgium	1,330	158			535	
Bolivia	4,883	2,475		47	1,589	
Bosnia and Herzegovina	170	9				
Brazil	9,892	4,517	353	47	4,465	
Brunei	37	18				
Bulgaria	2,926	339			156	
Burma	2,228	558				
Cameroon		889		5		
Canada 2006		23,564		74,980		
Chad		250				
Chile	2,676	892		519	769	
China	28,132	20,204			9,746	
Colombia	4,560	6,094			3,383	
Congo, Democratic Republic of the	37	39			756	
Congo, Republic of the	7	207				
Costa Rica					796	
Côte d'Ivoire	180	92	86			
Croatia	1,327	583				
Cuba	41	230				
Czech Republic	7,010	547			94	
Denmark	2,858	107	11			2
Dominican Republic 2006		96			8	
Ecuador	5	1,809			1,301	
Egypt	5,586	4,314	333	957	895	71
Equatorial Guinea	38					
Estonia	859					
Finland	694					
France	14,688	3,036			5,080	
Gabon	240	723				
Georgia	1,591	1,253				
Germany	24,364	3,379			3,843	
Ghana		5			309	
Gibraltar	0	0				
Greece	1,197	75				
Guatemala		480				
Hungary	4,407	987			335	
India	6,061	7,678	2	2,156	6,876	
Indonesia	5,797	5,721	808		1,370	56
Iran	19,246	7,018	19	570	7,936	
Iraq	2,501	5,418		918	1,637	
Ireland	1,550					

TABLE 1-1. (Continued)

Country	Gas	Oil	Condensate-gas	LPG/LNG	Refine Product	Water/Oily Water/others
Israel	176	442			261	
Italy	17,544	1,241				
Jamaica						
Japan	3,862	167				53
Jordan	439	49				
Kazakhstan	11,146	10,376	658		1,095	1,465
Kenya		4			928	
Korea, North		154				
Korea, South	1,423				827	
Kuwait	269	540				57
Kyrgzstan	254	16				
Laos					540	
Latvia	948				415	
Lebanon	43					
Libya	2,860	6,987	776			
Liechtenstein	20					
Lithuania	1,695				114	
Luxembourg	155					
Macedonia, Republic of	268	120				
Malaysia	1,965	31	3		114	19
Mexico	22,705	8,688		1,875	6,520	228
Midway Islands						
Moldova	1,906					
Morocco	830	439				
Mozambique	918				278	
Netherlands	3,816	365			716	
New Zealand	1,838	288	331	172	198	
Nicaragua		54				
Nigeria	2,560	3,396	21	97	4,090	
Norway 2006	6,529	2,444	31			457
Oman	4,126	3,558			263	
Pakistan	10,402	2,076			792	
Papua New Guinea		195				
Peru	1,078	1,551		654	15	
Philippines	565	107			112	
Poland	13,631	1,384			777	
Portugal	1,098	11			188	
Qatar	978	382	90	87		
Romania	3,588	2,424				
Russia	158,767	74,285	122	127	13,658	23
Saudi Arabia	1,880	4,139	212	1,183	1,148	
Senegal	43				8	
Serbia and Montenegro	1,921	323				
Singapore	106					
Slovakia	6,769	416				
Slovenia	840	11				
Somalia	0	0				
South Africa	908	980	11		1,379	
Spain	7,738	560			3,445	
Sudan	156	4,070			1,613	
Suriname		50				
Sweden	786					
Switzerland	1,662	94			7	
Syria	2,900	2,000				
Taiwan	406					
Tajikistan	549	38				

TABLE 1-1. (Continued)

Country	Gas	Oil	Condensate-gas	LPG/LNG	Refine Product	Water/Oily Water/others
Tanzania	253	888			8	
Thailand	1,348				323	
Trinidad and Tobago	659	336				
Tunisia	2,102	1,195			372	
Turkey	7,555	3,636				
Turkmenistan	6,417	1,457				
Ukraine	33,327	4,514			4,211	
United Arab Emirates	2,129	1,310	458	220	212	90
United Kingdom 2006	7,541	699	43	59	4,417	
United States 2006	548,665				244,620	
Uruguay	226	155				
Uzbekistan	9,706	868				
Venezuela	5,036	7,675			1,484	141
Vietnam	66		42		206	
Yemen	96	1,367		22		
Zambia		771				
Zimbabwe					270	
WORLD TOTAL KM	1,211,839	298,982	10,098	87,586	345,481	1,707

- Liquid pipelines
    - o Transmission lines
    - o Distribution lines
- Natural gas pipelines
    - o Gathering lines
    - o Transmission lines
    - o Distribution systems

Trends in pipeline development thus encompasses all related issues of technology developments for all pipelines, but specifically as related to liquid and natural gas pipelines.

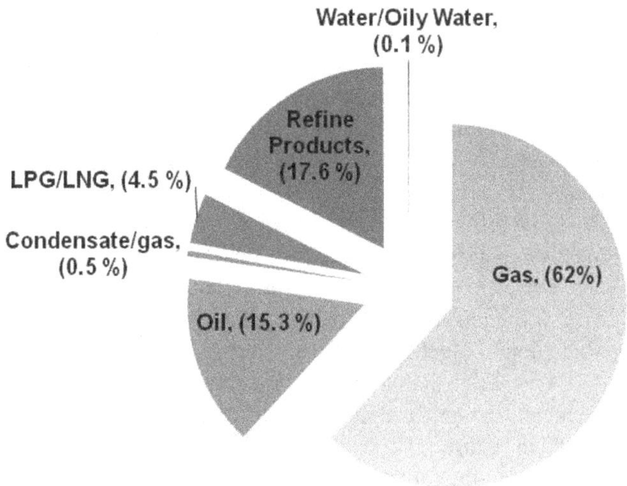

Figure 1-2. Percentage breakdown of world pipelines by product transportation, 2008

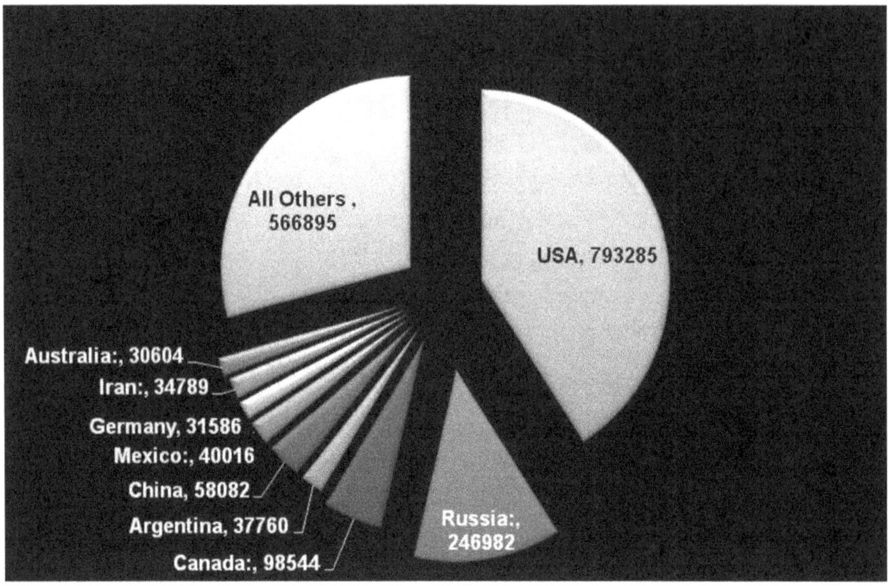

Figure 1-3. Market share breakdown of world pipelines (countries with >30,000-km pipelines)

Trends in Technology

Pressure and Material

Historically, the safe design of a pipeline has been related to a limitation of the hoop stress as some function of the material yield stress. The hoop stress is derived from a relationship using the Barlow formula, which related pipe stress to internal pressure, pipe diameter, and wall thickness. In the early part of the 1900s, pipe was low strength and provided limited economic capabilities in terms of pipeline technology. This is illustrated in Table 1-3, which shows the development of pipeline gas transportation through the 1900s.

Entering the 21st century, there are now projects with pressures up to 3000 PSI, 21 MPa, with diameters up to 48 in., 1220 mm. The big change in technology occurred with the introduction of high-strength low-alloy steels (HSLAs) in the 1960s.

The historical development and use of high-strength steel for pipeline application is illustrated in Figure 1-10 (Asahi et al. 2004). ExxonMobil and Nippon Steel Corporation jointly developed an ultra-high-strength large-diameter line pipe, American Petroleum Institute (API) X120. Nippon Steel Corporation developed the API X120 line pipe while pursuing the development and commercial application of X100 line pipe. The development of the X120 pipe accelerated the improvement in the strength of the line-pipe material (Akasaki 2005). The advent of this highest grade of line pipe thus occurred earlier than market forecasts had predicted.

The prime impetus for the increase in pressure and the increase in the material properties is economics (Figure 1-11). For the same design pressure, over 30% material savings and corresponding percentage of material and welding costs can be realized by the utilization of higher-grade X120 (Grade 931 MPa) versus X70 (Grade 483 MPa). It may be noted that Nippon Steel Corporation was the first heralding the development of X120 Steel.

On a large diameter pipeline project, 40% to 50% of the cost is related to material, and reducing the material cost therefore has a significant effect on project costs. This is illustrated in Figure 1-12, which clearly illustrates the benefit of using higher strength materials and the driving force for increasing the strengths to even higher values (Glover et al. 1999). Figure 1-12 provides a comparison of costs using Grade 483 as the base value for different pipe yield strengths at different operating pressures up to 10 MPa.

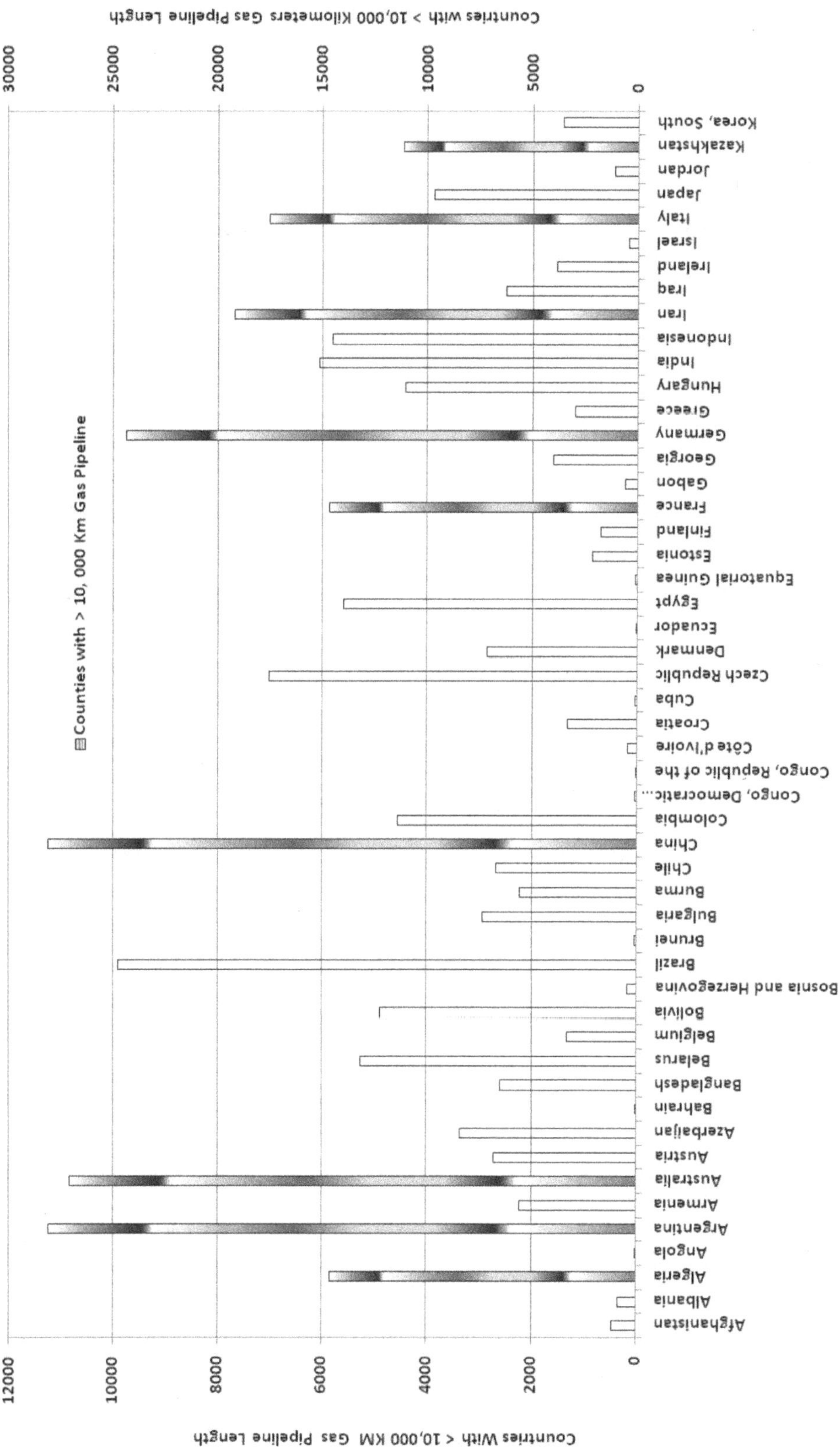

Figure 1-4. (A) Breakdown of onshore natural gas pipelines market share (CIA 2009) Note: Canada not included

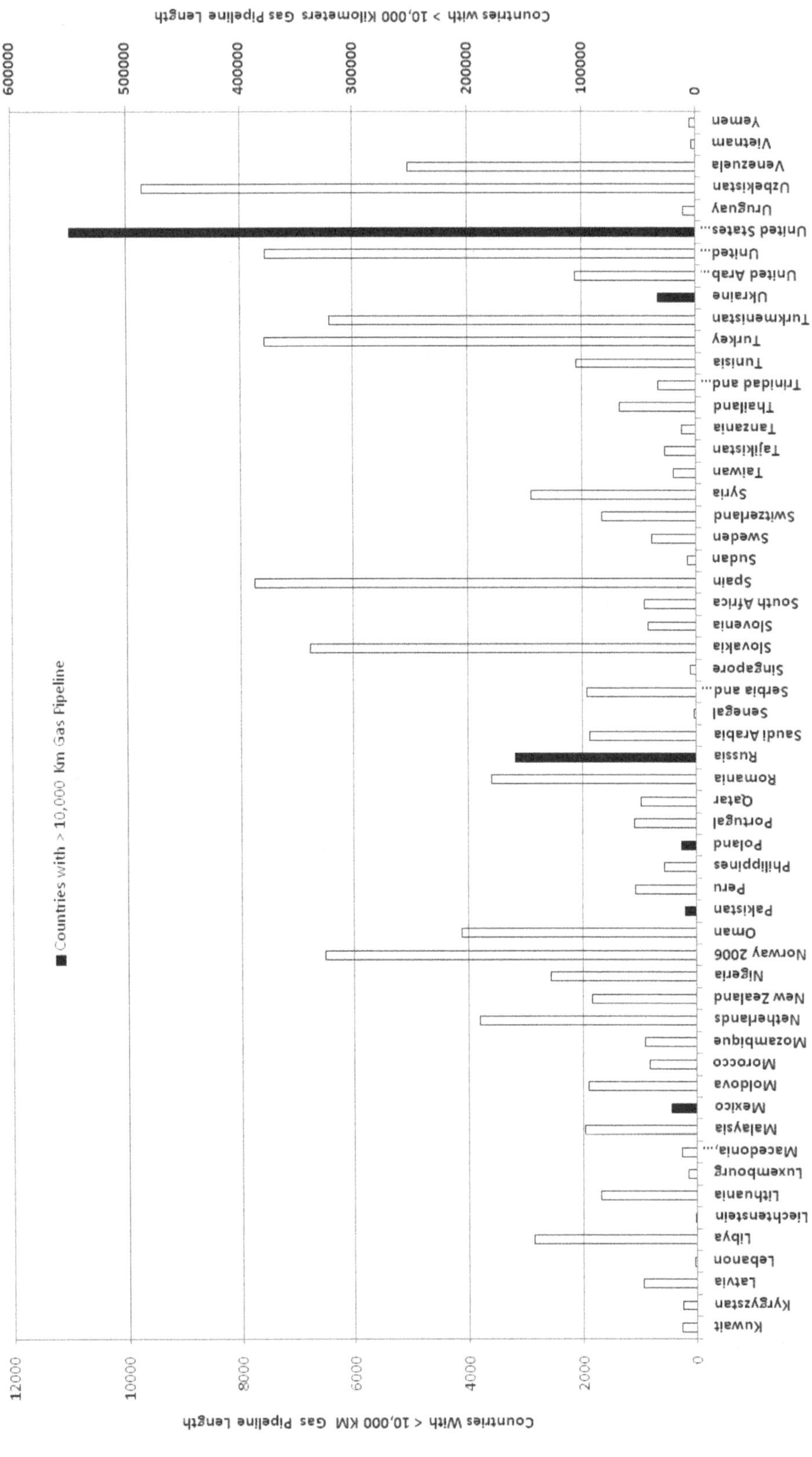

Figure 1-4. (B) Breakdown of onshore natural gas pipelines market share (BP 2009)

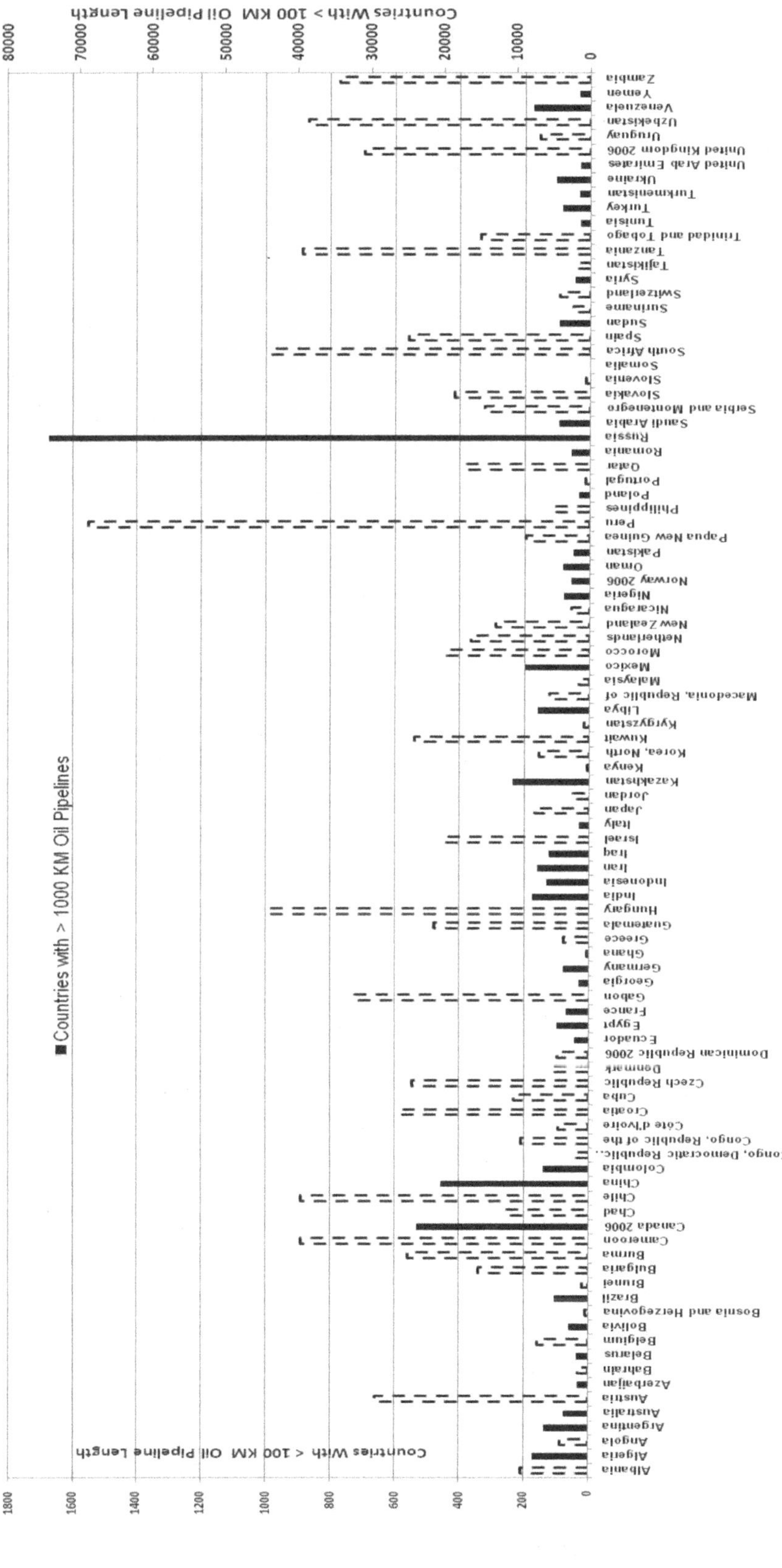

Figure 1-5. Breakdown of onshore oil pipelines (including heavy oil) market share (CIA 2009)

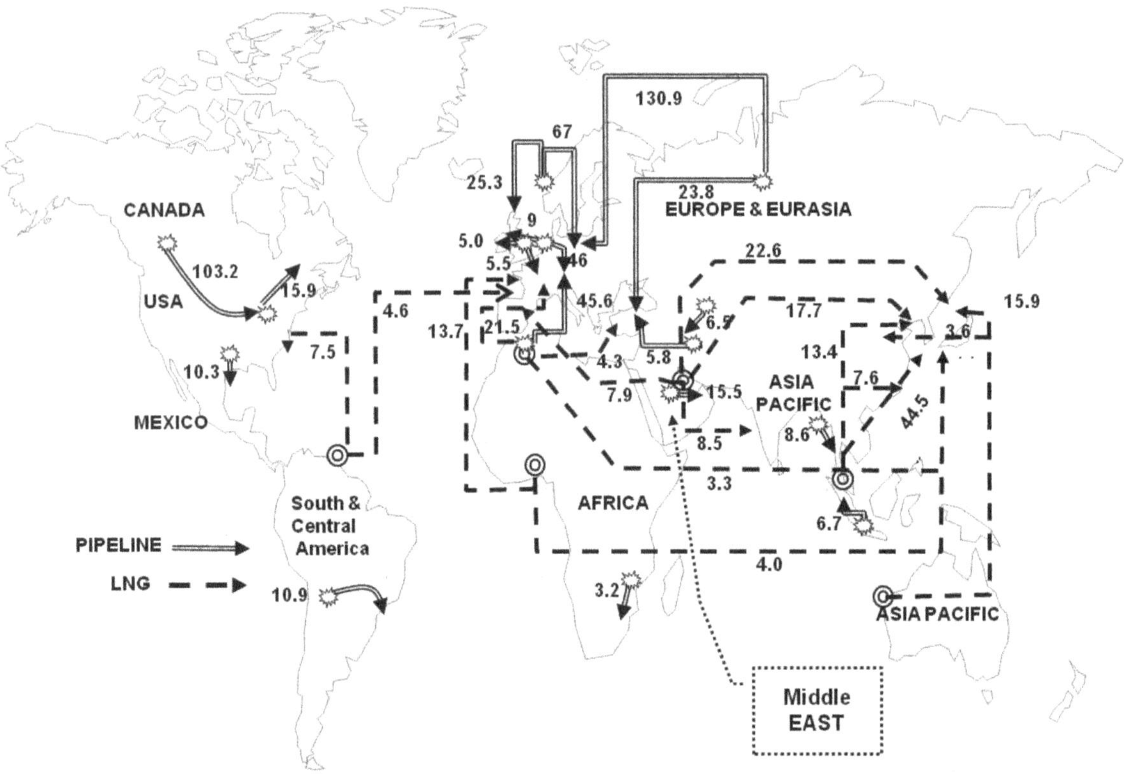

Figure 1-6. Major natural gas trade movements 2008 (in BCM; Source: BP 2009)

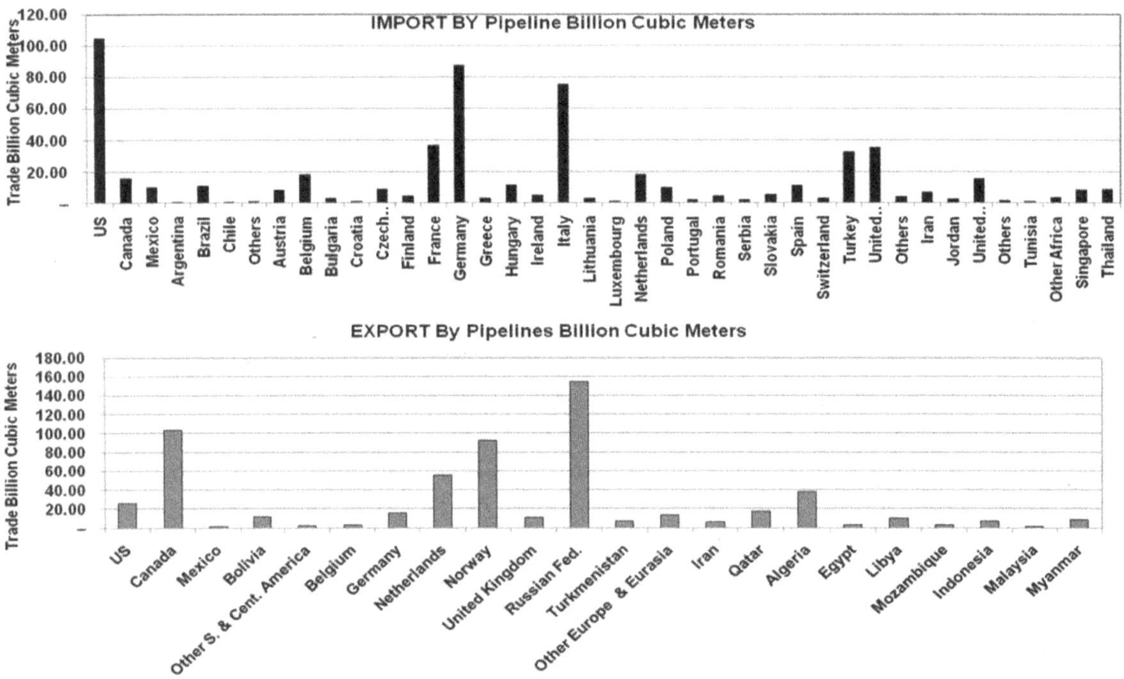

Figure 1-7. Global gas trade by pipelines, 2008 (in BCM; Source: BP 2009)

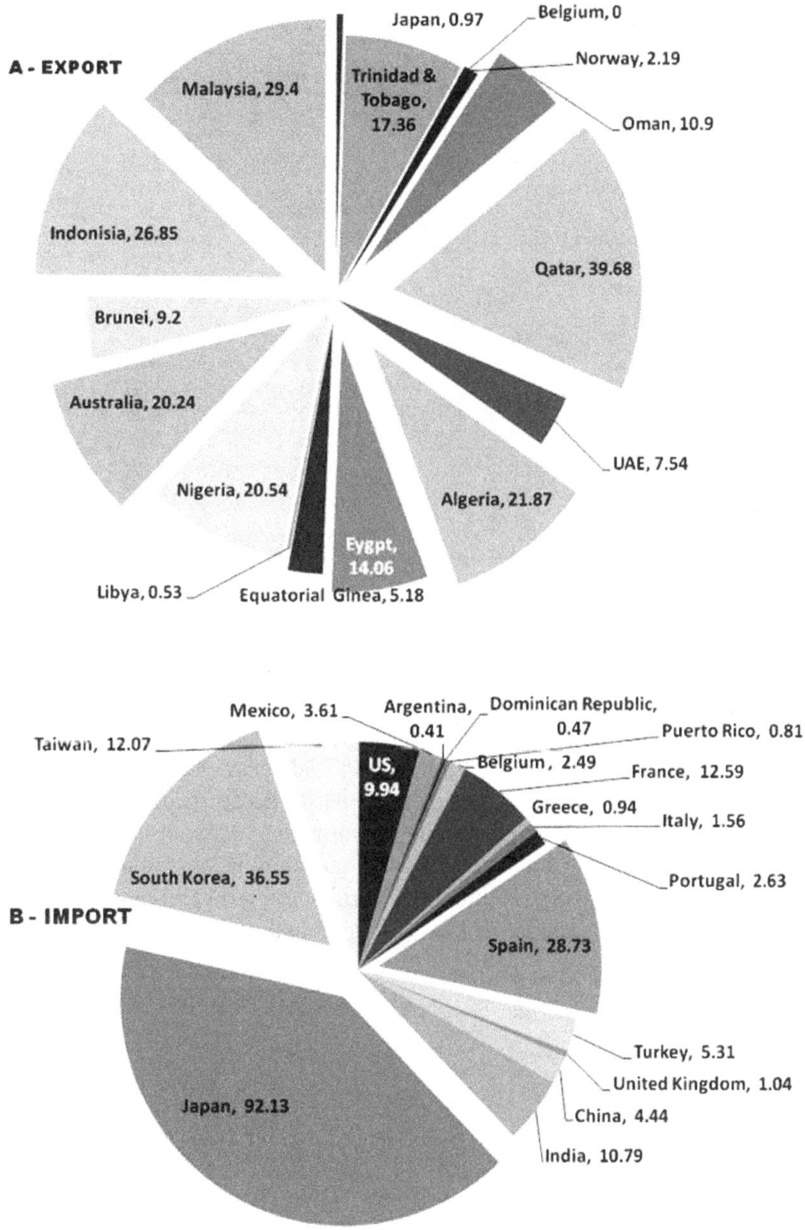

Figure 1-8. Global LNG trade movement, 2008 (in BCM; Source: BP 2009)

Accompanying this increases in material yield strength also increases the operating pressures. Although Figure 1-12 shows the benefit in terms of cost, it must be balanced with the design of the pipeline and consideration of *D*/*t* ratios together with strain limits and constructability issues such as pipe bending.

Construction Technologies

The major changes that have occurred in the construction technologies relate primarily to the continued expansion of mechanized and automatic techniques.

Mechanized welding is now a standard technology and has seen considerable advances in the previous few years including the use of one-sided welding, multiple arc technologies, and alternative welding technologies (Glover et al. 1990). The thrust has been to move

TABLE 1-2. World major daily oil imports and exports, 2008 in thousand barrels (BP, 2009)

Country	Crude Imports	Product Imports	Crude Exports	Product Exports
US	9,756	3,116	139	1,829
Canada	693	288	1,931	566
Mexico	40	500	1,459	150
South and Central America	610	864	2,479	1,136
Europe	10,859	2,892	281	1,741
Former Soviet Union	1	149	6,235	1,948
Middle East	220	202	17,925	2,203
North Africa	329	170	2,618	642
West Africa	18	281	4,466	121
East and Southern Africa	509	189	340	17
Australasia	489	400	274	37
China	3,581	812	75	313
India	2,557	458	^	718
Japan	4,069	856	–	356
Singapore	1,092	1,593	40	1,542
Other Asia Pacific	4,627	2,390	931	1,460
Unidentified*	–	17	257	394
Total World	39,451	15,175	39,451	15,175

the mechanization toward automation. The changes in welding technology have also been accompanied by changes in the inspection technology. Mechanized ultrasonic inspection techniques are also now commonplace for both mechanized welding and manual welding. Changes have occurred in this field relate to the introduction of phase array technology and improved sensitivity with respect to defect detection. Accompanying these changes have been the widespread use of alternative acceptance criteria based on engineering analyses and quality control from design through to material supply, fabrication, construction, commissioning, and operation while maintaining structural integrity. An example of this is the use of Alternative Integrity Verification (AIV) in lieu of pipeline hydrostatic testing by ensuring material and construction quality meeting the specified requirements. The AIV

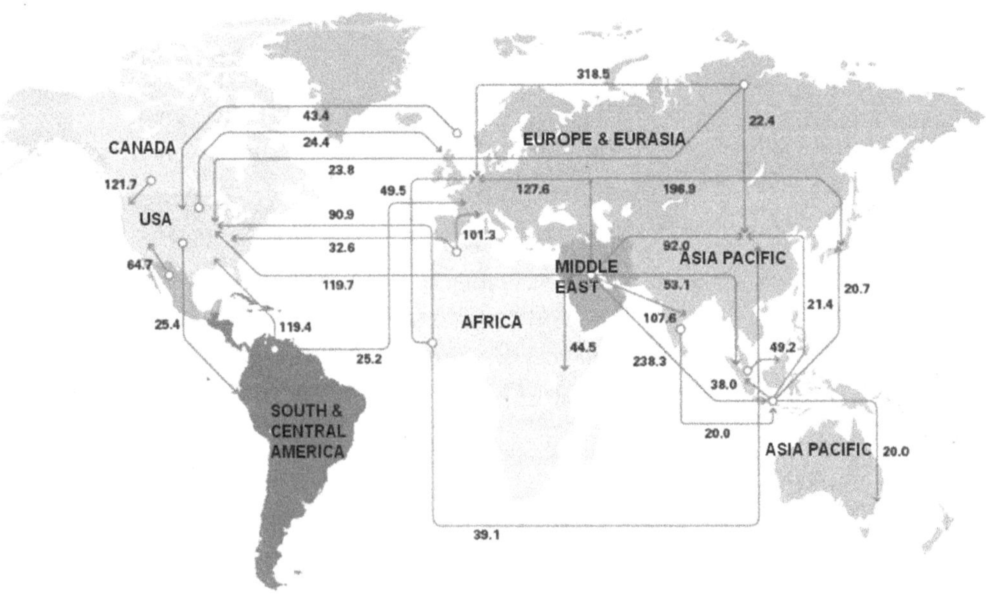

Figure 1-9. World major oil trade movements, 2008 (in million tons; Source: BP 2009)

TABLE 1-3. Historical gas pipeline developments

Year	Operating Pressure kPa	Operating Pressure PSI	Pipeline Diameter mm	Pipeline Diameter in.	Throughput/ Fuel Gas Ratio
1910	200	30	406	16	51/49
1930	2,000	290	505	20	68/32
1965	6,600	1,000	914	36	86/14
1980	8,000–10,000	1,160–1,440	1,220	48	89/11
1990s	12,000–15,300	1,740–2,220	1,220–1,420	48–56	93/7

quality management process generally includes all actions which are necessary to be performed to provide confidence and evidence of integrity and reliability in the constructed pipeline without hydrostatic testing. An AIV quality management process and typical associated AIV enablers which can provide the quality assurance/quality control at each pipeline project cycle point is shown in Figure 1-13. Refer to Chapter 10 for a similar information.

These technologies have resulted in considerable economic benefits for the pipeline industry while improving the structural reliability of pipeline systems. Emerging trends relate to the combination of these technologies into the pipe-laying sequence that will permit a single station for pipeline construction. These technologies are being used for both onshore and offshore pipeline.

Environmental concerns with river crossings, sensitive habitat, and crowded pipeline right of ways have led to increasing applications of directional drilling installations, and the trend has been to larger diameters and larger pulls. Related to the environment are continued efforts to minimize right-of-way aspect and reclamation techniques, all of which have seen considerable changes in the last decade (Mohitpour et al. 2007).

The assessment of pipeline safety and security continues to be a major topic, and the understanding of the risk and reliability and security related to system operation has led to many innovative approaches. The industry utilizes strain-based designs and is working toward the incorporation of reliability-based designs. Associated with integrity has been the introduction of a whole new era of pipeline inspection technologies based on crack detection including electromagnetic acoustic transducer (EMAT) technologies and strict quality control. These developments have primarily been in response to the occurrence of stress corrosion cracking (SCC) on major pipeline systems.

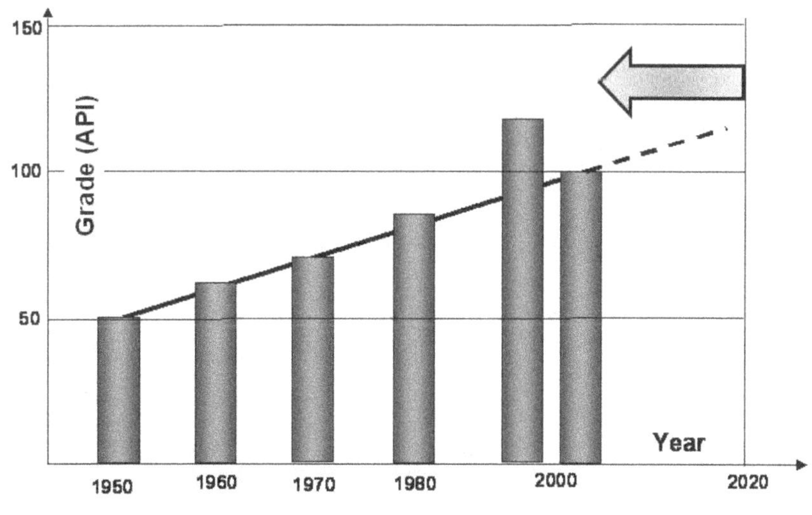

Figure 1-10. Historical development of line-pipe technology (after Asahi et al. 2004 and Terada 2005)

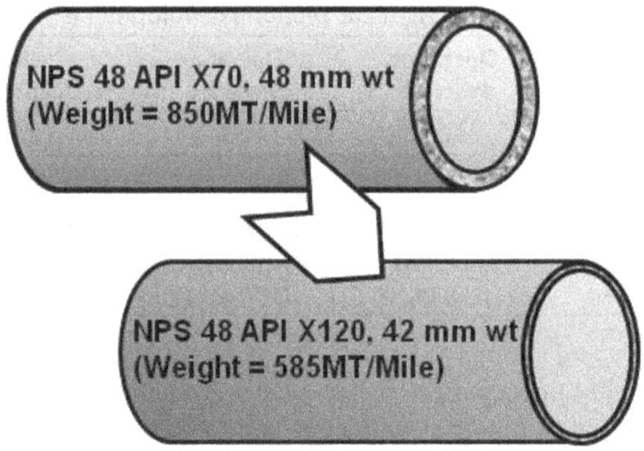

Figure 1-11. Comparison of wall thickness between conventional line-pipe API X70 and higher strength API X120 (after Ayukawa, 2005)

Current Development

Material. The major change that occurred in steel pipeline technology was the introduction of HSLAs. The essential characteristics of these steels are as follows:

- Fine grain size for good yield strength and toughness
- Low carbon and carbon equivalent for good toughness and weldability
- Chemistry control including micro-alloying with niobium, vanadium, titanium, and aluminum
- Process control and controlled rolling

These changes led to

- Higher strengths
- Improved toughness at low temperatures
- Improved weldability
- Improved resistance to sour service

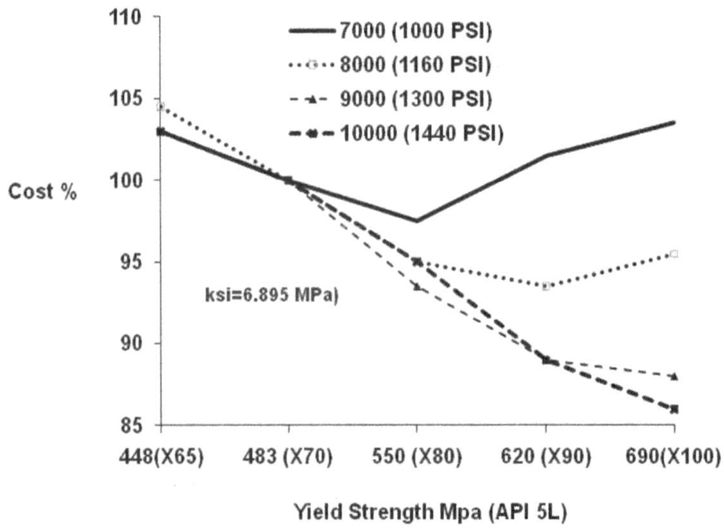

Figure 1-12. Cost-benefit using high-grade steel (Glover 1996)

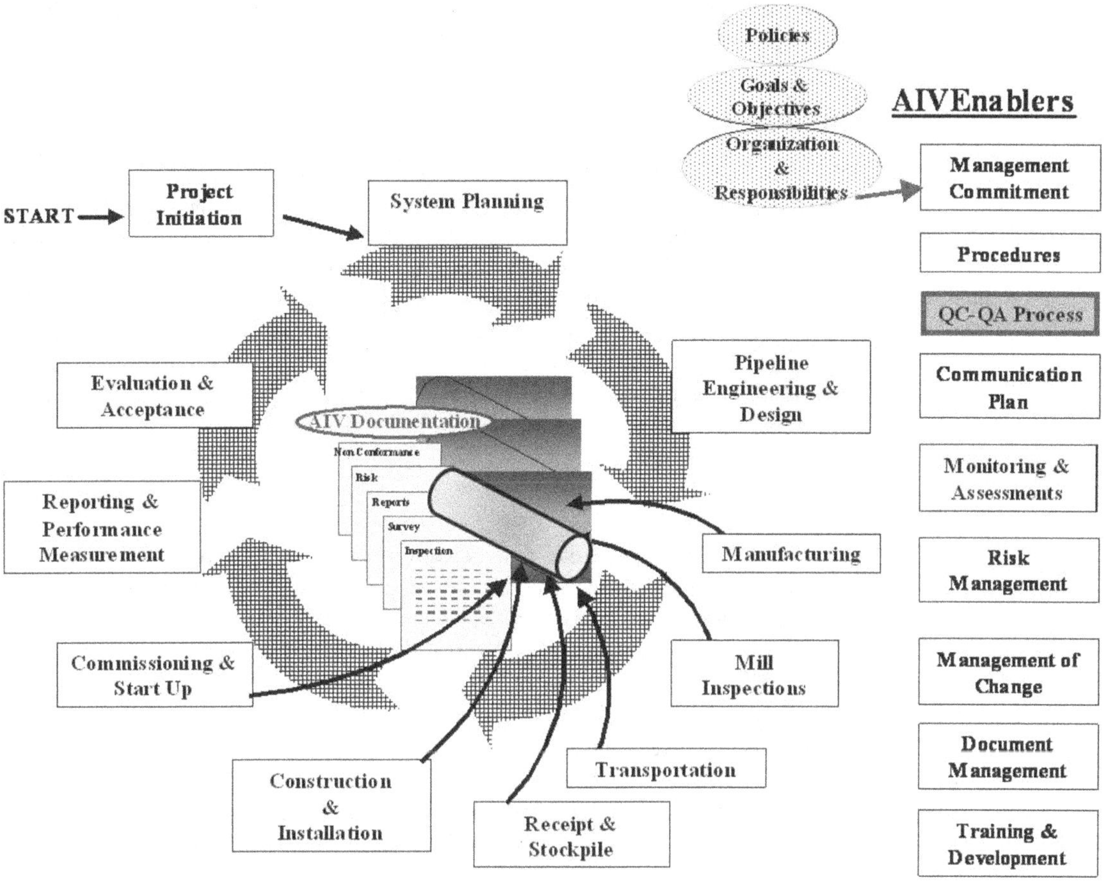

Figure 1-13. AIV quality management process and enablers (Refer to Chapter 10)

- Improved ductile fracture resistance
- Cost effectiveness.

These changes have now been incorporated into a new generation of pipeline steels, Grade 690-Grade 931 (API X100-API X120), which have been used in recent years for high-pressure systems (Akasaki 2005). These steels utilize modifications to the low alloys somewhat but also change to the process control on the controlled rolling and accelerated cooling.

In the development of extra high-strength X120 pipe, two pressing yet contrary concerns must be satisfied: High strength (120 ksi) and high toughness (>250 J) must be combined with appropriate corrosion resistance and weldability properties to meet different application environments.

The primary requirement of obtaining both high strength and high toughness is achieved by building a fine microstructure (lower bainite structure) into the steel material. The steel manufactures use "boron" as rare metal along with other elements.

The super-high heat-affected zone (HAZ) toughness is achieved with fine microstructures imparted by fine particles (Kojima 2005). This technology called HTUFF developed by Nippon Steel Corporation has solved the problem of brittleness that occurs in the HAZ during the welding of steel plates. Even when high heat-input welding that employs single-stroke welding at high temperatures is applied, welds remain impregnable to fracture cracks. The comparison of microstructure of conventionally produced steel and the high-toughness/high-strength steel produced through the HTUFF process is provided in Figure 1-14.

Composite Reinforced Line Pipe

Increased interest has been applied to the use of composite reinforced line pipe (CRLP) as a potential application for transmission pipeline systems and for gas transport/storage modules. CRLP uses a patented glass/resin reinforcing technology developed by NCF Industries. Composite reinforced gas transportation products are based on the concept of using composites to add hoop strength to steel line pipe or vessels (Figure 1-15). The steel and composite overwrap work in unison creating a hybrid product that carries the applied load of high-pressure gas. The advantage of the combined system is that high strengths can be achieved that can be operated at high pressures, while still offering enhanced pipeline integrity in a cost-effective way.

Composite

Fiber-reinforced polymeric composites that are currently used in pressure-retaining structures are seen as an alternative to conventional materials for pipeline systems. Some of the benefits of these composites relate to their corrosion resistance and their high-strength-to-weight ratio. The resistance in adopting composite materials, however, is primarily due to a limited understanding of the material behavior under a variety of loading and environmental conditions and a lack of qualified design methodologies. Ongoing research at the University of Alberta and other Canadian universities aims to overcome these limitations (Ellyin et al. 2000).

The advantage of FRP composites over conventional materials is their relative high strength, lightweight, corrosion resistance, and durability. The weight reduction translates into a direct cost saving due to reduced handling and installation costs, while improved corrosion and durability results in a long-term cost benefit. Furthermore, composite pipe technology has other advantages:

- The strength properties can be "tailored" for a particular loading application (i.e., designed and fabricated for characteristics of a particular region)

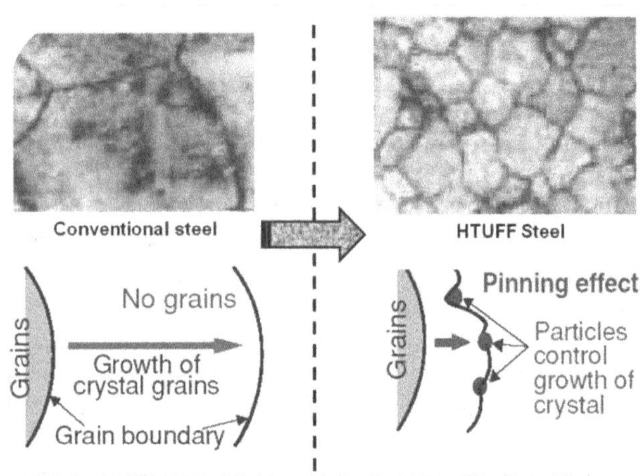

Figure 1-14. Production of high grade (API X120 steel) through control of crystal grain growth at extra high temperatures by means of nanoparticles (Kojima 2005).

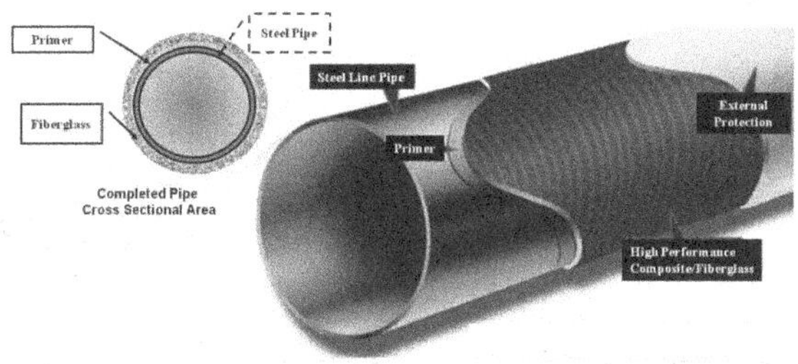

Figure 1-15. Composite reinforced line pipe **CRLP™** (Courtesy NCF Industries Inc., CA. Canadian, US and Foreign Patents issued and pending)

- "smart technologies" can be incorporated into composite structures, which can provide for remote monitoring of structural integrity or possible indications of failure and
- composite structures can be designed to "leak before break," which can be incorporated into a structural reliability approach. In addition, well-established manufacturing technologies (e.g., filament winding) exist for fabrication of composite pipe.

FRP composites are beginning to be used extensively by the offshore oil and gas transportation industry (currently confined to small diameter pipes, primarily due to their high-strength-to-weight and corrosion resistance properties (Vennett et al. 1999; McDonnell et al. 1995).

TRENDS IN PIPELINE CORROSION PROTECTION

Corrosion consists of an oxidation reaction and a reduction reaction at the surface of the corroding material. The oxidation reaction generates metal ions and electrons; the electrons are then consumed in the reduction reaction. For example, in environments where water is present (including moisture in the air), the electrons are consumed by converting oxygen and water to hydroxide ions. In iron and many iron alloys, these hydroxide ions, in turn, combine with iron ions to form a hydrated oxide (Fe(OH)2). Subsequent reactions form a mix of magnetite (Fe_3O_4) and hematite (Fe_2O_3). This red-brown mixture of iron oxides is rust. This basic oxidation/reduction reaction behind corrosion is illustrated in Figure 1-16 (Guthrie et al. 2001).

ASTM generalized forms of corrosion (Scully and Taylor 1987) include the following:

1. General corrosion shown as corrosive attack dominated by uniform thinning includes:

    - Atmospheric corrosion
    - Galvanic corrosion
    - Stray-current corrosion
    - General biological corrosion
    - Molten salt corrosion
    - Corrosion in liquid metals
    - High temperature

2. Localized corrosion indicated by high rates of metal penetration at specific sites include:

    - Crevice corrosion
    - Filiform corrosion

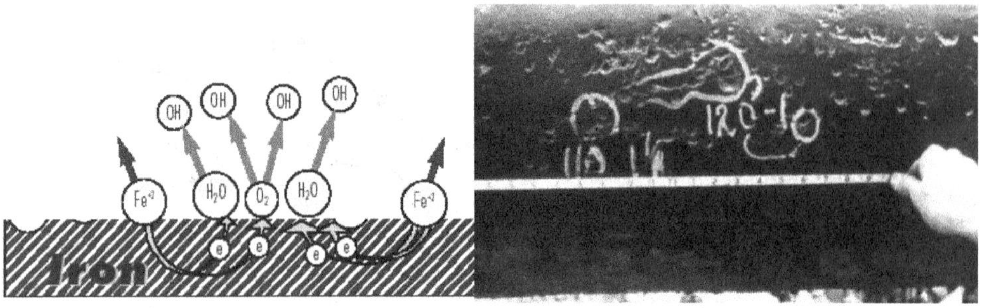

Figure 1-16. Oxidation/reduction reaction behind corrosion

- Pitting corrosion
- Localized biological corrosion

3. Metallurgically influenced corrosion affected by alloy chemistry and heat treatment include:

- Intergranular corrosion
- Dealloying corrosion

4. Mechanically assisted degradation indicated as corrosion with a mechanical component include:

- Erosion corrosion
- Fretting corrosion
- Cavitation and water drop impingement
- Corrosion fatigue

5. Environmentally induced cracking indicated by cracking produced by corrosion, in the presence of stress, include:

- SCC
- Hydrogen damage
- Liquid metal embrittlement
- Solid metal-induced embrittlement

The most effective method of mitigating corrosion on the external surface of a buried or submerged facility is by a dual system of protective coating supplemented by cathodic protection (CP) (NACE 2002).

The necessity to protect steel and iron against corrosion was recognized in the 18th century; however, the first coating application to pipelines appears to be between 1830 and 1850 when, in England, directions have been found on general use and application of coal tar and asphalt tar together with other material. Historical record indicates that in 1884, an English company was the first to use mineral filling materials for the manufacture of asphalt tar and mastic cladding.

While first time report on CP dates to 1875, the construction of first CP installation for pipeline is around 1906.

Coating Design

The primary method of pipeline protection is the use of a barrier coating. Failure of this protective layer can have a dramatic impact on the subsequent cost to operate the pipeline through either increased CP demand, pipeline rehabilitation, and, in some instances, the need for pipeline replacement or repair.

Coatings are effective in protecting pipeline against external corrosion over most of the area. It has been demonstrated that modern coatings do not fail in realistic simulated environments and that any local failures typically result from nonideal coating applications, which signifies requirements for proper application methods and procedures in shop or field particularly in field repair coatings (Moghissi 2003).

What constitutes a "Pipeline Coating Failure" (Figure 1-17) is a failure that can cause corrosion leading to pipeline leak or rupture or operational risks.

There are well over 100 different coating candidates that are available and used for pipeline applications. Some of the generic types of coatings that have been fully tested and are used for pipeline applications (Figure 1-18) (Coulson and Temple 1983) include:

- Coal tar enamel
- Coal tar/polyester
- Extruded polyethylene
- Fusion bonded epoxy
- Polyethylene tape (pressure sensitive)
- Polyethylene tape (butyl backing)
- Polyethylene tape (for high-heat application)
- Polyvinyl chloride tape
- Polyurethane
- Polyurethane/tar

Figure 1-17. Pipeline coating failures leading to operational interruptions

Figure 1-18. Typical pipeline coatings candidates

The first and most known pipe protection material was invented by Angus Smith in England and consisted on a mixture of coal tar and linseed oil. The first application of this coal tar paint and varnish in the USA dates back to 1885. The following list indicates systematic developments in coating systems and application to pipelines:

1822	First report on painting steel and iron against corrosion
1847	Basic essential of good painting techniques
1860	General use of coal tar for coating
1863	First use of asphalt paper for water pipeline protection in Crailsheim, Germany
1867	First appearance of bitumenized paper-wrapped pressurized pipes (to 20 bar) in Paris
1864	Zinc plating for protection of wrought-iron pipes
1875	Use of mineral wool as insulation of tarred/asphalted pipes
1885	First use of red lead as primary coating
1900	Use of jute, soaked in coal tar pitch for thicker coating
1910	Field coating of pipeline welds using straw and jute with a greasy material (later replaced with petroleum jelly)
End 1920	Application of bitumenized wool felt
1928	Use of hot-applied bitumen bands
1930	Electrical measurements for pipe covering — Karlsruhe Gas Institute
1950	First use of polyvinyl chloride tape — continuously laid pipe by wrapping machine
1953	Use of bitumenized glass wool
After 1960	Application of extruded polyethylene on pipes
1967	First use of fusion bonded epoxy (FBE) coating

CP Design

The term corrosion is generally used to describe corrosion reaction, corrosion effects, and corrosion damage. However, the term corrosion is used to specifically describe the reaction of a material with its surrounding environment that produces measurable changes and can lead to damage.

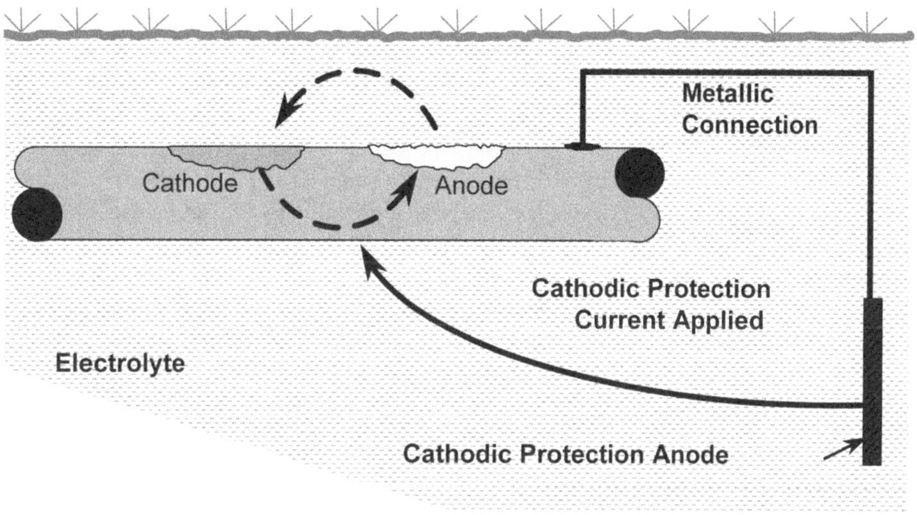

Figure 1-19. Macroscopic view of CP on a pipeline

Electrical energy added to metals when manufactured will leave when it is placed in a corrosive environment. This process causes the metal to corrode. CP is the application of an external direct current source to reverse this current flow to cause it to flow to the structure, thus to protect the structure. Therefore, CP electrically shifts the entire structure more negative than its original state and chemically creates a microscopic film on the surface of the metal making it passive so that it does not corrode. CP does not eliminate corrosion, but corrosion is simply transferred from the structure to anode(s) (Figure 1-19); that is, a cathode condition is replaced by an anodic one.

The concept of corrosion process first appeared in 1667 (Leierzapf 1985). It is referred to by Sir Humphry Davy on the CP of iron in seawater (Davy 1824). Davy (following Galvani's experiments and the discovery of the voltaic pile) was the first to establish that when an electrical current flows through some substances, these substances decomposed (a process later Davy called electrolysis). However, while the active and passive electrochemical processes were known in the 19th century, the reliable form of pipeline protection was only developed at the turn of the 20th century. The following lists the milestones achieved in the development of pipeline CP:

1667	Concept of corrosion process
1761	Discovery of CP of copper by zinc/iron
1787	Manufacture of white lead — Humphry Davy
1825–25	Practical experimentation on protection of copper-clad warship using zinc-cast iron plates — Humphry Davy
1834	Establishment of quantitative connection between corrosion material weight loss and electric current — Faraday
1836	First protection of iron in seawater — Humpry Davy
1875	First-time report on CP — F. Fischer
1890	First use of Impressed current — Thomas Edison
1892	First warning on gas pipeline damage by aggressive soil
1902	First practical CP using indirect current system — K. Cohen
1905	Use of impressed current CP in the USA — E. Cumberland
1906	Scientific study of CP fundamentals — F. Hubert and L. Goldsmith

1906	Construction of first CP installation for pipeline — Herbert Geppert, Karlsruhe, Germany
1910–18	Determination of current density requirements — O. Bauer and O. Vogel
1913	First use of term electrochemical protection referring to consumable anodes — Congress of Institute of Metals, Geneva
1920	Commencement wide use of CP in Europe
1928	First practical application of CP for pipelines
1928	First installation of CP rectifier on long distance gas line in USA — Rober J. Khoun
1930	Discovery of electro-potential as important variable in corrosion reaction
1952	First installation of deep anodes
1961	First anodic protection of products to prevent SCC in caustic soda electroplating
1970	First CP award — Rober J. Kuhn (father of CP in North America)

TRENDS IN PIPELINE INTEGRITY INSPECTION AND REHABILITATION

Pipeline integrity inspection has developed from a mere gauging of the internal size of a pipeline (hence, the possible acronym PIG [Pipeline Inspection Gauge]) to usage of modern

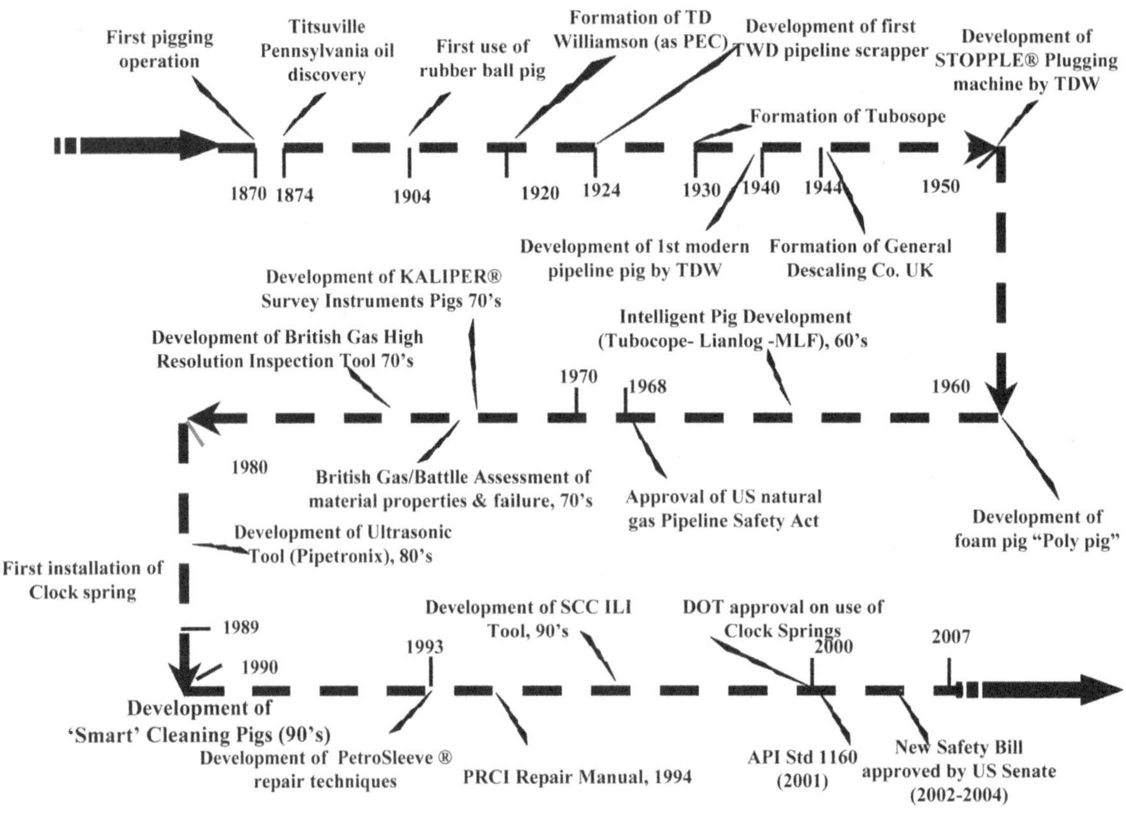

Figure 1-20. Historical milestones in inspection and rehabilitation techniques (Mohitpour et al. 2002)

development of high-resolution tools. This includes magnetic flux leakage (MFL), inertial, ultrasonic, and crack detection tools, etc. Equally and along with the development of pigging technology has been the development of repair techniques from simple grinding of defects or cutouts to modern sleeving techniques and the development of composite sleeves. Figure 1-20 represents some of the key historical developments and milestones in pipeline inspection technology and rehabilitation techniques.

The developments in pipeline integrity inspection and rehabilitation techniques are mostly a consequence of aging pipelines and safety-related issues and incidents that have been overshadowing the industry since the 1980s. While statistical incidents and fatalities as a consequence of pipeline operation are insignificant in relation to other forms of transportation mishaps and do not appear in any incident ranking statistical tabulation, they leave the severest impression and impact on public minds and governments.

Inspection Technologies

There are several techniques available to assess the integrity of pipelines. The most common of these are summarized as follows:

- Visual inspection
- Depth of cover survey
- External nondestructive testing (NDT)
  1. Radiography
  2. Magnetic particle testing
  3. Dye penetrant inspection
  4. Ultrasonic inspection
- CP monitoring
- Coating disbondment and damage survey
- Hydrostatic testing
- Geometry inline inspection (ILI) tools (low and high resolution)
  - o Caliper pig (low resolution)
  - o "Smart" cleaning pig (Short and Fletcher 1997)
  - o *x*, *y*, *z* geometry (inertial guidance) tool (Figure 1-21)
    Examples include:
    - ✓ High-resolution geometry tool (BJ GEOPIG™) for measurement of dent, ovality, and pipe movement, and
    - ✓ GPS position analysis

Figure 1-21. Low- and high-resolution geometry tools

- Metal loss ILI tools
  - o Conventional magnetic flux
  - o High-resolution MFL
  - o Transverse field inspection (TFI)
  - o Ultrasonic
- SCC tool

Use of high-resolution MFL (Figure 1-22A) involves induction of a magnetic field in the pipe wall. When the magnetic field travels through defects such as metal loss, the magnetic field is altered. This change in the magnetic field is detected by onboard sensors mounted in the tool. The data is downloaded and analyzed by computer algorithms and data analyst professionals. This technology, with modifications to the tool design, can also be used to find some types of cracks.

Unlike MFL tools, ultrasonics will require a liquid or a jell medium to be effective. In ultrasonic tools (UTs; Figure 1-22B), sensors from the equipment transmit an ultrasonic wave through the fluid (oil) and into the pipe wall. The UT signals are reflected off features within the pipe, and depending on the type of feature (i.e., weld, corrosion, crack, lamination), the reflected signal has different characteristics that are recorded by the tool. Computers and analysts assess the reflected signals and provide a final report regarding potential defects. UT technology is sensitive to cleanliness and speed of travel within the pipeline.

While each technique brings with it scope of applications, ILI and SCC tools are mostly used as inspection tools for performing various tasks including the following (Uzelac 2000):

- Metal loss
- Wall thickness measurement
- Cracking
- Geometry measurement
- Bend measurement
- Curvature monitoring
- Pipe movements and profiling

MFL tools are excellent in detecting metal loss within pipe wall, pipe property changes, and ferrous materials that are located adjacent to pipe wall. On the other hand, TFI tools reliably and accurately detect axial anomalies such as longitudinal seam weld cracks and corrosion (Yasinko et al. 2000).

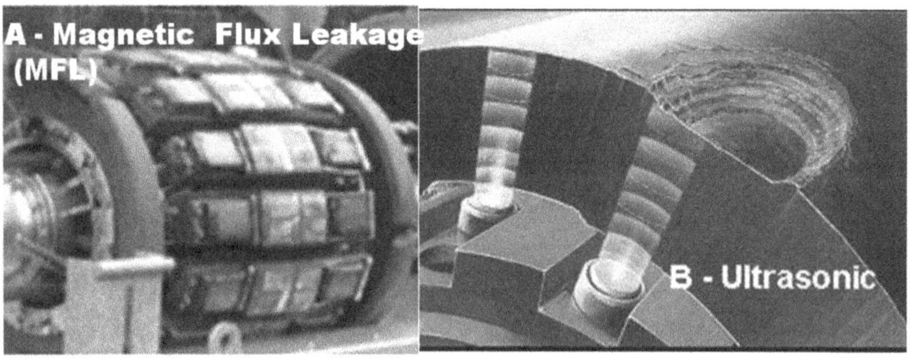

Figure 1-22. ILI tools: high-resolution MFL (A) and ultrasonic (B)

The development of ILI and SCC technologies have brought changing roles for pipeline inspection from a mere caliper/internal diameter measurement to detecting corrosion anomalies and cracking (Holden 1997).

Maximum benefits can be accrued from the utilization of each technique if the extent of the integrity problem is recognized and the scope of inspection to be performed is well defined. It is only then that ILI inspection can provide meaningful results in assessing the implication of defects if defect specifications and inspection requirements are matched accordingly.

There are also tools that are used for direct assessment of external corrosion defects [referred to as External Corrosion Direct Assessment (ECDA)] (NACE 2002). These tools are:

- Closed Interval Survey (CIS)
- Current–Voltage Gradient Survey [ACVG and Direct Current Voltage Gradient (DCVG)]
- Pearson's tool (electrical instrument and measurement in CP)
- Electromagnetic
- AC current attenuation survey

The tools can generally be used for various coating defect assessment, but cautions are given for selecting inspection tools not to be practical or reliable for some applications (NACE 2002). However, conditions for application of these tools include:

- Coating holidays
- Anodic zone on bare pipe
- Near river or water crossing
- Under frozen ground
- Stray currents
- Shield corrosion activity
- Adjacent metallic structures
- Near parallel pipelines
- High-voltage alternating current (HVAC) overhead electric transmission lines
- Shorted pipeline casings
- Under paved roads
- Uncased crossings
- Cased pipelines/piping
- At deep burial locations
- Wetlands
- Rocky terrain/rock ledges/rock backfill

Defects and Failures

Defects appear in the pipeline in many forms (Figure 1-23).

It is the operator's and the industry's standard that dictates, which defects need to be repaired and to what extent. Such defects are summarized below (Hewing 1999):

- Corrosion
  - ✓ General
  - ✓ Pitting
  - ✓ Erosion
- Gouge (possibly caused by mechanically induced damage)
- Weld defects
- Mechanical damage (dent with a gouge)
- Full rupture

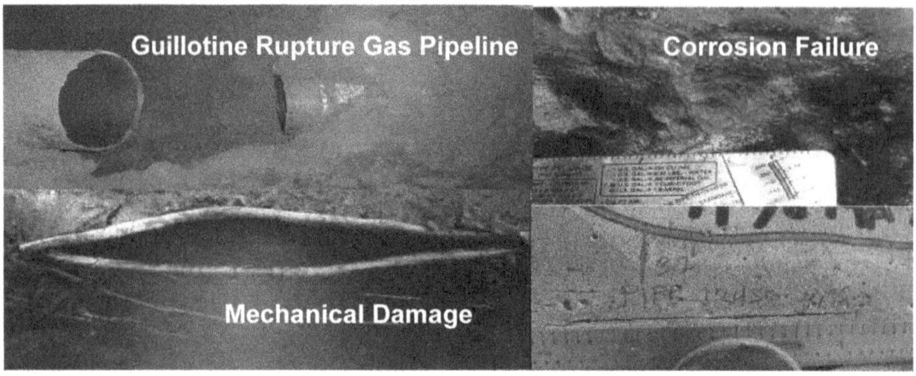

Figure 1-23. Some pipeline defects/failures

- Dents
  - ✓ Smooth (localized distortion)
  - ✓ Kinked (rapid change of contour)
- Cracks (planar two-dimensional defect with fracture)
- Lamination (manufacturing defect)
- Spalling (severe abrasion)
- SCC and hydrogen-induced cracking (HIC)
- Perforations (causing rupture/leaks)

The following describes some of the defects causing leading pipeline failures
Internal and External Corrosion. Corrosion is the disintegration of the pipeline material into its constituent atoms due to chemical reactions with its surrounding environment such as soil. It is the electrochemical oxidation of *metals* in reaction with an oxidant such as oxygen.

Unprotected pipelines, whether buried in the ground, exposed to the atmosphere, or submerged in water, are susceptible to corrosion (Figure 1-24). Without proper maintenance, every pipeline system will eventually deteriorate.

Corrosion can weaken the structural integrity of a pipeline and make it an unsafe vehicle for transporting potentially hazardous materials.

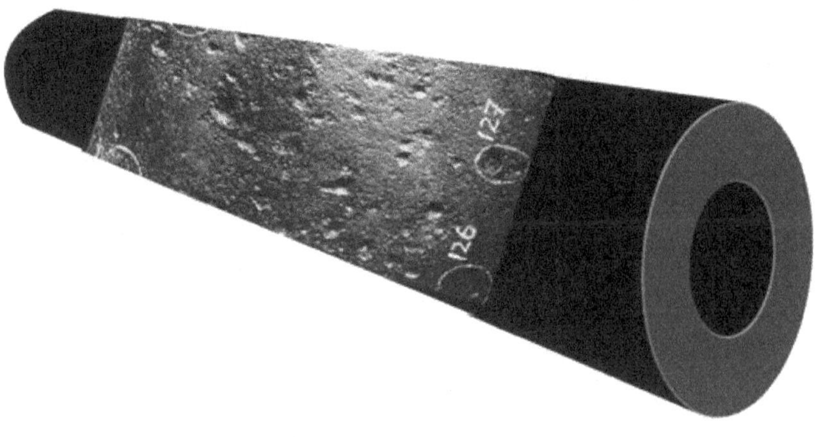

Figure 1-24. Externally corroded pipe

SCC. is a process involving the initiation of cracks and their propagation in metals, due to the combined action of tensile mechanical loading and a corrosive environment. It is thus a form of environmentally assisted cracking (EAC) that is of significance to oil and gas pipelines. SCC can develop when groundwater penetrates under or through the pipe coating, and given alignment with environmental, stress, and metallurgical conditions that are conducive to the formation of SCC, cracks can develop and grow to failure.

Indeed, it is the presence of tensile stresses that is dangerous, compressive stresses exerting a protective influence.

Two primary forms of SCC have been documented in oil and natural gas transmission pipelines, namely:

- high pH (range of 8.5 to 10.5) and
- near-neutral pH SCC (pH close to 7).

High pH cracks tend to develop in high-pH carbonate–bicarbonate solutions, and near-neutral pH cracking occurs in CO_2 solutions. These cracks are associated with a transgranular crack morphology (Figure 1-25).

High-pH SCC is a classical SCC, which was originally noted in gas transmission pipelines, more commonly found within 20 km downstream of the compressor station. High-pH SCC normally occurs in a relatively narrow cathodic potential range (−600 mV to −750 mV $Cu/CuSO_4$) in the presence of a carbonate/bicarbonate environment in a pH window from 9 to 13. Temperatures greater than 40°C are necessary for high-pH SCC susceptibility; growth rates decrease exponentially with temperature.

Intergranular cracking mode generally represents high-pH SCC. A thin oxide layer is formed in the concentrated carbonate–bicarbonate environment, which around the crack surfaces provides protection. However, due to changes in loading or cyclic loading, there can be crack tip strain resulting in breakage of oxide film. This results in crack extensions due to corrosion. Because of such a stringent environmental requirement for SCC initiation, this is not as prevalent as the near-neutral pH SCC. This type of SCC has been primarily noted in gas transmission lines.

Near-neutral pH SCC is a transgranular mode of cracking. The environment primarily responsible is diluted groundwater containing dissolved CO_2. The CO_2 originates (like in high pH) from the decay of organic matter. Cracking is further exacerbated by the presence of sulfate-reducing bacteria. This occurs primarily due to disbonded coatings, which shields the cathodic current that could reach the pipe surface. There is a free corrosion condition below the coating that results in an environment with a pH around 5 to 7.

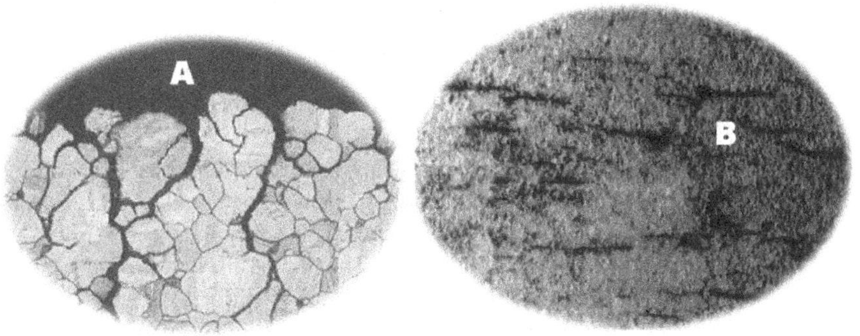

Figure 1-25. Typical high-pH SCC (A: intergranular SCC in a copper alloy and B: SCC in typical pipeline environment)

A cyclical load is critical for crack initiation and growth. There is field data that indicate that with a decreasing stress ratio, there is an increased propensity for cracking. Hydrogen is considered a key player in this SCC mechanism, where it reduces the cohesive strength at the crack tip.

There has been no correlation to a clear threshold for SCC initiation or growth. The morphology of the cracks is wide with evidence of substantial corrosion on the crack side wall.

A characteristic of both types of cracks is the presence of colonies of surface cracks in the body of the pipe that can link up to form larger more dominant cracks.

SCC frequently occurs in media that are little or nonaggressive toward the metal or alloy concerned in the absence of tensile loading (e.g., austenitic stainless steels in high-temperature water and steam). The associated weight losses are generally very small and even insignificant compared to the extent of the overall damage that can incur. This form of corrosion is of great practical importance and represents a permanent risk in numerous industrial installations, in terms of both the economic consequences and the safety considerations involved (personnel, equipment reliability, respect of the environment). There is no known category of commercial metals and alloys that is fully immune to SCC. Even materials such as glasses, plastics, and rubbers can also be prone to this type of attack in certain conditions.

It may be noted that the time necessary for a steel pipe or part to fail by SCC can vary from a few minutes to several years.

Manufacturing Defects. Manufacturing defects includes seam defects, hard spots, and laminations as well as defective pipe.

Hard Spots. Hard spots are usually developed during the hot rolling of a steel plate when an uncontrolled jet of water locally cools a portion of the plate too quickly. Hard spots are physically more brittle than the parent metal and therefore have less resistance toward cracking and crack propagation. Failures at hard spots are due to hydrogen stress cracking, which develops in areas of the pipe where all of the following three factors are present:

- A hard spot exists
- The pipe is exposed to sufficient atomic hydrogen
- The pipe operating stress is above 60% SMYS

Laminations. Pipe mid-wall laminations are a by-product of ingot steel production. Lamination is a metal separation within the pipe wall. Laminations are not uncommon in older pipelines and generally pose no integrity concerns unless they contribute to the formation of a blister. However, modern steel making practices use continuous casting, and these types of anomalies have essentially disappeared.

Pipe laminations, themselves, are unlikely to precipitate pipeline failure, as this is not a failure mechanism that is associated with historical failures as evidenced in industry incident reporting databases. The underlying reason for this is the fact that laminations lie within the plane of primary operating stresses.

Hydrogen blistering occurs when atomic hydrogen penetrates the pipe steel to a lamination and forms molar hydrogen, which cannot then diffuse through the steel. This sets up local stresses, and a continued buildup of hydrogen pressure can separate the layers of steel at the laminations, causing a visible bulging inside and outside the pipe, (Mohitpour et al. 2007). Hydrogen blistering at laminations is a potential cause of failures when there is a large presence of hydrogen (e.g., in sour gas or sour crude oil) pipeline service (Muhlbauer 2004).

Third Party/Mechanical Damage. Third party/mechanical damage is the result of random non-time-dependent events that are best addressed through the implementation of various preventative and mitigation (P&M) measures. It may be realized that many third-party damage incidents may lead to an immediate failure including a full rupture of a pipeline (Figure 1-26).

The third-party strikes that do not result in an immediate failure may go undetected until the damage is discovered through an integrity assessment of the pipeline. These third-party strikes

Figure 1-26. Typical third-party damage to a pipeline causing full-length rupture propagation in a gas pipeline

can cause dents, gouge in dents, and coating damage that could lead to external corrosion.

Incorrect Operations. The incorrect operations category applies to incorrect operations by operator personnel.

Threat of incorrect operations must be considered to be a potential threat to the operation of pipelines. This threat, however, can be effectively managed through the pipeline controls philosophy and the development of effective operating practices and the implementation of effective training programs.

Construction Damage: Construction damage usually includes dents and deformations (Figure 1-27) and other welding defects (Figure 1-28).

Weld defects depend on welding technique for joining pipes together (e.g., gas metal arc welding, shield metal arc welding, etc.). Such defects can include the following depending on the weld method:

- Cracks
- Misfire
- Bead misalignment
- Incomplete penetration and lack of weld root or side wall fusion
- Hollow bead
- Burn through
- Elongated and isolated slag inclusion
- Spherical porosity

Figure 1-27. Construction damage and dented/deformed pipe

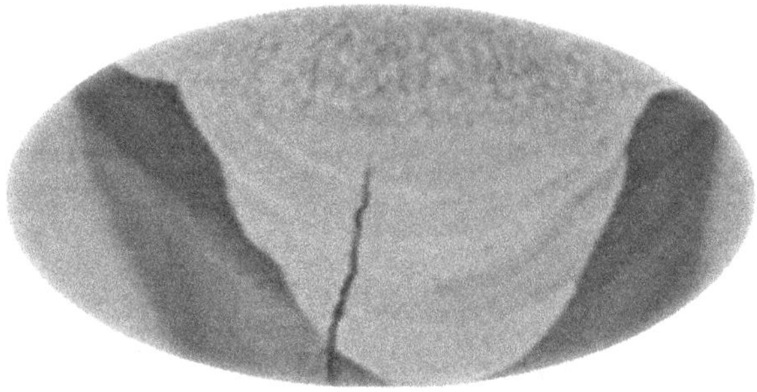

Figure 1-28. Typical weld defect (Hot crack)

- Internal and external undercut
- Arc burn
- Low cap and cap misalignment
- Shrinkage cavity

The industry's best practice for detecting dents in a pipeline (after construction) is through the use of ILI tools specifically designed to detect geometrical abnormalities.

Geotechnical, Weather, and Outside Forces. Geohazards in a pipeline can include many hazards but predominantly include hydrotechnical hazards (bank erosion, degradation, and encroachment), heaving and settlements, slope movements, landslides, and debris flows (Figure 1-29) (Esford et al. 2004).

The hazards presented to a pipeline are not directly related to the geotechnical condition encountered but are related to stresses induced by the geotechnical condition onto the pipeline. The amount of stresses transferred from a geotechnically induced condition is a function of many factors, which may include the pipe coating, the soil surrounding the pipeline, and the length of the pipeline affected by the geotechnical condition, pipe curvature, presence of other weights (such as buoyancy control weights, valves, etc.), and the rate of imposed movement.

Miscellaneous Equipment and Pipe Defects. Such failures are generally due to the following:

- Gasket/o-ring failure
- Stripped threads/broken pipe/coupling failure
- Control/relief equipment malfunction
- Seal/pump packing failure
- Wrinkle bend or buckle
- Miscellaneous

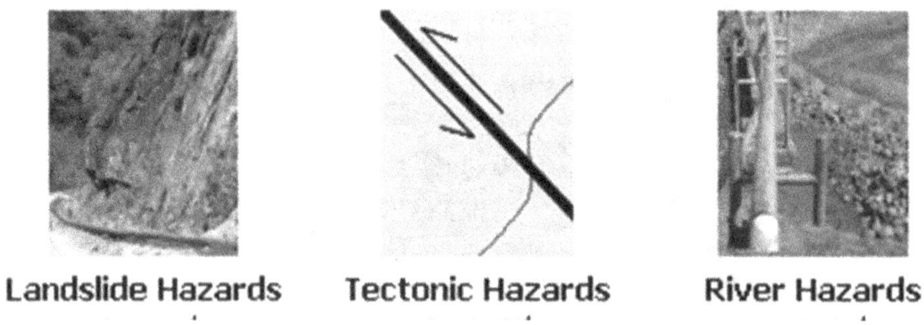

Figure 1-29. Typical geotechnical hazard along a pipeline route (Esford et al. 2004)

The miscellaneous failures include those occurring in pipe pumping/compression equipment and failure of pipe components (i.e., valves, flanges), plus other equipment such as tubing, gauges, etc. that resulted in a fluid release. Many of these failures occur inside pumping/compressor stations or processing plants.

DEFECT ASSESSMENT

Damage and corrosion may reduce the capability of a pipe to contain internal pressures, and if such damage is allowed to proceed, the pipeline may eventually leak or rupture. While corrosion or other damage in pipelines can be detected/observed by various methods, such defects (Figure 1-30) need to be assessed to determine pipeline capability for the intended operating pressure and stress level. A number of methods are available for such an assessment, some of which are expressed below:

1. *Corrosion defects:* A number of analytical techniques have been developed to determine if a defect will affect the pipeline's capability to operate at the maximum allowable operating pressure (MAOP). The best known method of assessing the remaining strength of corroded pipelines is that of the ANSI/American Society of Mechanical Engineers (ASME) B31G manual, first published in 1973.

The pipeline industry had used the "B31G" criterion to evaluate corroded pipe for removal or repair or for leaving it in service if the metal loss was within safe size limits as defined in the B31G criterion. A further approach to more accurately define the corroded area was developed known as the effective area method, which is probably better known by the term RSTRENG. The methods for assessment of corrosion defects are summarized in Table 1-4 below.

The table provides a summary of terms used in various equations of assessments and the type of corrosion shape. SMYS or SMTS refers to specified minimum yield (*Y*) tensile (*T*) strength. Folias factors considers the lipping effect (i.e., the expansion of pipe diameters during rupture), thus allowing for higher stress that is experienced by pipe at rupture location as a result of diameter enlargement.

Figure 1-30. Corrosion defects and assessment

TABLE 1-4. Summary of corrosion defect assessment methodologies

Method	Date Developed/ Published	Flow Strength Basis	Folias Factor	Area Shape
B31G	1970–1973	1.1 SMYS	2 terms	2/3 Ld Parabola
Modified B31G	1996	SMYS + 10 ksi	3 terms	Arbitrary 0.85Ld
NG 18	1981	SMYS + 10 ksi	3 terms	Defect area A/Ao; or rectangle Ld
RSTRENG 85% Area	1989	SMYS + 10 ksi	3 terms	0.85Ld
RSTRENG effective area	1990	SMYS + 10 ksi	3 terms	River bed profile iterative calculation
DnVRP-F101	1998/99	SMTS	2 term	Rectangle Ld
CorLas™ (Jaske et. al., 2002)	2002	SMYS + 10 ksi SMTS + (SMTS – SMYS)	J factor	Elliptical

Comparison of some of the methodologies for metal loss assessment as most used by the industry is summarized in Figure 1-31. The three methods compared in this figure allow a maximum defect depth of 80% of nominal wall thickness and predict failure stress based on an assumed flow stress (1.1 SMYS for B31G and SMYS plus 10 ksi for modified B31G) and the ratio of area of metal loss to original area with an applied geometry correction factor (Folias Bulging Factor). A defect is considered acceptable if the predicted failure stress level is greater than or equal to SMYS. Figure 1-32 presents a general description of the acceptable application of these methods (Baker 2003).

Figure 1-32 schematically illustrates the progression of defects with size: from "cracks," then to "grooves," then to "general or areal corrosion."

2. *Gouges:* Gouges arise as a result of a scraping action on the surface of the pipe and could be the result of the pipe having been dragged over a rock in the ditch, or by mechanical impact. The assessment of gouges originates with the assessment of part wall defects referred to as the NG18 surface flaw equations (Eiber et al. 1981).

3. *Weld defects:* The acceptability of flaws contained within gas or arc welded carbon and low-alloy steel pipe can be undertaken using the workmanship standards set out in several standard including API 1104 (API 2005).

REPAIR AND REHABILITATION TECHNIQUES

Throughout the history of the pipeline industry, pipeline repairs and rehabilitations have been traditionally conducted by simple grinding, mechanical sleeve, coating, or cutouts. Historical development includes the following:

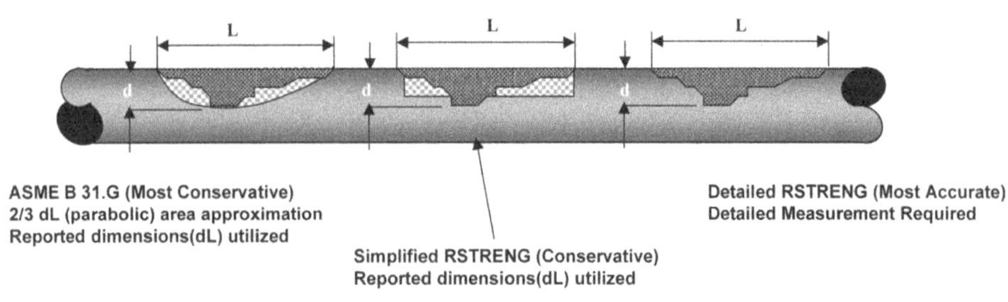

Figure 1-31. Comparison of major methodologies used for metal loss assessment

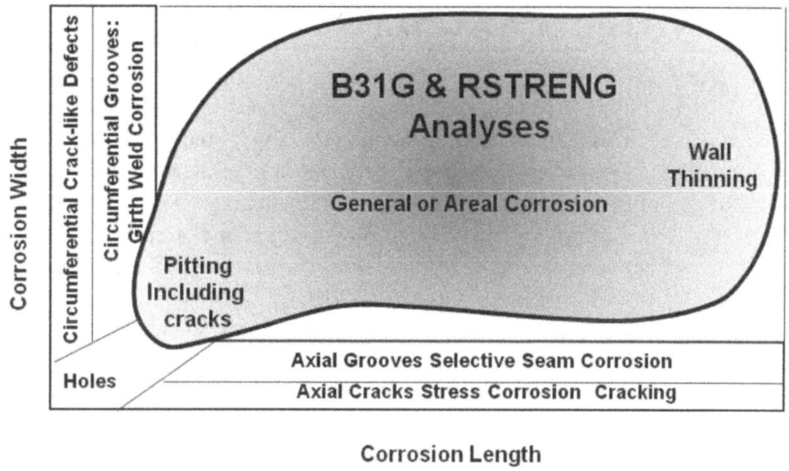

Figure 1-32. Applications area of B31G and RSTRENG (Baker 2003)

- Early years — Simple grinding
- 1970 and 1980 — Development in coating and composite material
- 1960–1980 — Increased R&D activities
- 1980s — Application of R&D result
- 1990s — Advancement in coating and composite repair techniques
- 2000 — Pipeline Research Council International (PRCI) Comprehensive Manual "P/L Repair Manual"
- 2000 — New Millennium Environmentally cost-effective alternatives for repairing damaged pipelines
- The rehabilitation techniques covering the traditional and newer alternative repair methods include the followings (Hewing 1999; Keifner et al.1994):
- Grinding repair
- Full encirclement split sleeve repair
- Pipe replacement or cut out repair
- Stopple and bypass
- Epoxy composition filled sleeve repairs
- Clock Spring® repair
- PetroSleeve® repair
- StrongBack composite wrap repair
- Armor Plate® pipe wrap repair

Relevant codes and standards for making repairs are summarized below:

Gas Pipelines	Oil Pipelines
o 49 Code of Federal Regulations (CFR) 192	o 49 CFR 195
o ASME B31.8	o ASME B31.8
o ASME B31.8S	o API 1160
o Canadian Standard Association [CSA] Z 662	o CSA Z 662
o PRCI Pipeline repair manual	o PRCI Pipeline repair manual
o Company operating procedures	o Company-operating procedures

PIPELINE INTEGRITY AND SAFETY

Pipeline operators manage the integrity, safety, and security of pipeline systems. It is this self-management and the technology development in continual condition monitoring of the pipeline that has led to improvement in pipeline structural integrity and operation and lessoning of operational accidents and generally lessening of injuries and fatalities globally. These improvements can be observed as an example from the US statistical information (Figures 1-33 and 1-34) (Pipeline Hazardous Material Safety Administration [PHMSA], Office of Pipeline Safety, 2005-2006).

Pipeline structural integrity in general is due to the following improvements:

- Pipeline protection systems (coating and CP facilities and monitoring)
- Corrosion protection design, detection, prediction and protection
- Inline inspection techniques
    - o Smart pigging
    - o Collapsible pigs
    - o ILI of unpiggable lines
    - o ILI mechanical damage detection
    - o Crack detection and SCC inspection
    - o Magneto-strictive sensor technology for nondestructive testing
    - o Real-time pigging operation monitory
    - o MFL technology
    - o Real-time inspection monitoring
    - o Crack detection technology
- Machinery heath monitoring and diagnostic
- Records/data management
    - o Advances in pipeline data management
    - o Computerized incident reporting and tracking systems
    - o Pipeline data management software
    - o Online data management systems
    - o Computerized data gathering and analysis
- Spatial analysis techniques/technology
- SCADA and cellular communications for remote operation and monitoring

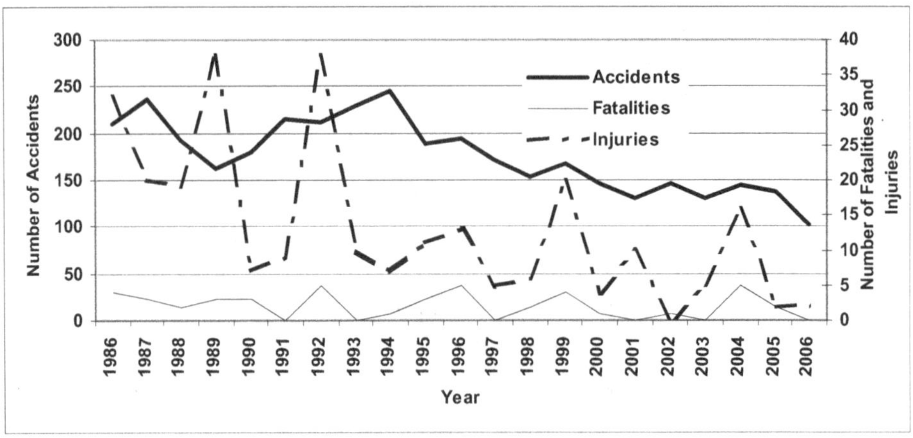

Figure 1-33. Hazardous liquid pipeline accident summary (PHMSA-OPS, 2005-2006)

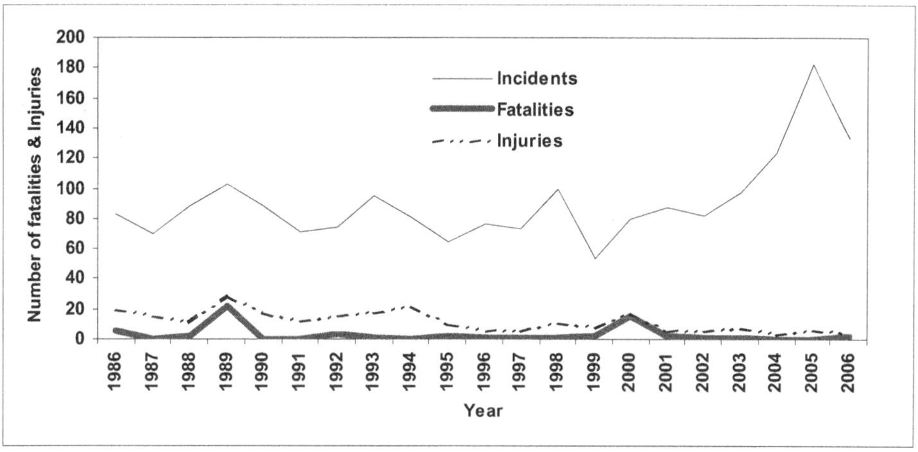

Figure 1-34. Natural gas incidents summary statistics (PHMSA-OPS, 2005)

- Pipeline security management
- Aerial-borne ROW encroachment monitoring
- Pipeline integrity condition/encroachment techniques

PIPELINE OPERATIONAL ISSUES/REGULATORY REQUIREMENTS

Operation of pipeline activities requires as a minimum taking into account all issues related to public and personnel safety protection of the environment, properties, and facilities security while transporting the products through the pipeline to the market.

Regulatory requirements and standards include the following

- United States DOT 49 CFR Parts 192 and 195
- National Energy Board of Canada
- British and European Regulatory
- Other international regulatory requirements
- ASME B31.4, B31.8, B31.8S, B31G, API 1160, National Association of Corrosion Engineers (NACE)

The operation of a modern pipeline system is governed trough a SCADA system, from a Pipeline Control Center (PCC). The operation is usually based on the hydrocarbon receipts and delivery schedules prepared and actualized timely by the pipeline-operating company.

Liquid pipeline companies usually rely on a computational pipeline monitoring (CPM) system installed for leak detection purposes for monitoring leaks/ruptures on a hazardous liquid transportation environment as per API 1130 (for operation, maintenance, testing, record keeping, and dispatcher training of the system).

Maintenance Issues

The ASME pipeline codes B31.4 and B31.8 contain extensive sections on operating and maintenance including detailed acceptance criteria for certain types of in-construction and in-service defects. The development of the criteria and appropriate validation owes much to the extensive research efforts sponsored by individual pipeline operators and industry consortia such as the American Gas Association (AGA), the Gas Research Institute (GRI), and the PRCI.

The ASME B31.8S standard sets out 22 possible threats to the integrity of a pipeline. The threats are either time dependent or time independent or stable hazards.

Time-dependent hazards include:

- External corrosion
- Internal corrosion
- SCC

Time-independent threats include:

- Damage caused by first, second, or third parties (instantaneous/immediate failure or delayed failure mode)
- Vandalism
- Incorrect operations
- Cold weather
- Lightning

Some of the stable threats that are time independent but are man-made or due to operational/construction/fabrication efforts include:

- Defective pipe seam
- Defective pipe
- Defective pipe girth weld and fabrication weld
- Wrinkle bend or buckle
- Stripped threads/broken pipe/coupling failure
- Gasket o-ring failure control/relief equipment malfunction
- Seal/pump packing failure

However, statistics show that the majority of threats are due to third-party damage or corrosion (Figure 1-35), as indicated in the analysis of recorded European and North American threats related to pipelines (Figure 1-36).

A similar trend may be noted when yearly breakdown comparison is made of various incidence facing transmission pipelines. An example is the yearly breakdown of failure

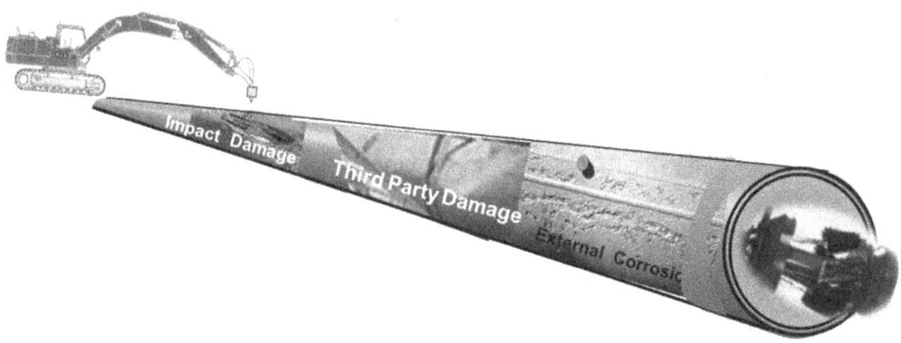

Figure 1-35. Major pipeline threats

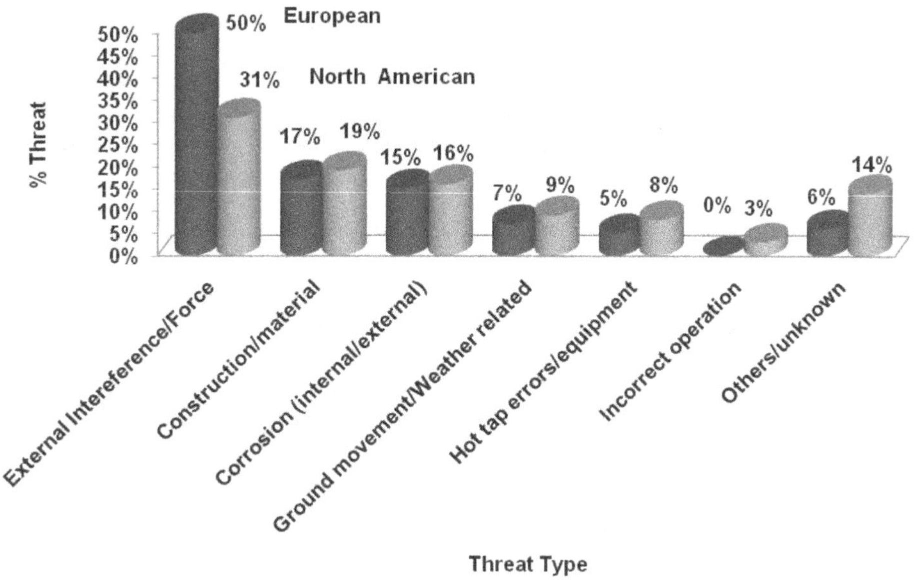

Figure 1-36. Comparison of onshore reportable threats associated with pipeline operation (European versus North American)

causes for Gas transmission pipelines in the USA (Figure 1-37) (HMSA 2007). As can be inferred from Figure 1-37, causes of pipeline Incidents are many and varied, but they do appear to be trending downward. Table 1-5 provides a summary of pipeline failure modes that are generally experienced by transmission pipelines.

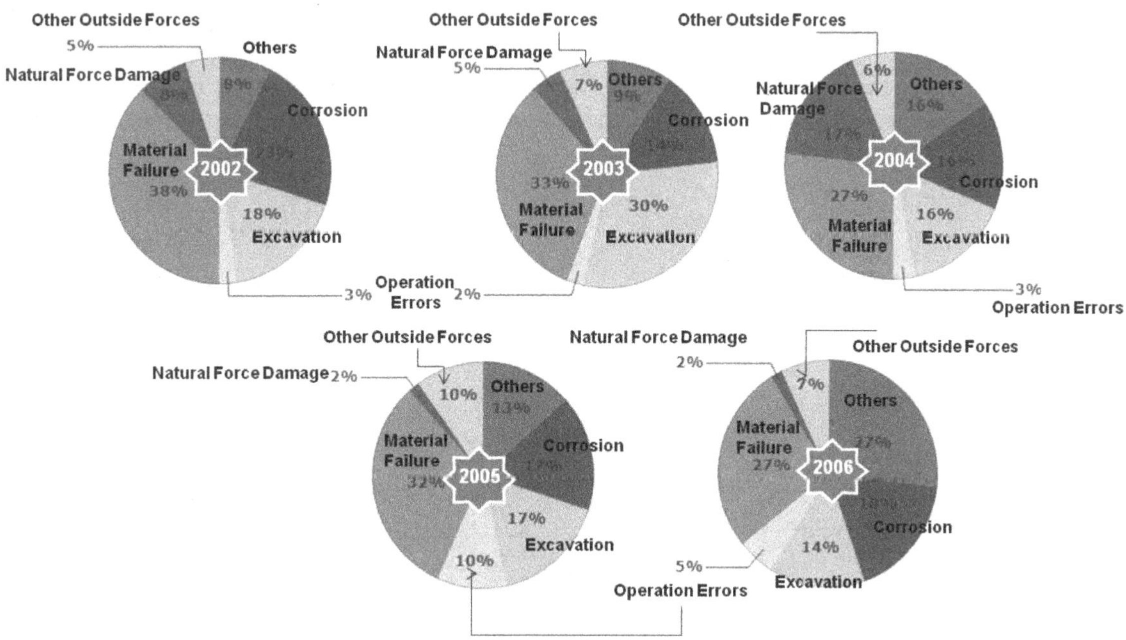

Figure 1-37. US gas transmission failures cause breakdown by year (HMSA 2007). Note: numbers are rounded up to nearest figures

TABLE 1-5. Summary of pipeline failure modes

Metal Loss	Cracking	Geotechnical Failure	External Damage	Material/ Construction Defect	Others
External corrosion	Hydrogen induced	Washout Erosion	Company First-party damage	Defective long seam weld	Fire
Internal corrosion	Stress corrosion	Slope movement	Contractor Second-party damage	Defective girth weld	Over pressuring
Gouging	Delayed cracking (mechanical damage)	Undermining Subsidence	Third-party damage	Defective spiral seam weld	SCADA malfunction
	Immediate cracking (mechanical damage)	Earthquake	Vandalism	Wrinkle Buckle	Lightning
	Corrosion fatigue		Terrorism	Defective pipe body	

Mechanical damage and cracks found in pipelines generally include weld cracks, SCC (Figure 1-38). Corrosion cracking found in pipelines include the following (Figure 1-39):

- Internal and external corrosion (both interacting and noninteracting)
- Internal/external pits
- Internal/external weld seam corrosion
- Internal preferential "trough" corrosion
- Weld cracks

PIPELINE INTEGRITY MANAGEMENT (PIM) PROGRAMS

Historically, pipeline integrity programs have been managed internally by individual pipeline companies and tailored to their unique systems. Historical development includes the following:

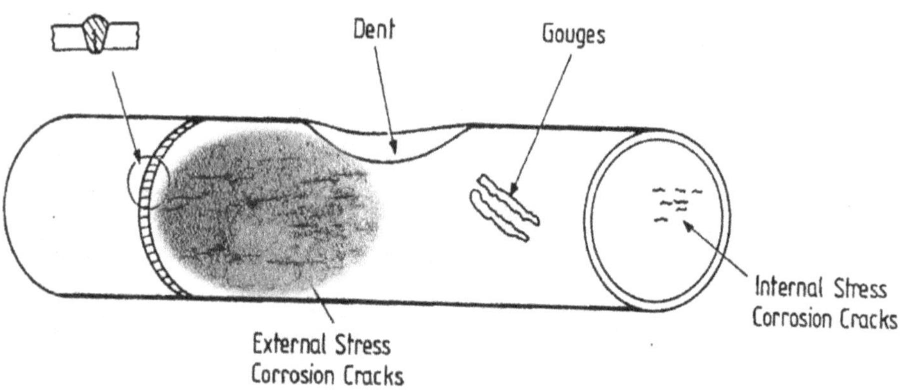

Figure 1-38. General summary of pipeline mechanical damage

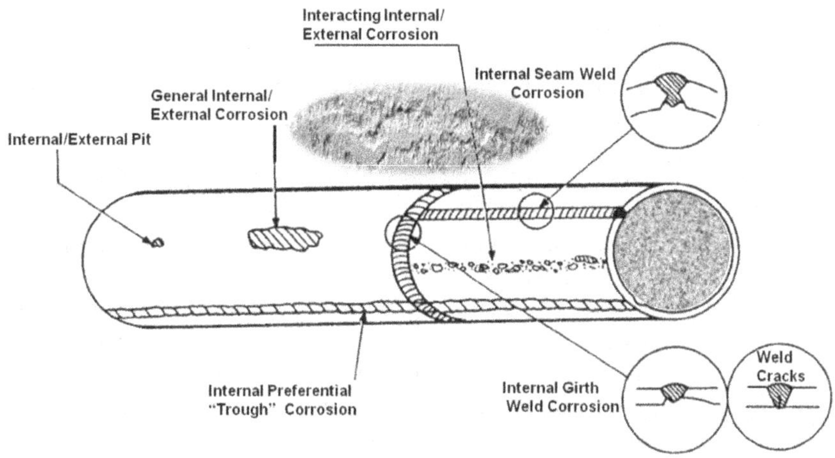

Figure 1-39. Typical pipeline corrosion and cracks

• Early years	Integrity Management Program (IMP) managed by individual companies
• 1980s to 1990s	More companies involved in IMP advancement of technology tools
• Late 1990s	DOT initiative on IMP
• November 2001	New API Standard 1160

Throughout the 1980s and 1990s, IMPs became more evolved in pipeline companies, and technology tools became more advanced in assisting companies with the various processes in their IMPs.

In the late 1990s and todate, the United States Department of Transportation (DOT) has taken initiative to ensure US liquid pipeline companies have IMPs in place that define the processes that a company utilizes for the purpose of improving pipeline safety and striving for incident-free operation.

API Standard 1160, Managing System Integrity for Hazardous Liquid Pipelines, First Edition, November 2001, was developed to assist operators to comply with newly established federal rules in the United States regarding pipeline integrity management in high-consequence areas (HCAs) (Jaske 2002).

API 1160 includes guidance for compliance with these US requirements, and it can be applied to other non-US-operated pipelines as well. The guiding principles used to establish API 1160 can be applicable to all pipeline systems and are as follows:

- Integrity is built in from the beginning
- Use trained people and defined processes to operate maintained facilities
- Integrity Management program must be flexible
- Integrate information
- Risk assessment is key and continuous
- Understand and utilize new technology
- Third-party evaluation of Integrity Management systems

An IMP, as defined by the US DOT (2007), is a documented set of policies, processes, and procedures that includes, at a minimum, addressing the following elements:

- a process for determining which pipeline segments could affect an HCA
- a Baseline Assessment Plan
- a process for continual integrity assessment and evaluation
- an analytical process that integrates all available information about pipeline integrity and the consequences of a failure
- repair criteria to address issues identified by the integrity assessment method and data analysis
- a process to identify and evaluate preventative and mitigative measures to protect HCAs
- methods to measure the IMP's effectiveness and
- a process for review of integrity assessment results and data analysis by a qualified individual

The key to any IMP is to establish a Baseline Assessment Plan with the following elements addressed:

- identify all pipeline segments that could affect HCAs,
- specify the integrity assessment method(s) for these segments (acceptable methods include internal inspection, hydrostatic pressure testing, or equivalent methodology)
- prioritize based on risk assessment
- provide a schedule for assessment and
- explain the technical basis for integrity assessment method(s) selection and risk factors used in scheduling the assessments

IMPs should be continual and also include methods to measure the program's effectiveness in assessing and evaluating integrity. The risks must be understood, managed, and controlled.

Risk assessment is an integrity management tool, and its purpose is to identify and quantify the risks associated with pipeline operation, such that remedial action can be performed in a timely manner. This is achieved through the ranking of potential risk to safety, environment, and operations.

Although the conducting of extensive risk assessments is somewhat new to the pipeline industry, other industries, such as the chemical plant industry, have had extensive practice in using risk assessments as a part of their management programs. Many of the practices used in assessing risk in other industries can be applied to the pipeline industry. Several risk assessment methods are used by industry, and the most common are failure probability methods and ranking systems. The most appropriate method depends on several factors, including system complexity, availability of historical data, and rigor required by the analysis. Pipeline integrity management decisions are made much easier by the risk assessment and prioritization process, which establishes a firm, documented basis for determining expenditures and schedules.

Technology is assisting pipeline companies to answer many complex questions with respect to safety, reliability, and integrity. It is evident that safety, reliability, and integrity will continue to be a top priority for pipeline companies and regulators worldwide for the years to come. We can be encouraged that our industry will continue to make significant progress in meeting the challenges faced with the development of new or with the management of aging infrastructure through our advancements in pipeline IMPs and technology.

Pipeline Industry Adopted PIM Program

PIM program adopted by the pipeline industry addresses the most relevant issues concerning prevention of potential risks affecting pipelines and facilities integrity, preserving health, and safety of communities and the protection of the environment along the pipeline right of way and from its facilities.

In order to implement and evaluate pipeline integrity and to assess its effectiveness, generally, pipeline companies put in place an IMP.

IMP is prepared in compliance with the following codes, standards, and regulations, taking into account laws and regulations concerning operation and environmental protection and applying all relevant parameters and requirements

- ASME B31.4 (1998), Pipeline Transportation Systems for Liquid Hydrocarbons and Other Fluids
- ASME B31.G (1991), Manual for Determining the Remaining Strength of Corroded Pipelines
- API 1160 "Managing System Integrity for Hazardous Liquid Pipelines"
- DOT Code of Federal Regulations (CFR) 49 Part 195 sections F and H
- Canadian National Energy Board (NEB) OPR
- CSA Z662 Oil and Gas Pipeline System standard

As well in preparation of IMP, standard, guideline and evaluations methods from the following organizations are also consulted:

- NACE International — The Corrosion Society (NACE)
  - o Corrosion control, CP, specific document consulted is NACE RP-0169 "Recommended Practice: Control of External Corrosion on Underground or Submerged Metallic Piping Systems"
- AP
  - o Tanks, line pipe, inspection, specific standards consulted are:
    - API 1104 "Welding of Pipelines and Related Facilities"
    - API 1107 "Pipeline Maintenance Welding Practices"
- PRCI
  - o Corrosion evaluations (RSTRENG) PRCI PR-03-805 "Modified Criterion for Evaluating the Remaining Strength of Corroded Pipe"
- ASTM International
  - o Materials, testing
- NEB of Canada — MH-2-95 "Stress Corrosion Cracking on Canadian Oil and Gas Pipelines"
- DNV OS-F101 "Corroded Pipe"
- ISA "Standards and Practices for Instrumentation"

Pipeline companies organize IMP document into four major framework components:

- Objectives
- Plans and programs
- Procedures and specifications and
- Management systems

The IMP also addresses prevention, monitoring, mitigation, and trending/analysis as subprograms.

IMP usually includes performance metrics to measure its effectiveness and identifying any requirement for improvements.

Typical industry's IMP manual generally contains the following:

1. Introduction and PIM Plan Framework
2. HCA Identification Plan
3. Risk Assessment Plan
4. Integrity Assessment Plan
5. Pipeline Repair and Remediation Plan
6. Integrity Reassessment Plan
7. Preventative and Mitigative Measures Plan
8. Management of Change Plan
9. Performance Measure Plan
10. Conformity Direct Assessment Plan
11. Communications Plan
12. Quality Assurance Plan
13. Documentation Plan
14. IMP Review and Revision Plan
15. Glossary of Terms

APPENDICES typically Included are:

- Assessment Guidelines, Procedures, Reports, and Forms
- Pipeline Inspection Guidelines
- Station Piping Inspection Guidelines
- Internal Inspection, For mainline
- ILI Data Prioritization
- Excavation Guidelines
- Site Observation Report
- Excavation Summary Report
- Coating Inspection Report
- Assessment of Localized Corrosion Procedure
- Assessment of General Corrosion Procedure
- Nondestructive Testing Report
- Excavation Inspection and Repair Report
- Assessment/Reassessment Schedules

Typical IMP contents are described below:

Introduction

This gives an introduction to the system and the reason for pipeline IMP program. The introduction describes the basis of the IMP and discusses standards and regulations regarding integrity management. Also, the introduction will include a discussion on human and financial resource considerations, regarding IMP implementation and administration.

HCA Identification

This provides an outline of considerations regarding the identification of HCAs. It also provides methods for defining HCAs, geo-referencing data, calculating dispersion distances (applicable to both liquid and gas lines), HCA documentation, managing the identification of new HCAs, and verification and removal of HCAs.

Risk Assessment Plan

Risk of Failure = Likelihood of Failure × Consequence of Failure

A risk management plan identifies these variables. The results of this analysis typically support the development of an assessment plan, risk prioritization, resource allocation, and other risk management objectives.

The risk management plan describes elements of a Risk Assessment Plan and considerations that would be made when developing and implementing it. It describes various risk methodologies and direct user in selecting one that could be best used to assess the threats to its pipelines and consequences of failure. It also describes processes for identifying the data sources, a data collection process, a quality assurance program, validation of the results of the risk assessment, and documentation requirements for risk assessment plan.

Integrity Assessment Plan

This describes the processes and considerations that may be made regarding the development and implementation of such a plan.

The consideration may generally include the development/formalization of the following procedures:

- Examination of aboveground pipelines
- Examination of buried pipelines
- Assessment of crack-like indications
- Assessment of clusters of crack-like indications
- Assessment of longitudinal weld seams
- Assessment of girth welds
- Assessment of arc burns
- Prioritization of defects for excavation and evaluation
- Placement of aboveground markers
- Launch and receipt of ILI tools
- Validation of ILI results
- Establishment of a standard ILI specification report
- Establishment of ILI tool selection criteria

A comprehensive ILI tool selection criteria for selecting each of the many different types of ILI tools available including MFL, ultrasonic, and geo pigs is generally referred to, and any of the following hydrostatic test procedures are also covered:

- General requirements and specifications for pressure testing, e.g., certification of stock pipe
- Test section plan and profile
- Test log, yield test data
- Test section data sheet and log
- Test failure log

This also covers planning tool for integrity management based on engineering assessments.

Pipeline Repair and Remediation Plan

Pipeline assessment utilizing inline inspection tools or other equivalent technologies may likely generate lists of pipeline conditions that have the potential to affect HCAs. Many conditions must be monitored in order to help ensure protection of the public and environment. Also, this plan specifies which pipeline conditions must be repaired defining a timeframe for such activities.

This plan, therefore provides considerations that may be made when developing and implementing a repair and remediation plan. Having a formalized program with procedures will allow personnel to take more efficient decisions regarding the evaluation and repair of pipeline defects.

This may also include the following procedures/reports:

- Application of relevant pipeline repair techniques
- Defect removal by grinding
- Full encirclement sleeve repair
- Cutout repair
- Operating pressure limitations during excavation and repair
- Standard excavation, inspection, and repair report
- Standard nondestructive examination (NDE) report
- Establish documentation requirements for pipeline repairs and remediation

Integrity Reassessment Plan

This plan provides consideration that is generally made when developing and implementing a reassessment plan. Integrity evaluations are periodically required on the pipeline. Pipeline segments may be re-ranked for assessment based on the integrity evaluation. The re-ranking will determine if a re-assessment is warranted at that time. Data gathered as a result of integrity activities needs to be trended and analyzed in order to improve the effectiveness of the assessment plans. Key trends will provide guidance for plan development and will assist in the scheduling of integrity management activities.

Depending on information available from ILI, it may be possible to develop and implement a formal reassessment plan based on risk analysis, which includes preestablished corrosion and crack growth rates. The plan would include establishment of reassessment intervals based on the integration of identified threats, results from the last integrity assessment, prioritization of data and results from the latest risk assessment, and documentation requirements.

Preventative and Mitigative Measures Plan

This plan generally discusses considerations that maybe made when developing and implementing a preventative and mitigative measures plan. The plan typically is designed to mitigate risk by reducing the likelihood of failure and/or consequence of failure.

It also helps with identifying preventative and mitigative measures, testing, and analyzing possible scenarios, establishing an interval for conducting preventative and mitigative analysis, defining cost/benefit for each measure, and then perform the cost-benefit analysis, establishing documentation requirements for the preventative and mitigative plan, and developing a depth-of-cover survey procedure for river/lake crossings.

Management of Change Plan

This plan discusses considerations that may be made when developing and implementing a management of change plan. The plan identifies and considers the impact of changes on the pipeline systems and their integrity. A management of change process includes the reason for the change, the authority for approving changes, analysis of the implications of the change, acquisition of required permits, communication of the change, time limitations, and qualification of staff.

The plan may include applicability; minimum information requirements, management of change process flow diagram, and documentation requirements.

Performance Measures Plan

This plan describes elements of a performance measures plan and considerations that may be made when developing and implementing a performance measures plan. The plan evaluates IMP performance and helps asking the following questions:

- Were IMP objectives accomplished?
- Were pipeline integrity and safety effectively improved through the IMP?

The plan might include establishment of attributes and a performance metrics and establishment of an interval for evaluating the effectiveness of the metrics in measuring the performance and documentation requirements.

Confirmatory Direct Assessment (CDA)

This usually discusses elements of a CDA plan and considerations that should be made when developing and implementing a plan.

Communications Plan

This plan is an important part of IMP and generally describes elements of a communications plan and considerations that may be made when developing and implementing this plan. The communications plan effectively keeps the appropriate company personnel, jurisdictional authorities, regulators, and the public informed about the company's efforts to help ensure the safety of the public and protection of the environment by its IMP activities.

The plan may include the following:

- Identification of forms and types of communication
- Identification of persons/groups needing to be involved in communication
- Establishment of a process for communicating implementation of the IMP and changes to the IMP
- Establishment of documentation requirements regarding communication
- Development of a Public Awareness Program
- Development of a Web-based version of the IMP and placement of it on an intranet Web site to effectively communicate to key employees the company's policies and procedures on the integrity management plan

Quality Assurance Plan

Elements of a Quality Assurance Plan and considerations that should be made when developing and implementing a plan are typically described here. The plan may provide a documented proof that a company is meeting the requirements of its IMP. Requirements include documentation, implementation, and maintenance that the company follows.

The plan may include:

- Establishment of qualification requirements for office personnel
- Establishment of qualification requirements for field personnel

- Establishment of training requirements for personnel
- Establishment of internal and external audit protocols to ensure that the IMP is delivering the desired results
- Development of master service agreements (MSAs)
- Development of an Operator Qualification Plan

Documentation Plan

The elements of a Documentation Plan and considerations that should be made when developing and implementing a plan are discussed here. The plan compiles a list of the documentation required to fully implement the IMP, including type of documentation required, the location(s) of the records, responsibility for completion of documentation, and frequency for updates (if required).

The Plan might include:

- Establishment of a list of documentation needed to fully implement an IMP
- Establishment of documentation requirements
- Locations of records, responsibilities for completion of documentation
- Frequency of updates

Review and Revision Plan

Review and revision plan and considerations that should be made when developing and implementing a plan are discussed in this section. The review and revision process is ongoing. Policies and procedures require continuous maintenance or they begin to deteriorate. It is important to notify manual holders of revisions in a clear and timely way. Similarly, manual holders have obligations in the revision process to (1) acknowledge that they have received the revisions and updated their manual and (2) to communicate the revisions to applicable manual users.

The plan may include:

- Identification of the persons to be involved in the review process
- Establishment of a review period, e.g., annually, semiannually, etc.
- Establishment of a revision record
- Establishment of a revision request form
- Establishment of a procedure for requesting a revision to the IMP, reviewing the request, and updating the IMP

Glossary of Terms

This may contain abbreviations and acronyms, terms and definitions, references, cautions, warnings, and other notes.

APPENDIX Guidelines, Procedures, and Forms

Industry-type IMP Appendix may contain the following guidelines, procedures, and forms regarding the assessment and prioritization of pipeline anomalies:

- Pipeline Inspection Guidelines
- Station Piping Inspection Guidelines
- Engineering Standard Internal Inspection, Mainline

- ILI Data Prioritization
- Excavation Guidelines
- Site Observation Report
- Excavation Summary Report
- Coating Inspection Form
- Assessment of Localized Corrosion Procedure
- Assessment of General Corrosion Procedure
- Nondestructive Testing Report
- Excavation Inspection and Repair Report

PIPELINE RISK ASSESSMENT AND SECURITY

Risk Assessment and Management

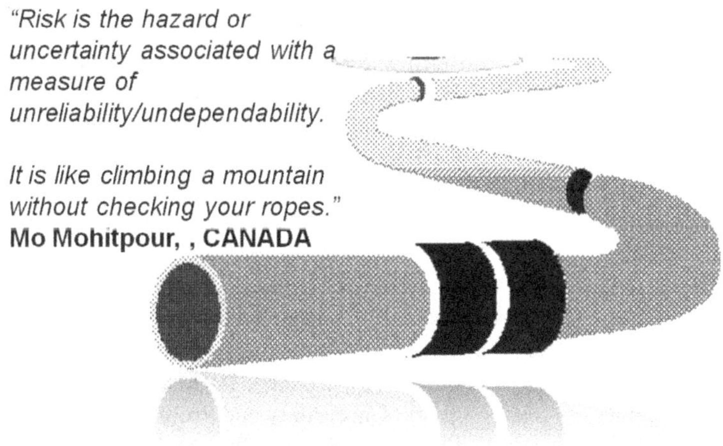

Going to work on foot, by car, or bicycle would render taking a risk. Traveling by bus, train, or plane will have associated risks. Depositing money in a bank, or in stocks, or piggy bank, under a mattress, will render other types of risks. Buying lottery ticket at the newsstand or gambling at a casino both hoping for a win engages one in activities that involve an element of chance — something intimately connected with risk (Vesper 2006).

The introductory remark, however, says it all.

Risk assessment has a long history in many industries and field including:

- Petrochemical
- Transportation including hydrocarbon and pipelines
- Nuclear
- Military
- Aviation and space
- Dam building and civil
- Mining
- Business and investment
- Health and safety

Ultimately, the scope and intent of a risk assessment depend on the needs of the user.

Each has applied various formal risk assessment (and ultimately risk management) techniques over the years and particularly over the past four decades (Ferguson 2009). All have a proactive approach to improving risks, as opposed to a reactive "fix-it-when-it-breaks" approach.

Historical Milestones/Chronology

Historically, risk assessment dates back about 3500 BC and is related to games of chance and gambling as depicted in Egyptian tomb paintings. It was not until the Renaissance (1450 to 600 AD), however, that a "scientific" or statistical basis for gambling was presented and thus a formal risk assessment method devised. The following provides a historical record of events that have lead to the risk assessment and management as we see it today as an inherent part of our pipeline business, most of which have been triggered by major loss events or public disasters (U of T 2002-PDA 2006):

600 BC	Book of "Game of Chance"
1630s	Galileo's essay "On Playing Dice"
1654	Creation of probability theory (Blaise Pascal and Pierre de Fermat)
1697	"Probability Theory — Normal Distribution," Abraham de Moivre
1687	Formation of "Lloyds of London"
1700 to 1782	"Systematic Process for Making Choices and Decision," Daniel Bernoulli
1800	Recognition of Voyage Insurance for travels by ship
1812	"Analytical Theory of Probability," Pierre Simon de Laplace
1838	1st US, Law Regulating "Steam Engine Inspection"
1852	US "Steamboat Inspection Act"
1881	1st notion of Workers' Compensation," by Bismark in Germany
1863	Recognition of the word "Risk" or "Risque" as element of chance
1905 to 1912	Workers' Compensation Laws, USA
1920	Formation of Tanker Insurance Company, Ltd. By BP
1921	"Risk, Uncertainty and Profit" by Frank Knight
1922	Probability Treatise by J.M. Keynes
1926	Theory of Games and Strategy by John von Neumann University of Göttingen
1930s	Formalized Approach to Risk Assessment in Engineering and Construction Industries, a framework for analysis of decision making (Ramsey, 1931 and Di Finetti, 1937)
1950s	Development of Chemical Risk Assessment, The Thalidomide disaster
1952	1st Demonstration of Unacceptable Risk Strategy," Harry Markowitz
1954 to 1960	"Risk Management Spectrum," 1st use of "Threshold Value^R (TLV ^{R})
1956	"Associating Risk Management and Cost Control," The *Harvard Business Review*
1962	Notion of "cost-of-risk" by D. Barlow, Massey Ferguson Toronto
1965	The Corvair unmasked! Ralph Nader's *Unsafe at Any Speed*
1966	Formation of "The Insurance Institute of America" and the designation "Associate in Risk Management"
1970	US Congress established Environmental Protection Agency, EPA
1971	Creation of Environment Canada
1972	First conference on risk analysis, US, the National Academy of Engineering and the National Academy of Sciences
1974	"Creation of Risk Management Circles," Gustav Hamilton, Sweden
1974	Formation of US Coast Guard

1975	"1st Use of Probabilistic Risk Analysis (PRS)" by US Regulatory Commission
1979	Three Mile Island Incident
1979	US Interagency Government Guidelines "Identifying and Assessing Chemical Risks"
1980	US Legislation directing EPA to clean up Hazardous Waste Sites
1980	Formation of "The Society for Risk Analysis" in Washington
1983	The Union Carbide Fertilizer Plant Bhopal disaster
1983	Establishment of the four steps paradigm for risk assessment (hazard identification, hazard characterization, exposure assessment, and risk characterization).
1986	Chernobyl Nuclear Disaster
1986	Formation of The Institute for Risk Management in London, UK
1989	The Exxon Valdez Oil Tanker Spill disaster
1990	UN Report "Natural Disaster Management," outlining nature of hazards, social and community vulnerability, risk assessment, forecasting, emergency management, prevention, science, communication, politics, financial investment, partnerships, and the challenge for the 21st Century.
1993	Fist use of the title "Chief Risk Officer," James Lam, at GE Capital
1995	Publication of 1st risk management Standard AS/NZS 4360:1995 (since revised in 1999)
1996	The Global Association of Risk Professionals
1998	First ISO Standard 14971 "Application of Risk Assessment and Risk Management" applied to medical devices
1998	First association of Risk and Mathematical Numbering Systems
2000	Y2K Issue
2001	September 11 Terrorism
2001 to 2002	2001 US law Enactment "Liquid Integrity Management Program (LIMP)," Oversight by PHMSA (Pipeline Hazardous Material Safety Authority)
2003 to 2004	US law "Transmission Integrity Management Program (TIMP)"
2004	Australia Standards *AS/NZS 4360*:2004 — *Risk Management*
2005	Integration of formal risk management into quality systems — and consequence modeling, Coburn
2008	US "Distribution Integrity Management Program (DIMP)"

Pipeline Risk Assessment

The risk associated with the pipeline, in terms of the safety of people, damage to the environment, and loss of income, depends on the expected failure frequency and the associated consequence. Such a risk is directly related to the type of fluids transported and the sensitivity of locations of the pipeline. In this context, pipeline failures are defined as loss of containment.

The consequences of the risks associated with accidents and emergencies that may occur in a pipeline transmission system can be most serious and alarming because, in general, they influence on personnel, public, and environment. It is therefore important to have in place an effective risk management plan, from assessment to mitigation, which will require a reliable analysis and prediction tool as well as response.

For pipeline industry, risk management is the set of actions taken to control risk. It implies the process for assessing this risk through execution of an action plan to control and reduce future risks.

Pipeline industry, which considers the enterprise-wide risk management (Meier 2000), generally focuses on pipeline operational risks (Figure 1-40) to assure continued safe and reliable operation benefiting the industry, specifically the owning company, its personnel, and the public with least impact on the environment.

Risk assessment and management is a decision support tool providing a structured approach, which identifies and documents possible cost-effective actions to improve facilities integrity and thus public safety and environmental protection. It is a decision framework that captures existing pipeline system information, its condition, and where funds could be effectively allocated to improve system reliability. It is a potential alternative to prescriptive regulations and, thus, is an "evolving" or "living" process for the reasons identified below:

- To address public safety and environmental concerns while managing expectations of greater efficiencies and cost control
- Rapid pace of change — past does not always predict the future
- Increasingly complex web of interrelated factors
- Increasingly high stakes with incidents
- Change from being reactive to proactive — concentrating on incident prevention and reduction in incident frequency
- To increase consistency and level of objectiveness in overall pipeline risk assessments
- To adjust and focus operation, maintenance, and rehabilitation activities on the risks inherent to pipeline operations
- To effectively communicate and formally demonstrate that risks are systematically identified, analyzed, and managed

Risk Assessment for Pipeline Transmission System

There are number of methods that are used in the industry to assess risks associated with pipeline operation and the enterprise in general. Two methods that are typically used are "decision matrix or qualitative" and "probabilistic or quantitative" models (D'Atri et al. 2008).

Figure 1-40. Interrelationship of enterprise-wide risk management

The decision matrix or qualitative model provides an evaluation of a qualitative (rather than quantitative) approach to risk assessment. It relates each pipeline risk according to the likelihood and the potential consequences of an event by a simple scale, such as high, medium, or low. This method focuses on the most crucial threats (Muhlbauer 2004). This qualitative approach, of combining probability and consequences, often uses numerical scoring methods to generate a relative risk ranking of various pipeline segments, of various lengths, along a pipeline route.

The probabilistic or Probabilistic Risk Assessment/Quantitative Risk Assessment (PRA/QRA) uses the fault tree analysis and techniques of the events and failures causing the event (Mohitpour et al. 2005). In line with the failure definition, the top event for the fault tree can be defined as "a loss of service in the pipeline system." The analysis then concentrates on identifying and developing the various contributing events and causes, which could lead to such a failure. The analysis considers all modes of failure, which could lead to a service outage. Thus, the mechanism starts on event that causes a loss in the system, then the tree goes forward in search of all possible consequences and subsequently backward, in search of all possible causes or initiators of the events. All possible paths are quantified based on the probability of each branch of the tree and identifies the initiator of the event possible damage (D'Atri et al. 2008).

The indexed method, or a probabilistic approach to risk assessments, compares risks associated with section of pipe to the other sections. For each pipe segment, risk is calculated as a product of the probability by the consequence of failure. However, for both methods, the probability of failure is calculated as an algebraic sum of the threats. Consequence of failure is calculated as the algebraic sum of the consequences. The relative weights of each variable in the algebraic sums represent the relative importance of each in contributing to total risk. This is algebraically defined in the following equations:

$$R = P \times C$$
$$P = \Sigma_{i=1}^{n} a_i A_i$$
$$C = \Sigma_{i=1}^{n} b_i B_i$$

where

R = risk
P = probability of failure
C = consequence of failure
A_i = threat
a_i = relative weight of threat i
B_i = consequence
b_i = relative weight of consequence i
n = number, consequence of failures

The index n, in turn, depends on a number of conditions, including the variables that influence the threat or consequence. For example, the threat of corrosion is dependent on soil type, the type of pipe coating, the efficiency of CP, etc. The model is completed by assigning values to each of these variables depending on the properties of each segment of the pipe. For example, Figure 1-41 shows the six threats (among many threats as identified under this figure) that contribute to the probability of failure and the six effects contributing to the consequence of failure.

Other factors to be considered are the maximum failure probability and the target reliability in a pipeline system:

Maximum failure probability = (tolerable risk failure consequences) and
Target reliability = 1 – (maximum failure probability)

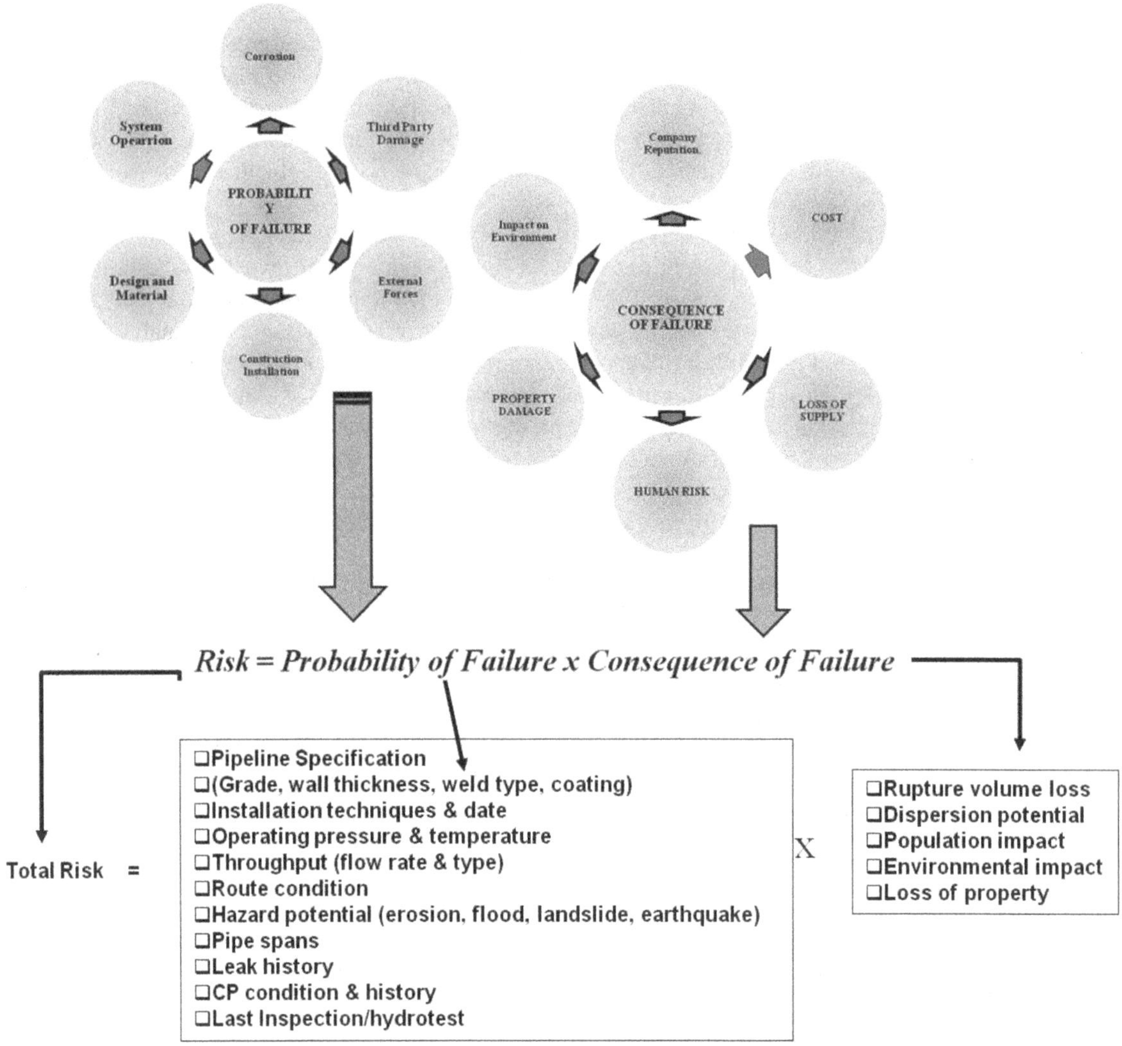

Figure 1-41. General scheme of threats and consequences, with examples affecting total risk

Tolerable risk is based on risk tolerance of general public; however, in general, tolerable risk is based on current safety performance of pipelines industry, which consider:

- Is safety associated with current codes widely accepted?
- Is average risk associated with current codes tolerable?
- the suggested approach
- define reliability targets to ensure *average* risk for all pipelines same or lower compared to current code
- same risk level for all pipelines

The general steps in a pipeline risk analysis are as follows:

1. Data gathering — identifying the location and characteristics of the pipeline and pipeline site of interest
2. Hazard identification — characterization of the pipeline route in sufficient detail to formulate potential accident scenarios and to permit subsequent evaluation of

accident probability, likely release amount, and nature and magnitude of resulting impacts

3. Probability analysis — probability analysis determining the likelihood of an event, expressed in relative (typically referred to as likelihood) or quantitative terms (typically referred to as probability)
4. Consequence analysis — examining the potential physical impacts and derivative consequences (e.g., harm to people or the environment) of a pipeline failure and accidental release of product
5. Risk calculations — generating a numerical combination of both the probability of an event and its consequences
6. Risk control — risk control consists of prevention and mitigation measures respectively to reduce the probability that a release of pipeline product will occur and to minimize the impacts of any release that might occur
7. Risk monitoring — consisting of a method monitoring the effectiveness of the risk assessment and management

Pipeline Security

Comparison of safety issues related to different modes of transportation (Figure 1-42) will indicate that pipelines are the safest means of transportation.

However, pipeline facilities, like any physical asset, are vulnerable to vandalism and terrorist attack. The means of attack may be physical, in the form of mechanical damage to equipment or the pipeline, causing disruption of flow or loss of containment or alternatively "cyber attack" on computer facilities SCADA system resulting in shut downs or malfunctions. It is possible for both forms of attack to be coordinated so as to occur simultaneously.

Some examples of sabotage against oil and gas pipeline as reported by press in the recent years are listed below:

1986 to present	Colombia, Cano Limon pipeline bombed over 950 separate occasions since 1986, disrupting flows for months at a time and costing an estimated \$2.5 billion in lost revenue

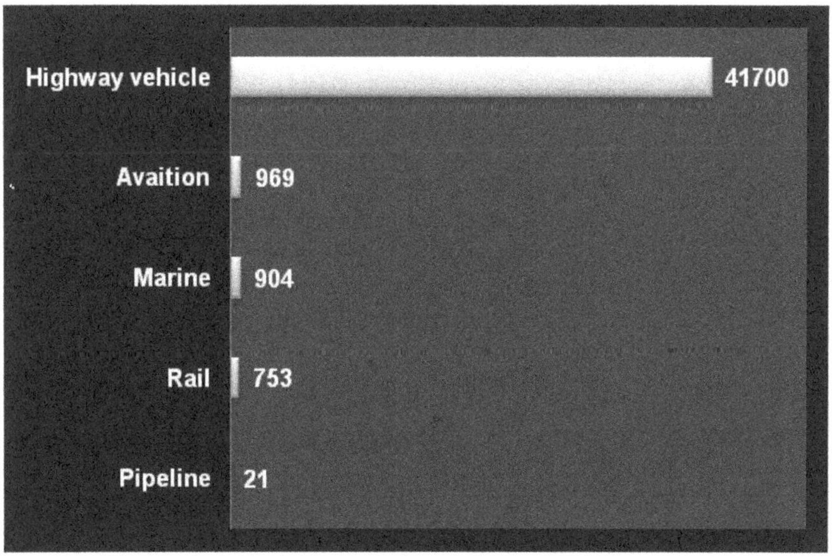

Figure 1-42. Pipeline safety comparison with other form of transport (Allegro)

October 1998	Nigeria, Niger Delta in Nigeria, scavenging gasoline leading to pipeline explosion, 1200 killed
June 2003	Baghdad, Iraq, sabotage and vandalism on 1000-km crude oil line from northern fields near Kirkuk to Turkey's port of Ceyhan on the Mediterranean Sea
July 2004	Ghislenghien, Belgium, near Ath gas pipeline leak/explosions, 23 people killed
Jan 2006	North Ossetia, Russia, Mozdok-Tbilisi pipeline sabotage
August 2006	Turkey, Agri province gas pipeline explosion sabotaged by Kurdish Workers' Party (PKK)
December 2006	Lagos, Nigeria, vandalized oil pipeline, 500 people killed
October 2007	Pakistan's Baluchistan province, unknown saboteurs blew up gas pipeline Och power plant coming from Och gas field, five killed (APP, 2007)
Oct 2008 Aug 2009	Sabotage to sour gas pipelines in the Tomslake area in the province of British Columbia

Risks to pipeline cannot be completely eliminated or prevented; however, the intent of pipeline security is to bring a risk-based approach to the application of the security measures throughout the entire pipeline industry (TSA 2008). US Transportation Safety Administration (TSA 2008) guideline thus provides criteria, which operators can follow to assess and determine criticality of each of their facilities. In addition, the guidelines identify baseline security risk reduction measures that can be implemented at each facility, as well as enhanced measures that can be implemented at facilities determined to be critical.

Since September 11, 2001, the pipeline industry has strived toward a security management program, which must be systematic, comprehensive, and proactive in managing security risks.

An effective pipeline industry security management program must generally contain the following elements:

1. A security policy setting out the senior management commitment
2. Well-defined roles, responsibilities, and authorities
3. A communication plan supportive of the pipeline security management program that is linked to a national security threat assessment system
4. Procedures to:
   a. identify and rank critical facilities and operations
   b. evaluate vulnerabilities of facilities and operations
   c. identify interdependencies of facilities and operations
   d. manage physical security
   e. manage secure information, which, if compromised, would impact the operation of the pipeline system
   f. manage the security of process control and SCADA systems
   g. protect cyber information and supporting infrastructure
   h. control physical access to security sensitive areas
   i. provide well established hiring and termination procedures for security sensitive positions
   j. screen visitors to facilities and operations
5. Effective training programs
6. Performance monitoring to continually assess conformance with the security management program including mechanisms for corrective action and preventive measures in event of nonconformance

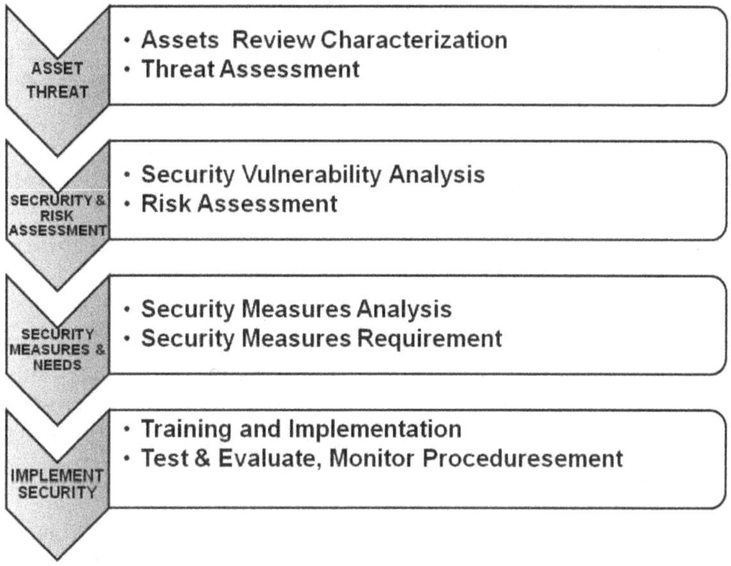

Figure 1-43. Systematic security assessment and management program

Besides the security plan, are the availability of trained personnel, guidelines, and appropriate tools that protect the pipelines. Examples of tools are:

- API 1164 "SCADA Security"
- API "Security Guidance for the Petroleum Industry"
- API "Security Vulnerability Assessment Methodology for the Petroleum and Petrochemical Industries" and
- US Transportation Security Administration, 2008 "Pipeline Security Guidelines

Asset security assessment needs through to Implementation of an appropriate security program is summarized in Figure 1-43.

Security Risk Assessment

The purpose of security risk assessment is to understand the present and future system security risks, access the security threats and the degree of influence probably engendered from such risks, and provide the basis for security strategy identification, establishment, and safe operation of facilities.

The following provides a method of calculating risk for information systems (Xiaolin et al., 2008) and can be similarly used in enterprise asset security assessment.

$$\text{Security risk} = (RAT,V) = R(L(T,V), F(\text{la}, \text{Va}))$$

where

R = function of security risk calculation
A, T, V = asset, threat, and vulnerability, respectively
la = value of asset acted security event
Va = harm extent of vulnerability
L = prob.ability of security event-induced threats, which utilize the vulnerability of asset
F = loss took place after safety time

REFERENCES

Akasaki, H., 2005, "Nippon Steel News," No. 326 Feb. Nippon Steel Corp., Pipe and Tube Sales Div., Pipe and Tube Division, http://www.nsc.co.jp/shinnihon_english/nsnews/pdf/20050208131406l5973.pdf.

Allegro, 2001 "Pipeline Safety — How Safe Are They". Allegro Energy Group, http://www.pipeline101.org/HSSE/safety.html.

API (American Petroleum Institute), 2004, *Security Vulnerability Assessment Methodology for the Petroleum and Petrochemical Industries.*

API (American Petroleum Institute), 2005, *Security Guidance for the Petroleum and Petrochemical Industries*, April.

API (American Petroleum Institute), 2005, *API 1104 — Welding Pipelines and Related Facilities*, 20th ed.

API (American Petroleum Institute), 2009, *API 1164 — Pipeline SCADA Security.*

APP (Associated Press of Pakistan), 2007, "Gas Pipeline Blown up," http://www.redorbit.com/news/business/1120838/gas_pipeline_blown_up_in_pakistans_baluchistan_province/index.html.

Asahi, H., Tsuru, E., Ohkita, S., Koyama, K., Akasaki, H., Murata, M., and Miyazak, H., 2004, "Development of Ultra-high-strength Linepipe, X120," Nippon Steel Technical Report No. 90, July, http://www.nsc.co.jp/en/tech/report/pdf/n9015.pdf.

ASME (American Society of Mechanical Engineers) B31.8S, 2001, "Managing System Integrity of Gas Pipelines," http://www.asme.org.

Ayukawa, N., 2005, "Comprehensive Capabilities Put X120 into Practical Use" Nippon Steel News, No 326 Feb., http://www.nsc.co.jp/en/nsnews/pdf/20061218135225l2203.pdf.

Baker M., 2003, "Stress Corrosion Cracking Study," Final Report for Office of Pipeline Safety (OPS), TTO #8 Integrity Management Program, Delivery Program DTRS56-02-D-70036, http://primis.phmsa.dot.gov/docs/sccReport/SCC%20REPORT%20FULL%20TEXT.PDF.

BP, 2009, "Statistical Review of World Energy," June http://www.bp.com/multipleimagesection.do?categoryId=9023755&contentId=7044552.

Coulson, K. E. W., and Temple, D. G., 1983, "An Independent laboratory Evaluation of External Pipeline Coatings" 5th Int. Conf. on the Internal and External Protection of Pipes, October, Innsbruck, Austria, BHRD Fluid, Eng., Cranfield, UK

CIA, 2009, "The World Fact Book — Field Listing," https://www.cia.gov/library/publications/the-world-factbook/fields/2117.html?countryName=&countryCode=&regionCode=%C5%A1.

D'Atri, M. F., Rodriguez, D., and García-Martínez, R., 2008, "Improving Pipeline Risk Models by Using Data Mining Techniques," www.igu.org/html/wgc2009/papers/docs/wgcFinal00663.pdf.

Davy, H., 1824, "Philosophical Transaction," 144, 151, http://www.corrosion-doctors.org/Biographies/DavyBio.htm.

Eiber, J., Maxey, W. A., Bert, C. W., and McClure, G. M., 1981, "The Effects of Dents on the Failure Characteristics of Linepipe," NG18, Report No. 125, Battelle Columbus Laboratories, May, 1981.

Ellyin, F., Wolodko, J., Dorling, D. V., Glover, A. G., and Jack, T., 2000, "Fibre Reinforced Composites in Pipeline Applications: Design Issues and Current Research," IPC 2000, ASME, NY.

Esford, F., Porter, M., Savigny, K. W., Muhlbauer W. K ., and Dunlop, C., 2004, "A Risk Assessment Model for Pipelines Exposed to Geohazards" Proceedings of the ASME International Pipeline Conference (IPC), Calgary, Alberta, Canada October 4–8.

Ferguson G., 2009, "Risk 101," Cameco Community Liaison Forum Monday, Saskatoon, Saskatchewan, Canada, April 6.

Glover A. G., 1996, "Use of High Strength Steel," NOVA Gas Technology Seminar, Nova Corporation (Now TransCanada Pipelines), Calgary, Alberta Canada, December.

Glover, A. G., Hodgkinson, D., and Dorling, D. V., 1990. "The Application of Mechanized Ultrasonic Inspection and Alternative Acceptance Criteria to Pipeline Girth Welds," Pipeline Technology Conference, October 1990, Ostende.

Glover, A. G., Horsley, D. J., and Dorling, D. V., 1999, "High-Strength Steel becomes Standard on Alberta Gas System," Oil and Gas Journal, Vol. 97, pp. 44–49, January.

Guthrie, J., Battat, B., and Grethlein, C., 2001, "Accelerated Corrosin Testing" The Advanced Materials and Processes Technology (AMPTIAC Quarterly), Vol. 6, p. 3.

Hewing, M., 1999, "Differing Repair Methods — Options and Benefits," Second Latin American Pipeline Integrity Conference, Puerto La Cruz, Venezuela, August.

Holden, E., 1997, "The Changing Role of Inspection," PII Publication, http://www.piigroup.com/reference_library/papers/papers.html

Jaske, C. E., 2002, "Managing Pipeline System Integrity: Introduction to API Standard 1160," Presented to the Pipe Liners Club of Tulsa, Tulsa, OK, January, 2002.

Jaske, C. E., Beavers, J. A., and Thompson, N. G., 2002, "Corrosion Prevention and Control", Vol. 49, p. 3. 2. Publisher: Kane, R. D. and Cayard, M. S., Chem.

Jaske, C. E., Vieth, P. H., and Beavers, J. A., 2002, "Assessment of Crack-Like Flaws in Pipelines," Corrosion NACExpo Denver, Colorado, April, Paper No. 02089.

Keifner, J. F., Bruce, W. A., and Stephens, D. R., 1994, "Pipeline Repair manual," PRCI Catalog L 51716e. USA, http://www.prci.com/publications/L51716e.cfm

Kojima, A., 2005, "Technology for Tuning Seeds and Needs Opens the Way to Higher-quality Products," Nippon Steel Corporation News, No. 328, April http://www.nsc.co.jp/en/nsnews/pdf/20061218135037 11981.pdf.

Leierzapf, H., 1985, "Workstoffe und Korrosion," Vol. 36, p. 88. Wiley VCH Verlag GmbH, Germany.

McDonnell, P., and McNamara, J. F., 1995, "Evaluation of GRP Composite Pipelines for High Performance Offshore Applications," Key Engineering Materials, Vol. 99–100, pp. 11–18.

McKnight, J., 1998, "Transportation in the 21st Century," *I. Mech. Eng. Presidential Address*, Latimer Trend & Company Ltd, Plymouth, England.

Meier, R. L, 2000, "Integrating Enterprise-Wide Risk Management Concepts Into Industrial Technology Curricula," Journal of Industrial Technology, Vol. 16, p. 4, August, www.nait.org.

Moghissi, O., 2003, "Identifying Locations Susceptible to Accelerated Corrosion: Finding a Needle in a Haystack," PRCI Conference, Calgary, Canada, April 8.

Mohitpour, M., McManus, M., and Trefanenko, W., 2002, "Trends in Pipeline Integrity Inspection and Rehabilitation Techniques," Proceedings of ASME International Pipeline Conference, Calgary, Alberta Canada, September 29–October 3.

Mohitpour, M., Szabo, J., and Van Hardeveld, T., 2005, *Pipeline Operation and Maintenance — A Practical Approach*, 1st ed., ASME Press, New York, NY.

Mohitpour, M., Golshan, H., and Murry, A., 2007, *Pipeline Design and Construction — A Practical Approach*, 3rd ed., ASME Press, New York.

Muhlbauer, W. K., 2004, *Pipeline Risk Management Manual Ideas, Techniques and Resources*, Elsevier Inc, UK.

NACE, 2002, "Control of External Corrosion on Underground Submerged Metallic Piping Systems," NACE Standard RP 0169, NACE, Houston, TX.

PDA, 2006, "An Incomplete History of Risk Management," https://store.pda.org/bookstore/TableOfContents/Risk_Assessment_Ch01.pdf.

PHMSA, 2007, "Significant Incidents Files" Nation Wide Gas Transmission Onshore Failures Causes, March 14.

Scully, J. R., and Taylor, D. W., 1987, "Electrochemical Methods of Corrosion Testing," *Metals Handbook*, Vol. 13, ASM, pp. 212–228.

Short, G. À., and Fletcher, M. Á., 1997, "The role of Smart Cleaning Pigs in Pipeline Rehabilitation and Maintenance," http://www.rstprojects.co.uk/pages/065399r0.pdf.

Terada, Y., 2005, "X120: Challenging Pipe-making Know-how," Nippon Steel News, No. 326 Feb., http://www.nsc.co.jp/en/nsnews/pdf/20061218135225 12203.pdf.

TSA (US Transportation Security Administration), 2008, "Pipeline Security Guidelines — A Draft," Revision 2a.APL, September 26, http://www.ilta.org/WhatsNew/2008/DRAFT_TSA%20Pipeline%20Security%20Guidelines.pdf.

U of T @ Austin (University of Texas at Austin), 2002, "A Short History of Risk Management: 1900 to 2002" Adapted from Risk Management Reports, Vol. 26, No. 12, December 1999, www.mccombs.utexas.edu/dept/irom/bba/risk/rmi/arnold/DOWNLOADS/Hist_of_RM_2002.pdf - 2009-02-09.

US DOT, 2007, "Gas Integrity Management Performance Measure Reports" http://primis.phmsa.dot.gov/gasimp/PerformanceMeasures.htm.

Uzelac, N. I., 2000, "In-Line Inspection Tool to Achieve Pipeline Integrity," The First Internet Conference on Pipeline Reliability (PICon), PICon Publication March 13, http://www.nrcan.gc. ca/mms/picon-e.htm.

Vennett, R. M., Williams, J. G., Lo, K. H., and Ganguly, P., 1999, "Economic Benefits of Using Composites for Offshore Development and Operations," Composite Materials for Offshore Operations, Vol. 2, ABS.

Vesper, J. L., 2006, "Risk Assessment and Risk Management in the Pharmaceutical Industry Clear and Simple. A Complete History of Risk Management." PDA Books, https://store.pda.org/bookstore/Advanced-Search.aspx/Risk_Assessment_Ch01.pdf.

Xiaolin, C., Xiaobin, T., Yong, Z., and Hongsheng, X., 2008, "A Markov Game Theory-Based Risk Assessment Model for Network Information System," International Conference on Computer Science and Software Engineering, http://staff.ustc.edu.cn/~xbtan/papers/A%20Markov%20Game%20Theory-based%20Risk%20Assessment%20Model%20for%20Network%20Information%20System.pdf.

Yasinko, E. A., Vieth, P. H., Dick, D. D., Nidd, G., and Pierantoni, T. D., 2000, "Platte TFI Program: Innovative Solutions to Hydrostatic Testing," ASME IPC 2000 Conference Calgary, Alberta, Canada, October 1–5, 2002.

Yergin, D., 1992, *The Prize: The Epic Quest for Oil, Money and Power*, Free Press, New York, p. 33.

Chapter 2

PIPELINE INTEGRITY MANAGEMENT PROGRAMS

INTRODUCTION

Pipeline integrity management comprises all the activities required to anticipate and prevent a pipeline failure. In most modern pipeline companies, these activities are a subset of general operations and maintenance measures and are related by an identifiable framework that describes the integrity management program (IMP). The objective of the IMP — executed in conjunction with other plans and programs — is to make the likelihood of a pipeline failure so remote within the life cycle of the pipeline system, that the risk of failure can be considered to be controlled.

The principal aim of Pipeline Integrity Management is to efficiently allocate operating and maintenance resources so as to assure the safety of personnel (members of the public, company employees, and contractors), environmental protection, and system reliability. If these three aims are met, operating costs and financial impact will be minimized, return on investment will be maximized, and company image will be enhanced. These requirements can be directed as part of a company's diligent operations and maintenance programs, and/or as part of rules or regulations that address the need to assess, evaluate, repair, and validate through comprehensive analysis of the pipeline segments.

Throughout the life cycle of a pipeline, the potential for pipeline failures can be due to various threats and causes, some of which are illustrated below in Figure 2-1.

Managing the potential for these threats and failures in a proactive manner throughout the life cycle of the pipeline is essential for all owners and operators of hazardous liquid or natural gas pipelines. The following high-level activities occur during a typical pipeline life cycle:

- Planning and design engineering
- Construction
- Monitoring, inspection, and leak detection
- Protection
- Maintenance and repair
- Operating philosophy
- Emergency procedures
- Decommissioning and disposal

Pipeline Integrity Management begins at the Planning and Design phase of the pipeline life cycle. Each of these phases or activities has its own unique measures and associated threats that need to be addressed by the integrity personnel. Examples of integrity measures used during the design phase can be summarized as follows in Table 2-1.

An Integrity Management system is intended to ensure that all aspects of a pipeline's operation throughout its life cycle are appropriate to mitigate the risks inherent to that pipeline, the hazards/threats presented externally, and the sensitivity of the environment in

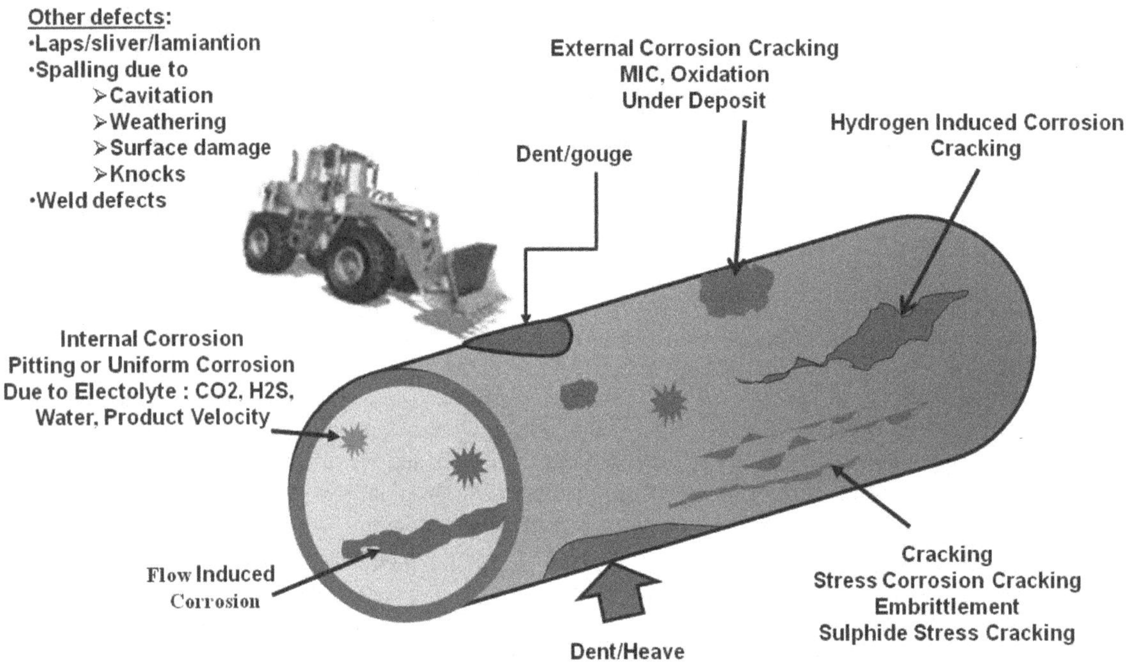

Figure 2-1. Illustrative examples of pipeline failure causes (Mohitpour 2010)

which it is located. This process begins with input from integrity personnel at the planning and design phase of the pipeline system and carried throughout the remaining construction, operations, and maintenance phases of the life cycle.

The goals of an IMP are to effectively maintain the integrity of the pipeline facilities through an active program of monitoring the condition of the pipelines and mitigating the adverse effect of potential defects and the prevention of any potential leaks or ruptures. In addition, the objectives of the IMP are to use risk management techniques to develop such a program in order to efficiently and effectively extend the life of the system.

TABLE 2-1. Example of design considerations with corresponding threat

Design Measure	Threat Addressed
Adequate wall thickness	All threats
External coating	External corrosion and SCC
CP system	External corrosion in the case of coating damage or degradation
Adequate depth of cover	Third-party interference, mechanical damage
Right-of-way signage	Third-party interference, mechanical damage
Correct selection of seam weld type and adequate quality assurance	Manufacturing (seam weld) defect
Adequate pipe mill quality control regarding metallurgy, dimension stability, and surface finish.	Defective pipe
Suitable, written welding procedure, inspection, and 100% radiography of girth welds	Construction/girth weld defects
Tight controls on and inspection of shop hot bends and field cold bends	Wrinkling
Route selection to avoid geohazards	Weather related and outside force
Internal coating	Internal corrosion
Hydraulic calculations and profile/flow specification to prevent water holdup and water dropout	Internal corrosion

An IMP, as defined by the United States DOT (U.S. Department of Transportation 2005; U.S. Department of Transportation 2006; U.S. Department of Transportation Pipeline and Hazardous Materials Safety Administration (PHMSA)), is a documented set of policies, processes, and procedures that as a minimum, address the following elements:

- a process for determining which pipeline segments could affect an HCAs
- a baseline assessment
- a process for continual integrity assessment and evaluation
- an analytical process that integrates all available information about pipeline integrity and the consequences of a failure
- repair criteria to address issues identified by the integrity assessment method and data analysis
- a process to identify and evaluate and evaluate preventative and mitigative measures to protect HCAs
- methods to measure the IMP's effectiveness and
- a process for review of integrity assessment results and data analysis by a qualified individual

Regardless of the type of program the company has developed, the IMP will be ever changing. The plans that comprised within the program will be continually improved as additional assessments are completed, new processes are developed, existing processes refined and expanded, operating experience is gained, and further advancements in technology becomes available.

The IMP should include the information and practices necessary to ensure that the pipeline facilities continue to provide safe, reliable, and environmentally responsible transportation. The IMP should also address the integrity of all pipeline facilities operated by the pipeline company.

INDUSTRY STANDARDS AND DOCUMENTS ADDRESSING INTEGRITY

Internationally recognized standards are commonly used for the design, construction, operation, and maintenance of facilities. In relation to pipeline integrity, there are a number of standards and documents that set out/outline pipeline integrity assessment requirements. Some of the more common integrity-related documents are listed below (Mora 2007):

- ASME B31.8S — Managing System Integrity of Gas Pipelines
- ASME B31G — Manual for Determining the Remaining Strength of Corroded Pipelines: A Supplement to ASME B31 Code for Pressure Piping
- API Std 1160 — Managing System Integrity for Hazardous Liquids Pipelines
- API RP 1162 — Public Awareness Programs for Pipeline Operators
- API Std 1163 — In-Line Inspection Systems Qualification Standard
- API Std 5T1 — Imperfection Terminology
- API RP 1110 — Pressure Testing Liquid Pipelines
- API 1156 — Effects of Smooth and Rock Dents on Liquid Petroleum Pipelines
- API RP 579 — Fitness for Service
- API Std 653 — Tank Inspection, Repair, Alteration, and Reconstruction
- API 570 — Piping Inspection Code: Inspection, Repair, Alteration, and Rerating of In-service Piping Systems
- CSA Z662 Annex N — Guidelines for Pipeline Integrity Management Programs

- NACE 35100 — Inline Nondestructive Inspection of Pipelines
- NACE RP0102 — Inline Inspection of Pipelines
- NACE RP 0169 — Control of External Corrosion on Metallic Piping Systems
- NACE SP0502 — Pipeline External Corrosion Direct Assessment Methodology
- NACE SP0204 — Stress Corrosion Cracking Direct Assessment Methodology
- CEN/TS 15173:2006 Gas supply systems — Frame of reference regarding Pipeline Integrity Management System (PIMS)
- CEN TS 15174 — A guideline for Safety Management Systems for natural gas transmission pipelines
- US Department of Transportation (DOT) Title 49 Code of Federal Regulations (CFR) Part 195, Integrity Management for Hazardous Liquid Pipeline Operators
- US Department of Transportation (DOT) Title 49 CFR Part 192 Pipeline Safety: Pipeline Integrity Management in High Consequence Areas (Gas Transmission Pipelines)
- Pipeline Library http://primis.phmsa.dot.gov/comm/PipelineLibrary.htm

Comparisons of various global codes addressing pipeline integrity issues including the requirement for an IMP are summarized in Table 2-2.

PIPELINE INTEGRITY REGULATIONS

Countries and regions throughout the world typically have some form of government oversight into the pipeline industry, and they have varying degrees of IMP requirements. Further comparisons of regulatory requirements outside of North America have not been included within this book; however, companies operating pipelines, regardless of location, need to access the country or region-specific requirements for Integrity Management in order to be compliant with applicable requirements.

The following section provides an overview of pipeline IMP requirements in North America from a regulatory perspective. It will mainly focus on the Federal pipeline Integrity Management requirements as stated in the "*Pipeline Safety Code of Federal Regulations (CFR) Title 49 Parts 192 and 195, Pipeline Safety: Pipeline Integrity Management in High Consequence Areas,*" used in the United States, and the "*NEB Onshore Pipeline Regulations, 1999 (OPR-99),*" used in Canada for federally regulated pipelines (U.S. Department of Transportation 2005; U.S. Department of Transportation 2006; U.S. Department of Transportation Pipeline and Hazardous Materials Safety Administration (PHMSA); Canada National Energy Board 1999; National Energy Board of Canada Web site, http://www.neb.gc.ca/).

Comparison of Regulatory Styles

Historically, the regulatory perspective has been highly reactive with a heavy reliance placed on prescriptive Industry Codes and Standards. The death of 298 schoolchildren in an explosion in New London Texas in 1937 led that State and eventually others to require odorant to be added to natural gas. In the United States, it was not until 1968 under President Johnston that Federal Pipeline Safety Legislation was enacted — The National Gas Pipeline Safety Act — which was largely in response to a pipeline failure in Louisiana killing 17 people. Prior to 1968, major interstate gas pipelines were designed, operated, and maintained in accordance with ANSI (later AMSE) Standard B31.8.

Strict adherence to the prescriptive requirement does not guarantee safety. "Prescriptive Regulations" require companies to conform to specific standards coupled with an inspection and penalty system to ensure compliance. Government retains primary responsibility for developing Regulations and for ensuring compliance. Regulations often specify what is to be done, how it is to be done, what will be inspected, when it will be inspected,

TABLE 2-2. Summary of global codes addressing pipeline integrity (updated after Healy et al. 2004)

Code/Reg. / Integrity Required	ASME B31.4 (Liquids)	ASME B31.8 (Gases)	ASME 31.8 (S) (nonmandatory)	IGE/TD/1 (UK)	AS 2885 (Australia)	CSA Z662–2007 (Canada)	DOT Part 195 (USA)	DOT part 192 (USA)	Norway, DNV OS-F101
Assessment Method	B31.G, safety factor (SF) = 0.72	B31.G, class location dependent	B31.G or similar —class location dependent	To be selected by operator	B31.G, RSTRENG or approved method	B31.G + engineering assessment	B31.G and RSTRENG	B31.G and RSTRENG and B31.8S	Per DNV Rules Sections C and D
Repair method	Various; grinding, steel or composite sleeve, cutout, etc.	Same as B31.4	As per 31.8. Additional guidance provided	Damage and operating pressure-dependent frinding, full	Encirclement sleeve or pipe replacement recommended	Similar to B31.4, must	Comply with 195.422	To comply with: 192.711 through to 192.719	Per DNV Section D
Integrity Plan	Written plans, which should be modified based on experience	As per B31.4	Systematic process/approach defined	Risk based, considering age and operating history	Risk based, written safety and operating plan	Must operate and maintain system to documented procedures	Mandatory, see rule 195.452	Mandatory, plan to B31.8 s, rule 192.617	Mandatory, as per section 11 C-101
Risk Analysis	Not considered	Not considered	Systematic process/approach defined	Should take into account frequency and consequences of all pipeline failure modes	Required for design and operation (Code AS 2885.1)	Can be used but not mandatory	Mandatory, see rule 195.452	Mandatory, see rule 192.917 or 192.7 (follows B31.8S)	Required as per Section D 600
Spill Prevention/ Blowdown	Regular patrols required to monitor activity near pipeline	Operators must have a leak survey plan	As per 31.8	Risk based	Risk based, leak detection system advised	Periodic line balance + other guidance (App. E)	Leak detection system mandatory, rule 195.452	As per B 31.8	Risk based

TABLE 2-2. (Continued)

Code/Reg. / Integrity Required	ASME B31.4 (Liquids)	ASME B31.8 (Gases)	ASME 31.8 (S) (nonmandatory)	IGE/TD/1 (UK)	AS 2885 (Australia)	CSA Z662–2007 (Canada)	DOT Part 195 (USA)	DOT part 192 (USA)	Norway, DNV OS-F101
External corrosion prevention	Based on effective coating and CP. CP monitoring <15 months, powers sources <2 months	Similar to 31.4	As per 31.8	Based on effective coating and CP, regular monitoring, Pearson and CIPS	Based on effective coating and CP. Special surveys (Pearson, DCVG) recommended	Based on effective coating and CP and regular monitoring	As per B31.4	As per B 31.8	As per Sections D 500 and D700
ILI	Not considered	Not considered	Guidance provided	Periodic inspection using intelligent pigs, frequency <10 yr	Previous intelligent pigs results should be considered, together with RBI	Should be considered, guidance given in Appendix D	See rule 195.452	Required as per rule 192.617	Required to confirm integrity as per Section 11, D400
Third-party damage prevention	Any failure cause should be established to prevent further incidents	As per B31.4	As per 31.8. Additional "performance measures" identified	Detailed records of inspection, surveillance, and pressures to be kept	All incidents must be recorded and investigated	All incidents must be recorded and investigated	As per B31.4	As per B 31.8s, rule 192.614	Per DNV-RP-F113 pipeline subsea repair
Failure history	Any failure cause should be established to prevent further incidents	As per B31.4	As per 31.8. Additional "performance measures" identified	Detailed records of inspection, surveillance, and pressures to be kept	All incidents must be recorded and investigated	All incidents must be recorded and investigated	As per B31.4	As per B 31.8 required	To be documented

and who will conduct the inspection. A prescriptive requirement thus describes the means to achieve the objective in terms of material and technology.

With "performance-based" or "self-regulating" regulatory regimes, governments can continue to safeguard the safety of citizens; however, it devolves to industry, in various degrees, responsibilities for establishing standards and inspection programs for ensuring compliance, and for providing performance reports to assure the government of the integrity of the system and the achievement of goals. A performance requirement thus describes "ends" in terms of performance that will assure meeting the objective. Government retains a capacity to verify the information being received and to conduct selective audits.

As indicated in Figure 2-2, "Goal-oriented" regulatory regimes are an evolution between prescriptive and performance-based Regulations. They ensure the onus for performance remains properly with the pipeline companies, and they encourage companies to develop proactive management systems that address the unique risks of their pipelines. At the same time, acceptance criteria that have been proven over time (such as ASME B31.8/ ASME B31.4, or CSA Z662 or) are required to be used.

Over time and pending no major incidents, regulators may desire to move towards a higher degree of performance and goal-based regulations. However, the pipeline industry and regulators need a common understanding of what the goals are and how these goals will be met. As well, this means the pipeline industry must take more responsibility in meeting the intent of regulations. The following issues need to be resolved before a higher degree of goal-based regulations can be realized:

- How do we define goal-based regulation?
- How do we develop meaningful performance measures?
- How can compliance and performance be measured?

Regulatory Style	Defines	Risk Approach	Compliance Obtained by
Prescriptive	means	deterministic	check list
Goal-oriented	goals	risk informed	professional judgement + acceptance criteria
Performance - based	measures	risk based	objective criteria
Self regulation	outcomes	company determined	self audit

Figure 2-2. Toward self-regulation (Vollman 2000)

- How do we incorporate the public perspective/input into goal-based regulation?
- And finally, are companies really prepared to take on more responsibility?

United States Pipeline Regulations

There have been some significant pipeline ruptures/failure incidents that have occurred on US pipelines in recent history that have brought a lot of attention to the pipeline industry. They have caught the interest of the general public and government agencies responsible for oversight. In March of 1993, a pipeline ruptured near Reston, Virginia, just outside of Washington, DC. Almost 10,000 barrels of fuel oil were spilled and eventually reached the Potomac River. Just a few years later, a pipeline owned by this same company spilled over 20,000 barrels of fuel oil into a river in South Carolina. In March of 1994, a natural gas line in Edison, New Jersey, ruptured and ignited near an apartment complex, sending a fireball into the sky that could be seen 30 miles away in New York City. This failure resulted in the evacuation of 1500 residents and property damage in excess of $25 million.

These incidents produced significant media coverage that led to considerable pressure for improved oversight from Congress, other Federal and State agencies, and the public. The federal Office of Pipeline Safety (OPS, see later) and industry leaders jointly arranged a Pipeline Safety Summit in June 1994 to discuss new ideas and potential paths. A wide variety of stakeholders were invited to join the discussion. The approach conceived at that summit has evolved and is now referred to as Integrity Management has subsequently been codified into several recent integrity-related rules.

In June 1999, a pipeline ruptured in Bellingham, Washington, spilling over 5000 barrels of gasoline into two creeks in a popular city park. The release and its subsequent ignition resulted in the deaths of two 10-year-old boys and a teenager, and environmental damage along 1.5 mi of a salmon stream. In June 2003, three people from this pipeline's operations and management were sentenced for criminal violations arising from this rupture — two of the three people served time in jail. Fines and penalties were issued in excess of $36 million. In 2000, a natural gas transmission line, weakened by internal corrosion, ruptured killing 12 people near Carlsbad, New Mexico. OPS has carefully studied the causes of these incidents.

In the United States, pipelines are operated under a variety of Federal and State Regulations and industry standards. The DOT OPS, within the PHMSA, is the federally regulated agency responsible for pipeline safety. The mandate of the OPS is to "ensure the safe, reliable, and environmentally sound operation of the nation's pipeline transportation system." The OPS publishes regulations on the design, construction, operation, maintenance, and repair of all regulated hydrocarbon and carbon dioxide pipelines. It also requires operating companies to develop and follow O&M procedures and emergency response plans.

The governing code for all gas and hazardous liquid pipelines under the Federal jurisdiction of the OPS is the *"Pipeline Safety Code of Federal Regulations (CFR) Title 49, parts 186-199."* The OPS, within the US DOT, PHMSA, has overall regulatory responsibility for hazardous liquid and gas pipelines under its jurisdiction in the United States.

Pipeline operators in the US are inspected by both Federal and State pipeline safety inspectors to ensure they meet or exceed the regulatory requirements and standards. Additionally, there are various government organizations, such as the National Transportation Safety Board (NTSB) and private citizens groups that monitor the operation of pipelines and make recommendations for improvements and changes. Ultimately, it is the responsibility of the pipeline operator to ensure the safety of its pipelines.

OPS is authorized to delegate to the States all or part of the responsibility for regulation of "intrastate" pipelines (pipelines inside the State boundaries). Through certification by OPS, States regulate, inspect, and enforce intrastate gas and liquid pipeline safety requirements. By signed agreement with the OPS, some State agencies may also inspect "interstate" pipeline safety requirements on behalf of the OPS. Some example of US State regulatory agencies includes:

- California Office of the State Fire Marshall (hazardous liquid pipelines)
- California Public Utilities Commission (gas pipelines)
- Florida Public Service Commission (gas pipelines)
- Florida Department of Agriculture and Consumer Services (LPG)
- Minnesota OPS
- Michigan Public Service Commission (gas pipelines)
- Oklahoma Corporate Commission
- Texas Railroad Commission
- Others

OPS Requirements for Pipeline Integrity

The previously mentioned incidents drove a new way of managing safety and a new approach to Regulation pipeline operations in the United States. The first major step in implementing the new approach was the issuing of the first Integrity Management Rule, applicable to large hazardous liquid pipeline operators, in December 2000, followed by a comparable rule for smaller liquid operators in January 2002. The Integrity Management Rule for natural gas operators was issued in December of 2003. Through these rules, OPS is attempting to answer the question of how to balance prescriptive-based, performance-based (which state what must be accomplished, but leave the specific details of the execution to the operating company), and goal t-based regulations.

Effective February 15, 2002 for liquid pipelines and February 14, 2004 for gas pipelines, all operators *must* have an IMPs in place for their companies. Central to the law are the Integrity Management requirements, which, among other things, requires operators to identify so-called "high-consequence areas (HCAs)" on their systems, conduct risk analyses of these areas, perform baseline integrity assessments of each pipeline segment, and inspect the entire pipeline system according to a prescribed schedule and using prescribed methods.

An HCA is identified as pipeline segments located in populated areas, in unusually sensitive areas of the environment, and in commercially navigable waterways. Companies are required to identify all HCAs to the OPS and submit specific IMPs to the OPS Research and Special Projects Administration, US DOT. All pipeline segments within HCAs must be inspected and remediation plans completed within a specified timeframe. For hazardous liquid pipelines, the baseline assessment and reassessment interval are 7 and 5 years, respectively. For gas pipelines, these times are 10 and 7 years, respectively.

Other provisions of the law include:

- Participation in planned-excavation one-call notification programs
- Increased penalties for violations of safety standards
- "Whistle-blower" protection for pipeline system employees
- Qualification programs for employees who perform sensitive tasks
- Authorization of some State participation in interstate pipeline oversight
- A required multi-agency program of research, development, demonstration and standardization to enhance the integrity of pipelines
- An inter agency task force to expedite environmental reviews when necessary to expedite pipeline repairs
- Government mapping of the pipeline system and assembling pipeline operator contact information for public dissemination

Liquid Pipelines

The hazardous liquid pipeline regulations are presented in the *U.S. Code of Federal Regulations (CFR) Title 49, Part 195.452*, and they address pipeline Integrity Management in HCAs. Methods identified by the liquid rule for evaluating the pipeline integrity include the use of ILI, hydrotesting, and other proven technologies. It requires a 5-year reassessment interval. It also provides prescriptive responses and timeframes for responding to ILI results. For example, the rule requires immediate response and a pressure reduction for metal loss indications that exceed 80% of the pipeline wall thickness.

The pipeline industry has also developed a recommended practice for pipeline operators to use when developing their IMPs. The API supported the development of *API 1160, "Managing System Integrity for Hazardous Liquid Pipelines"* (American Petroleum Institute 2001). This document provides guidance to the pipeline industry for managing integrity and is derived from proven industry practices for managing pipeline integrity.

There are eight essential elements that must be included in the operators of IMPs for hazardous liquid pipelines:

1. Identification of all HCAs
2. Baseline assessment program
3. A process for continual evaluation and assessment
4. An analytical process that integrates all available information about integrity and failure consequences
5. Repair criteria
6. A process to identify and evaluate the measures chosen to protect HCAs
7. Methods to measure the IMP's effectiveness
8. A process for review of results and data by a qualified individual

US Gas Pipelines

The natural gas pipeline regulations are presented in the *U.S. Code of Federal Regulations (CFR) Title 49, Part 192*, and they also address pipeline Integrity Management in HCAs.

ASME International also supports the development of standards for natural gas pipeline operators. ASME has produced a supplement to B31.8, "Managing System Integrity for Gas Pipelines" (American Society of Mechanical Engineers 2004), which addresses requirements for both prescriptive and performance-based IMPs, risk assessment methods, integrity assessment techniques and response plans, and performance measures.

There are 16 essential elements that must be included in the operators IMP for gas pipelines:

1. Identification of all HCAs
2. Baseline assessment program
3. Identification of possible threats to each segment
4. A direct assessment plan
5. Provisions for remediating conditions found during assessments
6. A process for continual evaluation and assessment
7. A CDA plan
8. A process to identify and implement mitigation programs
9. A performance plan using specific target measures
10. Record keeping provisions
11. Management of change process
12. Quality Assurance process
13. Communication plan
14. Procedures for providing regulators with copies of risk assessments or the IMP

15. Procedures to minimize environmental or safety risks during integrity assessments
16. A process to ensure continued assessment of possible HCAs

While the purpose and many of the provisions of the gas Integrity Management Rule are the same as those of the hazardous liquid pipeline rule, several differences can be distinguished in the gas Integrity Management Rule including:

- In the gas Integrity Management rule, OPS has "specified" the approach that an operator must apply to identify the areas near their pipelines where HCAs are possible. This approach involves calculating a potential impact radius, based on the diameter and MAOP of the pipe.
- The second significant difference is that the gas Integrity Management Rule allows the use of a technology called "Direct Assessment (DA)," in addition to pressure testing and ILI, to assess the integrity of the pipeline.
- The third significant difference is that the gas Integrity Management Rule has a strong reliance on the use of standards developed by ASME and NACE to support the identification of Integrity Management requirements. For example ASME B31.8S uses a threat-based approach that requires the operators to identify which threats apply to each segment subject to the Integrity Management Rule. The identified threats then serves as the basis for selection of integrity assessment methods.

The fourth significant difference is the time allowed for establishing the baseline assessment and reassessment interval. For hazardous liquid pipelines, the baseline assessment and reassessment interval are 7 and 5 years, respectively. For gas pipelines, these times are 10 and 7 years, respectively.

Canadian Pipeline Regulations

The NEB is a Canadian Federal regulatory agency that was established in 1959. Their mandate, under the NEB Act, is to regulate the construction and operation of interprovincial and international (cross-border) oil, gas, and commodity pipelines; the construction and operation of interprovincial and cross-border power lines; the setting of tolls and tariffs for oil and gas pipelines under its jurisdiction; the export of oil, natural gas, and electricity; and the import of natural gas. Interprovincial and cross-border pipelines are subject to the statutory requirements of the NEB Act and Regulations made by the NEB. The Regulations typically incorporate consensus standards such as the Canadian Standards Association (*CSA Z662 — Oil and Gas Pipeline Systems*).

The NEB tracks and monitors compliance to ensure that companies fulfill the requirements of the Act and Regulations as well as any conditions imposed at the time of approval of the facilities. In carrying out its mandate, the NEB promotes the safety of people, the protection of property and the environment, and economic efficiency in the Canadian public interest.

The NEB shares responsibility with the Transportation Safety Board (TSB) of Canada for incident investigation of pipelines. Since September 1, 1999, all companies must provide incident notifications and reports directly to the TSB. The TSB forwards all relevant information to the NEB, thus avoiding duplication of efforts. The TSB may choose to investigate only certain incidents, typically significant ruptures, in which case they are the lead investigator with the NEB providing assistance as required. Should the TSB choose not to investigate, then the NEB will assume responsibility for the file. Both the NEB and the TSB have the legal authority to seize pipe as evidence, close the site, interview witnesses, etc. The TSB determines the cause, cannot attribute blame ,and issues an "Occurrence

Report," The NEB has the statutory authority to adopt recommendations from the TSB report if sufficient information has been supplied to do so and reviews the corrective action taken by the company. The NEB responds to TSB reports and the NEB prepares a "Close-out Report" for the particular incident or rupture.

Canada also has provincial regulatory bodies that regulate any intra-provincial pipelines (pipelines inside the provincial boundary). Intra-provincial pipelines do not fall under NEB Regulations, thus are the responsibility of provincial regulators. Each province has a pipeline-related Act and Regulations applicable to their intra-provincial pipelines; these typically reference the NEB Regulations as a minimum. Example of provincial pipeline regulatory bodies includes:

- Energy Utility Board of Alberta ("EUB")
- Oil and Gas Commission of British Columbia ("OGC")
- Ministère des Resources Naturelles de Québec
- Ontario Energy Board ("OEB")
- Others

OPR-99

On August 1, 1999, a revised version of the NEB's OPR came into effect. The NEB undertook the development of the OPR-99 with the view that regulations should be less prescriptive and more goal-oriented. The shift to goal-oriented regulations gives companies more discretion in the selection of methods employed to meet NEB requirements. The onus is on each company to demonstrate the adequacy and effectiveness of the methods used.

The NEB's approach to goal-orientated regulations in OPR-99 includes the use of companion Guidance Notes that provide assistance in complying with the requirements of OPR-99. The Guidance Notes are considered by the NEB to be examples of acceptable practices and were developed cooperatively with NEB-regulated companies to provide clarity, practical advice, and suggestions to facilitate compliance and to promote shared responsibilities for safety and the protection of the environment. The Guidance Notes are essentially explanatory in nature.

Through the investigation of significant pipeline incidents, regulations have been modified. Recent investigations, such as the NEB Stress Corrosion Cracking Inquiry, resulted in new additions to the pipeline regulations in the area of pipeline Integrity Management.

The NEB intends to continue to ensure compliance with the OPR-99 by auditing systems, specifications, procedures, records, and training programs developed by companies and by performing inspections during the construction and operation of pipelines.

NEB Requirements for Pipeline Integrity

Canadian Federal pipeline Integrity Management requirements are stated in the "*NEB On-shore Pipeline Regulations, 1999 (OPR-99),*" Section 40 and 41:

<u>Section 40</u>: *A pipeline company shall develop a pipeline* IMP.

<u>Section 41</u>: *(1) If a company finds a level of defect in excess of that allowed by CSA Z662 on its pipeline, the company shall document the particulars of the defect, its likely cause and the corrective action taken or planned to be taken.*
(2) A company shall submit the documentation referred to in subsection (1) to the Board when required to do so under section 7.

As mentioned previously, the NEB has also published Guidance Notes to help provide assistance to interested parties in understanding the requirements of Sections 40 and 41 for pipeline integrity. The Guidance Notes for Sections 40 and 41 mention that an IMP may include the following four components:

1. A Management System

The IMP's "management system" should consider:

- the program scope, including a description of pipelines, goals, and objectives;
- the organizational lines of responsibility and reporting requirements to senior management for the IMP;
- the training and qualifications of management and staff required to develop and execute the IMP;
- the qualifications of consultants and contractors;
- the methods of keeping abreast of industry practices and research activities;
- the methods used to manage change in respect to design, construction, and operation of the pipeline;
- the methods used to measure the effectiveness of the program (audits).

2. Records Management System

The IMP's "records management system" or working records system (WRS) should consider:

- the requirements of applicable Standards such as CSA Z662;
- the maintenance of a WRS that allows access to historical pipeline records;
- the identification of applicable Standards and Codes;
- the documentation of condition monitoring and mitigation programs; and
- reviews of the IMP's effectiveness

3. Condition Management Program

Condition monitoring is intended to be a comprehensive, proactive, consistent process where the various components of the process are reviewed on an ongoing basis and updated as necessary. The IMP's "condition monitoring system" should consider:

- the use of periodic internal ILI tools (frequency of inspections may be based on engineering principles) and data;
- an engineering assessment of facilities to address pipeline integrity (results of ILI, hydrotests, investigative digs, etc.);
- the use of risk assessment methods to assist in assigning priorities for integrity evaluation of facilities or line pipe segments;
- where appropriate, monitoring and surveillance programs for slope movements, river crossings, depth of cover, etc.;
- a program to minimize third-party damage including line patrols; and
- the procedures used to track, analyze and trend the condition of the pipeline.

4. Mitigation program

The IMP's "mitigation program" should consider:

- the criteria and procedures for defect assessment and prioritization of repairs;
- the criteria and procedures used for:
  - pipe replacement (cutouts)
  - repair (grinding, sleeve, re-coating)
  - hot work, hot taps, welding, and maintenance welding
  - excavation procedures
  - hydrostatic testing and
  - operating pressure reductions (temporary or permanent)

- An outline of short-term (1 to 3 years) and long-term (4 to 10 years) mitigation program plans and priorities.

Based on what is mentioned in Sections 40 and 41 (in association with the Guidance Notes), the onus is on companies to demonstrate "adequacy and effectiveness" of the methods employed. As such, they are free to innovate and take full advantage of their intimate knowledge of their own pipeline system. Each company conducts its own audits and inspection program. All information contained within the program is open to verification by the NEB auditors.

ELEMENTS OF AN INTEGRITY MANAGEMENT PROGRAM (IMP)

An IMP is a set of safety management, analytical, operations, and maintenance processes that are implemented in an integrated and rigorous manner to assure operators provide protection to the facilities, public, and the environment. While most of the elements within an IMP provide some flexibility for an operator to develop a program best suited for its pipeline system(s) and operations, there are certain required features that each IMP must have in order to be effective. IMP Principles can be summarized as follows:

- Integrity must be built into pipeline systems from initial planning, design, and construction.
- Program is continuously evolving and must be flexible.
- Information integration is a key component.
- Risk assessment is a key element.
- Assessing risk to pipeline integrity is a continuous process.
- New technology should be evaluated and implemented.
- Performance measurement of the system and program must be an integral to the IMP.
- Senior management commitment is required.

Key Elements of a Basic IMP

Basic IMP requirements can be summarized as follows:

- Develop a *written Integrity Management Plan* that both identifies and addresses the *risks* on each segment of your pipeline system.
- Complete data gathering and identifying locations where a pipeline failure may lead to a significant impact and risk consequence.
- Develop a framework for implementing the Integrity Management Plan elements.
- Perform a risk assessment identifying threats, and integrate and analyze risk information. The risk analysis should identify the most significant pipeline threats in segments that can affect the pipeline. Examples of pipeline threats include corrosion, excavation-induced third-party damage, material defects, and operator errors.
- Establish a Baseline Assessment Plan (BAP).
- Perform the required Inspection assessments. Integrity assessments can be performed by ILIs (also referred to as "smart pigging"), hydrostatic pressure testing, direct assessment, or other technology that the operator demonstrates can provide an equivalent understanding of the condition of the line pipe. Integrating the assessment results with other relevant information to improve the understanding of the pipe's condition.
- Identify preventive and mitigative actions; take the necessary preventive/mitigative measures. Repair pipeline defects identified through the integrated analysis of the assessment results.
- Create an evaluation program to regularly review and update the Plan (the IMP should be considered iterative).

- Measure the program performance. Periodically evaluating the effectiveness of the IMP and identifying improvements to enhance the level of protection.
- Manage the changes through good documentation and records keeping subjected to a quality assessment/quality control process.

The data gathering begins with assessing all historical and current design information, operating processes, and procedures available, including all issues surrounding the current integrity status of the pipelines. The following is a typical sample of particular areas of data gathering needed for review and inputs to the program:

- Pipeline asset information including pipe details, alignment sheets, design basis (material, grade, coating type, etc. and schematics), etc.
- Information on the current and historical operation of the pipelines
- Historical and current ILI data
- Present methods currently employed for mitigating all pipeline defects
- All reports on corrosion repair programs
- All procedures and reports on inline cleaning programs
- If applicable, all procedures and applications of chemical inhibition (wax control and biocides) programs
- Details of the current method used for monitoring corrosion
- Risk reports, operation reports, and audit reports
- ROW and river crossing inspection reports and any other relevant monitoring/inspection information on the pipeline system

The integrity assessment should identify the specific assessment method(s) for each segment and the corresponding hazards. These methods must be based on the identification of the most significant integrity threats for the specific segment. The assessment plan must also include a schedule indicating when the assessments of each segment will be performed. The schedule must be risk-based, meaning that higher-risk segments are scheduled before lower-risk segments. Operators also need to document the technical basis for the assessment methods they select and the risk analysis performed to establish the schedule.

IMPs should be continual and also include methods to measure the program's effectiveness in assessing and evaluating integrity. The risks must be understood, managed and controlled. Risk assessment is an Integrity Management tool, and its purpose is to identify and quantify the risks associated with pipeline operation, such that remedial action can be performed in a timely manner. This is achieved through the ranking of potential risk to safety, environment, and operations.

Although the conduct of extensive risk assessments is somewhat new to the pipeline industry, other industries, such as the chemical plant industry have had extensive practice in using risk assessments as a part of their management programs. Many of the practices used in assessing risk in other industries can be applied to the pipeline industry. Several risk assessment methods are used by industry, and the most common are failure probability methods and ranking systems. The most appropriate method depends on several factors, including system complexity, availability of historical data, and rigor required by the analysis. Pipeline Integrity Management decisions are made much easier by the risk assessment and prioritization process, which establishes a firm, documented basis for determining expenditures and schedules.

Technology is assisting pipeline companies to answer many complex questions with respect to safety, reliability, and integrity. It is evident that safety, reliability, and integrity will continue to be a top priority for pipeline companies and regulators worldwide for the years to come. We can be encouraged that our industry will continue to make significant progress in meeting the challenges faced with the development of new or with the management of aging infrastructure through our advancements in pipeline IMPs and technology.

The IMP approaches set out in API 1160 and ASME B31.8S move towards developing a program based principally by conducting physical integrity assessments on individual pipeline segments. The order in which the integrity assessments are carried out is determined by the outcome of a system-wide risk assessment — those segments presenting higher risks being addressed first. Risk-based integrity assessment planning is widely used to prioritize the order in which assessments are performed; however, it is seldom used alone to determine the re-assessment intervals.

ASME B31.8S was published as a supplement to the gas pipeline design code, B31.8, and API 1160 was used as a model in the development of B31.8S. However, whereas API 1160's scope addresses all pipeline (and therefore addresses offshore, submerged lines), the scope of B31.8S is restricted to onshore pipeline systems. B31.8S specifies the elements that an IMP must have and B31.8 S addresses the 22 threats to pipeline integrity.

API 1160 — Managing System Integrity for Hazardous Liquids Pipeline (American Petroleum Institute 2001)

API 1160 describes how a written IMP is compiled that includes identification of all pipeline segments that could affect an HCA in the event of a pipeline failure, a plan for conducting baseline assessments of the line pipe in the aforementioned segments (rules are given for the incorporation of existing integrity assessment results), and includes a framework that addresses how each element of the operator's IMP will be implemented. API 1160 states the framework must address the following:

- The process for determining which pipeline segments could affect HCA's
- A baseline assessment plan (setting out the schedule of integrity assessments, the threats addressed, and the integrity assessment methods to be used)
- A process for future, periodic integrity assessments and evaluation of segments that could affect HCAs
- An analytical process that integrates all available information about pipeline integrity and the consequences of a failure
- Repair and mitigation measures to address issues identified by integrity assessments
- A process to identify and evaluate preventative and mitigative measures to protect HCAs. Methods to measure the IMP's effectiveness
- Methods to measure the IMP's effectiveness
- A process for review of integrity assessment results and data analysis by a qualified individual
- Maintenance of identified records
- A written IMP description
- Documentation supporting the decisions and analyses performed as part of the IMP

The framework shown in Figure 2-3 is taken from API 1160, and it provides a common structure upon which to develop an operator-specific IMP. In developing an IMP, pipeline operators should consider their unique Integrity Management goals and objectives and then use existing approaches or develop new processes to assure these goals are achieved.

There are many different approaches to implementing the different elements identified, ranging along a continuum from relatively simple to highly sophisticated and complex. There is no "best" approach that is applicable to all pipeline systems for all situations. However, it is important to recognize that an IMP should be a highly integrated and iterative process.

A brief overview of some of the key individual framework elements is provided in API 1160 as follows:

Initial data gathering, review, and integration. The first step in understanding the potential integrity threats both in HCAs and elsewhere along the pipeline system is to assemble information about potential risks. In this element, the operator performs the initial collection, review, and integration of data that is needed to understand the condition of the pipe and identify the

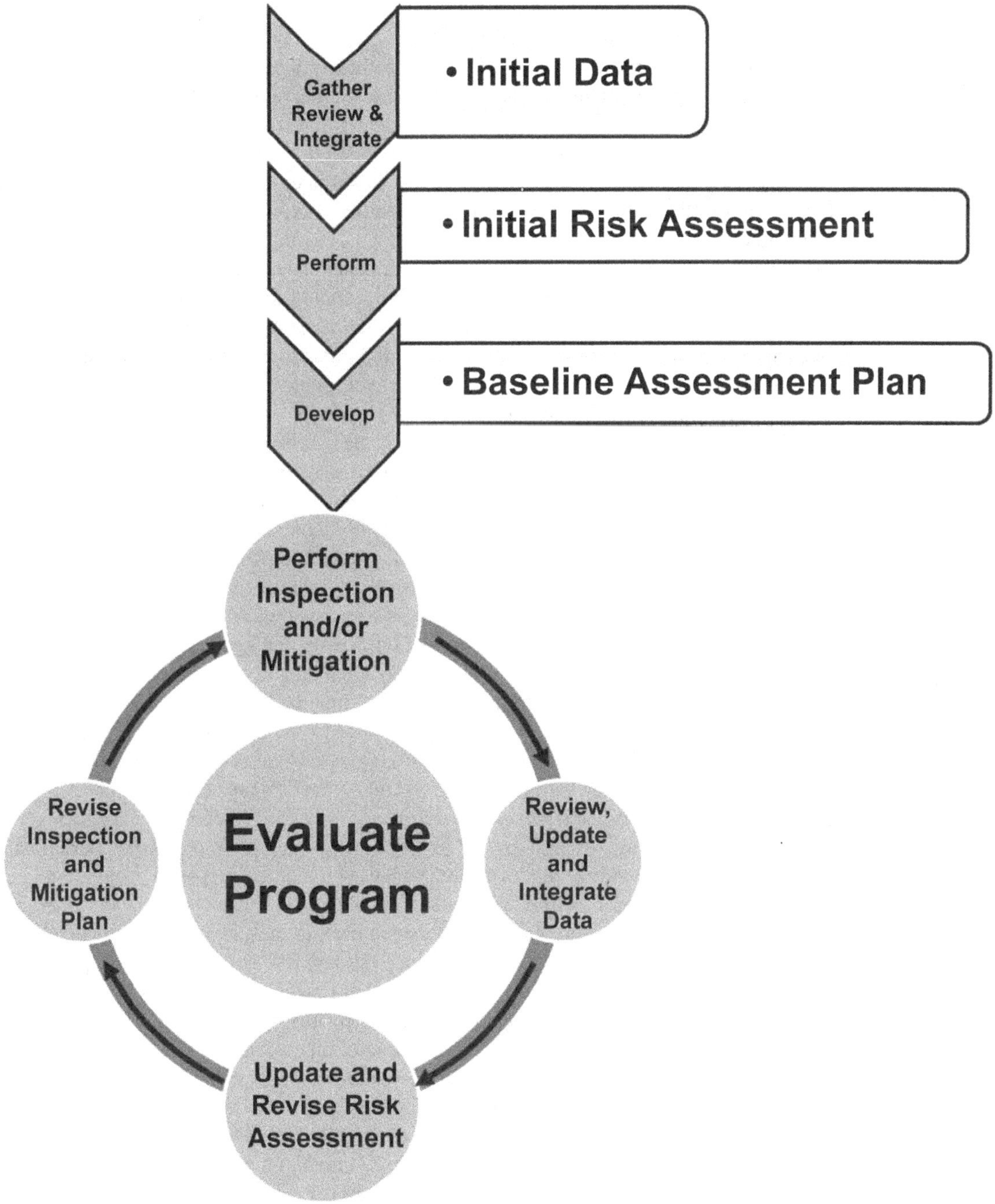

Figure 2-3. Integrity Management Plan (IMP) framework (Reproduced courtesy of the American Petroleum Institute)

location-specific threats to its integrity. The types of data to support a risk assessment include information on the operation, maintenance, and surveillance practices, pipeline design, operating history, and the specific failure modes and concerns that are unique for each system and segments within a system. For operators that are just formalizing an approach to Integrity Management, the initial data gathering may be focused on a limited number of parameters so that a screening for the most significant integrity threats can be readily identified.

Initial risk assessment. In this element, the data assembled from the previous step is used to conduct a risk assessment of the pipeline system. The risk assessment begins with a systematic and comprehensive search for possible threats to pipeline or facility integrity. The identification of potential threats should not be limited to a review of known risk categories, but should include steps to look for new or unique manifestations of risks. Through the integrated evaluation of the information and data collected in the previous step, the risk assessment process identifies the location-specific events or conditions, or combinations of events and conditions that could lead to loss of pipeline integrity, and provides an understanding of the likelihood and consequences of these events. The output of a risk assessment should include the nature and location of the most significant risks on the pipeline system.

Develop baseline assessment plan. Using the output of the risk assessment (or a screening assessment), a plan is developed to address the most significant risks and assess the integrity of the pipeline system. This plan should address integrity assessment activities (e.g., internal inspection or pressure testing), as well as any preventive and mitigative risk control actions identified in the risk assessment process.

Inspection and/or mitigation. In this element, the baseline assessment plan activities are implemented, the results are evaluated, and the necessary repairs are made to assure defects that might lead to pipeline failure are eliminated. As noted previously, a risk assessment may identify other risks that should be addressed. For example, if damage from excavation was identified as a significant risk in a particular area, the operator may elect to conduct additional patrolling, increase public communication, improve line marking, improve right-of-way clearance, actively engage local planning commissions, and/or enhance its excavator awareness program to reduce the likelihood of third-party damage to their line.

Review, update, and integrate data. After the initial integrity assessments have been performed, the operator has improved and updated information about the condition of the line. This information is retained and added to the database of information used to support future risk assessments and integrity evaluations. Furthermore, as the system continues to operate, additional operating, maintenance, surveillance, and other data are collected, thus expanding and improving the historical database of information to support Integrity Management.

Update and revise, reassess risk. Risk assessments should be performed periodically to factor in recent operating data, consider changes to the pipeline system design (e.g., new valves, newly replaced pipeline segments or rehabilitation projects, etc.) and operation (e.g., a change in flow or the hydraulic pressure profile), and analyze the impact of any external changes that may have occurred since the last risk assessment (e.g., population encroachment in new areas). The results of integrity assessments, such as internal inspection or pressure testing, should also be factored into future risk assessments, to assure the analytical process reflects the latest understanding of pipe condition.

Revise mitigation, and inspection plan. The baseline assessment plan should be transformed into an ongoing integrity assessment plan that is periodically updated to reflect new information and the current understanding of integrity threats. As new risks or new manifestations of previously known risks are identified, additional preventive or mitigative actions to address these risks should be performed, as appropriate. Furthermore, the updated risk assessment results should also be used to support scheduling of future integrity assessments.

Evaluate program. The operator should collect performance information and periodically evaluate the success of its integrity assessment techniques, pipeline repair activities, and other preventive and mitigative risk control activities. The operator should also evaluate the effectiveness of its management systems and processes in supporting sound Integrity Management decisions. Section 13 provides guidance for developing performance measures to evaluate program effectiveness, and guidance for conducting audits of IMPs.

Manage change. Pipeline systems and the environment in which they operate are never static. A systematic process should be used to ensure that changes to the pipeline system design, operation, or maintenance are evaluated for their potential risk impacts prior to implementation, and to ensure that changes in the environment in which the pipeline operates are evaluated. Furthermore, after these changes have been made, they should be incorporated, as appropriate, into future risk assessments to be sure the risk assessment process addresses the system as it is currently configured, operated, and maintained. As this final element indicates, managing pipeline integrity is not a one-time process. An IMP involves a continuous cycle of monitoring pipeline condition, identifying and assessing risks, and taking action to minimize the most significant threats.

ASME B31.8S — Managing System Integrity of Gas Pipelines (American Society of Mechanical Engineers 2004)

The program elements depicted above in Figure 2-4 are required for all IMPs and collectively provide the basis for a comprehensive, systematic, and integrated IMP. The B31.8S Standard requires that the operator document how its IMP will address the key program elements. This Standard utilizes recognized industry practices for developing an IMP. The process shown in Figure 2-5 provides a common basis to develop (and periodically reevaluate) an operator-specific program.

In developing the program, pipeline operators should consider their companies' specific Integrity Management goals and objectives, and then apply the processes to assure that these goals are achieved. This standard details two approaches to Integrity Management: a prescriptive plan or method and a performance-based plan or method.

- *Prescriptive plans* in which the re-assessment interval is prescribed by a code or standard and
- *Performance-based plans* in which the intervals are determined using criteria based on system performance measures (e.g., damage found per excavation, number of significant anomalies, number of unauthorized activities on pipeline right-of-way, and number and frequency of product releases.)

Prescriptive is very conservative but easier to implement. Performance is more flexible but requires significantly more data and analysis. You can select either program for entire system, each segment, or each threat within a segment.

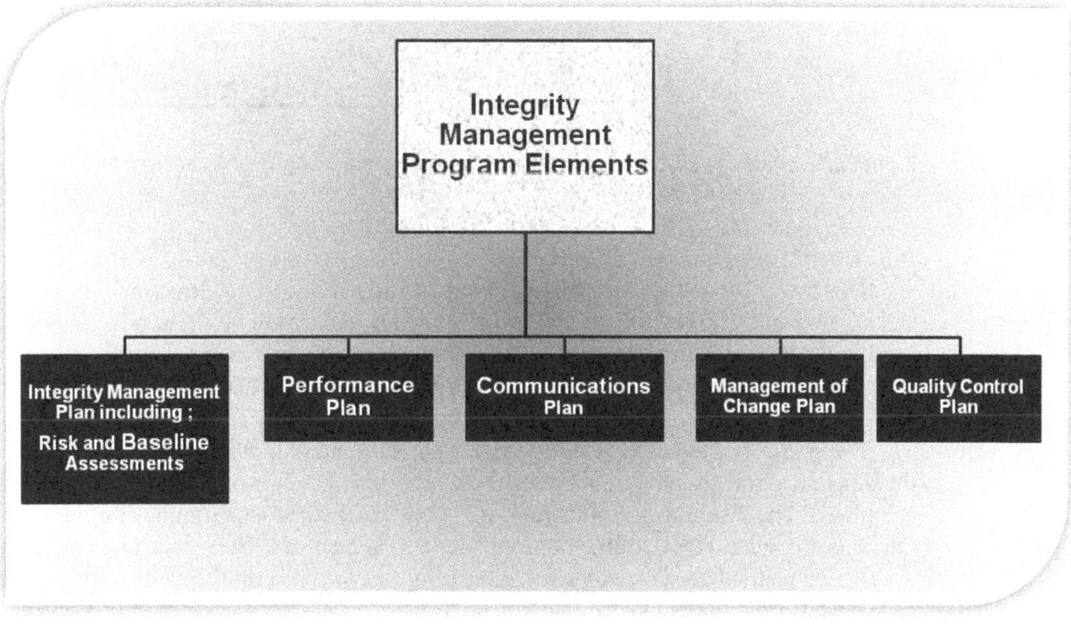

Figure 2-4. IMP element from ASME B31.8S (2004)

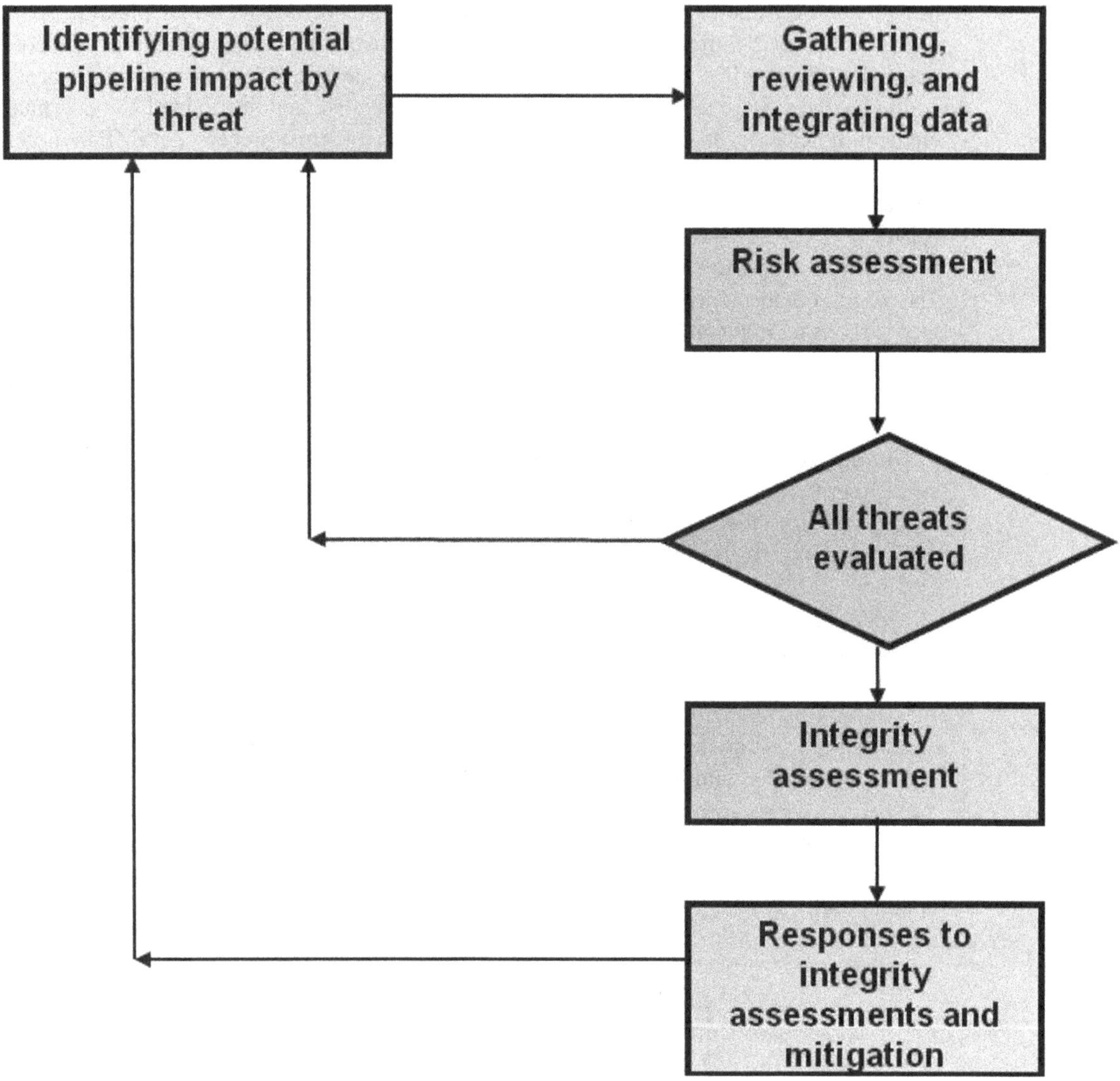

Figure 2-5. Integrity Management plan process flow diagram (American Society of Mechanical Engineers, 2004)

The prescriptive Integrity Management method requires the least amount of data and analysis and can be successfully implemented by following the steps provided in this Standard. The prescriptive method incorporates expected worst-case indication growth to establish intervals between successive integrity assessments in exchange for reduced data requirements and less-extensive analysis.

The performance-based Integrity Management method requires more knowledge of the pipeline and consequently more data-intensive risk assessments, and analyses can be completed. The resulting performance-based IMP can contain more options for inspection intervals, inspection tools, mitigation, and prevention methods. The results of the performance-based method must meet or exceed the results of the prescriptive method. A performance-based program cannot be implemented until the operator has performed adequate integrity assessments that provide the data for a performance-based program. A performance-based IMP shall include the following in the Integrity Management plan:

- a description of the risk analysis method employed
- documentation of all of the applicable data for each segment and where it was obtained
- a documented analysis for determining integrity assessment intervals and mitigation (repair and prevention) methods
- a documented performance matrix that, in time, will confirm the performance-based options chosen by the operator

The processes for developing and implementing a performance-based IMP are included in the Standard. There is no single "best" approach that is applicable to all pipeline systems for all situations. The Standard recognizes the importance of flexibility in designing IMPs and provides alternatives commensurate with this need. Operators may choose either a prescriptive- or a performance-based approach for their entire system, individual lines, segments, or individual threats.

The process of managing integrity is an integrated and iterative process. Although the steps depicted above in Figure 2-5 are shown sequentially for ease of illustration, there is a significant amount of information flow and interaction among the different steps. For example, the selection of a risk assessment approach depends in part on what integrity-related data and information is available. While performing a risk assessment, additional data needs may be identified to more accurately evaluate potential threats. Thus, the data gathering and risk assessment steps are tightly coupled and may require several iterations until an operator has confidence that a satisfactory assessment has been achieved.

ROLES AND RESPONSIBILITIES

This section (Murray 2010) of an IMP typically contains numbered paragraphs referring to:

- Statement of authority
- Introduction to the program
- Purpose and scope of the IMP
- The organizational structure for decision making
- Key functional roles and responsibilities
- Documentation of personnel qualification
- Training and hiring requirements

Typical contents of each of these sections are described as follows:

Statement of Authority

This is a signed and dated statement by the officer of the company who is accountable to the senior management for the efficient execution of the IMP. The statement expressly states the objectives of the program and management support for it. It provides certainty as to who is ultimately responsible for deciding any pipeline integrity issue, which cannot be resolved satisfactorily by staff.

Introduction to the Program

In this section, one typically finds a written description of the overall Integrity Management process and its alignment with the company's internal management systems and the appropriate regulatory requirements that must be satisfied. Often this section will describe the link between the conduct of the IMP to the Company's pipeline maintenance planning and budgeting process. (i.e., the IMP can provide a basis for a 5-year projection of integrity related activities and expenditures).

It should state the frequency with which the program will be reviewed. For example, "This document is under continuous review and is updated as required and at least once per year." Ideally the annual review should occur at the same time as the development of the Annual Maintenance Plan and budget.

The Introduction will also contain:

- A process flow diagram (i.e., A map of the program) for the management of the integrity of the system
- The goal(s) of the program are stated, for example, "To ensure that all pipeline facilities are suitable for safe reliable and environmentally responsible service"
- This section can also be used to describe the physical assets of the pipeline system and their geographical location.

Purpose and Scope of the IMP

The purpose of the entire document is to communicate the company's pipeline Integrity Management process and states its objective, e.g., "a comprehensive program targeted to achieve safe effective reliable, operation of the pressure components of the company's pipeline systems," The scope states the geographical and physical extent of the facilities/ operations covered by the IMP.

Organizational Structure for Decision Making

This section contains a set of organizational charts, which clearly show the lines of responsibility for the entire Integrity Management Plan. It further describes the amount of oversight provided by management to ensure that adequate resources are provided and that common approaches are being followed throughout the various geographical operating regions. All executive personnel and IMP personnel are identified by function (and often by name) on the charts.

Key Functional Roles and Responsibilities

This section must describe the organizational positions responsible for the Company's IMP, their respective functions, and specific and overall responsibilities for defined parts of the program. An organizational chart clearly setting out reporting relationships is essential. It is good practice to provide a description of each job responsibilities.

Typical functional responsibilities are shown in the following (note that job titles will vary in different Organizations):

Vice-President Engineering Responsibilities Typically, this Corporate officer is accountable to the Chief Executive Officer for ensuring IMP performance.

- Ensures program has adequate technical and budgetary support
- Ensures the requisite amount of training of IMP takes place (through monitoring of quarterly/semi-annual training reports)
- Reviews the management evaluation of IMP performance and takes appropriate action

Director Field Operations Responsibilities

- Ensures requisite training of field personnel
- Ensures program management, supervisors, and field staff have adequate support
- Reviews and approves changes to operating procedures
- Reviews and approves all repairs and preventative maintenance measures and procedures
- Attends Program Management review of IMP performance

Director of Engineering Responsibilities

- Corporate officer responsible for IMP performance
- Attends Program Management review of IMP performance and reports results to Vice President
- Ensures requisite training of technical support staff and reports on training to Vice President

- Reviews and approves assessment methods, integrity plans and budgets, data management needs, and changes to technical procedures
- Reviews changes to operating procedures and ensures that these are communicated through the Management of Change Process so that the requisite engineering assessment and planning procedures are updated

The IMP Oversight and Review Team

- This Review Team is responsible for periodically performing a quality assurance review of the IMP results and procedures and reports its findings to the Vice President Engineering.
- The IMP Review Team is identified in this section of the manual and usually consists of the same representatives from across the company that comprise the Preventative Maintenance Management Team plus additional persons that may be appointed by the Director of Engineering in advance of each meeting.
- The Review meetings are normally co-chaired by the IMP Team leader and the IMP Program Engineer.
- IMP Review Team meetings occur at least once a year, or more frequently if the need arises. Typical topics reviewed at each meeting include:
  - Assessment results
  - Regulatory Issues — noncompliances
  - Causes of any failures resulting from hydrostatic tests
  - Inspection tool performance and tool vendor performance
  - DA performance, recommendations, and conclusions
  - Repair methods used
  - Potential preventative measures
  - Staffing for Assessments and Repairs
  - Improvements to the IMP
  - Field activities operational information potentially affecting or important to IMP
  - Performance in identifying potential HCAs
  - Additional training requirements necessary to support IMP
  - Threat identification review
  - Additional items as necessary to aid in the success of the IMP program
  - Cost Reporting

It is good practice for audit and quality assessment purposes to include in this section of the Integrity Management Manual the frequency and topics discussed at the meetings of this team.

IMP Team Leader Responsibilities

- Develops and Maintains the IMP
- Ensures the program is regularly updated to reflect current integrity findings (e.g., new threats to HCAs), use of evolving technology, and changes to regulatory requirements
- Ensures team members are informed of all pertinent issues and are competent in the tasks they are asked to perform
- Ensures the pipeline integrity documentation is appropriately recorded and filed for ease of access in reviews or audits
- Participates in pipeline defect assessment decision making with respect to need for and type of repairs
- Prepares the assessment of the results from previous year's monitoring activities, ensuring all learnings are carried forward in to the planning process

- Prepares assessment of Integrity Management program performance for review by management team
- Leads the development and budget preparation of long-term pipeline integrity plans
- Provides regular summary reports to management that:
    - outline regulatory noncompliances that effect safety or the pressure containing operating conditions
    - have significant integrity initiatives with budget impacts

Preventive Maintenance Management Team

- Under the direction of the Directors of Engineering and Operations, this team develops inspection and preventative maintenance programs for all pipeline facilities and equipment, which could affect an HCA or pose a threat to the company's asset base.

Integrity Management Specialist Responsibilities

- Develops and maintains baseline data and re-assessment plan schedule
- Keeps current with Regulatory changes and initiates with the IMP team leader required program changes as a result
- Maintains risk assessment database
- Performs risk assessment modeling of pipeline segments, etc.
- Oversees Integrity Management assessments.
- Reviews ILI data for errors and deficiencies
- Oversees and recommends anomaly repairs and methods
- Reviews safe operating conditions for conducting immediate repairs
- Develops criteria for Integrity Management vendor/contractor selection

IMP Program Engineer Responsibilities

- Assists in developing and maintaining baseline and continual re-assessment plan schedule maintains Risk Assessment Database/performs Risk assessment modeling of pipeline segments
- Ensures procurement and good record keeping of Integrity Management data, including assessment and repair, and preventative/mitigative data. Incorporates regional Integrity Management data into centralized Risk Assessment Database
- Ensures integration of pipeline inspection data with all other pipeline Integrity Management data
- Prepares agenda for Program Management meeting for evaluation of IMP performance and selection of recommended Preventative and Mitigative options
- Specifies and compiles Integrity Management data, including DA, ILI, pressure testing, DA surveys, corrosion surveys, and nonpipe facility testing and inspection
- Recommends preventative maintenance programs for line pipe and nonpipe facilities
- Oversee ILI runs, hydrotesting, and other Integrity Management field activities
- Review hydrostatic test failures

Corrosion Specialist Responsibilities

- Field Corrosion Surveys
- Visual inspection of investigative digs of corrosion sites and third party damage
- Pipe coating repairs
- Develops and maintains database of all Integrity Management data including: HCA locations and type, as-built information, inspections, repair, preventative maintenance, third-party activities, surveillance, corrosion management, and CP systems

- Identifies necessary changes in type of data collected and/or methodology of integrity data collection
- Performs threat identification/risk analyses identifying HCAs

Pipeline Technician Responsibilities

- Performs field activities set out in integrity annual plan
- Marking and locating of buried structures
- Ensures complete documentation of Integrity Management activities.
- Ensures contractors have required qualifications and training
- Participates in review of regional annual results
- Participates in development of annual pipeline integrity plan

Documentation of Personnel Qualifications

This section will set out the qualifications and training deemed necessary by the company to perform each of the various functions identified within the IMP. Training records must be retained for Audit purposes. A typical example follows in Figures 2-6 and 2-7.

Training and Hiring Requirements

It is useful to provide a paragraph describing who is responsible for training and hiring Integrity Management personnel and updating the necessary requirements. The qualification and hiring of contractor personnel are generally handled in the technical qualification portion of the Contracted Services process. A statement should be made to the effect that the company uses prequalified contractor personnel.

HIGH CONSEQUENCE AREA (HCA) IDENTIFICATION

Although the identification of what is considered a "high-consequence area," or HCA, is not typically required for pipeline systems operating outside the jurisdiction of the United States Code of Federal Regulations, the following section shall briefly discuss the definition of HCAs, types of HCAs, and the identification of HCAs, as applicable to the *OPS Regulations, Code of Federal Regulations (CFR), Title 49 Parts 190 to 199*, as published by the US DOT PHMSA.

For Liquid Pipelines

Effective May 29, 2001, the *CFR* governing hazardous liquid pipeline operation and maintenance was amended to establish new requirements for *Pipeline Integrity Management in High Consequence Areas* (49 *CFR* 195.452). The purpose of these new requirements is to enhance and validate pipeline integrity and provide improved protection for HCAs that could be affected by an unintended release of hazardous liquids from a pipeline system. For hazardous liquids pipelines, HCAs are defined in 49 *CFR* 195.450 as:

1. A *high population area*, which means an urbanized area, as defined and delineated by the U.S. Census Bureau, that contains 50,000 or more people and has a population density of at least 1000 people per square mile
2. Another *populated area*, which means a place, as defined and delineated by the U.S. Census Bureau, that contains a concentrated population, such as an incorporated or unincorporated city, town, village, or other designated residential or commercial area.
3. A *commercially navigable waterway*, which means a waterway where a substantial likelihood of commercial navigation exists. These waterways are identified in the National Waterways Network, a geographic database created by the National Waterways GIS Design Committee.
4. An *area of the environment that has been designated as unusually sensitive to oil spills* (an "unusually sensitive area" or USA). USAs are defined in 49 *CFR* 195.6.

Job title	Job Skills	Education, Training and Experience
IM Program Engineer	Technical writing and verbal communication. Interpretation and application of published engineering calculation procedures and regulatory standards for pipeline systems. Application of metal corrosion principles and prevention practices. Supervision of technical staff / interface with senior management.	BS degree in Engineering, or equivalent experience. Experience in the design, inspection and repair of pipeline systems including: • Design and specification of pipe for procurement and installation utilizing ASME and ANSI standards. • Calculation of safe operating pressure for pipe anomalies using 831.G,modified B31.G, RSTRENG or similar methodology. • Direct participation in inspection, repair and preventative maintenance programs for line pipe and non-pipe facilities (e.g. pump stations).

Figure 2-6. Typical documentation of qualifications

For Natural Gas Pipelines

For natural gas pipelines, HCAs are defined in 49 *CFR* 192.903 as:
HCA means an area established by one of the methods described in paragraphs (1) or (2) as follows:

1. An area defined as:

    a. A Class 3 location under §192.5 or
    b. A Class 4 location under §192.5 or
    c. Any area in a Class 1 or Class 2 location where the potential impact radius (Figure 2-8) is greater than 660 ft (200 m), and the area within a potential impact circle contains 20 or more buildings intended for human occupancy or
    d. Any area in a Class 1 or Class 2 location where the potential impact circle contains an identified site

2. The area within a potential impact circle containing:

    a. 20 or more buildings intended for human occupancy, unless the exception in paragraph (4) applies; or
    b. An identified site

EMPLOYEE NAME:- . . . .

(first) (*last)*

A. - Integrity Management Program (IMP) Function

IMP Position Title :. . . . .

IMP Functionalresponsibilities

Responsibilities : ***per job description (Check all that apply) exceptions or additions (specify)*** :

B. - Relevant Education and Experience

List education and experience relevant to the IMP Functional Responsibilities that are listed in Job Skills Analysis. table Attach records to file as appropriate

Education

Education and Company training Yr Completed

-- ----------------

-- ----------------

Experience

No. Company *Relevant Experience* *Experience Time Yrs*

---------------- ---

Figure 2-7. Typical training record (kept for audit)

Where a potential impact circle is calculated under either method (1) or (2) to establish an HCA (see also Figure 2-9), the length of the HCA extends axially along the length of the pipeline from the outermost edge of the first potential impact circle that contains either an identified site or 20 or more buildings intended for human occupancy to the outermost edge of the last contiguous potential impact circle that contains either an identified site or 20 or more buildings intended for human occupancy.

Class Location Unit

A "class location unit" is an onshore area that extends 220 yards (660 ft) on either side of the centerline of any continuous 1-mile length of pipeline.

Dwelling Unit I Building

Each separate dwelling unit in a multiple dwelling unit building is counted as a separate building intended for human occupancy.

Class 1

A Class 1 location is:

1. An offshore area or
2. Any class location unit that has 10 or fewer buildings intended for human occupancy

Class 2

A Class 2 location is any class location unit that has more than 10 but fewer than 46 buildings intended for human occupancy. When a cluster of buildings intended for human occupancy requires a Class 2 location, the class location ends 220 yd (660 ft) from the nearest building in the cluster.

Class 3

A Class 3 location is:

1. Any class location unit that has 46 or more buildings intended for human occupancy; or
2. An area where the pipeline lies within 100 yd (300 ft) of either a building or a small, well-defined outside area (such as a playground, recreation area, outdoor theater, or

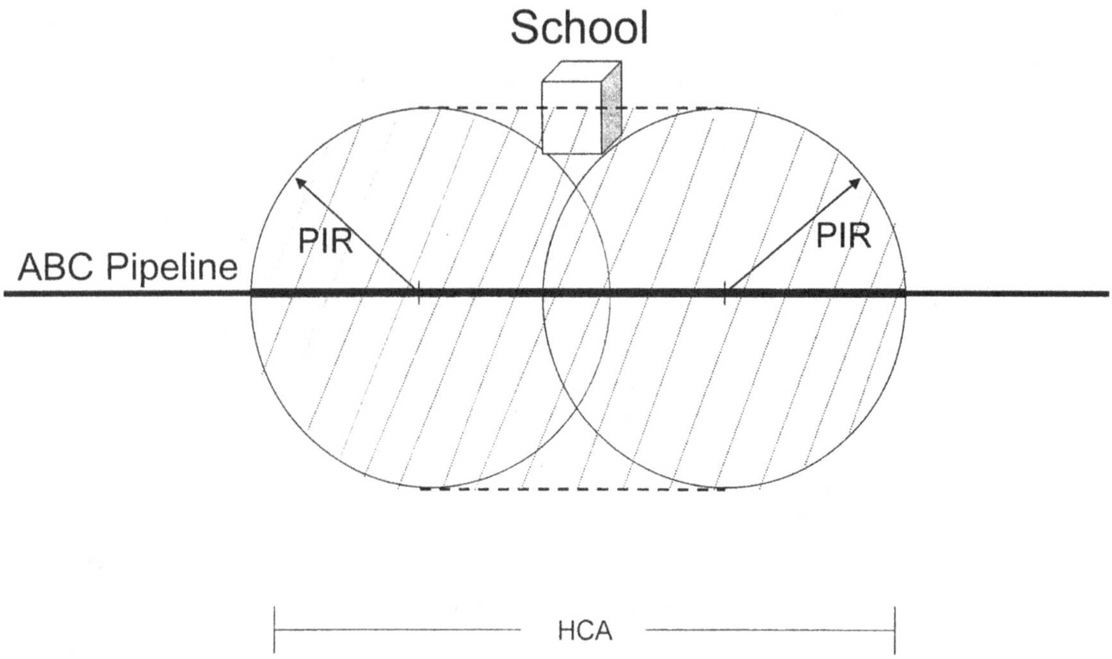

Figure 2-8. Determining HCA (American Society of Mechanical Engineers 2004)

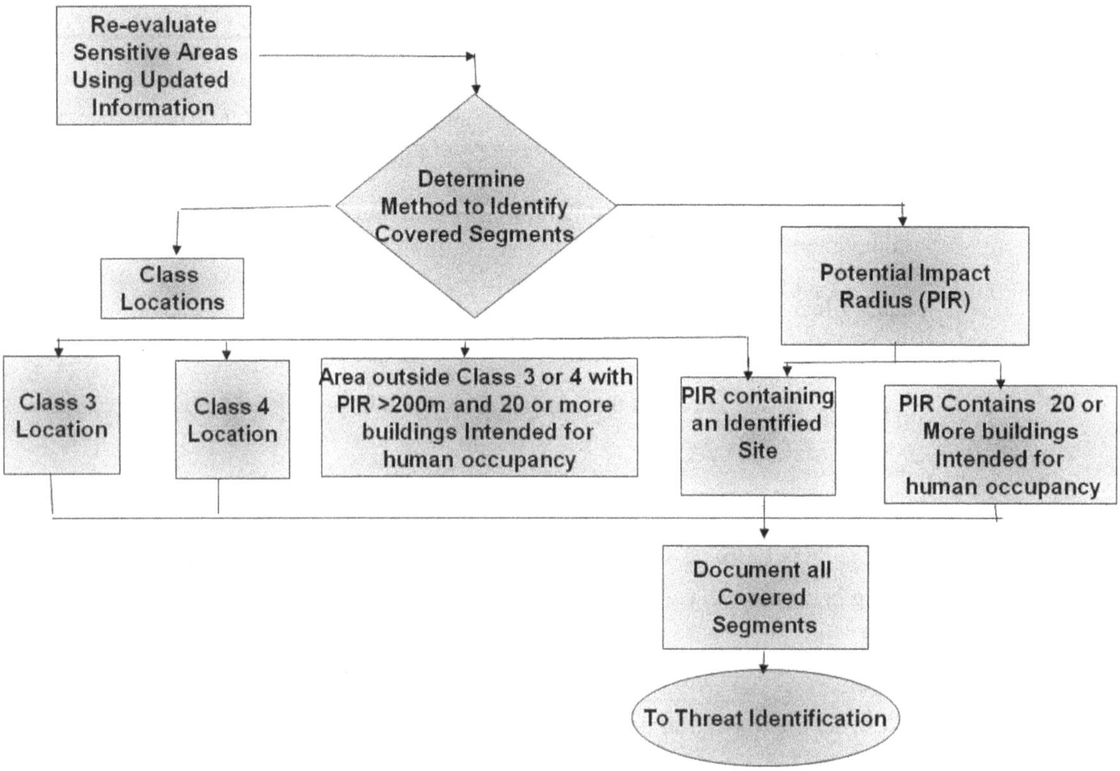

Figure 2-9. Area classification approach (Murray 2010)

other place of public assembly) that is occupied by 20 or more persons on at least 5 days a week for 10 weeks in any 12-month period (The days and weeks need not be consecutive.)

When a cluster of buildings intended for human occupancy requires a Class 3 location, the class location ends 220 yards (660 ft) from the nearest building in the cluster.

Class 4

A Class 4 location is any class location unit where buildings with four or more stories above ground are prevalent.

Identifying HCAs

HCAs, commonly called HCAs, are those locations where a pipeline spill might have significant adverse impacts to population, the environment, or commercial navigation. The definitions for various types of HCAs are periodically reevaluated and operators should be familiar with the current regulations for managing pipeline system integrity and the definitions of HCAs. The US federal government, through the OPS, makes maps and databases describing HCAs available to the pipeline industry and to the public through the OPS Web site. An operator must consider the federally prescribed HCAs in developing a pipeline IMP.

The physical location of some HCAs will change over time as new population and environmental resource data becomes available. Accordingly, the maps delineating the locations of HCAs are expected to be updated. Hence, it is important that the operator periodically ensure that its IMP considers the most recent HCA location information provided by the government.

Information about HCAs is used in several key elements of an IMP.

- Data gathering
- Risk assessment
- Inspection and mitigation

HCA information should also be incorporated into an operator's response plan.

As part of the data gathering and integration, US pipeline operators must determine the likelihood that a particular pipeline segment, segments, or system will affect an HCA in the event of a leak or spill. Operators must look at HCAs that the pipeline segment crosses, as well as those HCAs in proximity to the segment or system. In addition to the pipeline right-of-way, operators need to evaluate any impact zones for the pipeline and associated pipeline station and delivery locations. Impact zones will vary in size and complexity. When determining a potential impact zone, operators should consider:

1. The health and safety consequences of a release, including the possible need for evacuation
2. The nature and characteristics of the product or products transported (gas, refined products, crude oil, highly volatile liquids, etc.)
3. The operating conditions of the pipeline (pressure, temperature, flow rate, etc.)
4. The topography of the land associated with the HCA and the pipeline segment
5. The hydraulic gradient of the pipeline
6. The diameter of the pipeline, the potential release volume, and the distance between isolation points
7. The type and characteristics of the HCA crossed or in proximity to the segment
8. Potential physical pathways between the pipeline and the HCA
9. Potential natural forces inherent to the area (flood zones, earthquake zones, subsidence areas, etc.)
10. Response capability (time to detect, confirm and locate a release; time to respond; nature of response; etc.)

Developing an understanding of the potential impact zone ensures that operators include the appropriate HCA information for use in the risk assessment. HCA maps and databases provided by the OPS will include information about the quality of the HCA map or data. Commonly called "meta data" or data about the data, operators should document the meta-data for the HCA information at the time they acquire the HCA information from the OPS or other sources.

In addition to HCA information obtained from OPS, operators must document any HCA areas identified during routine operator conducted risk assessments, right-of-way surveillance, or other activity related to HCAs.

THREAT IDENTIFICATION AND RISK ANALYSIS PLAN

Identifying the potential threats to pipeline integrity and adopting a risk-based approached to pipeline Integrity Management are very important processes to have within the overall IMP. The Risk Analysis Plan consists of the systematic process in which potential hazards from facility operation are identified, and the likelihood and consequences of potential adverse events are estimated. Each pipeline segment is then given a numerical score based on the estimated risk. The ultimate goal of the Risk Analysis Plan is for the operator to be able to identify and prioritize the risks in a given pipeline segment so that adequate risk mitigation can be implemented to ultimately lower the risks and improve the integrity of the pipeline segment. More detailed and specific information on conducting pipeline risk assessments can be found in Chapter 10 of this book.

Threat Identification

In order to begin the risk analysis for the pipeline system and segments, the pipeline operator first needs to consider all the potential threats faced by the pipeline. All threats to the pipeline need to be considered. The threat identification information about the pipeline segments can then be used to assess the levels of risks within the Risk Analysis Plan.

The PRCI has extensively studied gas pipeline incident data and has identified 22 root causes that represent a threat to pipeline integrity and thus need to be managed. One of the root causes reported by operators is "unknown" (no root cause was identified or determined). The remaining 21 root causes or threats were grouped into nine categories more related failure types and further defined by three time-related defect types (time-dependant, stable, time-independent). The following is a breakdown of the nine categories and the corresponding 21 root causes identified:

1. Third-party damage
   - Damage inflicted by first, second or third parties
   - Previously damaged pipe
   - Vandalism
2. Corrosion-related
   - External corrosion
   - Internal corrosion
3. Equipment
   - Gasket or o-ring failure
   - Stripped thread/broken pipe/coupling failure

- Control/relief equipment malfunction
- Seal/pump packing failure

4. Incorrect operation

- Incorrect operational procedure

5. Weather related

- Cold weather
- Lightning
- Heavy rains or floods

6. Manufacturing-related defects

- Defective pipe seam
- Defective pipe

7. Construction

- Defective pipe girth weld
- Defective fabrication weld
- Coupling failure
- Wrinkle bend or buckle

8. Outside force

- Earth movement

9. Environment

- Stress corrosion cracking

The operator should consider each of these threats individually or in combination when assessing the potential threats for each pipeline system or segment. That is, the pipeline operator should consider the interactive nature of threats (i.e., more than one threat occurring on a section of pipeline at the same time). An example of such an interaction is corrosion at a location that also has third-party damage.

Risk Assessment and Analysis

It is important to distinguish between a risk management process and a risk analysis method. Risk analysis and assessment is the estimation of risk for the purposes of decision making. Risk management is the overall process that includes the results of the risk analysis and assessment, maintenance activity, and reintegration of data into subsequent risk assessments. Risk assessment methods can be very powerful analytical tools to integrate data and information and help understand the nature and locations of risks along a pipeline. However, risk assessment methods alone should not be relied upon to establish risk, nor solely determine decisions about how risks should be addressed. Risk assessment methods should be used a part of a process that involves knowledgeable, experienced personnel that critically review the input, assumptions, and results.

The risk assessment and analysis integrates available information about the threats to its pipelines and the consequences of failure. In general terms, the data gathered in the initial assessment of the pipeline and threat identification can be used to conduct a risk assessment for the various pipeline systems or segments. The data used should be as complete, as consistent, and as accurate as possible across the given pipeline segment. Actual data should be used for the analysis, and the use of global data assumptions should be avoided where possible.

There continues to be a number of different methods and products available for performing risk assessments and the operator should tailor the method used to meet the needs of the specific system. It is the operator's responsibility to apply the level of integrity/risk analysis methods that meets the needs of the operator's IMP. More than one type of model may be used throughout an operator's system. A thorough understanding of the strengths and limitations of each risk assessment method is necessary before a long-term strategy is adopted. Regardless, all risk assessment approaches have the following common components:

- They identify potential events or conditions that could threaten system integrity.
- They evaluate likelihood of failure and consequences.
- They permit risk ranking and identification of specific threats that primarily influence or drive the risk.
- They lead to the identification of integrity assessment and/or mitigation options.
- They provide for a data feedback loop mechanism.
- They provide structure and continuous updating for risk reassessments.

BASELINE ASSESSMENT PLAN (BAP)

The Baseline Assessment Plan is developed as a result of initial data gathering and risk assessment and consists of an initial inspection plan and possibly some mitigation activities including a schedule for these activities. To develop the Baseline Assessment Plan, the most appropriate inspection technique(s) must be identified for each pipeline segment and the work must be prioritized and scheduled. Assessment here means inspection (e.g., direct assessment, ILI, pressure testing). Inspection of each segment could be accomplished by hydrostatic testing, ILI, other equivalent technologies, or a combination of these techniques. The previously completed risk analysis and assessment can provide guidance to what factors need to be considered (see Figures 2-10 and 2-11).

The Baseline Assessment Plan, once developed, should identify what to inspect, how to inspect, and when to inspect. The baseline plan should also include a list of mitigation activities.

The operator should consider the following factors in developing the baseline plan:

1. Pipeline anomalies that can adversely affect pipeline integrity
2. Various inspection techniques typically used for underground pipelines
3. Methodology for evaluation of ILI data
4. Pipeline repair methodologies and other mitigation activities that can improve pipeline integrity.

The IMP program engineer is responsible for generating, maintaining and updating the Baseline Assessment Plan as necessary. This responsibility includes maintaining records of all revisions/updates made to the baseline plan. The IMP program engineer is responsible for ensuring that any prior assessments used in the baseline assessment plan meets the current assessment requirements and that required remediation actions have been taken.

The operator's IMP documentation should explicitly describe a process that requires that baseline assessment is performed on each pipeline segment that could affect an HCA. It should reference the results of the segment identification process and require that the baseline assessment plan include, individually, each segment determined to affect an HCA.

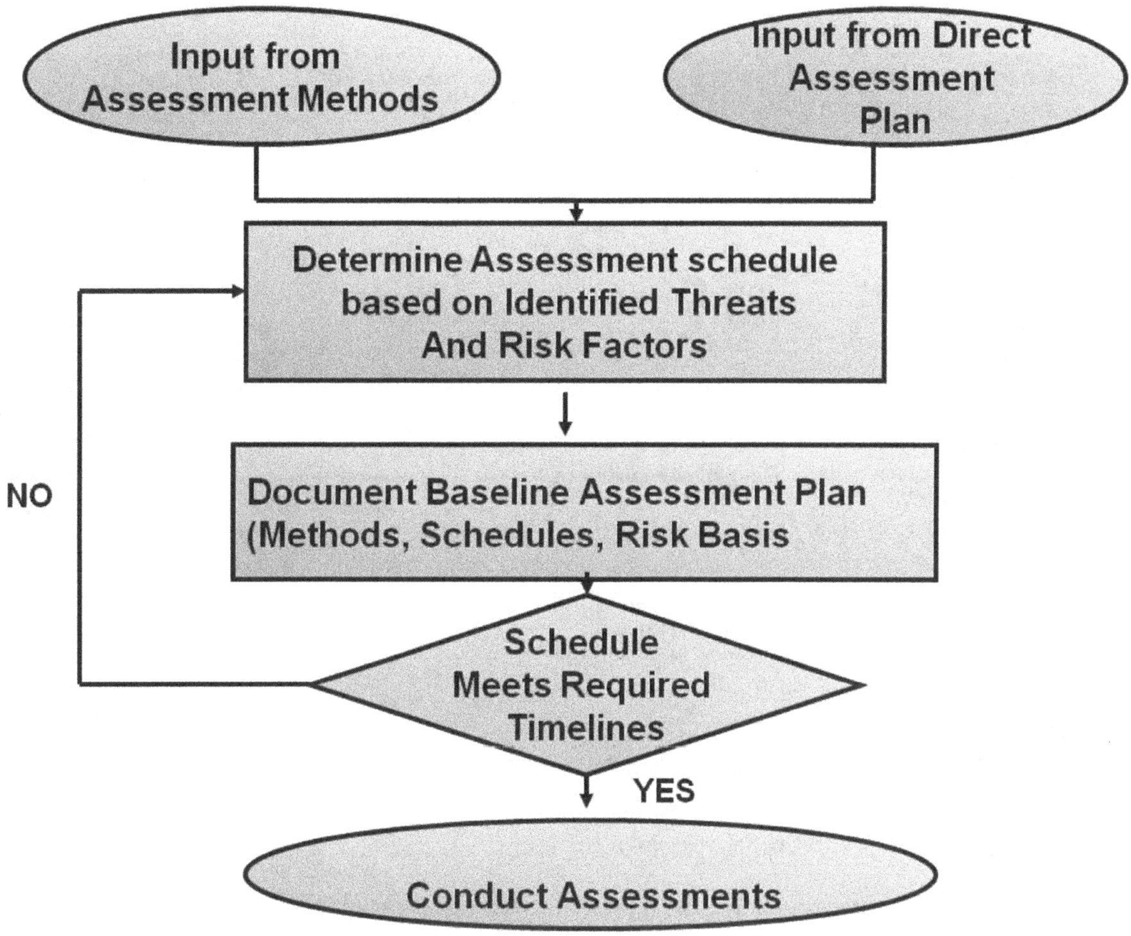

Figure 2-10. Baseline assessment

The Baseline Assessment Plan:

- must uniquely identify all of the pipeline segments to be assessed;
- must describe how inspections will be made (i.e., it must include the specific assessment techniques to be used for each segment covered by the plan, including specific tool types for segments to be assessed using ILI tools); and
- must describe when to assess (i.e., it must contain a specific schedule for when each pipeline segment covered by the plan is to be assessed).

It may consist of a single plan or a series of plans depending on the operator's preferences and business decisions. For example, the company may own and operate several distinct pipelines, or when managing the work of conducting assessments, these may be tracked more easily based on sections of line pipe that are identified by predetermined demarcation points such as launcher/receiver stations, valves, metering stations, pump stations, line connections, and line size transitions. However, each segment that could affect an HCA must be addressed *somewhere* in a baseline assessment plan. If an operator chooses to develop multiple plans, extra inspection time may need to be devoted to ensuring completeness.

The Baseline Assessment Plan often references other implementing documents or sections of the Integrity Management documentation such as segment identification results or risk analy-

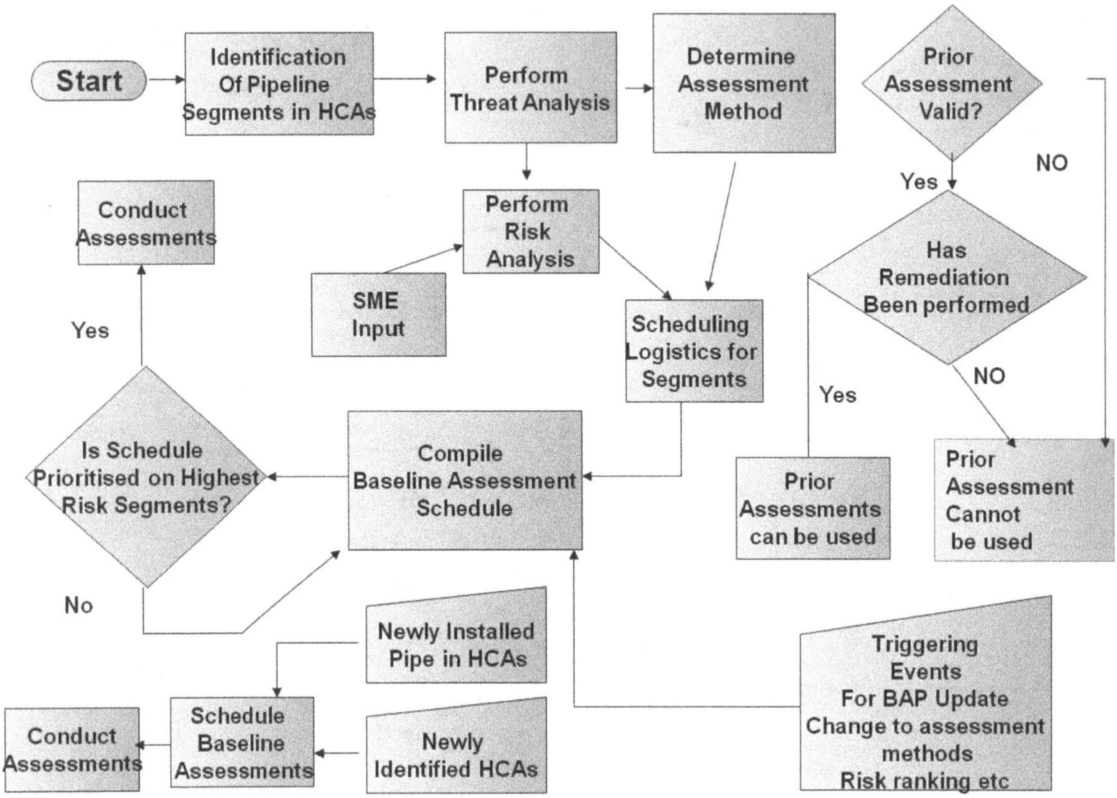

Figure 2-11. General overview of processes

sis documentation. It is often possible to be compliant through the use of prior assessment method(s), which address one or more of the threats identified for the covered segment as part of the current IMP. However remedial actions to address anomalies identified by such an assessment have to have been completed in accordance with the remediation section of the IMP. A commitment to complete a re-assessment within a set time period from the prior assessment is generally required. The re-assessment method for external and internal corrosion can be a CDA.

The operator can prioritize his baseline assessment schedule using a risk model (see Chapter 10).

The operator should develop and describe procedures to ensure that any assessments are conducted in a manner that minimizes environmental and safety risks. These procedures consider the following factors:

- Operational conditions of the pipeline, including operating pressure
- Type of assessment to be performed
- Cleanup and waste disposal
- Worker protection

Process Documentation

The IMP Program Engineer is responsible for documentation of the baseline assessment process. This documentation consists of:

- The database and results of the risk analyses used in establishing the Baseline Assessment Plan schedule priorities

- The database and results of assessment method selections for each HCA segment
- The procedure to determine the assessment schedule and maintenance of that schedule.

Events that could trigger the need for updating the Baseline Assessment Plan include:

- Ongoing results of assessments indicate the need to modify or change the assessment method for future inspections.
- Identification of new threats to the integrity of a pipeline segment requires additional inspection methods for a covered pipe segment.
- Identified changes in risk-ranking data (or risk analysis method) significantly changes the relative risk (and schedule priority) for covered pipeline segments.
- New or improved technologies are developed and proven to provide better evaluation of potential pipeline anomalies with equivalent cost structures.
- Acquisition of new pipeline systems.
- New HCA areas are identified along an existing pipeline route. This event requires baseline assessment of the line pipe in the newly identified area within a defined period of its discovery.
- Newly installed pipe is located in an HCA area. This requires a baseline assessment of the newly installed line within a defined time period (OPS sets the time limit within 10 years of its installation).
- A hydrostatic pressure test under preset conditions can be used to satisfy this baseline assessment requirement.

The entire Integrity Management team is responsible for recognizing information that could result in a change to the Baseline Assessment Plan. If a triggering event for a Baseline Assessment Plan change is observed, this must be directly communicated to the IMP program engineer. The IMP program engineer is also responsible for reviewing requested changes to the Baseline Assessment Plan, determining a need for changes to the Baseline Assessment Plan, and implementing those changes. All changes to the Plan should be recorded on the Baseline Assessment Plan Revision Log (Table 2-3) that is maintained by the IMP program engineer.

The Baseline Assessment Plan change log documents each change to the plan accurately and completely.

The Baseline Assessment Plan should be reviewed frequently *(usually annually)* by the IMP Review Team (program management meetings for evaluation of the IMP). This review will also assure the Baseline Assessment Plan is being kept up to date by the IMP program engineer.

INTEGRITY ASSESSMENT/MITIGATION PLANS

Integrity Assessment and Mitigation Plans describe how the pipe segments will be assessed according to the schedule and methods identified in the Baseline Assessment Plan. There are many tools and other proven methods available to assess the integrity of the pipeline system, with the three primary assessment methods being:

1. ILI — an inspection tool, often called a "smart pig," is run through the pipeline to evaluate the pipe's condition.
2. Pressure Test — the pipe is pressured to at least one and one half its normal operating limit to test the strength of the pipe. Water is usually used to pressure the pipe during the test.
3. Direct Assessment — a structured, multi-step evaluation is conducted to identify potential problem areas.

TABLE 2-3. Example of a Baseline Assessment Plan revisions log

Revision Date	Rev Code	Pipeline Description			Revision Description	Rev Name
		Area	Line No.	KP to KP	HCA Segment Numbers	

Revision Codes. List multiple codes for each revision entry as appropriate.
1. Change in Risk Ranking and/or Assessment
2. Change in Assessment Method Schedule
3. Addition of New Pipelines

Many of the most commonly used integrity assessment methods, such as ILI technologies, are constantly evolving, and the only reliable way to determine the state-of-the-art is to keep in touch with vendors, participate in industry-related technical exchanges, research centers, and correspondence with other pipeline operators.

More detailed and specific information on conducting *integrity assessments* and inspections can be found in Chapter 5 of this book. More detailed and specific information on conducting *pipeline repairs and mitigation* can be found in Chapter 7 of this book. Both of these chapters are reflective of the requirements within an operator's Integrity Assessment and Mitigation Plan.

From the information and the results of the risk assessment, as well as from understanding the type of threat faced by the pipeline, the pipeline operator should conduct an integrity assessment of the highest priority sections of the pipeline system in order to reduce and mitigate any known or potential risks and threats. Prior to selecting an assessment method, consideration should also be given to the 21 types of threats previously mentioned as part of the Threat Identification and Risk Analysis Plan and as indicated by the following (Table 2-4):

The assessment method selected should be documented as part of an Integrity Assessment Plan. More than one integrity assessment method may be required to address all the threats to a pipeline segment. Consideration should be given to any required or new regulatory requirements and should consider any Company-related or industry information on assessment method selection. The assessment method selected should also consider a number of items that need to be reviewed and validated by the operator's personnel and may include a number of the following items for consideration (Table 2-5):

The following describes the various assessment techniques and risk mitigation options that can be effective (to a varying degree) for some of the more common integrity threats faced by pipeline operators:

External Corrosion

- ILI
  - MFL
  - Ultrasonic
- Hydrotesting
- Direct assessment (ECDA)
  - Corrosion digs

Internal Corrosion

- ILI Tools
  - MFL
  - Ultrasonic
- Hydrotesting
- Direct assessment (Internal Corrosion Direct Assessment, or ICDA)
- Corrosion digs

TABLE 2-4. Identification and risk analysis plan

Type of Threat	Examples of Integrity Threat
Time dependent: corrosion/cracking	External corrosion Internal corrosion Stress corrosion cracking
Stable: manufacturing-related defects	Defective pipe seam, e.g., Pre-1970 low-frequency electric resistance welded (ERW) pipe Defective pipe
Stable: welding or fabrication related	Defective pipe girth weld Defective fabrication weld Wrinkle bend or buckle Stripped threads, broken couple
Stable: equipment	Gasket o-ring failure Control/relief equip. failure Previously damaged pipe Seal/Pump packing failure
Time independent: mechanical damage	Damage by third party Previously damaged pipe Vandalism
Time independent: incorrect operations	Incorrect operations
Time independent: weather-related, outside force	Cold weather Lightning Heavy rain or flood Earth movement

TABLE 2-5. Assessment method considerations

Assessment Method	Examples of Items for Consideration
ILI	Potential long seam failure concerns Pipeline has continuous diameter, no abrupt changes or telescoping Full-bore valves installed Pipeline does not have reduced diameter valve. Launcher and receiving barrels in-place or line is capable of accepting temporary barrels. Line has steady product/gas flow rate or may be temporarily capable of such flows. Pipe has greater than 3D bends? If no, tools avail. in limited diameters for 1½D bends. Line has bell and spigot or dresser couplings. Internal weld slag suspected
Hydrostatic testing	Customer(s) or end user(s) have alternative source for service or is/are flexible concerning interruption of service Potential long seam failure concerns Line is not piggable due to consideration listed above. Preassessment criteria results preclude use of external corrosion direct assessment for pipeline segment or a portion of the pipeline segment. Water source is available.
Direct assessment	Pipe is constructed of ferrous materials? ECDA cannot be applied to plastic pipe. Pipe has coating and coating is not tape. Rocky soil conditions? If yes, this can preclude some ECDA inspection methods. Concrete structures over pipeline? Roads, parking lots, etc. Limit access to pipeline with some inspection methods. Ground is not frozen? Frozen ground can affect readings from some inspection methods. Underwater sections; river crossings? Alternate assessment method may be required to assess this portion of the pipeline segment. Look at guided wave technology. Shallow cover or exposed pipe (influences effectiveness of some types of inspection methods)?

Stress Corrosion Cracking

- ILI Tools
  - Ultrasonic
  - EMAT
  - Elastic Wave
- Hydrotest
- Direct assessment
- SCC digs
- Recoating/rehabilitation plan
- Pipe replacement plan

Third-Party Damage/Mechanical

- ILI
  - Geometry and caliper tools
  - Inertia tools
- Aerial line patrols
- Public education programs
- Slabs, markers, signage

Geotechnical

- ILI
  - Inertia
- Slope strain measurement and monitoring
- Aerial line patrols
- Depth of cover survey
- River scour survey
- Pipeline strain relief

An understanding of the various pipeline anomalies and the conditions under which they can occur is essential in order to select the most appropriate inspection and mitigation techniques. API 1160 provides an excellent summary of the different types of anomalies and corresponding various types of ILI tools that can be used to detect them as part of an Integrity Assessment Plan (Table 2-6).

Table 2-7 includes a summary of some prevention and mitigation methods and their applicability to each of the 21 threats as given in ASME B31.8S.

Mitigation consists of two parts. The first part is the repair of the pipeline. Repair activities shall be made in accordance with ASME B31.8 or ASME B31.4 and/or other accepted industry repair techniques. Repair may include replacing defective piping with new pipe, installation of sleeves, coating repair, or other rehabilitation. These activities shall be identified, prioritized, and scheduled as part of the Mitigation Plan.

Once the repair activities are determined, the next part requires the operator to evaluate prevention techniques that prevent future deterioration of the pipeline. These techniques may include providing additional CP, injecting corrosion inhibitors and pipeline cleaning, or changing the operating conditions. Prevention plays a major role in reducing or eliminating the threats from third-party damage, external corrosion, internal corrosion, stress corrosion cracking, cold weather-related failures, earth movement failures, problems caused by heavy rains and floods, and failures caused by incorrect operations.

All threats cannot be dealt with through inspection and repair; therefore, prevention for these threats is a key element in the plan. These activities may include, for example, prevention of third-party damage and monitoring for outside force damage.

Data collected as part of the Integrity Assessment/Mitigation Plans inspection and mitigation activities needs to be analyzed and integrated with previously collected data. This is in addition to other types of Integrity Management-related data that is constantly being gathered through normal operations and maintenance activities. The addition of this new data is a continuous process that, over time, will improve the accuracy of future risk assessments via its integration. The Plans need to remain flexible and incorporate any new information that becomes known or available.

An integrity evaluation for each pipeline segment should be conducted at specified time intervals. This evaluation with accompanying data is used for re-ranking of the pipe segments. The re-ranking will determine if a re-assessment is warranted at that time. This evaluation determines whether an integrity assessment is required for each pipeline in the forthcoming months. Unscheduled integrity evaluations may need to be carried out if unusual or unexpected conditions arise that warrant such a review. The completed integrity evaluation needs to be documented and is to be retained for the lifetime of the pipeline.

Upon completion of a pipeline's baseline or integrity assessment, the reassessment interval must then be determined and included in the overall assessment plan. Each pipeline segment capable of affecting an HCA should be reassessed at intervals not exceeding a specified number of years, with any regulatory and/or code requirements being adhered to as well.

Once an assessment has been completed and a reassessment interval determined, the next reassessment year is to be documented as part of the Assessment Plan. The Assessment Plan will be updated based upon the results of integrity assessments and reassessment interval determinations.

The following two tables (Tables 2-8 and 2-9) are taken from ASME B31.8S and are examples of (1) Integrity Assessment Plan Summary and (2) Mitigation Plan Summary, for a hypothetical pipeline segment:

PERFORMANCE MEASURES PLAN

Measurement of performance is an integral part of a Pipeline IMP. Pipeline companies should conduct evaluations of their IMPs on an ongoing basis, and the performance should be reviewed at least annually and issues addressed. This will also provide the company with a continued measure of the IMP's effectiveness over time. Each operator should develop or choose significant performance measures at the beginning of the program and then periodically measure the effectiveness of the measures. Reports that summarize the results of these performance reviews should then be issued, evaluated, and discussed in order to continuously improve the program.

The performance information collected is used to periodically evaluate the success of its integrity assessment techniques, pipeline maintenance and repair activities, and the company's mitigative risk control activities. Examples of performance measures are given in the following Table 2-10:

For operators implementing more prescriptive programs, the performance metrics suitable for evaluating threat-specific conditions are given below in Table 2-11.

For operators implementing more performance-based programs, Table 2-12 below is from ASME B31.8S and gives a suggested list of performance metrics that can be used to measure the effectiveness for a performance based program. The operator can also develop their own metrics, and evaluations should be performed on at least an annual basis.

Audits

Internal and external audits should also be conducted to provide the operator with another source of information that can be used to measure and improve the effectiveness of the IMP. The performance measurement and audit results should be factored into future risk assessments as well. Audits may be performed by personnel with the company (self-assessments),

TABLE 2-6. Anomaly type and tools to detect them (reproduced courtesy of the American Petroleum Institute)

	Metal Loss Tools			Crack Detection Tools		Geometry Tools	
	Magnetic Flux Leakage (MFL)						
LI Purpose	Standard Resolution (SR) MFL	High Resolution (HR) MFL	Ultrasonic (Compression wave)	Ultrasonic (Shear wave)	Transverse MFL	Caliper	Mapping
METAL LOSS (Corrosion) External Internal	Detection[1], Sizing[3,10] No ID/OD Discrimination	Detection[2], Sizing[3]	Detection[2], Sizing[3]	Detection[2], Sizing[3]	Detection[2], Sizing[3]	No detection	No detection
NARRPW AXIAL EXTERNAL CORROSION (NAEC)	No Detection	No Detection[4]	Detection[2], Sizing[3]	Detection[2], Sizing[3]	Detection[2], Sizing[3]	No Detection	No Detection
CRACKS AND CRACK-LIKE DEFECT(axial) Stress Corrosion Cracking Fatigue Cracks, Longitudinal Seam weld Imperfections, Incomplete Fusion (Lack of Fusion) Toe-cracks	No Detection	No Detection	No Detection	Detection[2], Sizing[3]	Detection[2], Sizing[3]	No Detection	No Detection
CIRCUMFERENTIAL CRACKING	No Detection	Detection[5] and Sizing[5]	No Detection	Detection[2] & Sizing[3] if Modified[6]	No Detection	No Detection	No Detection
DENTS WRINKLE BENDS	Detection	Detection[7] & Sizing, Not Reliable	Detection[7] & Sizing, Not Reliable	Detection[7] & Sizing, Not Reliable	Detection[7] & Sizing, Not Reliable	Detection[8,10] Sizing,	Detection & Sizing, Not Reliable
BUCKLES	In Case of Detection Circumferential Position is Provided						
GOUGES	Detection[2], But No Discrimination as Gouges						No Detection

LAMINATION OR INCLUSION	Limited Detection	Limited Detection	Detection & Sizing[3]	Detection & Sizing[3]	Limited Detection	No Detection	
PREVIOUS REPAIRS	Detection of Steel Sleeves and Patches, Others only With ferrous markers		Detection Only of Steel Sleeves and Patches Welded to Pipe	Detection Only of Steel Sleeves and Patches Welded to Pipe	Detection of Steel Sleeves and Patches, Others only With ferrous markers	No Detection	No Detection
MILL RELATED ANOMALIES	Limited Detection	Limited Detection	Detection	Detection	Limited Detection	No Detection	No Detection
BENDS	No detection	No Detection	No detection	No Detection	No detection	Detection & Sizing[3]	Detection & Sizing[3]
OVALITIES	No detection	No Detection	No detection	No Detection	No detection	Detection & Sizing[3,11]	Detection & Sizing[3,9]
PIPELINE COORDINARTES	No detection	No Detection	No detection	No Detection	No detection	No Detection	Detection & Sizing[3]

NOTES

1. Limited by the minimum detectable metal loss
2. Limited by the minimum detectable depth, length, and width of the defects
3. Defined by the specified sizing accuracy of the tool
4. If the width is smaller than the minimum detectable defect width for the tool
5. Reduced Probability of Detection (POD) for the tight cracks
6. Transducers to be rotated by 90 deg
7. Reduced reliability depending on the size and shape of the dent
8. Depending on the configuration of the tool, also circumferential position
9. If equipped for ovality measurement
10. Available in tethered tool
11. If equipped for bend measurements

TABLE 2-7. Acceptable threat prevention and repair method (from ASME B31G (2004))

Prevention, Detection, and Repair Methods	TPD	Third-Party Damage		Corrosion Related		Equipment				Incorrect Operation	Weather Related			Manufacture Pipe			Construction Fab			O-Force	Environment
	(IF)	PDP	Vand	Ext	Int	Gask/ Oring	Strip/ BP	Cont/ Rel	Seal/ Pack	IO	CW	L	HR/F	Seam	Pipe	Gweld	Weld	Coup	WB/B	EM	SCC
Prevention/Detection																					
Aerial patrol	X	X	X	...	...	...	...	...	...	X	X	X	...	...	...	...	X	...	X	...	
Foot patrol	X	X	X	X	...	...	...	...	...	X	X	X	...	...	...	...	X	...	X	...	
Visual/mechanical inspection	...	...	...	...	X	X	X	X	...	X	...	.	...	...	X	...	...	...	...	...	
One-call system	X	X	X	...	...	...	...	...	...	...	...	.	...	...	...	...	...	...	...	...	
Compliance audit	...	...	...	...	...	...	...	...	X	...	...	.	...	...	...	...	...	...	...	...	
Design specifications	...	...	...	X	X	X	X	X	X	...	...	.	...	...	X	...	X	X	X	X	
Materials specifications	...	...	...	...	X	X	X	X	...	...	...	.	X	X	...	X	...	...	...	...	
Manufacturer inspection	...	X	...	...	...	...	X	X	...	...	...		X	X	...	X	...	...	...	...	
Transportation inspection	...	X	...	...	...	...	...	...	...	...	...	.	X	X	...	...	...	...	...	...	
Construction inspection	...	X	...	...	X	X	X	X	...	...	...	.	...	X	X	X	X	X	...	X	
Preservice hydrotest	...	X	...	...	...	...	...	...	...	...	...	.	X	X	X	X	X	X	...	..	
Public education	X	...	...	...	...	...	...	...	...	...	...	.	...	...	...	...	...	...	...	...	
O&M procedures	...	X	X	X	X	X	X	X	X	X	...	X	...	...	...	...	X	X	X	X	
Operator training	...	...	...	...	...	...	...	...	X	...	...	.	...	...	...	...	...	...	...	...	
Increase marker frequency	X	X	...	...	...	...	...	...	...	...	...	.	...	...	...	...	...	...	...	...	
Strain monitoring	...	...	...	...	...	...	...	...	...	...	...	X	...	...	...	...	...	...	X	...	
External protection	X	X	X	...	...	...	...	...	...	...	...	.	...	...	...	...	...	...	X	...	
Maintain ROW	X	X	...	...	...	...	...	...	...	...	...	.	...	...	...	...	...	...	X	...	
Increased wall thickness	X	X	X	X	...	...	...	...	...	...	...	.	...	...	...	...	...	...	X	...	
Warning tape mesh	X	X	...	...	...	...	...	...	...	...	...	.	...	...	...	...	...	...	...	...	
CP monitor/maintain	...	...	...	X	...	..	..	..	..	..	...	...	...	...	...	...	...	...	...	X	
Internal cleaning	...	...	...	X	...	...	...	...	...	...	...	.	...	...	...	...	...	...	...	...	
Leakage control measures	...	X	X	X	X	X	X	X	...	...	...	.	...	...	...	...	X	...	...	...	

Pig-GPS/strain measurement	...	...	...	.....	..	...	...	...	...	X	...	X	...	...	...	...	...	...	X	...	
Reduce external stress	...	...	...	.....	..	X	...	...	...	...	...	.	...	...	...	...	X	X	X	X	
Install heat tracing	...	...	...	.....	..	...	...	...	...	X	...	.	...	...	...	...	...	...	...	...	
Line relocation	X	...	X	.....	..	...	...	...	...	X	...	X	...	...	...	...	...	...	X	...	
Rehabilitation	...	X	...	X	..	...	...	...	...	...	...	.	...	...	...	...	X	X	X	X	
Coating repair	...	...	...	X...	..	...	...	...	...	...	...	.	...	...	...	...	...	...	...	X	
Increase cover depth	X	...	X	.....	..	...	...	...	...	...	...	..	...	...	...	...	...	X	...	...	
Operating temperature reduction	...	....	.....	...	X	...	...	X	...	...	...	.	...	...	...	...	...	...	...		
Reduce moisture	...	...	...	...X	..	...	...	...	...	...	...	.	...	...	...	...	...	...	...	...	
Biocide/inhibiting injection	...	...	...	...X	..	...	...	...	...	...	...	.	...	...	...	...	...	...	...	...	
Install thermal protection	...	...	...	.....	..	...	...	...	...	X	...	.	...	...	...	...	...	...	...	...	
Repairs																					
Pressure reduction	...	X	...	X	X	...	...	...	...	...	...	.	...	X	X	X	X	X	...	...	X
Replacement	...	X	X	X	X	X	X	X	X	...	X	X	X	X	X	X	X	X	X	X	X
ECA, recoat	...	...	...	X	X	...	...	...	...	...	...	.	...	...	...	X	...	...	...	...	...
Grind repair/ECA	...	X	X	...	..	...	...	...	...	...	...	..	...	X	X	X	X	...	...	...	X
Direct deposition weld	...	...	X	X	..	...	...	...	...	...	...	.	...	...	...	...	...	...	...	...	...
Type B, pressurized sleeve	...	X	X	X	X	...	...	...	...	...	...	.	...	X	X	...	X	X	...	...	X
Type A, reinforcing sleeve	...	X	X	X	..	...	...	...	...	...	...	.	...	X	X	...	...	...	...	...	X
Composite sleeve	...	...	...	X	...	...	...	...	...	...	...	.	...	...	...	...	...	...	...	...	...
Epoxy-filled sleeve	...	X	X	X	..	...	...	...	...	...	...	.	...	X	X	X	X	X	X	...	...
Mechanical leak clamp	...	...	...	X	..	...	...	...	...	...	...	.	...	...	...	...	...	...	...	...	...

General Notes: The abbreviations found in this table relate to the 21 threats discussed in paragraph 5. Explanations of the abbreviations are as follows:
Cont/Rel, control/relief equipment malfunction; L, lightning; Coup, coupling failure; PDP, previously damaged pipe (delayed failure mode); CW, cold weather ; Pipe, defective pipe; EM, Earth movement ; Pipe Seam, defective pipe seam; Ext, external corrosion; SCC, stress corrosion cracking; Fab Weld, defective fabrication weld; Seal/Pack, seal/pump packing failure; Gask/O-ring, Gasket or O-Ring; Strip/BP = stripped thread/broken pipe; Gweld, defective pipe girth weld; TPD (IF), damage inflicted by first, second, or third parties; HR/F, heavy rains or floods; Vand, vandalism; Int, Internal Corrosion ; WB/B, wrinkle bend or buckle; IO, incorrect operations company procedure.

or by auditors or subject matter experts from outside the organization. As mentioned in API 1160, examples of questions that IMP audits should assess include:

- Are activities being performed as outlined in the operator's program documentation?
- Is someone assigned responsibility for each subject area?
- Are appropriate references available to those who need them?
- Are the people who do the work trained in the subject area?
- Are quailed people used when required by code or regulation?
- Are activities being performed using an appropriate Integrity Management framework as outlined in this standard (API 1160)?
- Are all required activities documented by the operator?
- Are action items followed-up?
- Is there a formal review of the rationale used for developing the risk criteria used by pipeline operator?
- Are there established criteria for repairing, re-rating, replacing, or rerouting damaged pipelines? Are criteria established for these activities stated above for terminals, pump stations, associated piping and relief systems?
- Is there a formal review of the data provided by federal data sources that may have changed and what is the frequency of that review?

Integrity Management Reporting

As part of the performance reporting, the operator should include the following:
Overall assessment of the IMPs effectiveness

- List of integrity assessments completed and results obtained
- Repair/remediation work completed in the previous year
- Process improvements identified and implemented

TABLE 2-8. Example of integrity assessment plan summary table for a hypothetical pipeline segment (from ASME B31G (2004))

Threat	Oriteria/ Risk Assessment	Integrity Assessment	Mitigation	Interval years
External corrosion	Some external corrosion history, no ILI	Conduct hydrostatic test, perform ILI, or perform direct assessment	Replace/ repair locations where OFP below 1.25 times the MAOP	10
Internal corrosion	No history of IC issues, no ILI	Conduct hydrostatic test, perform ILI, or perform direct assessment	Replace/ repair locations where OFP below 1.25 times the MAOP	10
SOC	Have found SOC of near critical dimension	Conduct hydrostatic test	Replace pipe at test failure locations	3–5
Manufacturing	ERW pipe, joint factor < 1.0, no hydrostatic test	Conduct hydrostatic test	Replace pipe at test failure locations	N/A
Construction/ fabrication	No construction issues	None required	N/A	N/A
Equipment	No equipment issues	None required	N/A	N/A
Third-party damage	No third-party damage issues	None required	N/A	N/A
Incorrect operations	No operations issues	None required	N/A	N/A
Weather and outside force	No weather or outside force related Issues	None required	N/A	N/A

TABLE 2-9. Example of mitigation plan summary table for a hypothetical pipeline segment (from ASME B31G)

Example	Description
Repair	Any hydrostatic test failure will be repaired by replacement of the entire joint of pipe.
Prevention	Prevention activities will include further monitoring for SCC at susceptible locations, review of the CP design and levels, and monitoring for selective seam corrosion when the pipeline is exposed.
Interval for Re-inspection	The interval for reinspection will be 3 years if there was a failure caused by SCC. The interval will be 5 years if the test was successful.
Data Integration	Test failures for reasons other than external or internal corrosion, SCC, or seam defect must be considered when performing risk assessment for the associated threat.

General Notes: For this pipeline segment, hydrostatic testing will be conducted. Selection of this method is appropriate due to its ability to address the internal and external corrosion threats as well as the manufacturing threat and the SCC threat. The test pressure will be at 1.39 times the MAOP.

- Performance against the previous year's summary and any identified trends
- Notice of any changes in pipeline operating parameters, e.g., operating pressure, flow rate
- Any other integrity-related activities carried out in the previous year
- Identification of any work scheduled in the previous year but not completed
- A plan to complete work previously scheduled but not completed

Performance Improvement

The results of the performance measurement and any audits should be reviewed by the operator and evaluated to determine the effectiveness of the program, and any recommendation for improvements or changes should be acted upon in order to have continuous improvement built into the program.

All results and any changes made will need to be documented and distributed throughout the company as per requirements in the Communications Plan and the Management of Change Plan.

COMMUNICATION PLAN

Communications Plan should be developed and implemented so that information related to the results of the integrity program activities is properly distributed to the various internal and external stakeholders in a consistent and orderly fashion.

Communications should be communicated periodically and as often as necessary to communicate any significant changes or developments to the IMP. Some of the information should be communicated more routinely, and tools such as information mailouts and the company Web sites are effective ways of facilitating the communications.

An example (Mikkola 2007) of the general structure for communication plan within a Company can be given as follows:

External Communications

External communications should be issued and updated regularly regarding the operator's IMP and developments. Key parties and audiences with vested interests to be informed include:

- The general public (including landowners and public along the right-of-way)
- Local and regional emergency officials
- Public officials other than emergency officials

To assist and facilitate external communications, the operator should consider having programs in place such as Public Awareness Programs, information mailout programs,

TABLE 2-10. Example of performance measures

Goal	Description	Action
Reduce total number of reportable releases	A reduction in the total number of reportable liquid pipeline releases over a 5-year timeframe to achieve a demonstrable improvement of at least 10% from the baseline condition. Compare results with published statistics	Track and trend release information
Reduce total volume of reportable releases	A reduction in the total volume of unintended reportable liquid releases by at least 10% from the baseline condition within 5 years of the initiation date of this plan. Compare results with published statistics	Track and trend release information
Integrity assessment schedule maintained	Integrity assessments, method and timing completed per schedule	Take appropriate action for maintaining schedule
Community awareness	Community awareness notices and/or meeting satisfied	Monitor activity for compliance
Improved pipeline information	Provide greater knowledge of the pipeline conditions as the IMP progresses with specific attention to data accuracy, quantity, age, and usefulness	Pipeline database being implemented
Increased risk model correlation and relevance	Identify predictive correlations to the risk model in relation to the surrounding environment of the pipeline. The model should provide the ability to intuitively predict similar instances of increased risk.	Algorithm development and improvement to be completed by Systems Integrity personnel based upon results and trends from assessments and improved knowledge of pipeline conditions
Continuous algorithm improvement	Develop timely and useful preventative and mitigative strategies and define additional variables within the risk management algorithm that would provide effective measurement of program performance	Algorithm development and improvement to be completed by Systems Integrity personnel based upon results and trends from assessments
Improved internal audit scores	Improvement in the scores of internal audits of the Company pipeline system	Systems Integrity Department to develop an internal IMP review program, including assessment documentation and implementation review protocols. Results of the review will be used to improve overall program implementation strategies and not be based on a "scoring" system
External audits	Improvement in the scores of external audits of the Company pipeline system and/or a reduction in the number of findings	
Operational events	No unplanned relief valve openings	Reported by field operations to Systems Integrity Department
Reduction in immediate repair conditions	Performance goals based on detecting and repairing pipeline anomalies before they become immediate hazards or threats to the operational integrity of the pipeline	Development and implementation of improved corrosion mitigation program being coordinated by Systems Integrity Department

town hall meetings, and where appropriate, face-to-face discussions with affected landowners and stakeholders. A Public Awareness Program describes the key components of messages, means of delivery and frequency of activities for communicating with key audiences identified, including what the Company does to prevent accidents, how to recognize a leak, a reminder to prevent excavation damage by calling state excavation one-call numbers before digging, and what to do in an emergency. There are several industry-related publications that address Public Awareness Plans, such as *API Standard 1160 Section 10.1.5 — Improved Public Education, and API Recommended Practice 1162 — Public Awareness Programs for Pipeline Operators.*

Communications for landowner and tenants along the right-of-way should include company name and contact name/information (both routine and emergency) and include

TABLE 2-11. Threat-specific performance metrics from ASME B31.8S (2004)

Threats	Performance Matrics (Prescriptive Program)
External corrosion	Number of hydrostatic test failures caused by external corrosion Number of repair actions taken due to ILI results Number of repair actions taken due to direct assessment results Number of external corrosion leaks
Internal corrosion	Number of hydrostatic test failures caused by internal corrosion Number of repair actions taken due to ILI results Number of repair actions taken due to direct assessment results Number of internal corrosion leaks
Stress corrosion cracking	Number of in-service leaks or failures due to SCC Number of repair replacements due to SCC Number of hydrostatic test failures due to SCC
Manufacturing	Number of hydrostatic test failures caused by manufacturing defects Number of leaks due to manufacturing defects
Construction	Number of leaks or failures due to construction defects Number of girth welds/couplings reinforced/removed Number of wrinkle bends removed Number of wrinkle bends inspected Number of fabrication welds repaired/removed
Equipment	Number of regulator valve failures Number of relief valve failures Number of gasket or O-ring failures Number of leaks due to equipment failures
Third-party damage	Number of leaks or failures caused by third-party damage Number of leaks or failures caused by previously damaged pipe Number of leaks or failures caused by vandalism Number of repairs implemented as a result of third-party damage prior to a leak or failure
Incorrect operations	Number of leaks or failures caused by incorrect operations Number of audits/reviews conducted Number of findings per audit/review, classified by severity Number of changes to procedures due to audits/reviews
Weather-related and outside forces	Number of leaks that are weather related or due to outside force Number of repair, replacement, or relocation actions due to weather-related or outside-force threats

all relevant details about any maintenance or integrity activities, location information or maps, commodity information, excavation notification (one-call) numbers, as well as general information about the various plans, and other information available (i.e., emergency response, IMP, damage prevention programs, etc.). The operator must also have clearly posted marker signs along the right-of-way that include the company name and emergency contact numbers.

Internal Communications

The operator should implement an internal communications plan so that management and other personnel understand and support the IMP. This should include results of the latest performance measures reviewed on a periodic basis and summary of any changes to the IMP. Any changes to the IMP will need to be communicated to the operating company personnel in more detail through the Management of Change process.

MANAGEMENT OF CHANGE PLAN

Any changes that affect the pipeline IMP should be documented, and the program should be modified accordingly to accommodate the change. In order to accompany this, the process needs to be clearly defined, consistent, and flexible enough to accommodate both major and minor changes.

TABLE 2-12. Overall performance measures from ASME B31.8S (2004)

Miles inspected versus IMP requirement
Number of IMP changes requested by jurisdictional authorities
Jurisdictional reportable incidents/safety-related conditions per unit of time
Amount of IMP required activities completed
Fraction of system included in the IMP
Number of actions completed that impact safety
Number of anomalies found requiring repair or mitigation
Number of leaks repaired
Number of hydrostatic test failures and test pressures
Number of third-party damage events, near misses, damage detected
Risk reduction achieved by IMP
Number of unauthorized crossings
Number of precursor events detected
Number of right-of-way encroachments:
Number of pipeline hits by third parties due to lack of notification as locate request through the one-call process
Aerial/ground patrol incursion detections
Number of excavation notifications received and their disposition
Number and types of public communications issued
Effectiveness of communications
Public confidence in IMP activities
Effectiveness of the feedback process
IMP costs
Integrity improvement through use of new technology
Unscheduled outages and impact on customers

Managing change within the department or organization involves understanding how any changes within the organization can impact other departments, people, or process. The management of change also involves developing strategies and actions to manage the impact of change by the process of review, approvals, communication, and documentation. Failure to effectively implement any of these elements can adversely affect the program.

Any changes need to be recognized either before or shortly after they occur. Examples of changes that could affect program elements within the pipeline IMP could include:

- Adding, deleting, or otherwise modifying the pipeline equipment
- Changes in the fluid transported and/or its operating conditions in the pipe that may also affect the risk prioritization and any spill control or other mitigation measures employed
- Changes to flow rate and/or operating pressure, including design pressure or MAOP, pressure derates
- Restarting equipment or systems that have been out of service for an extended time and/or systems that have not been maintained
- Changes to existing procedures, or addition of new procedures
- Changes along the right-of-way, such as changes in land use, including an increase in population near the pipeline
- Regulatory changes

In order to ensure the safety and integrity, the operator will need to be developed and maintained. The process should include the following components:

- Reason for change
- Authority for approving changes

- Analysis of implications
- Acquisition of required work permits
- Documentation
- Communication of change to affected parties
- Time limitations
- Qualification of staff

In some cases, certain changes may require qualification or re-qualification of operations personnel. Training may also need to be provided to personnel to ensure proper implementation and operation of equipment or revised processes.

Guidelines need to be developed to ensure information is recorded in a consistent manner. Cataloging the references/data tools and adhering to the criteria set forth will ensure quick and easy tracking to the document, process, or task.

The integrity department of the company maintains a comprehensive listing of all resource material, data tools, and reference documents available within the department. This listing is maintained and revised accordingly for the purpose of establishing task ownership as well as a mechanism to manage changes that occur to various process or exercises being undertaken within the department. Maintaining a current inventory of resources facilitates wchanges to a process, procedure or task. This also ensures ownership of these references and tools are maintained. In conjunction with change initiatives within the integrity department, inter-department or company-wide, Management of Change is accomplished through procedures such as Project Plan/Design Basis Memorandum process, Engineering Standards and Operating and Maintenance Procedures. These procedures establish lines of communication and change authorization for the company as a whole.

DOCUMENTATION PLAN

When considering the abundance of information, the pipeline integrity department would receive and generate on a daily basis, it is evident that a well-organized system should be in place to store this information and to allow timely retrieval of it. The pipeline integrity document management system (DMS) should be developed to meet these needs and should consider the following items as part of the process.

The DMS should be a relatively flexible document management system capable of indexing and storing any form of information.

Unfortunately, it is not enough for an efficient system to simply be in place. Document management procedures should be created to ensure documents are indexed and filed in a consistent manner in order to facilitate efficient retrieval. Each individual within the department should understand them, be comfortable with them, and use them regularly.

Index Scheme

The most crucial element of any successful document management system is strict adherence to the index scheme. As information is received or generated within the pipeline integrity department, it is the responsibility of the individual(s) receiving or generating the information to ensure that it has been properly indexed. In order to achieve this, the individual(s) must have a solid understanding of the index scheme within the DMS.

Filing Guidelines

It should be ensured that the information is filed in a consistent and correct manner. Although the process used to index information is not particular to the form the information takes, the processes used to file information are:

Search and Retrieval Guidelines

Thus far the processes required to properly indexing and file a piece of information have been presented; however, it is not enough for individuals to understand only these processes. A complete document management system should also provide a systematic means of searching for and retrieving information that has been indexed and filed. Although a properly structured system can lend itself to this process in many instances by making the search for information intuitive, additional searching tools should also be available to reduce searching efforts.

Document Retention

Operators should define the retention policy for integrity-related information, according with government and/or regulatory requirements and corporate directions. Retention policy definition should be defined, but not limited, to: policies, manuals, plans, standards and procedures, studies, reports, decisions, etc.

The documentation plan should incorporate a comprehensive description about how information or data moves within the department. Several key processes carried out within the department are required to ensure sufficient information is gathered and assessed for the purpose of monitoring and mitigating integrity concerns. Many exclusive tasks take place within many of the key processes further supporting the data creation and correlation for the department. The integrity department has the utmost importance for managing information flow to maintain organization, support training, and facilitate information storage and archival activities required for the overall IMP.

Pipeline Document Management Systems

As a starting point of any pipeline integrity program, it is important that a pipeline operator has a documented management system that shows the policies and procedures in place to deal with all the operation of the pipeline system.

A complete documented management system for a pipeline operator should contain the following:

- Description of pipeline system, legal, and statutory duties
- Organization and control
- Key personnel
- Stakeholders
- Documentation and communication systems
- Management of change
- Risk analysis, evaluation, and control through whole life
- Integrity Management
- Health and safety management
- Emergency planning
- Emergency procedures
- Performance measures
- Training and qualification
- Management system review
- Management review
- Assurance audit of all elements and processes

The management system is usually a document that will be self-contained, or direct the user to other company documents that give all the necessary information.

If a company does not have a formal documented pipeline management system, or its constituent parts, then any integrity review will be slow and difficult to conduct, and many assumptions about the past, present, and future state of the pipeline will have to be made. An automated retrieval system must be put in place such that information can be retrieved and updated very fast and accurately.

This documentation system should include original design details, material documents, hydrostatic test records, failure and repair records, etc.

Document management system must be such that simple integrity or a full-scale engineering integrity reviews (see figure below) can be documented and filed as appropriately.

A data management system is very important in an IMP. A data management system facilitates the integration and analysis of information. The design principles of a data management system include:

- Support pipeline integrity initiatives.
- Standardize business practices and processes.
- Share applications across pipes.
- Create one common data warehouse.
- Utilize shelf or productized solutions.
- Utilize industry standards where appropriate.
- Build tools that fit the customer.
- Implement cost-effective solutions.
- Design for systems scalability.
- Keep it simple.
- Focus on performance.

The benefits of a data management system are:

- Promote operations of safe pipeline
- Respond to regulatory compliance
- Increase efficiency
- Lower operating costs
- Continuous improvement of pipeline data
- Foundation for corporate systems scalability

Typical Pipeline Integrity Management System

The most crucial element of any successful document management system is structuring the system for the appropriate recoding of data collections, retrievals, and updating.

The information needed for an integrity review depends on its objective. Obviously, if the review is to cover all aspects of the pipeline system, then a large amount of data will be needed.

The quality, reliability, and accessibility of information are crucial to the review. If there are few data, or it is of poor quality or unreliable, then the review will need to make major assumptions, and this will affect the conclusions.

Design, Material, and Construction Data

- Pipeline segment name
- Pipeline route coordinates, mile posts
- Pipe diameter
- Pipe wall thickness
- Pipe grade
- Design operating pressure and SFs, maximum operating pressure (MOP)
- Pipe type:
    - - ERW
    - - high-frequency welds
    - DC or low-frequency welds
    - flash welds
    - submerged arc welded (SAW)
    - double submerged arc welds (DSAW)
    - single submerged arc welds (SSAW)
    - -seamless

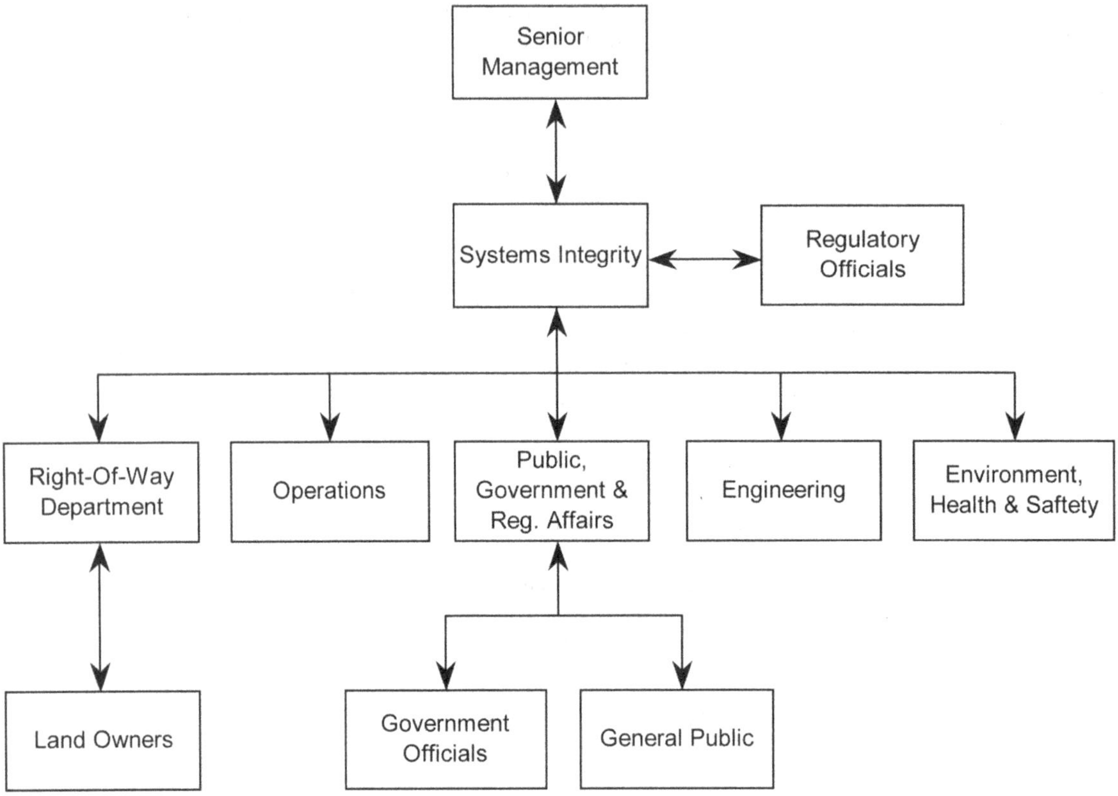

Figure 2-12. Example of communications plan overview (Mikkola 2007)

- Pipe manufacturer and production, historical manufacturing problems
- Construction date or age
- Weld quality and inspection
- Weld failure record
- Coating type
- Coating condition
- CP type
- CP condition
- Compressor station, booster station
- Valve locations, testing requirements, and closure times
- Type of soils (sand. rock, clay, etc.)
- Appurtenances, flanges, fittings, and instrumentation

Right-of-Way Data

- Width of right-of-ways
- Depth of burial
- Condition of right-of-way
- Frequency of patrolling, inspection records
  - -Aerial
  - Ground
- Encroachment check and mitigation

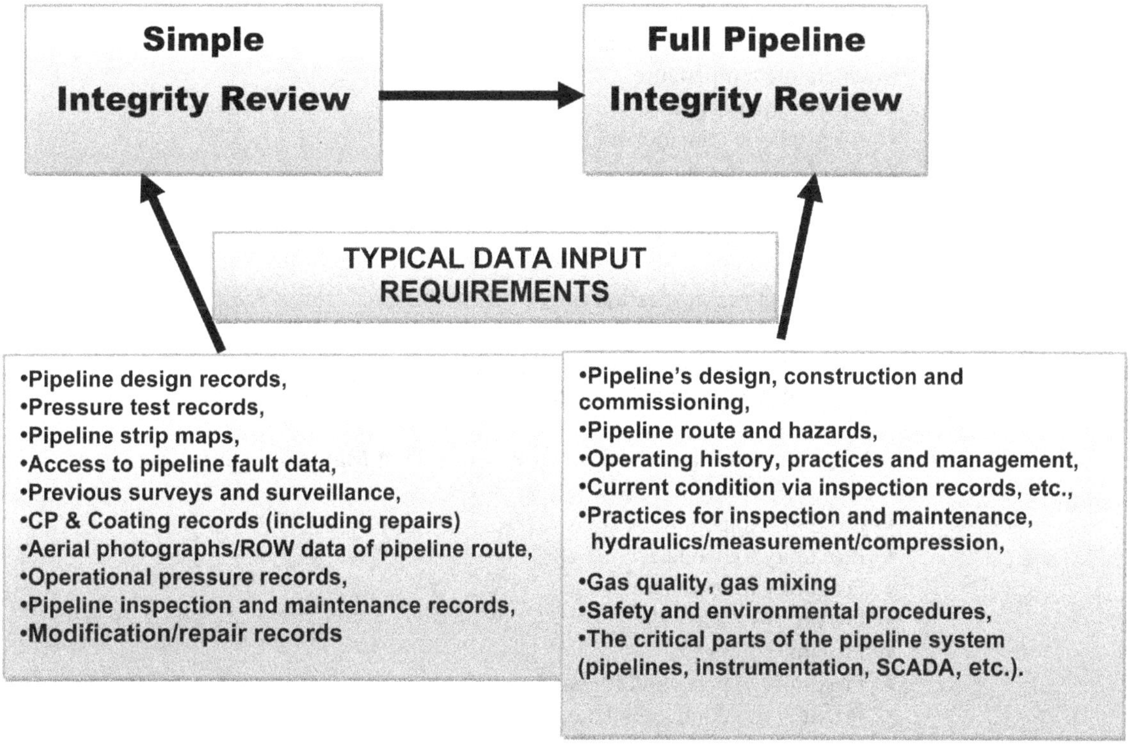

Figure 2-13. Types of integrity reviews and differing information requirements

- Pipeline markers and signage
- List of legal description and land owners
- Description of land use — rural, urban, farm, industrial
- Highway and railroad crossings — cased, uncased
- River, creek, and lake crossings
- Pipeline and other utility crossings, sharing right-of-way corridor Corrosion interference
- Community relationship
- Public awareness of pipeline
- One-call system use, effectiveness, and response timing
- Operator personnel presence during excavations
- Condition of pipe, pipeline exposure reports

Operation, Maintenance Inspection and Repair Data

- ILI results
- Results of ILI anomaly assessment
- Hydrostatic pressure testing data condition of pipe, failures during testing
- SCADA system
- Emergency response plan, drill, and training
- Emergency management plan
- Backup plan for communication and power failures
- Operators qualification and training plan

- Line pressure content or service (gas composition, gas mix, etc.)
- Pressure cycles and pressure profile
- Operating temperature
- Ambient temperature
- Atmospheric condition and data
- Pipe to soil readings
- Close interval survey
- Coating condition and inspection
- CP inspection
- Depth of burial inspection
- Re-route, replace section, line lowering
- Pipeline protection in river, creek, lakes and water ways
- Pipeline protection and monitoring in unstable ground

Records for Determining Portions of Pipeline That May Affect Sensitive Areas

- Proximity to populated areas
- Proximity to habitats
- Proximity to recreation areas
- Proximity to other HCA
- Proximity to farms
- Proximity to parks and forests
- Proximity to commercial areas
- Proximity to archaeological and historical sites
- Proximity to sensitive areas
- Proximity to other important areas

Incidents and Risk Data

- Past history of incidents, blow-downs, ruptures, and near misses, location
  - failure causes and root causes — consequences
  - remedial action — repair history
- encroachment history
- third-part activities
- Air, soil, and water sampling program
- Potential for human safety
- Potential for releases
- Potential for environmental impact (air, soil, and water)
- Potential for fire
- Potential for financial losses
  - human safety, injury, and fatality
  - damage to air, soil. and water
  - functional losses, legal expenses, fines, and punitive damages
- History on other company and industry systems for determining failure modes

The above only lists data that various pipeline operators have used for pipeline Integrity Management documentation. It only covers the pipeline and appurtenances. Other data collection will include those for measurements and compression. List of likely deficiencies should also be compiled.

Typical Reports Additionally, any document management system must provide an up-to-date report on the integrity condition of the pipeline, which will outline the following:

- Introduction outlining objectives, background, system description, data gathering system
- Metering/Compressor: Individual pipeline segment or station integrity reports to include observations and comments from all relevant disciplines, major equipment integrity reviews and comments on operations, maintenance, safety, and logistics, with station summaries and conclusions
- Pipeline Segments: Individual pipeline integrity reports covering ROW's/routes, ground movement, spanning, valves, ESDVs, etc., with summaries and conclusions
- CP System: An overall review of the CP System covering pipelines, associated stations, buried station/facilities, and piping
- Buildings, Control Center, Workshops and Stores: An integrity report to cover technical matters such as power supplies, air conditioning, workshop adequacy and utility stores operation, and control and adequacy of storage capacity
- Pipeline System Hydraulics: Hydraulic analysis data for current and future elements
- Safety and Environment: A comprehensive review reporting on company safety and environmental policies, strategies and procedures, and the implementation and monitoring of these
- Safety and Environmental Issues: safety and environmental issues and observations relating to individual station or other pipeline facilities
- Operations: An overall operations group review summarizing the individual facilities reports and assessing the organization, implementation, manning levels, logistics, training, etc.
- Maintenance: An overall maintenance group review summarizing the individual facilities reports and assessing the organization, implementation, manning levels, logistics, training, etc.
- Logistic Support: An assessment of road/water/air transport adequacy, availability, and reliability, in particular for operations and maintenance activities.
- Risk and integrity Assessment Summary: This would address all of areas of concern including adequacy of information gathering and recording, pipeline and station fitness for purpose and risk together with outline of risk assessment covered, based on compliance with governing codes.
- Recommendations: A summary of recommended corrective actions, using cross references as applicable

Review and Revision Plan

The review and revision process is ongoing. Policies and procedures require continuous maintenance or they begin to deteriorate. The integrity department is responsible for notifying manual holders of revisions in a clear and timely manner. Similarly, manual holders have obligations in the revision process to (1) acknowledge that they have received the revisions and updated their manual and, (2) to communicate the revisions to all applicable manual users (Mikkola 2007).

Requirements. A technical person or author should be involved from the beginning of the revision process and throughout the process to ensure proposed revisions are clear, concise, and appropriate for the subject and audience.

Revisions are issued when approved, or immediately, if required to enhance safety or required by changes in legislation.

All information in Integrity Management Plan manual should be reviewed periodically. Periodic reviews should be coordinated by the manual custodians. Revisions are to be clearly documented and compiled in a single master copy.

Records. To ensure manuals are kept up-to-date, insert all revisions as soon as they are received. Use the Revision Record at the front of the manual to keep a record of revisions received.

All Manual Holders are encouraged to contribute comments and suggestions for improvement. Use the Revision Request Form at the end of this chapter to submit proposed revisions. The Revision Request Form should be also available electronically.

Procedures. The following is an example of a process (Mikkola 2007) of what can be used for Revision Requests submitted throughout the year.

Manual User (or Technical Expert)

Submit Revision Request to manual custodian.

The Revision Request must be approved by the region or district before it is submitted to the manual custodian.

Manual Custodian

Determine whether the proposed revision is in regulatory compliance.

If the proposed revision is not in regulatory compliance:

deny the revision and

advise the originator.

If the proposed revision is in regulatory compliance, submit the Revision Request to the technical writer.

Technical Writer

Determine whether the proposed revision needs additional review.

Minor revisions (e.g., typos, wording or name changes) will be made without further review.

If the revision needs additional review:

assign tracking number and

consult with the appropriate technical expert(s) as required.

Technical Expert(s)

Approve or deny the proposed revision.

Technical Writer

Determine if proposed revision is approved or denied.

If approved by the technical expert(s), distribute the proposed revision to the regional/district field representatives.

If not approved by the technical expert(s), advise the originator.

Field Representatives

Coordinate and consolidate review comments from key regional/district staff as appropriate, then return consolidated regional/district response to technical writer.

Technical Writer

Consolidate comments from all regions/districts.

Technical Writer, Field Representatives

Assure revision has been finalized.

Revisions that cannot be resolved between regions or districts will be forwarded to operations management for resolution.

Technical Writer

Ensure final revision meets company documentation standards, then issue Revision.

Manual Holders

Insert revisions into the manual then update the Revision Record at the front of the manual (Figure 2-14).

Management Systems and Program Evaluation

A management system is a formal process or group of processes that, starting with a specified set of input information, direct a structured, and consistent set of steps, which result in specified outputs that are useful to the organization.

The Operator needs a management system because it is a way to formally provide instructions. It ensures consistent implementation and work products, as well as ensures appropriate dissemination of information and documenting program performance. Management systems are also important to and IMP because inadequate management systems have been shown to play a role in several recent accidents in North America and Europe.

Basic characteristics of a good management system can be outlined as follows (Murray 2010):

- Implementing instructions (procedures)
- Assignment of responsibility
- Description of required work products
- Schedule for product completion
- Documentation requirements
- Change control
- Performance monitoring and improvement

Implementing Instructions involves the proper implementation of approved procedures or other detailed guidance documents as part of the management system. This includes:

- Provision of any necessary detailed instructions
- Provision of some form of user-guide user through entire process
- Ability for an Auditor/Inspector to follow process from the procedure

Assignment of Responsibility involves the documentation of key tasks or assignments to a specific individual or job title. The Operator should be able to produce documentation of key assignments, describe qualifications and capabilities required for the position, and describe the experience and technical capability requirements.

Description of Required Work Products should include any reports, lists, and plans, etc., needed to complete the work or given tasks. This should be fitted to the needs of the users. It should include any forms to be filled out and, if appropriate, include a format and content guide.

Schedule for Product Completion should be clearly documented as this help ensure timely completion of the work and it helps ensure compliance with any rule schedule requirements. Auditors or Inspectors should look for a tracking system and ensure that regular reports on progress or status are distributed to key stakeholders.

Documentation Requirements:

- There should be program and process descriptions, including change documentation
- Auditor/Inspector should also look for meeting minutes, filled out forms, memos to file, etc.

- An Information dissemination should be conducted.
- Operator should require that program process information be promptly distributed to anyone who needs it.
- Each product should have a minimum distribution list.
- A timeframe for distribution should be established.

Change Control relates to the management of change process as defined and implemented within the organization.

Performance Monitoring and Improvement of an IMP relates to the operator's process for the continual evaluation and re-assessment of the integrity of pipeline segments that could affect HCAs documented with sufficient specificity and detail to provide assurance that it can be implemented in a consistent manner, and provides the ability for continuous improvement to the IMP program.

Process for periodic evaluation: The operator should have an adequate process for conducting periodic integrity evaluations of pipeline segments, including having the following defined:

- Program should prescribe a frequency for evaluations
- Procedures for performing evaluations
- Minimum information set to be reviewed
- Product to be developed

The frequency of evaluations needs to have a defined frequency and the period should be long enough for changes to input information to be meaningful. The period should be short enough to head off trends before failure occurs, and the process to re-examine evaluation frequency needs to be established and defined. The assessment interval selection process should consider:

- Interval should be re-evaluated regularly
- Should be in the Assessment Plan
- Every segment should be scheduled
- Should not default to any interval
- Should have a Documented Process to Consider and Choose an Interval
- Operator's process should consider things like:
    - Corrosion growth rate
    - Operator's CP performance record
    - Seam type
    - Potential for outside damage

Regular Review of Assessment Intervals should also consider other factors like:

- Risk analysis results
- Integrity assessment results
- Consequence analysis
- Pipe condition may have deteriorated

Implementing procedures for Pipeline Integrity Evaluations should also contain means to respond to changing conditions, such as, changes to HCAs, pipeline design and operation changes, consideration for operator industry experience, and the ability to perform unscheduled evaluations if required. The program should provide a procedure for integrity evaluations that are technically rigorous, provide conservative results, and provide for con-

Name:	Date:
Department:	Phone:
Manual Volume No.:	Subject Title:
Revision Date:	
Subject Page No.:	
Existing wording: ______	
Proposed wording: ______	
Reason for Change: ______	

Forward to the Manual Custodian

Prepared By: ______________________ Date: ____________

(Print Name)

Signature: ______________________

Figure 2-14. Revision process reference (Mikkola 2007)

sideration of new or emerging data. Information to be used required review materials for a given pipeline segment should include:

- Baseline assessment results
- Remediation history
- Previous assessment results

- Leak history
- Operating history

The end-product of the evaluations should be a report describing the actions taken (process) and the conclusions reached. It should include a description of overall pipeline integrity for the given segments, areas of concern, list of preventive and/or mitigative actions to be taken, and documentation of the distribution list. The program evaluation should also:

- Monitor integrity trends
- Identify and achieve integrity goals
- Monitor IMP implementation
- Identify IMP improvements
- Identify and communicate lessons learned
- Understand root cause of problems

Questions to be addressed as part of the conclusion of the evaluations should include:

- Is the actual progress in Baseline Assessments consistent with program requirements?
- Are data collection and integration efforts successful?
- Are remediation efforts meeting requirements?
- Is continual Assessment implementation consistent with program requirements?

CONCLUDING REMARKS ON INTEGRITY MANAGEMENT

As discussed within this chapter, pipeline Integrity Management is a process for assessing and mitigating the risks and threats faced by a pipeline in order to reduce both the likelihood and consequences of any incidents considered detrimental to the pipeline system. It is a systematic, comprehensive process designed to provide information to effectively allocate resources for appropriate prevention, detection, and mitigation activities.

Operating companies must proactively manage the integrity of their pipelines, and the commitment to do so must be firmly supported by the highest levels of management as well as by the directors and officers of the Company. One of the first requirements to manage an IMP is to develop and implement an Integrity Management plan. This is only an initial step in the process. Other aspects that need to be considered as part of the plan are:

- Develop the goals and objectives for the Program.
- Have management support these goals and objectives.
- Identify the roles and responsibilities of departments and individuals involved.
- Identify the threats to the pipeline system, including areas of high consequences.
- Complete a risk assessment of the pipeline segments.
- Develop a Baseline Assessment including schedules and activities.
- Perform the Integrity Assessment, including any inspections, analysis if data and mitigation/repair activities.
- Perform data management, document management, and record-keeping.
- Include plans for:
    - Management of change
    - Quality assurance

  - o Communications
  - o Training programs
- Measure the program performance and objectives, including ongoing assessment and re-assessment of the IMP.

The intent of a management system is to ensure that all aspects of pipeline activities throughout its life cycle are appropriate to mitigate the risks inherent to the pipeline and to its operation. As with any thorough and complete management system, the four major components of Plan, Implement, Measure, and Improve (Figure 2-15) need to be integrally built within the management process. The department or individuals responsible for the execution of the IMP need to ensure that this system is followed so that continuous improvement is inherent to transpire back into the process.

It is important to note that the Company's IMP will also be ever changing. Pipeline Integrity Management is a continuous process, and thus future information and risks need to be continually and properly evaluated. The established management systems need to ensure that the plans and programs are appropriately implemented and applied. The plans that are comprised within the program will then be continually improved upon as additional assessments are completed, new processes are developed, existing processes refined and expanded, operating experience is gained, and further advancements in technology become available.

Safe and reliable transportation is the primary goal and concern for all pipeline operators. Through an innovative pipeline integrity program, pipeline companies can achieve good standing in the community in which it operates, with those with whom it does business, and with all other stakeholders. Having a well-designed and properly implemented IMP is an important part in achieving this primary goal.

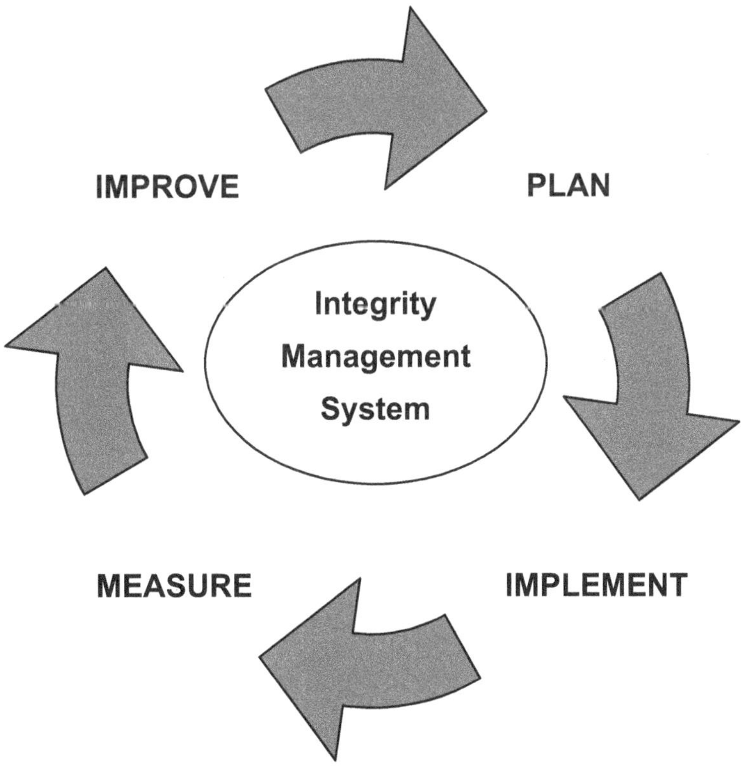

Figure 2-15. Components of a management system

REFERENCES

American Petroleum Institute (API), 2001, "Managing System Integrity for Hazardous Liquid Pipelines," Standard 1160.

American Society of Mechanical Engineers (ASME), B31.8S, 2004, "Managing System Integrity of Gas Pipelines," Supplement to ASME B31.8, ASME, New York, NY.

Canada National Energy Board (NEB), 1999, "On-shore Pipeline Regulations," http://www.neb-one.gc.ca/index_e.htm. National Energy Board Act http://laws.justice.gc.ca/en/N-7/text.html.

Healy, J., Jones, D. G., Clyne, A. J., Cazenave, P. B., and Alkazraji D., 2004, "Using Benchmarking to Optimise the Cost of Pipeline Integrity management," Pigging Products and Services Association (PPSA) London Seminar, http://www.ppsa-online.com/papers/2004-London-7-Healy.pdf.

Mikkola, C., 2007, Pipeline Integrity Management Program Overview - unpublished

Mohitpour M., 2010, Pipeline Integrity Overview, unpublished.

Mora, R. G, 2007, "Pipeline Integrity Management Program (IMP) — Key Documentation," http://www.tecnologiatotal.net/pipeline_integrity.pdf

Murray, A., 2010, Pipeline Integrity Management Systems Presentation Material – unpublished

National Energy Board of Canada Web site, http://www.neb.gc.ca/

U.S. Department of Transportation Pipeline and Hazardous Materials Safety Administration (PHMSA), http://www.phmsa.dot.gov/pipeline

U.S. Department of Transportation (DOT) Title 49 Code of Federal Regulations (CFR) Part 195, 2005, "Integrity Management for Hazardous Liquid Pipeline Operators."

U.S. Department of Transportation (DOT) Title 49 CFR Part 192, 2006, "Pipeline Safety: Pipeline Integrity Management in High Consequence Areas (Gas Transmission Pipelines), Final Rule."

Vollman, K.W., 2000, "Towards Goal-Oriented Regulation," International Pipeline Conference, Calgary Canada, October 4, 2000.

Chapter 3

PIPELINE COATING

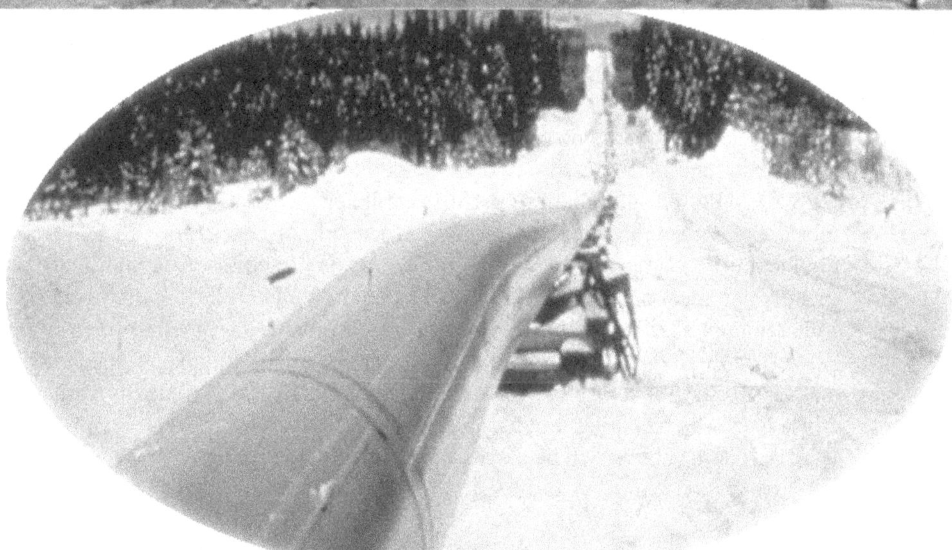

INTRODUCTION

Coatings are used to control the steel pipe corrosion both externally and internally. Coatings are the primary means of corrosion protection for a buried pipeline for new construction as well as for pipeline rehabilitation. While CP is applied to pipelines, coatings provide an assurance of defence against corrosion.

The exposed steel comprising the less than 1% of total surface on well-coated pipeline is generally due to small defects in the coating film. These defects (commonly called holidays) may result from:

- Skips by the coating machine
- Pinholes in the coating film as applied
- Cracks from excessive mechanical or thermal stresses
- Scrapes or gouges caused during subsequent handling of the coated pipe
- Penetration by rocks or other debris in the backfill surrounding the pipe
- Distorting stresses exerted on the coating by certain soils having a very high shrinkage rate upon drying.
- Penetration by growing roots
- Action of solvents in earth surrounding the pipeline (e.g., from leaks on a products pipeline)
- Action of bacteria in the soil surrounding the pipeline. Some coating materials are relatively inert in this respect, while others are sensitive to such damage — this subject has been given extensive research coverage in recent years.
- Damage from subsequent construction on other facilities making it necessary to uncover the pipeline.
- Any other action that will damage the coating film

Coatings isolate the metal from electrolytes in the soil, thus reducing the power requirements for CP; the CP protects the metal exposed to the environment at coating holidays (defects/holes, etc.). New pipe coatings may provide 99% of the pipe protection, but as they slowly deteriorate, the CP system provides more and more of the protection.

Therefore, coatings present a cost-effective means for protecting buried steel pipe from mildly corrosive environments. They are a protective system that physically shield the pipe steel from electrolytes in the soil.

Pipeline coatings have undergone dramatic technological changes over the past two decades. Therefore, today's coatings can provide better protection than their predecessors, continue to function under severe operating conditions, and be applied under less than ideal conditions. Present-day coatings perform at higher in-service operating temperatures, resist substantial damage in handling during construction or in operation (e.g., by soil stress or soil movement), and provide exceptional corrosion protection. Coatings also must be user-friendly and must be able to be applied in a coating/pipe mill or in the field. There are is a wide range of pipeline coating options that are available for pipeline corrosion protection. However, the diverse world of coating materials and techniques influence the decision-making process for selection — a situation compounded by the equally varied field conditions under which coatings are applied.

Each of the above will be discussed in detail in the sections below.

PIPELINE COATING

The primary method of pipeline protection is the use of a barrier coating. Failure of this protective layer can have a dramatic impact on the subsequent cost to operate the pipeline through either increased CP demand, pipeline rehabilitation, and, in some instances, the need for pipeline repair or replacement.

The primary function of a coating is thus to establish a permanent barrier between a structure and its immediate environment.

The performance of any coating system is directly related to the conditions encountered during the installation and operational life of a facility. Therefore, before any coating selection is initiated, it is imperative that the environmental and construction conditions of a facility are well understood.

Types of coating systems generally utilized for pipeline protection are:

- Coal tar/bituminous enamel
- Coal tar epoxy/asphalt mastic
- Coal tar polyester
- Extruded polyethylene/plastic (including Yellow Jacket YJ1, YJ2)
- Three-layered system (fusion bonded/polyethylene tape
- Fusion bond epoxy/powder coating (3M 206N, slow, standard, fast, and Napguard)
- Liquid epoxy
- 100% solids epoxy
- Pressure-sensitive polyethylene tape
- Butyl-backed polyethylene tape
- Heat-resistant polyethylene tape
- Poly vinyl chloride (PVC) tape
- Polyurethane
- Polyurethane tar
- Polyethylene tape (Tapecoat)
- Shrink sleeves
- Urethane coatings (multi components and block urethanes)
- High-performance composites (HPCC)

Some of the applicable coating standards, including the latest ISO standard (Coulson 2008), are:

- API RP 5L9 "External Fusion Bounded Epoxy Coating of Line Pipe"
- AS 3862 — Australian Std. "External fusion-bonded epoxy coating for steel pipes"
- Canadian Standard Association CSA Z245.20 "Fusion Bonded Epoxy" CSA Int.
- Canadian Standard Association CSA Z245.21 "Polyethylene tape" CSA Int.
- Canadian Standard Association CSA Z245.22 "External polyurethane foam insulation coating for pipe" CSA Int.
- DiN 30671 "Coating of Buried Pipelines with Polyethylene"
- EN 10289, Steel Pipes and Fittings for Onshore and Offshore Pipelines. External Liquid Applied Epoxy and Epoxy-Modified Coatings
- ISO 21809-2 Petroleum and Natural Gas Industries — External Coatings for Buried or Submerged Pipelines
- NACE RP0169 Control of External Corrosion on Underground or Submerged Metallic Piping Systems.
- NACE RP0185 Extruded, Polyolefin Resin Coating Systems With Soft Adhesives for Underground or Submerged Pipe
- NACE RP0190 External Protective Coatings for Joints, Fittings, and Valves on Metallic Underground or Submerged Pipelines and Piping Systems
- NACE RP0274 High-Voltage Electrical Inspection of Pipeline Coatings

- NACE RP0394 "Application, Performance, and Quality Control of Plant-Applied, Fusion-Bonded Epoxy External Pipe Coating"
- NACE RP0399 "Plant-Applied, External Coal Tar Enamel Pipe Coating Systems"
- NACE RP0602 "Field Applied Coal tar Coating"
- SP2888 "Liquid Polymer Coating"

The usage of some of the major coatings for pipeline applications, in the past several decades, is depicted in Figure 3-1. Comparison of conventional Fusion Bonded Epoxy (FBE) coating against an innovative powder coated multi-component coating HPCC is provided by Singh & Cox 2000.

Merits and demerits of some the coating systems indicated in Figure 3-1 are enumerated in Table 3-1. Characteristics and limitations of the coating are further clarified in Table 3-2 (Linchtenstein 2003). Application and operational limits of various coating listed above are summarized in Tables 3-3 and 3-4.

CODE REQUIREMENT FOR PIPELINE COATING PROPERTIES AND INSPECTION

Tables 3-5 summarizes the North American pipeline codes requirement for testing and inspection of pipeline coatings

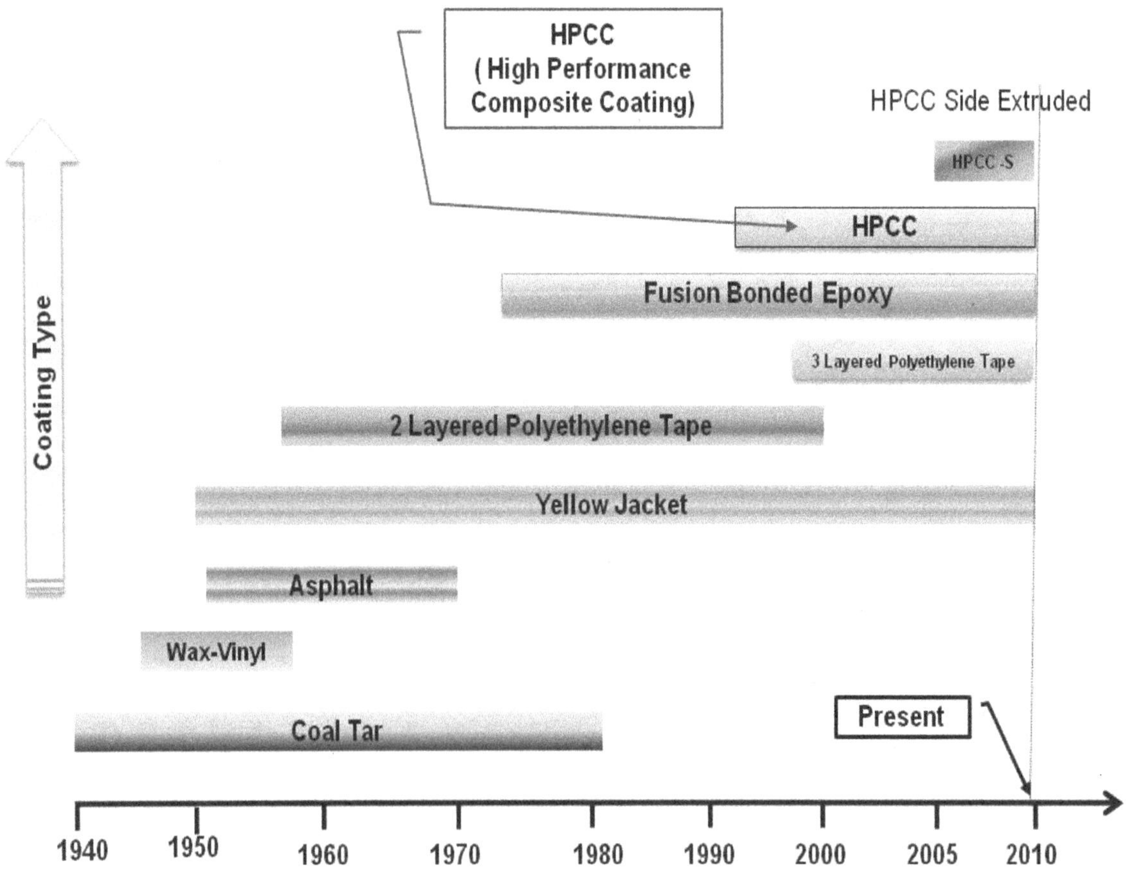

Figure 3-1. Pipeline coating timeline (Tailor et al. 2003; ShawPipe 2008)

TABLE 3-1. Coating systems advantages/disadvantage used in recent history

Advantages	Disadvantages
Asphalt-Coal tar: mostly from 1940 to 1981 (used from 1870)	
Easy to apply/good record	Subject to oxidation and cracking
Minimal surface preparation required	Soil stress has been an issue
Long track record in certain environments without failure	Limitations at low application temperatures
Excellent water/hydrocarbon resistance	Environmental and exposure concerns
Permeable to cathodic protection in event of failure	Associated with corrosion and stress crack corrosion failures
Tape Wrap (/double wrap- 2 layer): : 1960 -present	
Excellent adhesion chemical and corrosion resistance	Low-impact resistance/Thin film
Does not shield cathodic protection	High-moisture absorption and permeation
High operating temperatures possible	Expensive
FBE: 1975 to now	
Excellent adhesion chemical and corrosion resistance	Low-impact resistance/Thin film
Does not shield cathodic protection	High-moisture absorption and permeation
High operating temperature possible	Expensive
Two-layer extruded polyethylene: : 1960 -present	
Excellent track record	Limited temperature range
Good handling	Poor shear stress resistance
Low cost	Limited pipe sizes (<24 in [610 mm] OD)
	Can disbond due to low shear stress
Three-layer polyolefin: 1986 to now	
Excellent combination of properties	Best suited for electrical resistance welded pipes
	High thickness to eliminate weld tenting
Composite coating: 1990 to now	
Excellent combination of properties	Suitable only for large diameter pipes a
Conforms well to external raised weld profiles	Not designed for small diameter pipes (<406 mm OD)
Wax-Based Tape: 1945-1960	
Any configuration	Soil stress effects unknown
Very easy to apply	Low adhesion
Minimum surface preparation required	Bacterial attack
Easy to remove	Operating temperature limited
High Performance Composite Coating (HPCC)	
Excellent cathodic disbondment values	Expensive
Improved resistance to moisture permeation	
Can withstand high operating temperatures	
Good protection in severe environments	
Excellent conformity to raised welds	
Excellent field handling and flexibility characteristics	
Chemical resistant	
Good impact resistance	

From a regulatory point of view, regulations (such as *US DOT CFR 49, part 192 and 195*, 2009) generally require that whenever a pipeline is exposed for any reason, the coating and pipe must be inspected in the following situations:

- Inspection of buried valves and equipment
- Foreign line crossings
- Water run off removing the pipeline covering material
- Pipeline exposure due to natural hazards

PIPELINE COATING FAILURE MODES

Definition of failure: based on the concept of coatings and CP working together as a corrosion control system, coating failure on a steel structure ca be considered to have occurred

TABLE 3-2. Characteristics and limitations of pipeline coating (Linchtenstein 2003)

Pipe Coating (years in use)	Characteristics/Merits	Limitations
Coal tar enamels (80+ years of use)	• Least holiday susceptibility • Low-current requirements • Good resistance to cathodic disbondment • Good adhesion	• Limited number of manufacturers • Limited number of applicators • Health and safety concerns • Change in allowable reinforcements
Mill-applied tape systems (30+ years of use)	• Minimum holiday susceptibility • Ease of application • Good adhesion performance • Low-energy required for application	• Handling restrictions — shipping and installation • Ultraviolet (UV) and thermal blistering – storage potential • Shielding CP from soil • Soil Stress disbondment
Crosshead-extruded polyolefin with asphalt/butyl adhesive (40+ years of use)	• Minimum holiday susceptibility • Low-current requirements • Ease of application • Nonpolluting • Low-energy required for application	• Poor adhesion to steel • Limited storage capability (except with carbon black) • Tendency for tear to propagate along pipe length
Dual-side extruded polyolefin with butyl adhesive (25+ years of use)	• Minimum holiday susceptibility • Low-current requirements • Excellent resistance to cathodic disbondment • Good adhesion to steel • Ease of application • Nonpolluting • Low-energy required for application	• Difficult to remove coating • Limited number of applicators
Fusion bonded epoxy (FBE) (35+ years of use)	• Low-current requirements • Excellent resistance to cathodic disbondment • Excellent adhesion to steel • Excellent resistance to hydrocarbons	• Exacting application parameters • High application temperature • Subject to steel pipe surface imperfections • Lower impact and abrasion resistance • High-moisture absorption
Multilayer epoxy/extruded polyolefin systems (recent development)	• Lowest current requirements • Highest resistance to cathodic disbondment • Excellent adhesion to steel • Excellent resistance to hydrocarbons • High impact and abrasion resistance	• Limited number of applicators • Exacting application parameters • Higher initial cost • Possible shielding of CP current

when blistering or disbonding takes place. Invariably, under-film corrosion is visible. If the coating is blistered and disbonded and there is no corrosion, then it has not failed. If there is corrosion due to the coating shielding the cathodic current, it has failed (Neal 2000).

Werner et al. (1992) define failure as follows: "A buried pipeline coating has exceeded its useful life when adequate CP can no longer be economically maintained."

In summation, coating failures can be generalized as a failure that can cause corrosion leading to pipeline failure or operational risks. Coating failures depend on the type of coating but and condition that caused the failure. Some of typical coating failures are shown in Figure 3-2.

Major factors influencing pipeline external coating performance are summarized in Figure 3-3.

TABLE 3-3. Buried pipeline coating system applications (Wilmott, M., et. al 2000)

Coating System	Main Application
Polyethylene	Sub-sea flowlines, noninsulated
Polypropylene	Sub-sea flowlines, noninsulated
Polychloroprene	Risers, clamps, guides, and special flowlines
EPDM*	High-temperature flowlines and risers
Polypropylene (thick)	Oil: land and subsea
Polychloroprene (thick)	Oil: land and subsea
EPDM (thick)	Oil: land and subsea
Solid polyurethane	Oil: land and subsea
Foamed polypropylene	Multiphase land and subsea
Syntactic polypropylene	Multiphase land and subsea
Syntactic polyurethane Conventional	Multiphase land and subsea
Glass beads	Deep water
Glass syntactic epoxies	Deep water
Glass syntactic epoxies	Deep water
Pipe in pipe/jacket	Gas: land and sub-sea

Note * EPDM = ethylene propylene diene monome

External corrosion and environmentally assisted cracking problems such as SCC can threaten pipeline integrity. Most pipeline system s are protected from these problems by external coatings and impressed current CP. For a problem to develop, the coating must fail in such a way as to expose the pipe to a corrosive external environment and to prevent CP from reaching the steel surface. The mode of coating failure is the key to whether serious problems can occur.

Different coating systems under identical exposure conditions can fail in completely different ways. An example was recently encountered on a pipeline in Southern Alberta. The following photographs show coating conditions found at one site at which pipe was coated with polyethylene tape alongside FBE. The coatings were exposed to identical service conditions in identical soil environments. However, the way in which the coatings failed was completely different.

As can be seen in Figure 3-4, the polyethylene tape coating disbonded to form a wrinkle in the coating. By way of comparison, Figure 3-5 shows damage observed on the fusion

TABLE 3-4. Pipeline coating system operating range code requirement for pipeline coating properties and inspection

	Depth Limit (m) for offshore application	Temperature Range (°C)	U ($W/m^2/K$)
Anti-Corrosion Only			
Polyethylene	>1,000	60–80	
Polypropylene	>1,000	110–135	
Polychloroprene	>1,000	90–100	
EPDM	>1,000	135–150	
Insulation (High U)			
Polypropylene (thick)	>1,000	110–135	$U < 10$
Polychloroprene (thick)	>1,000	90–100	$U < 10$
EPDM (thick)	>1,000	135–150	$U < 10$
Solid polyurethane	>1,000	100–115	$U < 8$
Insulation (Medium U)			
Foamed polypropylene	<600	50–120	$4 < U < 10$
Syntactic polypropylene	>1,000	110	$5 < U < 10$
Syntactic polyurethane conventional	<300	100–115	$2.5 < U < 10$
Glass beads	>1,000	~100	$2 < U < 4$
Insulation (Low U)			
Glass syntactic epoxies	>1,000	~100	$2 < U < 4$
Pipe in pipe/jacket	various	100–150	U <1

TABLE 3-5. Code requirement for coatings

Code and Stipulation	Required Properties	Inspection Requirement
ASME ANSI B31.8 Transmission pipelines to be externally coated and CP applied. CP to be applied within 1 year of operation	• Test or experience must show that coating materials are corrosion resistant • Coating to be compatible with surface • Similar considerations as to CSA Code be given when selecting coating	• Coating inspected for defect prior to lowering in • Coating defects must be repaired
ASME ANSI B31.4 Pipelines to be externally coated. CP to be applied within 1 year of operation	• Mitigate corrosion • Adhesion • Ductility • Soil stress resistance • CP compatible • Low-moisture absorption and high electrical resistance for insulation type coating	• Pipeline coating to be inspected for holidays prior to lowering in • Coating defects to be repaired and inspected
CSA Z662-07 Requires pipelines be coated and CP applied	• Electrical isolation • Adhesion • Ductility • Soil stress resistance • CP compatibility • Degradation resistance • Different coatings must be compatible	• Inspection required before, during and after pipeline installation • Faults repaired and re-inspected

bond epoxy section of pipe exposed to identical field conditions. In contrast to the wrinkling associated with the tape system, the FBE coating has failed through localized blistering.

Examination of the water trapped under these coating failures showed that beneath the tape disbondment, the solution has a pH value of 6.5 to 7.0, whereas beneath the FBE blister, the pH of the trapped water was in excess of 10. These observations indicated that the tape formed an electrically shielding disbondment, while the FBE allowed penetration of the CP. and hence the generation of a high pH electrolyte next to the steel.

The failure mode of a pipeline coating has significant impact on whether corrosion and SCC will develop in the underlying metal. Due to the absence of effective CP, both localized corrosion and stress corrosion are generally observed beneath the disbonded tape coating, as shown in Figure 3-6. Cracks are grouped in colonies, often containing hundreds of individual cracks, and are usually found on the pipeline system in the body of the pipe. Research is ongoing to understand the mechanism and rate of crack growth (Figure 3-7).

In contrast, no localized corrosion or stress corrosion was found beneath the FBE blisters, suggesting that despite coating failure, the mode of failure allowed protection of the pipe surface by CP. These and other field observations suggest that FBE coatings fail in such a way as to allow penetration of CP to the steel surface. For this reason, most pipeline companies now use FBE coatings.

COATING PERFORMANCE

The ideal impermeable coating should be completely unaffected by the specific environment it is designed to block, be it most commonly humidity, water, or any other corrosive agent such as gases, ions, or electrons. The ideal impermeable coating should have a high dielectric constant and also have perfect adhesion to the underlying surface in order to avoid any entrapment of corrosive agents.

Figure 3-2. Pipeline coating failures

The performance of a pipeline coating is directly related to the conditions encountered during the installation and operational life of pipeline. Coating selection must thus consider environmental, construction, and operational conditions.

Thus, pipeline coatings must be compatible with the environment it traverses and indeed conditions that it will experience throughout a pipeline project life. Thus coatings are selected on the basis of their desired properties and performance characteristics under field and operating conditions (Figure 3-8).

Based on an evaluation of the relevant chemical, electrical, environmental, geotechnical, and mechanical conditions experienced by a pipe coating system during its service life, an optimum external pipeline coating for buried service should exhibit the following properties:

- Adhesion
- Peeling resistance
- Chemical resistance
- Electrical resistance
- Compatibility with CP
- Flexibility

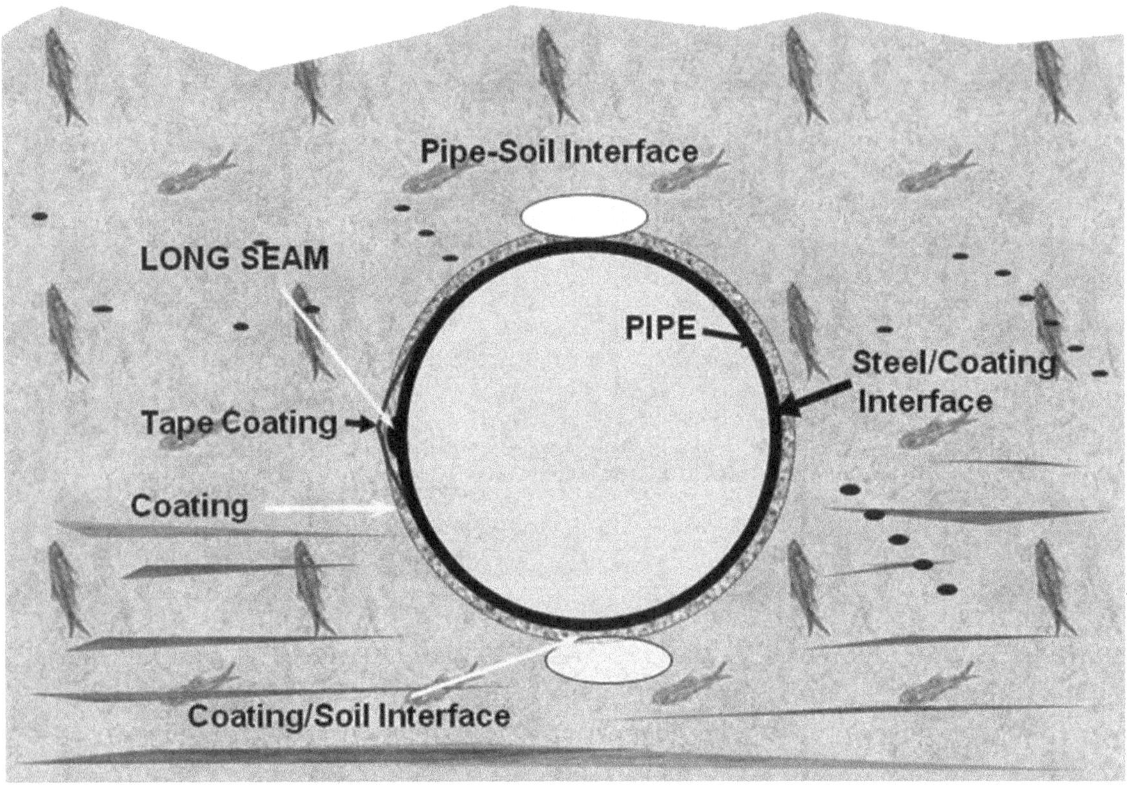

Figure 3-3. Factors influencing pipeline external coating performance

- Hardness/abrasion resistance
- Impact resistance
- Penetration resistance
- Soil stressing resistance
- Hot and cold water absorption resistance
- Stability at elevated temperatures
- Weathering resistance
- Compatible repair and girth weld coating material

The purpose of the tests is to assess the technical and potential financial benefits of adopting particular pipe coating technologies (e.g., thin versus thick film, FBE over three-layered polyethylene, etc.).

Table 3-6 lists the main standards that define the requirements and procedures for testing and performance evaluation of the above coating properties.

Other applicable and related standards that are reviewed for pipeline coating testing are listed in Table 3-7.

IMPORTANT COATING PROPERTIES FOR OPERATION

Adhesion

Permanent adhesion of the coating to pipe is essential, as a disbonded coating may permit moisture to accumulate between the pipe and coating, thereby creating an environment

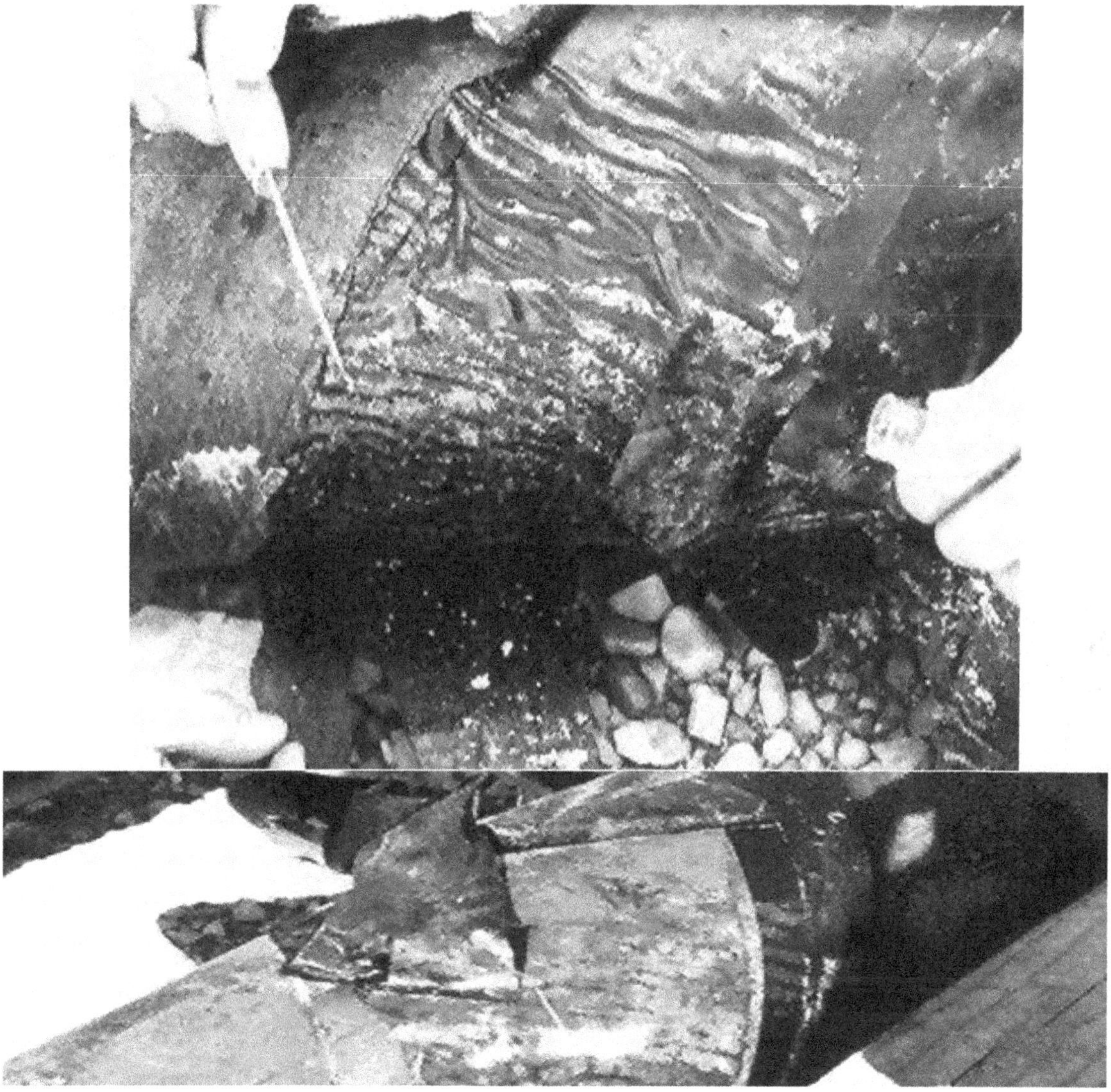

Figure 3-4. Disbonded tape coating showing longitudinal wrinkles where the tape has pulled away from the pipe

favorable to corrosion. A coating's adhesion reflects its ability to resist bending damage, cathodic disbondment, and soil stressing as well as the quality of coating application and pipe surface preparation.

Pull-off adhesion test and/or crosscut methods are usually employed to determine the adhesive strength of the coating materials (Malik et al. 2002).

In pull-off test, an adhesion tester consists of dollies made of aluminums, which are glued perpendicular onto the coated surface of the samples. After the curing of adhesive (glue), the testing apparatus is then attached to the loading fixture and aligned to apply the tension normal to the test surface. The force applied to the loading fixture is then gradually increased and monitored until either a segment of coating material is detached or a specified value is reached. The relative stress applied to each coating can be calculated as follows:

$$X = 4F/3.14(d)^2 \quad (3-1)$$

Figure 3-5. Blistered FBE coating (inset: blisters in girth weld area of a gas pipeline (Norsworthy 2009)

where

X = pull off strength achieved at failure; psi, kg/m^2
F = highest force applied to the test surface; pounds, kg
d = equivalent diameter of the original surface area stressed, in., m

The above is similar to tension/tearing test performed on a sample of tape coating material (Figure 3-9).

In crosscut adhesion test, a crosscut is typically made using a utility knife on the coated surface deep into the metal substrate. At the crosscut, the blade of the knife is inserted under the coating and with a levering action force applied to peel off the coating after the test. The peeled off/chipped off area is then observed to see the extent of removal of the coating from the substrate.

Adhesion testing consists of immersing a coating specimen in 75°C ± 3°C for 48 hours and/or at 95°C ± 3°C for 24 hours. While the specimen is still warm, a rectangle scribed down to the substrate, it is then air cooled to 20°C ± 3°C (maximum time of 1 hour) and a levering action is used to remove the coating until it demonstrates definite resistance various ratings are then assigned as follows:

Figure 3-6. Corrosion and SCC observed in the steel surface beneath dısbonded tape coating

- rating 1 — the coating cannot be removed cleanly
- rating 2 — less then 50% of the coating can be removed
- rating 3 — more than 50% of the coating can be removed, but it demonstrate definite resistance to the levering action
- rating 4 — the coating is easily removed in strips or large chips
- rating 5 — the coating can be completely removed as a single piece

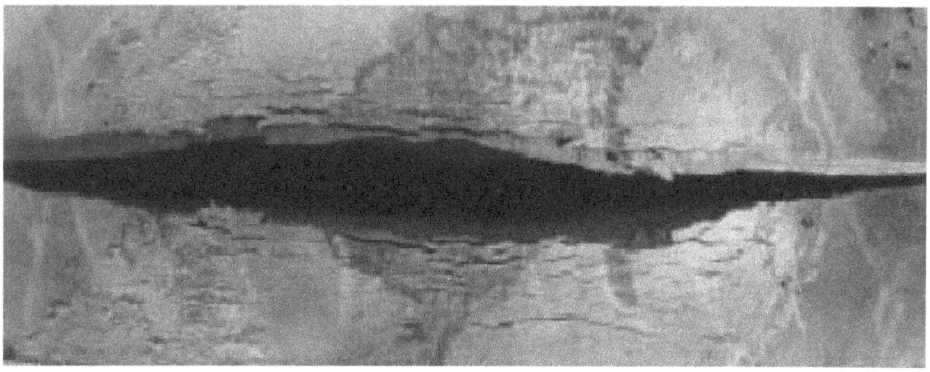

Figure 3-7. SCC colony on a large-diameter, high-pressure transmission gas pipeline (after Thomason 2001)

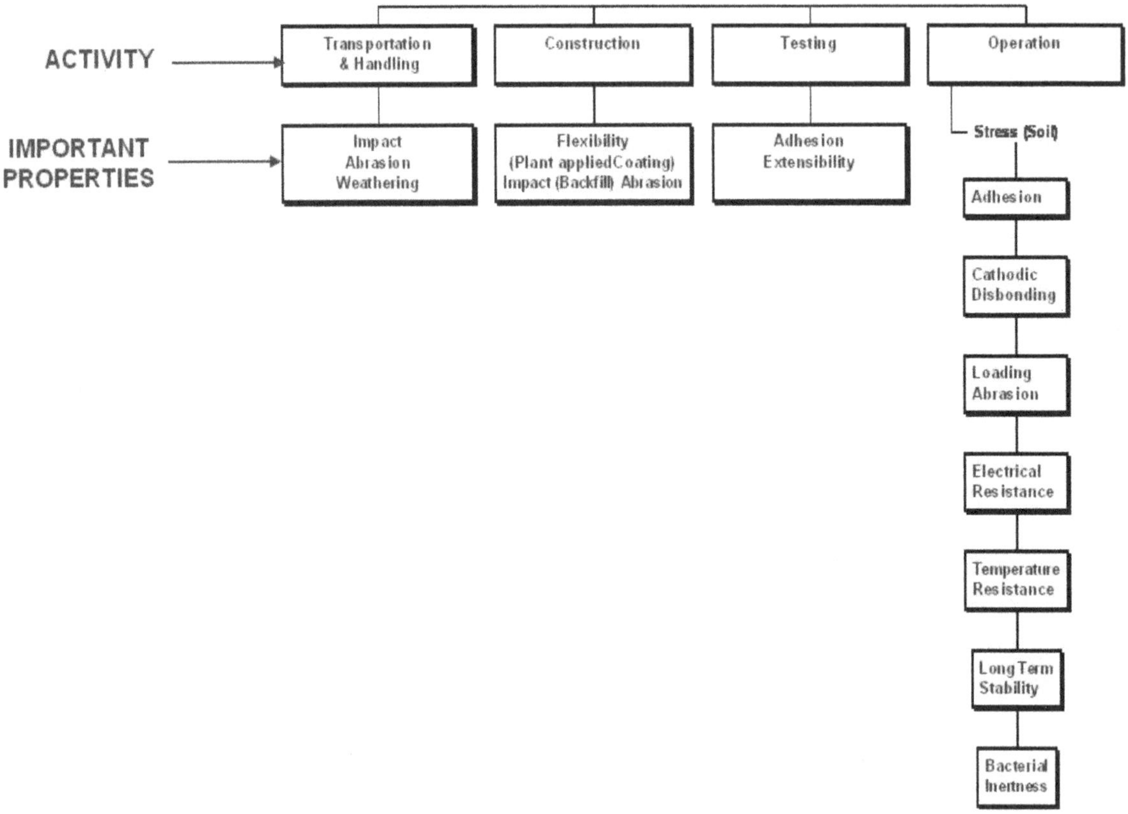

Figure 3-8. Desired property requirements

It is imperative therefore that the surface of the base pipe is properly prepared. This involves much more than simply removing surface dirt and rust. The presence of invisible contaminants such as soluble salts can have a very detrimental effect on the life of the coating (Neal 1999).

Another but least objective and easiest/fastest test method to perform is the X-cut knife adhesion test. In this test, an "X" is cut through the coating to the metal using a knife. The knifepoint is then used to pry at the point of the "X" in order to forcibly remove the coating from the substrate. The force used to remove the coating is subjectively assessed, and the location and extent of any disbonding is noted (Tator 2006) (Figure 3-10).

In order to perform well, coatings need to maintain a strong adhesive bond throughout their working life. High initial adhesive strength will be obtained from having a suitably prepared surface, good coating flow, and wetting during application. The longevity of the bond is related to the properties of the coating, its resistance to moisture penetration, soil strength, and cathodic disbondment, all of which are discussed later in the chapter.

Cathodic disbondment, long-term water immersion, and soil stress tests are commonly employed to evaluate the adhesion of a coating system.

Chemical Resistance

A coating system should be resistant to all chemicals such as hydrocarbons, acids, alkalis, and biological agents likely to come into contact with the coating. A typical chemical resistance test exposes the coating to the liquid and vapor phase of a chemical for a specified

TABLE 3-6. Coating property applicable test standards

Standard	Test Method	Application
ASTM C868-02(2008)	Standard Test Method for Chemical Resistance of Protective Linings	Chemical Resistance
ASTM D870-2002	Standard Practice for Testing Water Resistance of Coatings Using Water Immersion	Hot water soak tests
ASTM D1693-2008	Standard Test Method for Environmental Stress-Cracking of Ethylene Plastics	Stress Cracking
ASTM D2240-2005	Standard Test Method for Rubber Property — Durometer Hardness	Hardness Testing
ASTM D-2583-2007	Standard Test Method for Indentation Hardness of Rigid Plastics by Means of a Barcol Impressor	(Harness Testing)
ASTM D4060-2007	Standard Test Method for Abrasion Resistance of Organic Coatings by the Taber Abraser	Abrasion resistance (Taber test)
ASTM D4541-2009	Standard Test Method for Pull-Off Strength of Coatings Using Portable Adhesion Testers	(Peel test)
ASTM D5162-2008	Standard Practice for Discontinuity (Holiday) Testing of Nonconductive Protective Coating on Metallic Substrates	Holiday detection
ASTM G6-2007	Standard Test Method for Abrasion Resistance of Pipeline Coatings	Abrasion resistance
ASTM G-8-96(2003)e1	Standard Test Methods for Cathodic Disbonding (CD) of Pipeline Coatings	Cathodic disbondment
ASTM G9-2007	Standard Test Method for Water Penetration into Pipeline Coatings	Water permeability
ASTM G10-83(2002)	Standard Test Method for Specific Bendability of Pipeline Coatings	Flexibility
ASTM G11-04	Standard Test Method for Effects of Outdoor Weathering on Pipeline Coatings	
ASTM G12-07	Standard Test Method for Nondestructive Measurement of Film Thickness of Pipeline Coatings on Steel	
ASTM G14-2004	Standard Test Method for Impact Resistance of Pipeline Coatings (Falling Weight Test)	(Impact Resistance)
ASTM G17-1998	Standard Test Method for Penetration Resistance of Pipeline Coatings (Blunt Rod)	(Penetration Resistance)
ASTM G18-2007	Standard Test Method for Joints, Fittings, and Patches in Coated Pipelines	
ASTM G19-2004	Standard Test Method for Disbonding Characteristics of Pipeline Coatings by Direct Soil Burial	
ASTM G20-88(2002)	Standard Test Method for Chemical Resistance of Pipeline Coatings	Chemical permeation
ASTM G42-96(2003)	Standard Test Method for CD of Pipeline Coatings Subjected to Elevated Temperatures	Cathodic disbondment at high temperatures
ASTM G55-2007	Standard Test Method for Evaluating Pipeline Coating Patch Materials	Pipeline patching materials sealing abilities
ASTM G62-2007	Standard Test Methods for Holiday Detection in Pipeline Coatings	Holiday Detection
ASTM G70-2007	Standard Test Method for Ring Bendability of Pipeline Coatings (Squeeze Test)	Bendability/flexibility
ASTM G80-2007	Standard Test Method for Specific CD of Pipeline Coatings	Cathodic disbondment
ASTMG81-97a(2007)e1	Standard Test Method for Jaw Crusher Gouging Abrasion Test	
ASTM G85-02e1	Standard Practice for Modified Salt Spray (Fog) Testing	Salt spray
ASTM G90-2005	Standard Practice for Performing Accelerated Outdoor Weathering of Nonmetallic Materials Using Concentrated Natural Sunlight	Weathering

TABLE 3-6. (Continued)

Standard	Test Method	Application
ASTM G95-2007	Standard Test Method for Cathodic Disbondment Test of Pipeline Coatings (Attached Cell Method)	Cathodic disbondment
DIN 30670-92	Polyethylene Coatings of Steel Pipes and Fittings, Requirements and Testing	General testing requirements
DIN 30672	Softening Point. Lap Shear. Specific Gravity. Tensile Strength. Elongation. Hardness. Abrasion Resistance. Volume Resistivity Test	Peel Tests (For shrink sleeves and tapes)
ISO15711:2004	Paints and varnishes. Determination of resistance to CD of coatings exposed to sea water	Cathodic Disbondment
NACE RP0188-1999	Discontinuity (Holiday) Testing Of Protective Coatings	Holiday Detection
NACE RP0274-2004	High-Voltage Electrical Inspection of Pipeline Coatings	General Holiday Detection
NACE RP0490-2001	Holiday Detection of Fusion-Bonded Epoxy External Pipeline Coatings of 250 μm to 760 μm (10–30 mils)	Holiday Detection — FBE Coating
NACE TM0186-2002	Standard Test Method Holiday Detection of Internal Tubular Coatings of 250 m to 760 m (10–30 mils) Dry-Film Thickness	Holiday detection for internal coated pipe

period of time. At the end of a test period, the degree of coating degradation is evaluated (Figure 3-11) (Coulson and Temple 1983).

For this test, usually two holidays are drilled in the specimen. Half of he specimen with one hole) is immersed in a chemical reagent and the other half is exposed to the chemical vapor phase in a test bath for a specified period of time, typically up to 1 year. Loss of coating is then recorded for total failure, blistering, softening, disbanding, and penetration.

The test method must cover a procedure for evaluating the chemical resistance of a polymer-based protective coating, and the method closely approximates the service conditions, including the temperature differential between the external and internal surfaces of the pipeline, which may accelerate permeation of the lining by a corrosive media.

Electrical Resistance

An effective underground coating must have good dielectric strength to assure high electrical resistance. This resistance should not change appreciably with time or with exposure to water and/or other electrolytes.

Although coatings need not withstand continuous high-voltage stresses normally associated with electrical insulation, they must exhibit good electrical resistance. For example, coatings are exposed to 1500 V to 7500 V during holiday detection and 1 V to 3 V during the application of impressed current CP.

Also, the test must conclude on coating resistance in the electrolytes. In these test, the barrier coating resistance is typically calculated from the resistance and specimen (i.e., a coupon consisting of a steel rod coated with the appropriate coating) and geometry using the following equations (Coulson et al. 1991):

$$R_x = [P_c/(2\pi L)\ln(c/b)] + [P_e/(2\pi L)\ln(b/a)] \qquad (3-2)$$

where

R_x = coating resistance, Ω cm
P_c = coating resistivity, Ω cm
P_e = electrolyte resistivity, Ω cm
L = coating length, cm

TABLE 3-7. Other applicable coating testing standards

Description	Test Standard	Description	Test Standard
Adhesion Peel after Ageing	ASTM C836-89a	*Puncture resistance*	ASTM E154
Adhesion Test by Pell	ASTM C836	*PVC specification American*	CRDC 572
Adhesion Testing	ASTM D1000	*Resistance to Ageing (pipe)*	DIN 53370
Adhesive 180 deg. Peel	ASTM D1000	*Resistance to Oil*	ASTM D543
Adhesive strength	DIN 30672	*Salt Spray*	ASTM B117-94
Ashing	ASTM D2584	*Softening Point By Ring and Ball*	ASTM D36
Barcol Hardness	ASTM D2583	*Softening Point, Ring and Ball*	ASTM E28
Chemical resitance	ASTM D543	*Stress crack resistance*	ASTM D1693
Coefficient of friction	ASTM D1894	*Taber Abrasion*	ASTM C4060
Coefficient of Linear Expansion	ASTM D696	*Taber Wear resistance*	ASTM C501
Compression	ASTM D1751	*Taber Wear resistance*	ASTM C502
Cone penetration	ASTM D217	*Tear Resistance*	ASTM D412-92
Deflection Temperature	ASTM D648	*Tear Resistance*	DIN 53363
Density	ASTM D3575A	*Tear Resistance*	ASTM D1004
Density	ASTM D792	*Tear Strength*	ASTM D4533
Dielectric Strength	ASTM D1000	*Tear strength (trapezoidal)*	ASTM D1004
Elongation (High temp. 40 C)	ASTM D419-92	*Tensile Strength*	ASTM D3575T
Elongation and Recovery	ASTM D412	*Tensile Strength and Elongation*	ASTM D638
Environmental testing	ASTM D543	*Tensile Strength and, –30°C Elongation*	ASTM D412-92
Flammability	ASTM D635	*Tensile Strength and Elongation*	ASTM D412-92
Flexural strength and modulus	ASTM D790	*Tensile Strength and Mod*	ASTM D638
Hardness Shore A	ASTM D2240-91	*Tensile Strength after heat ageing*	ASTM D412-92
Impact (Charpy)	ASTM D256	*Tensile Testing*	ASTM D638(M)
Impact Resistance (seeved pipe)	DIN 30672	*Thermal Shrinkage*	ASTM D1204
Index of whiteness and yellowness	ASTM E313	*Thermal Stability*	ASTM D3895
Insulation Resistance	ASTM D256	*Thickness & Mass*	ASTM D751
Insulation Resistance	ASTM D149	*Water Absorbtion wt gain after 6 mo*	ASTM D471-95
Low-Temperature Brittleness	ASTM D746	*Water Absorption*	ASTM D570
Moisture Vapor Transmission Rate	ASTM E96	*Water Resistance*	ASTM D570
Penetration resistance	ASTM G17	*Weathering*	ASTM D2565

a = steel rod radius
b = coating thickness + steel rod radius
$c = b$ + the radial distance to the point of measurement in the electrolyte

With this technique, the resistive (barrier and inhibitive) properties of protective coatings can be ranked (Sonoke, 2008) as:

- $>10_8\ \Omega\ cm^2$ good
- $10_7\ \Omega\ cm^2$ to $10_8\ \Omega\ cm^2$ fair (standard)
- $10_6\ \Omega\ cm^2$ to $10_7\ \Omega\ cm^2$ doubtful
- $<10_6\ \Omega\ cm^2$ poor

Figure 3-9. Coating tension dollies

Compatibility/Cathodic Disbondment

In areas where the coating is damaged or is missing, CP (sacrificial anodes or impressed current) is generally used to supplement corrosion protection by coatings to protect the metal from corrosion.

However, if a cathodic over-voltage is applied, cathodic disbanding of the coating may result over time. Even in the absence of a cathodic over-voltage, cathodic disbanding of a coating can also occur. This is due to the increase of pH level, caused by an increase in hydroxyl ions at the cathode, which occurs at a holiday or discontinuity in the coating as a result of cathodic reduction reactions (Tator 2006).

As explained by Tator, these reactions are itemized below:

$$2H^+ + 2e^- \rightarrow H_2 \tag{3 – 3}$$

$$H_2O + ½O_2 + 2e^- \rightarrow 2OH^- \tag{3 – 4}$$

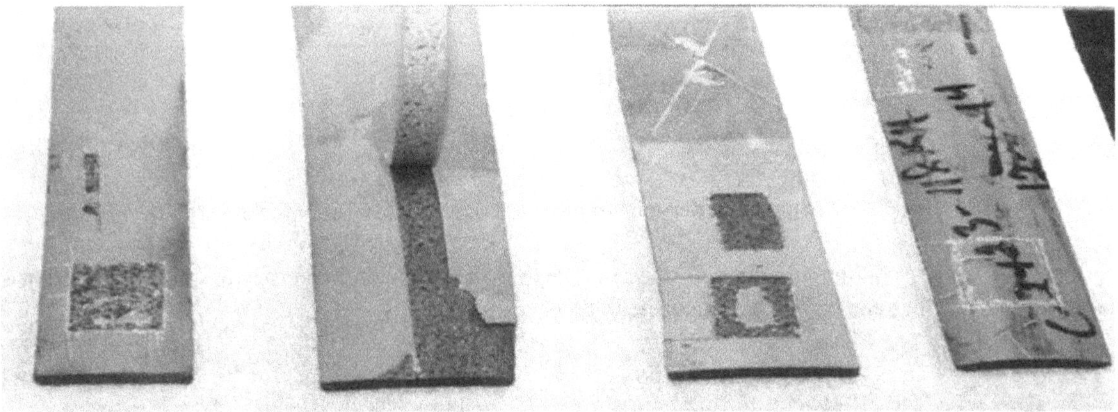

Figure 3-10. Knife adhesion testing of various FBE coating systems (Tator 2006)

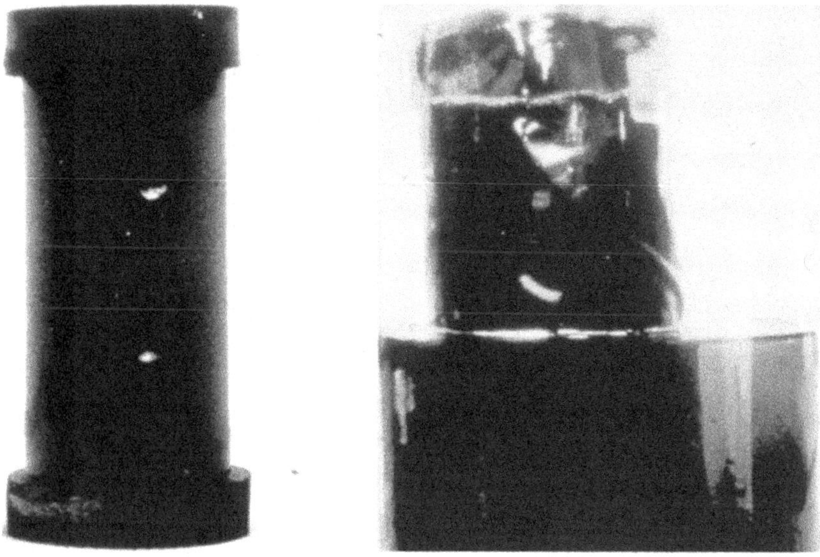

Figure 3-11. Typical chemical exposure test sample

$$Fe_2O_3 + 3H_2O + 2\,e^- \rightarrow 2Fe(OH)_2 + 2OH^- \qquad (3-5)$$

$$H_2O + e^- \rightarrow \tfrac{1}{2}H_2 + OH \qquad (3-6)$$

1. Reduction of hydrogen ions (Eq. 3-3) from the aqueous electrolyte (moisture in the soil) under the influence of an electrical current quickly increases pH level of soil (making the solution at the cathode more alkaline). As the pH rises, the reaction becomes unfavourable.
2. Increase in pH level through the reduction of oxygen from solution creates more hydroxyl ions (Eq. 3-4).
3. Ferric oxide reduction (Eq. 3-5) on the steel surface to form ferrous hydroxide results in a high pH environment in the crevice area (between the coating and the steel substrate) around the periphery of the holiday.
4. This is seen as the formation of a silver halo on the steel surface around the holiday and attributed to a loss of cathodic potential beneath the disbonded coating. This causes oxide growth to occur as a result of insufficient cathodic shielding.
5. Finally, water itself may be reduced, which will raise the local pH, as illustrated in Eq. (3-6).

Therefore, CP can induce the evolution of hydrogen gas at the exposed surface and increase the transportation of water to the coating defect site. These two effects can singularly or in combination increase the loss of coating adhesion at the edge of a "holiday," that is at a failure point in the coating.

Cathodic disbondment tests are used to assess a coating's compatibility with CP.

This test provides accelerated adhesion assessment and determines resistance of the coating to cathodic potential and current flow. Coated steel samples of are usually used for the tests. In the middle of the coated specimen, a hole is drilled through the coating to expose the substrate in this case pipe surface; Figure 3-12). Typically, a coating specimen with a predrilled holiday is immersed in a 3% salt solution at 20°C for 28 days. During the test, the coating samples are usually maintained at a nominal −1.5-V potential.

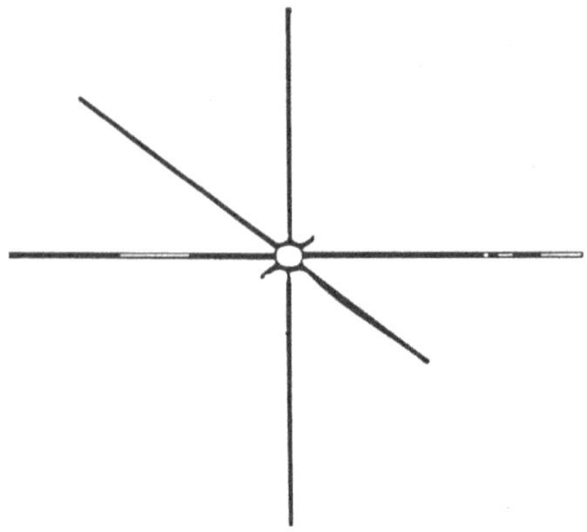

Figure 3-12. Example of radial cuts through coating

After completion of the test period (Figure 3-13), a levering action is used to lift the coating around the holiday, until the coating demonstrates a definite resistance to the levering action. The radius of disbondment is measured and reported.

Depending on the evaluated conditions that the coating may be subjected to test periods are as follows:

- 28 days, 20°C
- 14days, 35°C
- 14 days, 50°C
- 14 days, 65°C
- 48 hours, 65°C
- 24 hours, 65°C at −3.5 V

It may be noted that there are 22 cathodic disbondment test methods that are in use in many countries including US, European Union, France, Germany, Saudi Arabia, China, Canada, Australia, and New Zealand.

Table 3-8 provides a comparison of some standards for cathodic disbondment testing (Neal 1999). Only some testing parameters are included in this table.

Figure 3-13. Typical CD test result (80°C) (Banach 2004)

TABLE 3-8. Comparison of some standards test methods for cathodic disbondment

Standard	ASTM G8	ASTM G42	ASTM G80	ASTM G95	AS3862	CSA Z245.20	ISP15711: 2003(E)	NF A 49-711
Test Specimen (min. dimensions)	Suspended Pipe (O.D. > 19.05 mm) with sealed or plugged end	Suspended Pipe (O.D. > 19.05 mm) with sealed or plugged end	Schedule 40 Suspended Pipe (O.D. > 60 mm) with sealed or plugged end	Coupon, Panel, ring, In-Service Pipe	Panel (100 mm × 100 mm)	Panel, Pipe	4 Panels Method B only 300 × 150 × 2 mm	Pipe, 80 × 80 × thickness in mm
Temperature (°C)	Room Temperature	60 to Max. Service Temperature	Room Temperature	60 to Max. Service Temperature	R.T., 60, 110 or Max. Service Temperature	R.T., 60, 110 or Max. Service Temperature	Room Temperature	95, 110 or Max. Service Temperature
Duration (days)	30, 60, 90	30	60	90	1 to 28	1 to 28	182 (26 weeks)	2, 7, or 28
Holiday Size and Geometry	3 × 120° apart, 6.35 mm, Flat (small pipe), Cone (other)	3 × 120° apart, 6.35 mm, Flat (small pipe), Cone (other)	3 × 120° apart, 6.35 mm, Flat (small pipe), Cone (other)	3.2 mm, Cone	6 mm, Flat	3.2 mm, no geometry specified Cone	Method A – 10mm blasted Method B – 3mm, Flat	6 mm, Flat

It is shown that ASTM G8, ASTM G42, ASTM G80, and four of their generic modifications, as well as CSAZ245.20, EN 10289, DIN 30 671 (1987), API RP 5L7, and AWWA C-210 (old) and NF A49-711 will generate comparable results, when run under same temperature and duration conditions. Australian AS3862 and related AS/NZS 4352 will give approximately 15% to 20% smaller cathodic disbondments. ASTM G95 and ISO 15711 (E) test will give larger disbondments owing to the tests' durations (Holub et al. 2007).

Strain Polarization (see also disbondment)

Cathodic disbondment of strained coated pipe is performed to observe what effect field bending of pipe will have on coating compatibility with CP and resistance to cathodic disbondment.

The test method is the same as cathodic disbondment tests except that the shape panels allows for a 2.5° bend. The test panels are bent at lowest operating temperature that the pipe/coating is likely to experience and then put on test for 28 days at 20 °C to 30°C. They are then visually inspected for presence of any strain marks or cracks.

Penetration Resistance

During storage and operational service, coatings are subjected to concentrated pressures exerted by the weight of the structure and/or backfill. In situations where rocks or stones have not been cleared from the pipe bed and backfill, these pressures can be sufficient to penetrate and eventually perforate the coating.

A coating system must be able to resist penetration or deformation under conditions of static load. Resistance to penetration is evaluated by subjecting a known coating thickness to a constant static load. The percentage of coating thickness penetrated is evaluated.

This test determines the relative resistance of the coating system to penetration and deformation under static load (Figure 3-14). It is performed by placing an area of known thickness under a penetration rod for 14 days, the thickness is then measured and the percent penetration calculated, with a maximum allowance of 50%.

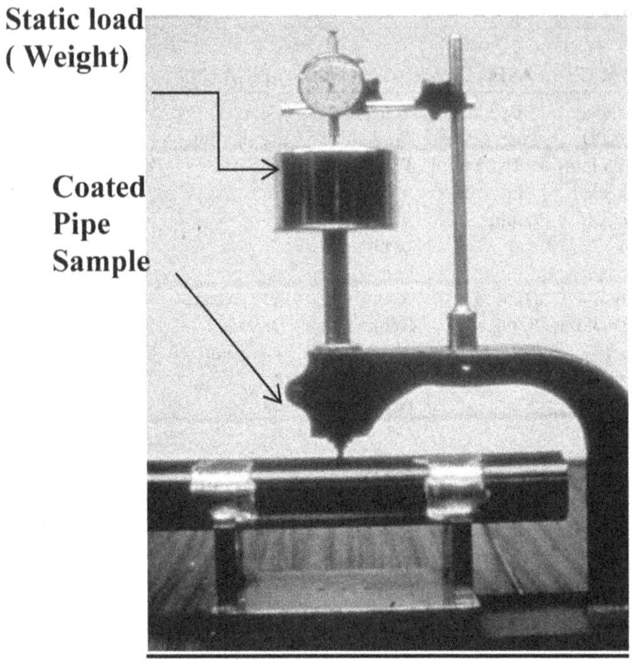

Figure 3-14. A penetration test apparatus

Soil Stressing Resistance

Coatings are subject to deformation and damage by soil stresses. The assessment on the influence of soil stressing and failing pipeline coating goes back as far as early 1970s (Argent 1977). Such a soil stresses can cause coating wrinkling and eventual disbondment and hence failure (Figures 3-15 and 3-16). The soil on top and side of the pipe causes the following and ensuing actions on the coating:

- Soil weight and pressure on top of pipe starts the movement of the coating.
- Significant downward soil force on the pipe sides, which start wrinkling and disbondment of coatings. This causes the coating to stretch or lose adhesion.
- The downward pressure also creates significant forces on the side of the pipe due to ditch walls restricting the path of soil movement to the sides. This causes more stress on the coating as the soil compacts around the pipe.
- However, there is little stresses on the underside of the pipe.
- Downward forces created above force the coating to wrinkle and disbondment on the undersides sides of the pipe where water/moisture from soil can penetrate causing pipe surface corrosion.

Soil stresses are more evident in clayish soil. Cohesive clays around the coated structure expand with absorption of moisture and subsequently contract during drying. These clays exhibit sufficient gripping action to remove the coating during contraction (Figure 3-17). Coatings must be able to withstand this type of soil movement, which can also be exacerbated by temperature cycling of the pipeline.

Typically, soil stress test consists of a soil box in which pipe samples are inserted to assess a coating's resistance to soil stress. Medium is circulated inside the coated samples

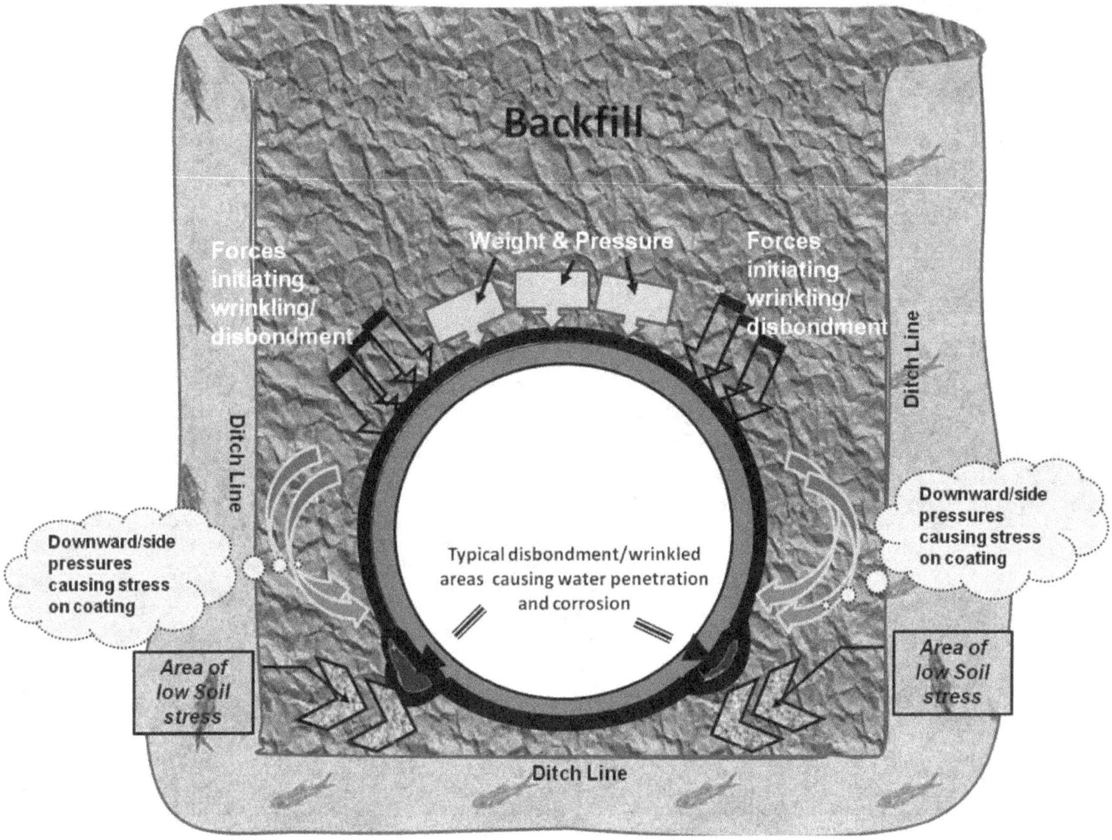

Figure 3-15. Soil stress forces (after Norsworthy 2009)

Figure 3-16. Degraded/wrinkled coating failure

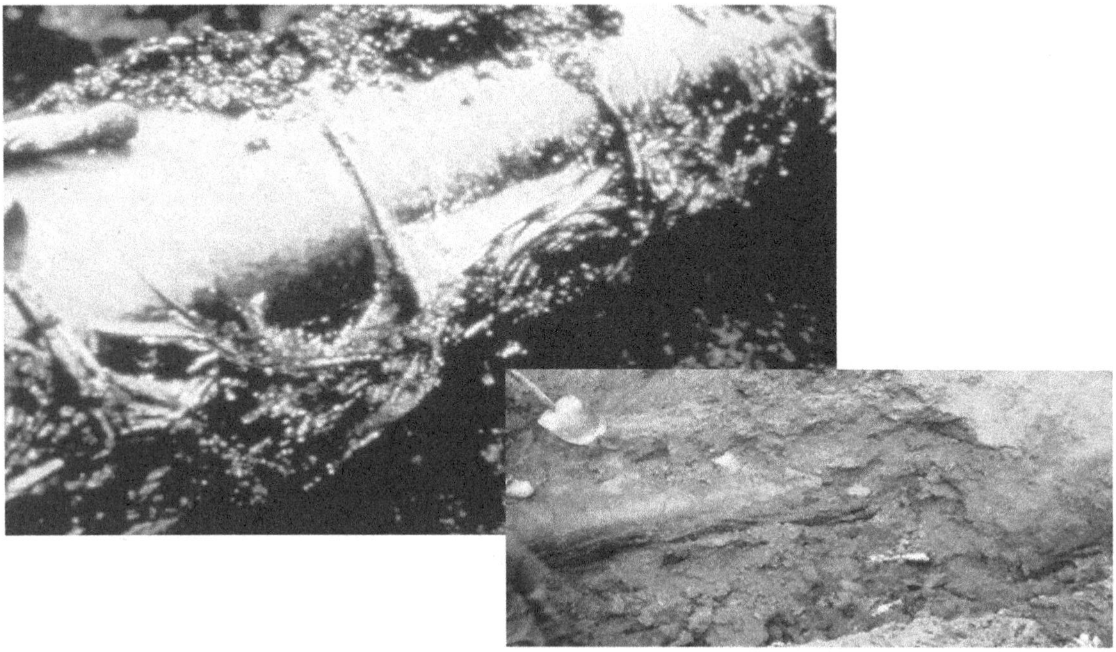

Figure 3-17. Soil stress — tape-coated pipe in clayish soil

at a given temperature. One cycle consists of wetting the clayish soil and allowing it to dry. After a predetermined number of cycles have elapsed, the clay is removed, and the coating is examined for adhesion loss.

Peel Adhesion

Usually performed on extruded polyethylene coatings, these adhesion tests determine if there is satisfactory adhesion between both the pipe and mastic and between the mastic and the polyethylene coating. They reflect the quality of materials as well as the caliber of application. Two test methods are employed:

1. Hanging Weight Adhesion Test

In this test (Figure 3-18), a known force is applied to the coating and the travel speed is calculated over a predetermined period of time.

The test specimen is a complete pipe test ring; cuts to the substrate are made 25 mm apart from the 90 zone to the 180 zone around the test ring and a tab is pulled away down to 45. The test is run from the 45 zone down to the 180 zone, and the time and distance peeled are measured (mm). The peeling may also be terminated when the time, calculated by the following formula, has been exceeded:

$$T = (0.125*3.13*D)/10 \qquad (3-7)$$

where

T = test time, minutes

D = Nominal pipe outside diameter, millimeters to one decimal

Peel speed is 10 mm/minute

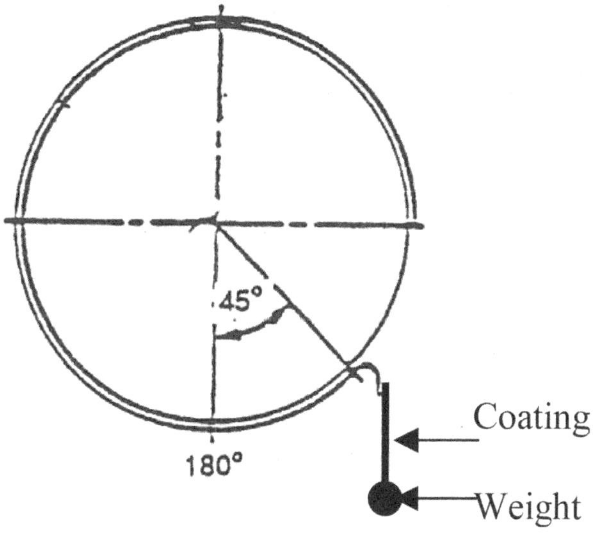

Figure 3-18. Hanging weight adhesion test

During the test, any increases in speed are noted as well as the probable cause for the increase such as an air pocket or bare spot on the pipe. The peeling speed is then calculated from:

$$PS = d/t \tag{3-8}$$

where

PS = peeling speed, mm/minute
d = distance peeled, mm
t = peeling time, min

2. Constant Rate of Peel Adhesion Test

In this peel test, the travel speed is constant and the force required to maintain that travel speed is measured over a set period of time.

The test specimen is prepared the same way as the Hanging Weight Adhesion Test, except that cuts to the substrate are made around the pipe so that the minimum peel length will be about 75 mm. An Instrument is used to perform the peel at a rate of 10±1 mm/minute. The test is continuously monitored to assess load drops (indicated by valleys on the graph), which may be due to elongation of the polyethylene, voids between the substrate and polyethylene, depletion of mastic and various other factors. The load required to peel the coating is then recorded.

The test is a failure if more than 5% of the total peeled length is below the specified minimum value. If the polyethylene breaks more than three times, the test is terminated and the minimum peel strength is calculated from:

$$P = L_p * 9.81 \tag{3-9}$$

where

P = peel strength, N
L_p = minimum peeling load, kg (excluding bottom 5%)

IMPORTANT PROPERTIES FOR CONSTRUCTION AND INSTALLATION

Flexibility

Coatings must be sufficiently flexible to resist cracking or disbondment during construction. For example, plant applied pipe coating requires sufficient flexibility for field bending between 1° and 2.5° per pipe diameter. Coating flexibility is determined by bending a coated strap over a fixed radius mandrel corresponding to the maximum amount of curvature expected (Figure 3-19). Tests are performed over a wide temperature range to determine the bending limits of a coating system.

A mandrel type bending machine are used to carry out the flexibility test by increasing the mandrel radius step by step until the coating fails (Malik et al. 2002). The performance of a coating depends on the actual amount of strain that it will encounter during field bending. The present strain is calculated as given bellow:

$$\%\text{Strain } \epsilon = (100t)/(2r + t) \qquad (3-10)$$

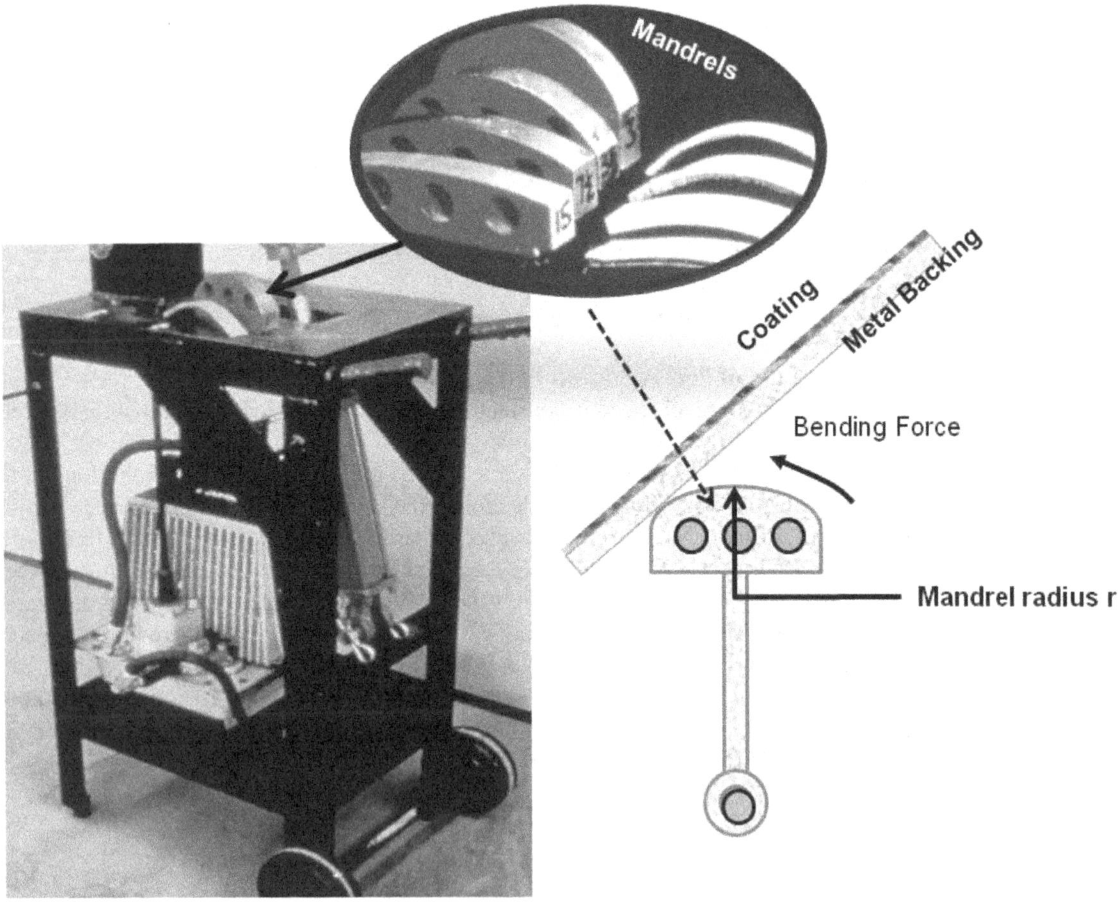

Figure 3-19. Typical flexibility test machine and mandrels

where

t = effective thickness of the specimens (DFT + metal)
r = radius of the mandrel shoe
DFT = dry film/coating thickness

The percent strain € is directly proportional to the effective thickness t of the coated specimen substrate).

Figure 3-20 (Coulson et al. 1983) represents the schematic of pipe bending strain notations from which it can be deducted that

$$€ = (L - D)/D \text{ or}$$

$$€ = [(R + r)\,\Theta - R\Theta]/D$$

Therefore:

$$€ = (r\Theta)/D = (r\Theta/2r) = \Theta/2 \quad (3-11)$$

where € is the maximum strain on outside of bend. As an example, a bend of 1.5 per pipe diameter length would result in a strain of € = Θ/2 = (1.5/2) x 3.14/180 = 0.0131.

Equally, the radius of a fixed mandrel required to induce a prescribed strain can be calculated from Eq. (3-6) as follows :

$$€ = t/(2r + t)$$

$$2€r + €t = t$$

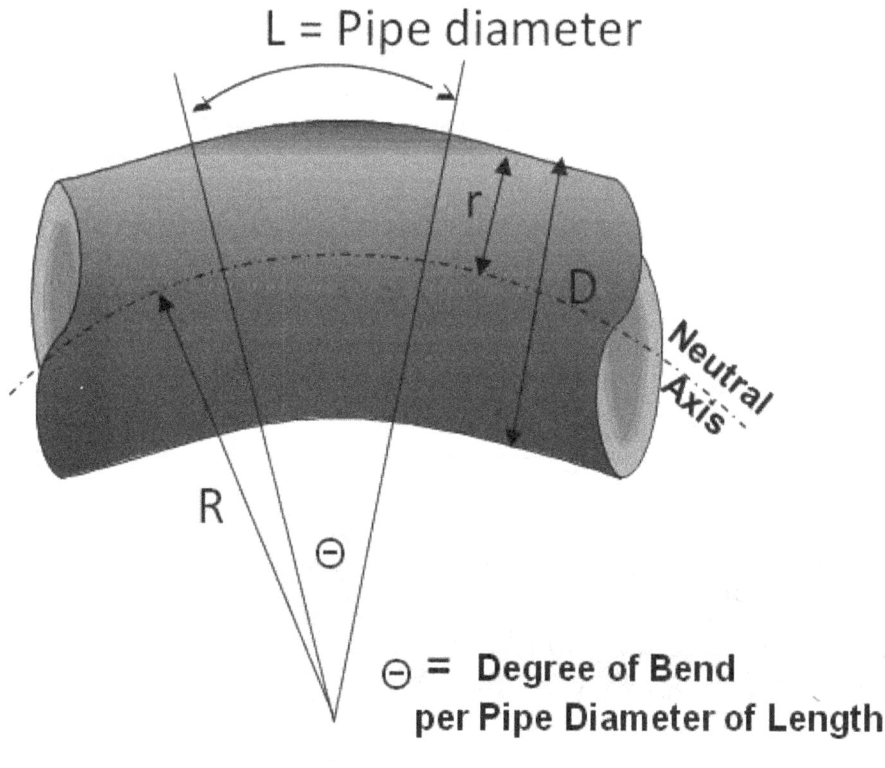

Figure 3-20. Schematic of pipe bending strain notations

Therefore:

$$r = t/2 \times [(1/€ - 1)] \qquad (3-12)$$

where, as previously indicated, t denotes total substrate thickness metal + coating thickness) and r denotes radius of curvature to the neutral axis also the mandrel radius). For a sample (substrate) thickness of 3.18 mm and 2.5° bend (€ = 0.0208), R = 70 mm approximately.

Flexibility evaluation is designed to provide a quantitative indication of the ability of a plant applied coating to resist cracking, tearing, disbanding, or any other mechanical damage as a result of field bending during construction.

In this test, typically 25 × 200-mm coated specimens, cut parallel to the axis of the pipe with smooth rounded edges to remove any potential stress rises, are typically subjected to a number of short radius bends over a series of mandrels, which are designed to simulate various degrees of pipe bending. Coating specimens are bent at a minimum angle of deflection of 2.5° per pipe diameter length using the following formula:

$$R = 22.43t \qquad (3-13)$$

where

R = required mandrel radius, mm

t = sample thickness, mm

Three specimens are usually each tested at −30°C, 0°C, and 20°C. They are then allowed to return to room temperature and are visually inspected for any form of damage such as white marks, stretch marks, or cracks.

Hardness Resistance

By the nature of construction, coated materials undergo a great deal of handling from the time the coating is applied until the coated material is lowered into the ditch and the backfilling operation is complete. A coating should have the ability to resist abrasion and any other damage during shipping, handling, backfilling, and in-service conditions. Soft coatings are more susceptible to damage than hard coatings (Figure 3-21).

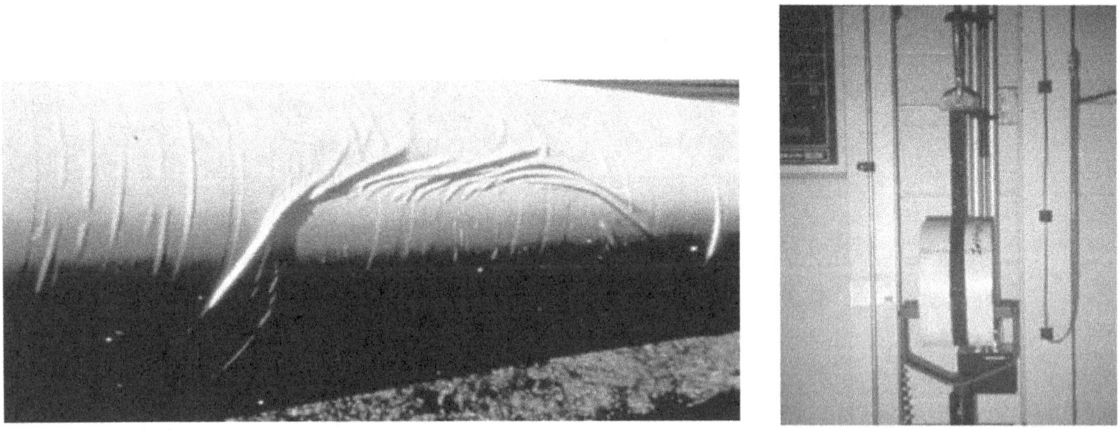

Figure 3-21. Wrinkling caused by lifting pipe by slings and typical hardness and abrasiveness — sling damage test

The shore scleroscope measures hardness in terms of the elasticity of the material. A diamond-tipped hammer in a graduated glass tube is allowed to fall from a known height on the specimen to be tested, and the hardness number depends on the height to which the hammer rebounds; the harder the material, the higher the rebound.

Shore hardness is a measure of the resistance of material to indentation by three spring-loaded indenter. The higher the number, the greater the resistance.

The hardness testing of coating is most commonly measured by the Shore (Durometer) test or Rockwell hardness test. Both methods measure the resistance of the coating toward indentation. Both scales provide an empirical hardness value that does not correlate to other properties or fundamental characteristics. Shore hardness, using either the Shore A or Shore D scale, is the preferred method for rubbers/elastomers and is also commonly used for "softer" plastics such as polyolefins, fluoropolymers, and vinyls. The Shore A scale is used for "softer" rubbers, while the Shore D scale is used for "harder" ones.

The coating resistance to handling and other factors listed previously is thus measured in terms of its hardness (ability to dissipate a dynamic load), which is measured using a Shore D Durometer at −50°C, −30°C, −10°C, 0°C, 20°C, 50°C, and 75°C in order to reflect various field conditions.

Conversion between Shore D hardness and Rockwell hardness is given in Figure 3-22, UM 2002.

It may be noted that both Shore D and Rockwell R are indexed scales and do not directly convert, but can be compared using the above captioned (Figure 3-22).

Abrasion Resistance

The abrasion test is intended to simulate pipeline movement in the ground during service. A flat bed of aggregates, similar to that used in actual backfill, is typically used to perform abrasion tests. Typically, a test section is placed on the aggregate bed after first checking the coating for damage (Espiner et al. 2003) (Figure 3-23).

A concrete saddle weights is then used to simulate the weight of the backfill to the top of the pipe. The pipe is then moved longitudinally over the aggregate in a 150-mm cycle at a rate of approximately 20 mm/min using a hydraulic cylinder attached to a mounting bracket clamped across the pipe end. Three cycles are performed before the weights were removed

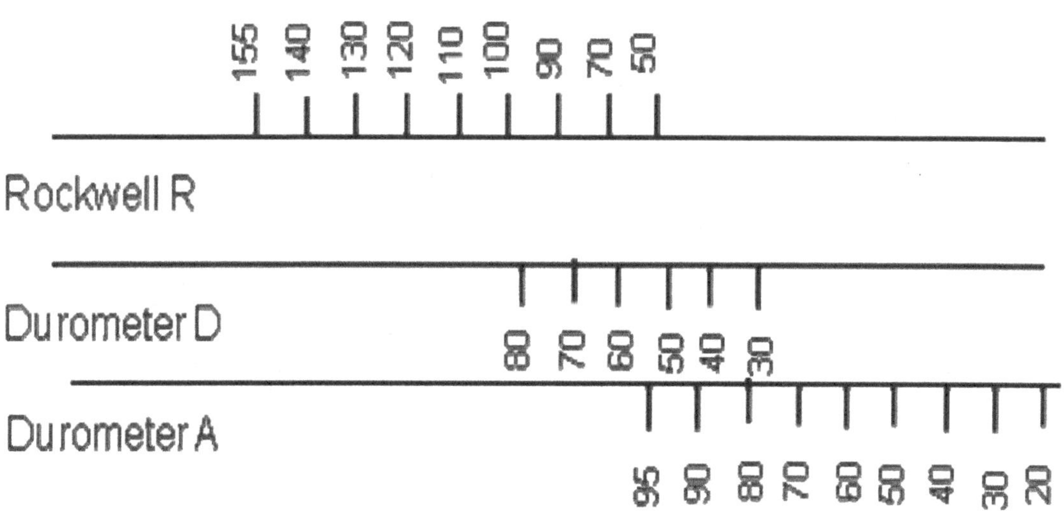

Figure 3-22. Hardness scales (after UM, 2001)

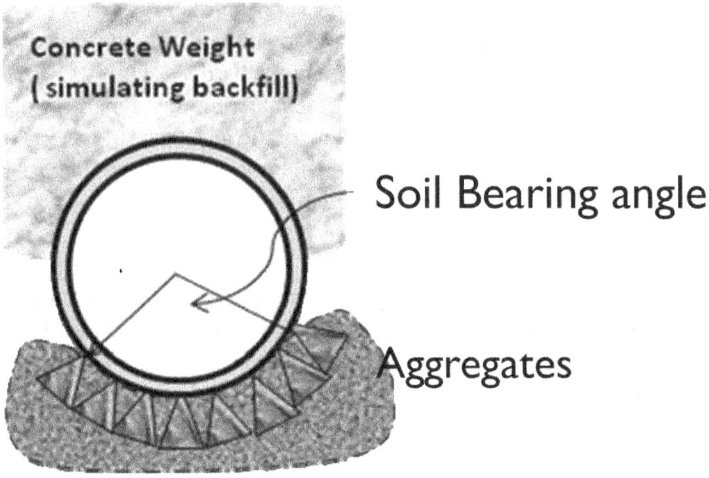

Figure 3-23. Abrasion test (after Espiner et al. 2003)

and the pipe area in contact with the aggregate inspected for coating damage. Hardness test as per above is also performed to confirm the abrasion resistance.

Sometimes abrasion and penetration by hard rocks) are measured qualitatively in a single field test. In such tests, production-coated steel line pipes are buried in different gradations of crushed hard rock, and a heavy load of about 2 tons (such as Caterpillar 966D wheel loader) is driven directly over the buried pipes for a number of times (usually twice daily for 12 consecutive days) (Williamson et al. 2000).

Subsequent excavation and inspection of the pipes are carried out to coating damage. Generally, coatings experienced progressively more damage (quantified by the number of holidays) with increase in rock size and progressively less damage with increase in coating thickness.

Impact Resistance

It is important for coatings to resist mechanical damage during installation and backfilling. Such damage to coating can particularly occur during pipeline trench backfill in rocky areas where backfill can contains rocks (Figure 3-24).

Figure 3-24. Backfill in rocky areas

The impact resistance properties of coatings are critical in determining the behavior under many of the conditions encountered in handling the pipe from shipping from the plant to being buried on the right of way. These include normal handling practices from:

- Stacking pipes in a stockpile or for transportation. Any sharp object such as stones or nails under or in between the pipes can cause damage. This effect is aggravated when there is relative movement between the pipes.
- Handling. In lifting pipes from stockpiles onto to transportation or vise versa, there is always the risk of contact with other objects. The most common impact scenarios are from pipe ends, dropped end hooks and tools.
- Bending/welding. Pipes have to be picked up from the right of way and placed into a bending machine. For welding, clamps have to be placed on the pipe ends.
- Burial and backfilling. In a rocky environment, both the backfill and ditch may contain rocks that are dropped on the pipe or impinge on the coating during burial.

There are several test methods used to determine the impact resistance of coatings, ranging from falling stones to falling weight (Figure 3-25) involving more precise and controlled impact scenarios. Open for discussion are the many factors that affect the measured impact resistance values as well as interpretations of results when there is evidence of substantial damage to the coating with no holiday during impact (Figure 3-26).

A method used for assessing impact resistance consists of an indentor of fixed weight, which is dropped from varying heights to produce a "point impact" on the coating. The failure point is then related to the height from which the falling weight ruptures the coating film and exposes the underlying steel surface. Impact resistance is the amount of energy required to cause the failure.

This test assesses the coatings ability to resist damage caused when rocks and frozen backfill are dropped onto the coated pipe during backfilling. For plant applied systems, it is

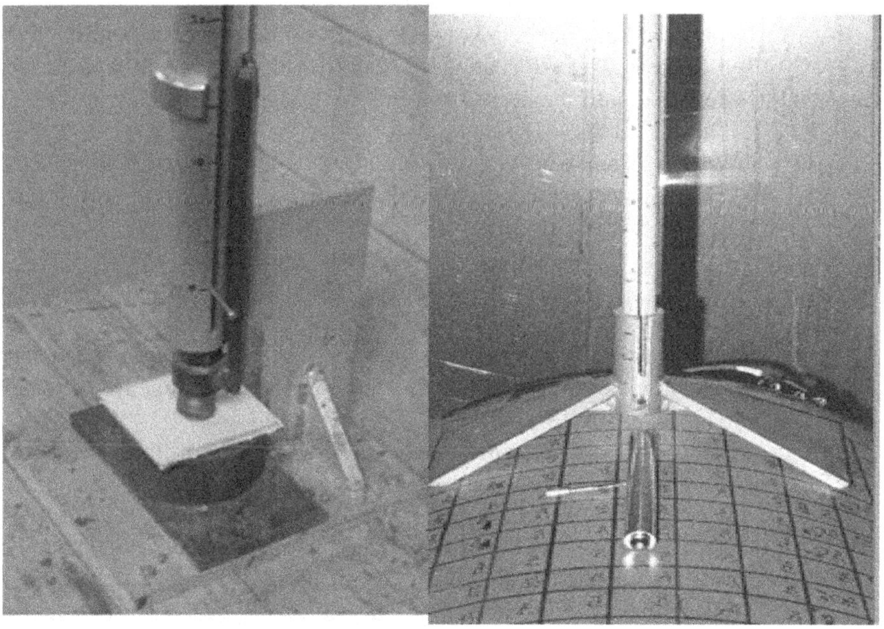

Figure 3-25. Typical impact resistance tests. a (left photo): falling weight on coating coupon (Menon 2004) and b (right photo): falling weight on coated pipe

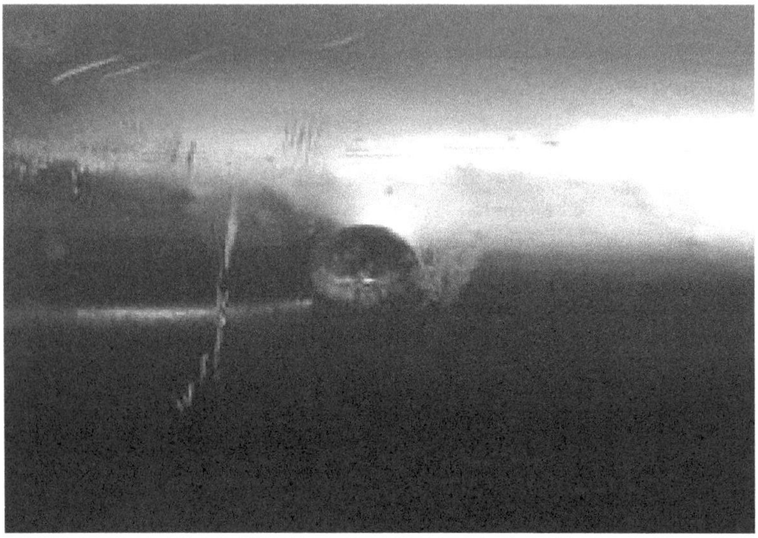

Figure 3-26. Typical impact test result (Banach 2004)

also an indication of the coatings resistance to mechanical damage caused during shipping, transportation, and handling on the construction site.

Impact on the coated pipe surface is produced by dropping a ball-bearing carrying a 1 kg mass down a slotted tube to produce a "point impact" on the surface of the pipe. The 1 kg mass is dropped from an initial height, and the specimen is checked for any breaks or cracks both visually and with a holiday detector at 1800 V to 2200 V for FBE and urethanes and at 10,000 V to 16,000 V for extruded polyethylene coatings. If, after three trials only one trial fails, one retest is performed and if the coating withstands the impact the height is increased; if the coating cracks or breaks, the height is decreased. This procedure is repeated until a constant maximum-impact-resistance energy value is obtained in joules (height * mass * 9.81 m/s^2).

Tests are done over a wide range of temperatures to anticipate the amount of damage, which may occur at a given construction temperature. Impact tests are performed at -30°C, −10°C, 0°C, 20°C, 50°C, and 75°C.

Weathering Resistance

Coating may be applied well in advance of proposed construction and/or in service dates. Therefore, coatings must be resistant to UV degradation and extreme temperature changes (Figure 3-27).

Also, some coatings chalk and lose thickness with exposure to sunlight during prolonged storage. This is a factor that needs to be seriously considered if the pipeline construction is delayed and the coated pipe will be stockpiled for an extended period. As a result, coated pipes have to be stockpiles and be protected with typically the use of tarps or whitewash paint to prevent UV degradation.

Whitewash, when used, is washed off the coating by rain and needs to be reapplied each year to maintain an adequate level of protection. In a typical application, the additional protection from UV degradation can typically range in cost, from 3% to 5% of the value of the coating per year.

Figure 3-27. Coating damage from sunlight

UV degradation and the effects of temperature can be evaluated by placing coating compounds in areas where coating storage may occur, exposing them to the elements. After a specified time period, the effects of exposure on the coating are determined by visual examination and by performing cathodic disbondment, flexibility, and impact resistance tests. The weathered coating sample test results are compared to the test results obtained from an un-weathered coating sample.

Compatible Repair and Girth Weld Coating Material

The girth weld coating system and repair coating should be compatible with the parent coating and possess properties, which meet the construction and operating conditions.

Coating Selection

Long-term coating performance is essential to prevent deterioration of a pipeline. Errors in coating selection may lead to failure and remedial repair costs, which can easily exceed the increased initial cost of a higher integrity coating. Factors used in coating selection are:

- Operating temperature
- Backfill/terrain characteristics
- Coating availability
- Overall coating cost

Operating Temperature

The pipeline operating temperature downstream of compressor stations and producing facilities is one of the criteria used for coating selection. These are the regions of pipe exposed to the highest temperatures. Laboratory coating tests are undertaken to assess coating properties at the operating temperature envisaged at a facility. From the test results, a coating can be selected, which will resist disbonding, melting, sagging, and softening over the period of the design life.

Backfill/Terrain Characteristics

Backfill/terrain characteristics, which need to be considered include:

- Rocky or frozen backfill — requires selecting a coating with excellent impact resistance, use of such a coating though relatively expensive can obviate the need to import fine backfill/padding material for the pipe
- Clay or glacial moraine — requires selecting a coating with excellent soil stressing resistance
- Undulating terrain — for plant applied systems, selecting a coating with flexibility sufficient for field bending under construction conditions is important

Coating Availability

Consideration should be given to the geographic source of the pipe, the pipeline location, the availability and location of coating applicator facilities, and probable method of transportation of the coated pipe. There are occasions where the final analysis of these factors leads to the selection of a coating system that is applied in the field with portable equipment or at a portable coating facility.

Factors that may eliminate the choice of an over-the-ditch applied coating system (using portable equipment or facilities) include:

- Coating manufacturer application requirements and construction temperatures
- The difficulty of providing and maintaining the optimum heat distribution in subzero conditions (e.g., variation of line heater speeds)
- The manufacturer's recommended application specification, which is difficult to meet for over-the-ditch systems during rainy periods and general winter construction
- Pipeline terrain
- Mountainous terrains may preclude the use of a field-applied coating system because it may be difficult to apply it properly under these conditions

Specifications

Once a coating system is selected, comprehensive specifications are required for coating application, handling, storage, joint coating, and repairs. Inspection of all phases of the coating operation is essential to ensure a uniform, high-quality protective coating.

Typical Performance Properties

As indicated previously, coatings must perform satisfactorily in all phases including transportation, stockpiling, construction, burial, and operation. These properties are re-emphasized as indicated in Figure 3-28 (Singh 2008).

There is no consensus as to the performance requirements among coating manufacture/suppliers and generally codes do not specify the requirements either. However, CSA Z245.20 and Z245.21 stipulate production coating test requirements as follows:

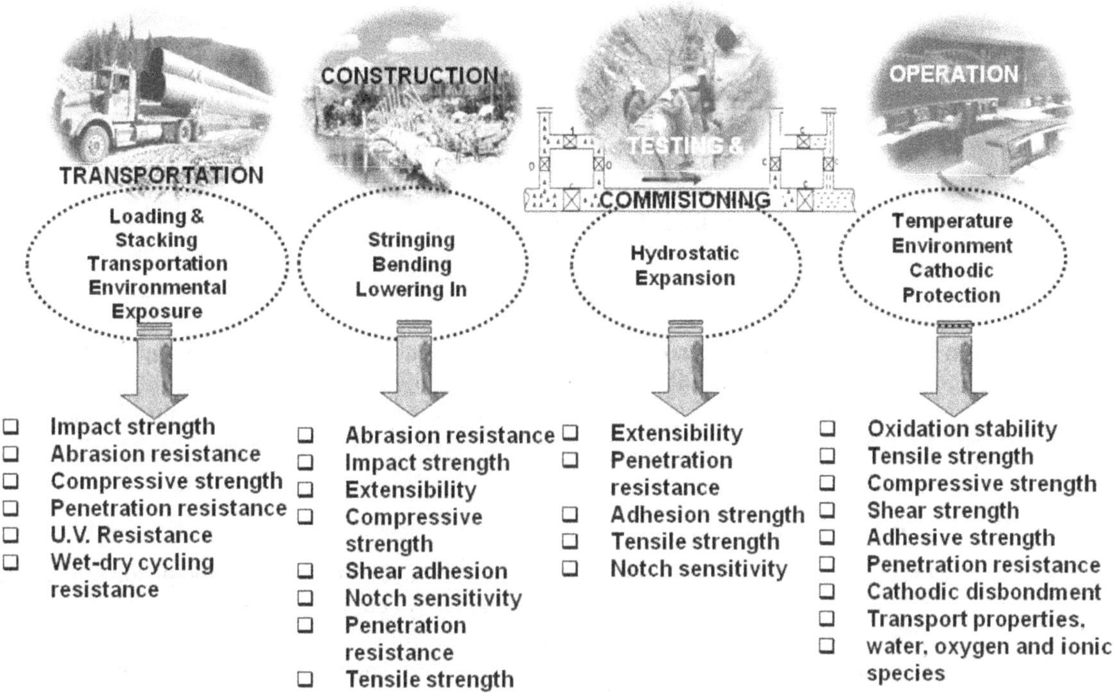

Figure 3-28. Coating Performance Properties Requirement (after Singh 2008)

1. Fusion Bond Epoxy Coating (CSA Z 245.20)

• 24 hours cathodic disbondment	11.5 mm max
• Cathodic disbondment (28 days, 65°C)	radius <20 mm
• Cross-section porosity rating	1-4
• Adhesion rating (hot water soak, 75°C)	CD radius <20 mm
• interface porosity rating	1-4
• 2.5-degree flexibility	no cracking
• 1.5 J impact	no holiday
• 24-hour adhesion rating	1-3
• Interface contamination	30% max

However typical FBE Qualification Requirements from RP0394-2002 provides the following guidance:

Test	Acceptance Criteria
• Cathodic disbndment (24 h)	Avg. radius: 6 mm (0.25 in.)
• Cathodic disbondment (28 d)	Avg. radius: 8 mm (0.3 in.)
• Cross-section porosity	Rating of 1–4
• Interface porosity	Rating of 1–4
• Flexibility (3°/PD at 0°C [32°F] or −30°C [-22°F])	No cracks, tears, or delamination
• Impact resistance	1.5 J (13 in.-lb) min
• Hot-water soak	A rating of 1–3 inclusive

2. External Polyethylene Coating

• Peel/adhesion	3.0, 19.6, 150 N
• Tensile stress at yield	–8.5 MPa for LDP/LLDP
	–11.0 MPa for MDP
	–17.0 MPa for HDP
• Elongation at break point	300% min
• 24-hour cathodic disbondment	12, 12, 7, 7 mm
• 24-hour adhesion test	1-2
• 2.5° flexibility	No cracking

Note: LDP/LLDP, low- or very low-density polyethylene tape; MDP, medium-density polyethylene tape; and HDP, high-density polyethylene tape

However, to assure pipeline coating quality, coating specifications must define minimum acceptable coating properties for acceptable performance on pipeline.

Field Joint Coatings

Once the most suitable mainline coating system has been chosen for the pipeline exterior, a compatible and appropriate field joint coating system must be selected. While mainline coating systems are typically applied under controlled conditions in a coating plant, the field joint coating must be installed in diverse environments, often under inclement weather conditions.

The selection of an appropriate field joint coating system depends not only on the variables and constraints listed above, but most importantly, the choice of the mainline coating system. Above all, the joint coating system must be compatible with and adhere to the mainline coating system.

While there are several coating standards and specifications covering mainline pipe coatings, such is not the case for field joint coatings. Instead individual users and contractors have developed their own specifications often with little bearing on the operating performance.

Surface cleanliness is the most important requirement in achieving a good joint coating, and in the sometimes difficult conditions found in the field, it can only be achieved by blast cleaning. Pipe preheat is required prior to cleaning and again before coating application. This ensures that there is no condensed surface moisture present. A good anchor profile is highly desirable to ensure the adhesion of coatings. Andrenacci et al. (1999) recommend a pattern at least 2.5 mi (90.025 mm) deep. They also have developed a weighted comparison of joint coatings (Table 3-9), which shows that by their reckoning a two-layer mastic sleeve is superior to a number of other methods.

Another more subjective comparison of various types of field joint completion systems is found in Table 3-10.

Pipeline coating choices for mainline and field joint protection are diverse. For onshore applications, mainline coatings are usually selected from a range of options including: single and dual layer FBE, three-layer polyethylene (3LPE) and more recently multilayer polypropylene coatings. A similarly broad range of field joint coatings exist with the most widely used technologies being FBE, liquid coatings, and heat shrink sleeves.

Selecting field joint systems that mirror the mainline coating represent a very logical and proven methodology. Field-applied FBE on FBE coatings and 3-layer Heat Shrink Sleeves on 3LPE systems are prime examples of this approach.

TABLE 3-9. Comparison of joint coatings (Andrenacci et al. 1999)

Factors	Weighting Factor	2-Layer Mastic Sleeve		2-Layer H-Melt Sleeve		3-Layer Sleeve		New Generation Sleeve		Cold-Applied Tapes		Epoxy Poly-urethane		FBE	
		R	W	R	W	R	W	R	W	R	W	R	W	R	W
1. Surface Preparation	1.2	1	1.2	4	4.8	3	3.6	3	3.6	1	1.2	5	6	5	6
2. Equipment requirements	1	1	1	3	3	4	4	3	3	3	3	2	2	5	5
3. Speed of installation	0.8	1	0.8	3	2.4	4	3.2	3.5	2.8	2	1.6	2	1.6	1	0.8
4. Operator skill set	1	2	2	3	3	4	4	3.5	3.5	1	1	2	2	5	5
5. Sensitivity to application	1.2	1	1.2	3	3.6	4	4.8	3.5	4.2	1	1.2	5	6	3.5	4.2
6. Crew size	1.2	1	1.2	2	2.4	4	4.8	3	3.6	3	3.6	2	2.4	5	6
7. Inspectability	0.7	2	1.4	2	1.4	2	1.4	2	1.4	4	2.8	5	3.5	5	3.5
8. Soil stress and mech. resistance	2	5	10	2	4	2	4	2	4	5	10	1	2	1	2
9. Shielding potential	0.8	3	2.4	4	3.2	3	2.4	3	2.4	5	4	1	0.8	1	0.8
10. Environmental concerns	0.7	1	0.7	1	0.7	4	2.8	3	2.1	4	2.8	5	3.5	4	2.8
11. Safety concerns	0.8	1	0.8	2	1.6	4	3.2	3	2.4	4	3.2	5	4	4	3.2
12. Cathodic disbondment	2	1	2	3	6	1	2	1	2	2	4	3	6	1	2
-low temp	2	5	10	5	10	3	6	3	6	5	10	3	6	2	4
-high temp															
Best overall system rating X weighting factor			1		6		7		2		4		5		3

Overall Coating Cost

The cost of pipeline coating is one component of the overall cost of a pipeline system. While in reality, coating cost is a fraction of the overall cost of a pipeline system; however, it accounts for about 7% to 10% of the cost of the pipe (Singh 2008), which in turn is about one-third cost of a pipeline project. Breakdown of a pipeline project cost is depicted in Figure 3-29.

When considering various coating options, the cheapest coating system for particular service may cause an overrun in cost for another application over the operational lifetime of the pipeline. In certain circumstances, a cheap coating, which has not performed well, may be the best choice such as gathering lines with a very short time in service), but for large transmission pipeline with anticipated long operation life), may cause unforeseen pipeline integrity problems causing increased operational and maintenance expenses.

There are four elements that make up the overall coating cost; these are:

- cost of the plant applied coating,
- joint completion costs,
- field repair costs, and
- the expected life of the pipeline (amortization costs)

The latter point is very important, since it can easily lead to overdesign or overspecify a coating with consequent cost increase. The choice of coating should therefore fully reflect the design life of the pipe.

TABLE 3-10. Comparison of common joint completion systems (Andrenacci et al. 1999)

System	Advantages	Disadvantages
Cold applied tapes	– Straight forward to apply-low skill	–must use a primer (typically solvent based)
	–few sizes	–spiral voids *bridging a major problem
	–adheres to most coatings	–shields cp system
	–effective in cool, low soil stress applications	–very soft, pressure sensitive adhesives
	–relatively insensitive to ambient temperatures	–low soil stress resistance
	–can tolerate a low degree of surface preparation	
Heat shrink sleeve 2 layer	–Simple & reliable installation-low skill	–limited resistance to soil stresses
	–excellent resistance to cathodic disbondment and undercutting	–requires a suitable torch
	–relatively insensitive to ambient temperatures	–installers must be trained
	–good track record	
	–can tolerate a low degree of surface preparation	
Heat shrink sleeve 2-layer hot melt	–Good resistance to soil stresses	–requires personnel & facilities to store, mix and dispense epoxy primer
	–effective at higher operating temperatures	–requires near-white surface preparation
	–excellent reisitance to cathodic disbondment, water soak, and undercutting	–epoxy is sensitive to temperature for proper cure. Force best in cold climates.
	–anti-corrosion effectiveness guaranteed if epoxy primer is properly applied and cured	–expensive
	–lower preheats than with 2-layer hot melts	
Field-applied FBE	–Excellent soil stress resistance	–requires induction heating in the field
	–effective for high temperature uses	–difficult to control application and cure temperature in cold, windy conditions
	–excellent cd resistance and hot water soak if the surface is properly prepared and the FBE is properly cured	–extremely sensitive to surface cleanliness
		–only useable with FBE coatings
Epoxy-urethane	–Excellent soil stress resistance	–extremely sensitive to surface cleanliness
	–simple to apply (brush, spray, etc)	–properties are very sensitive to cure temperature and surface dryness
	–relatively inexpensive	–accurate mixing ration and very good physical mixing are absolutely essential
	–good cd resistance and hot water soak if the surface is properly prepared and the FBE is properly cured	

Consequences of Incompatible Field Coating Choices

As indicated above, the field joint coating must offer compatible performance to that of the selected mainline coating. The coating application process is also very important in successful field joint coating. The following examples (Norman 2002) illustrate some of the issues with incompatible field coating application choices.

Heat shrink sleeves have been widely used in low ambient temperatures (under subzero temperatures). Although liquid products can work well on warm climate FBE pipe lines, they generally do not cure well below 10°C.

However, coating application techniques have been developed to allow for the application and cure of liquid epoxies on FBE-coated lines in sub-freezing temperatures. This

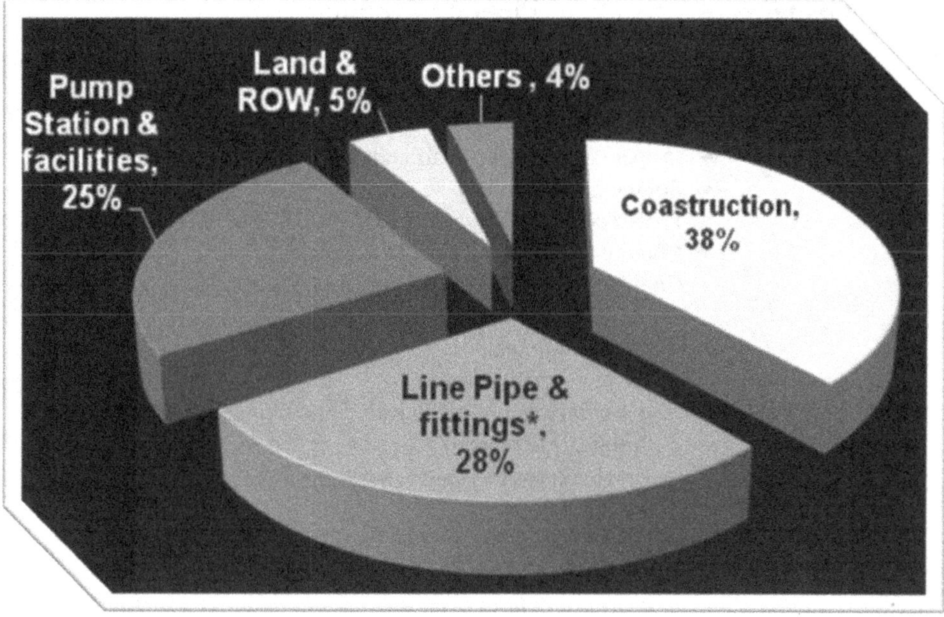

Figure 3-29. Typical pipeline cost allocation (oil pipeline) (note* includes coating 7% to 10%)

requires the bare steel pipeline joints to be preheated above 65°C prior to applying the epoxy.

Cold weather application also poses challenges for epoxies on 3PLE coatings. When the application described above for the FBE is used on 3LPE, the epoxy may cure on the steel portion; however, sufficient preheat may not be available to affect the cure throughout. Therefore, a secondary heating process would be needed to cure the epoxy on the 3PLE-coated situations for joint coating.

However, environmental conditions can often be made up for by preheating, postheating, and hoarding in. It is important to ensure that ambient conditions exist until a product is fully cured, not just during application.

Another practical consideration of incompatibility involves the holiday tests: If a different mainline and field joint coating systems were used, the holiday detector sweep done along the line would have to be adjusted at each joint. As an example, typical 3LPE holiday detection is conducted at over 10,000 V. Epoxy coatings require holiday testing at less than 5000 V. The implication is a re-calibration of the holiday test apparatus at each field joint, a logistics headache, which amounts to over 15,000 manual adjustments per 100 km of pipeline. Not only is this impractical, but there is danger of inadvertently subjecting the epoxy coating to high voltage required for the 3LPE, and thus causing damage, particularly at the steel/PE transition.

On lowering the coated pipeline into the trench, the bend forces must be maintained within the operating limits of the mainline and field joint coatings.

Another factor to consider during lowering-in is the transition zone (between the field and mainline) stresses. If epoxy is used for joint coating while the mainline has 3PLE coating, the large differences in thickness between 3LPE and thin epoxy film would create sharp transitions where the dynamic forces are intensified. Stresses due to the stinger roller during lowering in, or stresses due to pipe movement during service interacting at these sharp transition zones may damage the coating, particularly the cured thin brittle epoxy films.

Elements of Successful Pipeline Coating

Industries review of coating failures indicates that the following elements must be considered in having a very successful and trouble free pipeline coating (Wilmott 2007).

- Proper coating specification, including field joint coating
- Source of pipe supply
- Proper coating application procedure and qualifications
- Coating plant and inspection method
- Coating application control (quality control and quality assurance)
    - o Surface preparation
        - Pretreatment
    - o Coating application
        - Ambient conditions and application temperature
    - o Application method
        - Thickness
        - Cure
    - o Inspection and testing including holiday detection) and safety issues
- Construction technique and supervision
- Operational control

Some of the above elements singly or in combinations are described below:

1. Coating Specification and Qualification

Many of the in-ground coating problems are typically due to use of imprecise technical specifications (Norman 2002). Because of the nature of the terrain unique to each pipeline ROW, it is imperative that a coating specification be "project specific" relating directly to the specific type of coating to be used on the project. Specifications for fittings, field joints, and methods of repair should also be project specific.

All coating specifications should be completed and agreed prior to the award of a contract.

All coatings and coating systems should be procedurally qualified against the requirements of the technical specifications (Figure 3-30) (Tator 2006).

Coating plants should be qualified, preferably before the award of contract. The coatings should be qualified in both a testing laboratory and in the coating plant or field situation.

Applicators should be qualified by a three-step process. Firstly, the facility should be visited and examined in detail. The process should be examined with the question in mind: "what can go wrong." The expertise of the personnel and the company's training plans should also be examined. Secondly, a sample of production coating should be submitted for testing, and finally, a specification should be created, which addresses technical requirements and any areas of concern in the applicator's process.

Procedure qualification activities are necessary to ensure that the coating systems, coating plants, and contractors are competent to meet the requirements for corrosion protection of pipelines.

Coating specification as a minimum should aim for the following:

- Approved materials
- Coating storage
- Surface preparation
- Coating application
- Cure

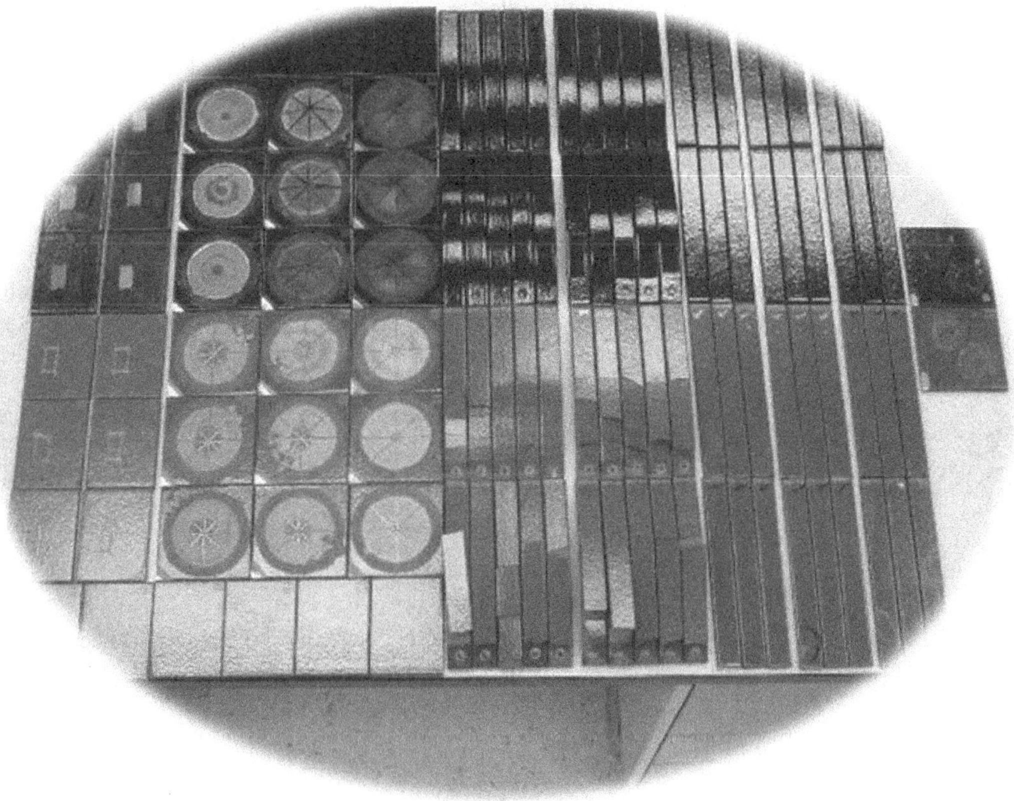

Figure 3-30. Typical coating qualification test panel series (Tator 2006)

- Holiday inspection
- Repair coating
- Reporting
- Safety
- Waste disposal
- Rights of inspection

Specifications must be objective and may be broadly categorized into performance specifications and process-oriented specifications. A performance specification aims at the performance characteristics of the finished product (e.g., an impact resistance of at least 3 J). A process oriented, on the other hand, directs the applicator as how to run the process (e.g., the pipe shall not travel faster than 4 m/min).

2. Source of Pipe Supply

Pipe is usually manufactured in a plant. The manufacturing processes may use chemical solutions for various purposes It is possible that the residues from some of these solutions could be present and detrimental to the adhesion of the coating to the steel substrate, even after some cleaning processes have been imparted to pipe surface.

Subsequent to the manufacture of pipe, temporary protective coatings (e.g., shellac or bitumen) have sometimes been applied to pipe and fittings in order to try to prevent corrosion during storage. This is an unacceptable procedure as it is usually extremely difficult to fully remove the temporary coating prior to the main pipe coating procedure. In addition, contamination of the abrasive blast cleaning materials usually results.

On some projects, bare pipe has been left on salt contaminated ground for lengthy periods. This may result to rust, corrosion, and possibly pitting.

All bare pipe and fittings should be stored on suitable wind-rows. The pipe should not be placed on the ground or in point contact with the "sand" wind-row, but polyethylene sheeting, or similar should be used to separate the pipe and the "sand." By this action contamination and corrosion should be minimized, and problems with the coating process should not accrue.

Inspection of Coating Plants

A thorough survey of the coating application plant should be undertaken prior to the award of the contract. The applicator's plant/machinery should be assessed; adequate quality assurance and quality control schemes should be in place. Inspection, laboratory facilities, and techniques should also be assessed.

Subsequent to award, there should be an ongoing dialogue between the coating plant and the end-product user and any other parties involved.

Independent, trained, and qualified coating inspectors should be utilized to achieving the desired coating standards.

Coating Applications

There are a number of steps involved in applying coating on the bare pipe successfully. These are listed and generalized in three categories:

A- Surface preparation, which generally includes:
- a. Cleaning
- b. Blasting
- c. Pretreatment

B- Application and
C- Testing/inspection
D- Transportation and storage

A. Surface Preparation

Undoubtedly, surface preparation is one of the areas in the pipe coating regime that is most problematical.

Without adequate surface preparation, the pipeline protective coating will not adhere correctly to the substrate. It is therefore imperative that all surface preparation processes are undertaken in a controlled manner in order to give a fully cleaned, correctly profiled substrate.

Many pipeline coating failures have occurred directly as a result of incorrectly cleaned substrates.

There are several standards that can be used for surface preparation; these are Steel Structures Painting Council (SSPC).

- SP-1 Solvent Cleaning
- SP-2 Hand Tool Cleaning
- SP-3 Power Tool Cleaning
- SP-4 Flame Cleaning
- SP-5 White Metal Blast Cleaning
- SP-6 Commercial Blast Cleaning
- SP-7 Brush-Off Blast Cleaning
- SP-8 Pickling
- SP-9 Weathering Followed By Blast Cleaning
- SP-10 Near-White Blast Cleaning

- SP-11 Power Tool Cleaning to White Metal
- SP-12 Water Jetting

NACE

- NACE 1 White Metal Blast Cleaning
- NACE 2 Near-White Blast Cleaning
- NACE 3 Commercial Blast Cleaning
- NACE 4 Brush-Off Blast Cleaning
- NACE 5 High- and ultrahigh-pressure water jetting

There are other blasting standards in addition to SSPC. There are:

BSI BS EN ISO 8501-1, 2007 "Preparation of steel substrates before application of paints and related products — Visual assessment of surface cleanliness — Part 1: Rust grades and preparation grades of uncoated steel substrates and of steel substrates after overall removal of previous coatings" and
BSI BS EN ISO 11124-2, 2002 "Preparation of Steel Substrates Before Application of Paints and Related Products — Specifications for Metallic Blast-Cleaning Abrasives Part 2: Chilled-Iron Grit"
Swedish Standard (St.Sa)

- St 2 Hand Tool Cleaning
- St 3 Power Tool Cleaning
- Sa 1 Brush-Off Blast Cleaning
- Sa 2 Commercial Blast Cleaning
- Sa 2 1/2 Near-White Blast Cleaning
- Sa 3 White Metal Blast Cleaning

The following list provides an outline of the main requirements in the surface preparation of pipes for coating:

1. Cleaning

Prior to blast cleaning, dirt, oil, and grease must be removed from the surface in accordance with a procedure such as SSPC SP-1 using wire brushes, scrapers, and a solvent such as xylene. Blast cleaning does not remove all of the grease from the surface, and this residual grease can lead to coating failure. If a continuous blaster, which recycles grit, becomes contaminated with oil and grease, all of the grit will have to be dumped.

2. Blasting

Most coatings require a blasted surface. Blasting performs two tasks: it cleans rust and mill scale from the surface and it provides a profile for the coating to adhere to. Most coatings require a given cleanliness as well as a given profile. The SSPC cleanliness standards for blasting in increasing order of cleanliness are:

- SSPC-SP 7 Brush-Off Blast Cleaning
- SSPC-SP 6 Commercial Blast Cleaning
- SSPC-SP 10 Near White Metal Blast Cleaning
- SSPC-SP 5 White Metal Blast Cleaning

There are other blasting standards in addition to SSPC. Table 3-11 provides a comparative listing of the different standards' cleanliness. The standards are not exactly equivalent.

TABLE 3-11. Comparison of standards for pipe cleaning

Cleaning Method	SSPC Standard #	NACE Standard #	ISO 8501-1 Std. #
Brush off	SP-7	4	Sa 1 (light blast cleaning)
Commercial	SP-6	3	Sa 2 (thorough blast cleaning)
Near-white metal	SP-10	2	Sa 2 1/2 (very thorough blast)
White metal	SP-5	1	Sa 3 (blast cleaning to visually)

Visual comparators can be use, generally, to check that a blasted piece of steel has achieved a given level of cleanliness. Comparators are either of the photographic variety, such as the ISO 8501-1 C1 and C2 "Surface Profile Comparators "or of the blasted metal sample variety, such as NACE Visual Comparators.

3. Surface Preparation — Pretreatment

Adhesion of some coatings will be enhanced with the application of a surface pretreatment after blasting and before coating. FBE is an example as it the main coating used in the oil and gas industry, which often has a pretreatment applied to the steel.

The two main pretreatments used are: chromate and phosphoric acid. Normally, just one of the two is applied to the surface. Chromate is applied to the surface and is allowed to dry on the surface to a final thickness of 501 m to 7511 m. Phosphoric acid is reduced in concentration with water and applied to the pipe at a given temperature for a given minimum time to etch the surface. The phosphoric acid is then washed off the surface with distilled water.

B. Coating Application

Following adequate preparation, the pipe must be prepared for coating applications. The coating system must be applied in a manner so as to at least meet the requirements of the technical specification. Some of the important points to consider are:

- Ambient conditions
- Application temperature
- Coating application parameters, such as speed, travels

Ambient Conditions. Ambient temperature, dew point, relative humidity, protection from rain/snow, and exposure to sunlight can all affect the finished coating. Different products will have different tolerances to ambient conditions.

The ambient temperature and the product's curing requirements will dictate whether or not a preheat must be used on the steel substrate. When a preheat is used, the time for the substrate to cool in the ambient air must be accounted for to ensure full cure.

The dew point is the temperature at which moisture in the air will precipitate as free water. Since many coatings are intolerant of water until they are cured, it is necessary to ensure that the surface to be coated is above the dew point, normally by 5°C, and remains so until cure is achieved.

Relative humidity is a measure of how much water is present in the air, expressed as a percentage of the maximum amount of water, which the air can hold at a given temperature. Many coatings consist of volatile components (e.g., solvent or water), which evaporate during the curing process. The lower the relative humidity, the easier it is for solvents to evaporate from the wet coating and the faster the coating will cure. Many coatings specify a maximum relative humidity above which application is not recommended.

Application Temperature. Most products have an application temperature window that will ensure complete cure. Application outside of the recommended temperature range could result in coating failure.

As an example and as indicated previously, application temperature for fast curing products such as FBE is critical because if the substrate temperature is too hot, the product will set up before it has fully flowed to all of the surfaces contours. If the substrate temperature is too low, the product will not fully cure and exhibit inferior properties.

Applications. There are many means to apply coating to steel surfaces. The application method must suit the form of the coating material as well as the size and shape of the area being covered. Liquid coatings, such as liquid epoxy, are commonly brushed onto girth welds and are sprayed onto larger components, often with a corresponding change in the product's thixotropy (properties).

FBE is a dry powder and in an FBE plant is fluidized, electrostatically charged, and sprayed onto the preheated pipe. FBE is flocked onto girth welds in the field by spraying nonfluidized, noncharged powder onto a preheated girth weld.

Two-layer plant-applied polyethylene uses two extruders to apply the mastic onto the steel and then the polyethylene over the mastic.

3LPE wrap the polyethylene onto the pipe from the side or sprinkle powdered polyethylene on from above the pipe.

C. Inspection, Testing, and Safety Procedures

On many projects, the inspection criteria and personnel and the testing regimes must be considered In order to obtain the required quality of work:

- All inspection and testing should be undertaken by qualified and trained personnel.
- All instrumentation should be calibrated regularly, and records of calibration should be maintained.
- All inspection and test results should be diligently recorded.

Coating Thickness. Thicker coatings generally exhibit better toughness, resistance to cathodic disbondment, impact ability, etc. Thicker coatings tend to crack more when bent. Since the coating cost is closely related to the thickness, it is a good idea to determine how much coating is required for a given application and specify this as a minimum thickness.

Coating thickness often varies over the coated surface. Thickness varies more with a viscous hand applied coating than with plant applied product. Coating specifications generally call for several thickness measurements to be made and the average of the measurements to be above a specified minimum. Additionally, a minimum thickness is often placed on the individual readings (the minimum of the individual readings will be less than the specified minimum average reading).

Cure Time. As indicated previously, coatings must cure before they are ready for use. The time to cure can be as short as a minute for FBE coating to many hours for a liquid epoxy or urethane. The times to cure are influenced by pipe preheat temperature and ambient conditions as well as the coating's own chemistry. The curing behavior of one brand of FBE might be different from another and even slightly different from batch to batch.

Manufacturers usually test how long their product takes to cure at a given temperature and provide this information with their product. The hotter the surface the product is applied to, the less time to cure.

The amount of cure, expressed as a percentage, can be calculated using a Differential Scanning Calorimeter (DSC) machine and a sample of virgin coating material and the set coating material. The DSC machine compares the thermal behavior of the virgin material, as it cures, to that of the set material.

Holiday Detection. Once a coating has cured and its temperature is close to ambient, the surface is checked for holes and imperfections in the coating with a DC voltage holiday detector. One side of the circuit is connected to the steel being inspected, the other end is made from a metallic brush or conductive rubber, which is run over the coating surface. If a hole exists in the coating, or there is a thin spot, a spark will jump the gap and a light or buzzer will alert the operator to the imperfection. When holidays are found, they should be fixed with an approved patch.

Coating should be stripped and the object recoated if the number of holidays is excessive.

Safety Considerations. There are a number of safety considerations, which are too numerous and detailed to cover in this brief introduction to coating. Some of the common safety concerns are:

- Abrasive blasting is bad for the respiratory system. It is important to use ventilated, sealed hoods and follow manufacturer's or equipment supplier's and regulatory requirements. Otherwise, one should not be exposed to blasting byproducts.
- Coatings are chemicals. It is imperative to read and follow the Manufacturer's Safety Data Sheet (MSDS) before handling and applying.

Testing Coatings. It important to review test results for selected coating for a pipeline project. The tests should reflect the service conditions expected and should be standardized tests where possible. Standard test techniques (ASTM, NACE, CSA provided previously).

Consideration must be give as to how laboratory results are obtained as this can affect the outcome. This includes the sample preparation, coating application, thickness, and test method. For example, the coating thickness will affect the cathodic disbondment radius; the thicker the coating, the less cathodic disbondment will occur. If two products are compared and their thicknesses are different, then that could explain the different cathodic disbondment results.

Coating testing should be carried out every time there is a formulation change, even if it is a small change to an inactive ingredient. Often small changes in chemistry will change the coating's properties in one or more key areas.

D. Transportation and Storage

Pipe storage can be critical if the coating is not protected from the environment. For transportation and storage, the coated pipe fitted with full-encirclement separators or similar suitable separators) to ensure that coating is not damage by stacked pipe weights in storage and through subsequent transportations (Figure 3-31). Coated pipes must be handles with padded forks and equipment.

All transportation and storage of coated pipe should be undertaken in a safe manner and so as not to damage the coating or the pipe. The guidelines on specifications, contractual agreements, prequalification of materials, and procedures and inspection levels should apply equally to field coating operations.

Proper storage of coated pipe is critical. FBE and other types of coated pipe placed in storage should be protected by a cover or a protective outer coating such as a white wash or tarp. If the coated pipe is stored over a long period (2 years), the coatings should be inspected by a qualified inspector.

Coating Failures and Cause

Coating failures may appear during application, at the stage of curing but most of after a certain period of service life of a pipeline. Various failures have been reported, and Table 3-12 summarizes some examples of common coating failures attributed to the external coatings used the most in the pipeline industry and the reason why they have occurred (Wilmot & Bacon, 2000, Wilmot et al., 2000, Wilmott 2007).

As indicated by Wilmott (2007), asphalt or coal tar was subject to oxidation and a loss of low molecular weight components through water dissolution and biodegradation. These effects caused coating to separates from pipe surface creating thin gap. Water penetration through these coatings would result in the development of corrosive conditions including microbial corrosion (MIC) and SCC failures.

Figure 3-31. Pipe storage awaiting coating applications (inset: Norsworthy (2009))

Tape coatings are based on polyethylene and polyvinyl chloride. PVC based products require addition of a plasticizer. Plastizers are low-molecular-weight organic materials such as dioctyl phthalate. The plasticizers can be lost from the coating through biodegradation, water dissolution, and sometimes local crystallization. The loss of plasticizer produces a brittle coating, which would make it prone to cracking in service. These coating require increased CP current demand, and shielding of CP and can ultimately cause localized corrosion.

Polyethylene tapes mostly fail due to development of shielding coating disbondments. The adhesive tapes can degrade through microbial action and does not have sufficient strength to withstand soil stress disbondment. The result is production of long narrow openings in the coating that can shield CP. Also, due to tenting at long seam and spiral welds, access of water to the steel pipe surface can occur resulting in localized corrosion as well as SCC and MIC.

Modern coating systems include stand-alone FBE, or FBE as the primer layer in a three-layer coating system.

The occurrence of localized corrosion leading to pipeline leak or rupture has been very small mostly due to improved application techniques and coating performance properties.

TABLE 3-12. Coating failure modes of typical pipeline coatings (Wilmott, 2007)

Coating Type	Failure mode	Failure cause
Asphalt & Coal Tar	Permeable embrittled coating	SCC major corrosive conditions
Polyethylene and polyvinyl chloride. (PVC)	Brittle/cracking	Corrosion and cathodic shielding
Polyethylene tapes	Coating Disbondment/tenting	CP shielding, SCC/MIC and corrosion at long seams/spiral welds
Fusion Bonded epoxy	Blister in coating & permeation	Localized corrosion
Three layer PE coating (Factory applied)	Disbondment	Some cathodic shielding
Three-layer PE coating (Over ditch)	Narrow disbondment, due to Soil stress	Sever cathodic shielding

For stand alone FBE coatings, a blister in the coating is the type of defect that is most often encountered. In continuously wet soil environments, water is able to migrate through the coating. But the pH level of water found under FBE coatings in such conditions indicates that it is in the region of 10 to 14 indicating an effective CP can reach the steel surface. In this scenario, the pH of the trapped electrolyte and the CP potential of the steel are outside the window where corrosion, MIC, or SCC can occur on pipeline steel.

Abrasive blast techniques are used in pipe surface cleaning and preparation. These techniques impart residual compressive stress in the outer layer of the pipe steel, which is beneficial in prevention of initiation of stress corrosion cracks. The application process for 3LPE ensures that a continuous sheath of polymer is extruded around the pipe. If the polymer is applied by side extrusion, the overlap between each revolution is fused to the next preparing an effective coating system.

On the other hand, mostly disbondments have been found in buried pipe lines coated with three 3LPE systems. The type of disbondments different in factory applied three-layer coatings and over the ditch applied tape coating. Wilmott (2007) indicates the long narrow shielding disbondments produced as a result of soil stress with tape coatings have not been observed with three-layer factory-applied coatings. This may imply that cathodic shielding is only one of the components that can contribute to pipeline failure.

Coating Failure Case Studies

Review of failures reported indicates the most recent trend including SCC in pipelines is attributed to coating failures. It is reported that Inadequate coating performance (NEB 1996) is a major contributing factor for increased SCC susceptibility of an underground pipeline. The majority of high-pH SCC failures have been associated with bituminous coatings (coal tar or asphalt), while the near-neutral pH-SCC failures have occurred most frequently on tape-coated pipelines (Beavers et al. 1993). The surface preparation conditions, degradation modes, and electrical behavior of these coatings are generally responsible for the type and prevalence of SCC on pipelines. It is reported that the effectiveness of a coating system in preventing SCC is related to three factors:

1. the resistance of a coating to disbondment
2. the ability to pass CP current should the coating fail and
3. the type of surface preparation used with the coating.

On the other hand, the reported cases of failure indicated that that most of all coating failures (over 90%) is a result of poor surface preparation, application, and important of all proper specification and QA/QC during coating applications (Table 3-13).

TABLE 3-13. Reported coating failures/case studies

Description	Coating System and Failures	Failure Cause	Location [Reference]
• NPS 30 (762 mm) pipe, stored 15 months • Temporary "protectives," shellac or bitumen, applied at pipe manufacturer plant • Marine environment Inadequate specification used	• FBE • Pitting corrosion prevalent on pipe (with 72% loss in wall thickness)	• Temporary "protectives" coating major source of problems • Stored pipes in marine environment, covered with a large amount of salts, not removed prior to abrasive blast cleaning	South America (Norman 2002)
• NPS 48 (1219 mm) Africa and NPS 30 (762 mm) India • 3 and 5 years in ground • Ground conditions vary: (sand to damp alluvial soil). India: through sharp stone/rock, to very wet clay • A CP applied • Operation temperature >60°C and 50°C (India)	• 3-layer low-density polyethylene (LDPE) coating system (50–75 μm of FBE powder, ~150 μm of a polyethylene co-polymer adhesive, overall total film thickness of 2–3 mm using an LPE) • Field joints protected with mastic, heat shrink, wrap-around sleeves • Corrosion under coating	• Lack of adhesion of coating to pipe • (LDPE) used very soft and easy to cut through. • No (FBE) remaining • Lack of proper coating specification	Africa and India (Norman 2002)
• NPS 12 to NPS 42 (304–1065 mm) • Wide ranging ground conditions • Operating temperature to >50°C	• FBE coating, thickness 400μm, First FBE-coated pipe used in 1977 • Lack of adhesion (onside of welds) • Pinholes in coating/leopard spotting • Blistering and foaming • Coating cracks	• Lack of adhesion: inadequate surface preparation, low-temperature application and moisture ingress under coating. Immediate application of joint coating right after welding material/technique. • Pinholes: inadequate surface preparation (salts). • Craters/blistering: extraneous material trapped in the coating while curing. Also moisture ingress contamination under the coating, or incorrect temperature profile during coating (cold ends) • Foaming: high moisture in the powder/too high an application temperature. • Strain lines/cracks: hydrostatic testing at very low temperatures. • Blistering of the coating soon after application over the circumferential weld — joint coating. Following cellulosic welding techniques hydrogen gas can be evolved. • Leopard spotting in ground formation of corrosion products underneath coating with adhesion loss (at pinholes)	UK (Norman 2002)
• NPS 36 900 mm) Pipe storage: in the center of a land mass • Operation: new project	• Coating system: 3LPE, coated in India. • Lack of adhesion at cut backs and pipe ends	• Lack of proper coating specification Improper allocation Poor Surface preparation • Corrosion visible under the coating at pipe ends. • The FBE first layer not wetted the substrate and not applied under the complete length of the PE	Iran (Norman 2003)

To remove the problematical corrosion situations, it follows that considerations must be given to:

1. Specifications and qualification
2. Pipe supply
3. Coating plant and inspection
4. Control of coating application
5. Control of site operations

Internal Coating

The presence of corrosive agents and conditions inside a pipeline can result in premature failure, loss of service, possible pollution, injury, property damage or fatalities. Common causes of internal corrosion include:

- Hydrogen sulfide, H_2S, which is a poisonous corrosive gas and forms sulfuric acid in the presence of water.
- Carbon dioxide, CO_2, which forms carbonic acid in the presence of water.
- Water vapor present in a pipeline will condense to liquid water when cooled sufficiently. Liquid water in a pipeline will contain dissolved corrosive gases as acids and may also contain bacteria responsible for some corrosion reactions. High velocity flow conditions can remove protective oxide layers on the inside of a pipeline and accelerate the corrosion attack. Low velocity flow conditions can allow wetting of the pipe wall to occur and near stagnant conditions, which are ideal for biological corrosion mechanisms.

Mitigating Internal Corrosion

Internally coating the pipeline is one way of mitigating internal corrosion besides improving the pipeline performance.

Internal coatings and linings can isolate the corrosive fluids from the pipe wall and as well make the pipe wall resistant to the type of attack expected to occur during pipeline operational life.

Internal coating will Improve energy efficiency, extend operating life, and reduce costs including the following:

- A reduction in surface roughness thus affecting/decreasing the friction coefficient at the pipe wall/fluid interface. This reduces the pressure drop in the pipeline causing an increase in the flow rate
- Capacity increase or pipeline diameter reduction (for new pipelines) for achieving the same flow capacity, thus an economic benefit
- A reduction in power requirement for pumping/compression and hence energy requirement cost
- Mitigation of the corrosion of the internal surface of the pipe during storage and transportation
- Provision of a clean smooth surface for ease of pipeline inline inspection
- Reduction in the effort and cost required for cleaning the pipeline after hydrostatic testing, and subsequent ease of disposal of water used in hydrostatic testing
- A decrease in pipeline maintenance through minimizing fouling and damage to valves
- A reduction in wear and erosion — internal coatings with adjustable-thickness layers extend pipeline life in erosion, sliding. and abrasion wear applications
- Lowers operational cost — internal coatings cut operational costs by reducing downtime, replacement materials and service expenses. They can also eliminate the need for expensive cleaning.

The most significant affect of internal coating is in the benefits that can be accrued from capacity capability of a pipeline system. As an example, comparison of internally coated with that noninternally coated NPS 34 gas pipeline shows a significant delivery increase assuming all intermediate deliveries and inlet pressure after compression remain constant (Figure 3-32).

The effect of increased delivery while decreasing the roughness is indicated in Figure 3-33. From this figure, over 20-fold in flow increase can be achieved if the line is internally coated.

Many of the internal corrosion problems can be taken care of by internally coating pipelines. An example is the pipeline transportation of sour gas. The types of corrosion that can occur in sour gas pipelines are:

- Hydrogen induced corrosion
  - HIC
  - Sulfide stress cracking (hydrogen embrittlement)
- Pitting corrosion
- General corrosion
- Erosion corrosion

HIC is an example of blistering that can occur in pipeline transportation of sour gas. It has been observed in both low and high yield strength steel pipes under both stressed and nonstressed conditions.

Hydrogen blistering/cracking results from the diffusion of atomic hydrogen produced by corrosive elements (such as H_2S) contained in a wet sour gas environment into steel

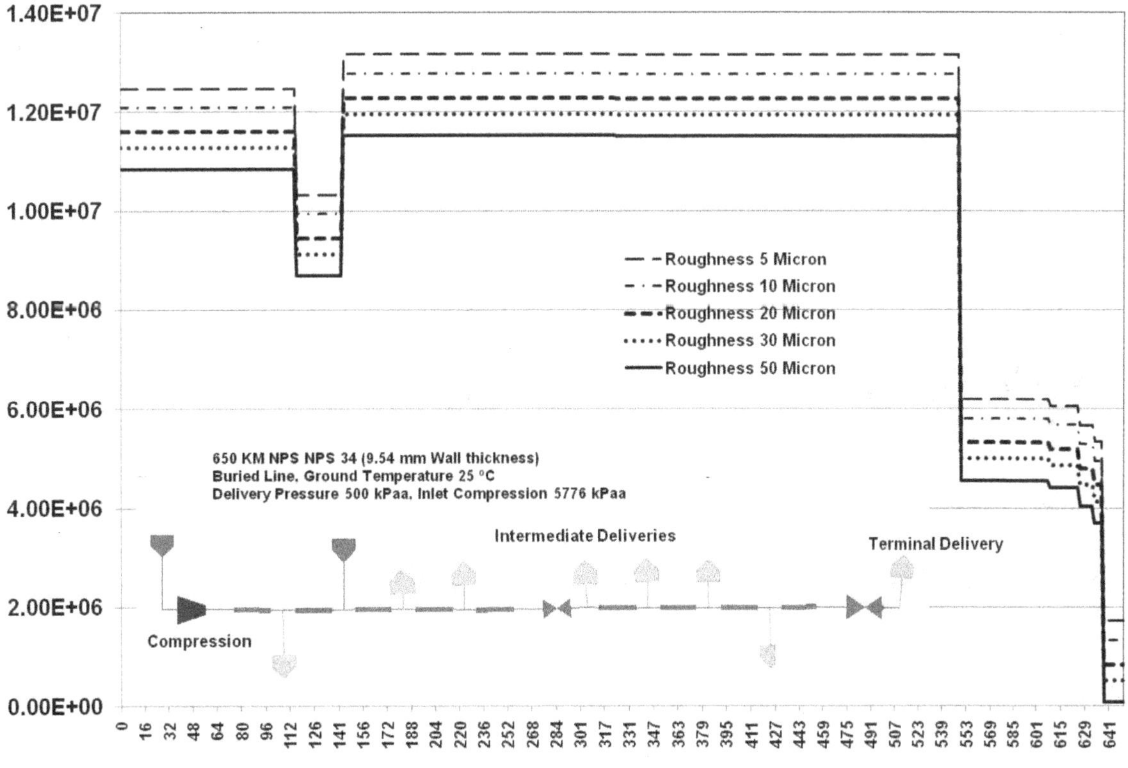

Figure 3-32. Effect of pipe internal roughness on flow capability (gas pipeline)

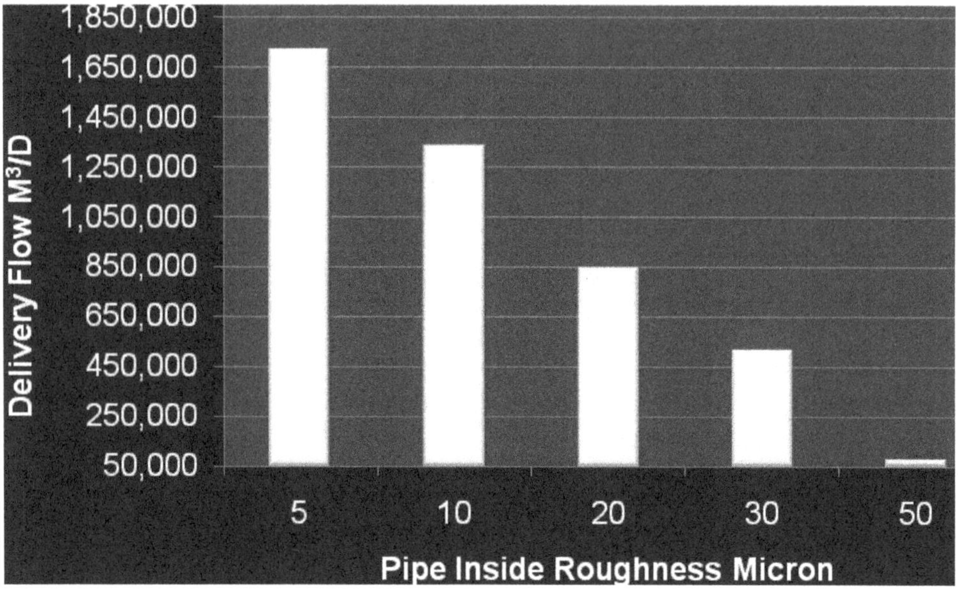

Figure 3-33. Gas pipeline flow capability increase with reduction in pipe roughness

where it is absorbed in laminations or inclusions contained in the pipe wall (Figure 3-34) (Mohitpour et al. 1990).

This atomic hydrogen then changes to nondiffusible molecular hydrogen, building up high localized pressures, which produce blisters or cracks (see inset in Figure 3-34) in the pipe wall.

Coating the Internal surfaces of the pipe with suitable coating material such as FBE powder will reduced the impact of corrosive elements intrusion as much as it improves flow efficiency by reducing flow drags in both liquid and gas pipelines.

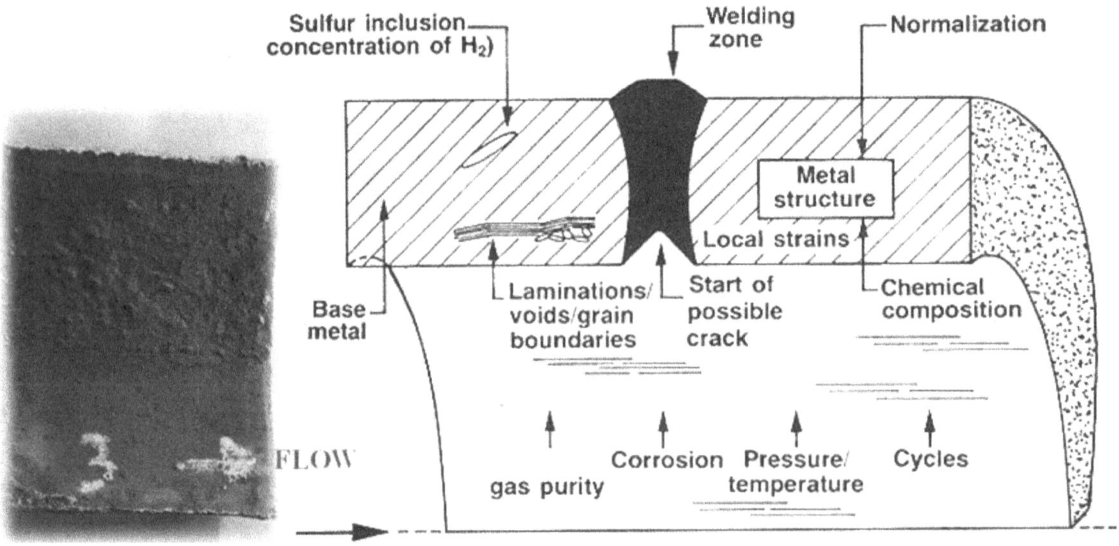

Figure 3-34. Factors influencing hydrogen embrittlement on internal surface of pipeline (Mohitpour et al. 1990)

Applicable Specifications for Internal Coating

Besides the standards listed previously, the following are also applicable for internally coating pipelines:

- API RP 5L2:2002: "Recommended practice for internal coating of line pipe for noncorrosive gas transmission service," 4th edition
- API RP 5L7: 1988: "Recommended Practices for Unprimed Internal Fusion Bonded Epoxy Coating of Line Pipe," 2nd Edition
- ASTM A 775: Standard Specification for Epoxy-Coated Reinforcing Bars
- ASTM D 3963: Standard Specification for Epoxy-Coated Reinforcing Steel

Factors to be considered during coating or lining selection should include:

- internal coating of field joints
- application methods
- availability of repair methods
- operating conditions
- long-term effects of the fluid(s) on the coating/lining
- resistance to pressure change
- influence of temperature gradients over the coating and
- compatibility with pigging operations

Internal Coating Application

There are a number of ways that inside of a pipeline can be coated in the coating plant or in the field. The application depends on varying wall thickness, valves, bends, and pipeline profiles.

The following indicted two applicable methods

1. *Coating between pigs*: The coating is loaded between two specially designed pigs. Using a predetermined driving pressure, the correct velocity is maintained to provide a controlled film thickness.
   Generally, a minimum of three coating runs would be required, with the coating train driven by completely dry and particulate-free compressed air. After each coating run, the same compressed air continues to slightly blow through the pipeline to ensure total evacuation of the released solvents and aid in drying between coats. The pot life of the coating is such to allow more than a reasonable time to perform the coating operation.
2. *Internal rotating sprayer*: This consists of a rotor-spray internal pipe coating and spray pump units, which travel inside pipe at a predetermined speed (Figure 3-35). Coating is usually FBE power and can be applied for varying the coating thickness by adjusting the spray nozzles.

Typical manufacturing process covers the following:

- Removal from stock yard
- Cleaning
- Rinsing
- Drying
- Internal coating
- Inspection
- Curing
- Inspection and repair of coating
- Marking

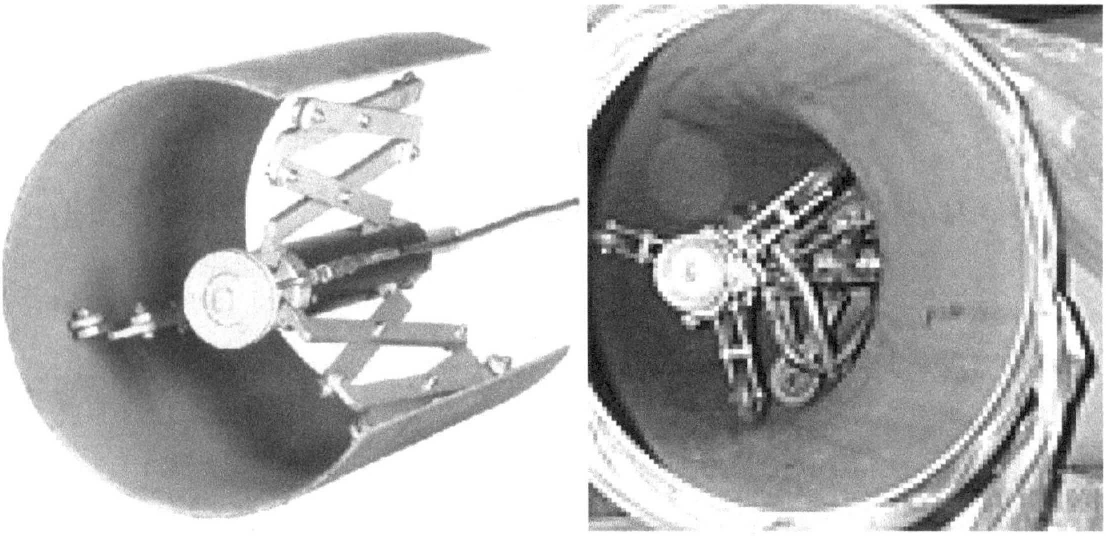

Figure 3-35. An internal coating rotor-spray — cut away and actual spray unit in action

Figure 3-36. Typical coated pipes and a bevel protector (Courtesy of JFE Steel Corporation)

- Attaching bevel protectors
- Marking and vinyl sheet cover on pipe
- Move finished product to stock yard
- Prepare for shipping

Typical finished product is shown in Figure 3-36. An important aspect of pipes, which are internally coated, is protecting the end bevels to assure not only minimum damage during transportation to work site but minimal bevel repair for welding purposes. This will reduce internal recoating of the pipe affected by welding.

The protector is made of a steel strip forming into a circular shape. The abutting ends of the ring are fitted with steel wedge clasps. Bevel protectors wedge, which has either machined reverse teeth or plain edges, is inserted into the claps and tapped into place to expand the protector inside the pipe end. This protects the vulnerable root face and bevel against damage during subsequent handling from mill shipment to arrival at the job site.

REFERENCES

American Society of Mechanical Engineers (ASME), B31.4, 1995, *Pipeline Transportation Systems for Liquid Hydrocarbons and Other Fluids*, ASME, New York.

American Society of Mechanical Engineers (ASME), B31.8, 1999, *Gas Transmission and Distribution Piping Systems,* ASME New, York.

Andrenacci, A., Wong, D., Mordaski, J. G., 1999, "New Developments in Joint Coating and Field Repair Technology," Materials Performance, Vol. 38, February. pp. 33–43.

Argent, C. J., 1977, "The Role of Soil Assessment Techniques in the Selection of Pipeline Coatings," 2nd Int Conference on Internal and External Protection of Pipes, BHRA Fluid Eng, Cranfield, Bedford, UK, September 7–9.

Banach, J., 2004, "Liquid Epoxy Coatings for Today's Pipeline Coating Challenges," NACE Northern Area Western Conference, Victoria, BC, February 16–19.

Beavers, J. A., Thompson, N. G., and Coulson, K. E. W., 1993, "Effects of Surface Preparation and Coatings on SCC Susceptibility of Line Pipe: Phase 1 — Laboratory Studies," Proc. ASME OMAE Conf (Conference on Offshore Mechanics and Arctic Engineering), New Orleans, LA, June, p. 226.

Canadian Standard Association CSA Z662-07, *Oil and Gas Pipeline Systems*, CSA Int., Ont. Canada.

Coulson, K., 2008, "Efforts Seek to Develop International Coating Standards," Pipeline and Gas Journal, August.

Coulson, K. E. W., and Temple, D. G., 1983, "An independent Laboratory Evaluation of External Pipeline Coating," 5th International Conf. on Internal and External Protection of Pipes, Innsbruk, Austria, August 25–27.

Coulson, K. E. W., Bario, T. J., and Werner. D. P., 1991," Test Show Barrier Coatings Do Not Block Cathodic Protection," Oil and Gas Journal. Oct, pp. 80–84.

Espiner, R., Thompson, I., and Barnett, J., 2003, "Coating/Backfill Interactions on High Pressure Transmission Pipelines — Technical and Financial Issues." NACE Corrosion 2003 Conf/, Paper # 3046, www.ceocor.lu/publications/italy/a/03_sector_a.pdf - 905k.

Holub, J., Wong, D. T., and Tan M., 2007, "Analysis of CDT Methods and Factors Affecting Cathodic Disbondment," Paper #0722, Corrosion 2007, Nashville, TN, March 11–15, 2007.

Linchtenstein, J., 2003, "Effectiveness of Pipeline Coatings for Corrosion Control," Journal of Material Performance, Vol. 42, number 6, June.

Malik, A. U., Andijani, I., Ahmed, S., and Al-Muaili, F., 2002, "Corrosion and Mechanical Behavior of Fusion Bonded Epoxy (FBE) in Aqueous Media," Desalination, Vol. 150, No. 3, November 10, pp. 247–254, http://www.swcc.gov.sa/files%5Cassets%5CResearch%5CTechnical%20Papers%5CCorrosion/CORROSION%20AND%20MECHANICAL%20BEHAVIOR%20OF%20%20FUSION%20BONDED%20EPOXY%20(F.pdf.

Menon, J., 2004, "One Single, Uniform Coating for the Entire Pipeline Using LPE — Liquid Polymer Coatings," Paper presented at the ACA (Australian Corrosion Association," Perth.

Mohitpour, M., Pierce, C., and Graham, P., 1990, "Design Basis Developed for H2 Service," Oil and Gas Journal, May 28, pp. 83–94.

Neal, D., 1999, "Good Pipe Coating Starts With Properly Prepared Steel Surface," Pipeline and Gas Industry, March 1999, pp. 38–43.

Neal, D., 2000, "Pipeline Coating Failure — Not Always What You Think It Is," Corrosion 2000, Paper #00755, Harding and Neal Inc, Houston, TX.

NEB (National Energy Board), 1996, "Stress Corrosion Cracking on Canadian Oil and Gas Pipelines," Report of the Inquiry, MH-2-95, Regulatory Support Office, National Energy Board, November 1996.

Norman, D., 2002, "Pipeline Coatings — the Good, the Bad and the Ugly," Paper presented at the Australasian Corrosion Association's Conference in Adelaide, Australia.

Norman, D., 2003, "Are We Protecting Our Assets?" http://www.davidnormancorrosioncontrol.com/html/papers/PDF/DavidNormanAreWeProtectingOurAssets.pdf.

Norsworthy, R., 2009, "Coatings Used in Conjunction With Cathodic Protection — Shielding vs Non-shielding Pipeline Coatings," 17th International Corrosion Congress, Paper #4017, http://www.polyguardproducts.com/products/pipeline/TechReference/NACE_4017.pdf.

Norsworthy, R., 2009, "Pipeline Coating Tutorials," Banff 2009 Pipeline Workshop, April, http://www.acr-alberta.com/BANFF09PIPELINEWORKSHOP/tabid/149/Default.aspx.

Tailor, D., Hodgins, W., and Gritis, N., 2003, "Field Joint Developments and Compatibility Considerations," http://www.shrinksleeves.com/literature/papers/TP_2003.pdf.

ShawPipe, 2008, "Pipeline Coating Solutions," Technical Literature, www.shawpipe.ca/literature/tech/ShawPipe_TL_R161008.pdf.

Singh, P., 2008, "Coating Standards CSA Z245.20/21," Second Canada–India Workshop on Pipeline Integrity, Calgary, Canada, September 25–26.

Singh, P., and Cox, J. J., 2000, "Development of a Cost Effective Powder Coated Multi Component Coating for Underground Pipeline," CORROSION 2000, March http://www.shawpipe.ca/literature/tech/TP_Underground.pdf.

Sonke, J., 2008, "Qualification of Protective Coatings Using EIS," http://www.sonke.eu/content/publications/2008eiseurocorr.pdf.

Steel Structures Painting Council (SSPC), 2003, *Fundamentals of Protective Coatings for Industrial Structures.*

Tator, K. B., 2006, "Qualification Testing of Fusion Bonded Epoxy Coatings for Pipeline Exteriors," http://www.kta.com/knowledge/PACE2006/FusionBondedEpoxy.pdf.

Thomason, N. G., 2001, "Gas and Liquid Transmission Pipelines," Appendix E, Report FHWA-RD-01-156, September by: Ruschau, G., and Al-Anezi, M.; Corrosion Costs and Preventive, http://www.corrosioncost.com/pdf/gasliquid.pdf.

UM (University of Maryland) 2001 "Material Hardness" http://www.calce.umd.edu/general/Facilities/Hardness_ad_.htm

US DOT, Code of Federal Regulations CFR 49, 2009, Part 192 and Part 195.

Werner, D. P., Lukezich, S. J., Hancock, J. R., and Yen, B. C., 1992, "Materials Performance, Survey Results on Pipeline Coatings Selection and Use," Materials Performance, Vol. 31, 1111, pp.16–21, National Association of Corrosion Engineers, November 1992.

Williamson, A. I., Singh, P., and Hancock, J. R., 2000, "The Resistance of Advanced Pipeline Coatings to Penetration and absorption by Hard Rock." Proceeding ASME International Pipeline Conference (IPC), Palliser Hotel, Calgary, AB, CND, Oct. 1-5, Paper# !PC 00-0177

Wilmott, M., 2007, "Choosing the Right Coating for Managing Pipeline Integrity," Technical Resources, Wasco Coating, Presented at 3rd Asian Pipeline Conference and Exhibition 2, http://ppsc.com.my/Petromine-Pipeliner.html.

Wilmott, M., and Bacon, T., 2000, "Economics and Potential Consequences of Improper Pipeline Coating Selection," AP Energy Business Publications Pte. Ltd, www.petromin.safan.com/mag/poct03/t52.pdf- 172k.

Wilmott, M., Highams, J., Ross, R., and Kopystinski, A., 2000, Journal of Protective Coatings and Linings (JCPL), Vol. 17, No. 11, pp. 47–54. http://www.corrosion-doctors.org/PaintCoatings/Buried.htm.

Chapter **4**

PIPELINE CORROSION AND CATHODIC PROTECTION

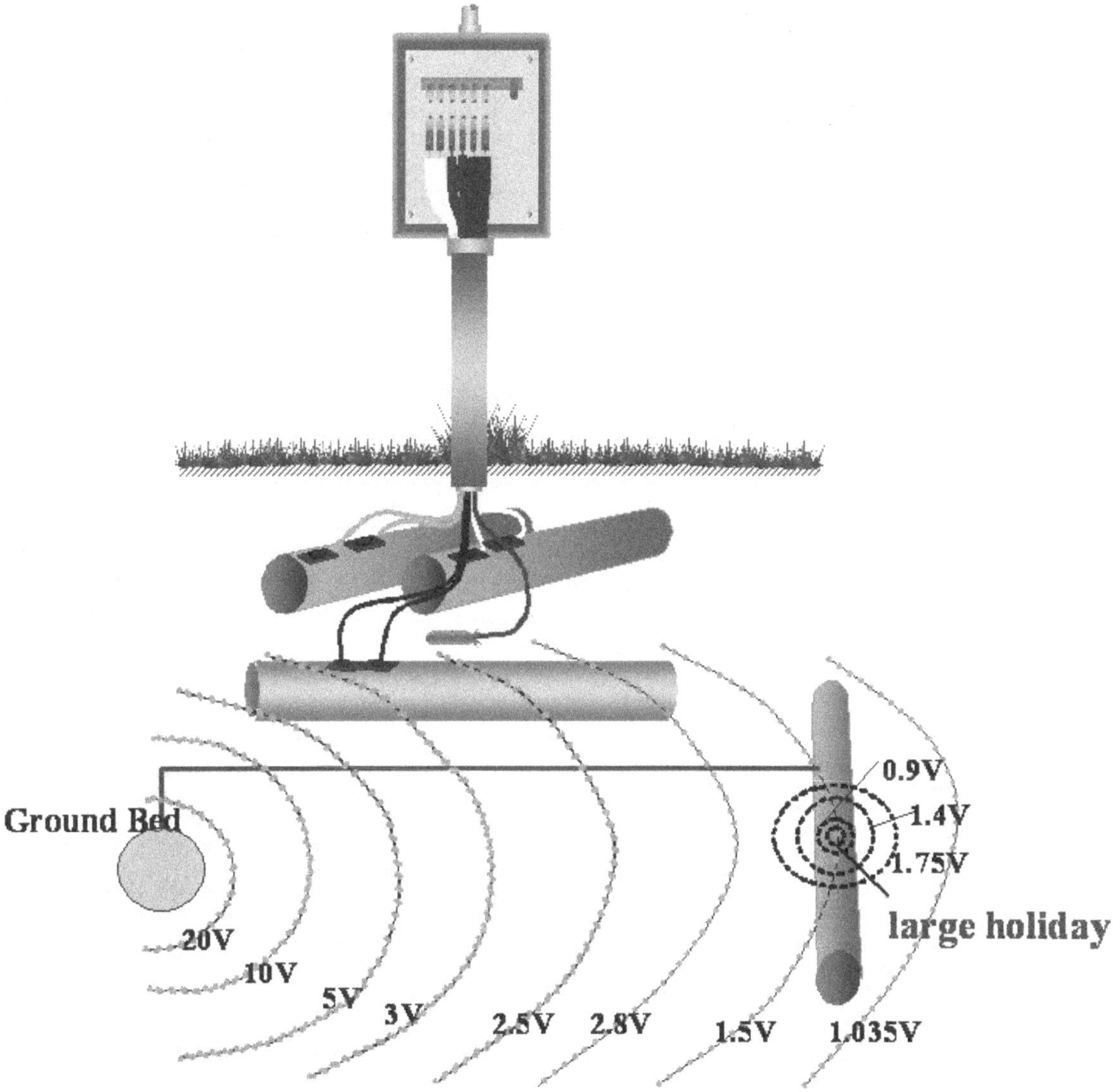

PIPELINE CORROSION AND CATHODIC PROTECTION (CP)

The prevention of external corrosion on underground structures is accomplished using a dual system of coatings and CP. To define corrosion, pipelines will be considered synonymous with all underground structures. All metallic corrosion is electrochemical in nature and is the result of direct electrical current flow.

DEFINITION OF CORROSION

An understanding of CP cannot be achieved without first understanding the mechanisms of corrosion. A simplified definition of corrosion, in its most general sense, is the degradation of a material due to a reaction with its environment.

NACE similarly defines corrosion as the deterioration of a substance (usually a metal), or its properties, as a result of its reaction with its environment. The environment in pipeline applications includes the soil and its many constituents (water, air, salts, bacteria, etc.).

For corrosion to occur, a number of conditions must be met. These are:

1. There must be an anode and cathode.
2. There must be an electrical potential difference between the anode and the cathode.
3. There must be a metallic connection between the anode and cathode.
4. The anode and cathode must be immersed in a common electrically conductive medium (an electrolyte, e.g., the ground).

Once all of these conditions have been met, corrosion will occur. The anode will corrode producing metal ions and electrons. Corrosion occurs where current leaves the surface of the pipeline (anode) via the surrounding soil Figure 4-1. Corrosion does not occur where current enters the pipeline (cathode).

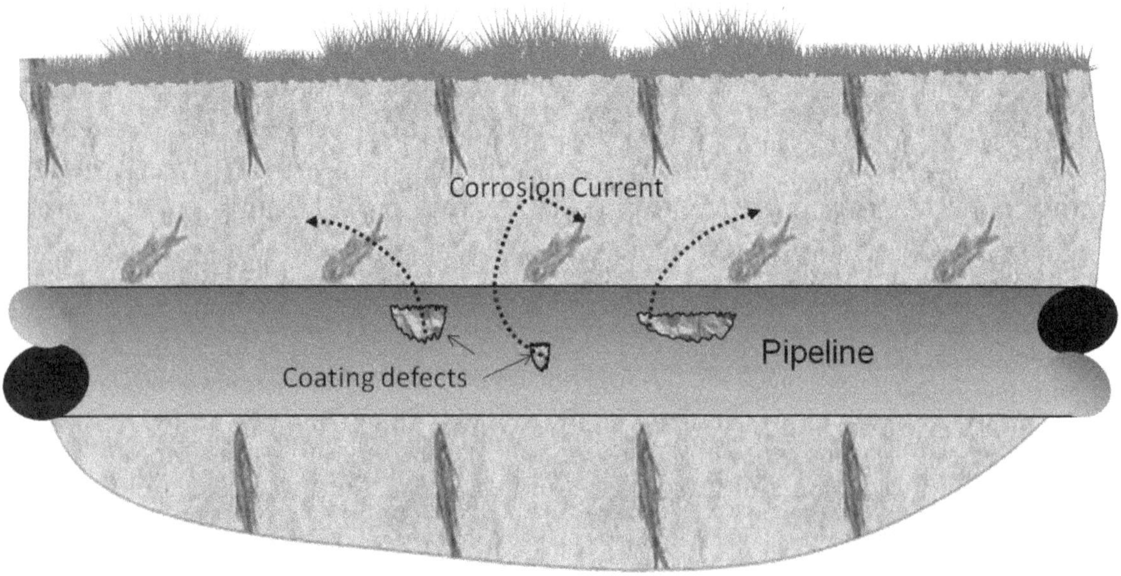

Figure 4-1. Microscopic view of a corrosion cell

MECHANISM OF CP

In understanding the mechanism of CP, the following attributes need to be applied:

Anode	Corroding structure — point where current is discharged into electrolyte
Cathode	Protected structure — point where current is picked up
Metallic Path	Electrical connection between anode and structure (electron path)
Electrolyte	Conductive medium between anode and structure (ion path)

Corrosion prevention is effected by negating one of the four conditions listed above. Generally, it is to make the surface of the pipe cathodic so that there is no current flow out and corrosion is stopped (Figure 4-2).

Therefore, CP is the cathodic polarization of all areas on the pipeline to be more active than the otherwise most active metallic site on the pipeline. The purpose of CP is to create an electrochemical arena within which the pipeline is protected from corrosion. Thus, in the case of CP of a pipeline system, the electrode potential difference between the anode and the cathode is eliminated by making all of the pipeline cathodic (Figure 4-3).

Thus, CP can simply be defined as the application of an external direct current (DC) source that causes current to flow to the structure being protected. In other words, an artificial potential gradient is created that will reverse the natural current flow. This causes the anode and cathode to exchange functions. An obvious problem is that something (the anode) will still be corroded. CP provides an alternative (sacrificial) anode, which is intended to corrode instead of the pipeline (the cathode).

CP REQUIREMENT

The typical components for CP systems (Figure 4-4), include anodes (impressed or galvanic), rectifiers, High-molecular-weight polyethylene (HMWPE) cable and a low-resistance backfill material typically referred to as coke breeze. The rectifiers are supplied with alternating current (AC) power. The AC is then converted to DC through a rectifying element (diodes or a selenium stack). The DC voltage is then directed to some type of anode configuration

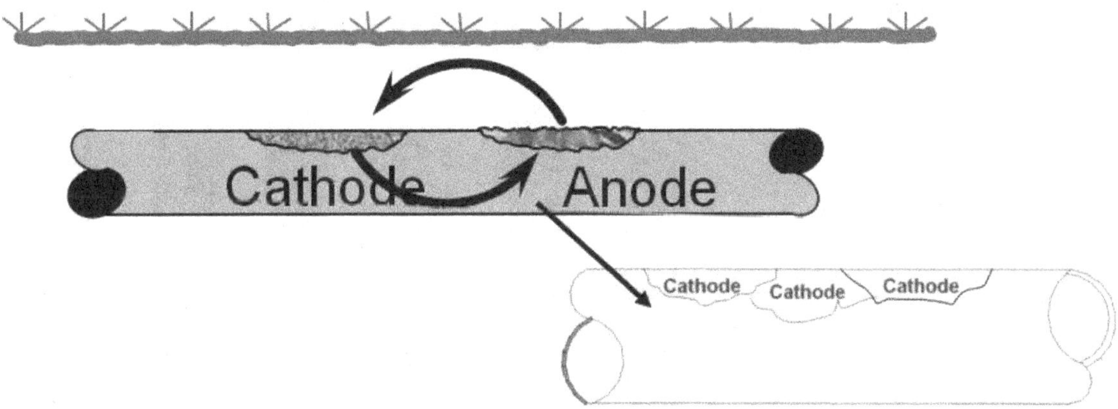

Figure 4-2. Corrosion protection

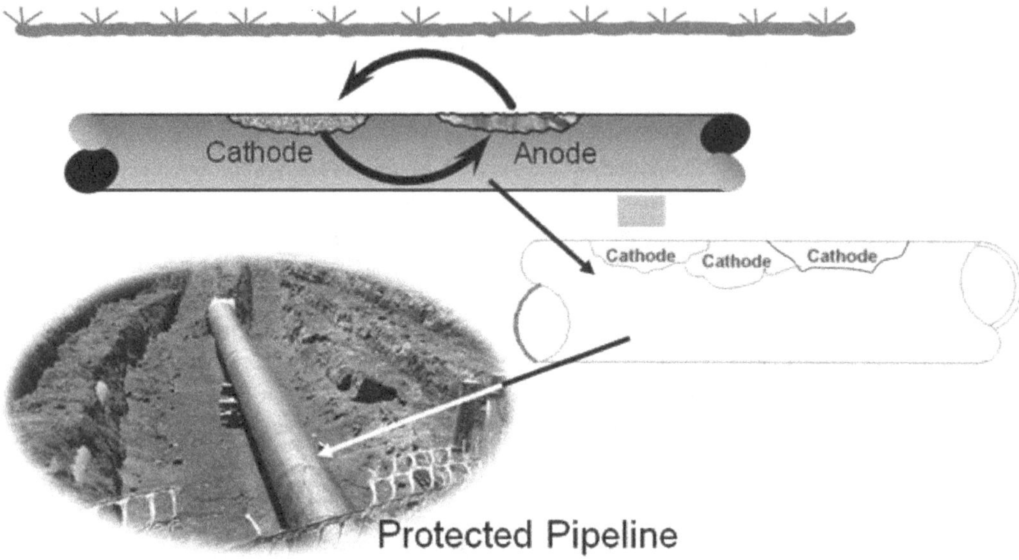

Figure 4-3. Corrosion mechanism and CP on a segment of protected buried steel pipeline

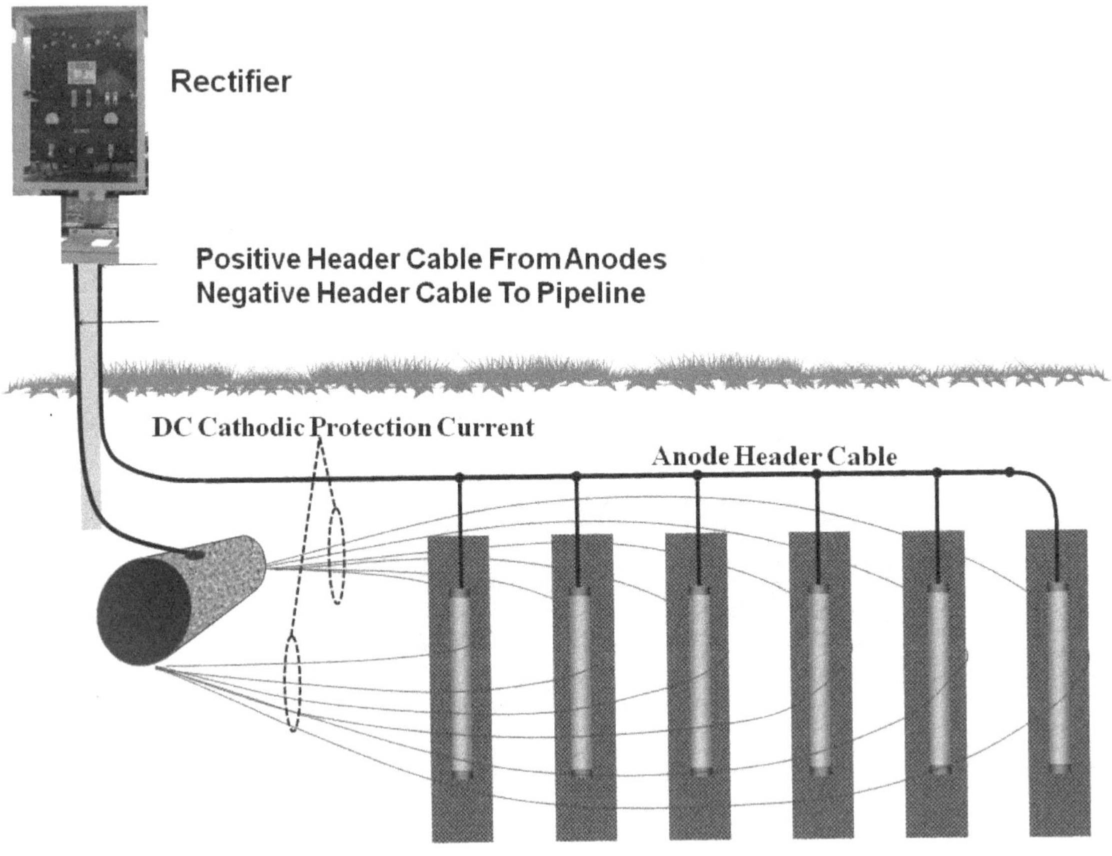

Figure 4-4. Typical CP system

that produces DC current (amps). The DC current then travels through the soil to areas along the pipeline that are exposed to the soil via holidays or holes in the coating. This DC current is what prevents the pipeline from corroding. The DC current then travels through the pipeline back to its source (the rectifier) to complete the circuit. Each rectifier has the ability to protect miles and miles of pipeline. The area of protection or influence may be dependent on the following factors:

- Coating condition — the better the condition of coating, the greater the area of influence from the CP system (DeGiorgi and Hamilton 1995).
- Soil resistivity — the lower the soil resistivity, the greater the area of influence from the CP system.
- Anode configuration — the number, arrangement (deep well or distributed), and placement of anodes affect the area of influence.
- Type of anode material — different anode materials and sizes affect the overall output of the ground beds (see later).
- Foreign structures — the lower number of foreign pipeline/structures in the vicinity of the ground bed, the greater the area of influence.

APPLICABLE STANDARDS AND PROCEDURES

The following lists the standards that are used by the pipeline industry and are applicable for the design, operation and maintenance of a CP system. The recommended practices describe the corrosion problems characteristic in below-ground steel pipes and structures and provide a general description of the methods currently used to provide CP against corrosion. However, it may be noted that since conditions that affect CP systems are subject to change with time, changes to the recommended practices or procedures, are often reviewed and implemented accordingly in the updated editions.

ANSI/NACE RP0104-2004 "The Use of Coupons for Cathodic Protection Monitoring Applications"

API RP 500 (R2002) "Recommended Practice for Classification of Locations for Electrical Installations at Petroleum Facilities"

API RP 651, 2007 "Cathodic Protection of Aboveground Petroleum Storage Tanks"

API RP 1632, 1996 "Cathodic Protection of Underground Petroleum Storage Tanks and Piping System"

API RP 2003, 2008 Protection Against Ignitions Arising Out of Static, Lightning, and Stray Currents

AS 2832.2, 2003, Cathodic Protection of Metals — Compact Buried Structures

ASTM G 57, 2006 "Method for Field Measurement of Soil Resistivity Using the Wenner Four Electrode Method"

BS 7361-1:1991 "Cathodic Protection. Code of Practice for Land and Marine Applications"

BSI BS DS/EN 13509 "Cathodic Protection Measurement Techniques"

BS EN 15257:2006 "Cathodic Protection. Competence Levels and Certification of Cathodic Protection Personnel"

CSA C22.3.6:1991 (R2003) "Principles and Practices of Electrical Coordination Between Pipelines and Electric Supply Lines"

OCC-1-1996. "Recommended Practice for the Control of External Corrosion on Buried or Submerged. Metallic Piping Systems"

DNV, RP-B 401 2005 "Cathodic Protection Design"

NACE Standard RP 072- 85- "Installation, Operation and Maintenance of Impressed

Current Ground Beds"
NACE Standard RP 0169-2007 "Control of External Corrosion on Underground Submerged Metallic Piping Systems"
NACE Standard RP0176-94 "Corrosion Control of Steel Fixed Offshore Platforms Associated With Petroleum Production"
NACE Standard RP0177-2007 "Mitigation of Alternating Current and Lighting Effects on Metallic Structures and Corrosion Control Systems"
NACE Standard RP0187-2005: "Internal Corrosion Direct Assessment Methodology for Liquid Petroleum Pipelines"
NACE Standard RP0193-2001 "External Cathodic Protection of On-Grade Carbon Steel Storage Tank Bottoms"
NACE Standard SP0207-2007 "Performing Close-Interval Potential Surveys and DC Surface Potential Gradient Surveys on Buried or Submerged Metallic Pipelines"
NACE Standard RP 0285-2002 "Corrosion Control of Underground Storage Tank Systems by Cathodic Protection"
NACE Standard RP 0286-2002 "Electrical Isolation of Cathodically Protected Pipelines"
NACE Standard RP 0502-2002 "Standard Recommended Practice — Pipeline External Corrosion Direct Assessment Methodology"
NACE Standard RP0572-2001 "Design, Installation, Operation, and Maintenance of Impressed Current Deep Anode Beds"
NACE Standard SP 0387-2006 "Metallurgical and Inspection Requirements for Cast Sacrificial Anodes for Offshore Applications"
NACE Standard SP0575-2007 "Internal Cathodic Protection (CP) Systems in Oil-Treating Vessels (Reaffirmed, formerly RP0575-2001)"
NACE Standard SP0607-2007 "Petroleum and natural gas industries—Cathodic protection of pipeline transportation systems"
ISO 15589-2:2004 (Modified) "Petroleum and natural gas industries — Cathodic protection of pipeline transportation systems — Part 2: Offshore pipelines"
NACE Standard TM 0101-2001 "Standard Test Method — Measurement Techniques Related to Criteria for Cathodic Protection on Underground or Submerged Metallic Tank Systems"
NACE Standard TM 0190-2006 "Impressed Current Laboratory Testing of Aluminum Alloy Anodes"
NACE Standard: Cathodic Protection of Pipelines/Tank
NACE Standard Specialty Package "Cathodic Protection of Pipelines/Tanks"
NFPA 70, 2008 "National Electrical Code"

PIPELINE CORROSION INFLUENCING FACTORS

The following factors influence the requirement for cathodic protection of a pipeline structure:

- Coating quality
  - No coating is perfect; however, the application of a good quality coating greatly reduces the amount of cathodic protection current required to protect a structure.
  - Cathodic protection protects the steel where there are coating defects.
- Differential Aeration Cells
  - different amounts of oxygen contacting the pipe in different sections
  - metal exposed to less oxygen becomes anodic to the steel with more oxygen Example: pipe under a road

- Galvanic Corrosion
  - two different types of metal in contact in an electrolyte, for example iron in contact with copper or stainless steel (SS) fittings. The iron will corrode to protect the copper or SS. This can work to an advantage for sacrificial anodes.
- Mill Scale Corrosion
  - Steel pipe is anodic to mill scale and will corrode severely in the right conditions.
- New and Old Pipe
  - New pipe is anodic to old pipe and will corrode if they are connected together.
- Dissimilar Soils
  - Examples: pipe in sand and clay or pipe passing through concrete and soil (sewage treatment main failure).
- Relative Size of Anodic and Cathodic Areas
  - A small anode area can cause a rapid failure versus a larger anode area when different metals are in contact. Example: small galvanized fitting on a steel line.

It may be noted that CP only works on the side of the metal where the anodes are located; therefore, only on outside or soil-side of the pipes and structures and not inside or above ground.

CP METHODS

CP can be applied by two methods — impressed current CP and galvanic current CP, ADEQ, 2002. Both systems function by causing current to flow from a sacrificial anode to the pipeline. The difference is the method by which the current is caused to flow.

Impressed Current System

An impressed current system requires an external AC power source, a rectifier (to convert AC current to DC current), a ground bed of anodes, a metallic connection from the anodes to the pipeline, and a common environment (the ground) (Figure 4-5).

These anodes are typically made of high silicon iron or graphite. Table 4-1 describes their use and type of backfill required.

The anode materials allow a large amount of current to flow from them in proportion to their own corrosion or consumption rate. The anodes cannot function without an external power source. The power source forces current to flow from the rectifier along the anode header cable to the anodes. The current then exits the anodes and travels through the ground to the pipeline. The current enters the pipeline at holidays or voids in the pipeline coating and then moves to a thermite-welded cathode cable. To complete the electrical circuit, the current flows back to the rectifier. The rectifier contains meters to monitor the current output and the voltage output (Figure 4-5).

The advantages and the limitations of an impressed current system are detailed below.

Advantages

- Can be designed for wide range of voltage and current
- High-ampere year output available from a single ground bed
- Large areas can be protected by single installation
- Variable voltage and current output
- Applicable in high-resistivity environments
- Effective in protecting uncoated and poorly-coated structures

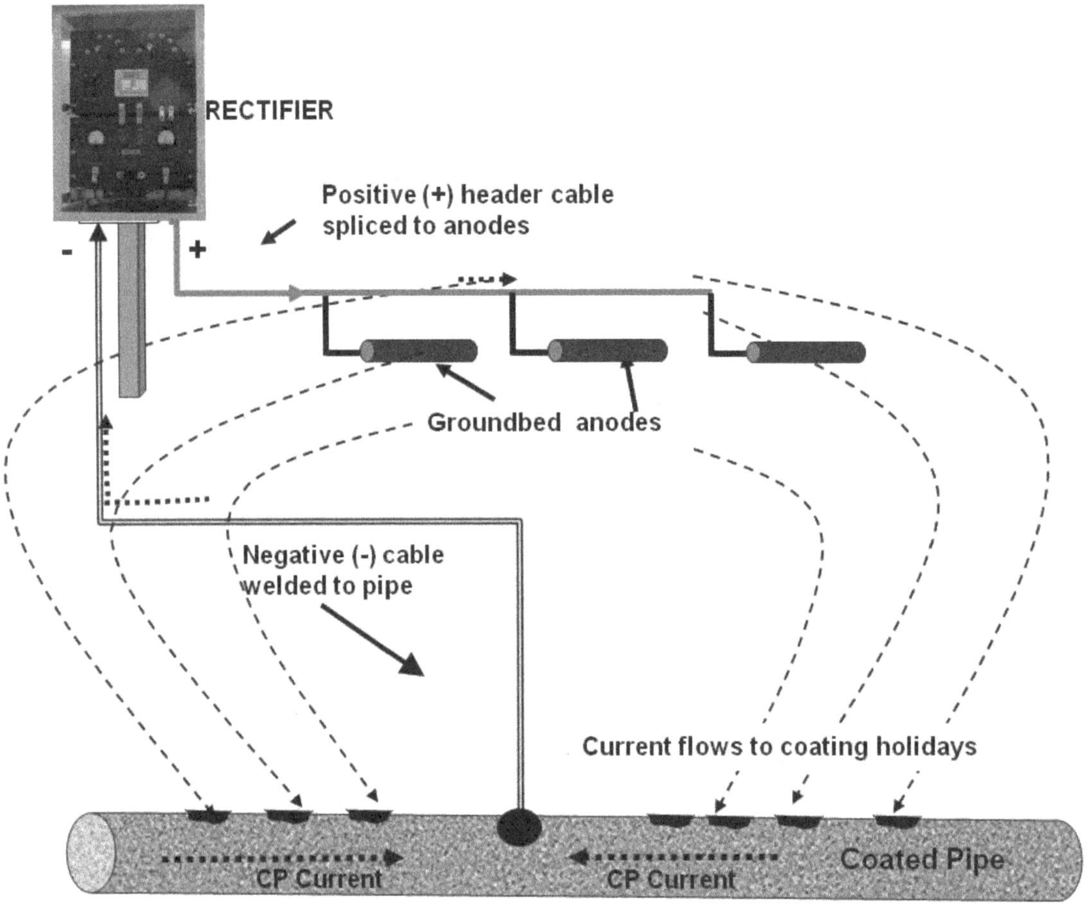

Figure 4-5. Impressed current CP system

Limitations

- Can cause cathodic interference problems with adjacent pipelines
- Subject to power failure and vandalism
- Wide range of voltage and current output

TABLE 4-1. Impressed current anodes

Type of Anode	Uses/Comments	Backfill
Sacrificial	Sacrificial anodes can be connected to an external power source and discharge current but their lifetime will be very short. They are not normally used for this application.	75% hydrated gypsum or bentonite
Steel	Any buried bare steel, such as old piping, or an automobile can be used.	None
High-silicon chromium iron	One of the most common anodes in use for pipelines and plants	Coke breeze
Graphite	Good life in well-drained soil	Coke breeze
Metal/metal oxide (ceramic)	Made from titanium and coated with a metal oxide. These anodes do not corrode away. They are designed to last a given time and stop working quite suddenly. Much lighter than most other anodes	Coke breeze or sea water or fresh water
Anodflex (continuous anode)	Good for long poorly coated structures, or very high soil resistivity locations	Coke breeze

- Requires periodic inspection and maintenance
- Requires external power that involves costs
- Overprotection can cause coating damage

Impressed current anodes include the following:
- Graphite
- High-silicon, chromium-bearing cast iron
- Platinum-coated materials
- Aluminum
- Magnetite
- Mixed metal oxide-coated titanium
- Conductive polymer
- Scrap iron or steel
- Lead-silver

Figure 4-6 shows a typical graphite anode used for CP applications. While largely superseded by silicon, iron graphite anodes are widely used for pipeline and tank bottom protection. Graphite anodes are relatively inexpensive and can be easily machined for cable connection. Normally supplied in diameters between 60 and 150mm and in lengths between 1200 mm and 1500 mm long, they are impregnated with linseed oil or resin to improve surface current distribution and give excellent protection.

When installing anodes in the ground, they are inserted in carbonaceous back fills. The backfill serves three purposes: (1) reduces anode-to-earth resistance, (2) increases current capacity by extending anode surface area and mass, and (3) reduces anode consumption through efficient electronic current transfer.

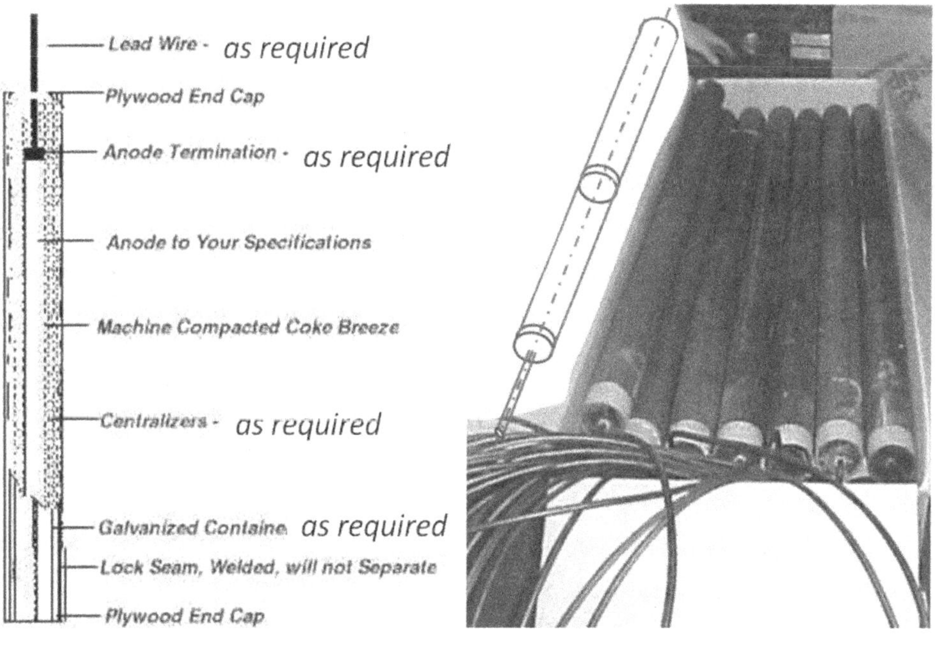

Figure 4-6. Typical anode (graphite)

Galvanic Current System

A galvanic current system requires anodes and a wire joining the pipeline to the anodes. The current in this system is provided by a natural reaction — a galvanic reaction — between the anodic material and the pipeline. It utilizes the same concept that provides electrical power in batteries (Figure 4-7). The voltage created by the galvanic reaction prevents current from leaving the pipeline and thus provides corrosion. The anodes are constructed of magnesium, zinc, or sometimes aluminum.

The current flows from the anodes to the pipeline via the ground. The current mainly enters the pipeline surface at voids in the coating (holidays). The potential gradient caused by the current flow effectively prevents current from leaving the surface and hence prevents corrosion.

Since all electrical circuits must be continuous (circuitous) to function, the current flows along the pipeline to the cathode cable connection — a thermic weld. The circuit is completed by joining the cathode and anode cables. The test lead is a convenient location for the joining to take place. It allows easy monitoring of the current flow (Figures 4-8 and 4-9).

Table 4-2 shows a galvanic series of metals in neutral soil referenced to a copper/copper sulfate reference cell (RC). Basically, the table indicates the potential difference that can be established between different metals. Current will flow from the more electronegative materials, or those materials with a more negative potential, to metals less electronegative. In other words, the current will flow from metals higher up on the table to those lower down on the table. For example, magnesium can be used to protect steel but graphite cannot. The greater the difference in the potentials, the greater will be the driving force (voltage) that will allow CP to be effected in lower soil resistivities. It should be noted that the potential of a material is significantly affected by the addition of other

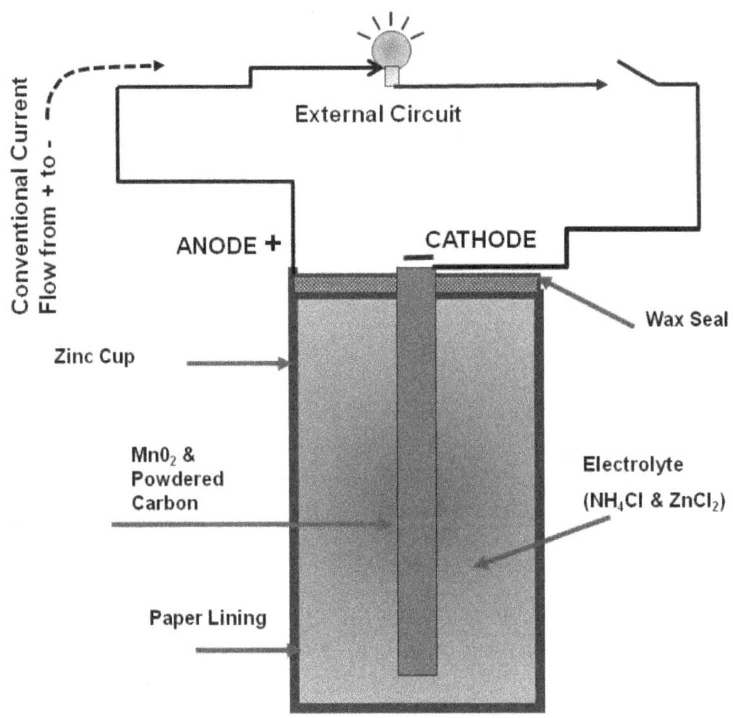

Figure 4-7. Typical arrangement in a dry cell battery

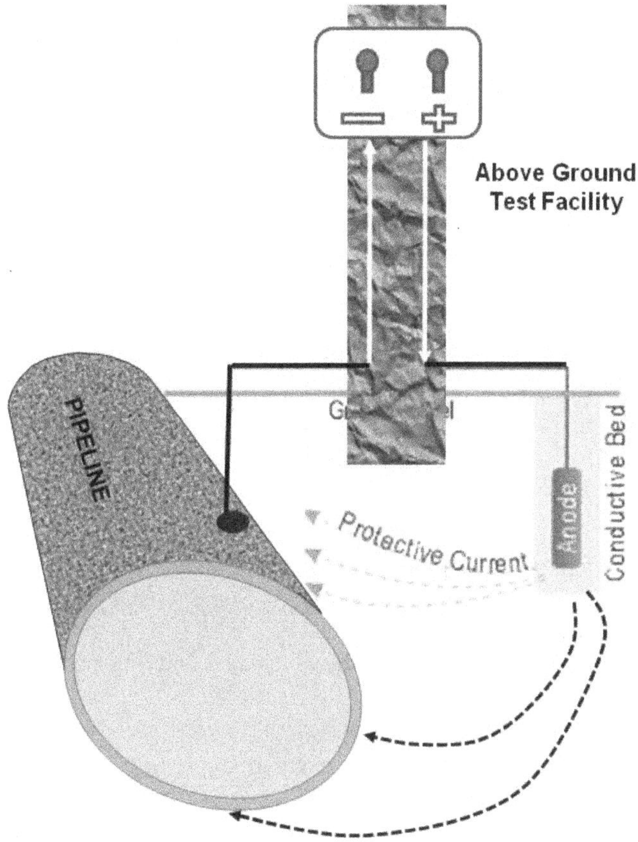

Figure 4-8. Sacrificial CP

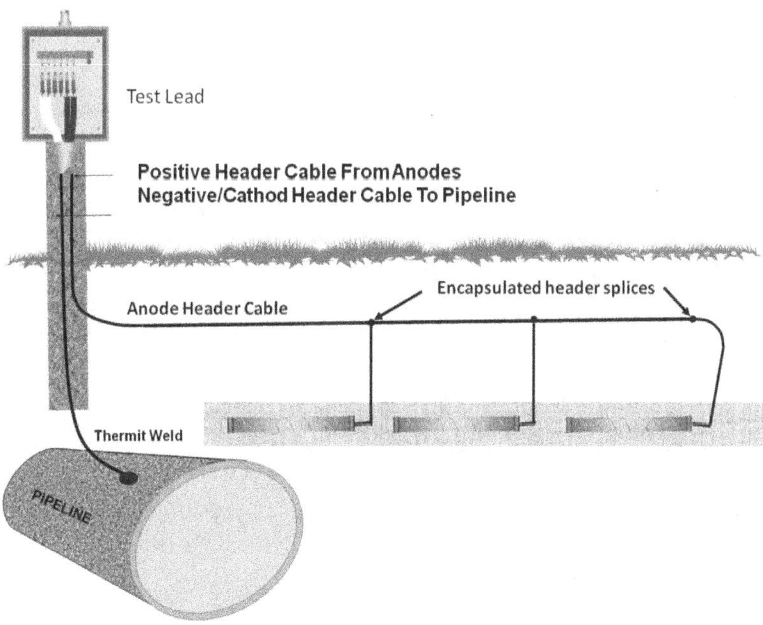

Figure 4-9. Typical sacrificial CP system arrangement

TABLE 4-2. Typical potentials for selected materials in neutral soil referenced to a $Cu/CuSO_4$ electrode

Material	Potential Volts
Magnesium	−1.75
91Mg-6A1-32n-0.15Mn	−1.55
Zinc	−1.1
95A1-5Zn	−1.05
Aluminum	−0.8
Steel (clean)	−0.5 to −0.8
Steel (rusted)	−0.2 to 0.5
Gray cast iron	−0.5
Lead	−0.5
Steel in concrete	−0.2
Copper, brass, bronze	−0.2
High-silicon cast iron	−0.2
Mill scale on steel	−0.2
Graphite (carbon)	+0.3

elements (e.g., pure magnesium has a potential of −1.75 V, while an alloy 91Mg-6A1-3Zn has a potential of only −1.55 V). Some typical sacrificial anodic materials and their uses are given in Table 4-3.

The following lists the advantages and limitations of galvanic systems.

Advantages

- No external power required
- No regulation required
- Easy to install
- Minimum of cathodic interference
- Anodes can be readily added
- Low maintenance
- Uniform distribution of current
- Installation (of anodes at time of construction) can be inexpensive
- Minimum right-of-way/easement costs
- Efficient use of protective current

Limitations

- Limited driving potential
- Low/limited current output/unit
- Installation (of anodes after construction) can be expensive
- Poorly coated structures require many anodes
- Can be ineffective in a high-resistivity environment

It might appear that galvanic systems would be the first choice for CP application since no external power sources or rectifiers are required. However, for extensive pipeline systems, the impressed current system is generally more practical. The major advantages of the impressed current system over the galvanic system are:

TABLE 4-3. Sacrificial anodes

Types of Anode	Uses/Comments	Backfill
Magnesium	Soil with resistivity <10,000 Ω cm	75% hydrated gypsum 20% bentonite 5% sodium sulfate
Zinc	Soil with resistivity <2,000 Ω cm	75% hydrated gypsum 20% bentonite 5% sodium sulfate
Aluminum	Sea water	None

- Its wide range of voltage and current output
- The large area covered from a single location
- It is effective in protecting poorly coated pipelines

SACRIFICIAL ANODE CHARACTERISTICS

Current Output

Current output is of primary importance in the application of sacrificial anodes as this determines the number of anodes required, and their rate of consumption, i.e., their service life.

Some properties of the metals used for sacrificial anodes are summarized in Table 4-4, below:

The theoretical ampere-hour supply per pound of metal is derived from the fact that the dissolution of one gram equivalent of metal transports one faraday of electricity, i.e., 96,490 coulombs or ampere-seconds.

Magnesium anodes typically are available in two alloys and packages as shown in Figure 4-10:

- High potential (−1.75 V sulfate reference electrode [CSE])
- Low potential (−1.55 V CSE)

Anode Efficiency

In practice, the theoretical ampere-hour output of sacrificial anodes is not all available for CP. While some of the anode material is consumed in protecting the structure, the remaining part of the material is consumed by the normal self-corrosion of the anode due to the action on it of the electrolyte.

The "anode efficiency" is the ratio of the ampere-hours actually supplied to the theoretical ampere-hour output per unit weight of metal consumed.

The efficiency of magnesium-alloy anodes is usually about 50%, i.e., the useful output is 50/100 × 1000 = 500 amp-hour/lb (1100 amp-hour/kg) of metal consumed.

To obtain this efficiency, the magnesium alloy should be of correct composition with a low-impurity content. The efficiency is also influenced by the environment. In water with a moderate to low salts content, the efficiency may be low because the current output is low, and consequently, the anode's own corrosion may be relatively high. The use of a special backfill (such as bentonite) around the anode gives a higher current output and a better anode efficiency (Figure 4-11).

Magnesium cylindrical anodes for vertical burial typically will have the weight and current densities as given in Table 4-5 below:

At increased temperatures, the self-corrosion of the anodes is higher and therefore the efficiency of the anode decreases. For this reason, magnesium-alloy anodes are generally not used when the temperature is higher than approximately 86°F (30°C) in brackish or saltwater or higher than approximately 113°F (45°C) in freshwater.

TABLE 4-4. Properties of some sacrificial anodes

Metal	Atomic Weight	Equivalent Weight	Theoretical Current Output Amp-Hour/kg
Magnesium	24.3	12.15	2,200
Zinc	65.4	32.7	820
Aluminum	27.0	13.5	2,970

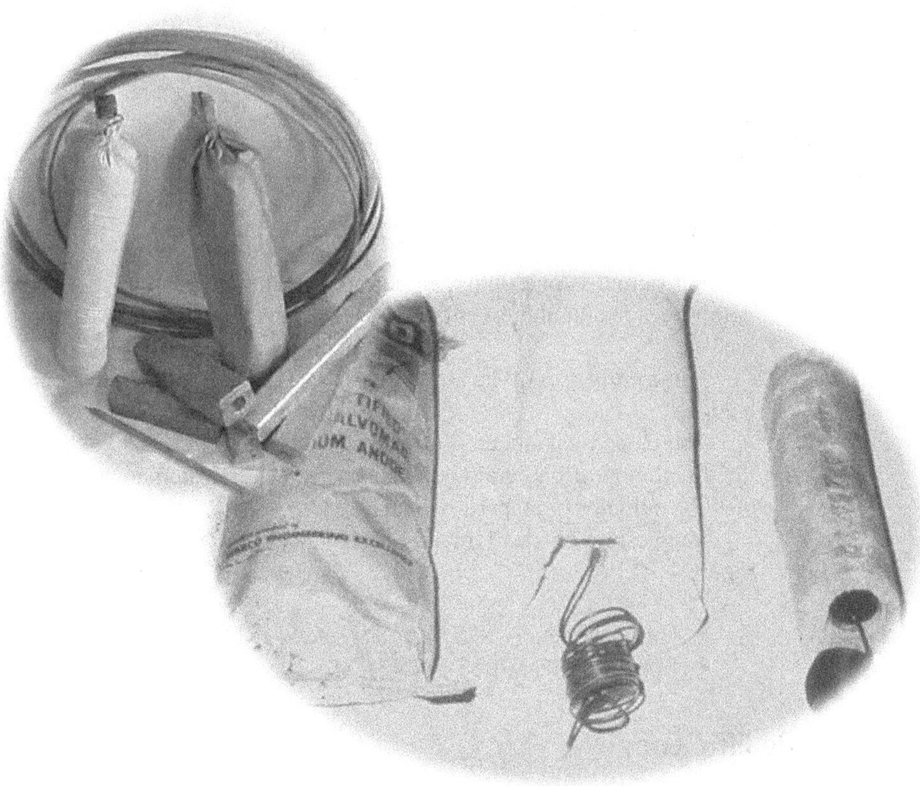

Figure 4-10. Typical cased magnesium anodes (inset: cased packaged and extruded)

Zinc anodes normally operate at an efficiency of about 90%, so that their ampere-hour output is approximately 0.90 × 372 = 335 amp-hour/lb (740 amp-hour/kg).

Aluminum anodes efficiency is about 75%, i.e., an ampere-hour output of 75/100 × 1350 = 1000 amp-hour/lb (2200 amp-hour/kg).

Current Output of Anodes

The current outputs of magnesium-alloy anodes and zinc anodes are as follows:

(a) Magnesium Alloy — The approximate output of magnesium, alloy anodes connected to a fully protected structure (–0.85 V relative to a standard copper/copper sulfate electrode) can be estimated using the data given in Table 4-5:

For example, the approximate current output of one 7.7-kg (17-lbs) magnesium-alloy anode buried in soil with a resistivity of 2000 Ω cm and connected to a buried steel structure, the steel-to-soil potential of which is maintained at –0.85 volt relative to a standard copper/copper sulfate electrode, is as follows:

$$\text{Current output} = (18 \times 10^4)/2000 = 90 \text{ mA}$$

Magnesium-alloy anodes are usually cylindrical, diameter 100 mm to 150 mm (4 in. to 6 in.), or 100 mm (4 in.) square blocks. For use in high-resistivity environments, or when a high current with a short life is required, anodes in the form of a bar or ribbon are used.

A typical vertical anode for pipeline protection is a 11-kg (22-lb) cylindrical anode of 110 mm (4.5 in.) diameter and 530 mm (21 in.) length buried in gypsum/bentonite backfill.

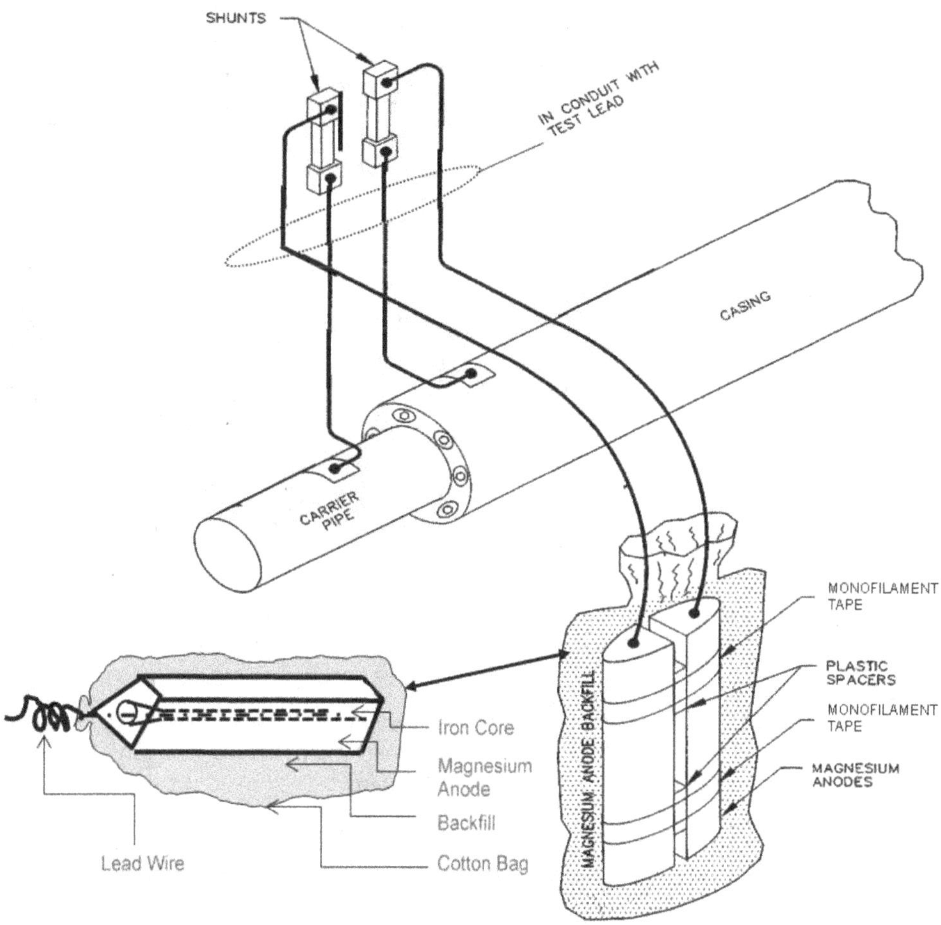

Figure 4-11. Typical vertical anode installation on a pipeline (with a cased crossing)

(b) Zinc — Zinc anodes have been used since the 19th century to protect steel structures from corrosion. These anodes are an effective choice for preventing corrosion in select soils and brackish waters. For these environments, an open-circuit potential of 1.1 V (with respect to a Cu/CuSO4 RC) would apply. They are made from 99.99% pure high-grade zinc and usually provide a 90% current efficiency and deliver a current capacity of about 854 amp-hour/kg (335 amp-hour/lb). However, zinc anodes generally have a lower current output than magnesium-alloy anodes, as zinc has a less negative potential (refer to Table 4-2). Zinc anodes are usually slabs 40 mm (1½ in.) in thickness, or bars, i.e., shapes with a high surface area to volume ratio to provide the maximum current for the weight of metal installed (Figure 4-12).

The type of zinc anode usually used for pipeline protection in soil has dimensions 40 × 40 × 1520 mm (1½ × 1½ × 60 in.) and weighs 15.5 kg (34 lbs). It is buried in gypsum/

TABLE 4-5. Magnesium anodes capacity

Magnesium Alloy (Cylindrical Anode)	Approx. Current Output (mA)
7–10 kg	(18×10^4)/Soil resistivity Ω cm
15 kg	(19.5×10^4)/Soil resistivity Ω cm
23 kg	(20.4×10^4)/Soil resistivity Ω cm

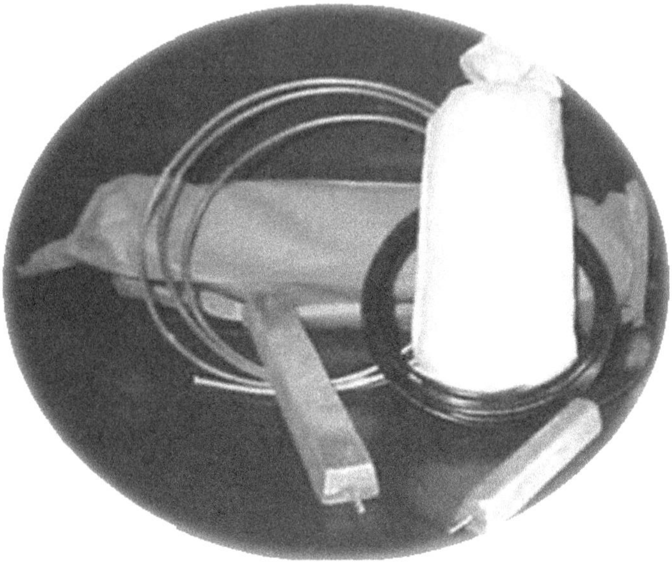

Figure 4-12. Typical zinc anode for pipeline protection application (soil and ribbon cast and extruded zinc anodes courtesy of CORRPRO)

bentonite backfill. This type of anode has a current output of approximately 50 mA in soil with a resistivity of 1000 Ω cm against a protected structure. Current output can be estimated from the following relationship:

$$\text{Current output} = 50{,}000/\text{soil resistivity in } \Omega \text{ cm}$$

(c) Aluminum Anodes — These anodes have very high current output and therefore are used for protection of massive structures and those that support facilities in a salty marine environment. These can typically deliver a current output of 2880 amp-hour/kg (1280 amp-hour/lb) with open circuit voltage of 1.05 (with respect to an Ag/AgCl RC) in saltwater.

Aluminum anodes deliver effective galvanic protection in situations where the anode will be buried in saline mud or exposed to water with lower chloride levels. Typical aluminum bracelet anode for protection steel pipeline in marine environment is depicted in Figure 4-13.

Ground Bed/Anode Backfill and Layout

Types of Ground Beds

The ground bed (anode bed) is the point of current discharge into the earth from the rectifier. Corrosion will take place at the anodes, which will eventually require replacement. Two types of ground beds are commonly used by the pipeline industry; the conventional horizontal type and the deep well type. With the conventional type, the anodes are installed horizontally in a straight line 2 m to 3 m below ground level. In the deep well type, the anodes are installed in a well 60 to 120 m below ground level. Each type has particular advantages and disadvantages, but both types discharge the rectifier current into the earth.

Conduction of current from the anode to the pipeline is most important. To achieve the highest current transfer, anodes are typically placed in backfill, which has low resistivity (<50 Ω cm), can easily absorb water, and will create a uniform environment. Typical anode backfill consists of the following:

- 75% gypsum
- 20% bentonite
- 5% sodium sulfate

Figure 4-13. Typical bracelet aluminum anodes for marine application

Horizontal Ground Bed. Horizontal ground beds for pipeline protection normally have the anodes buried about 2.5 m deep. The anodes form a continuous line, normally straight, but occasionally are curved (i.e., to follow the curvature of a stream or river/road).

To enhance current distribution up and down the pipeline, the first anode is generally not to be located closer than 100 m to the pipeline. The anode bed is often placed in low-resistivity soil and in the swampy/wet ground in low-lying areas. High-elevation locations are not ideal.

The anode spacing is normally 4.5-m center to center. Horizontal anodes are generally 20-kg anodes of high Si-Fe-Cr composition.

Their dimensions as previously indicated are about 1.5 m long with a diameter of 51 mm. Other common anode sizes for horizontal applications include 11 kg, 13 kg, 39 kg, and 50 kg.

Sometimes anodes are already prepackaged in a steel container whose dimensions are typically about 2.1 m long and 0.23 m in diameter. The inside of the steel container is usually filled with coke breeze. A short AWG #8 cable is attached to each anode contained in the container.

Each anode lead wire is spliced onto a larger header cable (typically #4 or #2). It is imperative that the splices be sealed since any contact between the positive copper header wire and the ground will quickly corrode the copper wire and the effect of anode(s) will be lost. It is also required that the cables be sheathed with a high-quality insulation.

The end of the header cable can be brought above ground and connected to a junction box. This will enable reconnection of the anodes if there is a break in the header cable over the life of the facility.

Where more than 20 anodes are used, it is sometimes advisable to centre tap the ground bed and to put two junction boxes at each end to allow for operational flexibility.

If the ground is dry, NPS 4-diameter plastic weeping tile are usually laid above the anodes prior to backfilling.

The ends of the ground bed are marked with a warning sign, which state that there is a buried CP ground bed and a contact number.

Deep Vertical Ground bed. Deep vertical ground beds normally have typically a 225-mm (about 9 in.) diameter hole drilled approximately 100 m to 120 m deep. The first anode begins 75 m below grade. The maximum number of anodes that the industry uses in hole is 12. For deep ground beds where a large mass of anodes is required, two or more holes may be drilled and spaced 10 m apart. Vertical anodes are typically 27 kg or 28 kg high Si-Fe-Cr anodes, which measures about 2 m in length. The anodes come with individual lead wires (AWG #8) brought to the surface (and used to lower the anodes down the hole). The anodes are spaced 1.5 m apart.

Prior to lowering the anodes in the deep well, centralizing tubes are connected to the top and bottom of each anode with a hose clamp to keep the anode from contacting the side of the hole.

After the anodes are lowered in, coke breeze is pumped into the bottom of the hole to 5 m above the top anode. As with the horizontal anode bed, it is critical that the anodes reside in the coke breeze backfill.

Strapped to the first anode will be a 20-mm PVC pipe, which is covered with a Nilex sock. The pipe and anode are together lowered down the hole. The purpose of the pipe is to vent any gas that may evolve away from the anode.

Without this vent pipe, the anodes would quickly become polarized and cease to function as intended. The vent pipe is terminated above ground (often in the junction box).

Prior to lowering the anodes, a resistance log is taken of the area where the anode will be installed. Any area that has significantly higher resistivity than surrounding areas will be noted and the anodes spaced around this region.

To provide operation and maintenance flexibility, individual anode leads are brought to the surface and connected in a junction box. A single cable (#2 or #4) will then be connected from the junction box to a rectifier as applicable.

When installing a deep ground bed on a pipeline, which is intended to be looped at some future date, the industry recommends the installation of the ground bed close enough to one pipe (or in between lines) so that any subsequent pipeline loop won't damage the anode lead wires.

As with the horizontal ground bed, a CP warning sign is located beside the facilities.

Vertical/Deep Versus Horizontal/Shallow Anodes/Ground Beds

1. Deep vertical anode ground beds can offer an advantage over surface ground beds when placed in low soil resistivity strata, underlying higher resistivity soil.
2. In soils of equal resistivities, deep anodes will have lower resistance to remote earth than horizontal surface anodes, since they more closely approach the conditions of being within an infinite conductor.
3. Surface potential gradients in the soil surrounding a deep anode installation will be less than for a surface anode system. The risk of interference to foreign structures is therefore correspondingly less with a deep anode system.
4. Deep anode ground beds are ideally suited where space is limited.
5. Deep anodes are essentially unaffected by seasonal soil moisture variations as surface anodes are sometimes inclined to be.
6. Deep anodes are less susceptible to accidental damage resulting from future excavating or trenching operations.
7. Deep anode ground beds are virtually inaccessible once installed. A break in the down hole cable would mean loss of the ground bed Surface-anode ground beds can be repaired should cable breaks occur.
8. For deep anode beds, individual anode lead wires protected in a plastic casing are an advantage.
9. Additional anodes could easily be added to a surface ground bed if required. This would not be possible with deep anode ground beds without a new hole.

10. Shielding effects are normally more pronounced with surface anodes. Deep ground beds generally exhibit more uniform current distribution.
11. The relative initial costs vary with the individual applications; however, the deep anodes are usually more expensive.

CP DESIGN CONSIDERATIONS

CP facilities may be required for new construction and the upgrading of old systems. In either case, similar factors must be considered prior to any design. These factors are:

- Existing CP
- Length and diameter of existing/new facilities
- Future plans for expansion

Existing CP in the area may have sufficient capacity to be able to protect the additional loads. This can be accomplished by simply resetting the rectifier output.

Current requirements for CP are calculated by using specific current densities and pipeline surface area (see Table 4-6). The specific current density varies with the pipeline coating quality and integrity and the soil conditions (Table 4-7). Improved coating requires less current density. Knowing the length and diameter of the pipe allows surface area to be calculated.

If future plans for pipeline expansion are known in advance, CP facilities can be designed to accommodate the requirement for protecting such facilities in advance. It costs much less to install one large ground bed than to install two smaller ones.

Ground Bed Location

Once the decision has been made as to the requirement for CP, a suitable location for the ground bed must be selected. Again, a number of factors must be considered. These are:

- Electrical power availability
- Soil resistivities
- Land availability
- Future plans for expansion

Cost elements/components affecting a CP system are identified in Table 4-8. The cost elements vary with type, size, and characteristic of pipelines and the right of way/terrain, location and type of power sources, as well as method of protection (anodic versus galvanic) (Koch 2002).

TABLE 4-6. Typical current density requirements for CP of bare steel

Environment	Current Density (mA/m^2)
Neutral soil	4–6
Well-aerated neutral soil	22–32
Wet soil	22–65
Highly acidic soil	32–160
Soil with active sulfate reducing bacteria	65–450
Stationary fresh water	11–65
Moving fresh water	54–65
Turbulent fresh water with entrained oxygen	54–160
Sea water	54–270

TABLE 4-7. Typical average current density requirements for CP of pipelines in soil

Coating Type	Average Current Density (mA/m^2)
Asphalt/coal tar enamel	300–1,500
Thin mil tapes	25–200
Extruded polyethylene	10–75
FBE	250–100

Electrical power supply cost depends on the location and availability of power versus the location of a ground beds. The cost of a ground bed installation can be very high if a poor location is selected. Expensive power sources such as thermal electric generators, solar power, wind power, or fuel driven generators may have to be considered if no suitable locations with electrical power can be found.

For ground bed location, generally, the wetter the location, the better the ground bed output. Since surface conditions do not always indicate the below ground conditions, soil resistivity measurements are used to indicate moist soil conditions. The lower the soil resistivity, the lower will be the required driving force (voltage) of the ground bed to effect CP over a set distance. In other words, the lower the soil resistivity, the greater the coverage for CP one can obtain from a single ground bed installation. Ponds, drainage ditches or natural water runoff courses are common locations for ground beds due to the low resistivity of their soils.

A ground bed radiates lines of equal potential (Figure 4-14), the potential decreasing with distance from the ground bed. Where a hole in the coating exists, the equal potential lines decrease as the current from the ground bed flows to the bare steel.

Just because a suitable site has been selected does not automatically mean the final ground bed site has been found. Some land owners do not allow pipeline companies to install facilities on their land, especially without a large monetary payment. These landowners are generally identified early before too much planning has been completed.

Once the above information has been obtained, calculations are performed to determine the parameters for the required CP installation (ground bed).

CP Installation

The installation and energizing of the ground bed is the final step in implementing a CP system for a pipeline (Figure 4-15). A number of procedures must be followed, these are:

TABLE 4-8. CP life cycle costs

Capital cost	• Design of facilities • Purchase of materials • Surveying land • Taking soil resistivity measurements • Purchase of land/rights of access • Paying for clearing/grading of mountainous/forested land • Installation costs • Cost of power (transmission line or generator)
Maintenance and monitoring costs	• Cost to verify CP meets an accepted criteria and is operating properly on a regular basis • Cost to repair CP facility when it fails to operate as required
Operating costs	• Cost of power • Annual payments for land access • Cost to water ground beds in dry soil

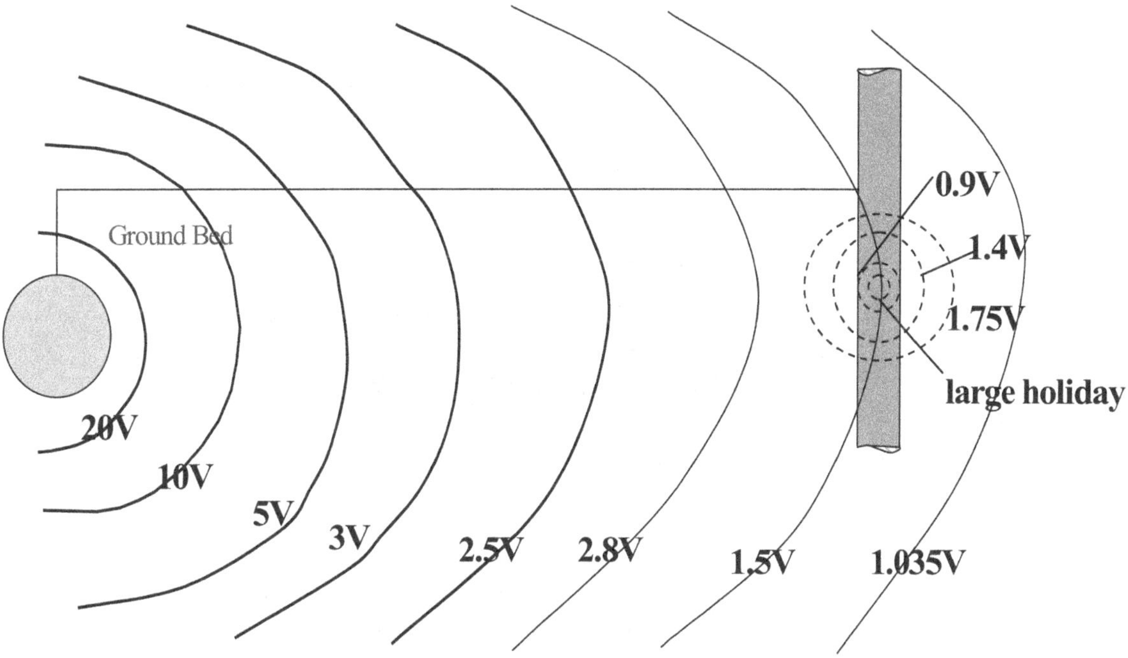

Figure 4-14. Equal potential lines radiating from a ground bed to a coated pipeline with a holiday

Determine the Location of Ground bed Easement. Often, the construction surveyor's stakes are still in the ground, which will make the exact determination of where the ground bed is to be installed very easy. If the stakes cannot be found, the legal survey plan of the ground bed must be consulted so that the correct location can be marked.

Laying Out of Anodes and Ground Bed Excavation. The ground bed materials are strung out along the proposed ditch to provide a marker, for the backhoe operator as to the length of the ditch, which is to be excavated (Figure 4-16). As indicated previously, horizontal anodes are usually buried at a minimum depth of 2.4 m, and the cables are buried a minimum depth of 0.5 m to 0.6 m. The anode depths will vary according to the soil resistivity

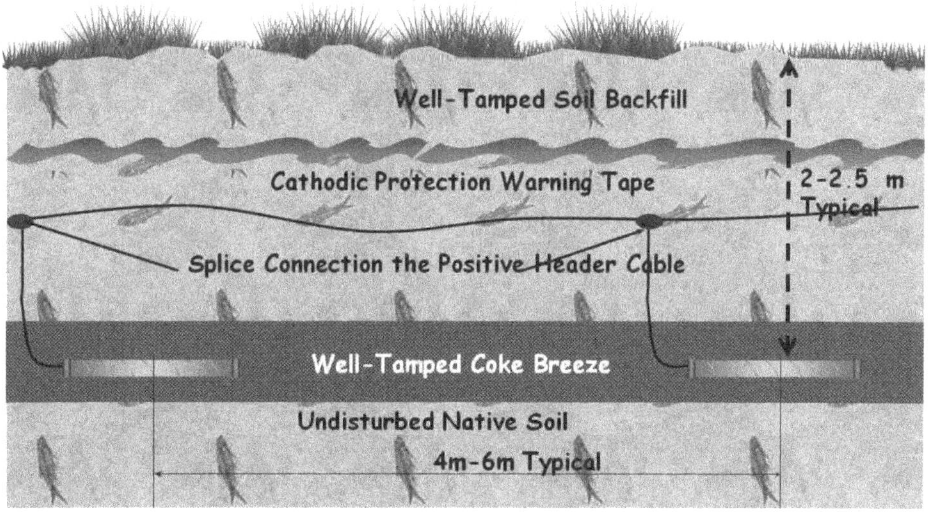

Figure 4-15. Typical installation of a horizontal ground bed

Figure 4-16. Anode marking and excavation

measurements taken earlier._The cable depths must be deep enough to prevent damage by farmers working their fields.

Splicing the Anode to the Header Cable. The anodes are spliced to the main anode cable that runs from the ground beds to a rectifier. This is done by means of a mechanical clamp and then coated. The splices, if not electrically sealed, would provide a spot for the current to leave, they therefore must be well taped/coated to ensure no current leakage.

Lowering Anodes Into the Ditch and Backfilling. Safety regulations generally do not allow human entry into the deep anode ditches without extensive modifications, which would require extra time and money. Therefore, the anodes are lowered into the ditch by a method that does not require personnel to enter the ditch (Figure 4-17). Ropes are often used as guides and constraints for the lowering in. Once the anodes are positioned, the backfilling can commence.

Figure 4-17. Lowering-in Anode into Anode Ditch

Attaching the Negative Cable to the Pipeline. The negative cable running from the rectifier to the pipeline is connected to the pipe using mechanical clamps or a thermite welding process. However, there are a number of ways that anode wires can be attached to the structure to be protected. These are:

- Direct thermite weld
- Thermite weld and wire through test station
- Clamp to structure and then thermite weld pigtail
- Direct weld anode core to structure

Figure 4-18 illustrates a typical clamping device and attachments that are used for attaching the anode wire to a pipeline outer surface.

Installing and Energizing the Rectifier. The last step to effect CP is to install the rectifier, connect the electrical power, and energize the ground bed.

The rectifier is the most common source of DC power that can be used for CP. This device simply converts AC power to low voltage DC power. It is available in a wide range of voltage and current ratings and is usually provided with a means of adjusting the DC output voltage to the required level. There are three types of rectifiers that are used by the industry. These are:

- *Constant voltage* — the DC voltage at the terminals remains constant for all current outputs up to the rated maximum rectifier current.
- *Constant current* — the current output remains constant over a wide range of circuit resistances up to the maximum rated output voltage.
- *Constant potential* — the current and voltage output vary to maintain a preselected structure-to-electrolyte potential.

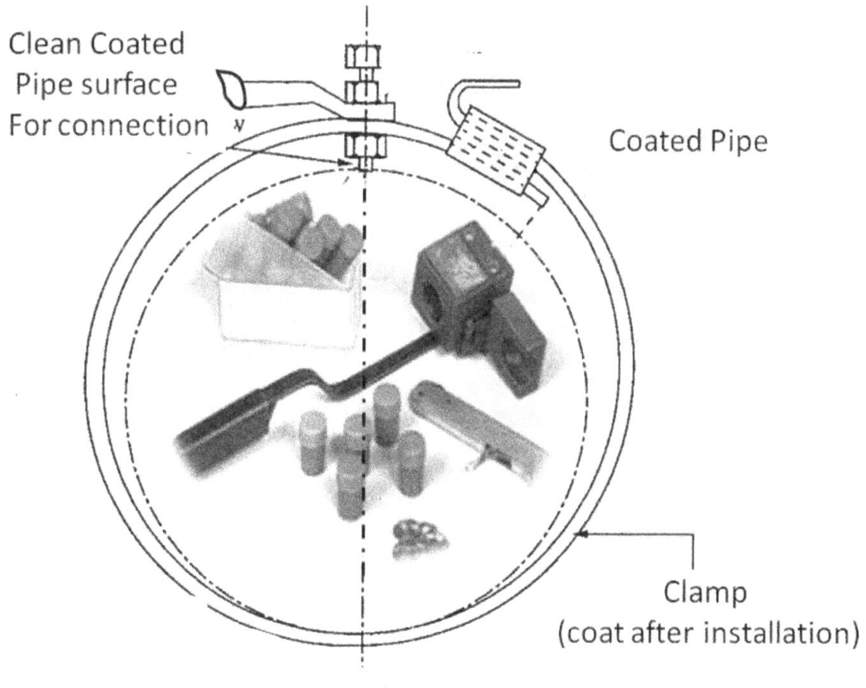

Figure 4-18. Typical clamp arrangement for attaching anode cable to pipe surface and for thermite welding (inset: clamping and splicing tools)

Electrical power to rectify can be supplied in a number of ways including:

- Third party power supply from a nearby electrical supply line via a transformer-rectifier unit
- Solar (photovoltaic) cells
- Engine generator sets
- Wind-powered generators
- Thermoelectric cells

A typical transformer rectifier arrangement is indicated in Figure 4-19.

Upon electrical connection, the CP will begin to work immediately. However, it will take several days before the pipe is polarized and stable before readings can be taken. Potential readings are then taken along the pipeline and compared to the readings obtained before the CP was activated (the natural readings). Depending on the readings, the output of the rectifier can be increased or decreased._In order to properly conclude the ground bed installation, the following requirements must be met; as-built drawings must be prepared, a detailed ground bed description developed, and a detailed rectifier description prepared.

These records enable precise details of the installations to be recorded. This enables future location of the ground bed and a method of monitoring what types of materials are being used, and their effectiveness as sacrificial anodes.

CP Monitoring

Determining the effectiveness of a CP system is accomplished by measurement of the voltage difference between the protected steel pipe (when the anode is in place and con-

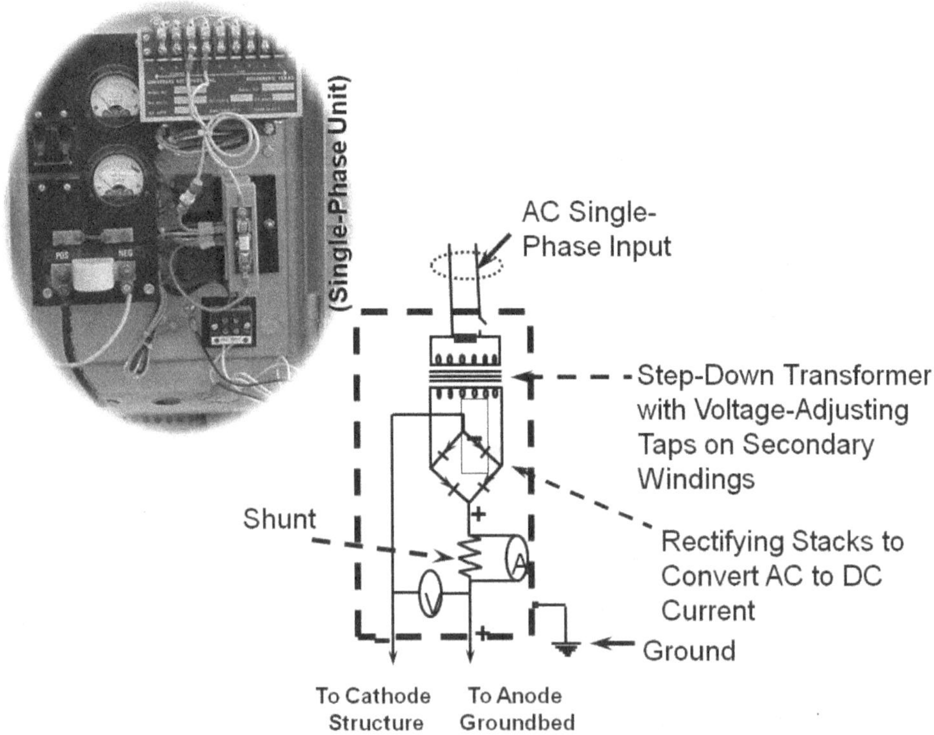

Figure 4-19. Typical cathodic protection transformer-rectifier

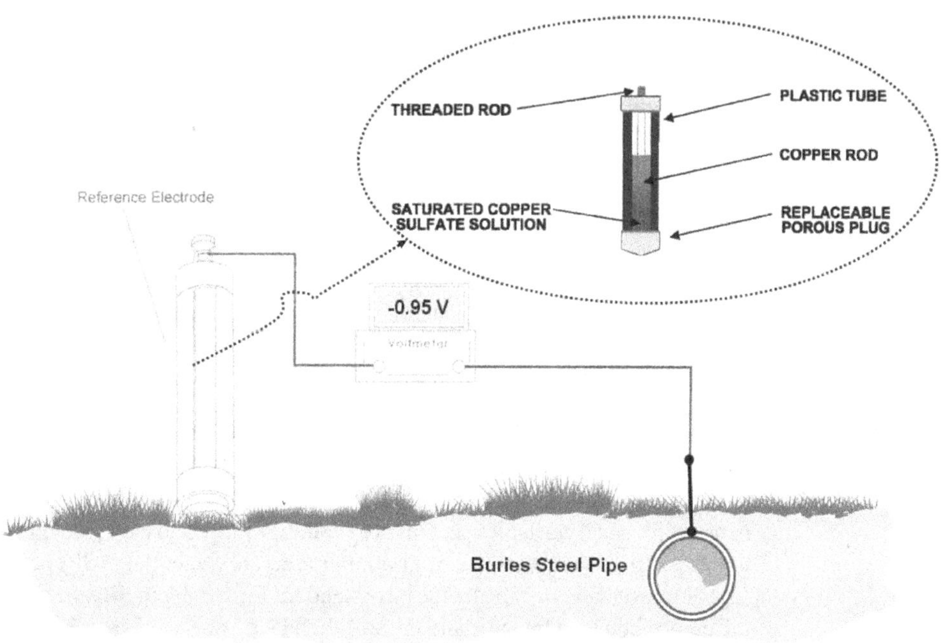

Figure 4-20. Copper–copper sulfate reference electrode

nected) and a Copper–copper sulfate ($Cu/CuSO_4$) electrode placed near the structure (Figure 4-20).

Arrangement for Measurement of CP Potential

A typical standard copper/copper sulfate reference electrode used for land based close interval survey (CIS) is made up of a central copper rod inserted in a saturated copper solution contained in a small plastic tube. The copper rod is connected to a porous plug at the measuring end. The other end of the rod is threaded for connection to a volt meter. This is shown as an inset in Figure 4-20. This reference electrode is practical and can be used in a variety of field applications and soil conditions due to its relative stability, (Pawson 2000 and 2003).

The industry has developed a number of procedures mostly for practical field application of CP monitoring; some of the typical ones noted are listed below (Alexander 2009; Bash 2006; NACE 1988):

- Use of the half cell
- Use of the "half cell" near an insulation joint
- Using thc half cell with the CP switching
- Determining remote earth
- Fluctuations during CIS time
- Two half cell close interval survey
- Close interval potential survey
- Two half cell coating survey
- Plotting a ground bed profile
- Plotting the position of a sacrificial anode
- Using a bright steel coupon

- Using a zinc coupon
- Using an isopotential cell
- Using the Alexander cell
- Interference testing and remedies
- Measuring CP current
- Continuity testing and testing an insulation joint
- Transformer rectifier monitoring and maintenance
- Monitoring offshore sacrificial anodes

Criteria for CP

The most highly anodic areas to be expected on a steel pipeline buried in most soils would have a potential of approximately –0.8 V as measured with respect to a copper sulfate electrode contacting the environment immediately adjacent to the anodic area. This is not generally practical to achieve because most pipelines are buried. The closest approach is therefore to place the reference electrode at the ground surface directly above the pipeline. To compensate for such variables as soil IR (voltage) drop and possible anaerobic bacteria, the industry strives to maintain a minimum pipe-to-soil potential (PSP) of –0.85 V to –1 V.

However, a widely accepted base standard for CP of pipelines requires that the potential difference between the pipe and the soil shall be no more electropositive than –950 mV for a standard saturated $Cu/CuS0_4$ RC. If the measurement is more negative than –0.950 V, then the structure is considered protected. If the measurement is more positive than –0.950 V, the structure is unprotected.

Also, a typical maximum allowable potential of –3000 mV (Cu/CuS04) is set to prevent adverse effects on some coating types. However, it may be noted that experimental results have shown that stress corrosion cracking is potential (i.e., voltage) dependant, and a potential more negative than –1.1 V may be detrimental where corrosion cracking is initiated. A highly negative potential can also produce hydrogen-induced cracking under some soil conditions. Also, it is known that this high negative potential can also increase the possibility of cathodic disbondment damage to protective coating. Therefore, a potential more negative than –3.0 V is not recommended without strong documentation of the electrochemical justification for a site-specific application (MUSTR 2001).

To assess the status of the CP along the pipeline, two wire test leads are placed at regular intervals, usually every 3 km, and at locations of possible concern such as pipeline crossings, cased crossings and river crossings. Two wires are attached as a safety precaution in case one of the wires becomes disconnected. The potential readings can be determined through an attachment to either wire.

Typically, a four-wire test lead is installed at insulation flanges, cased crossings, and pipeline crossings. Two wires are attached to each structure, which allow the monitoring of possible adverse effects from the foreign structure on the protected pipeline. If any effect is noticed, the test lead may be used to mitigate the problem.

Pipe-to-soil meters are permanently installed meters which monitor the CP. They are usually located at the ends of one pipeline system or in areas of high concern.

The meters allow nonspecialist technicians to easily monitor and record the CP status. All pipe-to-soil meters and rectifiers (volt and ampere output) are monitored once per month.

All test leads are monitored at least once a year to determine the electrical pipeline potentials. This survey provides a more accurate determination of the status of the CP along the entire length of pipeline than the monthly pipe-to-soil meter readings.

CP Records

For record keeping and repair purposes, the following records are usually kept by the industry and are required by regulatory bodies (MUSTR 2001):

- CP survey records and interpretations results
- Documentation of operation of corrosion protection equipment
- Documentation of CP system repairs

Typically, copies of all surveys and checks are relayed to the corrosion expert group, and these records are reviewed to ensure no facilities are below the criteria of protection.

Records of testing and inspection of CP systems must generally demonstrate compliance with the testing criteria specified by the corrosion expert or corrosion engineer that designed the CP system.

Steps that were followed in arriving at whether the CP system is or is not performing adequately to protect the pipeline system against corrosion are typically required to be included in the documentation. This documentation should provide enough information about the testing that any qualified CP tester would be able to perform the same steps and arrive at the same conclusion, when performing the test within a reasonably short period of time. These steps would provide an adequate number of measurements that will properly determine the adequacy of the protection provided.

For impressed current testing, the measurements typically include potential (voltage) measurements made at appropriate and adequate locations in both the current-on and current-off status. These measurements are made relative to a copper-copper CSE.

For galvanic systems, current-on measurements are the only option possible because the anodes are permanently attached. The current-on measurements indicate the distribution of current on the structure and where the weak spots in terms of protection may be located.

The following provide a definition of "corrosion expert" and a "corrosion tester" (MUSTR 2001):

- *Corrosion expert* — means a person who, by reason of thorough knowledge of the physical sciences and the principles of engineering and mathematics acquired by a professional education and related practical experience, is qualified to engage in the practice of corrosion control on buried or submerged metal piping systems and metal tanks. The person shall be certificated as being qualified by the NACE International as a senior corrosion technologist, a CP specialist, or a corrosion specialist or be a registered professional engineer who has certification or licensing that includes education and experience in corrosion control of buried or submerged metal piping systems and metal tanks.
- *CP tester* — means a person who can demonstrate an understanding of the principles and measurements of all common types of CP systems as applied to buried or submerged metal piping and tank systems and who has education and experience in soil resistivity, stray current, structure-to-soil potential, and component electrical isolation measurements of buried metal piping and tank systems. The person is typically required to be certificated as being qualified by NACE International, or equivalent.

CP Calculations for Buried Pipelines

On a coated pipeline that is cathodically protected, current flows from the anode of the CP system to holidays (or defects) in the coating. This current flow to the defects can be measured as a voltage between two electrodes on the surface of the ground. The voltage developed between two points on the surface of the ground obeys Ohms law, $V = IR$ where V is the voltage gradient and I is the current flow to the coating defect or holiday and R is the resistance of the soil between the two electrodes (Figure 4-21).

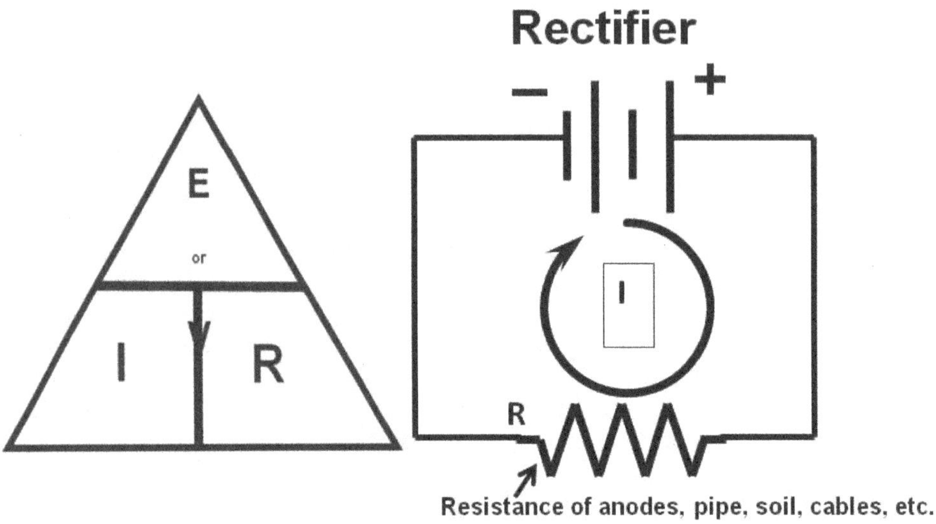

Figure 4-21. Ohm's law triangle and circuit resistance

The Ohm's law relationships can be written as follows:

$$E = IR \qquad I = \frac{E}{R} \qquad R = \frac{E}{I}$$

$$I = \frac{V}{R} \qquad R = \frac{V}{I}$$

The use of Ohm's law for corrosion measurement is also illustrated in Figure 4-21. For example, if the voltage applied through the rectifier is 5 V and measured current is 20 amps, the circuit resistance will be $R = E/I = 5/20 = 0.25\ \Omega$.

Typically, 1 amp of DC current discharging from steel into soil as it corrodes can remove about 9 kg (20 lbs) of steel in 1 year. One thousandth of an ampere can cause seven 6.35 mm (–¼ in.) diameter holes in an NPS 2 (2 in.) pipe in 1 year.

The purpose of CP current is to create an electrochemical arena within which the pipeline is protected from corrosion.

The following provides a detailed analyses of the electrochemical interconnection of a pipe, coating, and CP systems.

Attenuation (Figure 4-22) presents a general view of current and voltage along a pipeline with respect to a drain point (i.e., defect/holiday).

Symbols

R_S	linear pipe resistance, Ω/m
R_L	leakage (coating) resistance, Ω/m
α	attenuation constant, per kilometer
R_0	characteristic resistance, Ω
ρ	soil resistivity, μΩ cm
ρ^1	steel resistivity, μΩ cm
g	coating conductivity, $\mu\Omega^{-1}$ per m^2 (Siemens)
D	pipe diameter, mm
L	pipe length, km
t	pipe wall thickness, mm
a	pipe cross sectional area, cm^2

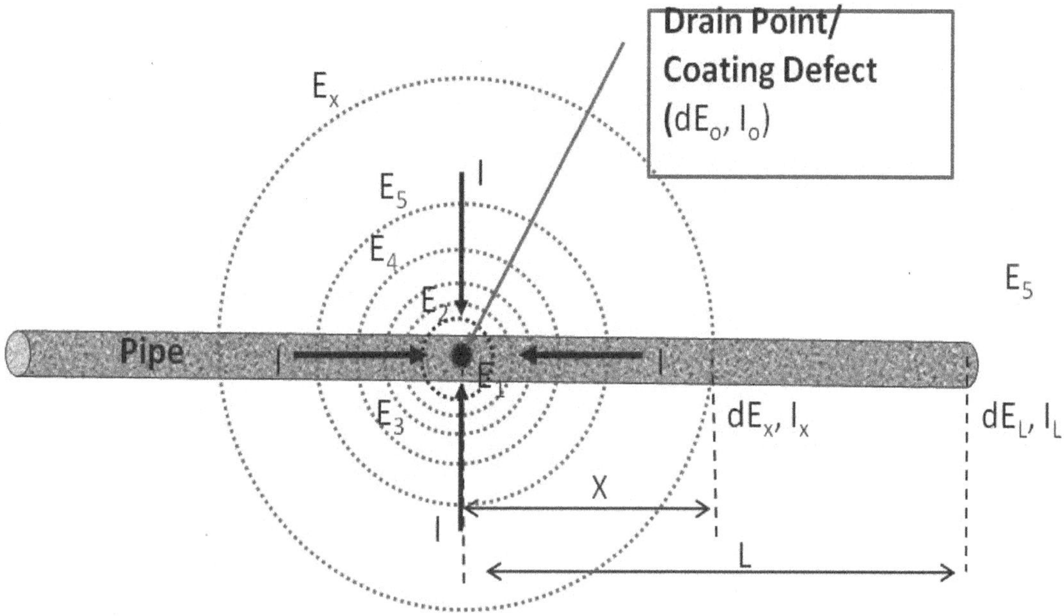

Figure 4-22. Voltage gradients around a coating defect (drain point)

A	surface area of pipe, m^2/m
W	weight of pipe, kg/m
dE_0	change in potential at drain point, V
dE_X	change in potential at point X, V
dE_T	change in potential at end of line, V
I_0	current at drain point, amps
I_X	current at point X on line, amps
I_T	current at end of line, amps

General Case

The equation for the general case can be written as follows:

a) Volts $$dE_X = dE \cosh\alpha X - I_T R_0 \sinh \alpha X$$

b) Amps $$I_X = I\cosh\alpha X - \frac{dE_0}{R_0}\sinh\alpha X \qquad (4-1)$$

OR:

c) Volts $$dE_0 = dE_x\cosh \alpha X + R_0 I_x \sinh(\alpha X)$$

d) Amps $$I_0 = I_X \cosh \alpha X + \frac{dE_X}{R_0} \sinh \alpha X$$

Finite Lines

The equation for finite lines can be written as follows:

a) Volts $$dE_0 = dE_T \cosh(\alpha L)$$

b) Amps $$I_0 = \frac{dE_T}{R_0} \times \sinh(\alpha L)$$

c) Volts $$dE_x = dE_T \cdot \cosh \alpha \left(L - X\right) \qquad (4-2)$$

d) Amps $$I_X = I_0 \cdot \frac{\sinh \alpha \left(L - X\right)}{\sinh (\alpha L)}$$

Infinite Lines

In almost all CP systems, there are no infinite lines. However, the equation for infinite lines can be written as follows:

a) Volts $$dE_x = dE_0\, e^{-\alpha X}$$

b) Amps $$I_X = I_0\, e^{\alpha X} \quad I_0 = \frac{dE_0}{\alpha R_L} = \frac{dE_0}{R_0}$$

c) Volts $$\operatorname{Ln} dE_0 = \operatorname{Ln} dE_X + \alpha X \qquad (4-3)$$

d) Amps $$\operatorname{Ln} I_0 = \operatorname{Ln} I_X + \alpha X$$

Linear Pipe Resistance

Steel resistivity varies from 16 μΩ cm to 23 μΩ cm. An average value is 18 μΩ cm. The equation for linear pipe resistance can be written as:

$$R_s = \rho^1 \times L/A \ (\Omega/\text{m}) \qquad (4-4)$$

Leakage (Coating Resistance)

For design purposes, the coating conductivity, g, of buried, nonfusion bonded epoxy coated landlines may be taken as 2000 μΩ/m^2. FBE coated lines can be considered to have a coating conductivity of 400 μΩ /m^2. This equation can be written as follows:

$$R_L = \frac{1}{g\,A\,L} (\Omega/\text{km}) \qquad (4-5)$$

Characteristic Resistance

The equation for characteristic resistance can be written as:

$$R_0 = \sqrt{R_S R_L}\ \Omega \qquad (4-6)$$

Attenuation Constant

The equation for attenuation constant can be written as any of the following:

a) $$R_0 = \sqrt{\frac{R_S}{R_L}} \text{ per km}$$

b) $$\alpha = \frac{R_0}{R_L}$$

c) $$\alpha = \frac{R_S}{R_0} = \frac{R_S I}{2E_a I} \qquad (4-7)$$

TABLE 4-9. Typical coating conductance in a 1000 Ω cm soil galvanic anodes calculations

	Average Coating Conductance	
Long Pipelines with Few Fittings	**Siemens/ft²**	**Siemens/m²**
Quality of Work		
Excellent	$<1 \times 10^{-5}$	$<1 \times 10^{-4}$
Good	1×10^{-5} to 5×10^{-5}	1×10^{-4} to 5×10^{-4}
Fair	5×10^{-5} to 1×10^{-4}	5×10^{-4} to 1×10^{-3}
Poor	$>1 \times 10^{-4}$	$>1 \times 10^{-3}$
Bare pipe (2" to 12") (5 cm to 30 cm)	4×10^{-3} to 2×10^{-2}	4×10^{-2} to 2×10^{-1}
Gas of Water Distribution with Many Fitting		
Quality of work		
Excellent	$<5 \times 10^{-5}$	$<5 \times 10^{-4}$
Good	5×10^{-5} to 1×10^{-4}	1×10^{-3} to 5×10^{-4}
Fair	1×10^{-4} to 5×10^{-4}	1×10^{-3} to 5×10^{-3}
Poor	$>5 \times 10^{-4}$	$>5 \times 10^{-3}$
Bare pipe (2" to 12") (5 cm to 30 cm)	4×10^{-3} to 2×10^{-2}	4×10^{-2} to 2×10^{-1}

Typical pipe to earth leakage conductance (g) for pipeline coatings in 1000 Ω-cm soil is indicated in (Table 4-9).

Symbols

I_a	anode current output, mA
R_a	anode resistance to earth, Ω
R_C	cathode resistance to earth, Ω
E_a	open-circuit potential between anode and reference electrode, V
E_C	open-circuit potential between cathode and reference electrode, V
E_O	open-circuit potential between anode and cathode = $E_a - E_c$, V
E_{cu}	closed circuit potential between cathode and reference electrode, V
E_d	driving voltage between anode and cathode for protected system = $E_a - E_{cu}$, V
W	weight per anode, kg
n	number of anodes
Q	weight of anode material consumed per unit current, kg/amp-year
I_{cd}	design current density for protection, mA/m²
Y	design life of system, years
L	length of anode, cm
D	diameter of anode, cm
W_m	weight of anode material required, kg
A	surface area of structure to be protection, m²
ρ	soil resistivity, Ω cm

Current Output

The equation for current output can be written as follows:

$$I_a = \frac{E_0}{R_a + R_c} = \frac{E_a - E_c}{R_a + R_c}\ \text{mA} \tag{4-8}$$

For negligible cathode resistance Eq. (4-8) can be written as:

$$I_a = \frac{E_d}{R_a} = \frac{E_a - E_{cu}}{R_a} \text{ mA}$$
$$= \frac{0.8}{R_a} \text{ for steel protected by Galvomag}$$
$$= \frac{0.6}{R_a} \text{ for steel protected by } H-1 \text{ alloy}$$

where

$$R_a = \frac{\rho}{2\pi L}\left(\ln\frac{8L}{D} - 1\right)$$

Anode Life

The following equation assumes a typical anode will be replaced when it is 85% dissolved. It can be written as follows:

$$\text{Life} = \frac{71.4\, W_m}{I_a} \text{ years} \quad (4-9a)$$

For magnesium anodes, the equation can be written as follows:

$$\text{Life} = \frac{108.7\, W_m}{I_a} \text{ years} \quad (4-9b)$$

Weight of Anode Material Required

The equation for the weight of anode material required is given by:

$$W_m = A \cdot I_{cd} \cdot Q \cdot Y \text{ (kg)} \quad (4-10)$$

where

I_{cd} = required current density, mA/m^2
Y = design life in years

Number of Anodes Required

The equation for the number of anodes required can be written as follows:

$$n = \frac{W_m}{\text{Weight of Anode}} \quad (4-11)$$

Impressed Current Anodes Calculations

Single Cylindrical Anode

The following formulae enable the anode to earth resistance of a single cylindrical anode to be determined:

$$R_v = \frac{\rho}{2\pi L}\left(\ln\frac{8L}{d} - 1\right)$$

$$R_h = \frac{\rho}{2\pi L}\left(\ln\frac{4L^2 + 4L\sqrt{s^2 + L^2}}{ds} + \frac{s}{L} - \sqrt{\frac{s^2 + L^2}{L}} - 1\right) \quad (4-12)$$

where

R_v = resistance of single vertical anode, Ω
R_h = resistance of single horizontal anode, Ω
L = length of anode including backfill, cm
d = diameter of anode including backfill, cm
ρ = electrolyte resistivity, Ω cm
s = twice the depth of anode, cm

For a single vertical anode, the following simplified expression can be used:

$$R_v = \frac{\rho K}{L} \tag{4-13}$$

where

R_v = resistance of single vertical anode, Ω
ρ = effective soil resistivity, Ω cm
L = length of anode, cm
K = shape function, representing ratio of anode length/anode diameter (from Table 4-10)
L/d = ratio of length to diameter of anode

Single Row Vertical Anode Group

The total anode to earth resistance of a group of vertical anodes in parallel and equally spaced in one row is given by the following equation:

$$R_n = \frac{1}{n} R_v + \frac{\rho}{S} \tag{4-14}$$

where

R_n = total anode to earth resistance of vertical anodes in parallel, Ω
R_v = anode to earth resistance of a single vertical anode, from Eq. (4-15), Ω
n = number of vertical anodes in parallel,

TABLE 4-10. The shape function

L/d	K
5	0.0140
6	0.0150
7	0.0158
8	0.0165
9	0.0171
10	0.0177
12	0.0186
14	0.0194
16	0.0201
18	0.0207
20	0.0213
25	0.0224
30	0.0234
35	0.0242
40	0.0249
45	0.0255
50	0.0261
55	0.0266
60	0.0270

TABLE 4-11. Paralleling factor

n	*P*
2	0.00261
3	0.00289
4	0.00283
5	0.00268
6	0.00252
7	0.00237
8	0.00224
9	0.00212
10	0.00201
12	0.00182
14	0.00168
16	0.00155
18	0.00145
20	0.00135
22	0.00128
24	0.00121
26	0.00114
28	0.00109
30	0.00104

ρ = soil resistivity, measured with pin spacing equal to S
P = paralleling factor, from Table 4-11
S = spacing between adjacent anodes, cm

Multiple Row Vertical Anode Group

An anode group composed of two or more rows of vertical anodes, separated by a distance substantially larger than that between the anodes within a single row, has a total resistance approximately equal to the total parallel resistance of all the rows. The formula generally used for paralleling resistance is:

$$\frac{1}{R} = \frac{1}{R_1} + \frac{1}{R_2} + \frac{1}{R_3} \times\times\times\times\times + \frac{1}{R_N} \qquad (4-15)$$

Typical CP Design Problem

Design Example 1

The following is an example of the design for an impressed current CP facility.

It is given that the estimated length of the pipeline is 38 km, the diameter is 500 mm, and the external coating will be fusion bond epoxy. Field experience and attenuation calculations confirm that in the area where the pipeline is to be installed, the maximum protective range is 44 km for the CP installation. If possible, the ground bed should be located at km post 19. A suitable site for a CP installation was established at km post 20.8, based on soil resistance measurements. The soil resistivity at anode burial depth and below the anodes is less than 5000 Ω cm.

The current required after 20 years of service is estimated to be 0.00007 amps/m^2. Therefore the total current requirement for this 38 km section of pipe is 4.2 amps, which is calculated as follows,

$$\begin{aligned}\text{Total Current Requirement} &= \text{Current Density} \times \text{Total Surface Area } (\pi DL)\\ &= 0.00007 \times \pi \times 0.5 \times 38{,}000\\ &= 4.2\text{ A}\end{aligned}$$

The standard size of rectifier is 8 A.

Silicon-chromium cast iron anodes will be used, each possessing an individual weight of 20 kg. From Eq. (4-11), the number of the anodes which will be required for a specified lifespan is:

$$N = \frac{(\text{Life, years}) \times (\text{Size of Rectifier, amps}) \times (\text{Consumption Rate, kg/amp-year})}{(\%\ \text{volume consumption}) \times (\text{Mass of anode, kg})} \qquad (4-16)$$

In this case, a consumption rate of 0.45 kg/amp-year and a 60% volume consumption were assumed for the 20-year lifespan. Eq. (4-16) indicates that six anodes will be required, i.e.,

$$N = \frac{20 \times 8 \times 0.45}{0.60 \times 20} = 6$$

The major components of the circuit resistance (R_T) are the anode-to-electrolyte resistance (R_A), the structure-to-electrolyte resistance (R_c) and the cable resistance (R_w).

There are various formulas available for determining R_A. The formula used is dependent on the ground bed installation configuration. In this case, the anodes are installed horizontally at a depth of 2.5 m, the center-to-center spacing will be 4.5 m, and the canned anode diameter is 0.3 m. The equation for determining R_A (derived from Eq. 4-12) in this example is as follows:

$$R_c = \frac{0.001588\rho}{L}\left[\ln\frac{4L^2+4L\sqrt{S^2+L^2}}{DS}+\frac{S}{L}-\frac{\sqrt{S^2+L^2}}{L}-1\right] \qquad (4-17)$$

where

R_A = anode-to-electrolyte resistance, Ω (2.0 Ω)
R_c = structure to electrolyte resistance Ω
ρ = soil resistivity, Ω-cm (5000 Ω cm)
L = length of installation, m (6 anodes × 4.5 m/anode = 27 m)
D = canned anode diameter, m (0.3 m)
S = twice the depth of anode, m (2.5 m × 2 = 5.0 m)
ln = natural logarithm

The structure-to-electrolyte resistance (R_c) is estimated to be .05 Ω after 20 years.

$$R_C = \frac{0.001588}{27} \times 50\left[\ln\frac{4\times27^2+4\times27\,(5^2+27^2)^{0.5}}{0.3\times5}+\frac{5}{27}-\frac{(5^2+27^2)^{0.5}}{27}-1\right]$$

$$= 0.05\ \Omega$$

The cable resistance (R_w) is calculated by determining the length of positive and negative cable to be used, the wire size and wire resistance.

Typical wire resistance are provided in the following table

	Diameter		Turns of Wire		Area		Copper Resistance[6]	
AWG	(in.)	(mm)	(per inch)	(per cm)	(kcmil)	(mm²)	(Ω/km)	(Ω/kFT)
2	0.2576	6.544	3.88	1.53	66.4	33.6	0.5127	0.1563
4	0.2043	5.189	4.89	1.93	41.7	21.2	0.8152	0.2485
6	0.162	4.115	6.17	2.43	26.3	13.3	1.296	0.3951
8	0.1285	3.264	7.78	3.06	16.5	8.37	2.061	0.6282

In this example, 130 m of positive and 50 m of negative (wire size is AWG #4 with resistance) is used. Therefore, R_W equals 0.15 Ω, approx. (180 m × 0.0008152 Ω/m).

Adding $R_A = 2.0\ \Omega$, $R_C = 0.05\ \Omega$, and $R_w = 0.15\ \Omega$, the total resistance (R_T) in this example is found to be 2.2 Ω.

The rectifier voltage can be calculated using Ohm's Law, where $V_{REC} = 2.2\ \Omega \times 8\ A \times 1.2$ (safety factor) = 21 V. The manufacturer's standard size is 24 V.

The easement length is determined as follows. The first anode needs to be remote from the structure, usually at least 100 m away. The installation was determined to require 6 anodes with a spacing of 4.5 m, so the total ground bed length must be 127 m (round to 130 m).

The results of the design are:

Size of rectifier (24 V, 8 A)
Number of anodes required (6)
Size of easement required (12 m × 130 m)
Location of ground bed (km post 20.8)

Before the ground bed can be installed, additional work must be completed:

A Legal survey and land title search must be conducted.
The required easement must be requested.
The necessary electrical power has to be acquired.

To ensure foreign companies, as well as the pipeline company, know where the ground bed is located; legal surveys and registered easements are essential. Their acquisition prevents many problems similar to those which have arisen in the past regarding new facilities being built through existing ground bed facilities.

The rectifier must have power to function. To obtain electric power in the field, take upward of 3 months or more. It is therefore imperative to apply for power as early as possible.

Design Example 2

Given the following information, for two pipelines constructed in parallel in the same trench, determine the CP current coverage requirement for Cases A and B. Also, calculate the number of anodes required for a 20-year protection life and select the appropriate rectifier amperage plus voltage for each case.

Pipeline A: NPS 4, wall thickness 0.237 in., length 4.5 mi.
NPS 12, wall thickness 0.25 in., length 4.5 mi.

Pipeline B: NPS 6, wall thickness 0.156 in., length 150 mi.
NPS 12, wall thickness 0.250 in., length 150 mi.

Assumptions:

Case A:

1. Coating quality = 500,000 Ω ft.^{2} = 1/g
2. Drain point shift potential, $E_0 = 2.0$ V
3. Mid-point shift potential, $E_x = 0.2$ V (1 rectifier)
4. Steel resistivity = 18×10^{-6} Ω/cm
5. Anode type = zinc

Case B:

1. Coating quality = 200,000 Ω ft.^{2}
2. Drain point shift potential, $E_0 = 2.0$ V
3. Mid-point shift potential, $E_v = 0.15$ V

4. Steel resistivity = 18×10^{-6} Ω/cm
5. Anode type = zinc

Solution for NPS 12 and Case A assumptions (NOTE: NPS 12 assume 12" diameter), single line only

a) Coverage

Steel pipe resistivity ρ_s= 18 μΩ/cm = 18×10^{-6} Ω/cm

For one mile $\rho_s = 1 \times 5280 \times 12 \times 2.54$ Ω

Using Eq. (4-4)

$$R_x = \rho_s \frac{L}{A} = \frac{\rho_s L}{\pi D t}$$

$$= 18 \times 10^{-6} \times \frac{1 \times 5280 \times 12 \times 2.54}{3.14 \times 12 \times 0.25 \times 2.54 \times 2.54}$$

$$= 0.0461 \ \Omega/\text{mile}$$

Using Eq. (4-5)

$$R_L = 1/(gAL) \ \Omega/\text{km}$$

But $1/g = 500{,}000 \ \Omega/\text{ft.}^2$ (coating conductivity)

Therefore, $R_L = 500{,}000 \times (1 \times 12/(3.14 \times 12 \times 5280)) = 30.16 \ \Omega/\text{mi}$

Using Eq. (4-7)

$$\alpha = \text{Attenuation factor}$$

$$= \sqrt{\frac{\text{Pipe resistance}}{\text{Coating resistance}}} = \sqrt{\frac{R_s}{R_L}}$$

$$= \sqrt{\frac{0.0461}{30.16}} = \sqrt{0.0015}$$

Using Eq. (4-6)

$$R_0 = \sqrt{R_s \cdot R_L}$$

$$\text{or } R_0^2 = R_s \cdot R_L$$

$$\text{and } \alpha = \sqrt{\frac{R_s}{R_L}} = \sqrt{\frac{R_s^2}{R_L^2}} = \frac{R_0}{R_L} = \frac{R_s I}{2E_A I}$$

where I = current at drain points, A

Using Eq. (4-3)

Voltage potential $E_x = E_0 \, e^{-\alpha x}$ where x = distance to be covered

Assuming voltage potential E_0 at drain point = 2 V, then;

$$\frac{E_x}{E_0} = \frac{0.2}{2} = e^{-\alpha x} = 0.1$$

$$\text{or } \alpha x = e^{0.1} = 2.3$$

$$\text{or } x = \frac{2.3}{\alpha} = \frac{2.3}{\sqrt{0.0015}}$$

= 60 miles in one direction or 120 miles in two directions

b) Rectifier Current Requirement

$$E_0 = IR_0 = I(R_s R_L)^{1/2}$$
$$\text{or } 2 = I\sqrt{0.0461 \times 30.16}$$

Therefore, I = 1.7 A in one direction
or I = 3.4 A in two directions (say 3)

Assuming interference of about 2 to 3 A, then
Total rectifier amperage requirement = 3 + 3 = 6 A

Select amplifier for 8 A; i.e., I = 8 A to allow for valves, interference abatements, etc.

Assuming zinc anode

W_m = anode weight, kg
I_a = anode current output

then using Eq. (4-9a) with anode life of 20 years;

Anode life = 20 = 71.4 × 20/I_a
or I_a = 71.4 mA

c) Number of Anode Required
Using Eq. (4-16),

$n = (I \times Y \times C)/(\text{weight of anode} \times \text{safety factor } (0.6))$
$= (8 \times 20 \times 0.5)/(20 \times 0.6) = 7$ anodes (approx.)

Design Example 3
Estimate resistance of horizontal anode ground bed 30.5-cm diameter, by 15.2 m long in soil of average resistivity = 4000 Ω cm. Assume depth of ground bed = 1.2 m.
Solution:

$$R_h = \frac{1 \times \rho}{2 \times 3.14 \times L}\left[\ln \frac{4L^2 + 4L\sqrt{S^2 + L^2}}{DS} + \frac{S}{L} - \frac{\sqrt{S^2 + L^2}}{L} - 1\right]$$

where
R_h = anode-to-electrolyte resistance, Ω
ρ = soil resistivity, Ω-cm (4000 Ω cm)
L = length of installation, m (1 anode = 15.2 m = 1520 cm)
D = canned anode diameter, m (0.305 m= 30.5 cm)
S = twice the depth of anode, m (1.2 m × 2 = 2.4 m = 240 cm)
ln = natural logarithm

Therefore:

$$R_h = \frac{0.1588}{1520} \times 4000\left[\ln \frac{4 \times 1520^2 + 4 \times 1520(240^2 + 1520^2)^{0.5}}{30.5 \times 240} + \frac{240}{1520} - \frac{(240^2 + 1520^2)^{0.5}}{1520} - 1\right]$$

$= 7.840 \qquad = 0.1579 \qquad = 1.012$

i.e., R_h= (0.1588 × 4000/1520) × (7.840 + 0.1579 – 1.012 – 1) = 2.5

Design Example 4

A gas utility will be installing 3049 m of 5.1-cm distribution line in a soil with average resistivity of 5000 Ω cm. Assuming a coating leakage conductance (g) of 2.14×10^{-3} Siemens/m^2 in a 1000 Ω cm soil and average potential shift of 0.5 V (measured to remote earth), calculate current to shift pipe potential for CP.

Solution:

$$R_L = \frac{1}{gAL}\text{(ohms)}$$

$$g = \mu\Omega/\text{m}^2\text{ (Siemens)}$$

1. Total surface area = (3.416 × 5.1/100) × 3049 = 488 m^2
2. Estimated leakage conductance (g) in 5000 Ω-cm soil will be

$$g = 2.14 \times 10^{-3} \times (1000/5000) = 0.428 \times 10^{-3}\text{ Siemens.}$$

Note $1/g = \Omega$

Total pipe resistance = $1/(g \times \text{surface area}) = 1/(0.428 \times 10^{-3} \times 488) = 4.8\ \Omega$, Therefore, estimated current to shift = $V/R = 0.25/4.8 = 0.052$ A

Design Example 5

A 120.7 km of NPS 16 (406.5 mm OD) lateral pipeline is to be laid between a mainline gas pipeline and a city gate to transport gas for city distribution. Isolation joints are to be installed at both ends of the pipeline. The lateral wall thickness will be 7.95 mm.

Assuming a soil resistivity of 4000 Ω cm and steel resistivity ρ' of $20.6 \times 10^{-6}\ \Omega$ cm, establish whether coverage can be provided with drain point shift of 2 V and mid point shift of 0.2 V (1 rectifier).

Assume a quality polyethylene coating with a leakage conductance (g) of 1×10^{-4} Siemens (in 1000 Ω cm soil).

Solution:

1. Steel resistance per km $R_s = \rho L/A$

$$R_s = 20.6 \times 10^{-6} \times (100{,}000)/[3.14(40.16^2 - 39.06^2)/4]$$
$$= 0.0207\ \Omega\text{ (per pipeline km)}$$

2. For coating

$$R_L = \frac{1}{gAL}\text{(ohms)}$$

$$g = 1/1 \times 10^{-4}\text{ Siemens in 1000 }\Omega\text{ cm soil,}$$
$$\text{or } g = 1/1 \times 10^{-4} \times (1000/4000)\text{ in 4000 }\Omega\text{-cm soil}$$

$$AL = 3.14 \times (40.65/100)1000 = 1276.41\text{ m}^2\text{/km of pipe length}$$

Therefore, $R_L = 1/[1 \times 10^{-4} \times (1000/4000) \times 1276.41] = 31.37\ \Omega/\text{km}$

3. Attenuation constant = $(R_s/R_L)^{0.5} = (0.0207/31.37)^{0.5} = 0.0257$

$$0.2 = 2 \times e^{-(0.0257X)}$$

Therefore, $0.1 = e^{-(0.0257X)}$ or $0.0257X = 2.3$, or $X = 2.3/0.0257 = 89$ km in one direction.

That is, with the rectifier in the middle, coverage can be provided for the pipeline.

CP Maintenance

CP does not necessarily eliminate corrosion. It does, however, remove corrosion from the structure being protected and concentrates the corrosion at another known location. At this known location, the current discharging structure (ground bed) can be designed for a reasonably long life. It is also easily tested and replaced at the end of its useful life without endangering the pipeline system being protected.

Maintaining and Fixing

Ground beds must be maintained and repaired as required. Shallow ground beds in soils with high resistivities, or which dry out during summer months, can be watered to increase their current output. This is commonly done by installing weeping tile (i.e., a 4-in. diameter plastic corrugated pipe with holes) directly over the anodes at the time of construction.

When ground beds are not working, they must be repaired immediately to minimize corrosion to the structure they are protecting. Common problems include broken rectifiers due to a failed stack or diode bridge, Thermoelectric Generator (TEG) not working due to gas supply freezing off at the regulator, ground beds not discharging current due to a nick in the header cable (which will quickly corrode). Broken test leads should be replaced and new test leads added where new foreign pipelines cross the protected facility.

Replacing the Ground Bed

The ground bed will normally be replaced when it no longer functions properly because the anodes have corroded away. Various calculation methods can be used to predict ground bed failure based on the current output and the time the ground bed has operated; however, this method only *predicts* failure. By definition, a ground bed fails when it no longer delivers the current (i.e., amps) to produce a structure-to-soil potential.

Typically, the voltage and amperage of a ground bed will be constant over the course of the ground bed's lifetime, assuming no substantial changes in soil conditions and/or addition of new structures. As the ground bed begins to deteriorate, the voltage will have to be raised in order to drive the same amount of current. With time, the amount of current discharged by the ground bed will decline.

Pipe/Soil Meters/Test Leads

Pipe/soil meters are another monthly recording system that monitor pipe soil potential (PSP).

Test leads (Figures 4.23–4.26) provide a quick and easy metallic connection for measuring potentials on buried structures such as pipelines, casings, and flanges. They also provide a location where the CP current can be measured and controlled if desired. At pipeline crossings, they provide a metallic connection to both pipelines to measure potentials or existing interfering currents if they exist.

Notes: 1. Terminal post as per drawing.
2. Use 2 black wires - +10 Str. Copper TWU
3. Thermite welds only to be used to attach leads to mainlines as follows:

Wall Thickness (mm)	Max. Line Press (kPa)
<2.8	Mechanical clamps only
2.8	Zero - 3445
3.2	4135
3.6	5170
4.0 and above	>5170 MOP

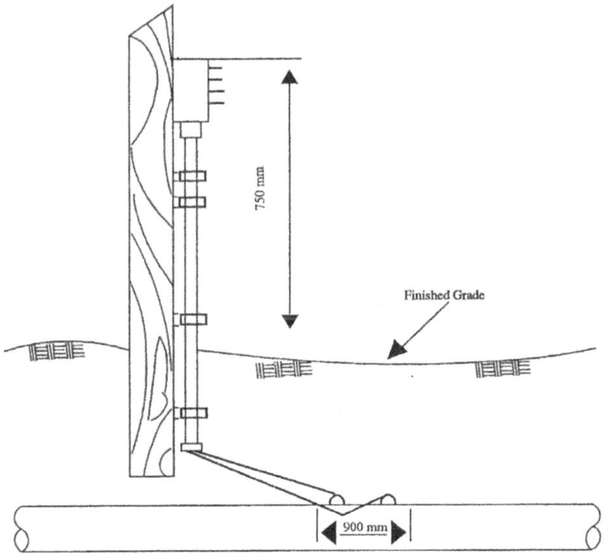

Figure 4-23. Typical test leads installations for mainlines

4. Coat thermite weld bead with compatible mastic costing and cover with a thermite weld cap.
5. Install terminal post at downstream road fence, cut line, etc.
6. If cannot reduce line pressure to specified limits prior to thermite welding, then only use mechanical clamps, see Figure 4.24.

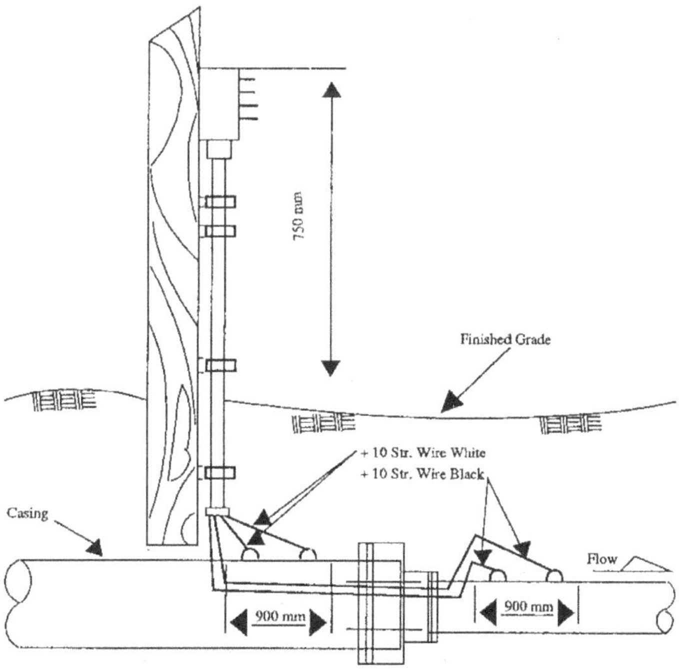

Figure 4-24. Typical test leads installations for cased crossings

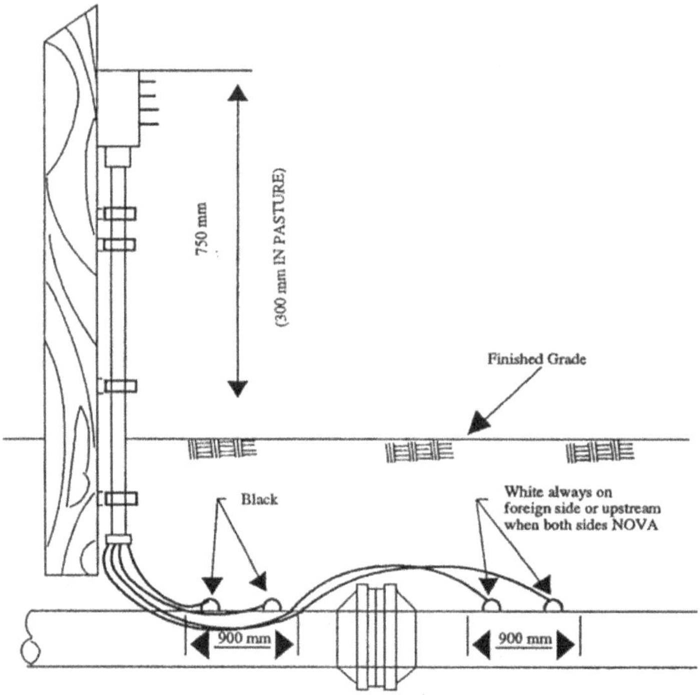

Figure 4-25. Typical test leads installations at insulation flanges

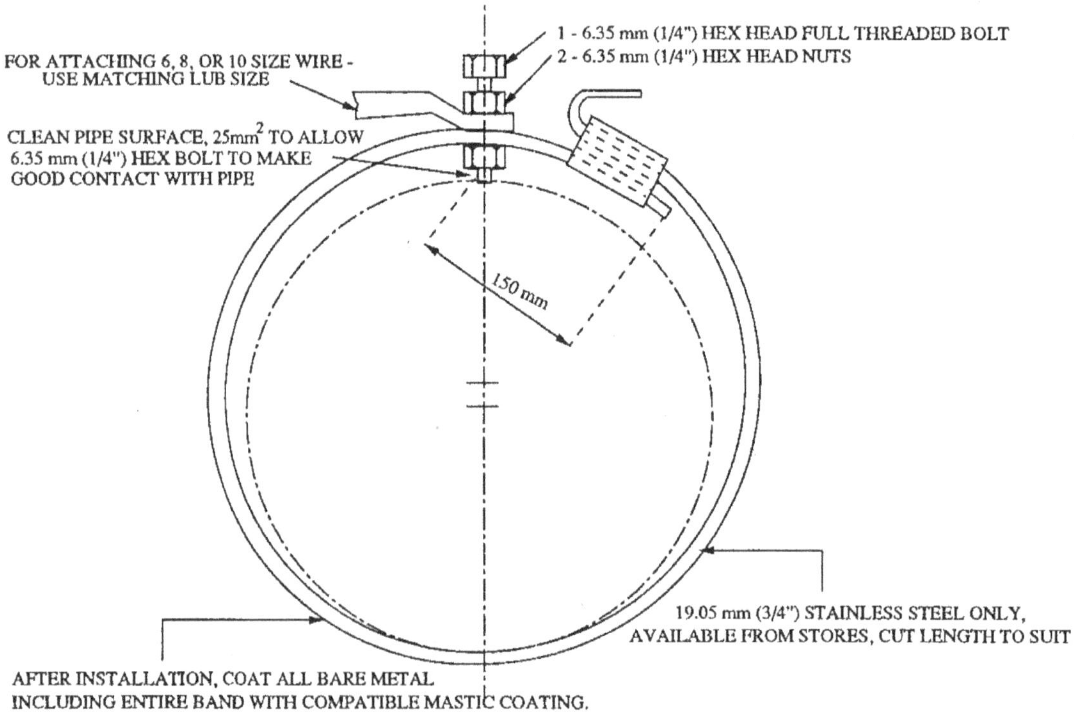

Figure 4-26. Typical mechanical clamp to attach test leads to pipelines

Operational Aspects CP Systems

Monitoring

CP facilities will require monitoring to ensure that CP is being achieved. Different companies will monitor different parameters with different frequencies. The following is a list of parameters that can be monitored along with typical monitoring frequencies.

CP Parameter	Typical Frequency
Rectifier/TEG volts and amps	Monthly
Structure-to-soil on potential	Annual
Structure-to-soil off potential	Annual
Structure-to-soil rest potential	Annual or as required
Close interval survey	Every 5 years or as required
Line current survey	As required
Soil resistivity survey	As required
Soil pH survey	As required
Concentration of sulfate reducing bacteria in soil	As required
Stray currents	As required
Bond current	Monthly (usually 2½ months) to annually
Leaks	As they occur

For new pipeline systems, a pipe to soil (PSP) survey is usually conducted after the CP system is energized. This initial survey should be performed within 18 months of the pipeline system installation.

The structure-to-electrolyte *on potential* is the best and easiest way to confirm that CP is being achieved. Care must be taken to ensure that this reading takes into account IR drop through the electrolyte. A structure-to-electrolyte *off potential* is required to satisfy the −850 mV to −1 V off criteria. A structure-to-electrolyte *rest potential* is required in addition to the off potential to satisfy the 100-mV polarization shifts. In this case, all rectifiers are turned off, resistance bonds disconnected, and the structure allowed to depolarize to its rest potential(s) before readings are collected.

The following electrical measurement are taken when conducting a periodic survey:

1. PSP s at test stations
2. Structure-to-soil and PSP s at compressor stations, meter stations, terminals and storage facilities
3. Impressed current rectifier output
4. Close interval PSP
5. Checks for carrier pipe to casing shorts
6. Isolation fitting and bond checks
7. Induced AC potentials and current levels in power line corridors
8. Telluric currents
9. Pipeline currents

A *close interval survey* is used to collect structure-to-electrolyte potentials at intervals shorter than the test lead spacing, and typically every 1m to 10m. The on, off, and rest potentials will all vary along the length of a pipeline or in various locations in a plant. These differences are due to many factors including: proximity to ground bed, soil chemistry and moisture content, presence of sulfate reducing bacteria, coating type and condition, interference from foreign structures, and depth of buried structure. A close interval survey can highlight regions of possible corrosion, which test leads may not show. A close interval survey is highly recommended, even if done infrequently.

A *line current survey* is used to measure how much current is flowing through a metallic structure. This survey is often used for bare or poorly coated pipelines to find regions of active corrosion. The line current survey plots the direction the current is flowing on the pipeline and its magnitude. Since corrosion occurs where current leaves the structure, it may occur where current flow in a pipeline flows to the same point (the point of possible corrosion) from both directions. Figure 4-19 illustrates where corrosion may occur on a pipeline. The dots represent the amount of current flowing on the pipeline while the arrows indicate the flow direction.

Frequency, Spacing of Readings, and Trending. Common industry practice is to monitor the effectiveness of the CP systems by completing PSP surveys at least once per year. If the PSP data on a pipeline can be taken each quarter, it will give a good indication of the seasonal and operational changes in protection on the effectiveness of the CP for the pipeline.

If the spacing of the test leads is relatively close, it gives a good representation of the protection levels; however, it may be noted that the readings only represent the protection at the location where the readings are taken.

The industry generally plots pipe to soil data to assess any trending in pipeline protection system. A typical trending plot is shown in Figure 4-27.

From this typical example, it can be seen that there are several areas that protection does not meet the –850-mV ON criteria. These areas are located roughly half way between rectifier locations. Since the readings dropped substantially in a relatively short distance from the rectifiers, the cause for the low readings was most likely coating degradation.

From the trending data, it can be seen that there are some seasonal changes as noted in the protection levels. This is often caused by change in ground bed moisture content and moisture around the pipeline.

Soil resistivity alone does not indicate soil corrosivity. Nevertheless, there is some relationship between the two properties as shown in Table 4-12 below:

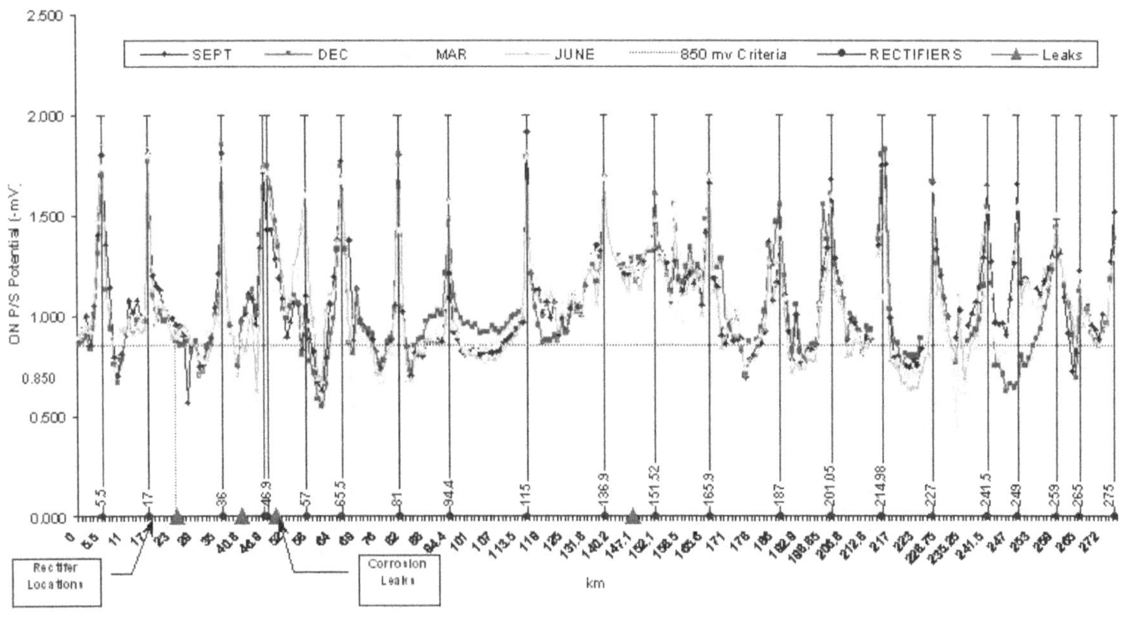

Figure 4-27. Typical CP trending plot

TABLE 4-12. Relationship between soil resistivity and corrosivity

Soil Resistivity (W cm)	Soil Corrosivity
<1,500	Very corrosive
1,500–3,500	Moderately corrosive
>3,500	Slightly corrosive

Some relatively high-resistivity soil may be very corrosive, e.g., acidic peaty soil. Corrosion can also occur in fairly high-resistivity soil if there is a considerable variation in resistivity at different points along the pipeline route, causing concentration cell effects.

However, soil resistivity is one of the best indicators of electrolyte corrosivity. Soil resistivity can be taken at intervals along a pipeline route and used in conjunction with other data to build a picture of possible locations of active corrosion. Such data, when considered together, can determine likely corrosion sites for excavation or further investigation. Typical Soil resistivities are shown in Table 4-13.

ELECTRICAL INTERFERENCE EFFECTS

Stray Currents

The stray currents are returning currents from large direct current systems employing an earth or with an un-insulated return path, such as a railway, tramway, welding generator, etc.; interruptions in the return path may multiply locally the strength of the stray currents.

An example where a stray current interference can occur is shown in Figure 4-28.

The most satisfactory method of dealing with stray currents would be to eliminate them at the source by suitable insulation or other means, but this is usually impossible and it is then necessary to allow for the interference of stray currents when designing the CP system.

Only general information can be given here on the design of CP systems to guard against stray currents, as conditions vary so widely.

The main consideration is that the final steel-to-soil potential after the system is installed should be as uniform as possible over the whole of the protected structure or pipeline. To achieve this, extra anodes (for those systems protected by galvanic anodes) must generally be provided in the vicinity of the points where the stray currents leave the steel to re-enter the soil. Part of the stray current effect can often be overcome by connecting the structure, via a resistor and a rectifier, to the negative terminal of the DC source, which produces the stray currents. This will make CP much easier to achieve, as it considerably reduces effects of peaks in the stray current.

TABLE 4-13. Typical soil resistivities

Soil Type	Resistivity Range (Ω cm)
Clay	200–15,000
Sand and gravel	5,000–100,000
Clay mixed with sand gravel	1,000–135,000
Marsh	200–400
Shale	1,000–50,000
Limestone	500–400,000
Rock	50,000–1,000,000
Sand saturated with sea water	100–200

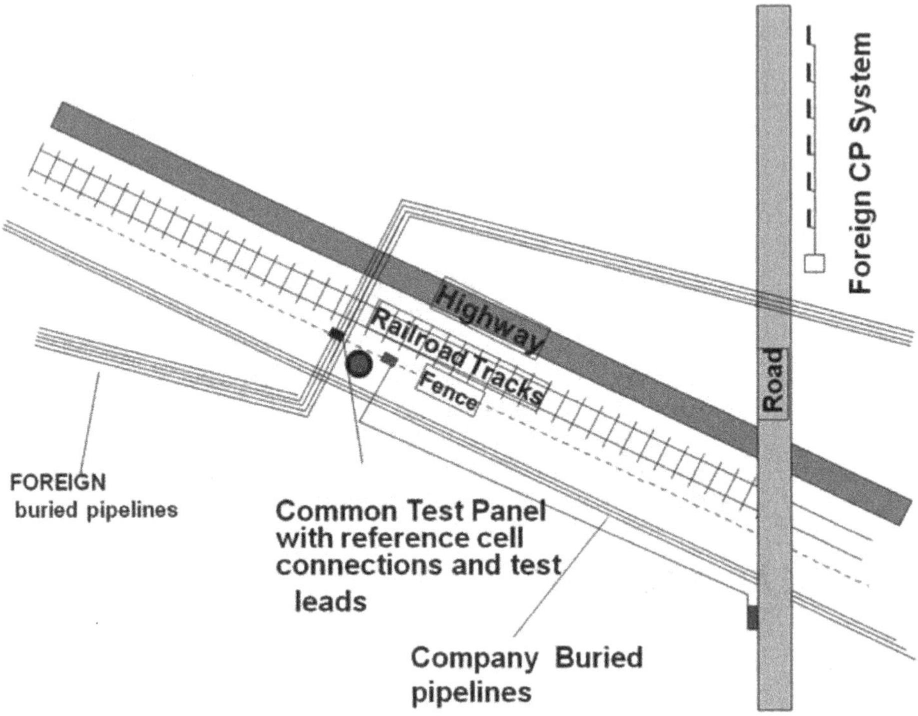

Figure 4-28. Typical sources of stray current interference at pipeline crossings

Effects from Other CP Systems

Electrical interference effects may arise from CP systems themselves, especially with impressed-current systems. Thus, the current applied to a protected structure or pipeline may be picked up by a neighboring and unprotected one, for example, where a protected pipeline crosses an unprotected one as shown in Figure 4-29. The unprotected pipeline becomes cathodic at the point where the current is picked up (A), and anodic at the point where the current eventually leaves the line (B), resulting in protection at "A" and corrosion at "B." Point B becomes anodic, so that the PSP measured with a copper/copper sulfate electrode becomes more positive when the rectifier is switched on. A positive change in potential of more than 20 mV when the rectifier is switched "off" and "on" indicates an objectionable degree of stray current corrosion.

Although electrical interference effects may be caused by rectifier installations, these effects are negligible with sacrificial-anode systems, as these anodes are placed much nearer to the protected structure than to any unprotected steelwork and because their driving voltage is usually much lower than that of impressed-current system ground beds.

Remedial Measures

As regards interference effects arising from the currents intentionally put into the earth by protection systems themselves, it must be ensured that neighboring structures are not adversely affected.

Sometimes interference can be prevented by bonding the neighboring structure or pipeline to the protected one, so that electrically the two become a single unit, and then applying CP to the combined unit. This is the recommended method for use where structures

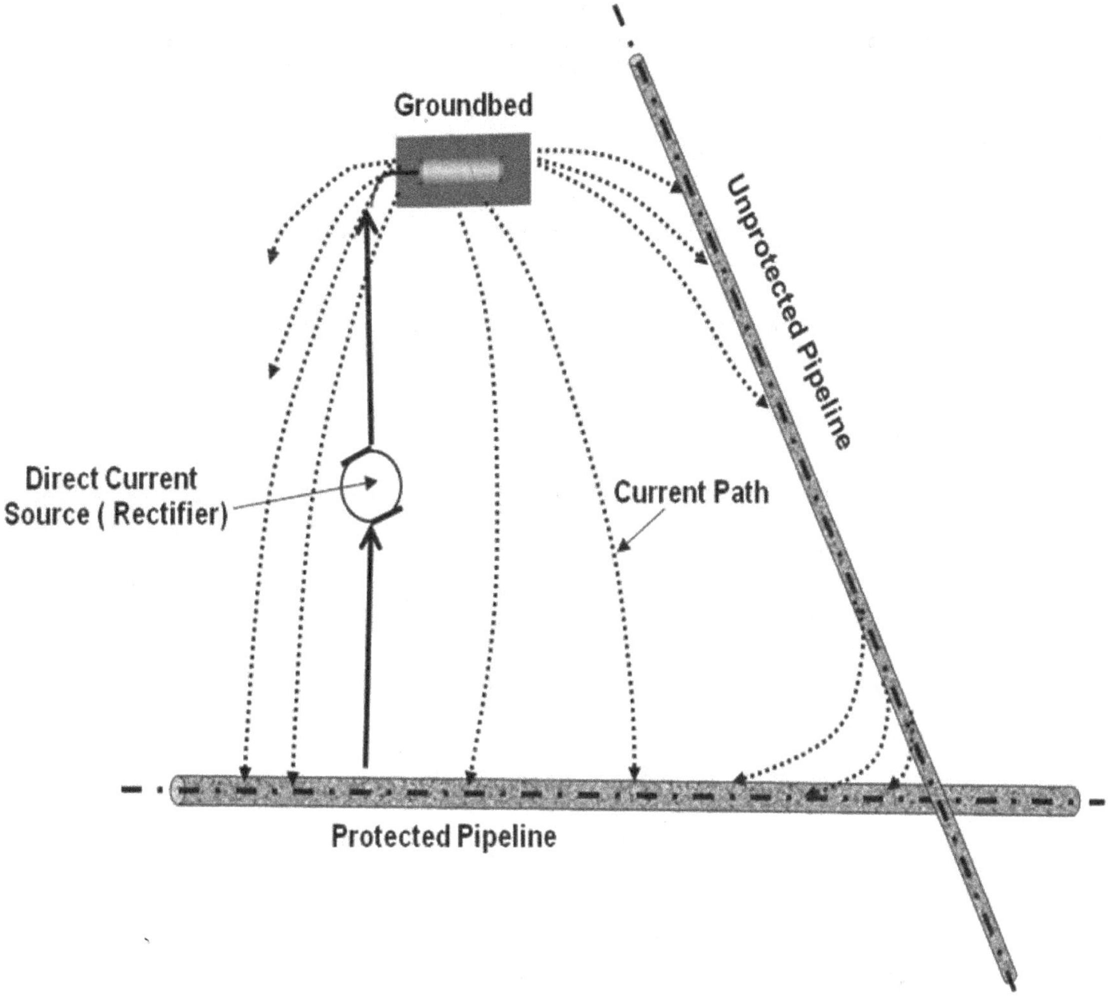

Figure 4-29. Current pick up by neighboring structures from CP system

are close together, as for example a pipeline, which is lying in a common right of way with neighboring pipelines. For a combined system of this type, a decision should be reached with other owners or operators before the CP system is designed and installed.

Many interference problems may have to be met by installing resistance/critical bonds rather than by making a combined CP system. The first step is to determine whether the second structure is adversely affected. This is done by noting if any change occurs in the steel-to-soil potential of the unprotected structure when the CP current is switched "off" and "on." Measurements should be taken where the current may be expected to leave the unprotected structure, which is the place where the structures are nearest to each other. If the steel-to-soil potential becomes more positive, i.e., anodic, by 20 mV or more when the CP is switched "on," the unprotected structure is liable to increased corrosion as a result of the operation of the protection system.

The application of resistance/critical bonds to eliminate corrosion by interference is based upon the principle that the current collected by the unprotected pipeline should be conducted via a metallic conductor to the protected structure. The electrical resistance of this bond is made high enough to prevent too much current from flowing to the unprotected structure. The resistance bonds are installed where the structures are nearest to each other and controlled so that no current leaves the unprotected structure to earth; this is checked by measuring the PSP with the rectifier current "off" and "on."

GUIDELINE FOR ANNUAL CP SURVEY/INSPECTION

The annual CP survey is generally undertaken by the industry in the evaluation of pipeline external corrosion. The following describes a general description of equipment and procedure necessary for performing such an Annual CP Survey/Inspection (Structure-to-Soil Potential) that generally meets the regulatory requirements. It also provides a brief description of the test equipment needed for the survey tasks.

Test Stations

In order to help inspect and maintain the integrity of the pipeline, test points are installed at approximately 2-km to 3-km intervals. In addition to the standard test stations (Figure 4-30), foreign line test stations, isolation flange (IF) test stations, casing test stations, block valves, permanent RCs, transitions and above-ground storage tank perimeters are also used. Test stations provide the same purpose of providing a means to monitor the CP system.

Cased Crossing Test Station

Cased crossing test stations are typically located at road crossings, river crossings and railroad crossings. Usually, one or two test leads from the casing and the carrier pipe are brought up into an above-ground test station. This test station helps determine if the casing

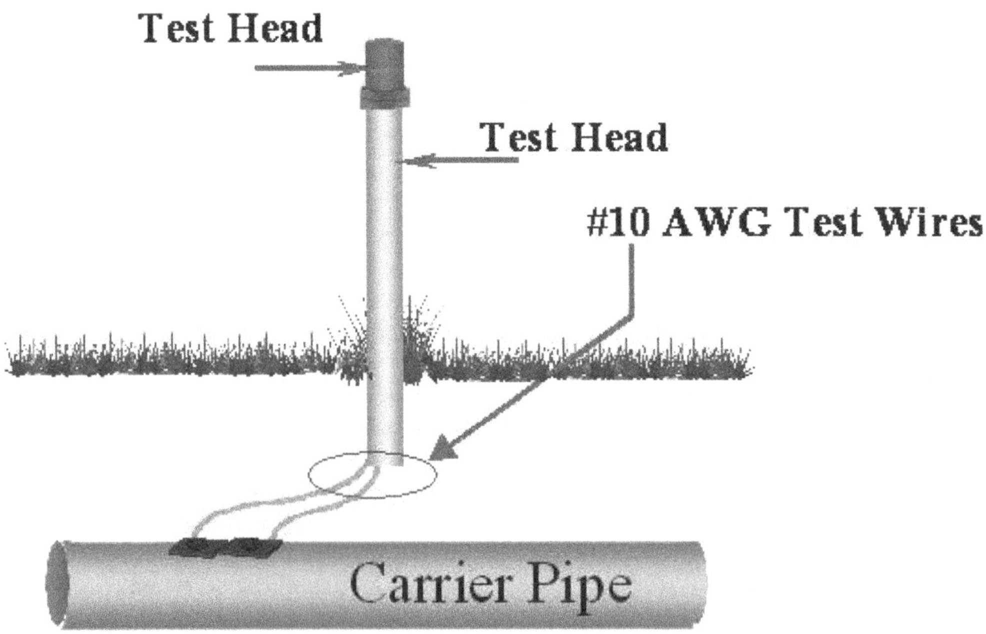

Figure 4-30. Typical pipeline test station

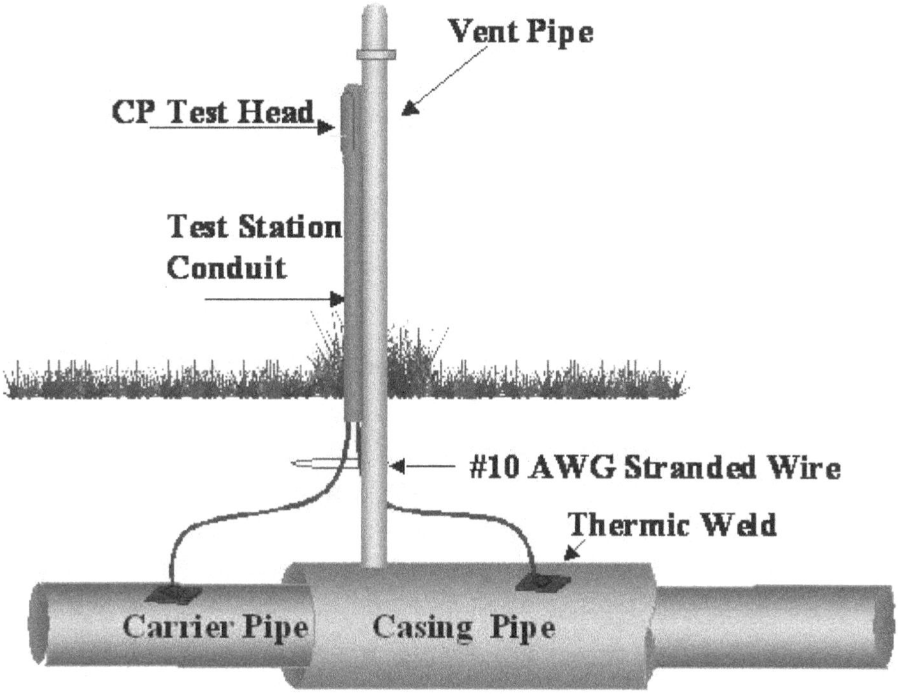

Figure 4-31. Typical cased crossing test station

and the carrier pipe are isolated from each other. A pipe-to-soil (P/S) potential for the carrier pipe and the casing-to-soil (C/S) potential need to be taken at this location. Figure 4-31 shows an example of a cased crossing test station.

Test Stations — Uncased Pipelines

Pipeline test stations are typically located at most crossings without casings, property lines, and at fence rows. Usually, one or two test leads from the carrier pipe have been brought up into an above-ground test station. This test station provides a connection point to the pipeline to help determine the levels of CP on the pipeline. A pipe to potentials are taken at these locations. The test stations are similar to that of mainline test stations (Figure 4-29).

Foreign Line Crossing Test Stations

Foreign line crossing test stations are typically located at most foreign pipeline crossings. Usually, one or two test leads from the owner company and the foreign facility are brought up into an above-ground test station, and in some locations, a buried permanent RC has been installed at the crossing. The test station provides a connection point to the pipeline that helps determine the levels of CP at the pipeline crossing. In addition, these test stations are sometimes used to bond the pipelines together. Bonding the two pipelines helps mitigate an interference condition that would probably damage one or even both facilities. When the pipelines have been bonded together, a shunt is usually installed to measure the current return. A P/S potential for each pipeline in the test station is taken at this location. Figure 4-32 below is an example of a typical foreign line crossing test station.

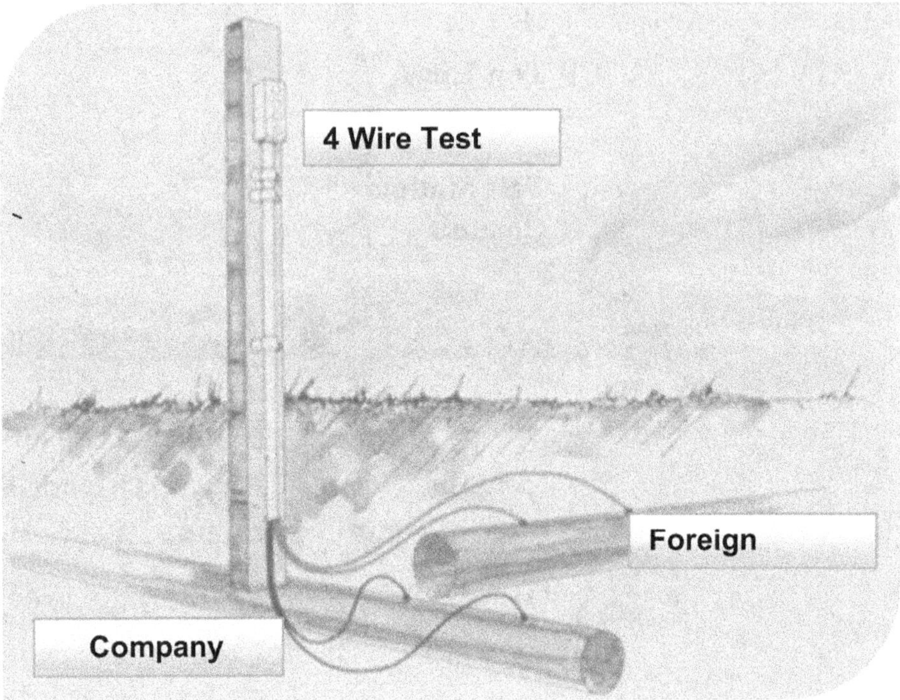

Figure 4-32. Typical foreign pipeline crossing test station

IF Test Station

Insulating flange test stations (Figure 4-33) are typically installed at block valves, compression and pumping stations, and measurement terminals. Two test wires from each side of the IF kit are usually brought above grade and inside a test station. Some of these locations may even have a shunt installed between the two sides to short the pipeline around the IF kit. P/S potentials are typically measured on each side of the flange. If a shunt has been installed, then bond current is measured at the test station.

CP Testing Equipment

Essentially, three pieces of test equipment (Figure 4-34) are required for a CP system when performing an annual survey. They are:

- Portable voltmeter
- Test leads
- Portable RC or electrode

Multimeter or Portable Voltmeter

The multimeter is one of the most used pieces of equipment in monitoring and inspecting the existing levels of CP. Multimeters generally require to have a high input impedance to eliminate errors in the data and can be used to measure the following:

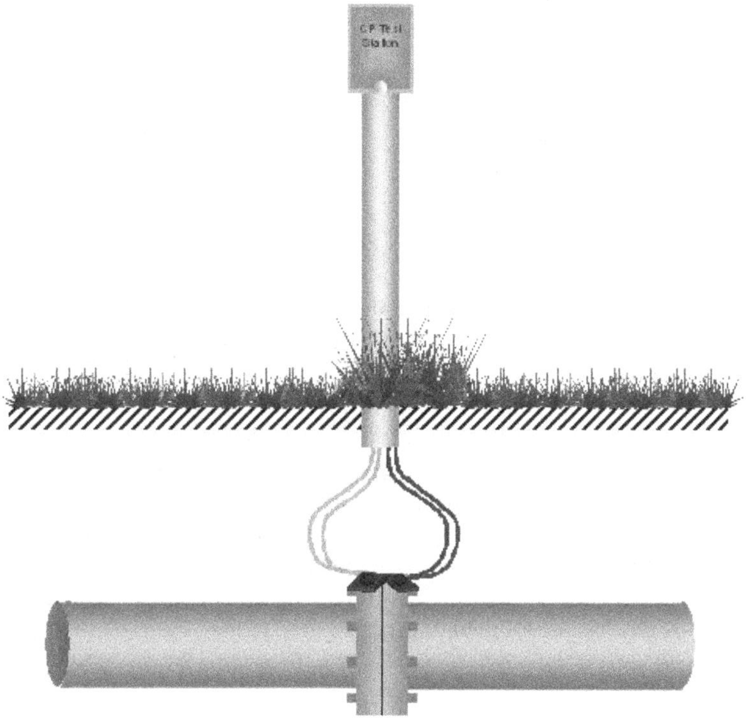

Figure 4-33. Typical IF test station

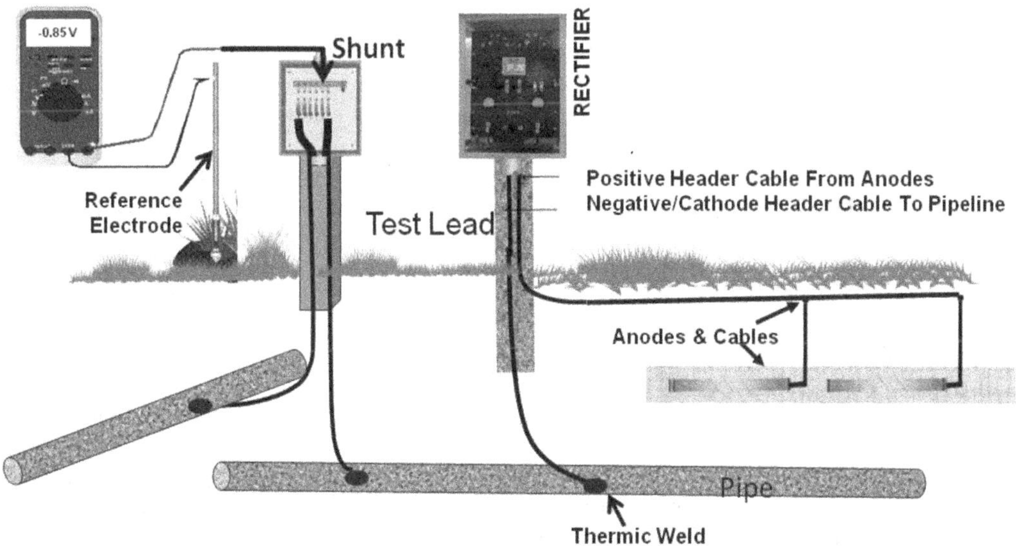

Figure 4-34. A CP system and test equipment

- AC volts
- DC volts
- DC millivolts
- Continuity
- Amps/milliamps
- Microamps

Multi-meters should be calibrated annually. Records of calibration are required to be kept for verification purposes. Test leads are an often over looked accessory to the multimeter. They should be inspected frequently and replaced when damaged. Physical damage, repetitive or 0.00 readings, are signs that the test leads will need replacing.

When conducting the annual survey, DC volts and millivolts are the primary settings used. Pipe-to-electrolyte potentials (P/S's) on company or foreign pipelines, casing-to-electrolyte potentials (C/S's) and rectifier voltage readings will require measuring DC voltage.

The DC millivolt range is generally used when measuring bond current, IF current, rectifier current output, and anode currents. It may be noted all currents are calculated by multiplying the millivolt drop across the shunt by that shunt's factor. A multimeter generally has the following features

- Display
- Diodes (protection and polarity)
- Ohms/residuals Functions

Reference Electrodes/Cells or Half Cells

CP testing depends largely on the proper use and interpretation of pipe-to-electrolyte measurements using either permanent or portable copper/copper-sulfate ($Cu/CuSO_4$) RC.

Precaution is necessary for the RC not to become contaminated, lost, or broken while in the field. As a result, two RCs are generally in use at all times. An RC contaminated from the environment can be cleaned and refilled or replaced. The copper rod inside the RC can be cleaned with a nonmetallic abrasive cloth. When refilling the RC, distilled water or a special antifreeze solution may be used along with copper sulfate crystals. Copper sulfate crystals need always be present in the solution. This ensures that a supersaturated solution is present in the RC.

Periodically, the cells need to be recharged due to evaporation of solution. The solution in the porous plug may also freeze in sub-zero temperatures even with antifreeze, which causes the cell to leak after thawing out. A typical $Cu/CuSO_4$ reference electrode is shown in Figure 4-35.

Performing CP Survey

An annual CP survey is generally carried out once every 12 months rarely exceeding 15 months. If an adequate levels of CP is not maintained on external pipeline system (i.e., minimum −850 mV) and monitoring and cleaning of the internal pipeline system is not carried out or chemical treatment program is not provided, it is likely that the pipeline will eventually have a corrosion related issue and/or failure. A failure because of unreliable data collection and analysis is not uncommon and must be avoided. Otherwise, the resulting costs could be too high.

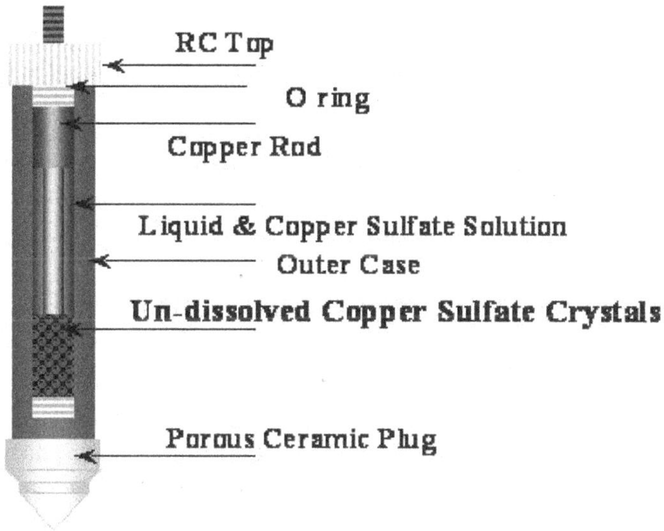

Figure 4-35. Typical reference electrode/cell

An annual survey will mostly require recording and documenting P/S potentials at about one 2-km to 3-km intervals along the pipeline. Also in urban areas, such as highly populated areas, gas plants, terminals etc., measurement of P/S potential is taken at closer intervals. The distance decreases as congestion, either people or other pipelines, increases. It is also usual to conduct potential measurements at every cased crossing and bond sites. In addition, all rectifiers will require to be checked prior to performing the survey to ensure that all units are functioning correctly.

PSP readings are necessary to be taken at foreign pipeline crossings and facilities to evaluate how other systems impact the owner company's pipeline. Typically, records of P/S potentials at each foreign line crossing that has an impressed current CP system whether such a system is bond with the own pipeline or not is taken. If a bond does exist, then the bond current needs also to be recorded.

Where IFs are installed a PSP on each side of the flange along with the current returned at each bonded flange is also necessary to recorded. In addition to bond currents, anode currents are also recorded at all locations where the individual anode cables are brought into a junction/shunt box. Similar procedures are generally used for reading critical bonds and rectifier current outputs.

CIPS Technique

The technique entails the measurement of the pipe or structure/electrolyte reference electrode potential (ON and instant OFF) at close intervals along the route of the selected pipeline section. Typically, P/S potential measurements are recorded at between 1-m and 2-m intervals along the pipeline route. The potential measurements are generally recorded using a CP measurement data logger from which a graphical profile of the pipe/soil potential data could be produced following the completion of the data collection survey.

The accuracy of potential measurements recorded is usually determined, in part, by the location of the reference electrode relative to the protected pipe/structure. Usually, a reference electrode is placed as close to the crown of the pipeline as reasonably possible while avoiding vegetation growth and other obstacles.

DCVG

This technique measures voltage gradients caused by the CP current flow through the soil to its length (without any additional or increased impressed current, which is sometimes employed in analogue DCVG measurements). Greater current density usually occurs at coating defects that expose bare steel with resulting increased voltage gradients in the soil. With this technique, the locations of defects in the external coating can be identified by peaks in the soil voltage gradient (Webb et al. 2006).

GUIDELINES FOR RECTIFIERS AND CRITICAL BONDS

A pipeline system can have many rectifiers, test points, and critical bonds installed along its length, at compression, measurement terminals, and pumping stations. As indicated previously, rectifier inspections are required to be undertaken monthly (not exceeding 6 weeks) and critical bond readings are usually taken at 2½ months interval to meet regulatory requirements.

Rectifiers have a singular purpose, to convert AC to DC. The converted DC voltage then powers and drives the anodes installed for that rectifier/ground bed.

As indicated previously, there are three types of rectifier units that are utilized by the industry to provide power to CP systems. These units include *Constant Voltage, Constant Current, and Constant Potential* units. In general, all rectifiers will have all of the same functional components in one form or another. Occasionally, some components may be constructed differently but have the same function (i.e., silicon diodes, selenium stacks or control cards). Units that are typically used in the industry can include tap adjusted air cooled or oil cooled rectifiers (Figure 4-36).

Standard/Tap-adjusted Air-cooled Units

Tap-adjusted air-cooled rectifiers, are the most common type of rectifier used by the pipeline industry. The rectifier outputs will vary for each unit depending on the size of transformer and the type of ground bed installed at that unit. Most of the components for the

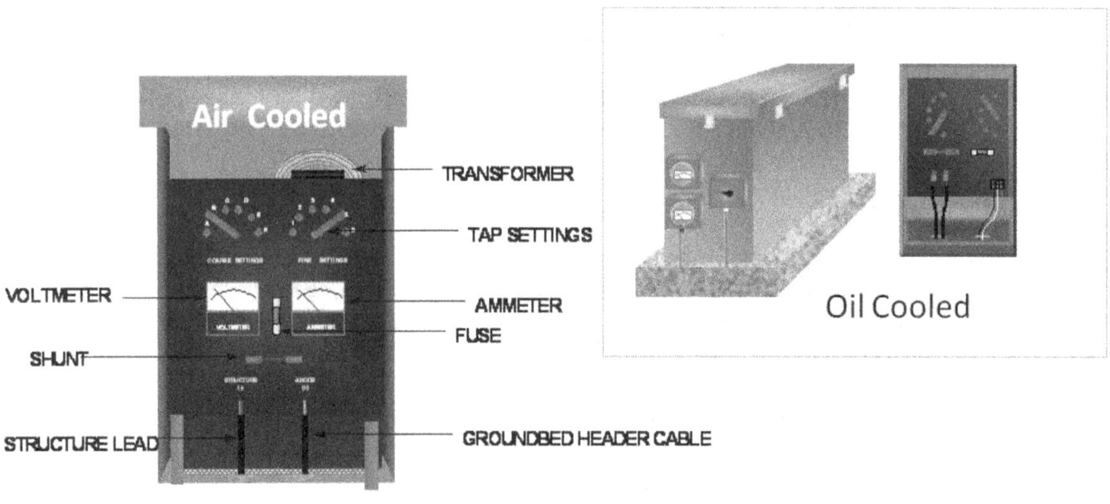

Figure 4-36. Air-cooled and oil-cooled transformer — rectifiers for pipeline cp applications

CPUs will be the same. CPUs are typically supplied with 120 VAC. However, the units may have 240 or 480 VAC.

Standard/Tap-adjusted Oil-immersed/Cooled Rectifier

Tap-adjusted oil-cooled units have generally components that are similar to air-cooled units except that all of the components are immersed in transformer oil. These units are typically used at pipeline compression, pumping/measurement stations, and terminals.

The majority of these rectifiers are designed for use in hazardous locations and hence to Class I Division 1 (of Electrical code). The rectifier outputs will vary for each unit depending on the size of transformer and the type of ground bed installed at that unit. In addition to eliminating or reducing the rectifiers' ability to produce a spark, the transformer oil helps to dissipate the heat of the rectifier through the sides of the unit and helps to keep the components clean.

Rectifiers Inspection/Maintenance Requirements

Regulatory requirements (including US DOT CFR Parts 192 and 195) require rectifiers be inspected once a month not exceeding 2½ months, but at least six times each calendar year to determine that the units are functioning correctly. The inspection includes visual as well as measurement readings.

Visual Inspection is generally undertaken to assess the following:

- Loose or disconnected wiring
- Obstructions for the ventilation of air-cooled rectifiers (bird or insect nests, manuals, paperwork, etc.)
- Discoloration of transformer oil (color and opaqueness)
- Blown fuses and lightening arrestors
- Discoloration or burnt areas inside rectifier cabinet (wiring, tap settings, circuit boards, etc.)
- A burnt or acrid smell when opening rectifier
- Physical damage to a Rectifier Unit (RU) cabinet
- Blistering or physical damage to the selenium stacks
- Similar readings between the rectifier's internal meters and the inspector's portable voltmeter

Typical inspection will include the following:

- Date of inspection
- Name of pipeline
- Line segment
- Watt-hour meter reading
- RU number
- Volts and amperes
- Tap or card controller settings (as applicable)
- P/S potential at the closest test point to the rectifier
- Initials of inspector

Meter Readings

After a rectifier has been visually inspected, the rectifier's output (volts and amperes) are also measured (using a multi-meter) and recorded on the most current RU inspection record

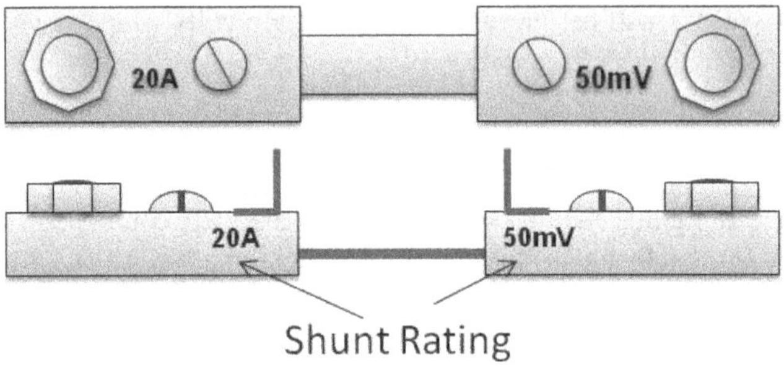

Figure 4-37. Typical rectifier shunt (Holloway type)

form. The multi-meter reading is used to verify the rectifier's meter readings and/or to make adjustments to the output.

Shunts

Shunts are extensively used to measures the voltage (usually in millivolts) drop across a known resistance. A typical shunt (Figure 4-37) is made of copper and comes in different sizes depending on the voltage/amperage requirements. The shunt's ampere and millivolt ratings are typically engraved on the shunt itself. The ratings will either be engraved on the top of or on the side of the shunt body.

Converting mVDC to Amperes

The conversion of mVDC from the portable voltmeter to amperes is dependent on the rectifier's shunt factor, which is dependent on the shut size. A shunt's resistance factor can be calculated by the following equation:

Shunt Factor (SF) = Shunt's Amp. (SA) divided by the Shunt's mV (SmV)

Example: Using the shunt in Figure 4-37, SF = SA ÷ SmV or SF = SA/SmV; therefore SF = 20 ÷ 50 = 0.4.

The millivolt reading from the portable voltmeter = 20 mVDC
To find the rectifier current/ampere reading, multiply the mVDC
from the voltmeter by the SF (i.e., amps. = 20 × 0.4)

Therefore, rectifier current = 8.0 A (record this in the amps. section of the RU Log Book).

Shunt Types

Shunt types include the Cott Shunt of varying resistance and SF (typically 0.001 Ω, 0.01 Ω, and 1 Ω, with respective SF of 1, 0.1, and 0.01), Holloway SS and RS shunts (SF 1 and 0.1, respectively) (Figure 4-38) shows a some of the shunts that are used by the industry:

Critical Bonds

Critical bonds can be found in numerous types of pipeline CP facilities and enclosures (i.e., Insulating flanges, junction boxes, test stations, steel enclosures, etc.). Regardless of the shunt type, every critical bond should be read and recorded in the same fashion.

Critical bonds can usually be found all along the pipeline. The most common type of critical bond is found at foreign pipeline crossings. Foreign pipelines companies have CP systems installed to protect their pipelines from external corrosion. Bonding is required

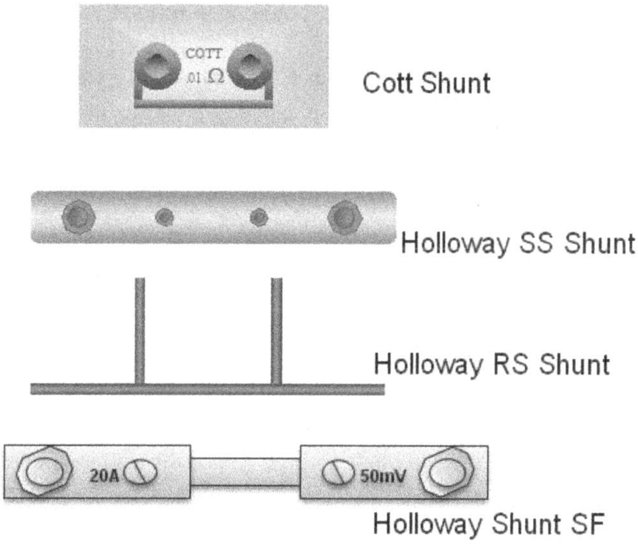

Figure 4-38. Typical shunts

when others cross these foreign lines to assure the integrity of pipeline systems. The bonds provide a metallic path for current return to the foreign pipeline. Without this bond, corrosion can occur at accelerated rates. One ampere of current return to a foreign pipeline without the bond can consume approximately 8 kg (20 lbs) of steel per year.

Regulatory bodies generally require critical bonds to be inspected at intervals not exceeding 2½ months, but at least six times each calendar year. These "critical bond" inspections include current magnitude, current direction, and company pipeline's PSP.

REMEDIAL PROGRAM

Remedial measures are typically carried out as soon as possible after the surveys, tests, or inspections indicate that corrosion protection is not adequate. These measures may include the following:

1. repair, replace, or adjust components of CP systems
2. provide additional CP facilities
3. repair or replace damaged or deteriorated coating
4. apply a protective coating to bare piping systems
5. repair, replace, or adjust continuity and interference bonds
6. remove injurious metallic contacts
7. repair defective electrical isolation devices
8. repair shorted casings by re-establishing electrical isolation filling the annular space with protective material or both

EQUIPMENT PROTECTION AND SAFETY ISSUES

Pipelines in parallel, in proximity, or crossing a power line can be affected by induced voltages, fault currents, and lightning strikes. These effects not only need to be mitigated in order to reduce the risk of damage to individuals and equipment but also precautions must be taken when maintaining nearby CP including rectifiers.

Rectifiers are generally equipped with lightening and surge protection facilities on both input and output circuits. Also, lightening arrestors may be installed between foreign structures and transmission pipelines, electrical tower footings or grounding systems and as well across pipeline insulating flanges/joints. Explosion proof construction methods are required where a combustible atmosphere is expected,

However, hazards are possible, and therefore, personnel maintaining rectifies or other CP facilities must be protected and appropriate safety steps implemented.

The degree of shock hazard and threshold levels of voltage and/or current that can be tolerated by human beings is dependent on many factors. The possibility of shock hazard depends on factors such as the voltage level and duration of human exposure, human body and skin conditions, and the path and magnitude of any current conducted by the human body. The magnitude of current conducted by the human body is a function of the internal impedance of the voltage source and of the voltage impressed across the human body and the electrical resistance of the body path. This resistance is also dependent on the contact resistance, i.e., wet or dry skin, standing on dry land or in water, and on the current path through the body, i.e., hand-to-foot, hand-to-hand, etc.

The industry considers that 15 V AC (RMS) open circuit and a source current of 5 mA or more under a short circuit condition constitutes shock hazard between parts of a structure or between structure and ground.

The beginning sensation of shock, which may occur at 1 to 8 mA, may not be painful or harmful to a human being but may lead to an accident by causing a rapid involuntary movement of a person.

Human Resistance to Electrical Current

To understand safety related issues due to exposure to electrical current hazards, the following provide the effect of human resistance to electrical current (NSC 1980):

Dry skin	100,000 Ω to 500,000 Ω
Wet skin	1000 Ω
Standing or lying in water	150 Ω
Internal body: hand-to-foot	400 to 500 Ω
Internal body: ear-to-ear	(about) 100 Ω

Current Values Affecting Human

The following table provides the effects due to exposure to and contact with electrical currents (OSHA 1997). The list indicates the consequences of electrical shocks when experienced due to faulty or inappropriate operation of electrical equipments.

Reading	**Effect**
1 mA or less	No sensation — not felt
1 mA to 8 mA	Sensation of shock — not painful; individual can let go after 48 hours — muscular control not lost
8 mA to 15 mA	Painful shock — individual can let go at will — muscular control is not lost
15 mA to 20 mA	Painful shock — muscular control lost — cannot let go
20 mA to 50 mA	Painful — severe muscular contractions — breathing difficult

50 mA to 100 mA (possible)	Ventricular fibrillation — death will result if prompt cardiac massage not affected
100 mA to 200 mA (certain)	Defibrillator shock must be applied to restore normal heart beat. Breathing probably stopped
200 mA and over	Severe burns — severe muscular contractions — chest muscles clamp heart and stop it during shock (Ventricular fibrillation is prevented) Breathing stopped — heart may start following shock or cardiac massage may be required

REFERENCES AND OTHER SOURCES CITED

Alexander, R., 2009, "Cathodic Protection of Steel Pipelines," Technical Information, http://freespace.virgin.net/roger.alexander/technic.htm.

ADEQ (Arizona Department of Environmental Quality), 2002, " Protecting underground storage (UST) From Corrosion.", http://www.azdeq.gov/environ/waste/ust/download/brochure/cp.pdf

Bash, 2006, "R. J. Kuhn's 0.85 V, CSE, Cathodic Protection Criterion for Buried Coated Steel Pipelines Is Scientifically Sound," *Corrosion 2006*, San Diego, CA, March 12–16.

*Beackmann, W., Schwenk, W., and Prinz, W., 1997, *Handbook of Cathodic Corrosion Protection: Theory and Practice of Electrochemical Protection Processes*, Gulf Publishing Company, Houston, TX, pp. 505–534.

DeGiorgi, V. G., and Hamilton, C. P., 1995, "Coating Integrity Effects on Impressed Current Cathodic Protection System Parameters," *Boundary Elements XVII*, Computational Mechanics Publications, Southampton, http://www.beasy.com/images/pdf/publications/papers/cint.pdf.

Koch, G. H., 2002, "Corrosion Cost and Preventive Strategies in the United States," U.S. DOT, Federal Highway Administration, Research, Development, and Technology, Turner-Fairbank Highway Research Center.

MUSTR (Michigan Underground Storage Tank Rules) DEQ, 2001, "Cathodic Protection Testing Criteria," January 18, http://www.michigan.gov/documents/deq/deq-std-opmemo18_250026_7.pdf.

*Jankowski, J., "Monitoring Methods of Cathodic Protection of Pipelines," Gdańsk University of Technology, http://www.pg.gda.pl/chem/Katedry/Korozja/html/JCM/8_02_02.pdf.

*Holtsbaum, W. B., "Performance of Deep Groundbeds in Western Canada," http://www.anotec.com/articles/ArticleFiles/Art-25-Performance-of-Deep-Groundbeds-Canada-Holtsbaum-47.pdf.

NACE, 1988, "Cathodic Protection Monitoring for Buried Pipelines," Report CEA 54276, NACE International.

NACE RP0169, 2002, "Control of External Corrosion on Underground or Submerged Metallic Piping Systems." NACE International, Houston.

NSC (National Safety Council), 1980, *Accident Prevention Manual for Industrial Operations*, National Safety Council, Chicago, IL.

*Neal, D., 1999, "Good Pipe Coating Starts With Properly Prepared Steel Surface," Pipeline and Gas Industry, Vol. 82, March, pp. 38–43.

OSHA (U.S. Department of Labor Occupational Safety and Health Administration), 1997, "Controlling Electrical Hazards," OSHA 3075, http://actrav.itcilo.org/actrav-english/telearn/osh/hazard/elect.htm.

Pawson, R. L., 2000, "Cathodic Protection Close Interval Surveys — Field Manual," http://www.delweg.com/library/cips_guide.pdf. or http://www.delweg.com/library/technical_papers.htm.

Pawson, R. L., 2003, "Preparing for a Close Interval Survey From OTL Survey Systems," Technical Papers from Around the Web, http://www.delweg.com/library/technical_papers.htm.

Pawson, R. L., 2008, "Analysis and Interpretation of CIS Data," *Corrosion*, NACE International, New Orleans, LA, March 16–20.

*Peabody, A. W., *Control of Pipeline Corrosion*, 2nd ed.

*Saskachewan, Ministry of Advanced Education, Employment and Labour? "Recognizing Hazards in High Risk Work Areas," Saskatoon, Regina, Canada, http://www.readyforwork.sk.ca/teachers/movingonline-2008/Hazard-Information-Supplement/RecognizingHazards.pdf.

*Shwehdi, M. M., and Johar, U. M., 2003, "Transmission Line EMF Interference With Buried Pipeline: Essential and Cautions," Proceedings of the International Conference on Non-Ionizing Radiation at UNITEN (ICNIR 2003), Electromagnetic Fields and Our Health, Malaysia, October 20–22, 2003.

Webb, N., Wyatt, B., Roche, M., and Thirkettle, 2006, "Combined External Coating and Cathodic Protection Assessment Techniques — A Case Study for a Pipeline in West Africa", Cefracor Conference in Aix, France, June.

*Wessling, D., 2002, "Capabilities and Limitations of Techniques for Assessing Coating Quality and Cathodic Protection on Buried Pipelines," Cathodic Protection Theory and Practice Conference, Sopot, Poland, September.

* sources cited for information

Chapter 5

DATA GATHERING TOOLS FOR INTEGRITY ASSESSMENT

There are a number of ways that data can be collected for the purpose of physical integrity assessment of pipelines. These include the tools and direct assessment methods, which are listed below:

- ILI tools
  - o MLF
  - o UT wall thickness and crack detection
  - o Eddy current
  - o EMAT
  - o Inertial and caliper
- Hydrostatic testing
- Direct assessment

INLINE INSPECTION (ILI)

ILI tools ("smart pigs") initially became available to the pipeline industry in the 1970s to detect corrosion and pipeline deformation. One of the initial conditions of the construction of the Trans-Alaska crude oil pipeline in the early 1970s included a technical stipulation in the right-of-way agreement between the pipeline's partners and the United States Department of the Interior, the federal agency with primary oversight for construction of the pipeline, required the development of a pipe deformation monitoring system. The stipulation was included in the agreement because of the government's concern about possible settlement of the buried, hot pipeline in areas containing unidentified thaw unstable ice-rich permafrost. The Superpig curvature pig was developed in response to the stipulation in the right-of-way agreement. The curvature pig was a three section-articulated ILI pig 14-ft long weighing 5000 lb. The center, or transducer, section contained 16 wheels that contacted the pipe through connections with spring-loaded arms. Differences in the distance traveled by the wheels determined the curvature of the pipe. It was run a number of times in the pipeline, but discontinued when it became lodged in a valve cavity and had to be removed in pieces through the valve bonnet (Ulrich 1996).

Over the years, as computer and electronic technologies became much more advanced and cost-effective, the tools have undergone significant improvements in their range, applicability, and accuracy and have made it possible to identify other anomalies such as cracks. Today, there are different tools available that specialize in the inspection and data gathering for different types of anomalies. Accuracy and reliability varies with each tool, pipeline condition, and other factors. In conducting an ILI program, the operator should evaluate the capabilities of the available tools for the intended inspection and formulate a plan to validate the results.

Today, the use of ILI tools is a standard procedure for the collection of pipeline data required for integrity assessment and fitness-for-purpose studies. Their major task is to provide accurate geometric information regarding the length, width, depth, orientation, and location of a flaw. The major advantage of ILI tools is their capability to survey the entire pipe circumference while the pipeline remains in operation. They are usually pumped through the line to be inspected (i.e., free-swimming tools) and do not require their own drive.

Tool selection and accuracy are vital to the success of this effort, and it is important to realize that inspection tools have their strengths and weaknesses, and we need to be aware of them when making them part of our defect assessment tool kit. In this section, we will briefly review some key aspects of the API 1163 standard (2005) for ILI, before examining the main types of inspection tools used and comparing typical accuracies and range of use. Comparing the data from successive inspection runs provides the means of uncovering and dating new anomalies as well monitoring the growth of existing features. This requires being able to overlay the data sets spatially and if possible incorporating them into a geographical information system (GIS). ILI tools must be used in conjunction with a range of other data and information (e.g., confirmatory digs and surface defect profiling) to obtain the greatest accuracy for use in defect assessment methods.

Code Requirement

Overview of API 1163

The API 1163 "ILI Systems Qualifications Standard" (2005) is an umbrella document for planning, conducting, and validating an inspection. It was developed by multiple stakeholders including pipeline operators, consultants, tool vendors, and government regulators, and it is written in performance-based language. The standard is not technology specific, and by deliberately requiring the pipeline operator and the ILI tool provider to share the responsibility for the desired end result, it contains the means of ensuring a continuous improvement feedback loop.

The process involves the three steps of preparation, execution and verification, and the tasks follow the flow chart shown in Figure 5-1. Before describing these various steps, it is important to consider the contracting considerations, which must precede them. Establishing the inspection acceptance criteria is critical, and this will entail considering those circumstances where partial loss of data, or data degradation, due to say speed excursions, occurs. The data analysis specifications such as Probability of Detection (POD), Probability of Identification (POI), and sizing must be clearly stated in the contract. The reporting criteria for defects and verification of anomalies will also be included as will the cost and timing conditions, pertaining to the need for a re-run, if necessary.

The pipeline operator must clearly articulate the specific threats that are being investigated so that the appropriate tool technology is selected. Since the operator has familiarity with both the physical and operational characteristics of the pipeline system, he has an understanding of any constraints it may pose. The purpose of performing the ILI forms part of the company's integrity management program, so it is important that the goals and objectives of this program are met. Some useful guidance may be found in documents such as API 1160 and ASME B31.8S.

The ILI service provider must identify the performance capabilities of the inspection tool and particulars regarding its proper application and use. This will take the form of a performance specification sheet that details the ability of the tool to detect, identify, and size anomalies. The tool performance must be qualified by the service provider using at least one of the following:

- Verified historical data
- Large-scale tests from real or artificial anomalies
- Small scale tests, modeling, and/or analysis

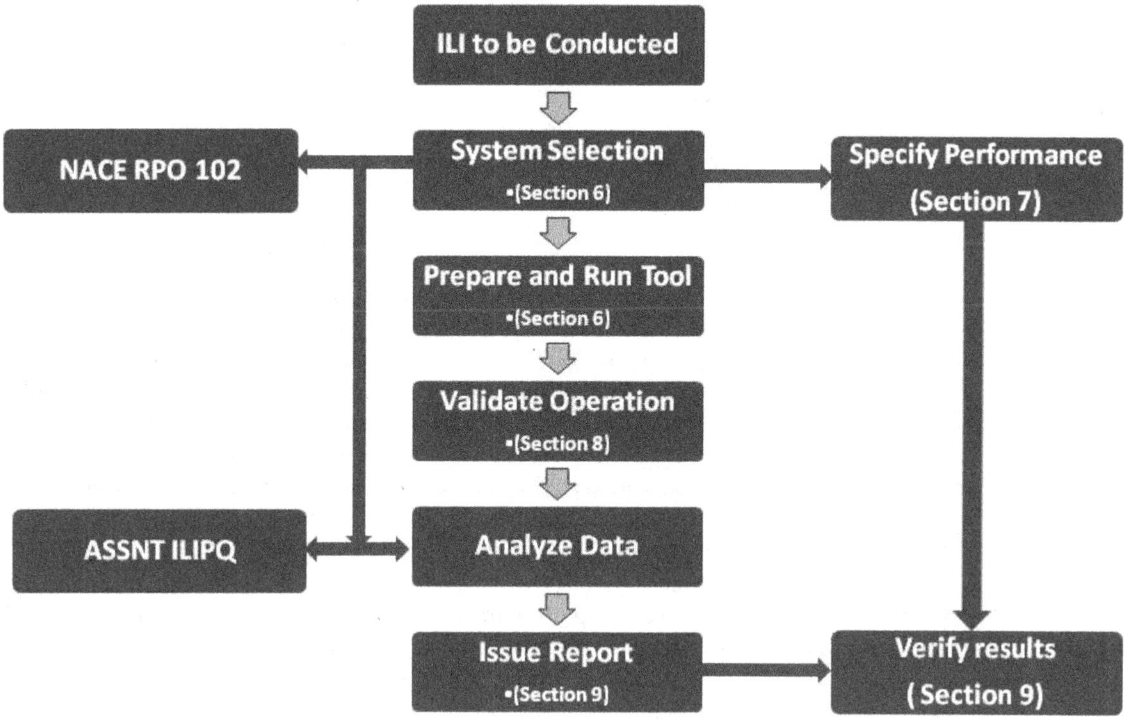

Figure 5-1. Flow chart of the API 1163 three-step process (courtesy of API)

Step 1 — Preparation Stage

Assuming that an ILI can provide the operator with what he needs to know, the first preparatory step involves documenting what is already known about the condition of the pipe. For example, what kind of damage and its severity may be expected to be present? Is the damage considered to be relatively stable or expected to grow quickly? Next one needs to identify the end use for the inspection results collected so that the appropriate level of tool accuracy is sought.

A high-level review of the available tool suitability should then be made matching the technology to the defect type(s) previously identified. In addition, the operator should consider the piggability of the pipeline.

Checking the piggability of the line is very important to ensure a clear passage for the ILI tool selected. This extends beyond determining whether or not the tool might stick, but also considers the possibility of damaging the tool — a potentially expensive proposition. The pipe must also have ready access to enable safe launching and receiving of the ILI tools.

In order to determine the level of tool accuracy possible, the effect on data quality of the cleanliness of the pipeline and speed excursions when going through fittings needs to be determined in consultation with the service provider/tool vendor. The length of the tool run is another factor needing careful consideration since it has an impact on the amount of onboard memory required for the tool.

Prior to agreeing upon a contract to conduct an ILI,, it is common for ILI vendors to request that a "Pipeline Questionnaire" be completed by the operator. The purpose of this questionnaire is to ascertain the piggability of the line, its operating parameters, and the performance criteria sought by the operator. In order to complete the questionnaire, the operator will need to draw from several sources of information including:

- As built drawings
- Bills of materials
- Previous inspection records
- Anecdotal information from field/construction staff

Step 2 — Execution Stage

It is highly desirable that the personnel and equipment used to perform ILIs and analyze the results are qualified to API 1163 and its companion documents, ASNT ILI-PQ (NACE 2002) and NACE RP0102 (Beller 2007). Combined, these three standards provide requirements and processes for the qualification of ILI systems, including the inspection tools, their software, and the personnel to operate the systems and analyze the results. Tool preparation (Figure 5-2), by the service provider, is a critical step at the execution stage.

Before running the inspection tool, it is useful to clarify with the pipeline operator the project requirements as well as the pre- and post inspection requirements. This clarification should include verifying the line operational parameters (product, velocity, configuration of traps, valves, 1.5D bends, etc,) as well as the cleaning method, which is related to the required cleanliness for the tool type. The tracking methods to be used, e.g., above-ground markers, GPS, acoustic sensors etc., should also be confirmed at this stage.

Pipeline cleaning (Figure 5-3) may be carried out by using scraper, brush, descaling, knife, or jetted pigs. The pigs may be run separately or in trains. It is good practice to run a gauging pig prior to launching a sophisticated ILI tool to reduce the possibility of it sticking or becoming damaged.

Figure 5-2. Preparing the inspection tool (courtesy of Rosen)

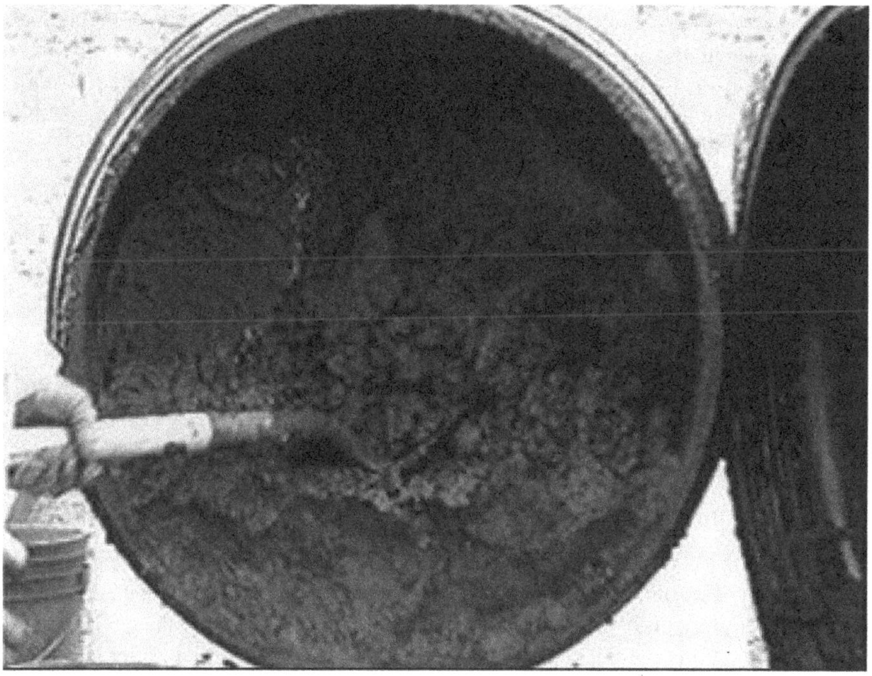

Figure 5-3. Removing debris after a pipe cleaning run

It is important too to analyze the debris emanating from the cleaning pig run in order to quantify the debris removed, estimate the amount of corrosion scale, and the measurable degree of cleanliness or dryness with in the pipe. This information can be subsequently used to optimize the cleaning of similar lines.

Scheduling is key to a successful ILI operation as it has a potential impact on customers, manpower, and budget. The logistics of the operation includes ensuring access for heavy equipment that may be needed to launch/receive tools plus the need for tracking sites for pig locating during the run. Both of these aspects could give rise to landowner issues, so it is important to consult and engage with these affected parties early in the project planning stage.

Depending on the tool requirements, seasonal flow variations in the pipeline may be important since this can affect tool speed. Hence, there is a need for operating procedures to ensure consistent tool speed and good communication between the field operators and the control center. Clearly, there is a requirement for sufficient skill levels on the part of the field operators and adequate staffing, since depending on the length of the tool run, the operation could last as long as 24 hours. The remaining logistical tasks include tracking the tool, and this may be facilitated through the use of an onboard transmitter — vitally important should the tool become stuck. There is also a need to plan for handling hazardous waste derived from the cleaning run or the inspection tool run.

Step 3 — Verification Stage

The results of the ILI must be verified after it is completed, and the reported results must meet or lie within the performance specification previously agreed upon for the pipeline being inspected. This requires a pre-agreement concerning the verification criteria, the type and extent of the verification activity, such as verification digs, and who will be responsible for undertaking these specific activities. When verification digs are performed, the information from the measurements taken should be given to the service provider so as to confirm and continuously refine the data analysis processes. Any discrepancies between the reported inspection results and those verification measurements that are outside of performance specifications should be documented. The methodology and data used for performance specification qualification must be clearly documented. The performance limitations of a particular system vary according to the technology being used. In addition, the following factors need to be considered:

- Anomaly orientation and proximity to other anomalies or pipeline components (welds, etc,)
- Anomaly shape and area
- Maximum and minimum wall thicknesses
- Tool velocity
- Pipeline cleanliness
- Pipe metallurgy
- Pipe curvature (i.e., bend) or deformation
- Pipe wall coverage
- Acceptable sensor loss or data degradation

Prior to undertaking field verification work, the data analysis process to be used must be confirmed. Having established that sound data has been collected during the inspection run, one of the first data validation steps will be to make a comparison of the reported locations and distinctive features of the pipeline section with their actual location and type, e.g., valve and girth weld locations and deliberate changes of wall thickness. When previous inspection runs have taken place, the most recently recorded data will be compared with historical results, including data from previous excavations and repairs. This supplemental information is important to the later step of determining defect growth rates. Where no previous ILIs have taken place on a section of pipeline, the API 1163 verification process allows for the experience gained using the inspection tool on a pipeline with similar characteristics to be used in lieu. However, the use of historical data from other lines provides confidence only for tools and technologies with a long and successful history. Sometimes the results of tool calibration test data will be used when no other data sources exist. The performance specifications to be verified include:

- POD
- POI
- Sizing accuracy (tolerance, certainty, confidence level)
- Location accuracy (linear, circumferential)

In order to verify the accuracy and demonstrate that the results are consistent with the performance specification, statistically valid comparisons are made between reported and measured characteristics. Inconsistencies found during the process verification must be evaluated and resolved. If the inconsistencies cannot be resolved, then the inspection results are not verified.

Figure 5-4 defines some of the tool reporting variables used in the verification process. Verification digs are generally conducted first at the reported location of the most severe features and measurements taken using some of the methods described below. The measured data is compared with that reported by the tool, and if necessary, a recalibration is made. If the site has previously been excavated, a comparison can be made with known features. This process works very well whenever the operator has many previous measurements covering a range of defect dimensions. A unity plot, such as that shown in Figure 5-5, is a convenient way of indicating tool accuracy including confidence intervals. Well-chosen inspection tools generally perform well in these unity plots. Interpretation of tool data is less reliable and consistent when involving the use of indirect methods, which require extensive analysis and skilled personnel to interpret the results.

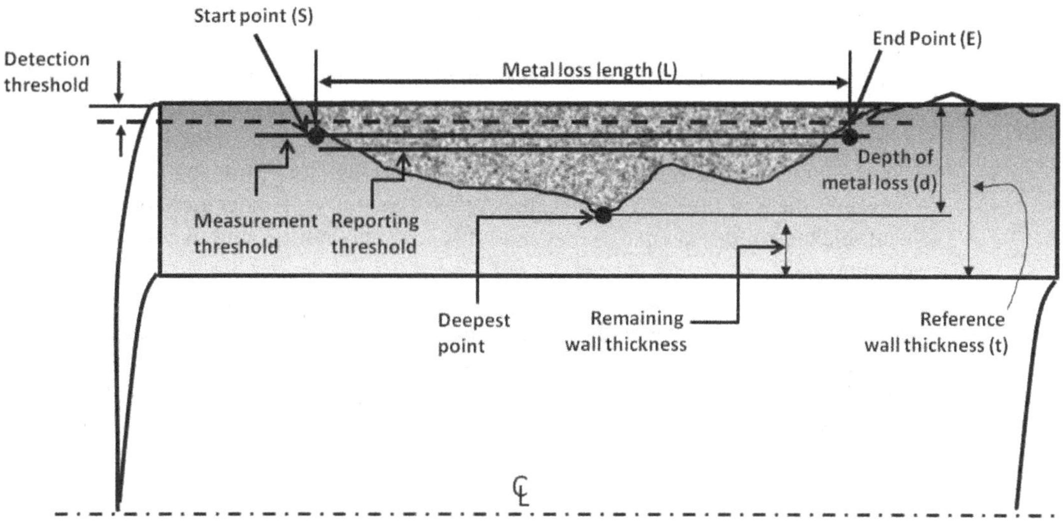

Figure 5-4. Tool reporting descriptors

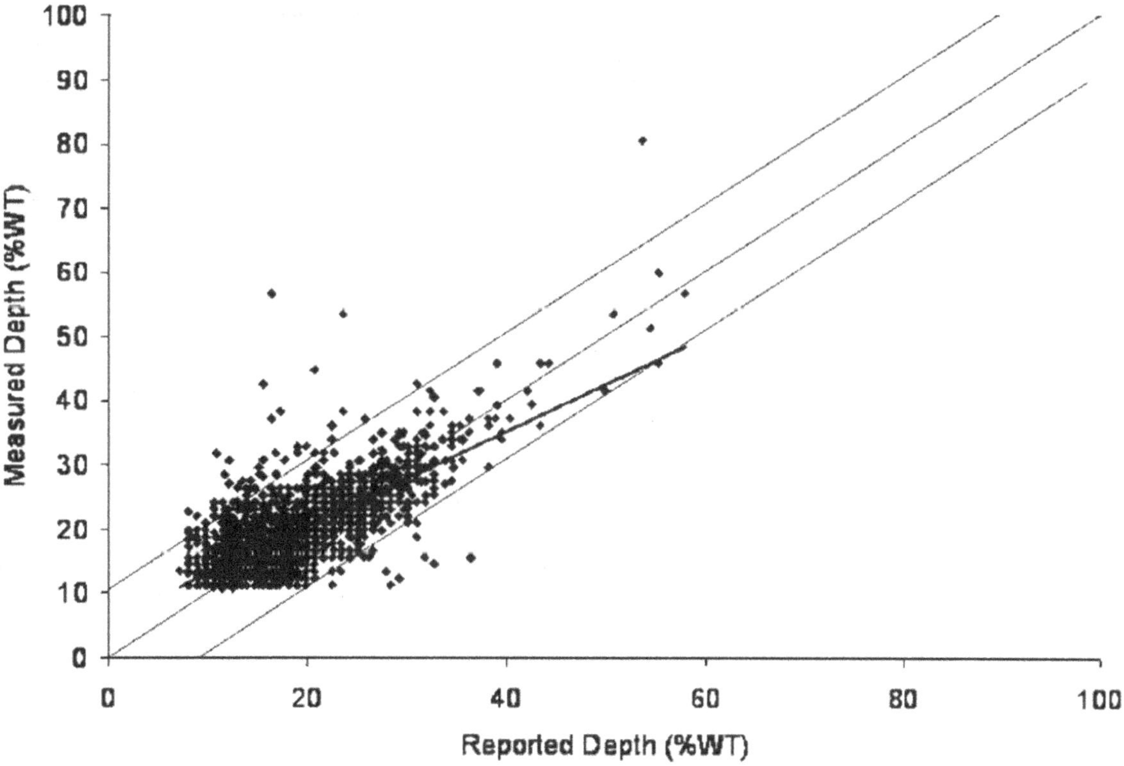

Figure 5-5. Unity plot of measured versus reported depth

The time required to complete an analysis is typically of the order of 2 to 3 months, although a preliminary report is generally provided much sooner. Ideally, the new inspection data will be compatible with that obtained from previous runs. If so, this will minimize errors in establishing reference positions and enable new anomalies to be identified and existing features to be compared. The validation dig data sometimes shows inconsistent sizing and erroneous indications and characterizations. These may be due calibration issues with the tool caused by travel speed, pipe cleanliness, temperature, or increased pipe wall thickness (for example, a magnetic flux tool may not be able to completely saturate the pipe wall).

The number of verification measurements needed and the method used to verify the performance specification will depend on the amount of historical data associated with the pipeline and/or the inspection system. The information obtained from previous excavations will be helpful to providing an understanding of the defect population on the line.

Standards, such as API 1160 or ASME B31.8S, give good guidance when selecting tools for detecting different types of anomalies.

ILI TOOLS

Having identified the integrity threats to the pipeline, these should be matched to the inspection tool capabilities. Different threats may mean the use of different tools. Table 5-1 provides a list of different types of tools and their broad range of capability. In general metal loss, due to internal or external corrosion, can be detected using magnetic flux leakage (MFL) or ultrasonic tools (UT). Crack detection, whether seam weld anomalies arising from manufacture, or stress corrosion cracking, requires the use of an ultrasonic or possibly a transverse flux magnetic leakage tool. Deformation and mechanical damage such as dents, gouges, and buckles can be detected to varying degrees, by caliper, MFL, UT, or eddy current tools. Movement of the pipe centre line is best detected using inertial tools.

The early ILI tools were used for various forms of geometry survey. Intelligent or smart pig tool development began in the mid-1960s primarily for the purpose of measuring corrosion. Tuboscope was one of the early pioneers with their Linalog MFL pigs. This provided pipeline operators with an early warning of any major problems. The inspection log was essentially qualitative providing a degree of defect severity grading. In the mid-1970s British Gas and Battelle had completed major investigative programs on pipeline material properties and failure mechanisms. This led to the development of the first high-resolution inspection tools and the provision of quantitative performance requirements. Some years later in the 1980s, Pipetronix and others developed the alternative ultra-sonic testing technique that requires fluid coupling between the tool and the pipe wall.

The industry classifies the ILI tools as geometry, metal loss, or crack detection tools and employs a variety of technologies to evaluate the geometry and condition of the pipeline.

TABLE 5-1. ILI tool capability — anomaly matrix

Tool Type	Removed Metal	Cracks	Metallurgical Changes	Geometry Changes	Others
MFL	X		X	*	*
Circumferential MFL	X	X	X	X	*
UT wall thickness	X			X	*
UT crack		X		*	*
Eddy current		X	*		*
EMAT		X			*
Inertial				X	*

X indicates capability; *, possible capability.

Figure 5-6. GE PII MagneScan™ MFL inspection tool

A caliper tool is a geometric inspection tool used to detect dents, wrinkles, field welds, bend radii, and pipe ovality. It is tolerant of moderate amounts of debris within the pipeline, so it is also useful in identifying obstructions harmful to other tools. When combined, with an onboard inertial device, it can provide the location, orientation, and depth measurement for each dent, which is extremely useful in threat assessment since dents in the lower half of

Figure 5-7. ROSEN CDP MFL inspection tool

Figure 5-8. BJ Pipeline Inspection Services VECTRA™ MFL inspection tool

the pipe are most likely to have been caused by rocks in the ditch, rather than by mechanical damage. ILI tool capability range is depicted in Table 5-1.

MFL Tools

Metal loss tools contain highly sophisticated sensors that measure the MFL, (Figures 5.6 to 5.8), that is detected from an induced axially oriented magnetic field in the pipe wall as the tool traverses the inside of the pipeline. The sensors record and measure any deviations in the flux density. Such deviations are an indicator of a change in wall thickness or other anomaly that causes a disturbance in the magnetic field. Today's high-resolution tools have more advanced capabilities to collect and store more precise length and depth data for each anomaly in comparison to older-generation MFL tools developed in the 1970s and 1980s. These high-resolution tools also have the capability to determine whether the metal loss is internal or external to the pipe wall. Sizing accuracy is best for geometrically simple defect shapes. Sizing accuracy can degrade where pits are present or defect geometry becomes complex, if defects are axially aligned, or if excessive tool speeds are experienced. In gen-

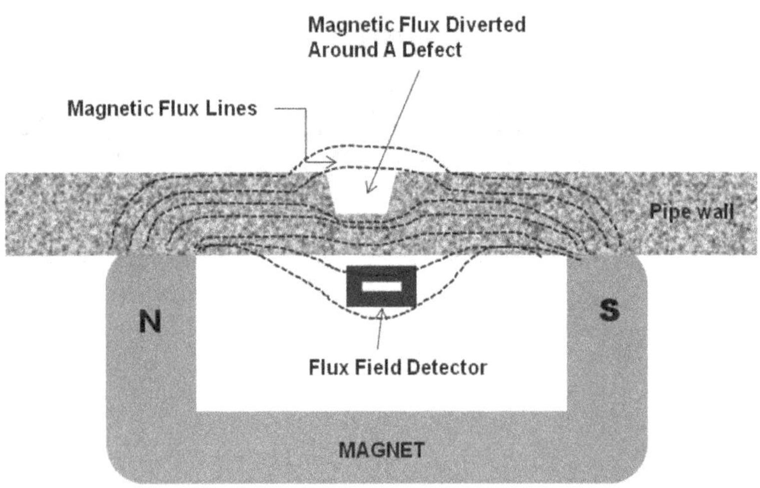

Figure 5-9. Principle of MFL

eral, MFL tools provide a cost-effective alternative for assessing the corrosion condition of a pipeline and are commonly used as a proven a capable assessment method by pipeline operators. There has been limited success identifying third-party damage using magnetic-flux leakage tools. MFL tools are not useful for accurately sizing deformations.

ILI has established itself as the primary and most efficient source of information regarding flaws that are present on a pipeline. As such, it has become an indispensable part of pipeline integrity management. Historically, there have been two major technologies used in ILI for corrosion, MFL and UT, each having their distinct properties and fields of application.

MFL tools (Figure 5-9) use the Hall effect to provide an indirect form of measurement of metal loss and are well suited to detect and characterize volumetric defects such as corrosion. The detectable defect depth depends on the length of the feature with short deep defects more susceptible to length inaccuracy. They are not suited to finding cracks or metallurgical anomalies such as laminations. Circumferentially placed magnets saturate the wall of the pipe, while single, bi-, or tri-axial sensors detect the amount of flux leakage (Figure 5-10). The number of sensors and their deployment markedly improve the tool accuracy (Figure 5-11) and ability to differentiate between metal loss on either the inside or outside diameter of a pipe. The signal obtained by the sensor depends on the length-to-width ratio of the defect (narrow axial anomalies cannot therefore be reliably detected. As with most tools, internal diameter line restrictions, e.g., sharp bends and heavy heavy-wall fittings, constrain their use while their detection and sizing accuracy is related to travel speed.

The ability to saturate the pipe wall places a limit on the maximum wall thickness, results may also be affected by the pipe steel characteristics. It is not feasible to construct tools to operate effectively in small diameter pipe <75 mm to 100 mm.

Figure 5-10. The disposition of magnets and sensors on an MFL tool

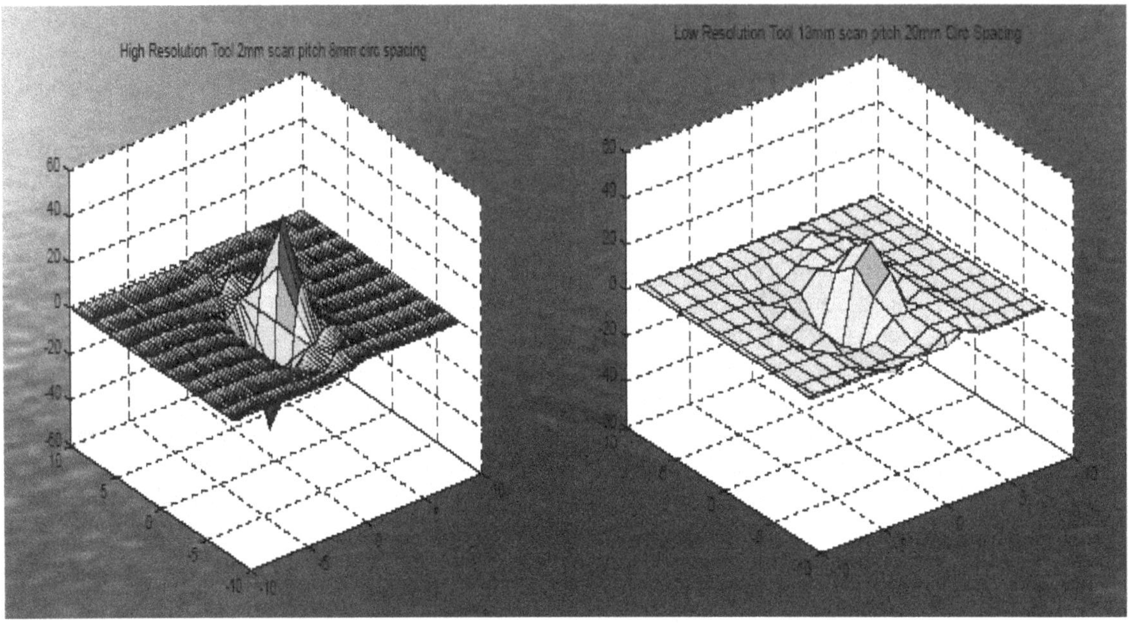

Figure 5-11. High and low defect resolution (courtesy of BJ Services)

An MFL tool can also find metal gain as, for example, on welds and patches. It may also be used to detect areas of cold working and debts and bends.

Figure 5-12 illustrates the process of calibrating the signals from the MFL tool against a number of features of known size and orientation. As noted previously, this is done during the tool run preparation stage and the data may also be used for verification purposes.

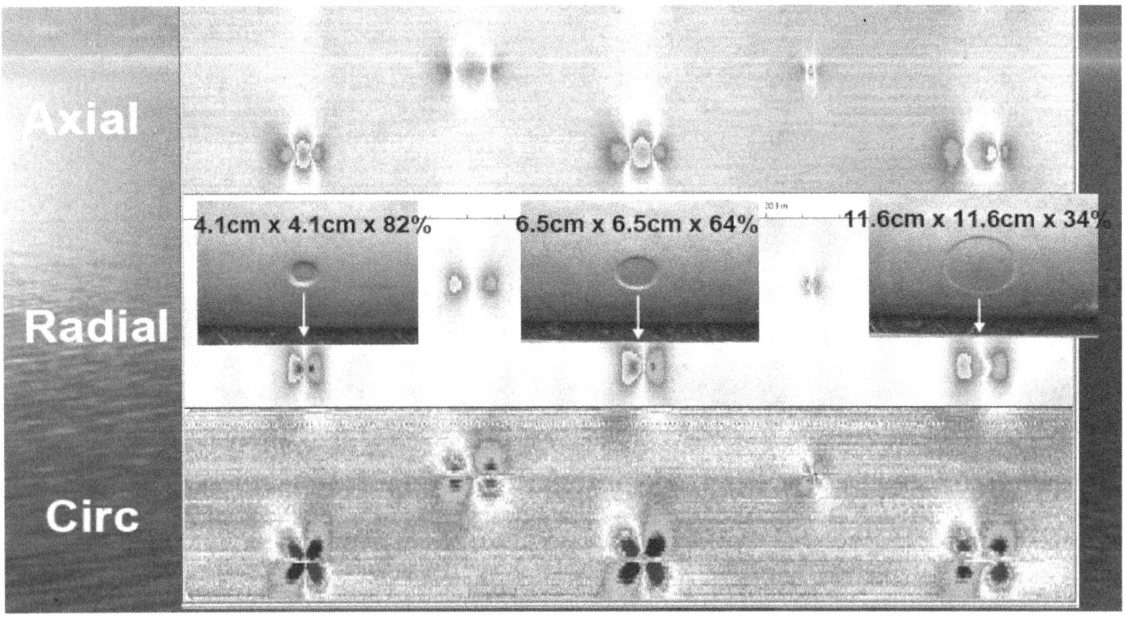

Figure 5-12. The Calibration of MFL signals with known defects (Courtesy of BJ Services)

TABLE 5-2. Typical MFL tool specifications

Parameter	Standard Resolution	High Resolution
Min detectable defect depth	20% wall thickness (WT) 80% Confidence	10% of wall thickness (WT) 80% Confidence
Accuracy of defect depth	±15% WT	±10% WT
Length sizing accuracy	±0.51 in. (13 mm)	±0.39 in. (10 mm)
Location accuracy axial	3.94 in. (0.1 m) (rel weld)	3.94 in. (0.1 m) (rel weld)
Location accuracy circumference	±30 deg	±5 deg

MLF tool specification detection capability and defect sizing are respectively indicated in Tables 5.2 to 5.4.

Ultrasonic Tools UTs (Corrosion/Cracks)

UT is used to measure wall thickness by using transducers to transmit an ultrasonic pulse into the pipe wall and record the difference in the reflection times to the internal and external walls. It is critical that the signal be acoustically coupled to the ID of the pipe; thus, a liquid couplant such as crude oil or several types of refined products can typically be used. This method allows for direct linear measurement of the local pipe wall thickness which can be used to approximate, with appropriate calculations, the remaining strength of the corroded pipe.

Ultrasound is a nondestructive testing technology that has been applied for a variety of inspection tasks for many years. A major advantage of ultrasound is the ability to provide quantitative measurements. This means that the actual wall thickness of a pipe section can be determined with high accuracy and reliability. The reporting accuracy regarding depth measurement for the latest generation of tools is around ±0.4 mm to 0.5 mm. The highest possible resolution that can be achieved today is 0.06 mm. Usually, thresholds for depth measurement of metal loss or cracks are set at 1 mm; however, lower thresholds are possible.

There are several different means available for introducing sound waves in to a pipe wall of which the most common is the use of piezo- electric transducers. Electro magnetic acoustic transmission is also used in circumstances where a liquid couplant is not present. The most widely used tools available from several vendors make use of piezoelectric transducers.

Figure 5-13 (Beller 2007) shows the principle applied for wall thickness measurement which is used for the detection and sizing of metal loss features, such as corrosion or goug ing. An added benefit is the ability to detect and identify mid-wall flaws such as laminations and inclusions and also certain categories of material separations and voids, such as HIC.

The important issue is that the sensors (transducers) are aligned at right angles to the wall to be inspected. The transducers used are operated in an impulse-echo mode, in other words they act as transmitters and receivers of the acoustic wave used for the measurement. The type of transducer chosen (i.e., dynamic range, focal point, etc.) and the characteristics

TABLE 5-3. MFL detection capabilities

(Excellent = 95%, Good = 85%, Fair = 70%)			
Parameter	**Std Res**	**Hi Res**	**Extra Hi Res**
Pitting corrosion	Good	Excellent	Excellent
General corrosion	Excellent	Excellent	Excellent
Mechanical damage	Poor	Fair	Fair

TABLE 5-4. MFL characterization/sizing capabilities

(Good = ±20% with 80% confidence, fair depends on severity)			
Parameter	**Std Res**	**Hi Res**	**Extra Hi Res**
Pitting corrosion	Poor	Good	Good
General corrosion	Fair	Good	Good
Mechanical damage	Poor	Poor	Poor

of the electronics used (i.e., pulse repetition frequency, sampling rate, etc.) has a major influence on the detection threshold, the accuracy, and depth and length resolution.

Ultrasound constitutes the only reliable technology currently available for the detection and sizing of cracks in pipelines.

Figure 5-14 (Beller 2007) depicts the crack inspection principle. Here the probes are placed not at right angles to the wall. The sensor carrier design must rather ensure that the incident ultrasound signals are refracted in a manner that they will propagate a shear wave at a 45 deg. angle inside the pipe.

Ultrasound tools are ideally suited for the inspection of thick wall line pipe, as often found in offshore pipelines. Wall thicknesses of up to 50 mm can be inspected with the same specifications as thinner wall.

Ultrasonic crack detection tools (Figure 5-15) operate by introducing an ultrasonic pulse into the pipe wall at an angle such that it generates a shear wave traveling circum-

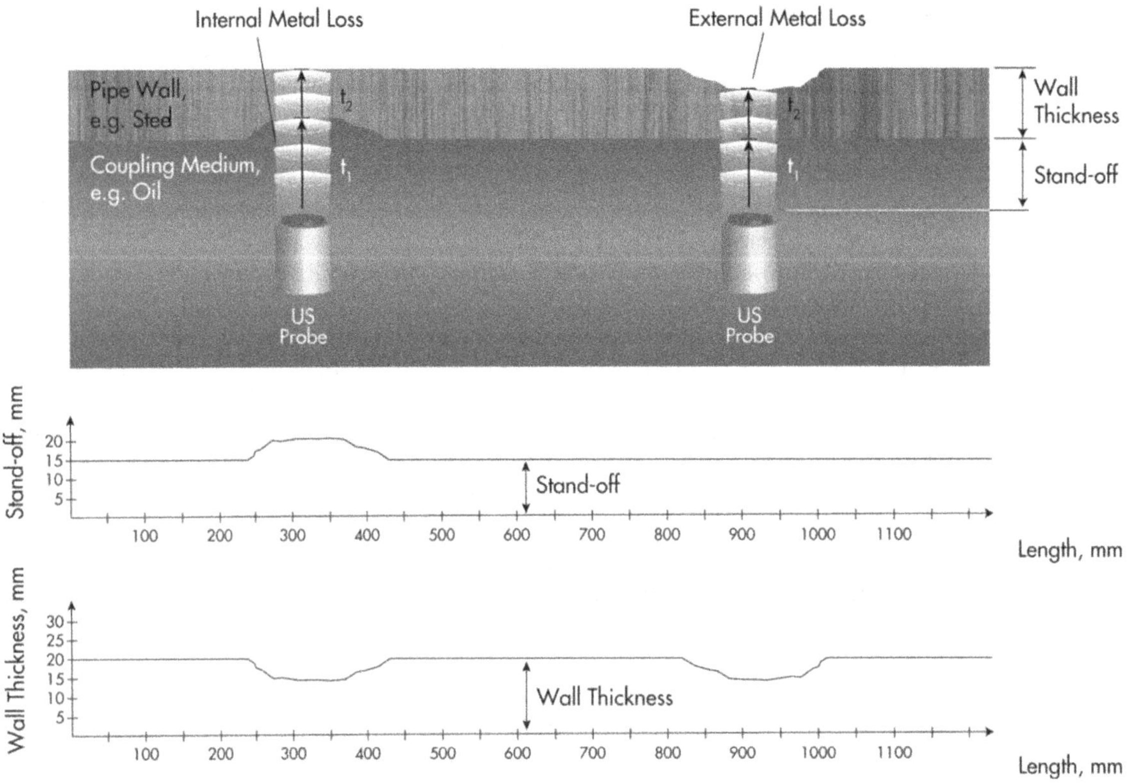

Figure 5-13. Ultrasound principle wall-thickness measurement (courtesy of Pigging Products and Services Association [PPSA] and NDT Systems & Services AG)

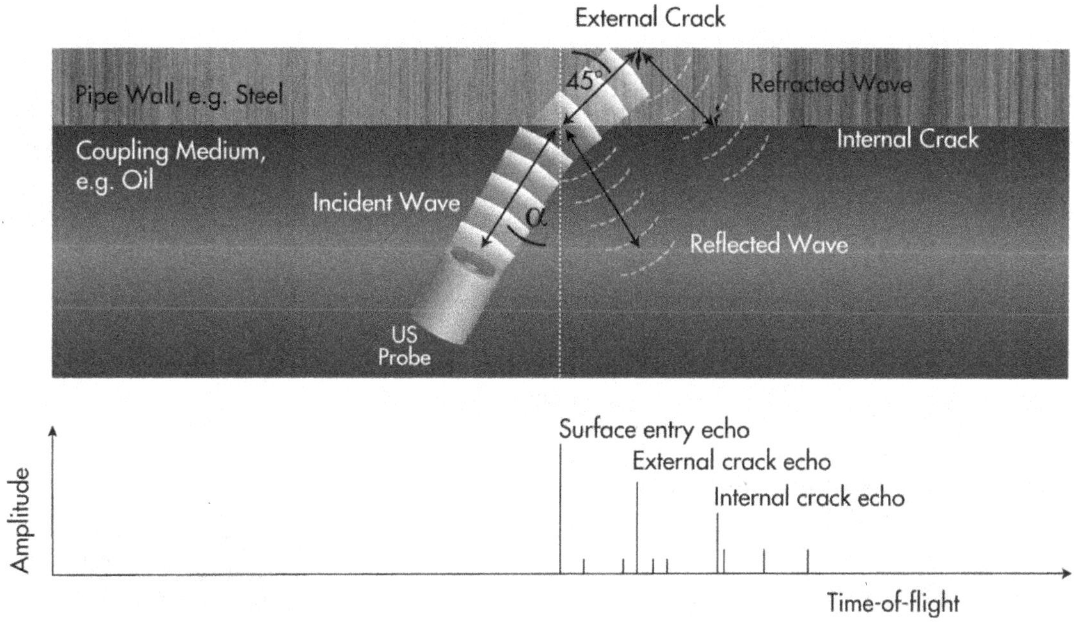

Figure 5-14. Ultrasound principle crack inspection (courtesy of PPSA, and NDT Systems and Services AG)

ferentially through the pipe wall as it is reflected off the internal and external surfaces. If the pulse encounters a crack, it is reflected back along the same path and is received at the transducer. The tool is capable of detecting cracks, stress corrosion cracking, hook cracks, lack of fusion, as well as narrow and axial corrosion.

Figure 5-16 shows the layout of a typical UT. The batteries needed for the energy supply and the electronics used, e.g., for controlling the ultrasound sensors and recording data, are housed in pressure vessels. A trailing sensor carrier always houses enough sensors to ensure full circumferential coverage of the pipeline being inspected. The pressure vessels and the sensor carrier are connected via universal joints to ensure that the tool can also negotiate bends. Odometer wheels are used for distance measurement.

Sound waves can be in the form of either compression or shear waves, the former being introduced perpendicular to the wall of the pipe, the latter at an angle to the surface. In both cases, the transducers in the head of the tool are comprised of small plates of piezoelectric material, which vibrate (resonate) whenever an alternating voltage is applied to their elec-

Figure 5-15. Ultrasonic crack detection tool (Courtesy of GE/PII)

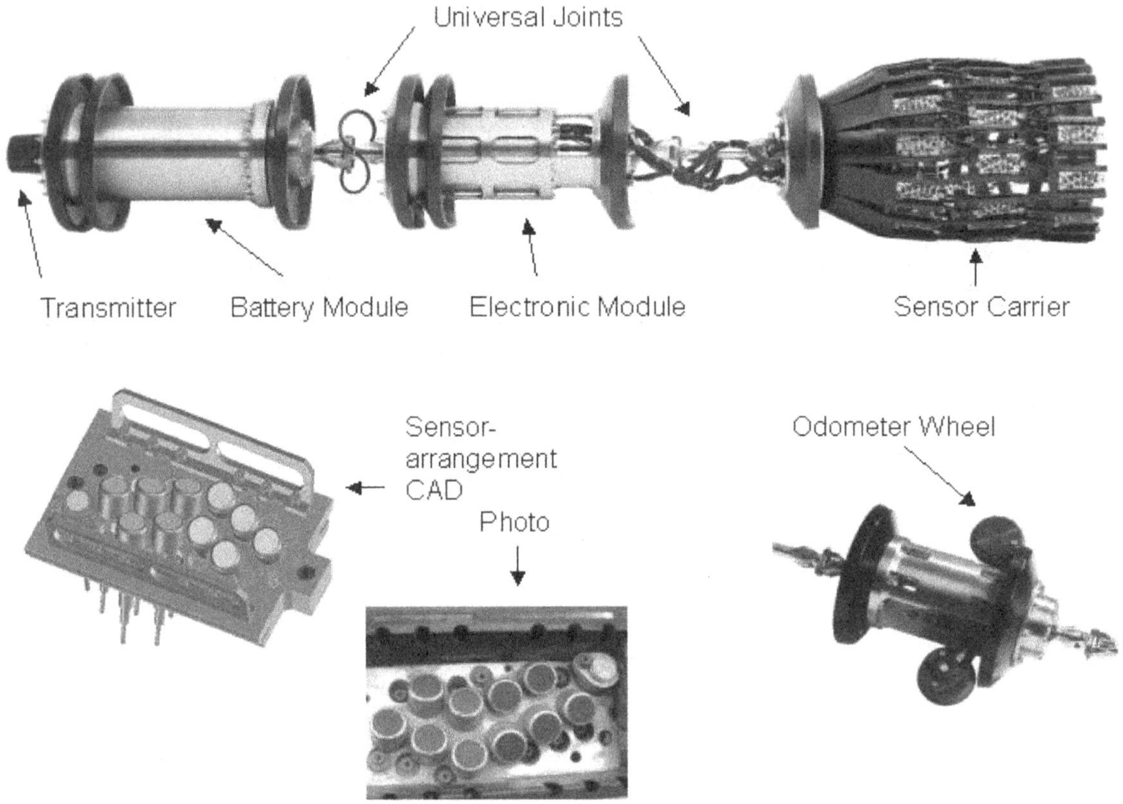

Figure 5-16. Layout of an ultrasonic ILI tool (crack detection; courtesy of PPSA, and NDT Systems & Services AG)

trodes. Similarly, incoming sound waves produce the inverse effect, that is, they cause the transducer to vibrate and thereby generate an alternating voltage that can be measured. The transducer signal "echoes" from the inside and the outside of the pipe wall are measured in order to determine the local wall thickness. The ultrasonic transducers are set at a fixed distance from the pipe wall known as the standoff distance (SO). Only entry echoes, which define the SO and the outer wall echoes are used to determine the wall thickness and other pipe wall defects.

The transducer signal reflection or echo is measured to determine the distance traveled (distance = sound velocity in the fluid medium × ½ propagation time). Clearly, a delay is required between the time of transmission and receipt of the return signal. This is achieved by pulsing the transducer so that it can act as both transmitter and receiver. The time of flight is altered when the sound wave encounters a defect.

Compression wave UTs are able to discriminate between defects located at mid-wall and on the internal or external diameter of the pipe. These tools are also useful in identifying welded attachments and sleeves, welds associated with many types of repairs, blisters and inclusions, HIC, laminations, and wall thinning on bends.

Unlike the MFL tool, there is no maximum wall thickness constraint, though there is a limit on the minimum wall thickness. A further limitation is it cannot be run in small diameter pipelines (typically less than 150 mm). The tool is sensitive to a larger number of features than

TABLE 5-5. Typical compression wave UT tool specifications

Parameter	Specification
Accuracy of depth measurements	±0.02 in. 80% Confidence
Accuracy of depth measurements for flat surfaces and wall thickness	±0.008 in. (0.2 mm)
Longitudinal resolution	0.12 in. (3 mm)
Min detectable corrosion depth	0.008 in. (0.2 mm)
Min size of pits detected: no depth measurement	0.39 in. diameter 0.06 in. depth

MFL, though it requires a higher degree of pipe cleanliness to be effective. It provides a direct, linear measurement of wall thickness and reliable depth predictions; hence, its accuracy supports a more accurate MOP calculation in the defect assessment phase.

A compression wave tool has its limitations (Tables 5.5 and 5.6) — it can only be run in a liquid due to its need for efficient sound transfer to the pipe wall. This also places demands on the cleanliness of the pipe prior to the inspection. It is not effective for use with thin walled pipe ($t < 4$ mm) and is only marginally effective for pipe in the range (4 mm $< t <$ 6 mm). Significant general corrosion can result in signal loss. It is unable to reliably detect the location of casings, or any repairs that are not welded to the pipe. The tool itself is long and this might require lengthening regular launching/receiving traps.

Shear wave UTs use the same technology as compression wave tools, the distinction being the angle of incidence at which the sound wave enters the pipe wall (Figure 5-17), Since significant cracks lie perpendicular to the main stress component, the angle of incidence of the signal is adjusted for the refraction effect from the fluid being transported, so as to introduce it at a propagation angle of 45 deg. to the pipe wall. The limitation of the tool is depicted in Tables 5.7 to 5.9.

The requirement of a fluid couplant, to improve the transfer of sound energy into the pipe wall, creates operational difficulties for operators of gas pipelines who are anxious to detect crack like features on their pipelines The use of an UT such as has just been described requires placing the tool in a liquid slug, usually water, which effectively takes the line out of normal service for the period of the inspection and the subsequent drying of the line. On large diameter pipelines, the slug may be around a kilometer in length so a considerable amount of water needs to enter and leave the pipe. Pig trains are run in advance of and behind the liquid slug, whose motion is controlled by the back pressure in the gas stream. It can be quite difficult to control the speed of the slug in hilly terrain, also as it passes features in the pipeline such as, crossovers, dead legs, valves, and changes in wall thickness. Since the tool accuracy is closely linked to tool speed, this can have a profound effect on the verification of the tool run. It was because of these shortcomings of running conventional UT in a gas line that a significant investment has been made by gas pipeline operators in the development of an alternative means of introducing sound waves into the wall of the pipe.

TABLE 5-6. Compression wave UT detection capabilities

(Good = ±20% with 80% confidence level; fair depends on severity)

Parameter	Capability
Pitting corrosion	Fair
General corrosion	Excellent
Mechanical damage	Excellent (Dents only)

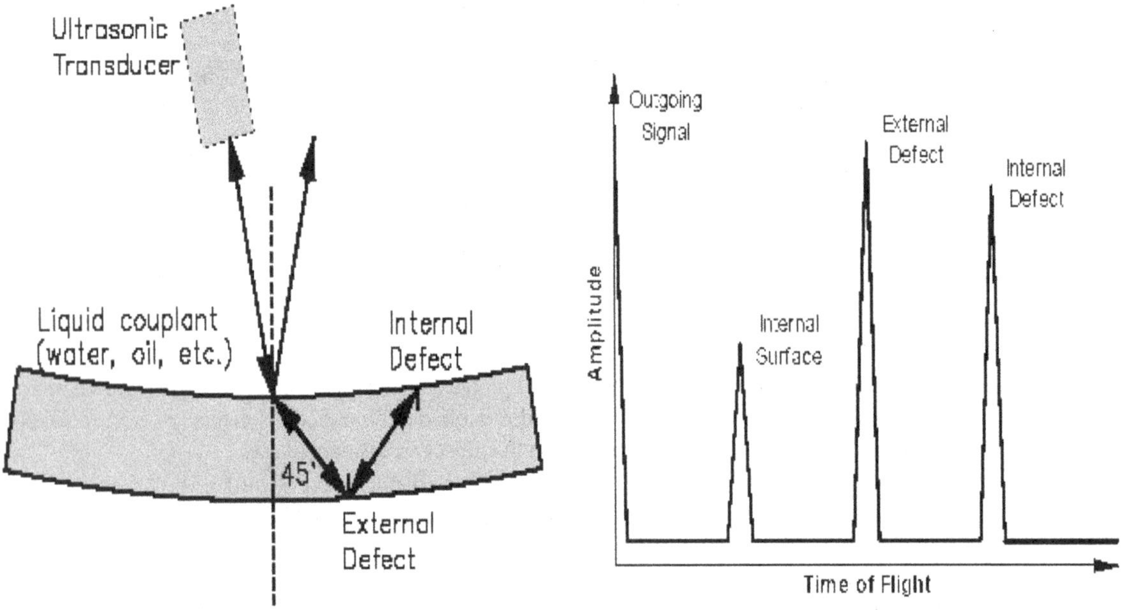

Figure 5-17. Shear wave UT

EMAT

The EMAT is one such development. Its principle of operation is shown alongside a conventional means of introducing sound waves into a pipe wall (Figure 5-18).

Transverse Flux Tools

This tool is more sensitive to axially aligned metal loss defects in comparison to more standard MFL tools because it magnetizes the pipe wall circumferentially. It is thus more of a specialized tool used to measure axially aligned metal loss defects such as narrow axial/longitudinally orientated seam corrosion. It is also used to detect cracks and lack of fusion. Tools' characterization and sizing capabilities are indicated in Table 5-10.

MFL tools have been widely used in gas pipelines for many years, since, unlike UTs, they do not need a liquid couplant to transmit signals to and from the pipe wall. Their short-

TABLE 5-7. Typical shear wave UT tool specifications

Parameter	Specification
Minimum detectable defect depth	0.04 in. (1.0 mm)
Accuracy of depth measurements	Classified in categories:
	<12.5% wall thickness (WT)
	12.5 to 25% WT
	25% to 40% WT
	>40% WT
Accuracy of length measurements	±10% WT for feature, >100 mm
Accuracy of width measurements	±1.97 in. (50 mm) for crack fields
Min detectable size of pits: with depth measurement	0.79 in. (20 mm) diameter
	0.04 in. (1 mm) depth
Location accuracy: axial	0.004 in. (0.1 m) (wrt girth)
Location accuracy: circumferential	±5 deg

TABLE 5-8. Shear wave UT detection capabilities

(Good = ±20% with 80% confidence; fair depends on severity)	
Parameter	**Capability**
Axial seam corrosion	Excellent
Axial cracks	Good
ERW seam defects	Fair

coming of not being able to reliably detect crack like features prompted the development of the transverse or circumferential magnetic flux tool shown in Figure 5-19. The magnets are arranged in the four quadrants as shown with the sensors aligned axially. Some operators have claimed to have found good success in detecting seam cracks and long axial corrosion features. Identifying an anomaly allows for follow-up; however, precise measurements of a defect are not yet possible. A typical tool is shown in Figure 5-20.

Geometry ILI Tools

Deformation or geometry tools are most often used for detecting damage to the line involving deformation of the pipe cross-section, which can be caused by construction damage, dents caused by the pipe settling onto rocks, third-party damage, and wrinkles or buckles caused by compressive loading or uneven settlement of the pipeline.

Standard Caliper Tools

Standard caliper tools (Figure 5-21) are tools that measure deviations in the geometry of a pipeline's diameter. They record a channel of data for each caliper arm which are typically spaced around the circumference of the tool, and these arms ride against the inside diameter of the pipeline. Any change to the geometry of the pipe diameter will cause a relative movement in the arm and will then be recorded. Changes in the pipe diameter geometry can be due to pipe bends, dents, buckles, valves, or changes in wall thickness. This type of tool can be used to discern deformation severity and overall shape aspects of the deformation. With some effort, it is possible to identify sharpness or estimate strains associated with the deformation using the standard caliper tool output.

High-Resolution Geometry and Mapping Tools

High-resolution tools (Figures 5.22 and 5.23) provide the most detailed information about the deformation, resulting in a more accurate measurement of smaller and a more complex shaped dents. Some also indicate slope or change in slope, which can be useful for identifying bending or settlement of the pipeline. Mapping tools are based on inertial navigation using built-in gyroscopes and accelerometers and establish the geographical coordinates of girth welds which can be useful for creating pipeline alignment maps, populating GIS information systems and determining pipeline ground movement.

TABLE 5-9. Shear wave UT characterization/sizing capabilities

(Good = ±20% with 80% confidence; fair depends on severity)	
Parameter	**Capability**
Axial seam corrosion	Good
Axial cracks	Fair
ERW seam defects	Poor

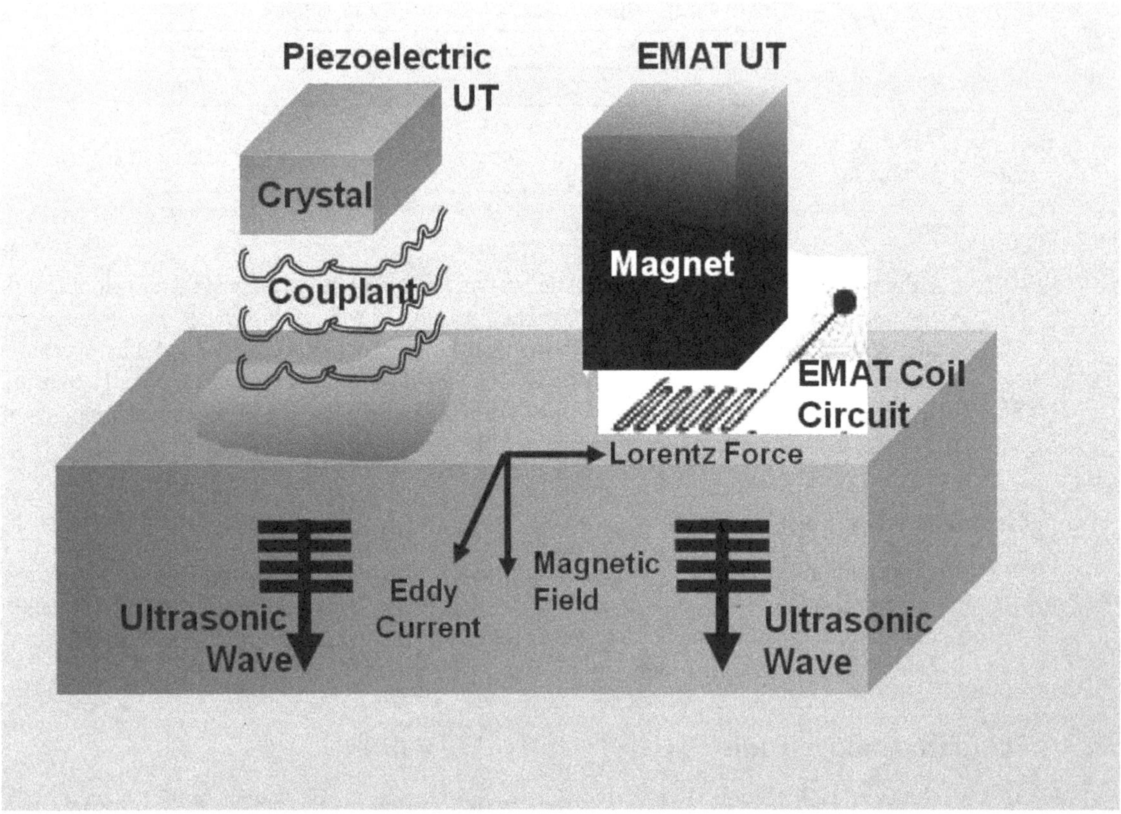

Figure 5-18. EMAT technology

TOOL ACCURACY

All the data used for defect assessment, including that obtained from ILI tools, have an inherent measurement error. Usually, the ILI service provider will state the accuracy of tools in data sheets and final reports. When using defect geometry for assessment purposes, errors of measurement are usually ignored and other safety factors are introduced into the assessment methods for compensation. Only the Det Norske Veritas Part A (Det Norske Veritas 1999) code allows for input of ILI tool accuracy and measurement technology, by introducing probabilistically calibrated coefficients. Table 5-11 summarizes the detection and sizing capabilities of various tools.

ILI results can be improved by a combination of effective validation and incorporating sizing uncertainties. The effect of tool error on depth and length of a detected defect

TABLE 5-10. Transverse MFL characterization/sizing capabilities

(Good = ±20% with 80% confidence; fair depends on severity)

Parameter	Capability
Pitting corrosion	Poor
General corrosion	Fair
Mechanical damage	Poor
Axial Seam corrosion	Good
Axial cracks	Fair
ERW seam defects	Poor

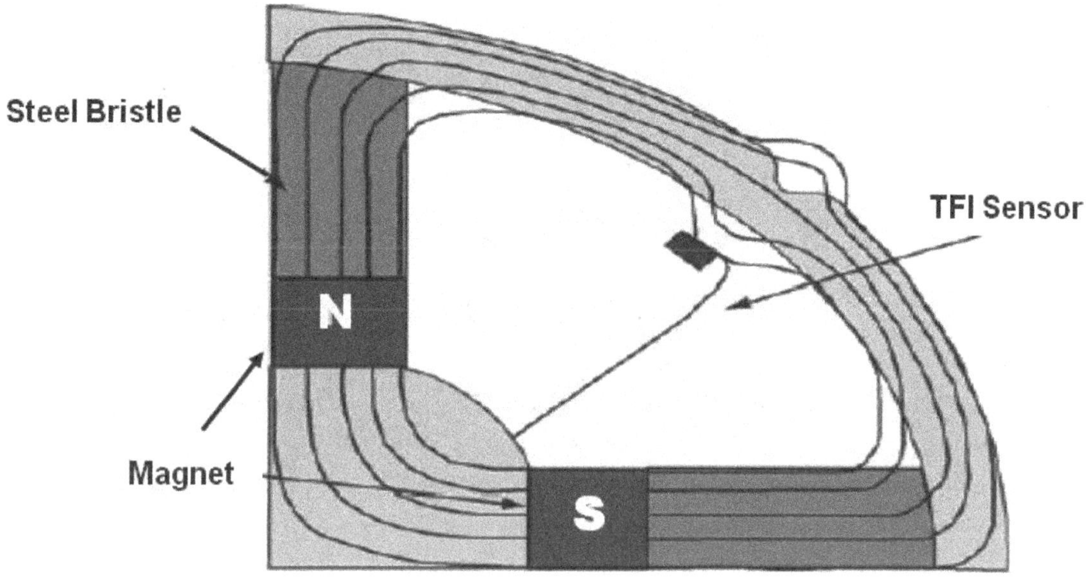

Figure 5-19. Transverse flux leakage

is shown in Figure 5-24. An "acceptable" defect could be "unacceptable" if we take lower bound inspection tolerances.

The use of aboveground markers, GPS, and inertial navigation systems on board the line inspection tools combine to provide a very accurate means of pipeline centre line positioning and accurate locations for excavations (Figure 5-25).

Control points are spaced approximately every 2 km with valves or other clear features providing reference points. An inspection tool with inertial capabilities can accurately map the area in between the survey control points. The location accuracy is highly dependent on a coordinate survey.

Figure 5-20. Transverse flux tool (axial flaw detection [AFD] tool; courtesy of Rosen)

Figure 5-21. BJ Pipeline Inspection Services PROFILE™ caliper tool inspection tool

LOCATION ACCURACY

Figure 5-26 shows a 500-mm error over a distance of 4000-m peaking in the middle, which is the furthest distance from a reference point. Placing a reference point in the middle reduces the error by half and pushes the error out to the ¼ point mark. This is the method used by service providers to check the accuracy of the survey. They will process the data using half the reference points and then use these reference points to establish the survey accuracy. Once it is known that they are within the limits required, they will reprocess the data with all the remaining control points.

Figure 5-22. BJ Pipeline Inspection Services GEOPIG™ caliper tool provides high-resolution caliper data and inertial pipeline mapping information

Figure 5-23. Geometry inspection tool (courtesy of Rosen)

TOOL DATA VALIDATION METHODS

Having determined the existence of anomalies during ILI runs, it is extremely important that they are investigated. This is done by carrying out investigative digs, exposing sections of the pipe and examining both the external surface and the pipe wall through thickness for

TABLE 5-11. Summary of detection and sizing capabilities for various tools (from Westwood and Hopkins (2004))

	Metal Loss Tools			Crack Detection Tools		Geometry Tools	
Defect	MFL -LR	MFL-HR	UT	UT	MFL	Caliper	Mapping
Corrosion	D and S1	D and S	D and S	D and S	D and S	NO	NO
Axial cracks	NO	NO	NO	D and S	D and S	NO	NO
Circumferential–Cracks	NO	d3 and s4	NO	D and S2	NO	NO	NO
Dents	d	d and s	d and s	d and s	d and s	D and S	D and s
Laminations	d	D	D and S	D and S	NO	NO	NO
Mill defects	d	D	D	D	d	NO	NO
Ovality	NO	NO	NO	NO	NO	D and S	D and S5

D or d indicates defects; S or s, sizing; 1, no discrimination on internal/external diameter; 2, sensors have to be rotated 90 deg.; 3, unreliable or reduced detection capability; 4, unreliable or reduced sizing capability; 5, tool is specially equipped to measure ovality; 6, SR, standard resolution; HR, high resolution; 7, MFL field is in transverse direction.

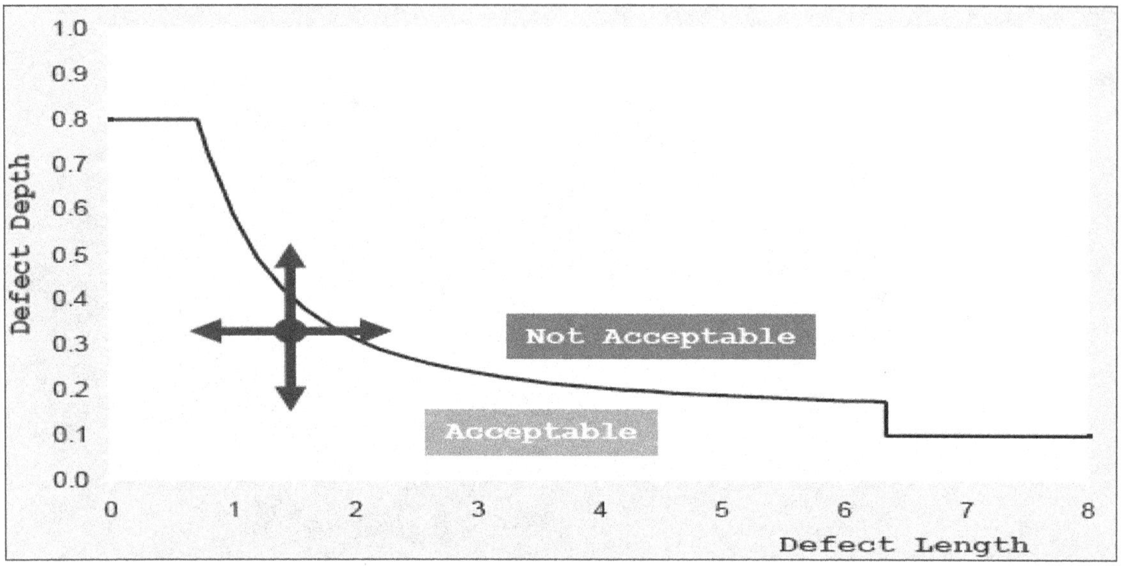

Figure 5-24. Defect assessment — probabilities and errors

defects This exercise provides both a validation (ground truthing) of the inspection tool's ability to identify defects and to size them appropriately.

An accurate description of the defect dimensions is an essential input to the assessment process. Dig outs can be very expensive (about US $80,000 on average for large diameter (900+ mm) pipes so validating tool accuracy is extremely important in order to keep maintenance costs manageable.

A suitable safe excavation must be made and the operating pressure for the pipeline reduced (often below 80% MOP) to facilitate safe working procedures around the pipeline.

The surface of the pipe must then be cleaned (Figure 5-27) before starting an inspection. A number of nondestructive testing techniques are available to assist in the detection and sizing of defects. The most common of these are:

- Magnetic particle and dye penetrants
- Ultrasonic inspection
- Eddy current inspection
- Laser profiling

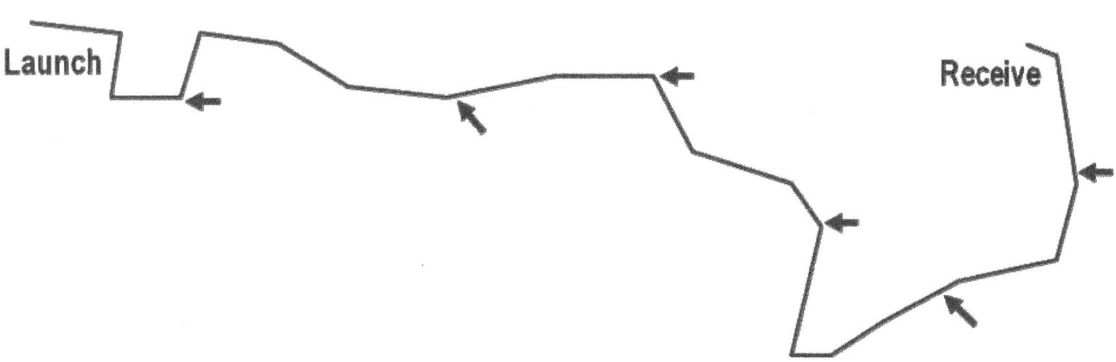

Figure 5-25. Mapping the location accuracy of the ILI survey

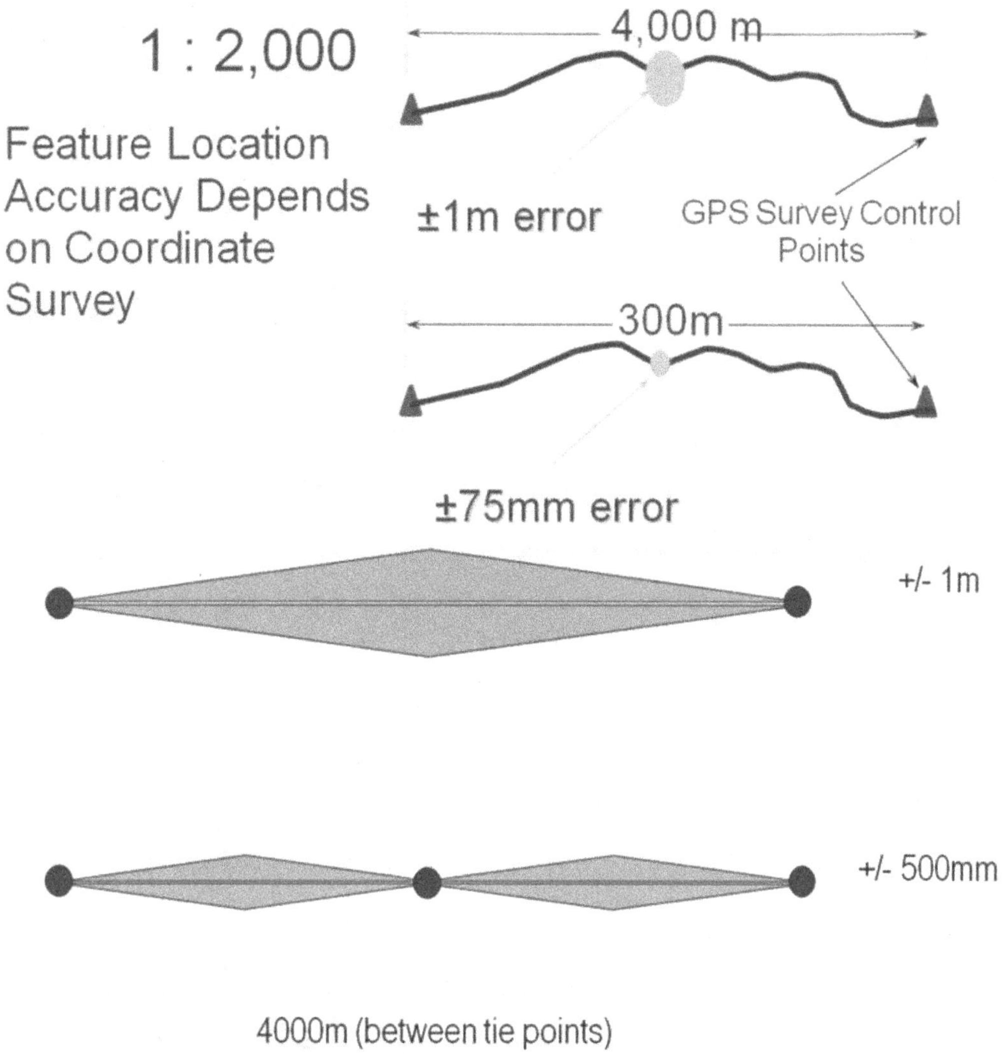

Figure 5-26. Mapping accuracy

Magnetic Particle Inspection (MPI)

MPI, (Figure 5-28) is an NDE method primarily used to detect surface-breaking flaws in ferromagnetic materials such as steel and iron. It can also be used to locate sub-surface flaws; however, its effectiveness quickly diminishes depending on the flaw depth and type. The MPI method, along with liquid penetrant inspection, is one of the oldest and most widely utilized forms of NDE.

Magnetic particle testing uses magnetic fields and small magnetic particles, such as iron filings, to detect flaws. The particles can be applied dry or wet by suspending them in a colored or fluorescent liquid.

The technique uses the principle that magnetic lines of force (flux) will be distorted by the presence of a flaw in a manner that will reveal the flaw's presence. The flaw (for example, a crack) is then located from the "flux leakage," following the application of fine iron particles, to the area under examination.

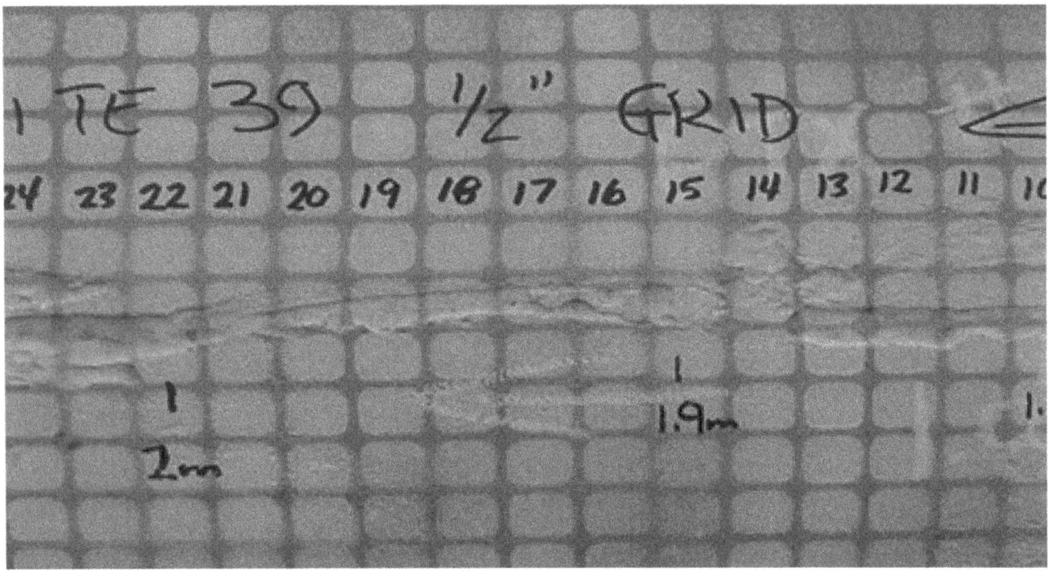

Figure 5-27. A cleaned, prepared surface for mapping of defects

Liquid penetrant inspection (LPI) is one of the most widely used NDE methods and is used to reveal surface breaking flaws by bleed out of a colored or fluorescent dye from the flaw. Its popularity can be attributed to its relative ease of use and its flexibility. The technique is based on the ability of a liquid to be drawn into a "clean" surface-breaking flaw by capillary action. After a period of time called the "dwell," the excess surface penetrant is removed and a developer is applied, which acts as a "blotter."

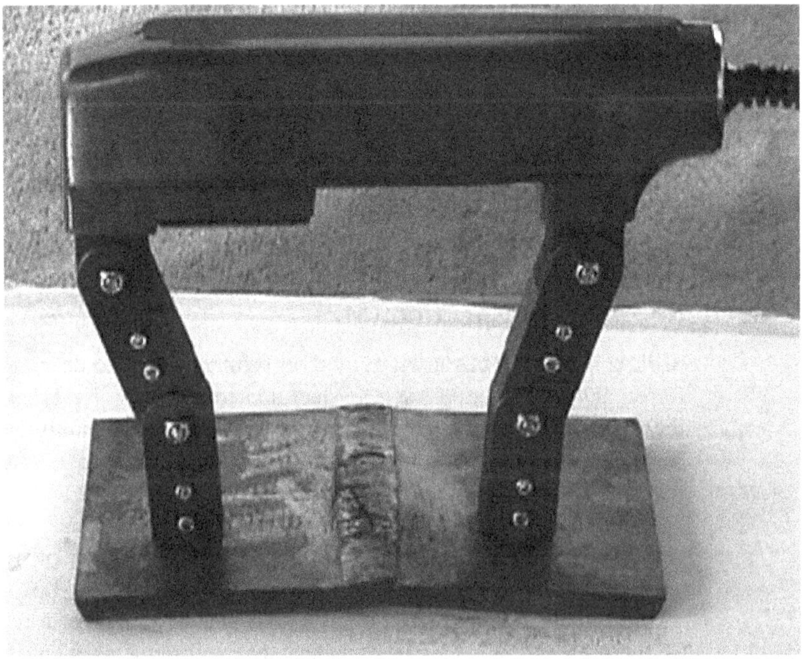

Figure 5-28. MPI

The developer draws the penetrant from the flaw to reveal its presence. Colored (contrast) penetrants require good white light, while fluorescent penetrants need to be used in darkened conditions with an ultraviolet "black light." Among the advantages of using dye penetrants we can list:

1. It has high sensitivity to small surface discontinuities.
2. Almost all types of materials may be inspected.
3. Parts with complex geometric shapes are routinely inspected.
4. Indications are produced directly on the surface of the part and provide a visual record.
5. Penetrant materials and associated equipment are relatively inexpensive. (Aerosol spray cans are commonly used.)
6. They can be used to accurately determine defect depth for removal by grinding.

However, the method also has some disadvantages:

1. Precleaning is critical as contaminants can mask defects.
2. The inspector must have direct access to the surface being inspected.
3. Surface finish and roughness can affect inspection sensitivity.
4. Multiple process operations must be performed and controlled.
5. Post cleaning of acceptable parts or materials is required.
6. Chemical handling and proper disposal is required.

Table 5-12 below provides a comparison between different MPI techniques.

Ultrasonic Testing

Hand-held wall thickness measurements are made with an ultrasonic transducer that transmits sound waves into the test material. These sound waves are reflected from the opposite or inside surface and travel back to the transducer. In essence, the gauge listens for the echo from the opposite side with ultrasonic testing, in order to reduce the energy input requirement, a coupling medium of liquid between the transducer and the test part is necessary. Ultrasound itself is a mechanical vibration or pressure wave. Vibrations above 15 kHz are generally referred to as ultrasound, but for nondestructive testing, the range is usually from 1 MHz to 30 MHz or higher. These sound waves can be highly directional and can be focused into a small spot or a thin line depending on the requirements. They can also be limited to a very short duration, which is important for fine longitudinal resolution or accurate thickness measurement.

UT inspection can be used to detect surface flaws such as cracks and internal flaws such as voids or inclusions of foreign material. Two methods are used for flaw detection: the through-transmission and the pulse-echo method. In the through-transmission test method, two transducers are used, one as a transmitter and the other as a receiver. The two transducers are located on opposite sides of the test part; clearly, this is not possible with a pipeline, so recourse is made to the pulse echo technique.

TABLE 5-12. MPI technique comparisons

Type of MPI	Sensitivity	Advantages	Disadvantages
Dry powder	2-5-mm-long cracks	Easiest to use	Lowest sensitivity
Wet fluorescent	1-mm-long cracks	Highest sensitivity	Requires UV lamp; can only be used in low light
Black on white contrast	1- to 2-mm-long cracks	Easily photographed	Requires application of white contrast paint

Ultrasonic Testing — Pulse Echo Technique

The transducer acts as both transmitter and receiver and produces a pressure wave in response to the electrical pulse, which has been applied to it (Figure 5-29). The pressure wave travels through the coupling medium, which is usually water, to the exposed pipe wall. At the interface of the coupling medium and the pipe, the ultrasonic pulse enters the pipe but a portion of the pressure wave is reflected back to the transducer. After entering the part, there is a partial reflection from the internal flaw. The remainder of the pulse travels to the back wall, where another partial reflection and transmission occurs.

The distances traveled in the pipe and the time segments of the electrical signal are proportional in the same material. Note, however, that the sound velocity is material dependent. By calibrating a particular specimen with a given sound velocity, the elapsed time between the front and back surfaces can be easily translated to a measure of pipe wall thickness.

The test described above uses a longitudinal wave inspection with normal incidence when the ultrasonic beam is perpendicular to the surface of the pipe. This method is used to detect internal flaws; however, it is not very suitable for detecting surface flaws such as cracks.

Shear Wave Transducer

Shear waves are used to detect both surface cracks and internal flaws (Figure 5-30). In a shear wave, the particle motion is perpendicular to the direction of wave propagation, while in a longitudinal wave, they are aligned. Shear waves are generated from longitudinal waves. When the angle of incidence of an ultrasonic beam is not 90 deg., refraction occurs and the beam splits into two parts, in a solid material: a longitudinal and a shear wave beam.

The longitudinal beam has a greater refraction angle than the shear wave. By increasing the angle of incidence at one point, the longitudinal beam ceases to exist and only the shear wave remains. This angle is referred to as the first critical angle.

A cross-section of a pipe is shown in Figure 5-30 with the detection of a crack on the inside surface of the pipe on the right hand side. A similar arrangement on the left side shows the geometry of flaw detection on the outside surface of the pipe.

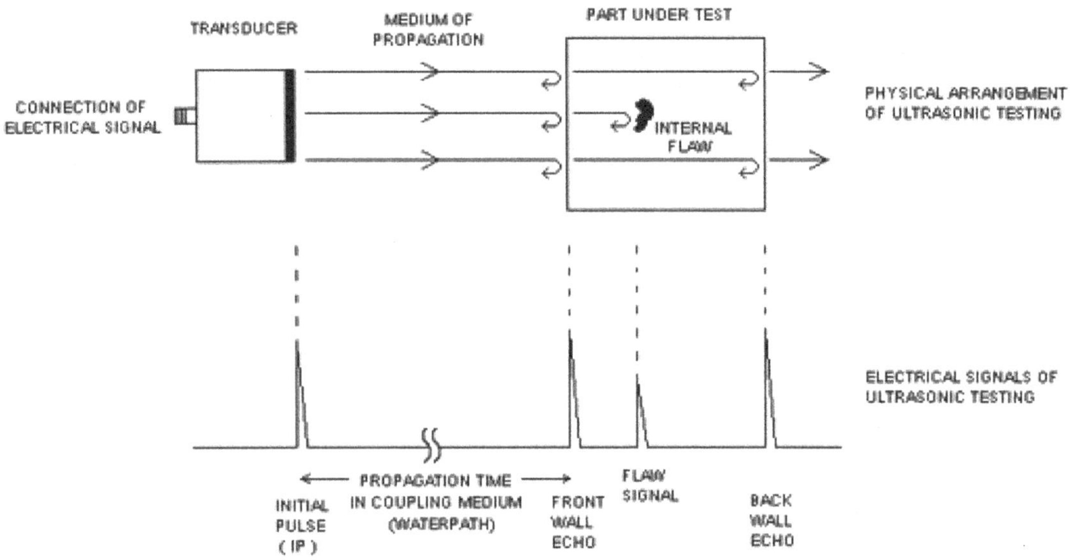

Figure 5-29. Pulsed echo technique

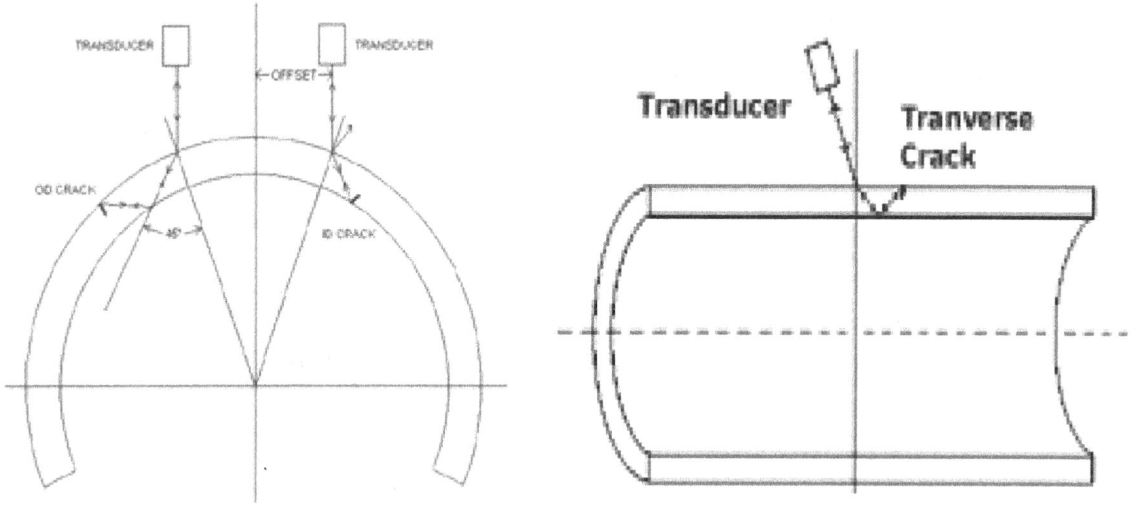

Figure 5-30. Shear wave probe

The shear wave can bounce several times between the I.D. and O.D. surfaces of the pipe, and any internal flaws in the path of the beam are detected as well. Notice that the right-hand transducer produces a shear wave that travels clockwise inside the pipe, while the left-hand transducer produces a shear wave traveling counter clockwise.

In order to detect transverse cracks, the transducer has to be located over the center of the pipe and angled in a plane containing the center line of the pipe.

Phased Array Testing

Phased array testing is a specialized type of ultrasonic testing that uses sophisticated multi-element array transducers and powerful software to steer high-frequency sound beams through the test piece and map returning echoes, producing detailed images of internal structures similar to medical ultrasound images. It is used for inspection of critical structural metals, pipeline welds, aerospace components, and similar applications where the additional information supplied by phased array inspection is valuable.

Eddy Current Testing

Eddy current testing utilizes the principle of electromagnetic induction to locate near-surface cracks, measure thickness, and categorize certain material properties in metals. An eddy current probe generates a magnetic field that induces currents that flow in a circular path in the test material. Changes in the integrity or thickness of the test piece will in turn affect current flow, the magnetic field, and ultimately the magnitude and phase of the voltage in the coil. The instrument monitors the probe output and displays information for analysis. When an energized coil is brought near to the surface of a metal component, eddy currents are induced into the specimen. These currents set up a magnetic field that tends to oppose the original magnetic field. The impedance of the coil in close proximity to the specimen is affected by the presence of the induced eddy currents in the specimen.

When the eddy currents in the specimen are distorted by the presence of the flaws or material variations, the impedance in the coil is altered. This change is measured and displayed in a manner that indicates the type of flaw or material condition.

Hydrostatic Testing

Hydrostatic testing of existing pipelines has been widely used to demonstrate or revalidate the pipeline integrity and serviceability (Kiefner 2001).

Hydrostatic testing involves isolating that portion of the pipeline undergoing testing, filling it with water, and then pressurizing the line to a specified pressure to check for leaks. US federal safety regulations for pipelines require that pipelines used to transport hazardous or highly volatile liquids be tested at a pressure equal to 125% of the MAOP for at least four continuous hours and for an additional four continuous hours at a pressure equal to 110% or more of the MAOP, if the line cannot also be visually inspected for leakage during the test.

This method of pressure testing also has long been an industry-accepted assessment method for validating a number of potential integrity threats to pipelines.

This integrity assessment method can be both a strength test and a leak test. Testing a pipeline above the operating pressure will demonstrate the absence of defects that could result in failure up to the test pressure at that given time. Not all anomalies will be removed during the test; only those reaching a critical size will be removed (Figure 5-31).

Several pipeline codes and standards, for example, ASME B31.8 (2007) and API RP 1110 (2007) — Pressure Testing of Steel Pipelines, contain details on conducting pressure tests for both post construction testing and for subsequent testing after a pipeline has been in service for a period of time. They specify the test pressure to be attained and the test duration in order to address certain threats. They also specify allowable test mediums and under what conditions the various test mediums can be used.

The operator should consider the results of the risk assessment and the expected types of anomalies to determine when to conduct inspections utilizing pressure testing. Gener-

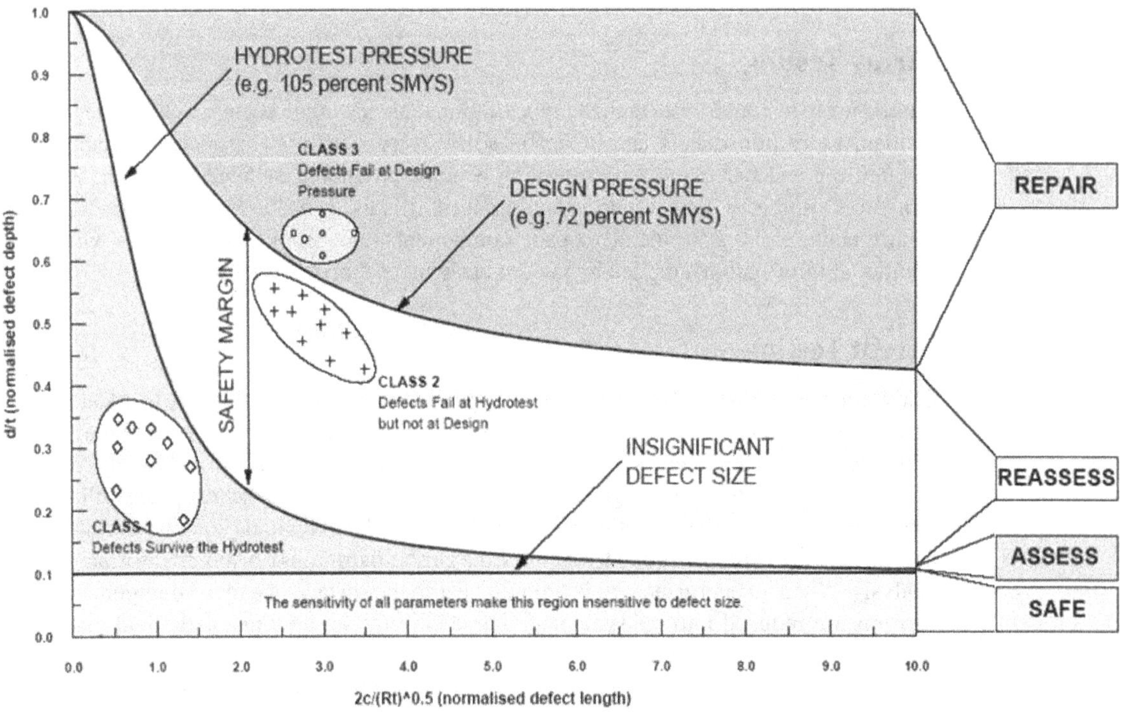

Figure 5-31. Effect of pressure on nondimensionalized defects

ally, defects such as corrosion-caused metal loss, dents, buckles, and some types of longitudinal cracks can be dealt with more effectively by using an appropriate ILI technology. If the type of defects suspected cannot be found reliably by means of an ILI tool, a hydrostatic test may be used to validate a safe operating pressure level for a specific period of time.

It may be noted that both field experience and full-scale laboratory tests have revealed the benefits and limitations of hydrotesting of a pipeline. However numerous ILI practices have also shown the superiority of ILI technologies over hydrostatic testing. For example, the MFL tool for the detection of metal loss caused by corrosion has shown that such ILI technologies can detect much smaller seam weld anomalies (Kariyawasam et al. 2006) than those found by hydrostatic testing of up to 110% pressure.

UT technology is capable of reliably detecting and sizing crack-like features larger than 30 mm (1.2 in.) in length and 1 mm (0.04 in.) in depth, including incipient of SCC This is significantly better than hydrotesting for crack detection in the critical regime of SCC (Figure 5-32).

Katz et al. (2005) indicate that another fundamental difference in crack detection and assessment between hydrostatic testing and ILI technologies is that the ILI provides detailed information on crack location, size, and distribution along the pipeline. Such data then can be used for the development of a comprehensive integrity management plan (Kariyawasam et al. 2006), whereas hydrostatic testing is a type of snapshot testing that removes all cracks greater than critical size at the test pressure but provides no information on the remaining sub-critical cracks.

In this regard, cost and long-term reliability of results would become a factor in deciding the choice between ILI technology and hydrostatic testing for revalidating the pipeline integrity. Needless to say, hydrostatic testing can have environmental implications due to discharging of hydrotest water, which in line inspection would not have.

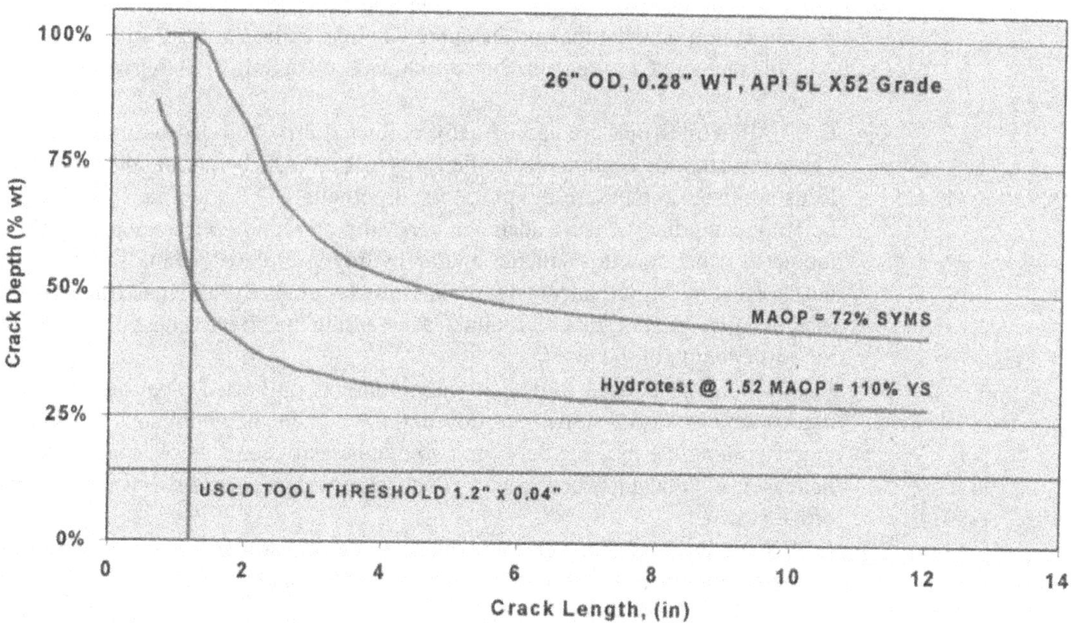

Figure 5-32. ILI and hydrostatic testing comparison (Katz et al. 2005)

DIRECT ASSESSMENT

As described within ASME B31.8S, direct assessment is an integrity assessment method utilizing a structured process through which the operator can utilize this assessment method in order to determine the integrity of the pipeline. This method is more commonly applicable to un-piggable natural gas pipelines and can be done by integrating knowledge of the physical characteristics and operating history of a pipeline system or segment with the results of inspection, examination, and evaluation, in order to determine the integrity.

B31.8S describes two types of assessment methods for direct assessment, namely, ECDA and ICDA.

EXTERNAL CORROSION DIRECT ASSESSMENT (ECDA)

ECDA is a formalized method of assessment of the integrity of a pipeline with respect to external corrosion. Pipeline operators have historically used a variety of electrical survey tools to manage external corrosion, and ECDA incorporates these tools in a formal process containing four steps. These four steps include:

1. Preassessment
2. Inspections
3. Examinations and evaluations
4. Postassessment

Preassessment includes compiling historic and current data to determine whether ECDA is feasible, defines ECDA regions, and selects indirect inspection tools.

Inspections are typically aboveground inspections to identify and define the severity of coating faults, other anomalies, and areas where corrosion activity may have or may be occurring. Two or more complementary indirect inspection tools are used over the entire pipeline segment under investigation to provide improved detection reliability under the wide variety of conditions that may be encountered along a pipeline right-of-way. Indirect inspection tools include: CIS, Pearson Survey, DCVG, and a Soil Resistivity Survey.

- *CIS* — Pipelines that are cathodically protected are equipped with test stations. These test stations contain electronic leads that are attached to the pipeline. These leads are used to measure the pipe-to-soil potential. The potential should be sufficiently cathodic to ensure adequate corrosion protection but not excessively cathodic to cause coating damage and/or hydrogen embrittlement. The potentials measured at these permanent test stations only originate from a small fraction of the total pipeline length. The CIS is aimed at assessing the effectiveness of the CP over the entire length of the pipeline.
- *Pearson Survey* is used to detect coating defects on buried pipelines. The technique employs an AC signal introduced to the pipeline and compares the potential gradient along the pipeline between two mobile earth contacts. At coating defects, increases in voltage gradient occur, which are noted on record sheets as the survey progresses.
- *DCVG* measures the direct current voltage gradients in the soil above a cathodically protected pipeline. This method is used for locating and sizing coating faults.
- *Soil Resistivity Surveys* are important when evaluating the severity of corrosive areas and helps in the selection of sites for CP installations. All soils conduct electrical current. However, while some soils have good conductivity, most soils have

poor conductivity. The resistivity of the soil is influenced by many factors, including soil type, moisture content, temperature, and electrolytes. The Soil Resistivity Survey helps identify the locations and severity of corrosive areas along the pipeline. Ideally, it should be undertaken at the same time each year so as to minimize the effect of changing moisture content.

Direct Examinations and Evaluations are performed at the most likely locations to contain coating or corrosion damage. This step includes prioritization of indications from indirect inspections; excavating locations where coating flaws and/or corrosion are most likely, measuring coating damage and corrosion defects, evaluation of damage severity, root cause analysis, and a process evaluation.

Post assessment defines reassessment intervals and assesses the overall effectiveness of the ECDA process.

INTERNAL CORROSION DIRECT ASSESSMENT (ICDA)

As described by ASME B31.8S, ICDA can be used for determining integrity for the internal corrosion threat on pipeline segments that normally carry dry gas but may suffer from short-term upsets of wet gas or free water (or other electrolytes). Examinations of low points or at inclines along a pipeline, which force an electrolyte such as water to first accumulate, provide information about the remaining length of pipe. If these low points have not corroded, then other locations further downstream are less likely to accumulate electrolytes and therefore can be considered free from corrosion. These downstream locations would not require examination.

Examinations are performed at locations where electrolyte accumulation is predicted. For most pipelines, it is expected that examination by radiography or ultrasonic NDE will be required to measure the remaining wall thickness at those locations. Once a site has been exposed, internal corrosion monitoring method(s) (e.g., coupon, probe, UT sensor) may allow an operator to extend the re-inspection interval and benefit from real-time monitoring in the locations most susceptible to internal corrosion.

For more information on both the ECDA and ICDA process as an integrity assessment method, see ASME B31.8S (2004), nonmandatory Appendix B, paragraphs B1 and B2. Also, NACE RP0502 — Standard Recommended Practice — Pipeline External Corrosion Direct Assessment Methodology, is recommended for guidance when conducting external corrosion direct assessment.

REFERENCES

API 1163, 2005, *In line Inspection Systems Qualification Standard*, 1st ed., American Petroleum Institute, August 2005, 37 pp.

API 1110, 2007, *Pressure Testing of Steel Pipelines for the Transportation of Gas, Petroleum Gas, Hazardous Liquids, Highly Volatile Liquids or Carbon Dioxide*, 5th ed., American Petroleum Institute, June 2007, 18 pp.

ASME B31.8S, 2004, *Managing System Integrity of Gas Pipelines*, American Society of Mechanical Engineers, January 14, 2005, 68 pp.

ASME B31.8, 2007, *Standard Gas Transmission and Distribution Systems*, American Society of Mechanical Engineers, November 30, 2007, 202 pp.

Beller, M., 2007, "Pipeline Inspection Utilizing Ultrasound Technology," Presented at PPSA Aberdeen Seminar, Germany, 2007.

Det Norske Veritas, 1999, *Recommended Practice RP-F101: Corroded Pipelines*, DNV, Norway.

ILI-PQ, 2005, *In-line Inspection Personnel Qualification and Certification*, American Society for Nondestructive Testing, January 1, 2005, 22 pp.

Kariyawasam, S., Colquhoun, I., and Larios, J., 2006, "Comparison of Technologies for SCC Threat Validation and Assessment." Paper number IPC2006-10434, Proceedings of 6th International Pipeline Conference (IPC 2006), Calgary, Alberta, Canada, September 25–29.

Katz, D., Gaom, M., Limon, S., and Krishnamurthy, R., 2005, "Advances in Crack Assessment for Pipeline Integrity," 11th International Conference of Fracture (ICF11), Paper 5068, Turin, Italy, http://www.icf11.com/proceeding/EXTENDED/5068.pdf.

Kiefner, J., F., 2001, "Role of Hydrostatic Testing in Pipeline Integrity Assessment," Northeast Pipeline Integrity Workshop, Albany, NY, June.

NACE RP0102, 2002, *In-Line Inspection of Pipelines*, National Association of Corrosion Engineers, December 1, 2000, 35 pp.

Westwood, S., and Hopkins, P., 2004, "Smart Pig Defect Tolerances: Qualifying the Benefits of Standard and High Resolution Pigs," Paper IPC04-0514, Proceedings of 5th International Pipeline Conference (IPC 2004), Calgary, Alberta, Canada, pp. 1181–1194. ASME, New York, NY.

Chapter 6

DEFECT ASSESSMENT

INTRODUCTION

The reliable operation of a pipeline is dependent on maintaining the structural integrity of the system. Defects, which could compromise structural integrity, can be introduced into the pipeline at any point throughout its life cycle, that is, during pipe manufacture, construction, and operation. It is vital therefore that the significance of these defects is assessed, and this process forms part of the regulatory framework for conducting engineering assessments in many jurisdictions. The word defect is used in a general sense here; however it has a specific meaning to those involved in pipeline integrity, so before progressing further, it will be worthwhile examining the definitions provided in API 1163 (2005).

Indication: A signal from an ILI system. An indication may be further classified, or characterized, as an anomaly or imperfection
Anomaly: An unexamined deviation from the norm in pipe material, coatings, or welds
Feature: Any physical object detected by an ILI system (i.e., anomalies, components, nearly metallic objects, etc.)
Imperfection: An anomaly with dimensions and characteristics that do not exceed acceptable limits
Defect: A physically examined anomaly with dimensions, or characteristics, that exceed acceptable limits

This chapter describes a number of quantitative methodologies that are available to the pipeline integrity specialist to evaluate the severity of a defect or damage to the pipeline. These methodologies can be broadly categorized as "fitness for service," or "fitness for purpose" assessments and their output is information, which is then used to decide whether to continue to operate the system, as is, in an altered state, or to repair, or replace the damaged pipe.

A fitness for service assessment may also be used to provide guidance on re-inspection or hydrostatic re-testing intervals as part of an integrity management program. In addition, it can be used to ensure that adequate management of change processes are in place when rehabilitating existing facilities, or undertaking a change of service. One needs to be aware that while some imperfections can remain benign over the entire life of the pipeline, others can grow over time into defects, depending on the operating conditions.

For the most part, the various defect assessment methodologies have a common basis — the branch of engineering science known as Fracture Mechanics, which links the stresses acting on the pipeline, its material properties, and the type and size of defect (Figure 6-1).

Hence, in order to conduct an effective defect assessment, it is important to have an understanding of each of these elements. Before discussing each of these, in turn, it will be useful to consider first the main types of defects found in pipelines and their cause.

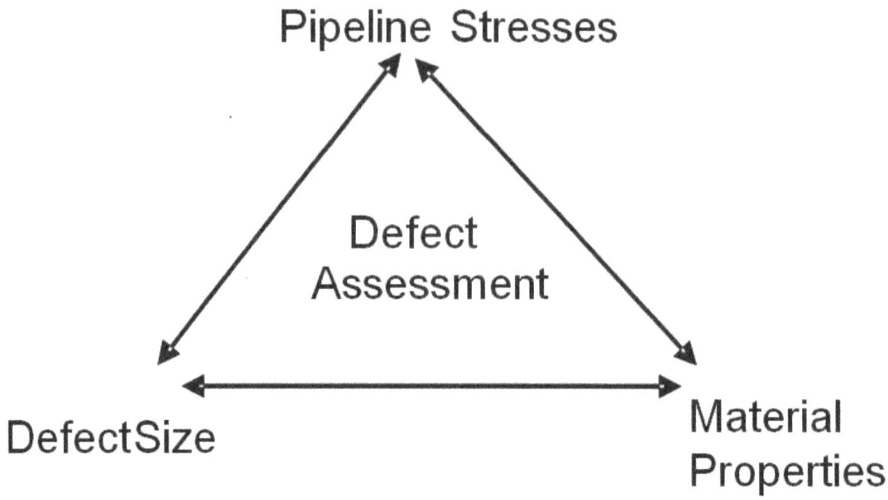

Figure 6-1. Defect assessment linkages

COMMON DEFECT TYPES

Imperfections and defects can be introduced into a pipeline at any point during manufacture, construction, and operation, with the manufacturing process including both steel production and pipe making (Table 6-1). Many of these imperfections will grow into time dependent defects as a consequence of cyclical loading or operating error such as over pressuring the pipeline. Misaligning pipe during welding, or damaging it by denting, can create stress concentrations, which become the sites for crack initiation.

Pipe Manufacture

Pipes are made from steel in one of three different ways:

1. broaching a cast steel ingot (e.g., seamless pipe)
2. forming from a rolled plate (e.g., SAW and spiral pipe) or
3. forming from a coiled strip (e.g., ERW pipe)

TABLE 6-1. Summary of failure modes

Metal loss	Cracking	Geotechnical Failure	External Damage	Material/Construction Defect	Other
External corrosion	Hydrogen induced	Washout/erosion	First-party damage	Defective long seam weld	Fire
Internal corrosion	Stress corrosion	Slope movement	Contractor/second-party damage	Defective girth weld	Over pressuring
Gouging	Delayed cracking (mechanical damage)	Undermining/ subsidence	Third-party damage	Defective spiral seam weld	SCADA malfunction
	Immediate cracking (mechanical damage)	Earthquake	Vandalism	Wrinkle/buckle	Lightning
	Corrosion fatigue		Terrorism	Defective pipe body	

Undetected imperfections present in the ingot, plate, or strip are transferred to the pipe in the manufacturing stage. It is more common these days in North America and Europe for steel to be continuously cast into slabs rather than poured into ingots. However, in the past and in some countries to day, ingots are still used, so it is important to realize how imperfections can be introduced through this process. Contraction of the steel during solidification results in a "pipe" forming in the ingot, and it should be removed prior to line pipe production as it could cause laminations in the finished pipe such as is shown in Figure 6-2.

The continuous casting process involves pouring the molten steel from the furnace on to a cooled chute where it forms a billet, which is withdrawn continuously to form either a coil strip or a plate. Three types of imperfection may be introduced during the casting process:

- alumina clouds (from the wall of the chute)
- oxidation of the metal during pouring
- center line segregation

Since seamless pipe is invariably used in aboveground facilities, such as process plants, rather than in pipeline construction, we will not discuss its method of manufacture. As noted earlier, there are three principal methods of manufacturing line pipe, the choice usually being dictated by pipe size and the grade of the steel. The electrical resistance welding process shown in Figure 6-3 does not involve the use of a welding consumable since the seam is produced by heating the edges of the rolled plate to a white heat using either a sliding contact DC resistance, or a high-frequency induction heater, and then pressing them together. The long seam is heat treated to ensure good grain refinement, and the resulting weld is tested ultrasonically as the continuous pipe travels along the production line. The entire pipe joint is hydrostatically tested, and since this process involves inserting sealing plugs into the ends of the pipe joint, it is essential that the pipe ends are subjected to ultrasonic testing. (This remains true for any method of pipe manufacture.)The edges of the coil strip from which the pipe is made have to be carefully machined as, historically, improper edge preparation as well as the use of low-frequency induction heating has been the cause of most of the defects introduced in this type of pipe. Figure 6-4 illustrates a hook crack that is most commonly found in low-frequency ERW pipe. In the CSA Z662 (2007) standard, "low frequency" means less than 1 kHz. The cause of hook cracking is metal separation around nonmetallic inclusions, and it is most commonly found in high-sulfur "dirty" steels, that is, steels with a sulfur content above 0.03%.

The presence of hard spots on the long seam has also been found, and these are attributable to intermittent cooling, while uneven edge heating and application of pressure from the sealing rollers results in stitching defects. This latter problem is confined to low-frequency ERW pipe manufacture and is directly related to the creation of variable material properties

Figure 6-2. Plate lamination defect

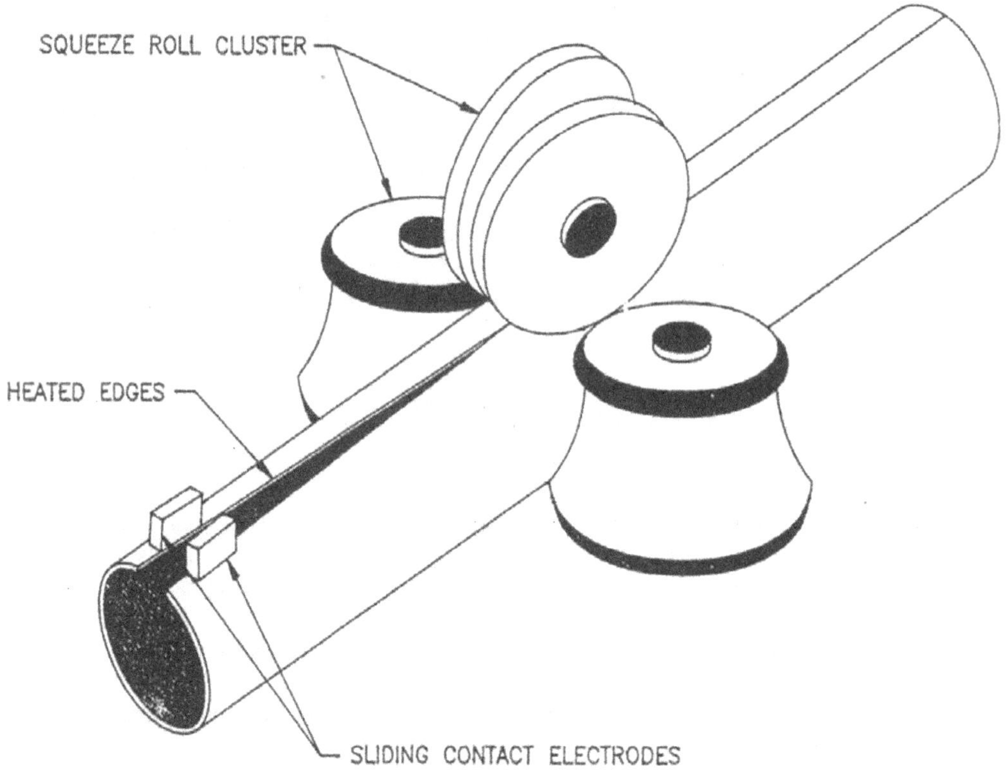

Figure 6-3. ERW line pipe manufacture using DC sliding contacts

at regular intervals along the length of the seam. There is a size restriction with this type of pipe manufacture, with the largest diameter being approximately 600 mm.

The submerged arc method of making the pipe long seam is used in the spiral welding process, where coil strip, with prepared edges, is drawn through straightening rollers at an

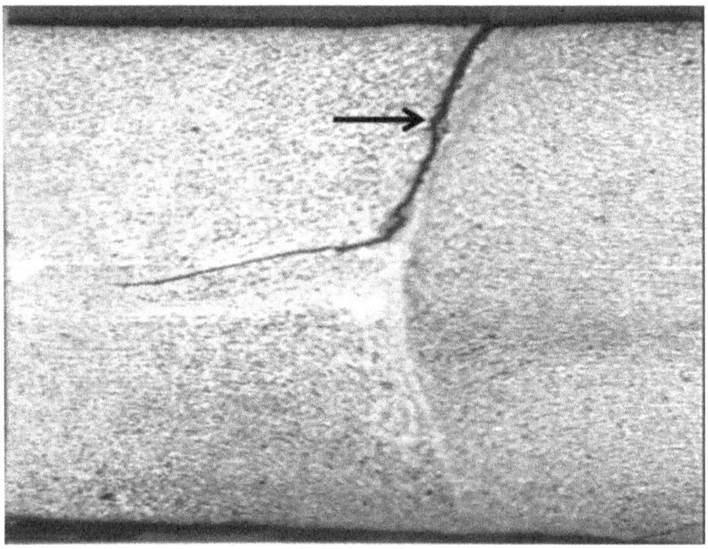

Figure 6-4. Hook crack in a low-frequency ERW pipe

angle to the direction of pipe manufacture, to form a helically wound cylinder. This process enables the manufacture of larger diameter pipe than the ERW process and has been used to produce pipe of X70 grade of relatively heavy wall thickness in diameters up to 1000 mm.

Higher grade steels and those of thicker wall can be made using the UOE process (see Figure 6-5). As with the other methods, the first step in manufacture is to inspect the stock material using ultrasonic probes. The plate is then edge milled to the correct width before being pressed into a "U" shape, then an "O" shape, before being tack welded internally. The "O" press is used to obtain uniformity in the circumferential length, maintain an appropriate seam gap, avoid low peaking, and provide uniform curvature. The long seam is completed using submerged arc welding, and the ends of the pipe are rounded and expanded hydraulically to meet the dimensional requirements. The amount of expansion is up to 1.5% (maximum) of the original diameter, meaning there is permanent set at the ends of the pipe. Another purpose in mechanically expanding the pipe ends is to remove residual stress produced by the thermal effects of submerged arc welding and the removal of local strains generated during UO forming. This action improves roundness, straightness and diameter sizing. Recovery of mechanical strength is lowered by the Bauschinger effect generated during the "UO" forming.

Figure 6-6 illustrates four types of weld defect that are found during submerged arc welding pipe manufacture:

1. Weld bead cracking associated with center line solidification. Side wall cracking is also possible, caused by incorrect welding parameters or center line segregation in the plate material.

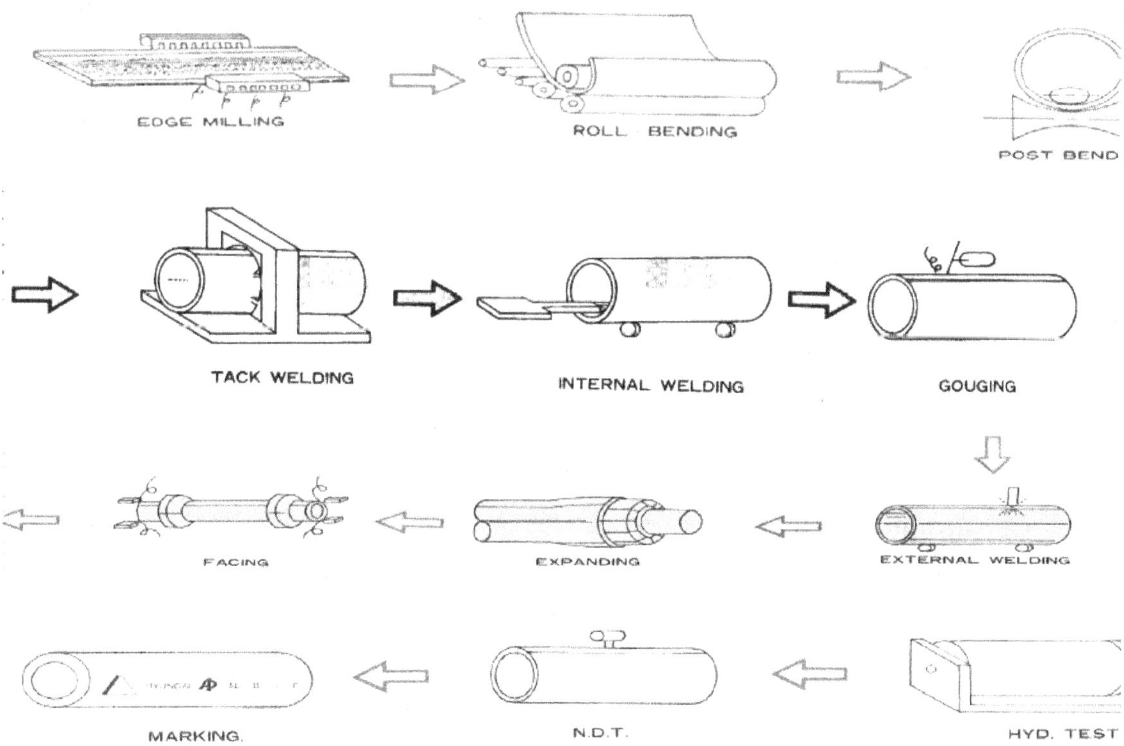

Figure 6-5. The UOE long-seam submerged arc welding line pipe manufacturing process (courtesy of Hyundai Heavy Industries)

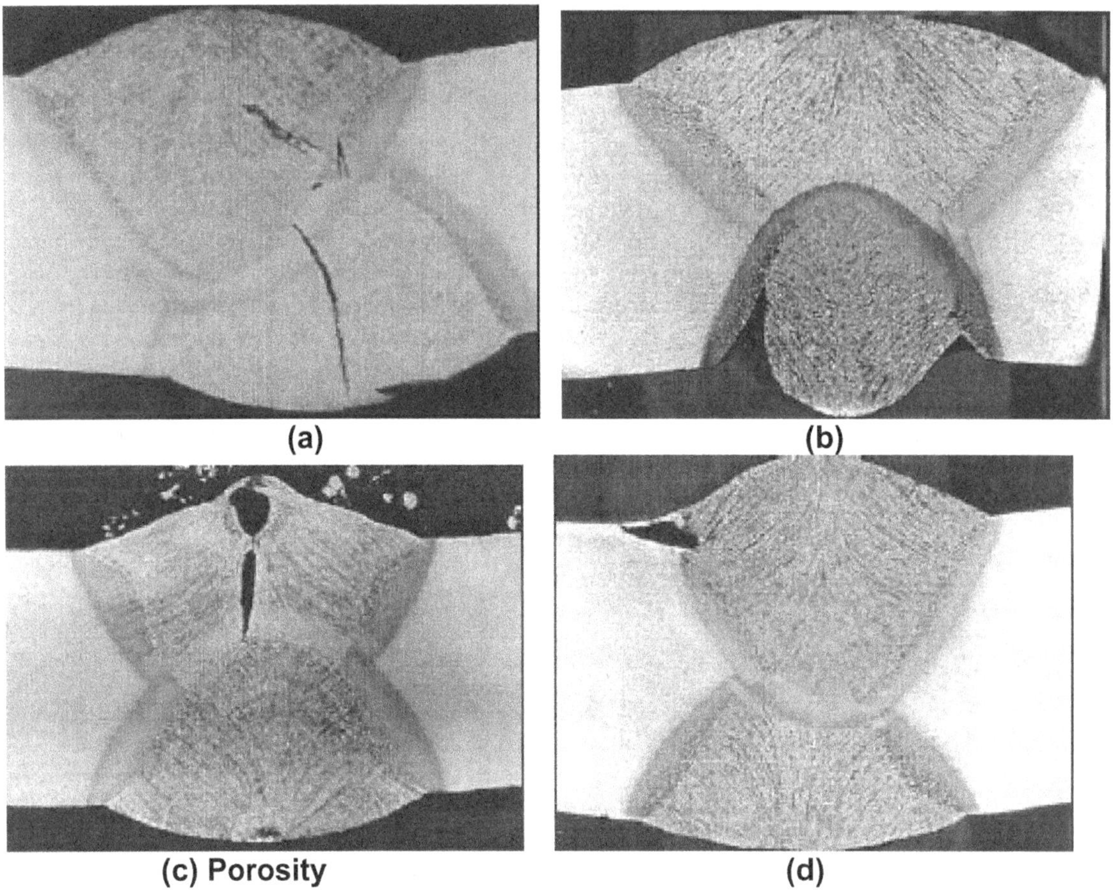

Figure 6-6. Submerged arc welding pipe making defects [courtesy of Alasdair Clyne (DnV)]

2. Misalignment or poor pipe fit up that can lead to lack of fusion defects.
3. Undercutting or lack of fill is caused by incorrect welding procedures; this type of defect can easily become the site of crack initiation.
4. Porosity due to the use of damp flux or incomplete flux cover. This is often found in conjunction with other types of weld defects.
5. Lamellar tearing and toe cracking due to non-metallic inclusions in the pipe material; it is possible for such cracks to grow as a result of cold expansion of the pipe ends.

Construction Defects

Construction defects can be caused through mishandling the pipe during transportation, lowering in, and backfilling operations. Detailed descriptions of these activities may be found in Chapter 9 of Mohitpour et al. (2007). If the pipe has been loaded properly, that is, in accordance with appropriate pipeline standards, then during transportation by sea, rail or truck the possibility of out of roundness will be reduced, as will the likelihood of damage to any pipe coating.

Denting of the pipe can arise either through poor padding practice, laying the pipe on boulders, as shown in Figure 6-7, boulders striking the pipe during backfill, or from external mechanical damage as shown in Figure 6-8.

Figure 6-7. External denting caused by poor laying practice

Defects may also be introduced through the girth welding process with cracking caused by improper welding procedures, HIC, or poor fit up (misalignment). Arc burns are caused by inadvertently making contact with the pipe and involve metal loss. Incorrect spacing of side booms, during lowering in, can lead to excessive bending stress and wrinkling or kinking on the compressive surface of the pipe.

Operational Defects

It is during operations that imperfections in the pipe body or welds can grow to become defects. This is particularly the case for cracks, which may grow as result of cyclical pressure loading. Incorrect operations, such as over pressuring, can occur if there is an over-reliance placed upon the logistical accuracy and capability of the SCADA system and no provision is made for pressure relief. A number of liquid pipeline ruptures, for example, have been traced directly to valve closures going undetected while pumps continue to operate. Corrosion, both internal and external, can present a continual integrity threat.

External damage can also occur during operations and is closely related to encroachment on to the right of way and the need on the part of the pipeline operator to be vigilant. Impact from farm ploughing implements, or a back hoe, can cause both denting and gouging and involve a geometric distortion of the pipe shell in the case of a dent, or metal loss from a gouge. It is also possible to have a combination of the two (Figure 6-9). The action of denting the pipe may cause permanent plastic deformation of the circular cross-section of the pipe as well as extensive cold working of the material adjacent to the dent. Indeed, as we shall see later, it is customary when analyzing a dent defect to make an increased depth allowance for possible cracking or metal loss at the base of the dent.

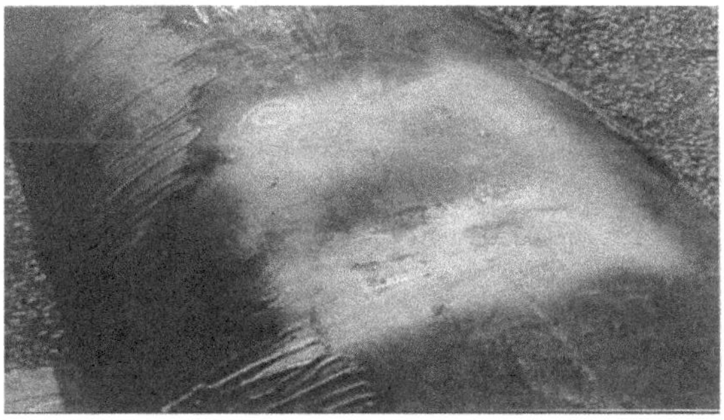

Figure 6-8. Dented pipe caused by mechanical damage

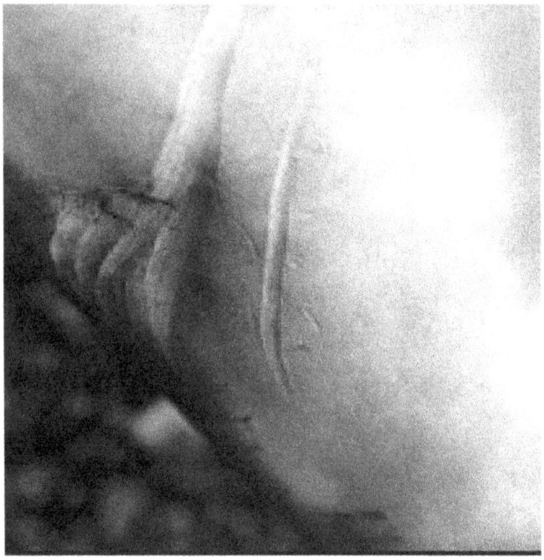

Figure 6-9. A dented pipe with a deep gouge

Driver and Zimmerman (1998) studied the likelihood of puncturing a pressurized pipe as a result of external third-party interference and were able to relate the puncture resistance to the tooth size (Figure 6-10) and capacity of the equipment. Whether a pipe is gouged, when it is struck, depends on a number of factors including:

- the shape and sharpness of the indenter
- the pipe material properties
- the frictional resistance between the surface of pipe and the indenter
- the pipe geometry (diameter and wall thickness)
- whether or not the pipeline is pressurized

Figure 6-10. Size of potential indenter (back hoe teeth)

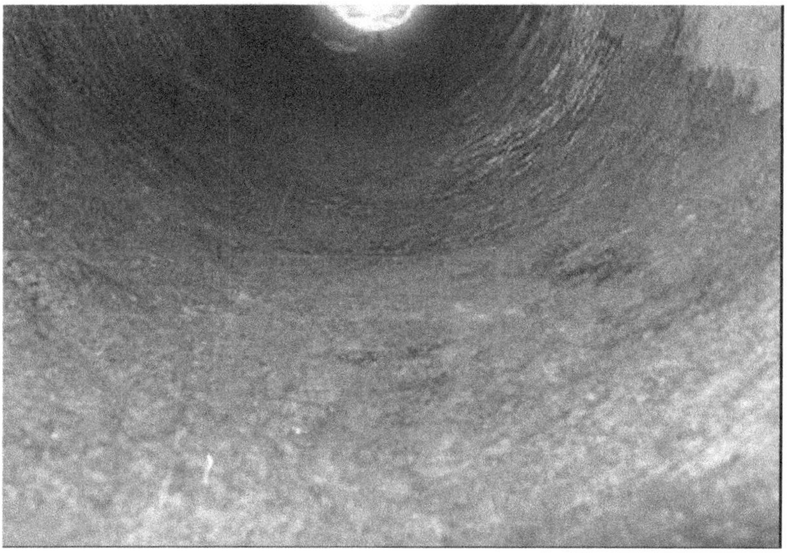

Figure 6-11. Microbial-induced internal corrosion

Corrosion is a major cause of pipe defects resulting in metal loss and can occur internally, depending on the type of product being transported, as well as externally. Microbial-induced corrosion shown in Figure 6-11 is a consequence of the presence of sulfur-reducing bacteria within the flowing product. External corrosion occurs as a result of a failure of the pipe coating — the pipeline's first line of defense. Ideally, when a coating fails, it should not shield the cathodic protection, rendering it ineffective.

However, this is not always the case, particularly with tape coatings. The corrosion may be generalized as shown in Figure 6-12(a), or very localized in the form of deep pitting

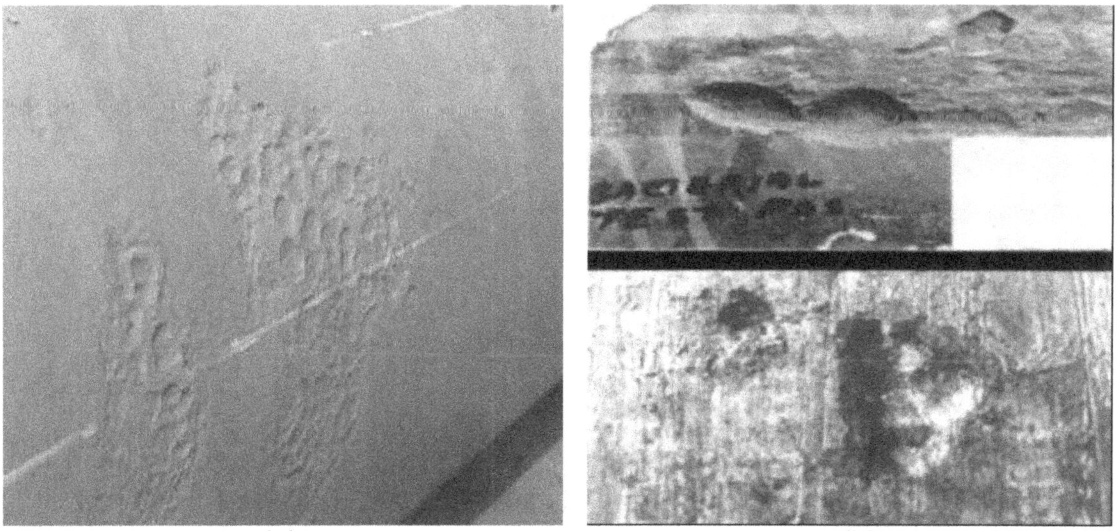

(a) (b)

Figure 6-12. Corrosion: external generalized (a) and pitting corrosion (b)

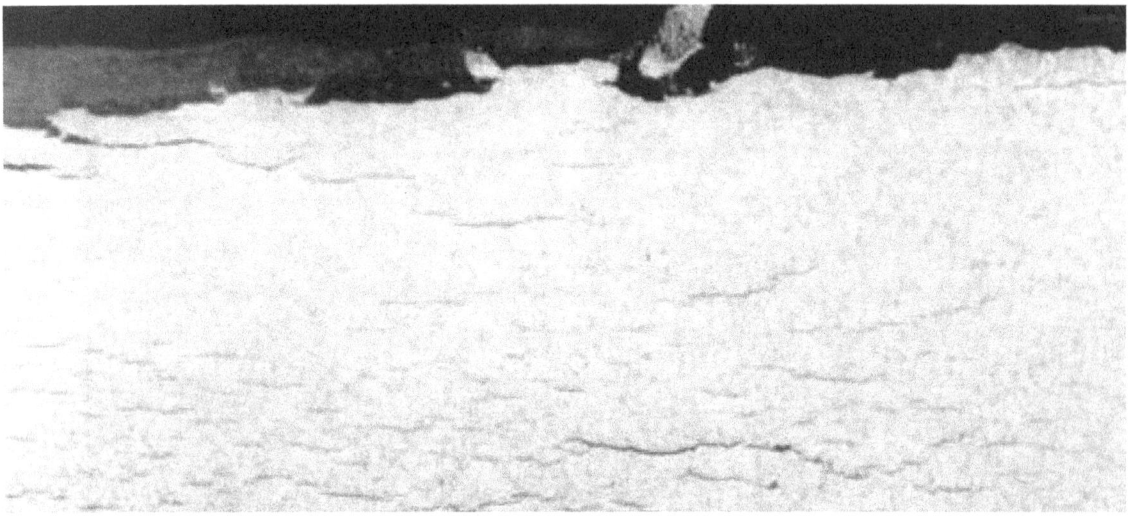

Figure 6-13. Stress corrosion cracking

as can be seen in Figure 6-12(b). Another potential defect threat, caused in part, by failure to adequately protect the pipe surface, is stress corrosion cracking (Figure 6-13). These surface breaking cracks are either inter-granular, (high pH), or transgranular (near neutral pH) in nature and can coalesce to form cracks of a critical length.

Slope movement and subsidence can impose significant longitudinal bending stresses on a buried pipe, which may result in buckling or wrinkling (Figure 6-14a). Both can cause a high localized radius of curvature with associated very high strains, perhaps sufficient to tear the pipe surface (Figure 6-14b).

Table 6-1 provides a summary of possible pipe failure modes based upon the threats identified in the ASME B31.8S pipeline integrity standard (2004).

(a) (b)

Figure 6-14. Pipe buckling and wrinkling (a) and localized tearing (b)

PIPELINE STRESSES

The second element in defect assessment is the determination of the stresses induced in a pipeline as a consequence of the applied loading and the type of restraint offered. Buried pipelines can be regarded as being fully restrained if their burial length is such as to mobilize sufficient soil frictional resistance as to restrict longitudinal movement. On the other hand, aboveground piping may be either restrained or unrestrained, depending on how it is supported and/or tied in to other facilities.

Almost all onshore pipelines have been designed in accordance with working stress design principles, whereby the worst case loading conditions include the effects of pressure and temperature; axial forces acting along the length of the pipeline; and secondary bending stresses. These bending stresses are caused, for example, by differential settlement and perhaps surface loads. The working stress design philosophy attempts to take account of any uncertainties in characterizing loads, and the structural response of the pipeline, through the use of safety factors. For example, the MOP, of the pipeline, referred to in the United States as MAOP, is determined by modifying the well-known Barlow formula used in determining the hoop stress in a thin walled pipe.

$$P = \frac{2t\sigma_{\text{smys}}}{DFJT} \tag{6-1}$$

where

σ_{smys} = specified minimum yield stress for the pipe material
t = wall thickness
D = pipe outside diameter
F = design factor including for gas pipelines a location factor
J = joint factor
T = temperature-derating factor

The factors F, J, and T may be found in the appropriate pipeline standard.

The effect of internal pressure will cause the pipe circumference to expand, and as a consequence, the pipe will try to contract longitudinally, due to the Poisson effect (i.e., conservation of material). If this contraction is restrained (by an anchor block, soil friction etc.), a longitudinal stress will develop in the pipe.

The magnitude of the longitudinal stress is approximately $\nu\sigma_{\text{h}}$, where σ_{h} is the hoop stress and ν is the material's Poisson ratio ($\nu \approx 0.3$ for steel). This induced stress may act in combination with other longitudinal stresses such as are caused by thermal effects or where a pipe end is capped or the pipe changes direction rising vertically out of the ground. In the latter case, axial stress is induced by the longitudinal force due to the internal pressure acting on the pipe ends.

By considering horizontal force equilibrium (shown in Figure 6-15), the resultant longitudinal force in the pipe wall, F, is approximately $P(\pi D_i^2/4)$, acting on a cross-sectional area of $\pi(D_o^2 - D_i^2)/4$. Thus, the axial stress σ_x is,

$$\sigma_x = \frac{F}{A} = \frac{P\pi D_i^2/4}{\pi(D_o^2 - D_i^2)/4} = \frac{PD_i^2}{D_o^2 - D_i^2} = \frac{PD_i^2}{(D_i + 2t)^2 - D_i^2}: \tag{6-2}$$

For a thin wall pipe, the axial stress found using Eq. (6-2) can be approximated with little error by writing it as:

$$\sigma_x = \frac{PD}{4t} \tag{6-3}$$

where D is the nominal (outside) diameter.

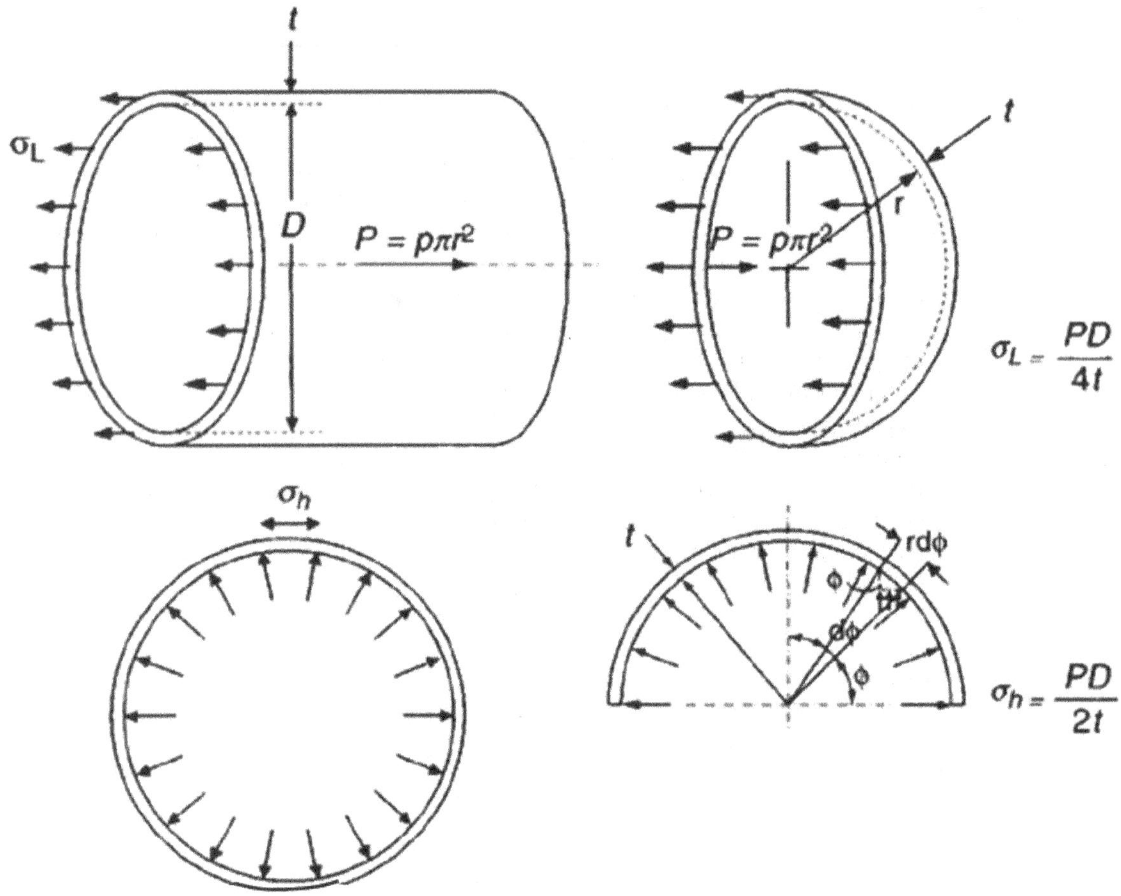

Figure 6-15. Equilibrium considerations for an internally pressurized pipe

Thus, the axial stress caused by internal pressure on the pipe end is equal to half of the hoop stress. It is because the longitudinal stress is comparatively smaller than the hoop or circumferential stress that we are primarily interested in those defects, which are aligned with the longitudinal axis. Combining the effect of both the longitudinal and circumferential stresses is important, however, in establishing the stress field within the pipe surface.

Longitudinal thermal stress is set up in the pipe whenever the flowing temperature of the process fluid differs from the temperature at installation, assuming conditions of full restraint. It is given by,

$$\sigma_{LT} = E\alpha(T_2 - T_1) = E\alpha\Delta T \qquad (6-4)$$

where

α = coefficient of linear expansion for the pipe material
E = elastic modulus
T_1 = initial temperature
T_2 = final temperature
$\Delta T = T_2 - T_1$

Hence, the longitudinal stress in a buried pipeline due to pressure and temperature is given by,

$$\sigma_L = \sigma_x + \sigma_{LT} = \frac{\nu PD}{2t} - E\alpha\Delta T \tag{6-5}$$

A negative value for σ_{LT} indicates a compressive stress due to temperature, which is the case when a buried pipeline is operated at a temperature greater than the installation temperature. When the temperature difference ΔT is such that the absolute value of the second term in Eq. (6-5) is greater than the first, then the pipe will be in compression. The maximum effective stress due to the combination of pressure and temperature is the algebraic sum,

$$\sigma_{eff} = \sigma_x - \sigma_L \tag{6-6}$$

In most pipeline standards, the value of σ_{eff} is limited to 0.9 of the yield stress, derated as required for temperature. As we shall see later in Theories of Failure, this limit is based upon a failure criterion known as the maximum shear stress, or Tresca criterion.

Bending stress can be introduced to the pipe during its service life through surface loading, differential settlement, or line lowering activity. Detailed descriptions of the methods available to determine the bending stresses from these conditions may be found in Mohitpour et al. (2007).

Bending stress may also come about as a result of lowering in during installation, indeed this condition may be the most onerous stress condition experienced by the girth weld and should form part of the weld defect acceptance assessment process. Since the pipe is suspended at a number of points along its lowering in length, it behaves essentially as a continuous beam over many supports. An analysis is further complicated by the fact that during lowering in the continuous pipe will not lie in a single plane (Figure 6-16); hence, a three-dimensional structural analysis will be required to determine the magnitude and locations of the largest stresses (Sen and Zhou 2008).

Another potential limiting stress condition, which may require investigation at some point, occurs when a liquid pipeline undergoes conditions of slack flow, where a free surface develops within the pipe. This could lead to the formation of a partial vacuum so the local buckling strength of the pipe should be checked using the following equation,

$$P_{cr} = \frac{2E}{1-\nu^2}\left(\frac{t}{D}\right)^3 \tag{6-7}$$

where

P_{cr} = critical pressure to cause buckling
ν = Poisson's ratio

Example:

An NPS 36 pipe (914-mm diameter), with a wall thickness of 6-mm experiences a slack flow condition; verify that structural collapse will not occur.

Solution:

$$P_{cr} = \frac{2\times 208\times 10^6}{1-0.3^2}\left(\frac{6}{914}\right)^3 = 129 \text{ kPa}$$

Since the negative pressure due to a vacuum cannot exceed 101.325 kPa (standard atmospheric pressure), there is no possibility of collapse.

Figure 6-16. Non-planar lowering in of strung pipe

Principal Stresses

It is clear from the foregoing that, in a pipeline system, stresses can act simultaneously in several different planes. Assuming for the moment that internal pressure is large when compared to the direct stresses in the longitudinal and tangential (circumferential) directions. (Although as we shall see later, this is generally NOT the case.) Figure 6-17 illustrates the normal and shear stresses acting on an element within the pipe wall thickness. The tri-axial state of stress acting on an element within the pipe wall can be written in matrix notation as,

$$\begin{bmatrix} \sigma_x & \tau_{xy} & \tau_{xz} \\ \tau_{xy} & \sigma_y & \tau_{yz} \\ \tau_{xz} & \tau_{yz} & \sigma_z \end{bmatrix} \qquad (6-8)$$

In a generalized stress system, by performing a number of axis rotation transformations to this matrix, one obtains a coordinate system x', y', z' such that the shear stresses disappear and the state of stress can be described purely in terms of the principal stresses σ_x', σ_y', and σ_z', represented on the diagonal of the matrix as shown in Eq. (6-9). The axisymmetric nature of an internally pressurized cylinder is such that its longitudinal axis and the two orthogonal diametral axes aligned in the x and y directions are principal axes, so that the primes indicated can be omitted and the stresses in these directions are principal stresses.

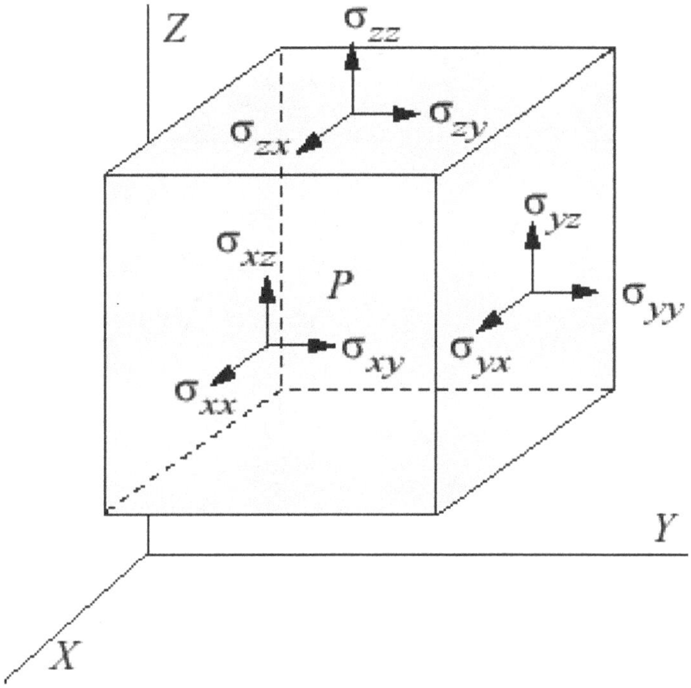

Figure 6-17. Three-dimensional stress state on a pipe wall element

$$\sigma_{PR} = \begin{bmatrix} \sigma_{x'} & 0 & 0 \\ 0 & \sigma_{y'} & 0_{yz} \\ 0 & 0 & \sigma_{z'} \end{bmatrix} \qquad (6-9)$$

For convenience, these principal stresses can be rewritten as σ_1, σ_2, σ_3 and they can be ranked in that order in terms of their relative size. The internal and/or the external pressure in a pipe, which corresponds to the radial stress in a thin wall, is negligibly small when compared to the stresses in the other two orthogonal directions, so the pipe surface can be regarded as being in a bi-axial stress state. This type of condition is described as plane stress and in matrix terms the relationship is as shown in Eq. (6-10).

$$\begin{bmatrix} \sigma_x & \tau_{xy} \\ \tau_{yx} & \sigma_y \end{bmatrix} \qquad (6-10)$$

where $\tau_{xy} = \tau_{yx}$ for static equilibrium.

Mohr's Circle

Mohr's circle is a graphical means of determining the principal stresses (strains) within a structural element if the stress (strain) state at any point on its surface is known (Figure 6-18). It was developed in 1882 by Otto Mohr and was widely used by engineers in the era before hand-held calculators and microcomputers. It is a highly intuitive way of showing the relationships between normal and shear stresses in a structure and for quickly estimating the maximum stresses. It is also useful as an aid when conducting failure investigations

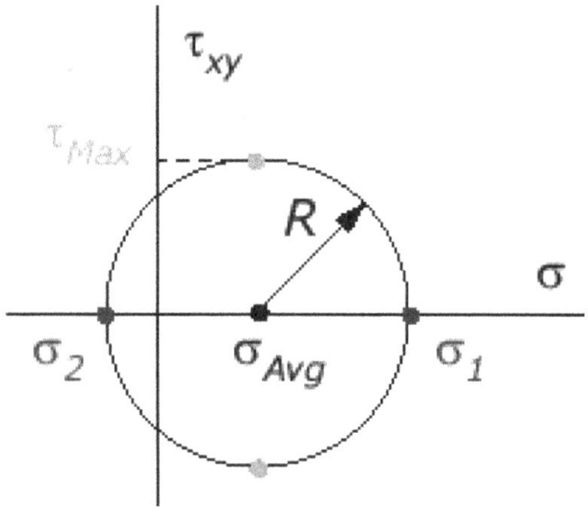

Figure 6-18. Mohr's Circle of stress

since one can often deduce from it the direction of the failure load. The interested reader is directed to any standard text on Strength of Materials for a complete description of the underlying equations relating the stress state at any point to the principal stresses. It is sufficient for the present purpose to accept that the normal and shear stress components acting on an element will always lie on Mohr's Circle. It was noted earlier that in a pipeline system the principal stresses align with the principal axes, and that they equal the maximum values of the longitudinal and circumferential stress. It is a subtlety of the Mohr's Circle method that the angles between the stresses on the surface of the element are doubled within the circle. It follows therefore that the orthogonal longitudinal and circumferential stresses must lie on the diameter of the Mohr's Circle. The maximum shear stress on the pipe surface occurs when the stress element is rotated 45° away from the principal directions and hence must be perpendicular to them when plotted on the circle. Thus, in Figure 6-18, we can construct the circle quickly by simply plotting the values σ_1 (say hoop stress) and σ_2 (longitudinal stress) along the abscissa. The maximum shear stress is thus given by the intercept on the vertical axis and numerically must be equal to,

$$\tau_{xy} = (\sigma_1 - \sigma_2)/2 = (\sigma_h - \sigma_l)/2 \qquad (6-11)$$

Theories of Failure

The purpose in expressing the longitudinal and circumferential stresses on a pipe surface in terms of principal stresses is to relate them to implicit failure criteria contained within the pipeline codes and standards. A considerable number of such failure theories exist; however, only the most common are discussed here. These theories, dependent on the nature of the material in question (i.e., brittle or ductile), are listed in Table 6-2.

The distinction between ductile and brittle material type behavior is largely a function of the strain level involved, with a 5% elongation criterion at failure being both arbitrary and reasonable. All of the stresses used in the above theories are principal stresses.

In the Rankine formulation, it is assumed that failure occurs when one of the three principal stresses equals the ultimate strength of the material in either tension σ_T or compression σ_c. Assuming $\sigma_1 > \sigma_2 > \sigma_3$ then failure occurs when $\sigma_1 = \sigma_T$, or $\sigma_3 = \sigma_C$.

This criterion is more commonly used with brittle materials, than with modern line pipe.

TABLE 6-2. Failure theories

Material Type	Failure Theory
Brittle	Rankine's maximum normal stress criterion
Ductile	Tresca maximum shear stress, Von Mises strain energy

The Tresca shear stress theory states that yielding will start when the maximum shear stress equals the value of the maximum shear stress in a tension test specimen of the same material, when that specimen starts to yield.

The theory also predicts that the yield strength in shear, $\tau_{xy} = 0.5\sigma_y$. Referring back to Eq. (6-11), we can include this equality to give,

$$\tau_{xy} = (\sigma_1 - \sigma_2)/2 = (\sigma_h - \sigma_l)/2 = \sigma_y/2 \qquad (6-12)$$

From which, it can seen $(\sigma_h - \sigma_l) = \sigma_y$. This is very similar to the formula given in B31.8 and other codes for limiting the value of combined stress, to $0.9\sigma_y$, when the longitudinal stress is compressive (negative), indicating that there is an implicit factor of safety against yielding.

Note that while the criterion relates the maximum shear stress to the difference in the principal stresses, the largest of these stresses must itself be less than the yield shear stress. i.e.,

$$|\sigma_h| \le \sigma_y; |\sigma_l)| \le \sigma_y; |\sigma_h - \sigma_l| \le \sigma_y \qquad (6-13)$$

The Tresca criterion is widely used with ductile materials but can be quite conservative for some load combinations. It requires that the two principal stresses lie within the shaded zone shown in Figure 6-19.

The Von Mises theory is also known as the maximum distortional energy theory. It states that yielding will occur when the distortional energy in a unit volume of material reaches the same energy for yield/failure as a uniaxial tension test specimen of the same material.

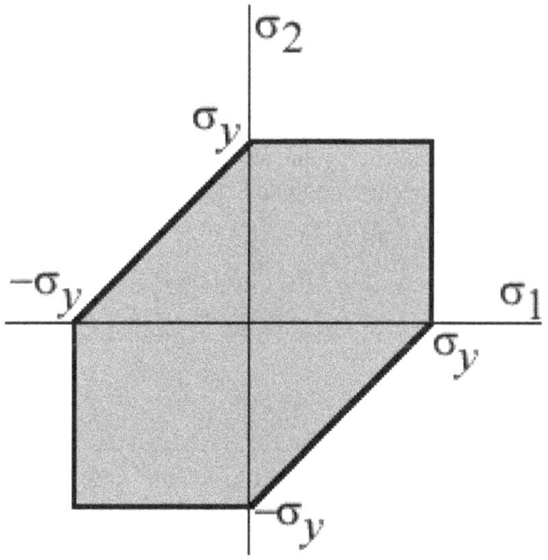

Figure 6-19. Tresca shear stress criterion yield envelope

Mathematically, this condition may be written as,

$$\frac{1}{2}\left[(\sigma_1-\sigma_2)^2+(\sigma_2-\sigma_3)^2+(\sigma_3-\sigma_1)^2\right]\le\sigma_y^2 \qquad (6-14)$$

For a plane stress condition $\sigma_3 = 0$ and Eq. (6-12) reduces to

$$\sigma_1^2-\sigma_1\sigma_2+\sigma_2^2\le\sigma_y^2 \qquad (6-15)$$

This is the equation of the ellipse that would circumscribe the Tresca Yield envelope.

Examples:

Determine the maximum shear stress on the outer surface of a fully restrained buried pipe initially pressurized to 7000 kPa. The outside diameter of the pipe is 400 mm and the wall thickness 8 mm. It may be assumed that there are no temperature effects present.

Solution:

The hoop stress is found from,

$$\sigma_h=\frac{Pd}{2t}=\frac{7000\times 400}{2\times 8}=175\ \text{MPa}$$

and the axial stress =

$$\sigma_A=\frac{\nu Pd}{2t}=0.3\frac{7000\times 400}{2\times 8}=52.5\ \text{MPa}$$

The stress in the radial direction is due the pressure differential between the inside and outside of the pipe; hence $\sigma_3 = 7$ MPa.

Since there are no shear stresses present in the radial, longitudinal, and tangential directions, the principal stresses are $\sigma_1 = 175$ MPa, $\sigma_2 = 52.5$ MPa, and $\sigma_3 = 7$ MPa.

The maximum shear stress is found from

$$\tau=(\sigma_1-\sigma_3)/2=\frac{175-7}{2}=84\ \text{MPa}$$

As discussed earlier, it is more usual and practical to consider a thin walled pipe as being in a two-dimensional stress state in which case the maximum shear stress is found by considering only the hoop and longitudinal stresses, that is,

$$\tau=(\sigma_1-\sigma_2)/2=\frac{175-52.5}{2}=61.25\ \text{MPa}$$

MATERIAL PROPERTIES

While aluminum, polyethylene, and fiber glass composites are used as pipeline materials, steel is invariably used for the transportation of hydrocarbons in large quantities and at high pressures. Consequently, it is the properties of line pipe steel that will be the focus of attention in this section. Most large diameter line pipe in use today is specified using the API 5L standard (2008) as the basis for requirements. Companies can, and often do, exceed these base requirements to suit their particular needs.

Figure 6-20 depicts the progression, over time, of one important material property, that of tensile strength. However, of itself, ultimate tensile strength provides only a portion of the material's resistance to in service failure, except in those circumstances where gross section yielding or highly ductile fracture occurs. This type of plastic collapse mode of failure will be addressed later. Of equal importance to tensile strength is the toughness property of the material, toughness being a measure of the ability of the material to withstand the presence of a sizeable crack-like defect. Often the original material test reports for in service pipelines have been mislaid, so it then becomes necessary to re-establish the pipe properties before conducting a defect assessment.

The increase in tensile strength depicted in Figure 6-20 has been achieved as a result of a number of initiatives taken in steel pipe manufacture. These include the use of cleaner steels, combined with alloying, and grain refinement due to processing techniques such as quenching and tempering, or thermo-mechanical rolling. The latter two techniques involve

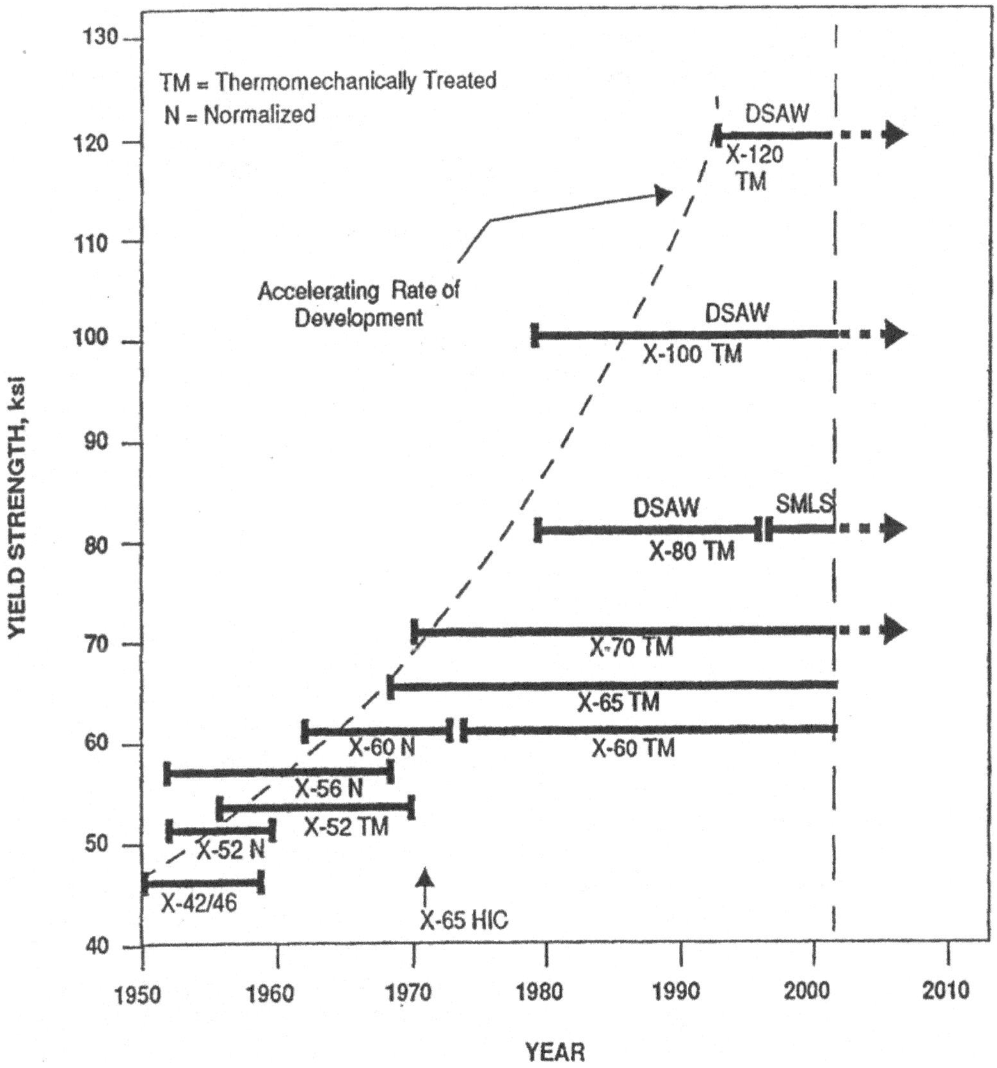

Figure 6-20. A development of the yield strength of pipeline steels over time (Courtesy Malcolm Gray, Glover 2004)

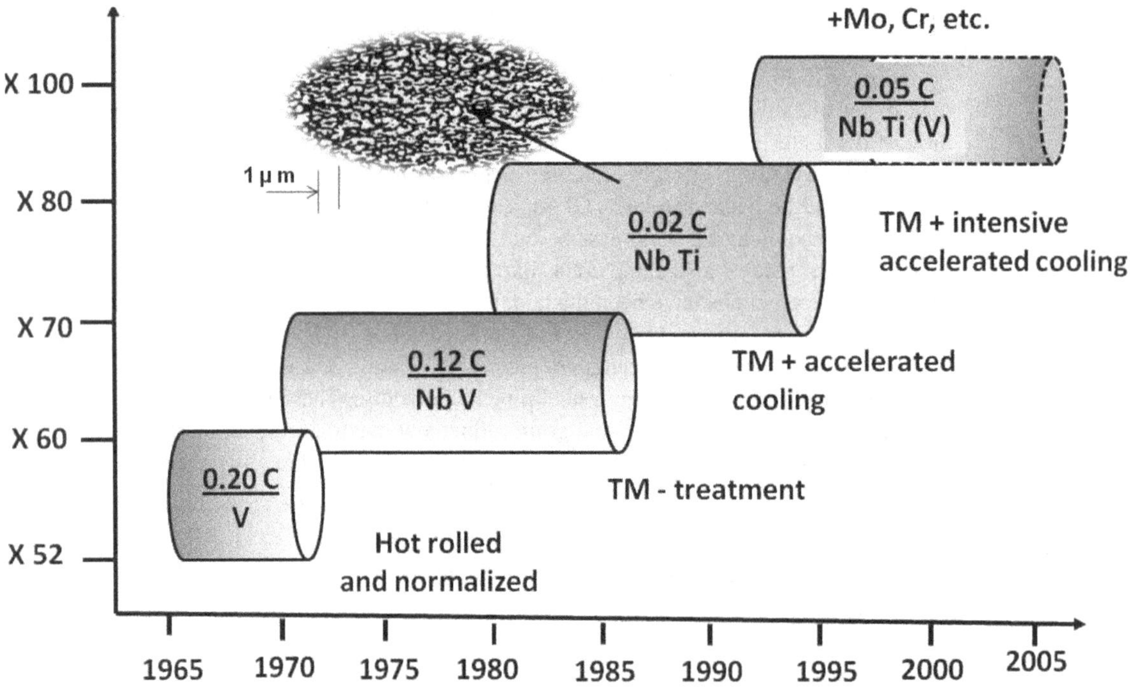

Figure 6-21. Effect of thermomechanical processing on steel strength (after Wang et al. 2006)

the skilled intervention of appropriate heat treatment during the plate manufacture. In the thermo-mechanical rolling process, supported by state-of-the-art process control, the plate is rolled at precisely defined temperatures and with exactly controlled amounts of thickness reduction. The required fine-grained, homogeneous microstructure, to obtain optimum strength and toughness, is obtained through recrystallization, transformation, and precipitation. Variations in the soaking temperature, time, deformation schedule, and final cooling rate, contribute to the evolution of the microstructure and the properties attained. Extensive rolling following the start of the austenite to ferrite transformation results in marked increase in strength due to subgrain formation.

Steel strength can be increased by increasing the carbon content as the interstitial carbon atoms add more frictional resistance to prevent dislocations. However, such an approach has the downside of reducing toughness while also increasing hardness. This makes welding more difficult and adds to strain aging; hence, the approach of adding rare earths such as indium and niobium as alloying elements instead of carbon.

It will be clear from Figure 6-20 that a wide variety of steel grades and pipe manufacturing techniques have been used in pipe manufacture and given the longevity of well maintained pipe line systems, much of these types of pipe are still in service. It follows therefore that any type of thorough defect assessment must take into account both the method of manufacture (see Common Defect Types), and the pipe material properties. For example, a sizeable number of in service failures in North American pipelines are traceable to hook cracking, which occurs on the long seam of low-frequency ERW pipe. While many of these hook cracks were uncovered during the pre-commissioning hydrotesting, a number may have survived and can grow through fatigue.

As noted in Common Defect Types, the current practice of continuous casting of the molten steel has generally replaced the use of ingot molding, thus reducing the probability of introducing inclusion defects. However, as will be addressed later in Defect Assessment

Methods, pipes manufactured in the mid-1970s and before may well contain a wide range of manufacturing defects. Pipes constructed in the period 1950 through to 1975 will very likely have relatively low toughness, less than 40 J, since minimum acceptable values were not specified at that time in the various standards. Material toughness can be determined both qualitatively and quantitatively. Charpy Vee notch (CVN) testing and the drop-weight tear test (DWTT) are both examples of qualitative measures, which are easily obtained and widely used in quality control. They are also used in the prevention of structural failures, although their values are only relative measures. By contrast, quantitative measures of toughness such as the stress intensity (K), surface energy criterion (G), crack tip opening displacement (CTOD), and crack tip opening angle (CTOA) have an absolute meaning and are used in the fracture mechanics calculations of Defect Assessment Methods to calculate failure pressure and/or critical crack defect size.

For more than a century, the Charpy test has been employed as an acceptance test characterizing the impact toughness of a material. The original parameter is the energy consumed while breaking a standard notched specimen having a net cross-section of 80 mm^2. The Vee notch impact test measures the impact energy of a prepared specimen in Joules or ft-lbs. Figure 6-22 illustrates the test set up, while Figure 6-23 shows an example of a failed test piece.

Mild steel, being a body-centered cubic material, will undergo a transition from ductile to brittle behavior under conditions of decreasing temperature, or increasing loading rate. Hence, if this impact energy is plotted versus temperature, for a number of samples, many steels will show comparatively low values at low temperatures, which rise in a transition regime and enter a plateau at elevated temperatures (upper-shelf toughness) as shown in Figure 6-24.

Charpy tests have the advantage of being simple and relatively inexpensive to conduct. However, even at full size, the specimen is small and for some pipe wall thicknesses even smaller samples may be required (Figure 6-25).

Size is in fact one of the limitations of the test, for thicker wall and higher strength pipe, a dynamic tear, or drop weight test could more accurately reflect the toughness of the

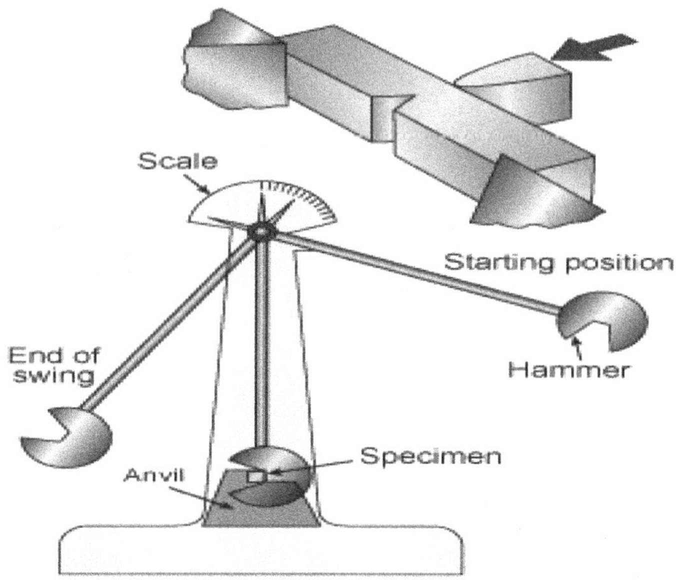

Figure 6-22. Charpy impact testing machine and sample

Figure 6-23. CVN test sample

material. For some studies, involved in predicting steady state running ductile fractures, the CVN specimen was considered too small to develop significant propagation resistance.

Accordingly, alternatives to the CVN test as a means to improve arrest toughness predictions were sought, with early work at the US Naval Research Laboratory and Batelle Memorial Institute showing that the DWTT was better at characterizing propagation resistance. In addition, the Batelle experiments showed that a transition temperature could be obtained from observation of the change in fracture surface appearance of a DWTT specimen.

The DWTT uses the kinetic energy of a falling weight to break a suitably notched specimen (Figure 6-26). The specimen is a plate measuring 10 × 4 × 0.5 in. thick and has a hard surfaced weld bead placed on one side, which is precracked by having a starter notch

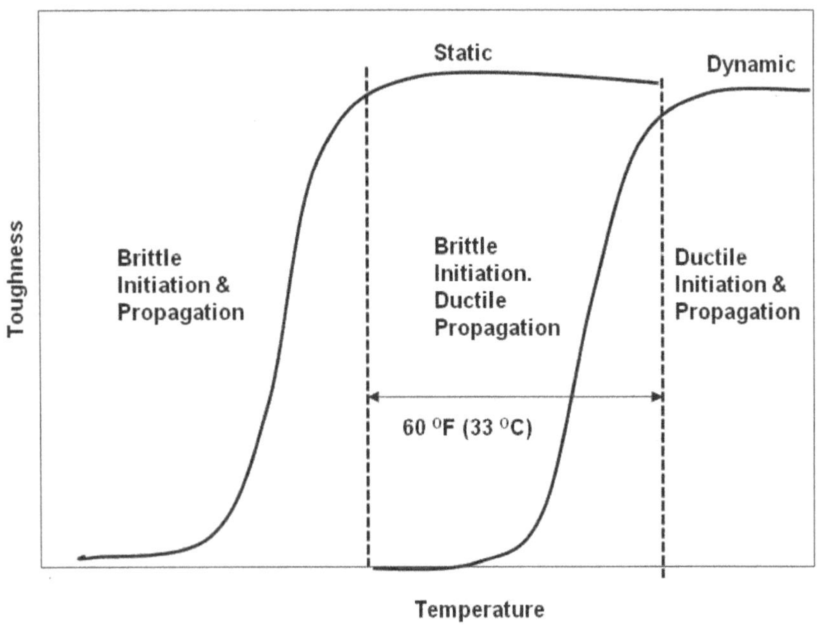

Figure 6-24. Typical brittle to ductile transition curve

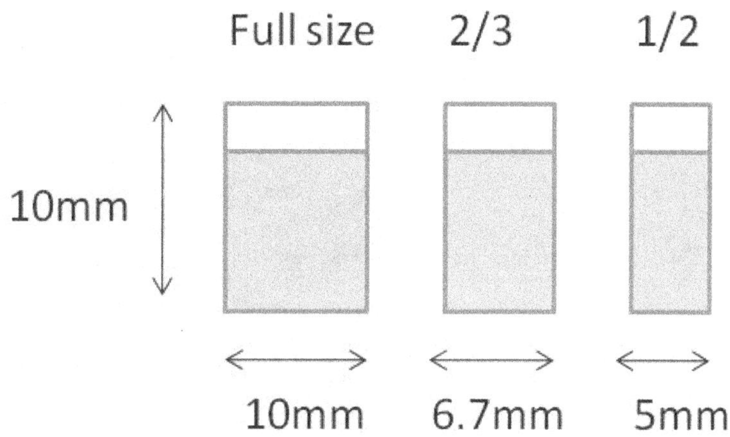

Figure 6-25. Different sized Charpy specimens

milled into it. As noted above, the DWTT provides information regarding the highest temperature at which any particular steel is likely to fracture in a brittle mode by cleavage. For this reason, it is known as the nil ductility temperature, since above it, fractures will exhibit some degree of plastic flow and have a fibrous appearance on the fracture surface.

The failure of a brittle material is preceded by little or no plastic deformation and is almost instantaneous, which is what makes it so potentially dangerous. A ductile failure on the other hand is associated with significant plastic deformation. Safe operation of a pipeline is only possible if the operating temperature is well above the transition regime. In cold climates, the lowest temperature and highest stress experienced by the pipeline may occur during hydrotesting, so one needs to take this into consideration when specifying the material properties.

The Canadian CSA Z 662-07 standard requires, "the minimum design temperature for notch toughness purposes to be taken at or below the lowest expected metal temperature when the pipe stress exceeds 50 MPa (7250 psi) during pressure testing and service under design conditions, having due regard to past recorded temperature data, the minimum fluid temperature that could occur, and the possible effects of lower and ground temperatures."

Figure 6-26. DWTT specimen

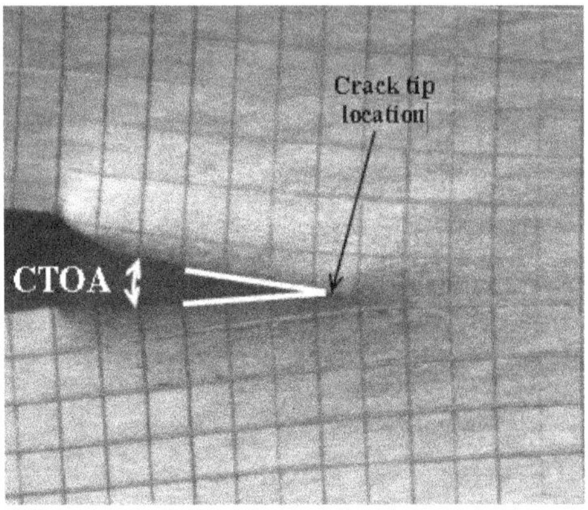

Figure 6-27. CTOA and tip location (Darcis et al. 1998)

Subsequent to the development of the DWTT, work has progressed on developing several other methods for measuring fracture resistance particularly where there is a high degree of stable tearing of the material during cracking. Among these is the CTOA approach, which is based on the CTOD method proposed by Wells (1961). Both are ductile fracture criteria and are becoming recognized as one of the more widely accepted properties for characterizing fully plastic fracture. As such, they can be used as an alternative to the absorbed fracture energy for the assessment of the toughness of pipeline steels.

The principal advantages of the CTOA method for pipeline steels are that it can be directly measured from the crack opening profile in addition to being descriptive of the geometry of the fractured pipe. Figure 6-27 shows the tearing action that takes place during the test, while Figures 6-28 and 6-29, respectively, show a double cantilever beam CTOA test specimen and a typical location on the pipe surface where it may be cut directly, without any flattening. The specimen and test procedures have been codified in ASTM E-1820 (2001).

In addition to the fracture toughness, one also requires knowledge of the tensile properties of the pipeline material as shown in Figure 6-30. The ring expansion test shown in Figure 6-31 is the most accurate method of determining the yield strength of the pipe material, although it is relatively expensive to perform. The test arrangement prevents no other form

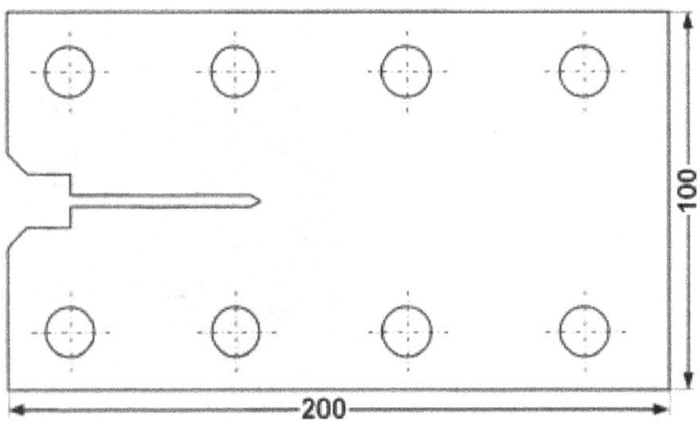

Figure 6-28. Double cantilevered beam CTOA specimen (Darcis et al. 1998)

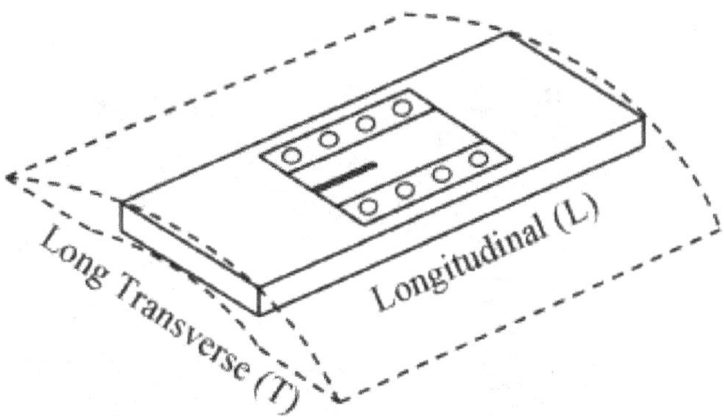

Figure 6-29. Orientation of the CTOA specimen in the pipe (Darcis et al. 1998)

of loading than radial pressure from taking place. The simplest approach to mechanical properties testing is to use a conventional tensile testing machine on the type of specimens shown in Figure 6-32. As can be seen, these consist of either flattened straps, or round bars, taken from the body of the pipe as shown in Figure 6-33.

Some observations are in order with respect to the results obtained from these tests. The plate rolling process, described in Common Defect Types, creates anisotropic properties, so the strength will differ in the transverse and longitudinal directions. Secondly while the test results from flattened, strap tests will be sufficient to confirm the grade of pipe material, this data does not truly represent the mechanical properties of a pipe in-service. The flattening process will affect the yield strength and ductility of the material, the difference being especially accentuated with high-strength steel pipe made by the thermo-mechanical controlled rolling process (TMCP). Glover and Rothwell (2004) noted a decrease of around 10% to 12% in the yield strength measured from transverse orientation flattened strap specimens

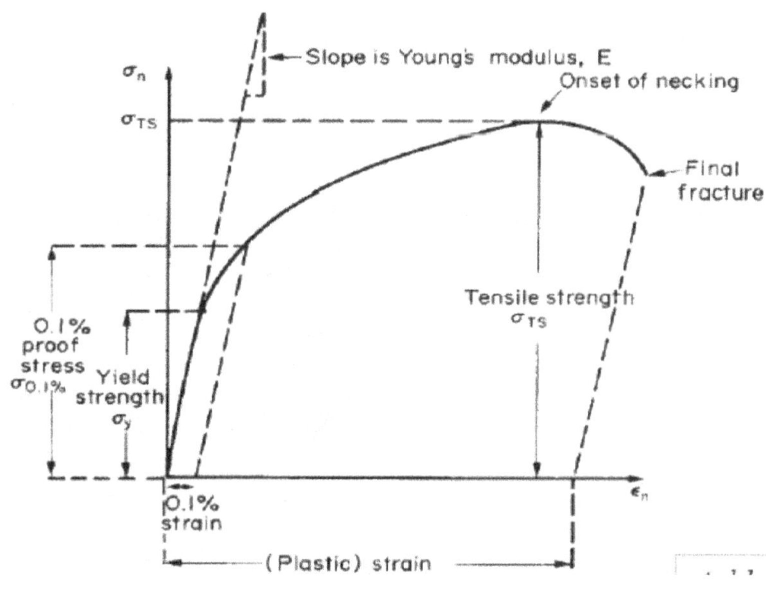

Figure 6-30. Tensile properties

Figure 6-31. Ring expansion test set-up (Chauhan et al. 2009)

compared to machined round specimens or ring expansion tests for high-strength pipeline steels. They also indicated that the yield strength and ultimate strength in the hoop direction were about 20% and 5% higher, respectively, than in the longitudinal direction.

Besides determining the tensile properties in the pipe body, it is also necessary to test the weld material as this information will be used in the assessment of weld defects. The size and orientation of the test specimens are shown in Figure 6-34.

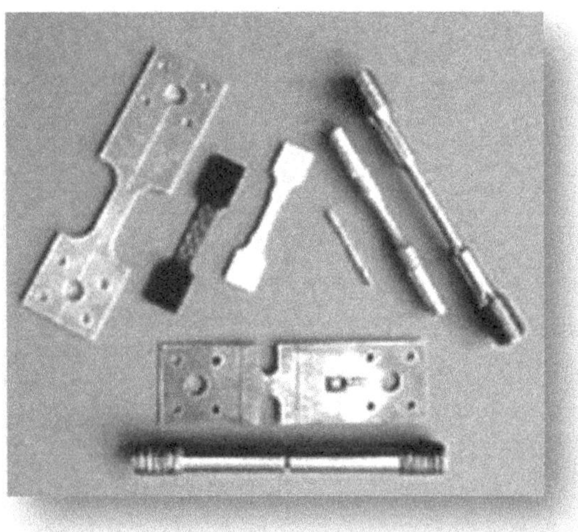

Figure 6-32. Tensile testing samples (Courtesy of Canmet–MTL)

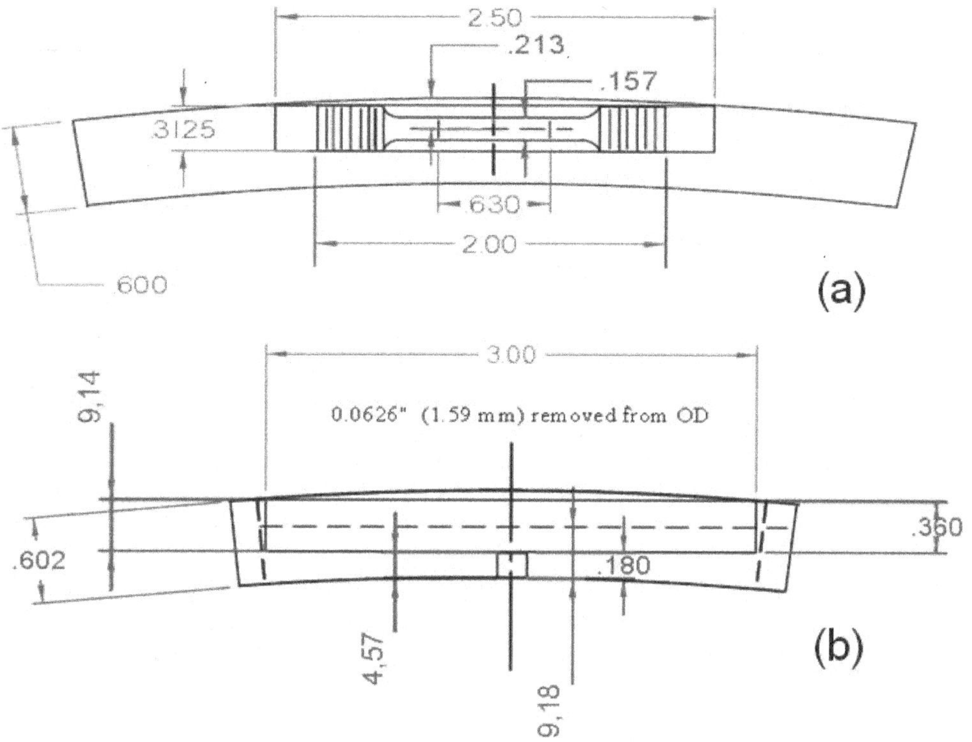

Figure 6-33. Location and sizes of round bar (a) and flattened strap test (b) samples (Courtesy of Canmet–MTL)

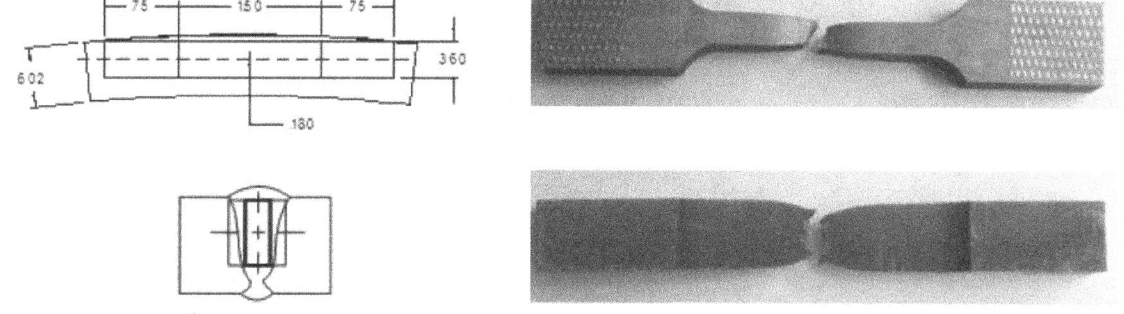

Figure 6-34. Typical weld metal tensile test specimens (Courtesy of Canmet–MTL)

FRACTURE MECHANICS IN DEFECT ASSESSMENT

The term "fracture mechanics" involves the evaluation of defects or discontinuities in a structure, which in the present context, means predicting the effect of inherent flaws on the pressure containing capacity of a pipeline. The approach used is based upon attempting to describe the stress field that surrounds any crack found in the body of the pipe. The earliest attempt at such a description was the seminal work conducted by Griffith on brittle materials in the early 1920s. In a series of experiments conducted on glass rods, he observed that the longer the rod, the lower its fracture strength. This led him to postulate that all materials contain a distribution of fine cracks and that specifically, longer rods had a greater probability of containing a large flaw. He reasoned that these cracks produced stress concentrations and that as the load increased, eventually, a limiting stress was reached in the vicinity of the crack tip. This physical insight led to the development of an energy balance criterion, which, in its simplest form, states that a crack will continue to grow when the amount of elastic strain energy released by the material is greater than, or at least equal to, the energy required to create the new crack surfaces. We can see how this instability comes about by considering Figure 6-35, which illustrates a rectangular thin plate of material containing an elliptical through thickness crack. It is conventional to define the major axis of the crack as being of length "2a." The applied load will create a far field tensile stress σ, acting in a direction perpendicular to the crack. The description of a thin plate means that we can make a reasonable assumption of plane stress behavior.

Surface Energy of a Crack

If the plate did not contain a crack, then the stress would be uniform at any cross-section. For this condition, the strain energy stored in an elastic body per unit volume is the area under the stress strain curve i.e.,

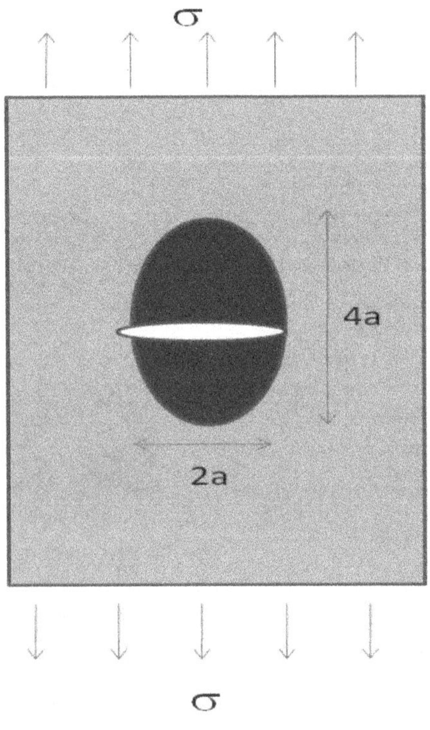

Figure 6-35. Classic Griffith crack

$$U_0 = -\frac{1}{2}\frac{\sigma^2}{E} \qquad (6-16)$$

As it is, load cannot be transferred across the face of the crack so that a stress concentration will occur in the remaining ligament and one can imagine the flow of stress around the newly appearing crack. Furthermore, there is a release of strain energy in the plate because of the inability of the crack surfaces to support the load. The amount of energy released is related to a specific volume of material surrounding the crack and Griffith assumed that its shape on the plate, would be elliptical in plan, with the major axis being twice the length of the minor axis, that is, $4a$ units long, and hence, its area is $2\pi a^2$.

Note that when a crack appears, two surfaces are formed, so the total surface area of a through thickness crack is $4at$, and if we define the term γ_s as being the specific surface energy, that is the energy per unit area, then the energy associated with the crack surface formation U_s is $4a\gamma_s t$.

We require the energy released per unit thickness rather than per unit volume so,

$$U_e = -\frac{1}{2}\frac{\sigma^2}{E}\left[2\pi a^2\right] \qquad (6-17)$$

Griffith supposed that all of this stored energy release was converted into surface energy U_s in creating a new crack per unit thickness such that,

$$U_s = 2a2\gamma_s = 4a\gamma_s \qquad (6-18)$$

The total energy U_T is the sum of the released and stored energies, that is,

$$U_T = U_s + U_e = 4a\gamma_s - \frac{\sigma^2\pi a^2}{E} \qquad (6-19)$$

Considering Eq. (6-19), it may be seen that if increasing the half crack length, a, lowers the total energy U_T, then the crack is unstable and it will grow. Conversely, a stable crack length represents a condition of minimum total energy. To find this length, we differentiate the total energy with respect to "a" and set the resulting value to zero.

$$\frac{dU_T}{da} = \frac{d}{da}\left[4a\gamma_s - \frac{\sigma^2\pi a^2}{E}\right] = 0 \qquad (6-20)$$

That is,

$$4\gamma_s - \frac{2\sigma^2\pi a}{E} = 0$$

Solving for stress, we obtain,

$$\sigma = \left[\frac{2E\gamma_s}{\pi a}\right]^{0.5} \qquad (6-21)$$

If the second derivative of Eq. (6-20) is a negative quantity, this would indicate that the crack is unstable and will continue to grow. One can also rearrange Eq. (6-21) to obtain

$$a_c = \left[\frac{2E\gamma_s}{\pi\sigma^2} \right] \tag{6-22}$$

and note that a_c is half the critical crack length for a given stress σ. It can be further noted that if,

$a > a_c$ the crack is unstable and will continue to grow
$a < a_c$ the crack is stable and hence will not grow.

Eq. (6-22) is known as the Griffith criterion, and it applies only to brittle materials, that is, materials, which do not deform plastically. With the exception of pipelines made from steels of very low toughness, one can expect the remainder to undergo some plastic deformation. Orowan (1945) and Irwin (1948) both built on the Griffith criterion by noting that the effective surface energy, including the work done by plastic deformation around the fracture surface, can be substituted for the specific surface energy in the Griffith equation. This is an empirical approach used to obtain agreement between observed fracture loads and crack lengths. It can be written as

$$\gamma_{eff} = \gamma_s + \gamma_p \tag{6-23}$$

where

γ_p = work done in plastic deformation
γ_s = specific surface energy
γ_{eff} = effective surface energy

Note that the value for γ_p is much larger than the specific surface energy it replaces, $2\gamma_s$, in Eq. (6-21)

$$\sigma = \left[\frac{E\gamma_p}{\pi a} \right]^{0.5} \tag{6-24}$$

Irwin used an experimental approach to relate strain energy release to crack extension and defined this as a strain energy release rate, G. The critical strain energy release rate, G_c for crack growth was found to be equal to twice the specific surface energy γ_s hence,

$$\sigma = \left[\frac{EG_c}{\pi a} \right]^{0.5} \tag{6-25}$$

He also supplemented this energy approach by introducing the concept of a stress intensity factor K, which he defined as

$$K = \left[2\gamma_s E\right]^{0.5} = \left[EG_c\right]^{0.5} = \sigma\sqrt{\pi a} \tag{6-26}$$

The critical value of K for crack growth is known as K_c

The stress intensity factor depends on the applied stress, the crack size, and the geometry, so Eq. (6-26) can be generalized by including a geometry factor Y, such that,

$$K = Y\sigma\sqrt{\pi a} \tag{6-27}$$

This geometry factor changes the crack tip stress field causing it to increase, and its determination, for even the simplest of conditions of relating crack geometry to the applied load, requires the use of elaborate numerical methods. Fortunately, these factors can be looked up in a reference text such as Tada and Paris (1985) for a very large number of conditions. Often Y will be expressed as a function, $Y = f(a/w)$, where a is the half crack length and w is the dimension over which the crack is growing.

Stress Analysis of Cracking

The significance of K can also be determined using stress analysis. There are three major modes of cracking as shown in Figure 6-36. Mode I, the opening or tensile mode, is the most common failure geometry for plane stress conditions, which is applicable to thin wall pipe.

The stress field in the vicinity of a crack is shown in Figure 6-37. The stresses at a distance from the crack tip may be written for all values of r and θ as,

$$\sigma_x = \sigma_{app}\left(\frac{a}{2r}\right)^{0.5}\cos\frac{\theta}{2}\left(1-\sin\frac{\theta}{2}\sin\frac{3\theta}{2}\right)$$

$$\sigma_y = \sigma_{app}\left(\frac{a}{2r}\right)^{0.5}\cos\frac{\theta}{2}\left(1+\sin\frac{\theta}{2}\sin\frac{3\theta}{2}\right) \qquad (6-28)$$

$$\tau_{xy} = \sigma_{app}\left(\frac{a}{2r}\right)^{0.5}\sin\frac{\theta}{2}\cos\frac{\theta}{2}\cos\frac{3\theta}{2}$$

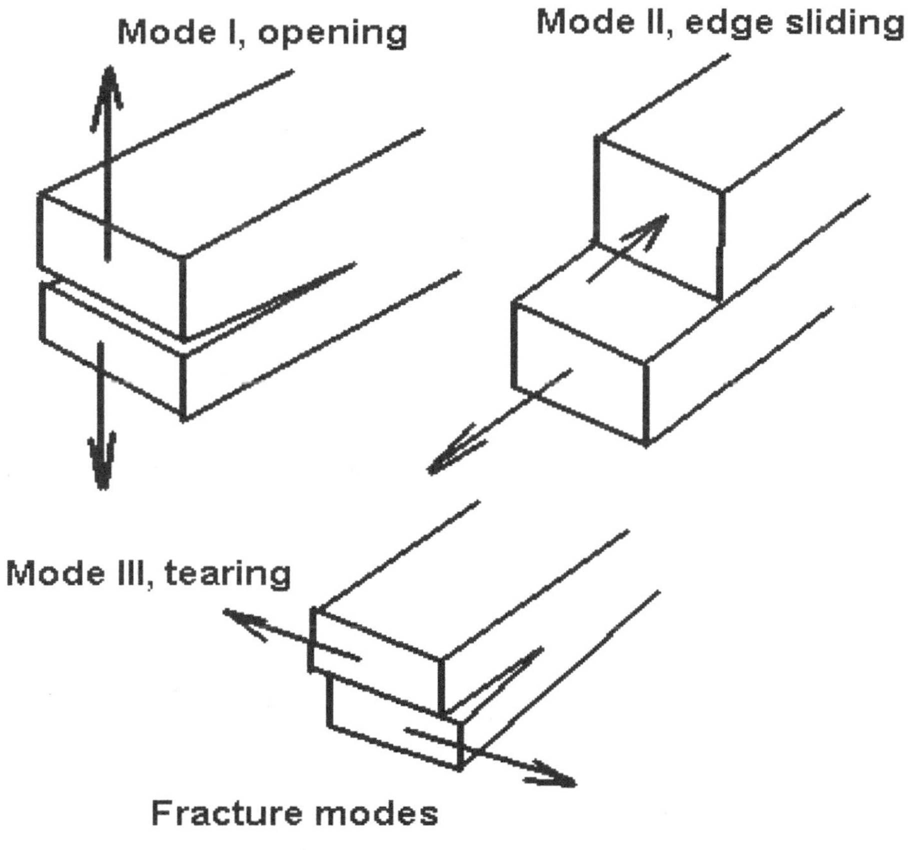

Figure 6-36. Fracture modes of failure

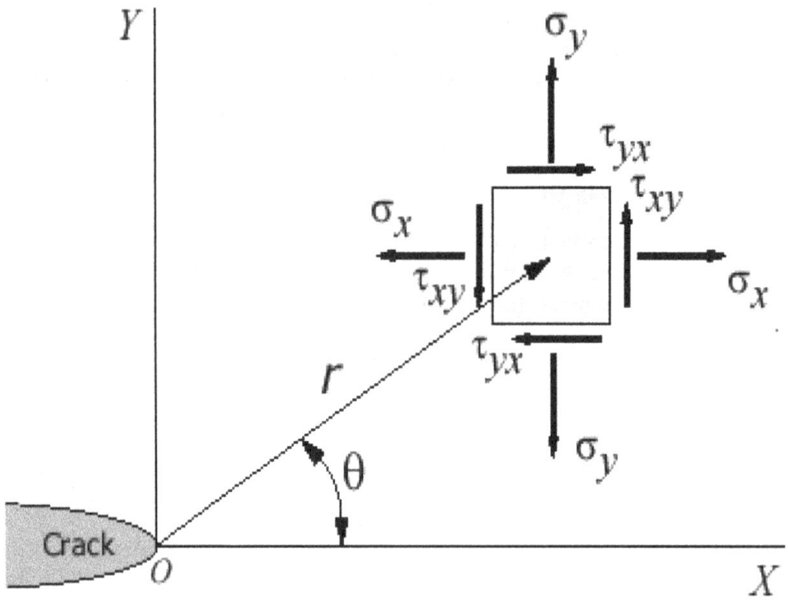

Figure 6-37. Stresses at a distance from the crack tip

If we use a scalar factor, $\sigma_{\text{app}}\sqrt{\pi a}$, which was defined in Eq. (6.26) as the stress intensity factor, and we can refer to here as K_{I}, then the magnitude of the stresses can be rewritten as,

$$\sigma_x = \frac{K_{\text{I}}}{(2\pi r)^{0.5}}\cos\frac{\theta}{2}\left(1 - \sin\frac{\theta}{2}\sin\frac{3\theta}{2}\right)$$

$$\sigma_y = \frac{K_{\text{I}}}{(2\pi r)^{0.5}}\cos\frac{\theta}{2}\left(1 + \sin\frac{\theta}{2}\sin\frac{3\theta}{2}\right) \qquad (6-29)$$

$$\tau_{xy} = \frac{K_{\text{I}}}{(2\pi r)^{0.5}}\sin\frac{\theta}{2}\cos\frac{\theta}{2}\cos\frac{3\theta}{2}$$

Of these stresses, the largest is in the σ_y direction along the x axis $\theta = 0$ and has the value,

$$\sigma_y = \frac{K_{\text{I}}}{(2r\pi)^{0.5}} \qquad (6-30)$$

As r tends to zero, both σ_y and σ_x tend to infinity, which is clearly not possible since plastic yielding will limit the stresses to the yield strength of the material (Figure 6-38). In this circumstance of small crack size, the safe operating stress is determined by calculating the net section yield of the ligament. The plastic deformation near the crack tip is suppressed due to the surrounding elastic material. It is noteworthy that the stress field arises from the presence of the crack and not directly from the background applied stress, indeed one has to move relatively far away from the crack tip for the applied stress to dominate.

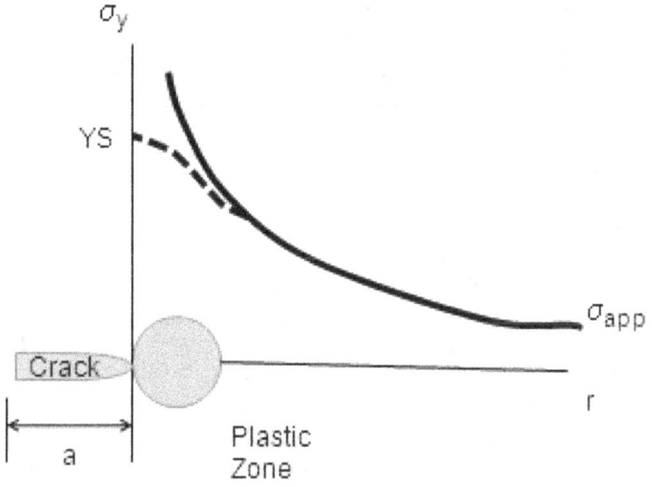

Figure 6-38. Limiting stress in the vicinity of a crack tip

The stress intensity factor K is a measure of the severity of the stress in the vicinity of the crack tip and it is the near tip stress field, which directly influences crack growth.

It will be apparent from Eq. (6.26) that increasing the half crack length "a" and increasing the applied stress will have the same effect — that of increasing the stress intensity. When the stress and/or the crack length increase to a level that causes unstable crack growth, this defines the *critical* stress intensity, K_c. If the loading condition is Mode I, then one can use the term K_{IC} as the Mode I fracture toughness, which, as indicated in Material Properties, is a property of the material that can be measured.

The analysis outlined above to determine K_{IC} is based upon linear elastic fracture mechanics, which is applicable if the plastic zone is small and the background stress remains within the elastic range. This is a reasonable assumption for modern pipeline steels where a damage tolerant design approach is used. Such an approach assumes the existence of preexisting imperfections such as cracks and ensures that, during the service life of the pipeline, the applied stress will not cause K_{IC} to be exceeded. It also adopts a "leak before break" philosophy, which we will consider in the following section.

Leak Before Break Philosophy

A leak before rupture is one possible outcome of a sequence of crack growth events. Inherent subcritical flaws in a ductile pipe material may grow in service as a result of fatigue, corrosion, accidental damage, or overpressure. The determination of whether a leak will develop into a rupture depends on:

1. the orientation of the flaw
2. the size and shape of the flaw
3. the type and intensity of the applied loads
4. the tensile and fracture toughness properties of the material
5. the presence and distribution of any residual stress

Kanninen and Popelar (1985) developed an approach to assess whether a pipe would leak before rupturing. They assumed the presence of a semi-elliptically shaped surface crack that had progressed part way through the pipe wall. Such a crack has a continually changing length; hence, the stress intensity factor will also vary along the crack front and this requires a correction to be made. This shape correction factor, Q, involves the evaluation of an elliptical integral ϕ, which is used to define the crack front.

Using a linear elastic fracture mechanics criteria for crack propagation, results in

$$K_{\mathrm{I}} = K_{\mathrm{IC}} = 1.12 P_0 \frac{R}{t}\left[\left[\frac{\pi a}{Q}\right] \sec\left(\frac{\pi a}{2t}\right)\right]^{0.5} \qquad (6-31)$$

where

K_{IC} = critical stress intensity factor for quasi static loading, MPa m$^{0.5}$
P_0 = the pressure at the time when the crack begins to propagate, MPa
R = the pipe radius, mm
t = the pipe wall thickness, mm
a = the depth of the crack, mm
Q = a geometric correction factor that takes into account the defect shape.

Q depends on the length and depth of the crack as well as the ratio of the applied stress, σ_a to the yield stress of the material, σ_y and is given by,

$$Q = \Phi^2 - 0.212\,(\sigma_a/\sigma_y)^2 \qquad (6-32)$$

where the elliptical integral

$$\Phi = (1 + 1.464(a/c)^{1.65})^{0.5} \qquad (6-33)$$

and c = the half length of the crack (mm).

Note the change in notation for crack length from the $2a$ used previously to $2c$.

Once the part through wall crack propagates in an unstable manner, it will eventually become a through wall crack at which point it is possible to write,

$$K_{\mathrm{I}} = K_{\mathrm{d}} = \frac{P_1 R}{t}\left[\pi c\left(1 + 1.61\left(\frac{c^2}{Rt}\right)\right)\right]^{0.5} \qquad (6-34)$$

where K_{d} is the dynamic stress intensity factor for the pipe material and is a measure of the resistance to growth of a running crack as opposed to K_{Ic}, which represents resistance to quasi static initiation.

P_1 is the pressure after the crack has progressed through the pipe wall and, for a gas pipeline, is steadily reducing as the product inside the pipe leaks out. The above equations have their basis in the linear elastic fracture mechanics analysis of wide flat plates with a central crack, so to be representative of an axial defect in a cylindrical pipe, a correction has to be made to take account of the bending effects that are now introduced. Folias (1965) developed such a correction factor, and it appears in parenthesis in Eq. (6-34).

By equating and then algebraically rearranging Eqs. (6-31) and (6-34), and introducing the artifice,

$$\frac{t}{t} = \left[\frac{K_{\mathrm{d}}}{K_{\mathrm{IC}}}\right]^2 = 1$$

into the equations, the resulting expression is

$$\left[\frac{c}{t}\left(1 + 1.61\frac{t}{R}\left(\frac{c}{t}\right)^2\right)\right] = 1.254\left[\frac{K_{\mathrm{d}}}{K_{\mathrm{Ic}}}\right]^2\left[\frac{P_0}{P_1}\right]^2 \frac{a}{t}\sec\left(\frac{\pi a}{2t}\right) \qquad (6-35)$$

This equation defines the conditions for which a part wall crack will grow unstably to a through wall defect and will continue to grow in an unstable manner in the axial direction rupturing the pipe. For a given pipe where R and t are known and $\left[\frac{K_d}{K_{1c}}\right]^2\left[\frac{P_0}{P_1}\right]^2$ are specified, a graphical representation of c/t versus a/t can be made (see Figure 6-39).

Values that lie below the resulting curve define the conditions for leak before break, while those values above the line indicate leak with rupture. If both of the relevant critical stress intensity factors are determined experimentally, then one can determine if a given crack measuring (a, $2c$) will leak before rupture. Equation (6-35) is a limiting state condition, and each of the curves plotted in Figure 6-39 may be thought of as a limiting boundary surface — a recurring concept we will return to later in the discussion of assessment methods.

Example:

An NPS36 pipeline is made of X65 grade steel with fracture toughness, $K_{IC,}$ of 55 ksi $\sqrt{\text{in}}$ and has a wall thickness of 0.5 in. Its MOP is 1200 psi. If a surface crack 1.0 in. long, or greater can be detected, determine if the pipe will leak before rupturing.

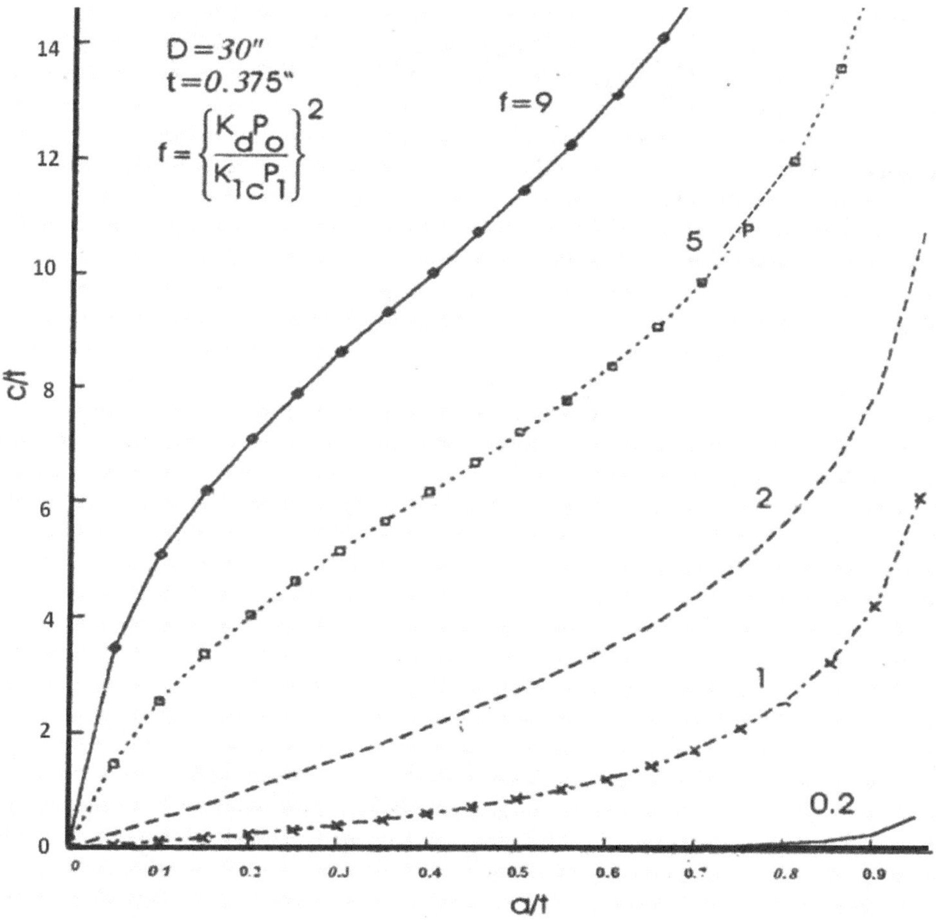

Figure 6-39. Results of a leak before rupture analysis for various ratios of dynamic and quasi static stress intensities and pressures (Popelar et al. 1996)

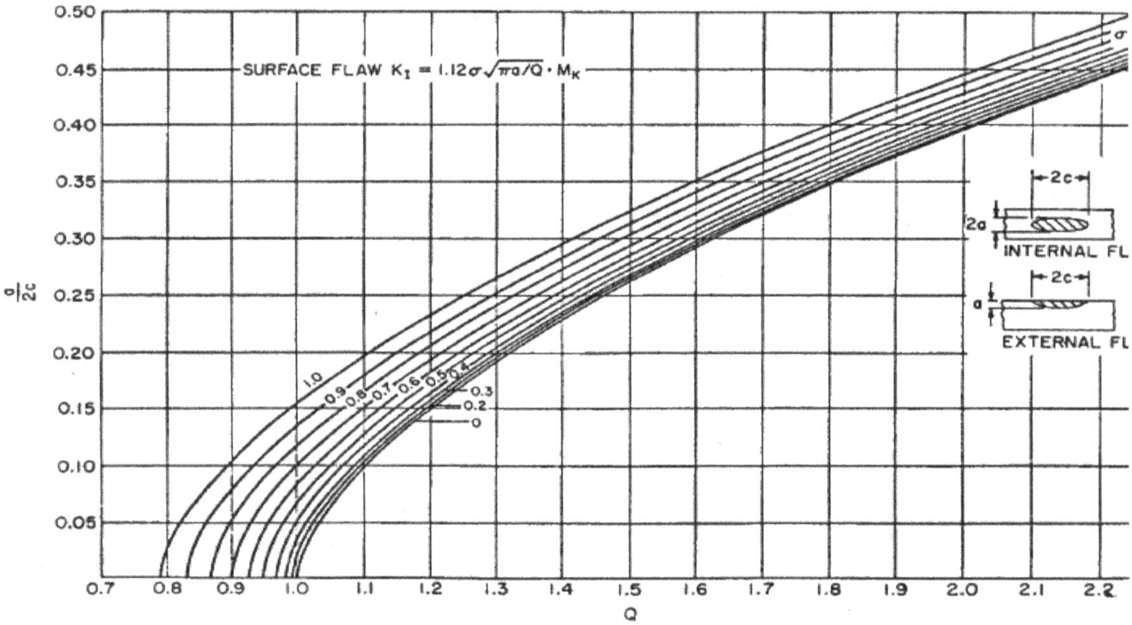

Figure 6-40. Effect of crack depth/length ratio $a/2c$ and σ/σ_y ratio on flaw shape parameter Q (Barsom and Rolfe, 1987, p. 47. Reprinted by permission of Prentice Hall Inc., Englewood Cliffs, NJ)

Solution:

It is appropriate to model the defect as a thumbnail crack under tension, in which case, we can make use of the following relationship

$$K_i = \frac{1.12}{\sqrt{Q}}\sigma\sqrt{\pi a}$$

where Q is the flaw shape parameter and the units of σ are ksi.

To ensure leaking before rupture, the crack must be stable so $K_{1C} > K$ for $a = 0.5$ in., the thickness of the pipe.

The hoop stress $\sigma = \dfrac{PD}{2t} = \dfrac{1200 \times 36}{2 \times 0.5} = 43200$ psi

Hence

$$55 = \frac{1.12 \times 43.2\sqrt{0.5\pi}}{\sqrt{Q}}$$

From which $Q = 1.215$.

We now need to make use of the graphical relationship shown in Figure 6-40 linking the flaw shape parameter Q to the ratios of crack depth/length ratio and σ/σ_y.

Since $\dfrac{\sigma}{\sigma_y} = \dfrac{43.2}{65} = 0.665$, then from the graph at a value of Q of 1.17, we obtain

$\dfrac{a}{2c} = 0.19$, hence

$2c = a/0.19 = 0.5/0.19 = 2.63$ in.

This is a detectable crack length.

Fatigue Loading

Until now we have only considered static conditions, we now need to consider crack growth under the influence of alternating stresses brought about by cyclic loading, that is, the fatigue process. Natural gas is a homogenous fluid, so gas may enter or leave the pipeline without a need to halt the stream flow. By contrast, liquid hydrocarbons are most often transported as batches within the same pipeline, requiring a frequent need for stopping and restarting flow. Hence of the two, liquid hydrocarbon pipelines are much more susceptible to fatigue damage. The structural design approach to pipelines reflects this reality. There is an acceptance that a distribution of cracks exists in the pipe body, from the beginning of its service life, with critical defects being removed during the precommissioning hydrotest. The remaining cracks, when subjected to repeated cyclic loading will continue to grow (Figure 6-41) with the expectation that they will be detected during inspection and removed by repair. In short, a damage tolerant philosophy is used to manage the integrity of the system to ensure adequate remaining strength is maintained between inspections so that the likelihood of a failure is acceptably low.

Irrespective of the shape of the actual pressure spectrum in a pipeline, which will be far from sinusoidal, for the purpose of defining some useful terms, consider the cyclical stress relationship shown in Figure 6-42.

It can be seen that the cyclic stresses have a maximum (σ_{max}), minimum (σ_{min}), and mean value (σ_m).

The stress range is defined as,

$$\sigma_r = \sigma_{max} - \sigma_{min} \qquad (6-36)$$

the amplitude being half the range $\sigma_a = (\sigma_{max} - \sigma_{min})/2$.

The mean value is defined as $\sigma_m = (\sigma_{max} + \sigma_{min})/2$, and the load or stress ratio R is defined as $\sigma_{max}/\sigma_{min} = P_{max}/P_{min} = K_{max}/K_{min}$, since the stress intensity factor has a linear relationship with stress.

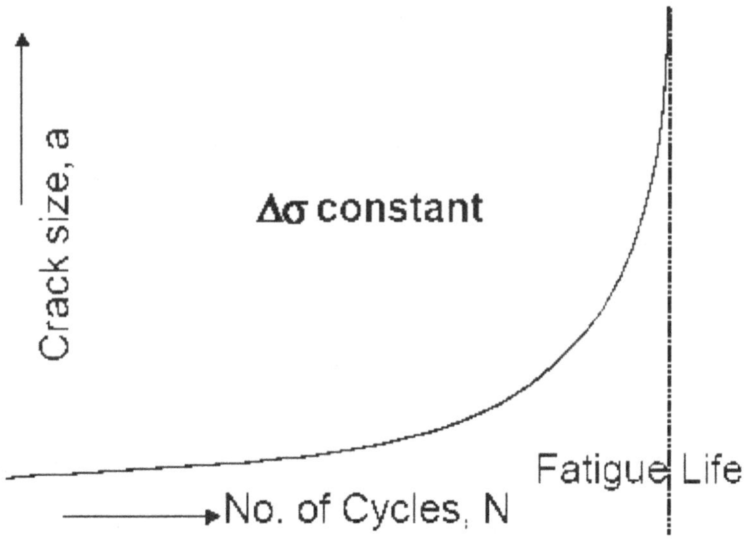

Figure 6-41. Crack growth to failure

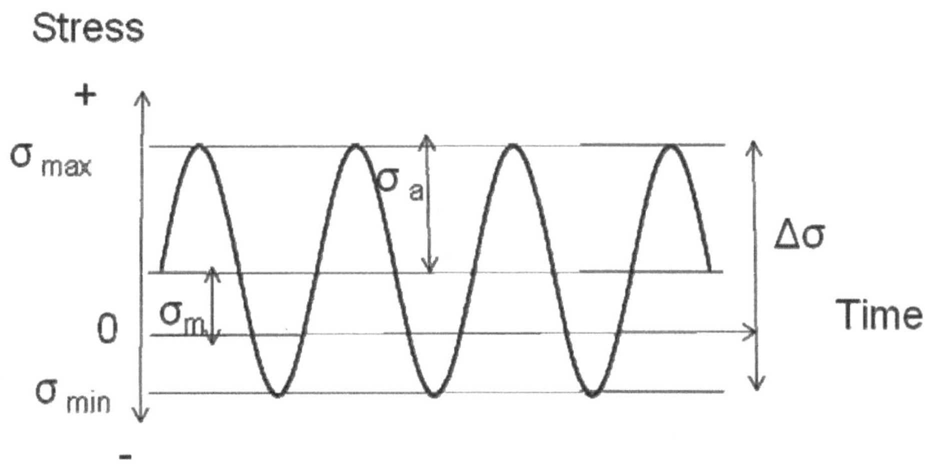

Figure 6-42. Cyclic stress

Stress Life Approach

Two approaches are commonly used to access fatigue damage and remaining life in pipelines. The first of these, the stress life approach, does not distinguish between crack initiation and propagation; rather it deals with the life to failure of the pipeline structure. In the case of pipelines, where the stresses are limited by the working stress design to lie within the elastic range, the predicted lifetime can be quite long (high cycle fatigue). The second approach adapts the principle of linear elastic fracture mechanics to cyclical loading and is used to predict the crack grown rate from the site of an original crack.

The stress life approach, also known as the damage accumulation method, provides a measure of the pipeline material performance under a high degree of cyclic loading. The number of cycles to failure is determined by using an S-N diagram, which is specific to the type of material and its environment. It is derived from conducting fatigue tests on a large number of smooth specimens for the material under consideration.

There are two ways in which the S-N relationship is generally plotted. In the first of these, the stress is plotted against a logarithmic scale of the number of cycles — a semi-logarithmic plot. Alternatively, both the stress and the number of cycles are plotted in the form of a log-log relationship — a logarithmic plot as shown in Figure 6.43. This figure illustrates a typical S-N relationship for steel showing a break point, known as the endurance limit, or fatigue limit, at around 1 million cycles. The corresponding stress value is about 0.5 of the ultimate tensile strength of the steel.

S-N diagrams are not always readily available so a good approximation of the S-N relationship can be obtained by establishing two points, on the log-log plot (Figure 6-43). The first point can be set using either the ultimate stress at 100 cycles or, 0.9 σ_{ult} at 1000 cycles. The second point corresponds to the assumed endurance limit, i.e., 0.5 σ_{ult} at 1 million cycles.

The straight line relationship shown in the figure is fully defined by the slope between the two points and the intercept of the line with the vertical axis. It has the form,

$$\text{Log}_{10}\,\sigma = a + b\,\text{Log}_{10}N \qquad (6-37)$$

So that between 100 and 1 million cycles, the S-N relationship can be written as,

$$\sigma = 10^{a} N^{b} \qquad (6-38)$$

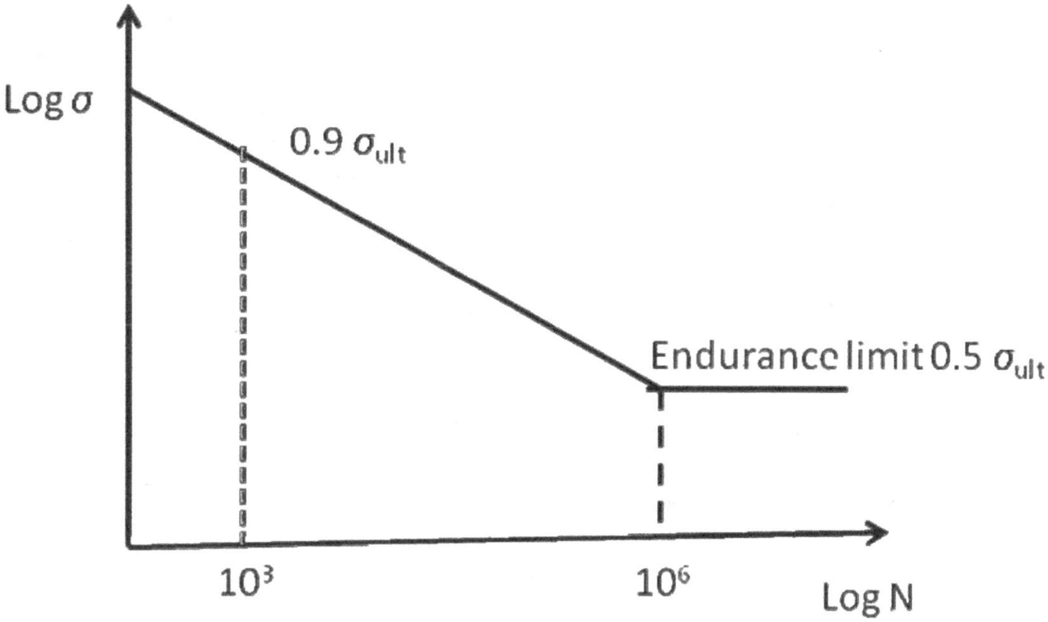

Figure 6-43. Typical S-N logarithmic relationship for steel

One important fact that should be noted from the S-N diagram is that below the endurance limit, failure will not occur, and therefore, the material would have an infinite life.

Where possible, rather than approximating the S-N relationship, a meaningful S-N curve can be constructed by the constant stress amplitude testing of at least 12 specimens. At each stress level, a minimum of three tests are conducted and the results averaged to obtain the point plotted on the S-N curve.

When a structure is subjected to a number of variable loading conditions, the total number of stress cycles required for failure can be determined through the use of the accumulated damage theory proposed by Palmgren and developed later by Miner. In this concept each individual loading condition contributes linearly to the total accumulated damage. The damage contribution for an individual load spectrum at a given stress level, is taken to be proportional to the number of cycles applied at that stress level, n_i, divided by the total number of cycles N_i that would be required to reach failure at the same stress level. Miner's rule describing fatigue failure is written as,

$$\frac{n_1}{N_1}+\frac{n_2}{N_2}+\frac{n_3}{N_3}+\frac{n_4}{N_4}+....=1 \qquad (6-39)$$

While Eq. (6-39) is a useful means of determining fatigue life, for a structure subject to a number of cyclical loads of varying stress magnitude, it contains the simplifying assumptions that the damage accumulation is linear and that it must reach a value of unity. It makes no allowance for the sequence of crack growth and may in some cases be un-conservative. For example, Miner's total life approach should predict a shorter life if the crack tip experiences high-stress cycles early in its life followed by smaller amplitude cycles later, than if the load sequence was reversed. Smaller stress cycles following large amplitude stress cycling are in fact more damaging to the structure (damage becomes more sensitive to load cycles as the crack grows).

In short, one should always apply a factor of safety to the predicted fatigue life predicted by this approach.

As noted earlier, the stress life approach does not differentiate between crack initiation and propagation while the experimental determination of the S-N relationship for the material is based on the use of smooth specimens. It was pointed out in Common Defect Types that pipelines contain imperfections, these inevitably cause stress concentrations, which become the sites for crack initiation. The presence of defects will have a marked effect on the growth of a crack and on its fatigue life. Stress concentrations occur at welds because of their shape and/or the introduction of a defect during the welding process. They may also contain residual stresses, which have the effect of ensuring that all cyclical stresses are tensile. The weld failure shown in Figure 6-44 was caused by a concentration of stress at the toe of the weld. Notwithstanding these factors, S-N diagrams may be used, with caution, to assess the fatigue life of welds provided they have been developed from weld tests in, which the weld is free from significant defects. Significance in this context will be those defects that are readily detected and accepted by the workmanship requirements set in the appropriate standards. The S-N curve chosen must take account of the scatter in test data, which is done by establishing a curve two standard deviations below the mean S-N curve (assuming a log normal distribution).

The S-N relationship for welds has the form $N\sigma^m$ = constant.

Therefore,

$$n_1\sigma_1^m = n_2\sigma_2^m = n_3\sigma_3^m \qquad (6-40)$$

This allows small stress cycles to be converted into one equivalent cyclic stress with the associated number of cycles.

If the welds contain defects that lie outside of workmanship standards, then the S-N approach is invalid and should NOT be used, instead we must use a fracture mechanics approach.

Fracture Mechanics and Fatigue

Fatigue crack growth of a damage tolerant pipeline must take account of the fact that an inherent imperfection will grow from a small initial flaw to reach a size that causes fast unstable crack propagation. Periodic inspections are conducted during this period so as to enable an intervention to be made before the critical size is reached. Figure 6-45 illustrates the three stages of fatigue crack growth; initiation, stable propagation, and fast fracture.

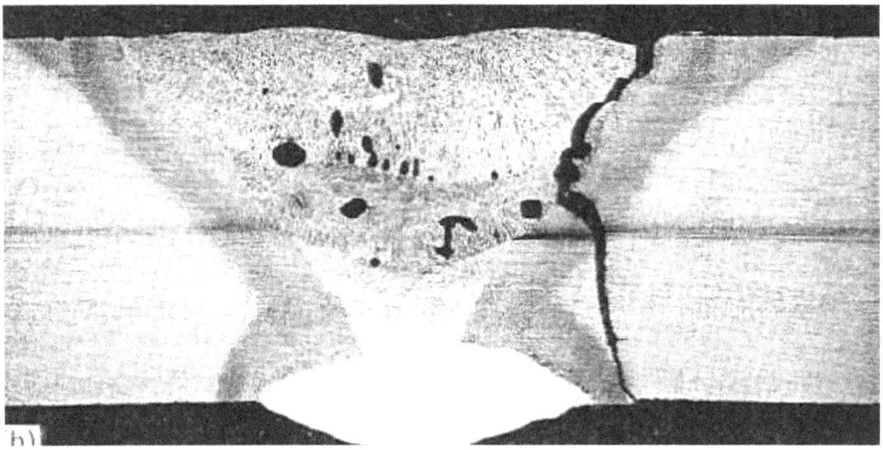

Figure 6-44. Fatigue crack growth from the weld toe

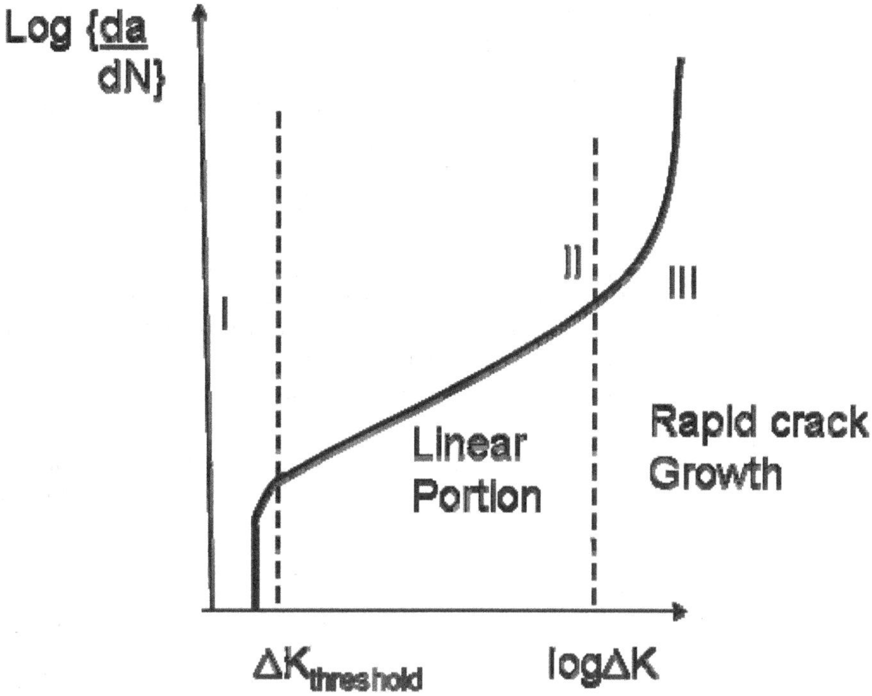

Figure 6-45. Stages of fatigue crack growth

For initiation to occur, the stress intensity range must exceed some threshold value, which is greatly influenced by the physical environment. This stage may be entirely bypassed in circumstances where the pipe has experienced some existing cracking prior to going into service.

Stable propagation describes that portion of the life cycle where small cracks, i.e., those greater than 0.1 mm long, start to grow in a stable fashion (stage II, the linear portion of Figure 6-45). At stage III, the critical crack size is being approached and the instability known as fast fracture occurs. This portion of the fatigue life is comparatively small.

The concept of fatigue crack growth analysis had its beginnings in the 1960s in the seminal work of Paris and Erdogan (1960). Fatigue crack growth is governed by the range of the stress intensity factor (ΔK) during a loading cycle. In general terms, the rate of crack propagation can be written as,

$$\frac{\mathrm{d}a}{\mathrm{d}N} = f(\Delta K, R) \qquad (6-41)$$

Within the linear portion of the logarithmic plot of crack growth rate versus stress intensity range in Figure 6-45, we can write the relationship

$$\frac{\mathrm{d}a}{\mathrm{d}N} = C\Delta K^m \qquad (6-42)$$

This is known as the Paris law and has been found to be applicable for a wide range of materials. It defines a threshold value for ΔK, below which crack growth will not occur. The Paris law is used for the entire range of da/dN, which is a conservative approach at low values of ΔK, and of no consequence at very high ΔK where only a small number of cycles occur.

Commonly accepted values found in the literature for C and m for steel are 0.42×10^{-11} and 3, respectively; however, the actual crack growth properties can be determined experimentally.

The simplest experiment uses a wide center crack test specimen for which the solution for K is already known, that is, derived from Eq. (6-26).

$$\Delta K = \Delta\sigma\sqrt{\pi a}$$

In this equation, $\Delta\sigma$ is the nominal stress range within the cycle. The test specimen undergoes cyclical loading providing the crack growth rate da/dN, which is then plotted as a function of ΔK. The resulting graph will be a straight line when plotted on log–log paper, enabling the coefficients c and m to be determined.

Great care should be taken when using the quoted values for C in the growth rate calculations since it has dimensional units such as psi $\sqrt{\text{in.}}$, ksi $\sqrt{\text{in.}}$, or MPa $\sqrt{m}$ depending on the units chosen to express the growth rate/cycle. The exponent m is nondimensional; however, it is critical in the unit conversion process when converting C from one system to another to multiply by the fraction:

$$\frac{(\text{Linear dimension conversion})}{(\text{Stress dimension conversion})^{m}}$$

To demonstrate the applicability of the Paris law, consider an axial crack in a pipeline, which because of its length can be considered from a stress intensity viewpoint to be infinitely long.

The initial crack is of length $2a_0$ and the pressure due to hoop stress is σ. The stress intensity factor K is identical to that for a wide plate with a center crack given previously, i.e.,

$$K = \sigma\sqrt{\pi a}$$

The total number of cycles for the crack to reach a critical length $2a_c$ can be obtained by integrating the equation defining the fatigue crack propagation, that is, Eq. (6-42) between the limits a_0 and a_c or a_{final} (a_f)

$$N_f = \int_{a0}^{a_f} \frac{da}{C(\Delta K)^m} \qquad (6-43)$$

where C and m = fatigue crack growth constants and may be found in documents such as BS7910 (2005).

Substituting in Eq. (6-43) for ΔK

$$N_f = \int_{a_o}^{a_f} \frac{da}{C(\Delta\sigma\sqrt{\pi a})^m} \qquad (6-44)$$

Integrating between initial and final crack sizes gives

$$N = \frac{a_f^{1-\frac{m}{2}} - a_0^{1-\frac{m}{2}}}{C\Delta\sigma\pi^{\frac{m}{2}}\left(1-\frac{m}{2}\right)} \text{ for } m \neq 2 \qquad (6-45)$$

and

$$N = \frac{a_0}{C\left(\sigma^2 \pi\right)} \ln \frac{a_f}{a_0} \text{ for m} = 2$$

where N is the number of cycles required for a defect to grow from an initial length; a_0, to a final crack length a_f.

Figure 6-45 showing the various stages of crack growth rate predicted by the Paris law does not indicate that in regions I and III, the growth rate is significantly affected by the stress ratio R and it has a moderate affect in region II. Consequently, modifications have been made to the Paris law, the one due to Forman et al. (1967) being given in Eq. 6.46.

$$\frac{\mathrm{d}a}{\mathrm{d}N} = \frac{C \Delta K^m}{(1-R)K_c - \Delta K} \qquad (6-46)$$

An example of a pipeline fatigue failure can be seen in the photograph of the damaged cross-section shown in Figure 6-46. The beach marks typical of fatigue can be clearly seen below the initial crack defect.

Example:

A pipe contains an initial defect of length $2a_o = 10$mm and is subjected to a constant amplitude cyclical stress ranging between a minimum value of 50 MPa and a maximum of 250 Mpa. Assume the fatigue crack growth rate is governed by the equation.

$$\frac{\mathrm{d}a}{\mathrm{d}N} = 0.42 \times 10^{-11} \Delta K^3$$

1. Calculate the crack growth rate when the crack has the following values 10 and 30 mm
2. Assuming that the relevant fracture toughness is 60 MPa $m^{0.5}$ estimate the number of cycles to failure

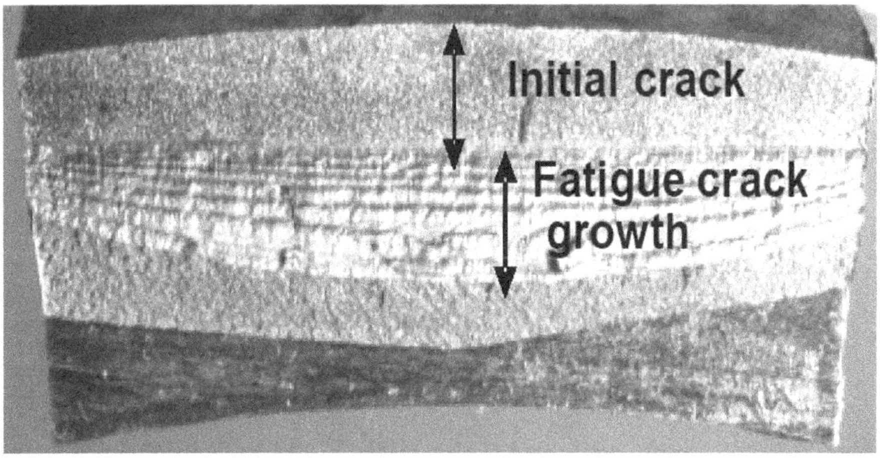

Figure 6-46. Failed pipe section showing fatigue growth

Solution:

Determine the critical crack size $2a_c$ by rearranging Eq. (6-26)

$$a_c = \frac{1}{\pi}\left(\frac{K_{Ic}}{\sigma_{max}}\right)^2 = \frac{1}{\pi}\left(\frac{60}{250}\right)^2 = 0.01833m$$

which means the critical crack size is 36.66 mm

To find crack growth rates,

$$2a = 10 \text{ mm}, \Delta K = \Delta\sigma\sqrt{\pi a} = 200\sqrt{\pi x5/1000} = 25.06 \text{ MPa}\sqrt{m}$$

$$\frac{da}{dN} = 0.42\times10^{-11}\times(25.06)^3 = 0.66\times10^{-7}\, m/\text{cycle}$$

$$2a = 30 \text{ mm}, \Delta K = \Delta\sigma\sqrt{\pi a} = 200\sqrt{\pi\times15/1000} = 43.41 \text{ MPa}\sqrt{m}$$

$$\frac{da}{dN} = 0.42\times10^{-11}\times(43.41)^3 = 0.344\times10^{-6}\text{m/cycle}$$

Fatigue life

Substituting in Eq. (6-45) for C, m, $\Delta\sigma$, a_f, and a_0

$$N = \frac{a_f^{1-\frac{m}{2}} - a_0^{1-\frac{m}{2}}}{C\Delta\sigma\pi^{\frac{m}{2}}\left(1-\frac{m}{2}\right)} = \frac{18.33^{-0.5} - 5^{-0.5}}{0.42\times10^{-11}\times200\pi^{1.5}(-0.5)} = 91.5\times10^6 \text{ cycles}$$

Life Prediction Example:

It was noted earlier in this section that as the flaw grows, it becomes necessary to conduct the crack growth analysis in stages. Consider a high-strength steel test specimen whose yield strength is nominally 100,000 psi. The specimen contains an edge notch with an initial length $a_0 = 0.3$ in. The tensile stresses on the specimen vary between 25,000 psi and 45,000 psi. Estimate the total number of cycles required to cause failure. The fracture toughness is 150 ksi $\sqrt{\text{in.}}$

Solution:

The procedure is as follows:

1. Obtain the stress intensity factor relationship
2. Estimate the initial flaw size
3. Knowing K_c, K_{Ic} and the nominal stress σ, calculate the critical flaw size $a_{crit} = a_f$
4. Obtain the relationship linking da/dN with ΔK, let us assume a Paris law,

$$\frac{da}{dN} = 6.6\times10^{-9}(\Delta K_i)^{2.25}$$

The units of da/dN are inches/cycle and ΔK_i in ksi $\sqrt{\text{in.}}$

5. Determine ΔK_i from step (2)
6. Integrate the expression for *da/dN* between the limits of a_o and a_f to obtain the fatigue life

For an edge crack in tension, the appropriate stress intensity function is,

$$K_{1c} = 1.12\sigma\sqrt{\pi a}$$

From which the critical crack size a_c is,

$$a_c = \left(\frac{K_{1c}}{1.12\sigma_{max}\sqrt{\pi}}\right)^2 = \left(\frac{150}{1.12 \times 45\sqrt{\pi}}\right)^2 = 2.8 \text{ in.}$$

Assume an interval of crack growth say Δa = 0.1in., (smaller intervals will provide slightly better accuracy).

We use the average crack size within the interval to determine ΔK_i and replace the derivative da/dN with the approximation $\Delta a/\Delta N$ to obtain the number of cycles ΔN for each interval in turn, starting with the growth from 0.3 in. to 0.4 in. Note σ is given in terms of ksi.

$$\Delta N = \frac{\Delta a}{6.6x10^{-9}\left(1.12 \times 20 \times \sqrt{\pi a_{avg}}\right)^{2.25}} = \frac{0.1}{6.6 \times 10^{-9}\left(1.982 \times 20 \times \sqrt{0.35}\right)^{2.25}}$$

$$\Delta N = 12500 \text{ cycles}$$

Table 6-3 below contains the results from repeating the above process indicating that a total of 86700 cycles are required to cause failure.

Example:

A pipe with a fracture toughness of 60,000 psi $\sqrt{\text{in.}}$ is subjected to a repeated stress cycle from zero to 20,000 psi. A small crack has started to grow in the pipe, which can be assumed to be quite long so that the stress intensity factor $K_{1C} = \sigma\sqrt{\pi a}$ applies. The constants C and m in the Forman equation for crack growth rate are 5 × 10^{-14} and 3, respectively, where *da/dN* is expressed in inches/cycle and the stress intensity is in psi $\sqrt{\text{in}}$. Determine

(a) The crack length at failure
(b) The crack growth rate when the crack is 2 in. shorter than the critical length
(c) The crack growth rate when the crack is 0.1 in. shorter than the critical length

Solution:

(a) $K_{1c} = \sigma\sqrt{\pi a}$ from which $a_f = \frac{1}{\pi}\left(\frac{60,000}{20,000}\right)^2 = 2.86$ in.

(b) Crack length $a = 2.86 - 2 = 0.86$ in.

$$\sigma_{min} = 0 \longrightarrow K_{min} = 0 \text{ and}$$

$$\Delta K = K_{max} = 20,000\sqrt{0.86\pi} = 32,874 \text{ psi}\sqrt{\text{in}}$$

TABLE 6-3. Crack growth over accumulated cyclical loading

a_0 (in)	a_f (in)	a_{avg} (in)	ΔK	ΔN	ΣN cycles
0.3	0.4	0.35	23.5	12,500	12,500
0.4	0.5	0.45	26.7	9,750	22,250
0.5	0.6	0.55	29.4	7,550	29,800
0.6	0.7	0.65	32.2	6,150	35,950
0.7	0.8	0.75	34.6	5,200	41,150
0.8	0.9	0.85	36.6	4,600	45,750
0.9	1.0	0.95	38.8	4,100	49,850
1.0	1.1	1.05	40.5	3,700	53,550
1.1	1.2	1.15	42.5	3,300	56,850
1.2	1.3	1.25	44.5	2,950	59,800
1.3	1.4	1.35	46.1	2,700	62,500
1.4	1.5	1.45	47.7	2,550	65,050
1.5	1.6	1.55	49.3	2,350	67,400
1.6	1.7	1.65	51.0	2,200	69,600
1.7	1.8	1.75	52.5	2,050	71,650
1.8	1.9	1.85	54.0	1,900	73,550
1.9	2.0	1.95	55.6	1,800	75,350
2.0	2.1	2.05	56.8	1,700	77,050
2.1	2.2	2.15	58.5	1,600	78,650
2.2	2.3	2.25	59.6	1,500	80,150
2.3	2.4	2.35	60.8	1,450	81,600
2.4	2.5	2.45	62.5	1,400	83,000
2.5	2.6	2.55	63.5	1,350	84,350
2.6	2.7	2.65	64.8	1,200	85,550
2.7	2.8	2.75	66.0	1,150	86,700

$$\frac{da}{dN} = \frac{C\Delta K^m}{(1-R)K_c - \Delta K} = \frac{5\times10^{-14}(32{,}874)^3}{(1-0)60{,}000 - 32{,}874} = 6.5\times10^{-5} \text{ in./cycle}$$

(c) Repeating step (b) with $a = 2.76$ in. results in $da/dN = 9.2 \times 10^{-3}$ in./cycle. That is, only 109 additional cycles i.e. $0.1/9.2 \times 10^{-3}$ will be required to reach 2.86 in. and failure.

Irregular Loading

The cyclical pressure loading, which occurs in a real pipeline operation, particularly a liquids line, is quite irregular; see for example Figure 6-47. Performing fatigue calculations using either, the stress life or fracture mechanics approach, requires a simplification of the load spectrum. This is achieved through the use of cycle counting methods, which characterize the irregular pressure, or stress, time history as a distribution of individual cycles of known amplitude. There are a number of different counting methods available, the most common in use with pipeline companies being the rain flow cycle counting approach. This was developed by Dowling and Sochie (1982) and is highly amenable to digital processing. It is a two-parameter approach providing information about the mean value as well as the amplitude of each cycle.

Descriptively one should imagine the stress time history spectrum rotated through 90° where it notionally looks like a pagoda style roof (Figure 6-48). A raindrop will flow down the roof until a peak or trough is encountered at which point it falls vertically until it either reaches another roof below, or strikes the ground. The beginning and end of each flow path defines the extremities of a half-cycle. A manual process for reducing the data from the load spectrum would look like:

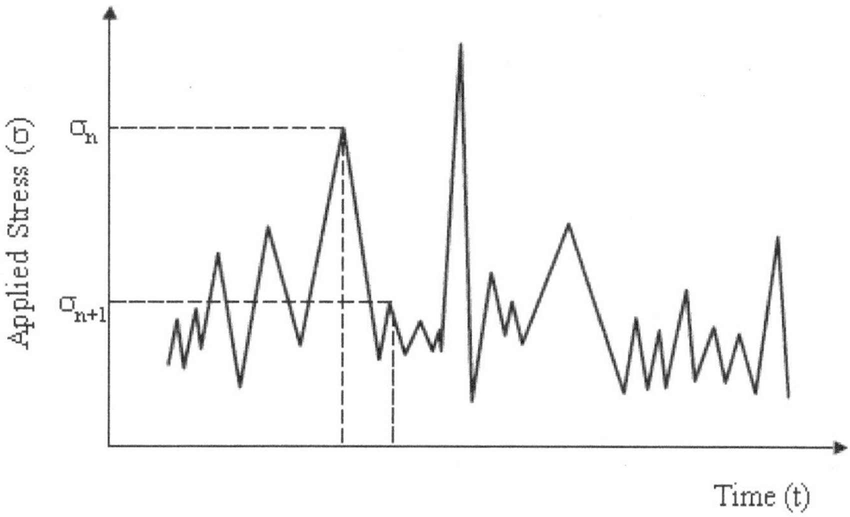

Figure 6-47. Irregular stress spectrum

1. Reduce the time history to a sequence of (tensile) peaks and (compressive) troughs.
2. Rotate the time history clockwise 90° so that the earliest time occurs at the top.
3. Each tensile peak is imagined as a source of water that "drips" down the pagoda.
4. Count the number of half-cycles by looking for terminations in the flow occurring when either:
   a. it reaches the end of the time history;
   b. it merges with a flow that started at an earlier tensile peak; or
   c. it flows opposite a tensile peak of greater magnitude.
5. Repeat step 4 for compressive troughs.
6. Assign a stress magnitude to each half-cycle equal to the difference between its start and termination point.

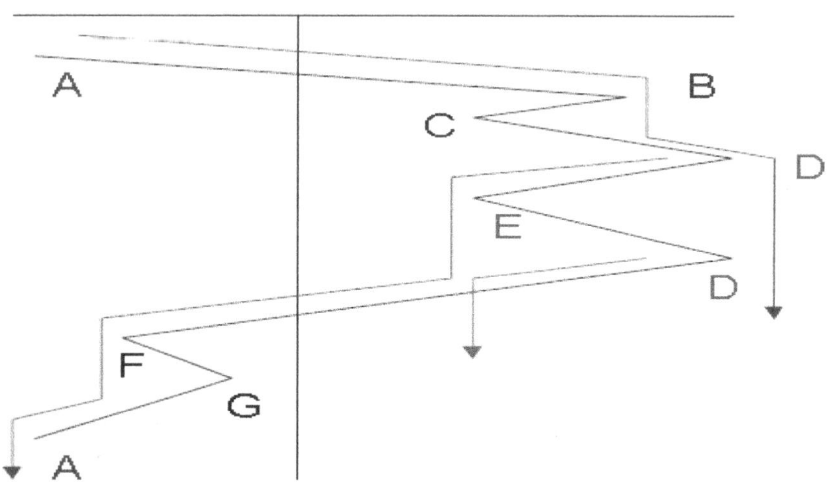

Figure 6-48. Rain flow on a Pagoda roof analogy for cycle counting

7. Pair up half-cycles of identical magnitude (but opposite sense) to count the number of full, or complete, cycles. Typically, there will be some residual half-cycles in the time history. For example, in Figure 6-48, there are four cycles, two full cycles A-D-A and D-E-D, and two half cycles B-C and F-G.

Modern computer-based algorithms no longer use the manual step-by-step procedure shown above, although the generic name "rain flow" still persists.

DEFECT SIZING

In Defect Assessment Methods of this chapter, it will be apparent that the accuracy of the assessment methods discussed is highly dependent on our knowledge of the size and type of defect found. Tool selection and accuracy are vital to the success of this effort, and it is important to realize that inspection tools have their strengths and weaknesses and we need to be aware of them when making them part of our defect assessment tool kit. In Chapter 5, we examined the ability of ILI tools to detect and size anomalies, which warrant more detailed scrutiny by means of excavation and in the ditch measurement. Mapping the exposed surface of the pipe provides the necessary detail required to establish its remaining strength. While a surface crack length may be measured directly using a rule, accurate crack depth is quite difficult to establish even using a hand-held UT. By comparison, metal loss, whether generalized, or local, is easier to measure. The quickest and easiest, but least accurate, way to accomplish this task is to use a pit gauge, which measures pit depth using a protruding needle that is calibrated on a flat surface before being placed in the defect. The depth is then read out on a scale, dial, or digital display.

A bridging bar and depth micrometer (see Figure 6-49) works in a similar manner to a pit gauge by measuring the pit depth from a reference plane, the bridging bar, to a point on the end of the micrometer. The bridging bar can be set up to operate in either of the two modes shown, spanning or cantilevered, depending on the location of the corroded area. It is held in place by magnets affixed close to its ends, freeing the operator to concentrate on taking depth measurements along the corroded length. The depth micrometer used on this equipment is more accurate than a pit gauge, but the procedure requires more time to set up. Pen probes are another means of finding the remaining wall thickness. They use a small transducer that transmits ultrasonic waves into the pipe wall. The probe is small enough to fit into small corrosion pits so as to accurately determine defect depth.

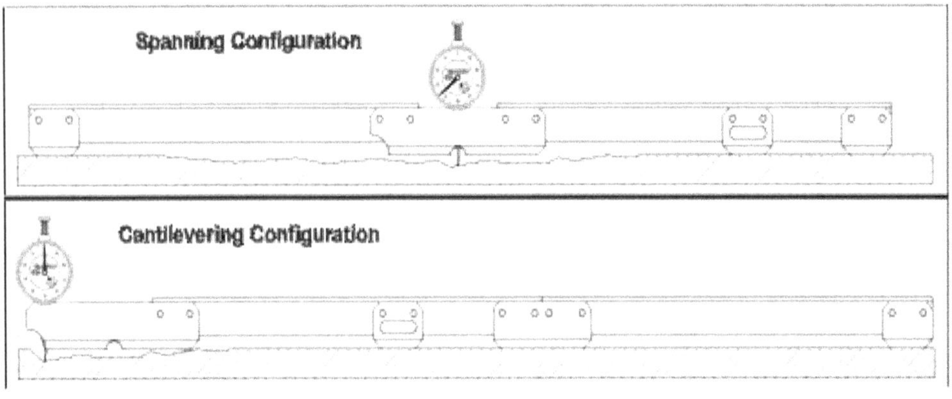

Figure 6-49. A bridging pit gauge (courtesy of Western Instruments Inc.)

Figure 6-50. Laser profile imaging tool (courtesy of RTD Inc.)

More refined methods are now available to measure and map an entire area, including the laser profiling inspection tool (LPIT) shown in Figure 6-50. It uses laser triangulation to measure and map the corrosion depths in their exact spatial orientations to accurately represent corrosion found on the pipe as a digital C-scan image. A representation of the accuracy, which can be achieved, may be seen from Figure 6-51, where a digital photographic image of a corroded pipe surface is compared with the C-scan image. The increase in depth is displayed by a change in the color intensity on the C-scan image.

Common to all of the profiling methods described above is the need that the corroded area must be thoroughly cleaned and all corrosion products removed so that the dimensions can be accurately measured.

DEFECT ASSESSMENT METHODS

When a defect is discovered in a pipe causing uncertainty as to its continued serviceability, an engineering critical assessment can be performed, if sufficient information is available to determine its fitness for service.

The necessary information relates to the nature and physical dimensions of the defect, the geometric and material properties of the pipe, i.e., its diameter and wall thickness, its fracture toughness at the service temperature, and the specified minimum yield stress. In addition, it is useful to know the recent operating pressure history so that maximum stresses can be calculated.

These parameters form the input to a model describing the interaction between the defect, the stress and the toughness of the material. The types of defects, which can be addressed in this manner include corrosion induced metal loss, and cracking or planar defects in the body and the long seam weld of the pipe. Crack like anomalies are dealt with quite differently in various jurisdictions. Whereas in the United States, crack-like features need to repaired or removed from pipelines, in Canada, it is permissible to perform an engineering critical assessment using some of the methods outlined in this Section. While mandatory in the United States, many operators of transportation pipelines in other jurisdictions already perform ILI, hydrostatic testing, and direct assessment methods to verify the safety

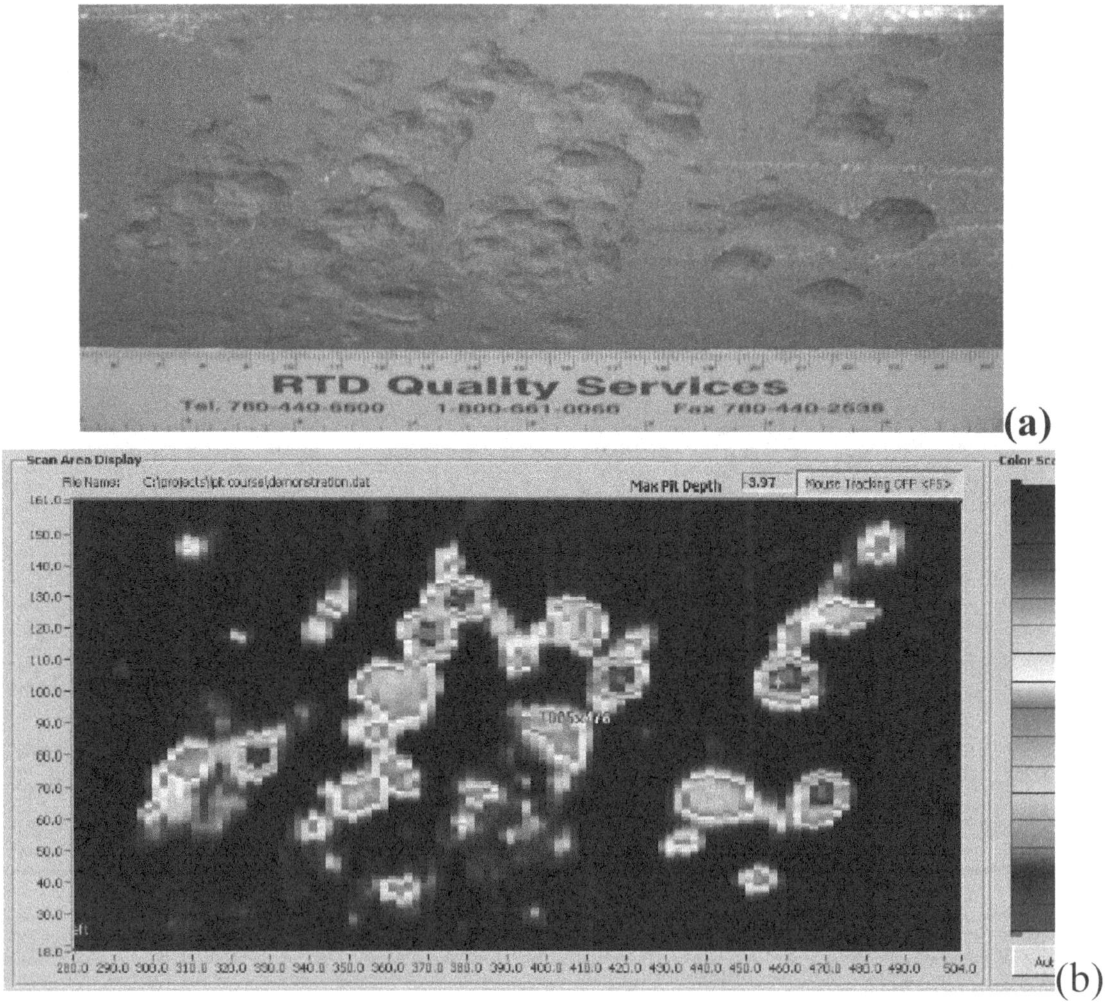

Figure 6-51. Accuracy representation LPIT (a: field and b: LPIT data; courtesy of RTD Inc.)

of their system. The ASME pipeline codes B31.4 and B31.8 contain extensive sections on operating and maintenance including detailed acceptance criteria for certain types of in service defects. The methods subsequently developed to meet these criteria have their origins for the most part, in classical fracture mechanics, adjusted to account for strain hardening and inelastic behavior. Their development and validation owes much to the extensive research efforts sponsored by individual pipeline operators and industry consortia such as the American Gas Association (AGA), the Gas Research Institute (GRI), and the Pipeline Research Council International (PRCI). Most of the methods are semi-empirical and involve simplifying assumptions about the characterization of the defect. Most pipeline steels may be regarded as being both tough and ductile; hence, the majority of defect assessment techniques consider plastic collapse as the limiting failure state. The major causes of pipeline failure throughout the world tend to be either metal loss due to corrosion, or as the result of external damage. Which of the two mechanisms found to be dominant, in a given jurisdiction, is highly dependent on the population density, and the types of pipeline system in place. For example, gathering systems are more prone to pitting corrosion failure than third-party damage. Hence, regulatory authorities in producing regions are more likely to place corrosion top of the list, whereas in Western Europe, external damage prevails. In the following section, methods are provided for assessing corrosion and other blunt defects, dents, gouges, and fatigue cracks. Several worked examples have been included to help provide clarity.

Gouges

Gouges arise as a result of a scraping action on the surface of the pipe and could be the result of the pipe having been dragged over a rock in the ditch, or by mechanical impact. The associated metal loss is often accompanied by local plastic deformation and the formation a thin layer of hardened material. This work hardened area may have very low ductility and may contain small cracks. Depending on the orientation of the gouge, several methods of assessment are available. However, if the mechanical interference is severe enough to cause a dent, then an alternative approach must be used.

The gouge reduces both the burst strength and fatigue strength of the pipe with a longitudinally orientated gouge considered a more severe defect than one in the circumferential direction. The assessment of gouges assumes that failure is controlled by plastic collapse. These assessment methods have been validated against the results of 48 full-scale vessel burst tests containing artificial, longitudinally orientated, machined V-shaped notches. It is recommended that the measured depth of the gouge be increased by 0.5 mm to account for the possibility of cracking at the base of the gouge.

The assessment of gouges originates with the assessment of part wall defects referred to as the NG18 surface flaw equations. Figure 6-52 (a and b) illustrate what is meant by a part wall and a through wall defect. For ductile pipe, subjected to internal pressure, the failure stress σ_f of an axially oriented gauge is given by:

$$\sigma_f = \bar{\sigma}\left[\frac{1 - d/t}{1 - M^{-1}\left(d/t\right)}\right] \quad (6-47)$$

where

d = defect depth

t = wall thickness

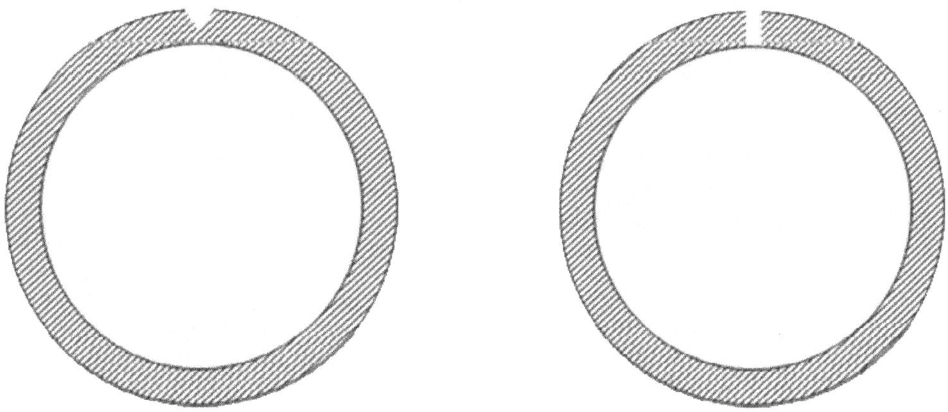

(a) (b)

Figure 6-52. Part-wall (a) and through-wall (b) defects

and the flow stress

$$\overline{\sigma} = \left(\frac{\sigma_y + \sigma_{\text{ult}}}{2} \right) \text{or } 1.15\ \sigma_y \qquad (6-48)$$

The Folias factor M is used to take account of the stress magnification as a consequence of bulging due to wall weakness. The analytical solution for this factor is an infinite series with three commonly used approximations given by,

$$M = \sqrt{1 + 0.52 \left({}^{2c}\!/\!_{\sqrt{Dt}} \right)^2} \qquad (6-49)$$

$$M = \sqrt{1 + 0.6275 \left({}^{2c}\!/\!_{\sqrt{Dt}} \right)^2 - 0.003375 \left({}^{2c}\!/\!_{\sqrt{Dt}} \right)^4} \qquad (6-50)$$

$$M = \sqrt{1 + 0.8 \left({}^{2c}\!/\!_{\sqrt{Dt}} \right)^2} \qquad (6-51)$$

The variables defect length $2c$, pipe diameter D, and wall thickness t are shown in Figure 6-53(a).

Kastner et al. (1981) developed the following plastic criterion for circumferential gouges, which are assumed to be rectangular in profile and oriented perpendicular to the axis of the pipe:

$$\frac{\sigma_f}{\overline{\sigma}} = \frac{\eta(\pi - \beta(1-\eta))}{\eta\pi + 2\sin\beta(1-\eta)} \qquad (6-52)$$

where $\beta = 2c/D$ and $\eta = 1 - {}^{d}\!/\!_{t}$

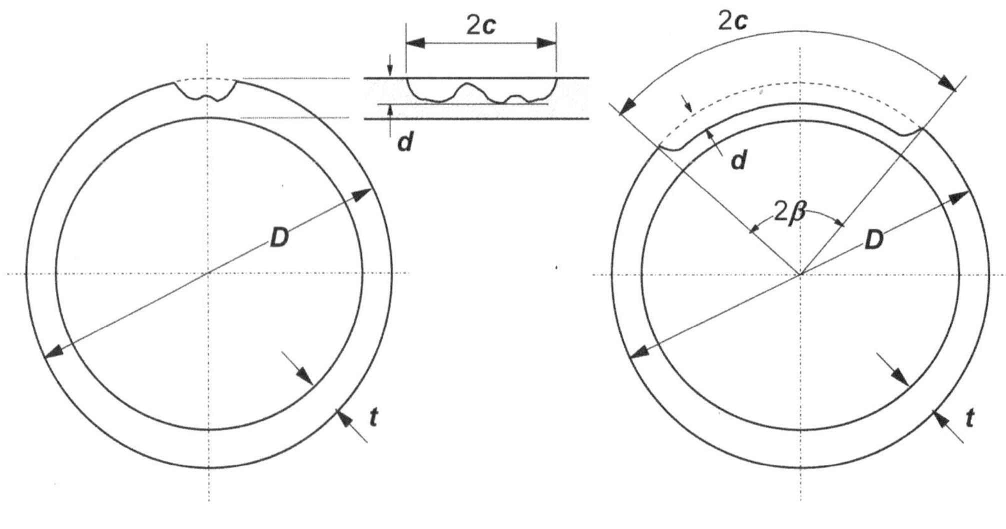

Figure 6-53. Dimensions of a longitudinal (a) and a circumferential (b) through-wall crack defect (Courtesy Penspen Integrity Inc.)

$2c$ is the maximum circumferential length of the defect [see Figure 6-53(b)], and the flow stress in this model is taken to be the average of the yield and tensile strength. The model was developed for a single part-wall defect subjected to internal pressure, axial, and bending loads. The failure criterion was developed solely for pressure loading; however, it has been shown to be applicable to any axial load by replacing the axial stress due to internal pressure (caused by the Poisson effect), with the total axial stress.

Dents

A dent is a depression caused by contact with an indenter that results in plastic deformation of the pipe wall. It causes a local stress and strain concentration as well as a local reduction in pipe diameter and produces a gross disturbance in the curvature of the pipe wall (see Figure 6-54). Dents are defined by their geometry (Figure 6-55); a smooth dent causes a smooth change in the curvature of the pipe wall, whereas a kinked dent causes an abrupt change. Plain dents are smooth and have no wall thickness reductions.

Spring back is defined as the reduction in dent depth due to elastic unloading that occurs, when the indenter is removed from the pipe. An unconstrained dent will spring back or return to its former shape and attempt to re-round when the indenter is removed. A constrained dent, however, is not free to re-round because the indenter remains in contact with the pipe.

Re-rounding occurs when the dent attempts to move outward upon re-pressuring of the pipe. The change in dent depth under internal pressure does not significantly reduce the burst strength of pipe — unless the dent is very deep. An empirical limit for a plain dent under static internal pressure loading has been derived from the results obtained from extensive full-scale testing. All dent depths in the full-scale tests were measured at zero pressure; hence, a correction for spring back was applied to dents measured while under pressure. Tests by Battelle, British Gas (BG Technology), and others suggest that plain dents less than 8% of the pipe diameter (measured at pressure) do not significantly affect the burst strength of pipe. Dents of 8% pipe diameter may, however, restrict product flow and passage of pigs and render the pipe unserviceable. In the British Gas testing, dents up

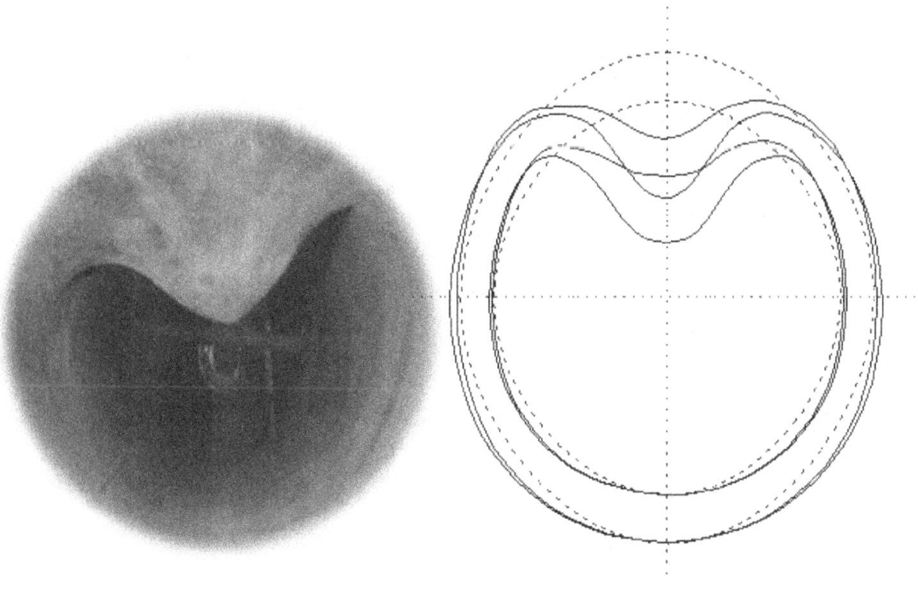

Figure 6-54. An extremely deep dent in ductile pipe

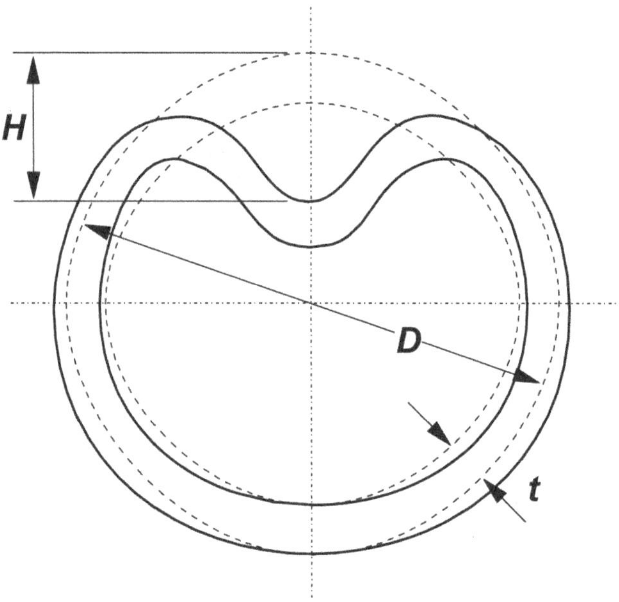

Figure 6-55. Dimensional definitions in a dent under pressure (Eiber et al. 1981)

to 24% pipe diameter (Figure 6-54) did not significantly reduce burst strength. These results would indicate that the code requirements are suitably conservative.

Cyclic pressure tests indicate that the fatigue life of a plain dent is less than that of an equivalent circular section of pipe. In general, the deeper the dent, the shorter will be the fatigue life, which also depends on the mean stress level. Higher mean stresses promote re-rounding of the pipe. The length of the dent also has an influence on the likely orientation and location of any fatigue cracking present. Long dents have a tendency to contain cracks, which are axially orientated and which usually occur in the center of the dent, whereas with shorter dents cracking is found around the flanks.

The European Pipeline Research Group (EPRG), among others, has developed an empirical model (Roovers et al. 2000) for predicting fatigue life of plain dents. It enables the fatigue life to be calculated using the S-N curves for plain pipe, modified to account for a stress concentration. In common with all such fatigue life models for denting, the EPRG recommends a model uncertainty factor of 13, to be applied to the predicted fatigue life (others use a factor of 10).

A constrained plain dent will have a fatigue life that is at least that of an unconstrained plain dent of the same depth. Consequently, the same proposed method can also be used. The equation for the EPRG fatigue life estimate is given by,

$$N = \frac{1000}{13}\left[\frac{\sigma_u - 50}{2\sigma_A K_s}\right]^{4.292} \qquad (6-53)$$

Where

N	= number of fatigue life cycles
σ_u	= ultimate tensile strength UTS
σ_A	= stress factor= $(\sigma_{max} - \sigma_{min})/2$
K_s	= dent factor = $2.871\,(1.43tH_r)^{0.5}$
$2\sigma_A$	= UTS $(B(4 + B^2)^{0.5} - B^2)$
H_r	= dent depth at pressure
t	= wall thickness

B = parameter that is a function of the cyclic stress amplitude σ_A
B = $(\sigma_A/\text{UTS})(1 - (\sigma_{max} - \sigma_A)/\text{UTS})$
Safety factor =13

To use the model, we count the number of cycles n over a specified period D_f using the rain flow cycle counting method described in Fracture Mechanics in Defect Assessment for each selected pressure range. The damage fraction n/N in each cycle is calculated and then summed to compute the fatigue damage over the period D_f. The minimum predicted fatigue life is then given by $1/D_f$.

An alternative to the EPRG model has been developed by Fowler (1993), referred to as the Stress Engineering Services, or SES, model. Fowler performed testing and analytical work to establish the effects of pipeline dents (without a gouge), under cyclic internal pressure loading. The results indicated that plain smooth dents that were less than 5% of the diameter for pipe with a diameter to thickness ratio of less than 30 are not a problem for normal pipeline service and unlikely to be a concern for normal gas line service. Fowler estimated the fatigue life based on the equivalent number of cycles for a pressure range from 0 to1200 psi using Miner's rule and by assuming the fatigue curve has a similar form to Eq. (6-40):

$$N = CS^{-3.74} \tag{6-54}$$

where

N = equivalent number of cycles
$C = 2.0E + 6$ cycles
S = stress, which is assumed linear with pressure

The S-N fatigue curve first used for this analysis was the API_RP2A curve X' with:

$$N = C\left(\frac{\Delta\sigma}{\Delta\sigma_{ref}}\right) \tag{6-55}$$

which has an endurance limit at 200 million cycles, at a stress of 3330 psi. σ = pressure fluctuation, $\sigma_{ref} = 11{,}400$ psi, $m = 3.74$.

Miner's rule was used to derive a linear damage factor since the pressure and stress amplitudes vary widely. By combining Eqs. (6-54) and (6-55), an equation for the fatigue life of a plain dent is given by:

$$N = 2\times 10^{6}\left(\left[\frac{\Delta\sigma}{\Delta p}\right]\Delta p\frac{1}{11{,}400}\right)^{-3.74} \tag{6-56}$$

where N is fatigue life cycles and Δp, cyclic pressure.

$\left[\frac{\Delta\sigma}{\Delta p}\right]$ = stress intensification factor given in Table 6-4 taken from Fowler (1993)

The spring back correction is given by the following equation,

$$\frac{D_0}{2R} = 1.43\frac{D_r}{2R} - 0.0023 \tag{6-57}$$

where

$2R$ = nominal pipe diameter
D_0 = dent depth measured at zero pressure
D_r = dent depth measured at pressure

TABLE 6-4. Values of stress intensification factor ($\Delta\sigma/\Delta P$) for an average operating pressure of 1000 psi (Fowler 1993)

D/t	18	20	25	30	35	40	45	50
$D_0/2R$								
1	13.94	13.25	15.00	20.93	29.80	38.68	47.55	56.43
2	19.64	21.18	25.50	32.53	39.87	47.20	54.53	61.87
3	25.35	29.11	38.34	44.14	49.93	55.72	61.51	67.31
4	31.05	37.04	51.49	55.74	59.99	64.24	68.50	72.75
5	36.76	44.97	64.63	67.34	70.05	72.77	75.48	78.19
6	42.46	52.89	77.78	78.95	81.29	81.29	82.46	83.63
7	48.17	60.82	90.92					
8	53.87	68.75	104.10					
9	59.58	76.68	117.20					
10	65.28	84.61	130.40					
11	70.99	92.53	143.50					

The assumption is made that the application of n_i cycles at a stress amplitude $\Delta\sigma$, for which the average number of cycles to failure is N_i, cause an amount of fatigue damage that is measured by the cumulative cycles ratio n_i/N_i. The estimated fatigue life X can be calculated as:

$$X = \frac{1}{\sum \frac{n_i}{N_i}} \qquad (6-58)$$

Example of a Plain Dent Fatigue Assessment

A 24-in. diameter pipeline, 0.6-in. wall thickness has been struck by an excavator. Inspection has shown a smooth dent with no associated metal loss or cracking. The dent is located in the body of the pipe and is 4.2% of the pipeline diameter when measured at pressure. The pipeline undergoes daily pressure cycling in the range 800 psi to1200 psi. The pipeline has an expected life of 20 more years. Determine if the dent is acceptable (a) under static loading and (b) the cyclic pressure loading. Use the SES Model.

The spring back correction is first determined from

$$\frac{D_0}{2R} = 1.43\frac{D_r}{2R} - 0.0023$$

$$= 1.43 \times 0.042 - 0.0023 = 0.05776$$

Hence, $D_0/2R = 5.776\%$ and $D/t = 24/0.6 = 40$

Table 6-4 provides values of $\Delta\sigma/\Delta P$ for an average operating pressure of 1000 psi and ratios of $D_0/2R$. By interpolation, the required $\Delta\sigma/\Delta P = 79.38$ $\Delta P = 400$ psi so the fatigue life of the plain dent is,

$$N = 2\times10^6\left(\left[\frac{\Delta\sigma}{\Delta p}\right]\Delta p\frac{1}{11{,}400}\right)^{-3.74} = 2\times10^6\left(79.38\times\frac{400}{11{,}400}\right)^{-3.74} = 43{,}371 \text{ cycles}$$

Since the pipeline is cycled once per day, the predicted fatigue life including a factor of safety of 13 is

$$43{,}371/(365 * 13) = 9.14 \text{ years}$$

the dent is not acceptable under cyclic loading.

Dents and Gouges

For a smooth dent containing a defect (Figure 6-56), the fracture model developed by Hopkins et al. (1989) is defined as follows in SI units

$$\frac{\sigma_h}{\bar{\sigma}} = \frac{2\cos^{-1}}{\pi}\left[\exp -\left\{113\frac{1.5\pi E\left[Y_1\left(1-\frac{1.8H_0}{D}\right)+Y_2\left(10.2\frac{H_0}{2t}\right)\right]^{-2}}{\bar{\sigma}^2 Ad}\exp\left[\frac{\ln\left(0.738C_{v2/3}-K_1\right)}{K_2}\right]\right\}\right] \quad (6-59)$$

and in Imperial units

$$\frac{\sigma_h}{\bar{\sigma}} = \frac{2\cos^{-1}}{\pi}\left[\exp -\left\{\frac{1.5\pi E\left[Y_1\left(1-\frac{1.8H_0}{D}\right)+Y_2\left(10.2\frac{H_0}{2t}\right)\right]^{-2}}{\bar{\sigma}^2 Ad}\exp\left[\frac{\ln\left(C_{v2/3}-K_1\right)}{K_2}\right]\right\}\right]$$

where

$$\bar{\sigma} = 1.15\sigma_y\left(1-\frac{d}{t}\right),\ K_1 = 1.9,\ K_2 = 0.57$$

and

$$Y_1 = 1.12 - 0.23\left(\frac{d}{t}\right) + 10.6\left(\frac{d}{t}\right)^2 - 21.7\left(\frac{d}{t}\right)^3 + 30.4\left(\frac{d}{t}\right)^4$$

$$Y_2 = 1.12 - 1.39\left(\frac{d}{t}\right) + 7.32\left(\frac{d}{t}\right)^2 - 13.1\left(\frac{d}{t}\right)^3 + 14\left(\frac{d}{t}\right)^4$$

Note the following are all in Imperial units. (K_1 and K_2 are nonlinear regression parameters.)

σ_f = hoop stress at failure, lbf/in.^{2}
σ_{bar} = plastic collapse stress of an infinitely long gouge, lbf/in.^{2}
A = fracture area of Charpy (0.083 in.^{2} for a 2/3 Charpy specimen)
E = Young's Modulus (30,000,000 lbf/in.^{2})

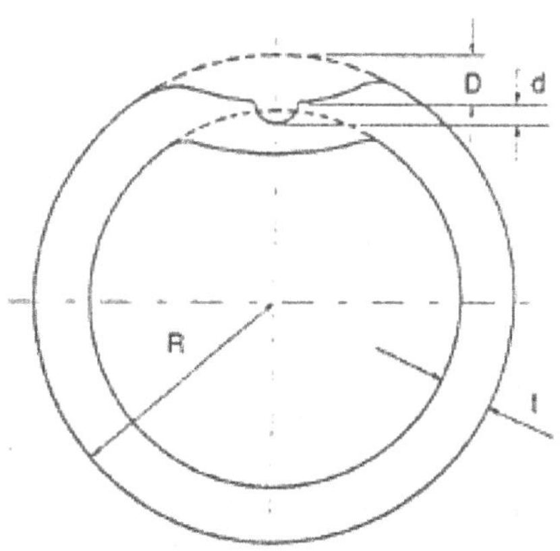

Figure 6-56. A dent containing a gouge

$C_{v2/3}$ = 2/3 Charpy toughness, ft lbs.
d = maximum or average depth of part wall defect, in.
H_0 = dent depth measured at zero pressure, in.

Eq. (6-59) takes into account the effects of both membrane action and bending due to the presence of a dent of depth H_0. The membrane and bending stresses are given, respectively, by the terms

$$\sigma_m = \sigma_h\left(1 - \frac{1.8H_0}{D}\right) \text{ and } \sigma_b = \sigma_h\left(10.2\frac{H_0}{2t}\right) \quad (6-60)$$

If the depth of a gouge exceeds a critical value, then a through wall defect results. If the length of such a through wall defect exceeds a critical length L_c, then rupture will occur. That is when,

$$L_{defect} \geq \sqrt{\left[\left(\frac{\sigma_h}{1.15\sigma_y}\right)^{-2} - 1\right]\left(\frac{Dt}{0.8}\right)} \quad (6-61)$$

The above model, while it may appear from its complexity to be fairly exact, has a lot of scatter when its predictions are compared to test data. It is more useful as a research tool than a practical model, and since it is NOT conservative, its use is not recommended in the absence of specialist advice (Rosenfeld et al. 2002).

Example:

An NPS 36-in diameter pipeline, 0.5-in wall thickness, and grade X70 has been dented and gouged (Figure 6-56) during construction. The dent is 2.5 in. deep, and the gouge is 20% of the wall thickness. The pipeline is to be operated at 80% specified minimum yield strength (SMYS). Determine whether or not it is necessary to repair the damage. The ⅔ size Charpy toughness value is 61 ft-lbs. and the area of the fracture surface of a ⅔ Charpy specimen is 0.083 in.^{2}. Young's modulus E is 30,000,000 lbf/in.^{2}.

$$\frac{\sigma_h}{\bar{\sigma}} = \frac{2\cos^{-1}}{\pi}\left[\exp - \left\{\frac{1.5\pi E\left[Y_1\left(1 - \frac{1.8H_0}{D}\right) + Y_2\left(10.2\frac{H_0}{2t}\right)\right]^{-2}}{\bar{\sigma}^2 Ad}\exp\left[\frac{\ln\left(C_{v2/3} - K_1\right)}{K_2}\right]\right\}\right]$$

Evaluating each of the terms in turn,

$\sigma_{bar} = 1.15\sigma_y\,(1 - d/t) = 1.15 * 70{,}000\,(1 - 0.2) = 64{,}400$ lbf/in.^{2}

$$\frac{1.5\pi E}{\sigma_{bar}^2 Ad} = \frac{1.5\pi\, 30{,}000{,}000}{64{,}400^2 \times 0.083 \times 0.1} = 4.106$$

$Y_1 = 1.12 - 0.23(d/t) + 10.6(d/t)^2 - 21.7(d/t)^3 + 30.4(d/t)^4$
$Y_1 = 1.12 - 0.23(0.2) + 10.6(0.2)^2 - 21.7(0.2)^3 + 30.4(0.2)^4$
$Y_1 = 1.373$
$Y_2 = 1.12 - 1.39(d/t) + 7.32(d/t)^2 - 13.1(d/t)^3 + 14(d/t)^4$
$Y_2 = 1.12 - 1.39(0.2) + 7.32(0.2)^2 - 13.1(0.2)^3 + 14(0.2)^4$
$Y_2 = 1.0524$
$K_1 = 1.9$
$K_2 = 0.57$

$(10.2\, H_0/2t) = 10.2(2.5)/2 \times 0.5 = 25.5$ and $(1 - 1.8H_0/D) = (1 - 1.8 \times 2.5/36) = 0.875$

$\sigma_f/\bar{\sigma} = 2/\pi\cos^{-1}[\exp - \{4.106[(1.373)(0.875) + 1.0524(25.5)]^{-2} \exp(3.887)\}]$
$\sigma_f/\bar{\sigma} = 2/\pi\cos^{-1}[\exp - \{4.106 \times (0.00127) \times 48.764)] = 2/\pi\cos^{-1}(\exp - 0.2542)$
$\sigma_f/\bar{\sigma} = 0.4345\ \bar{\sigma} = 0.4345 \times 64{,}400 = 28{,}011\ \text{lbf/in.}^2$

$$\frac{28{,}011 \times 100}{70000} = 40\%\ \text{SMYS}$$

This provides a factor of safety of 2 for an 80% SMYS limit on operating stress.

Corrosion Defects

Corrosion damage in pipelines can be detected by various methods. When it is discovered, the primary concern is whether the pipeline remains structurally sound at the MAOP. If not, the pipe may leak or rupture.

A number of analytical techniques have been developed to determine if a defect will affect the pipeline's capability to operate at the MAOP. The best known method of assessing the remaining strength of corroded pipelines is that of the ANSI/ ASME B31 G manual (2009), first published in 1973. It was later perceived to be conservative, so refinements to the underlying concepts were made leading to the development of a modified B31G and later the effective area method. All three approaches stem from the innovative work conducted by Kiefner and Maxey in the early 1970s to predict the failure stress level for a through wall axial defect in a pipe. They developed a semi empirical relationship based on the results obtained from a number of full-scale tests. This relationship, commonly referred to as the log secant model, accounts for the flaw length along the axis of the pipe, the flaw depth through the pipe wall, and the bulging that occurs with surface breaking flaws. It links the hoop stress at failure, to the geometry of the pipe and its toughness as follows.

$$\frac{K_c^2 \pi}{8c\bar{\sigma}^2} = \ln \sec \frac{\pi}{2}\left[\frac{M_p\, \sigma_p}{\bar{\sigma}}\right] \qquad (6-62)$$

where

$K_c{}^2 = \dfrac{12\, C_V\, E}{A_c}$ and C_V is the upper shelf Charpy V-notch impact energy, ft-lb

A_c = area of Charpy V-notch specimen, in.^{2}
E = elastic modulus, psi
$L = 2c$ = length of area affected by corrosion and/or stress corrosion cracking, in.
$\bar{\sigma}$ = flow stress, psi (SMYS + 10,000 psi)
σ_p = failure stress, psi
SMYS = specified minimum yield strength, psi
D = nominal outside diameter of the pipe, in.
t = nominal wall thickness of the pipe, in.
A = the area of metal loss due to either corrosion or stress corrosion cracking
A_0 = the area of metal, Lt, which would remain intact, if there were no flaws

An axial defect in a pressurized pipe will disrupt the flow of hoop stress causing it to be concentrated toward the ends of the defect. The Folias factor M takes account of the stress magnification as a consequence of bulging due to pipe wall weakness and its simplest form is given by:

$$M = \sqrt{1 + 0.8\left(L/\sqrt{Dt}\right)^2} \qquad (6-63)$$

where t is the wall thickness. As noted earlier, the factor M derives from an infinite series solution, with alternative forms provided by Eqs. 6-49 and 6-50 above.

The method used to determine the minimum toughness requirements is based on the Batelle/AGA relationship for ductile initiation developed by Kiefner and Duffy et al. (1973) from work originally described by Burdekin and Stone (1966).

They in turn had used the Dugdale (1960) elastic plastic strip yield model to derive the relationship between crack opening displacement and the applied stress in a biaxial stress field. They added the Folias factor M_T and the use of flow stress (to account for the strain hardening of typical pipeline steels).

From the foregoing, it can be seen that the maximum flaw size, which can be tolerated in a pipe, is a function of its diameter, pressure, wall thickness, and steel grade — in short its operating stress. The effect of toughness on the failure of a ductile material is readily demonstrated by considering the factors contained in Eq. (6-62). Figure 6-57 shows the result of plotting the logarithm of the normalized stress $M\sigma_h/\bar{\sigma}$ against the logarithm of the normalized toughness $K_c^2\pi/8c\bar{\sigma}^2$ from which it can be seen that as the fracture toughness increases the normalized stress approaches a value of unity. When the quantity $K_c^2\pi/8c\bar{\sigma}^2$ is greater than 3.0 pipe steel is generally referred to as being flow stress dependent, meaning that the failure stress for the defect is independent of toughness. Conversely, at values below 3.0, the steel is said to be toughness dependent. The flow stress or instability criterion for determining the failure of a through wall defect is found by equating the normalized flow stress to 1, from which,

$$\sigma_h = \frac{\bar{\sigma}}{M_T} \qquad (6-63)$$

By plotting the inverse of the Folias factor M for a range of non-dimensional crack lengths as shown in Figure 6-58, we can readily obtain the hoop stress at failure for a through thickness flaw. Equation 6-63 can also be used to define the boundary between surface flaws that fail as ruptures and those that will fail as leaks in flow stress dependent

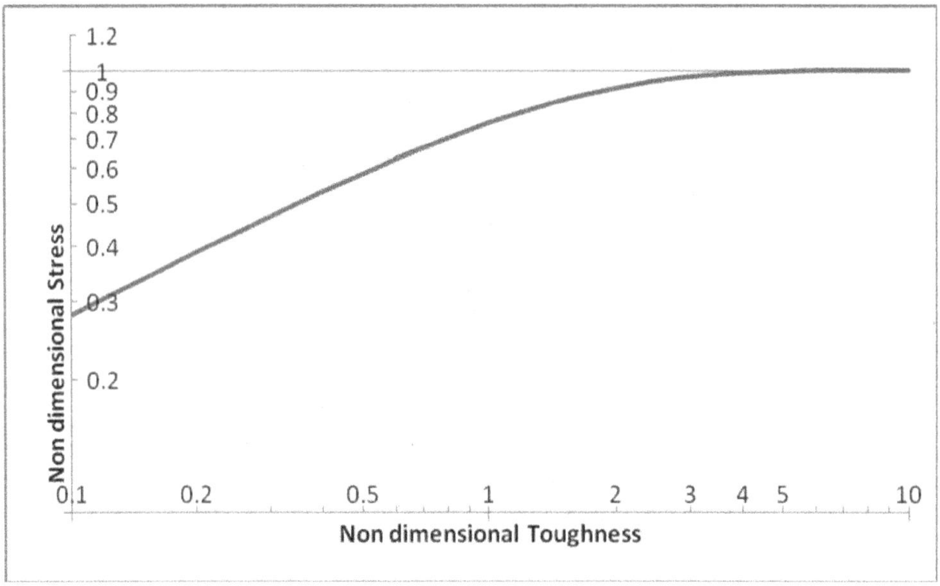

Figure 6-57. Relationship between failure stress and through wall flaw size

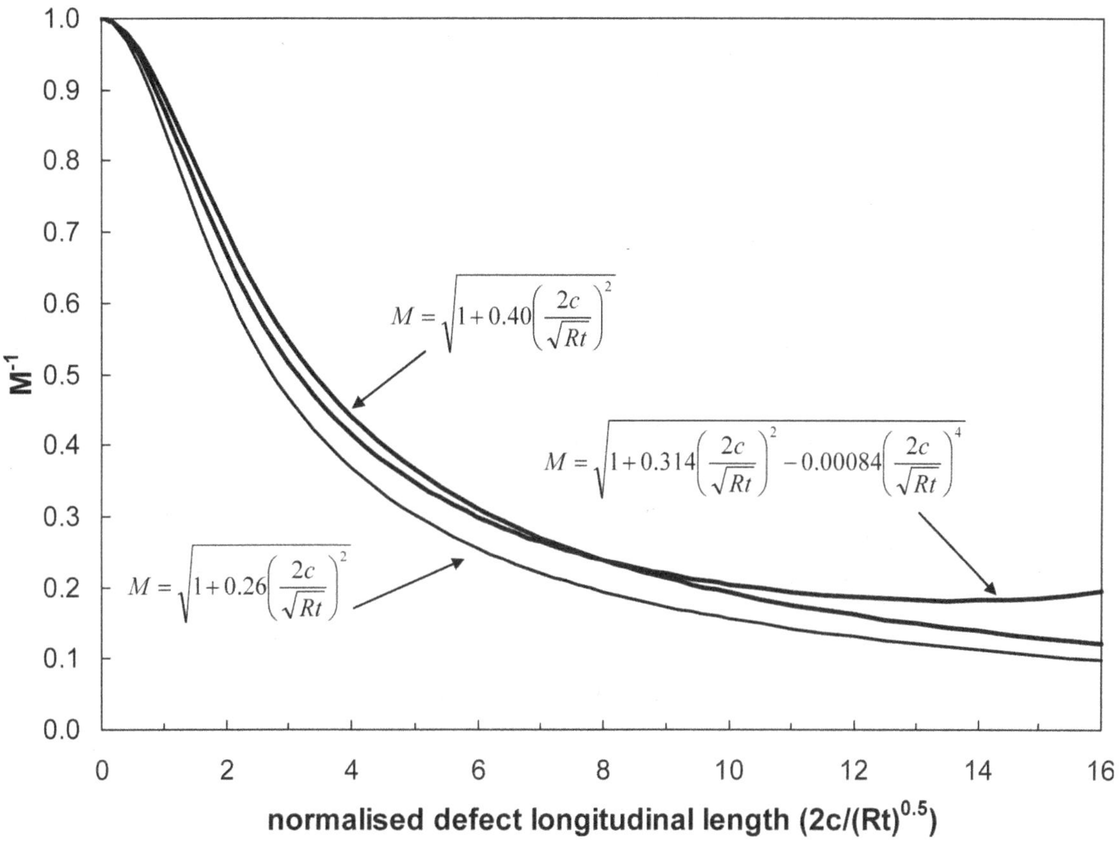

Figure 6-58. Flow stress-dependent failure criterion

materials. The flow stress-dependent failure criterion means that, graphically from Figure 6-58 for a given representation of Folias factor, a point that lies above the line will rupture whereas a point below the line will leak but not fail. Note that the ordinate in Figure 6-58 could also be written as $\frac{\sigma_h}{\bar{\sigma}}$.

Example:

A 36-in. (914.4 mm) diameter pipe of X65 grade steel and 12.7-mm wall thickness has a through wall defect 6 in. (152.4 mm) long. If the operating pressure is 850 psig, determine if the pipe will leak or rupture. Assume the flow stress is 110% of the specified minimum yield stress and a two-term Folias factor. $M = \sqrt{1+0.26\left(2c/\sqrt{Rt}\right)^2}$

Solution:

For failure to occur, the failure stress must be less than the hoop stress; otherwise, the pipe will continue to leak.

1. Determine the hoop stress at operating pressure

$$\sigma_h = \frac{PD}{2t} = \frac{850\times36}{2\times0.5} = 30{,}600 \text{ psi} = \frac{5844\times914.4}{2\times12.7} = 210.4 \text{ MPa}$$

2. Calculate the failure stress
   The non-dimensional defect length = $2c/(Rt)^{0.5} = 6/(18 \times 0.5)^{0.5}$
   $= 152.4/(457.2 \times 12.7)^{0.5} = 2$

$$M = \sqrt{1 + 0.26\left(2c/\sqrt{Rt}\right)^2} = \sqrt{1 + 0.26(2)^2} = 1.428$$

The flow stress $\bar{\sigma} = 1.1 \times 65{,}000 = 71{,}500$ psi $= 1.1 \times 446 = 491.6$ MPa

The failure hoop stress $= \bar{\sigma} \times M^{-1} = 491.6/1.428 = 344.2$ MPa $= 71500/1.428 = 50{,}070$psi.

Since the failure hoop stress is greater than operating hoop stress the pipeline will leak but not fail.

We can solve the problem graphically using Figure 6-58 by determining the intercept for the nondimensional length 2 since M^{-1} is identical to the ratio $\sigma_h / \bar{\sigma}$ hence, we can find σ_h directly.

The B31G (2009) assessment criterion was developed for smooth corrosion defects occurring on the body of pipelines operating at, or above 40%, of the SMYS. The pipeline material was assumed to be weldable carbon or high-strength low alloy steel. The smooth contour assumption is consistent with the requirement for no high-stress raisers to be present in the defect. It need only be used in cases where the depth of corrosion exceeds 10% of the pipe wall thickness. The assessment of localized corrosion up to a depth of 80% of wall thickness is permitted. The appropriate equation to be used depends upon the value of a non-dimensional factor, A, which is determined from the following equation using the length of the corroded area.

$$A = 0.893L/\sqrt{Dt} \qquad (6-64)$$

For values of A less than or equal to 4,

$$P_{\text{safe}} = 1.1\,P\left[\frac{1 - \frac{2d}{3t}}{1 - \left(\frac{2d}{3t\sqrt{A^2+1}}\right)}\right] \qquad (6-65)$$

and for values of A greater than 4.0

$$P_{\text{safe}} = 1.1\,P\left[1 - \frac{d}{t}\right] \qquad (6-66)$$

Here P_{safe} is the maximum safe operating pressure for the corroded pipeline and P is the greater of the established maximum allowable operating pressure or the design pressure, except that P_{safe} may not exceed P.

The distinction between Eqs. (6-65) and (6-66) arises from the assumption made regarding the corrosion defect shape. In the former instance, a parabolic shaped defect is assumed (hence the inclusion of the factor $^2/_3$). It is more appropriate however for long corrosion defects to be characterized as a rectangular shape, which accounts for the different form of Eq. (6-66). Two other important assumptions in the derivation of the B31G approach are:

1. The maximum hoop stress is equal to the material yield strength
2. The flow stress is simply 1.1 times the yield stress

One can present Eqs. (6-65) and (6-66) in terms of pipe hoop stress, the flow stress, and the Folias factor, given by Eq. (6-49) as follows,

$$\sigma_f = \bar{\sigma}\left[\frac{1-2d/3t}{1-M^{-1}\left(2d/3t\right)}\right] \tag{6-67}$$

The maximum length restriction on the validity of use of Eq. (6-67) corresponds to a value of Folias factor M of ≤ 4.12. The equivalent formulation for a long corrosion defect is given by

$$\sigma_f = \bar{\sigma}\left[1-d/t\right] \tag{6-68}$$

A contiguous corroded area whose maximum depth is greater than 10% but less than 80% of the wall thickness can be assessed provided that its length, as derived from Figure 6-57, does not exceed the value given by Eq. (6-69).

$$L = 1.12B\sqrt{Dt} \tag{6-69}$$

where L is the maximum allowable longitudinal length of the corroded area, mm, and B is a value equal either to 4.0 for maximum depths less than 17.5% of the nominal wall thickness, or can be obtained from Figure 6-59 for maximum depths between 17.5% and 80% of the nominal wall thickness. Note well, here "c" is the maximum depth of the corroded area.

Equation (6-70) provides an alternative means of determining the non-dimensional factor B.

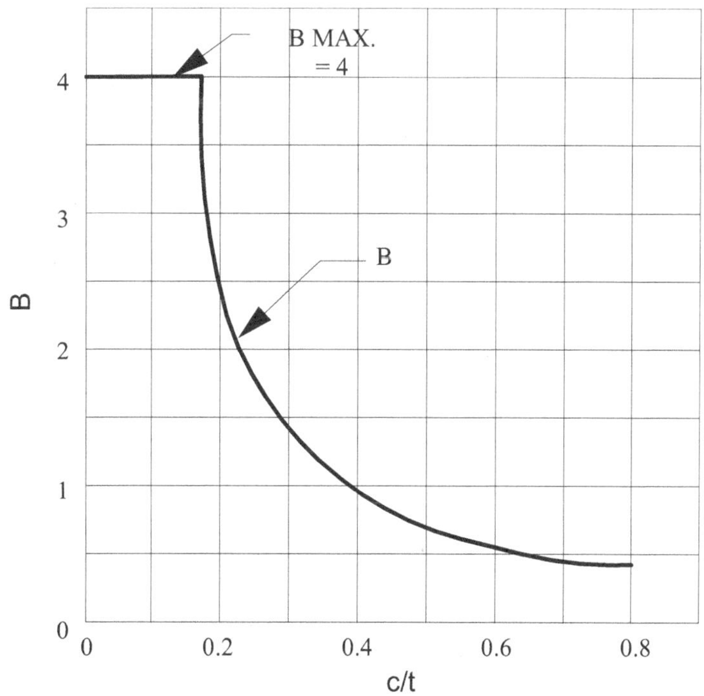

Figure 6-59. Determination of nondimensional variable B

$$B = \sqrt{\left[\frac{c/t}{1.1(c/t) - 0.15}\right]^2 - 1} \qquad (6-70)$$

Although widely used by the pipeline industry, a perception emerged that the formulations given above were unduly conservative essentially because of the underlying assumptions regarding flow stress, defect geometry and the representation of the Folias factor. Accordingly, all three were reworked in an effort to improve accuracy. The flow stress definition was changed to that originally proposed by Kiefner and Maxey in the research performed on behalf of the AGA, that is a simple addition of 10 ksi, or 68.94 MPa, to the specified minimum yield stress of the material. This was considered to be a better description of the post yield behavior of relatively low-strength steels and more in line with the conventional fracture mechanics description of flow stress (the average of the yield and ultimate tensile strength, $(\sigma_y + \sigma_{ult})/2$. The three-term expression originally proposed for the Folias factor was reinstated and an arbitrary geometric factor of 0.85 was substituted for the $^2\!/_3$ used previously. The modified B31G equations are given by:

$$\sigma_f = \bar{\sigma}\left[\frac{1 - 0.85\, d/t}{1 - 0.85\, d/t \; 1/M}\right] \qquad (6-71)$$

where

$\bar{\sigma}$ = SMYS + 10 ksi (68.94 MPa) and

$$M = \sqrt{1 + 0.6275\left(L/\sqrt{Dt}\right)^2 - 0.003375\left(L/\sqrt{Dt}\right)^4} \qquad (6-72)$$

$$\text{For}\left(L/\sqrt{Dt}\right)^2 \le 50.0$$

$$M = 0.032\left(L/\sqrt{Dt}\right)^2 + 3.3 \qquad (6-73)$$

$$\text{for}\left(L/\sqrt{Dt}\right)^2 > 50.0$$

A further approach to more accurately define the corroded area was developed known as the effective area method, which is probably better known by the term RSTRENG. In both the effective area and the RSTRENG approaches, the surface defect is expressed in terms of a generalized area of metal loss A so that the nominal failure stress is given by

$$\sigma_f = \bar{\sigma}\left[\frac{1 - A/A_0}{1 - A/A_0 \; 1/M}\right] \qquad (6-74)$$

The remaining strength approach, RSTRENG, uses an iterative means to establish the lowest safe maximum working stress. The river bottom, profile of the defect, (Figure 6-60) is projected on to the longitudinal axis and is accurately defined at number of intervals, which need not be regularly spaced. If N points are used to define the defect, then the number of possible combinations of defect length and depth, which need to be evaluated is

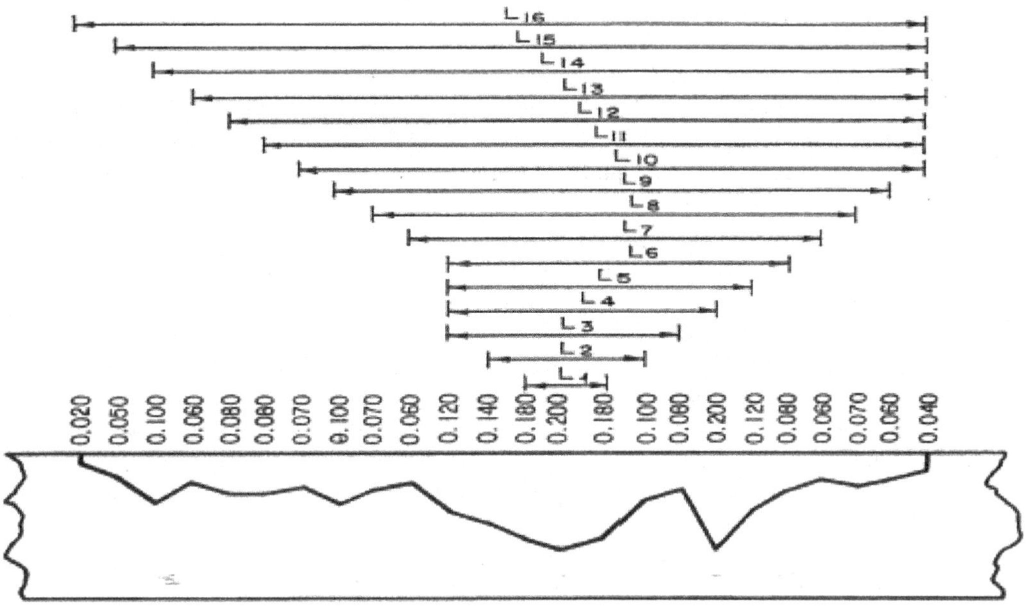

Figure 6-60. Profile of Corrosion depth along the "River bottom" path (Kiefner and Vieth 1989)

$N!/[2(N-2)!]$. Thus, for a 21-point profile, a total of 210 evaluations will be made from which the worst effective flaw is determined and the safe operating pressure calculated. The accuracy of the modified B31G and effective area methods have been validated using 86 tests performed on pipes containing corrosion defects of known size. Kiefner and Associates (2010) have made available a free download of a spreadsheet named KAPA, which performs these iterative calculations for both blunt corrosion defects and cracks.

Further enhancements to B31G were made in 1996 as a result of a joint industry study undertaken by BG Technology (now known as Advantica). The project produced guidance for the assessment of single and interacting defects as well as a means for determining the actual shape of the defect. Following the same underlying principles as B31G, the significant changes were those made to the Folias factor and the definition of flow stress. The latter is based on the ultimate tensile strength rather than the yield strength of the steel, which was considered appropriate for flaws failing by plastic collapse. The method was validated using the results obtained from 70 full-scale tests performed on fully restrained pipes containing machined single and interacting defects. In addition, three dimensional nonlinear finite element analyses were performed to further validate the approach.

The single defect equation is given by:

$$\sigma = \bar{\sigma}\left[\frac{1-d/t}{1-d/t\;1/Q}\right] \qquad (6-75)$$

where

$$Q = \sqrt{1+0.31\left(L/\sqrt{Dt}\right)^2} \qquad (6-76)$$

and

$\bar{\sigma} = 0.9\sigma_{ult}$

The process for determining the safe working pressure is like RSTRENG, iterative, but characterizes the defect as a number of pits contained within a corrosion patch. The assessment determines whether or not the defect behaves as a single irregular patch or as a series of single pits within each patch. Interaction between the pits is also considered and the failure pressure is taken to be the minimum from all of the analyses.

The results and methods obtained from the BG Technology study were subsequently incorporated into another joint industry study undertaken by Det Norske Veritas (DnV). (Bjornoy et al. 1997). In addition to simply considering the effect of internal pressure, this study also considered axial and bending loads, leading to the development of equations for assessing single defects under combined loading. The methodology for assessing corrosion under internal pressure only, has been validated against results obtained from 138 full-scale vessel tests, which included both real corrosion and machined defects. The methods developed in both joint industry studies have been incorporated into a DnV recommended practice (1999). To be consistent the Society's safety philosophy, a probabilistic calibration of the equations was performed taking into account both defect sizing and material properties uncertainty. This results in a partial safety factor approach based on the Load and Resistance Factor Design (LFRD) methodology. Within the recommended practice, partial safety factors are given for different types of inspection method and for differing levels of measurement accuracy. An alternative approach uses the allowable stress design format, and enables the calculated failure pressure to be multiplied by a single safety factor based on the original design factor, with the user making a judgment call on the effects of uncertainty associated with the defect sizing.

Table 6-5 summarizes the main differences in the various methods used to make a remaining strength assessment of corroded pipe, while Figure 6-61 illustrated the various representation of metal loss.

A result of 86 burst tests conducted by Kiefner and Vieth (1989) has shown that RSTRENG provides more accurate results, that is, less data scatter when compared to B31G and modified B31G as can be seen from Table 6-6.

Constructing Defect Assessment Curves

The results of all of the above methods for assessing defects can be depicted graphically by constructing acceptance curves using the basic equations. For example, Figure 6-62 shows plots of the loci of limiting values of normalized defect depth versus normalized defect length from Eq. (6-75) for two pressure conditions. The lower of the two plots applies to the pre-commissioning hydrotest pressure, while the upper plot is for results obtained at the maximum allowable operating pressure. As was pointed out earlier in the discussion of Figure 6-58, both these lines represent a "Failure" condition. A number of supposed part-wall metal loss defects have been plotted on the diagram from which it can be seen that two defects are predicted to fail at the MAOP condition - they would need to be repaired immediately.

TABLE 6-5. Summary of methods for assessing corrosion

Method	Flow Strength	Folias Factor	Areal Shape
B31G	1.1 SMYS	2 terms	2/3 Ld Parabola
Modified B31G	SMYS + 10 ksi	3 terms	Arbitrary 0.85Ld
NG 18	SMYS + 10 ksi	3 terms	Defect area A/A_0; or rectangle Ld
RSTRENG 85% area	SMYS + 10 ksi	3 terms	0.85Ld
RSTRENG effective area	SMYS + 10 ksi	3 terms	River bed profile iterative calculation
DnVRP-F101	SMTS	2 terms	Rectangle Ld

SMTS = specified minimum tensile strength.

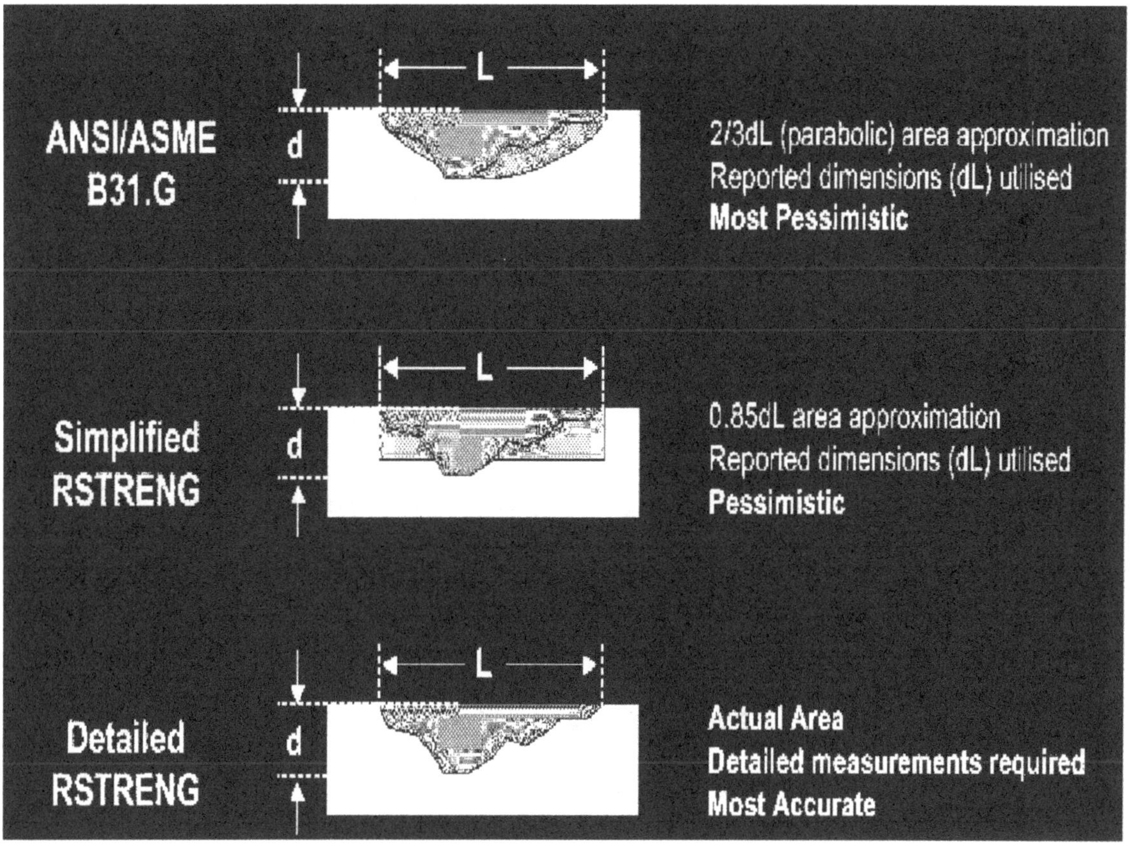

Figure 6-61. Remaining strength assessment representation of metal loss

Two other defects are small enough that they would survive a hydrotest pressure. The remaining defects falling between both plots will need to be reassessed using either more accurate defect measurements or more detailed analysis.

Interacting Defects

The methods described above all assume a single defect. We need to consider how to deal with multiple defects. Figure 6-63 shows one approach to determining the longitudinal extent and interaction distances within a region of localized corrosion. Several empirical interaction rules have been developed from full scale testing and/or the use of finite element modeling to enable a corrosion colony to be assessed. They are, however, limited in their coverage of pipeline size and defect geometries. One such rule is that inline defects are

TABLE 6-6. Statistics of failure pressure ratio from different prediction methods

Prediction Method	Mean	Standard Deviation	Maximum	Minimum
Elastic limit	1.591	0.322	2.847	1.16
B31G	1.989	1.099	6.367	0.966
Modified B31G	1.666	0.531	3.287	0.854
RSTRENG	1.102	0.183	1.613	0.691

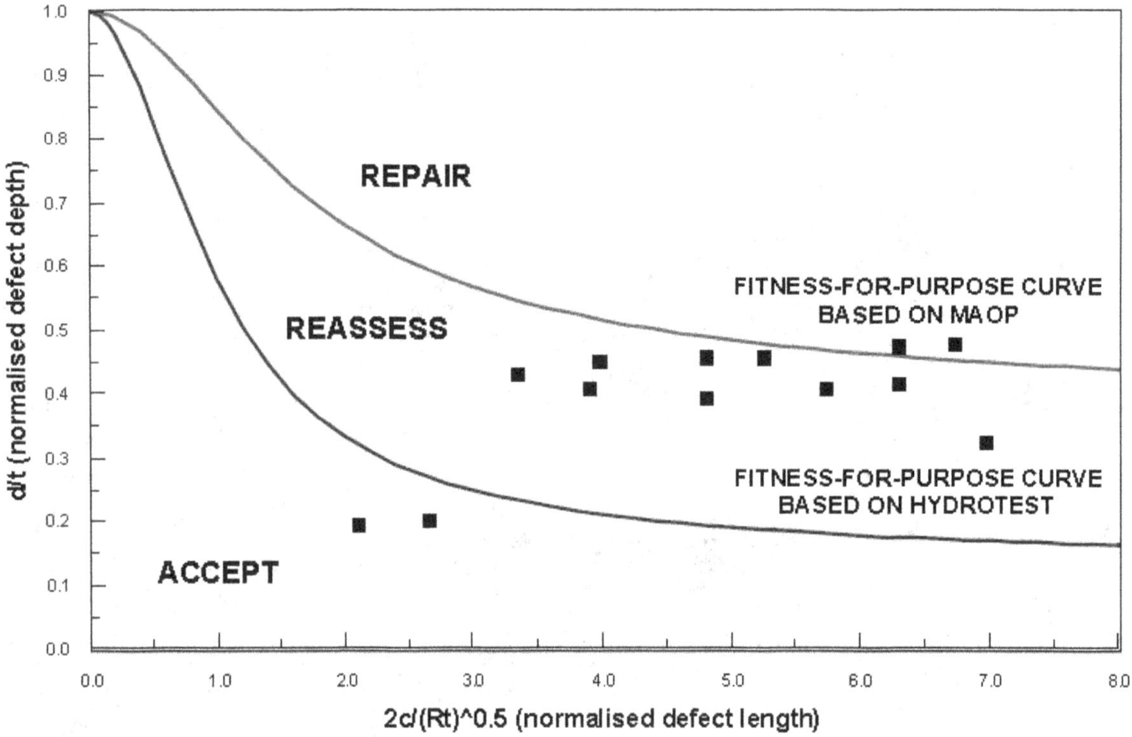

Figure 6-62. A typical defect assessment curve

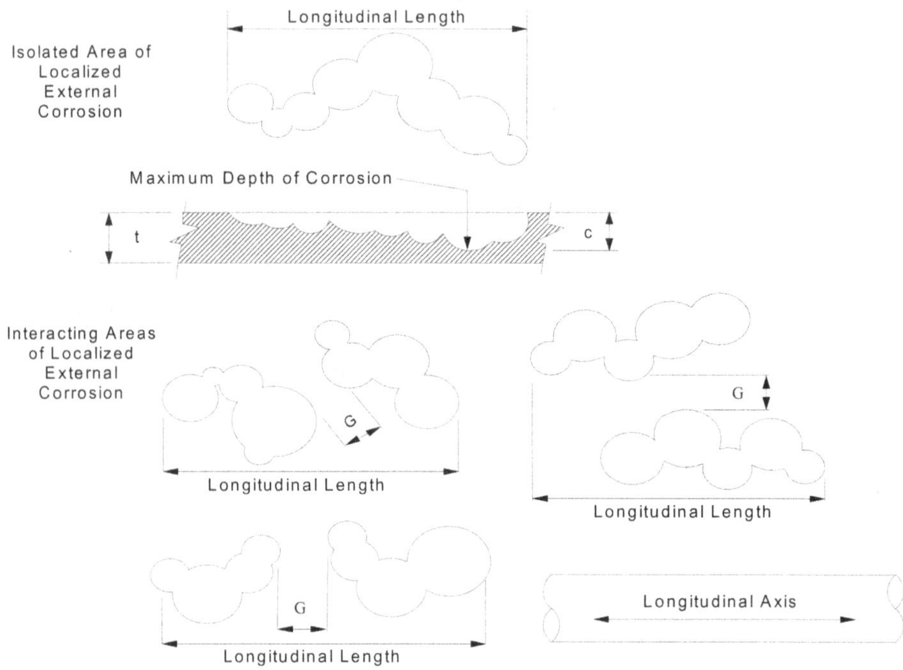

Figure 6-63. Method of determining longitudinal extent of localized corrosion and interaction distances (ASME B31.8)

assumed to interact if the axial distance between them is less than the length of the shortest defect. In this circumstance, the combined length of the defect is the sum of the individual defect lengths *and* their separation distance(s). For circumferentially spaced defects that overlap in the axial direction, interaction is assumed to occur if their circumferential separation is less than, or equal to, the full wall thickness. The combined defect length is taken to be the overall longitudinal extent of the corroded area. The reader is referred to O'Grady et al. (1992) and/or the Dnv Guidance Document for Assessing Corrosion (1998) for specific details.

Weld Defects

Pipeline girth welds are not a major cause of pipeline failures; however, defects are sometimes detected during service, and will require assessment. The acceptability of flaws contained within gas or arc welded carbon and low alloy steel pipe can be undertaken using the workmanship standards set out in several standards. API 1104 (2008), for example, covers the acceptance standards to be applied to production welds tested either to destruction or non-destructively inspected by radiography, magnetic particle, liquid penetrant, and ultrasonic test methods. These acceptance standards are based on empirical criteria and place primary importance on flaw length. Such criteria have been proven to have an excellent record in practice. They are, however, quite conservative and require for example the repair of non-planar defects such as slag and porosity, which are generally accepted as being benign, Similarly, many planar defects are also considered to be unacceptable, despite full scale tests, which show high failure stresses even under the most severe loadings at low temperatures (Cosham and Kirkwood 2000).

As a consequence, Appendix A of API 1104 (2008) "Alternative Girth Weld Acceptance Criteria," allows the use of an alternative approach, based on fracture mechanics analysis, which incorporates the significance of flaw depth as well as length. The fitness for purpose criteria used provides for more generous allowable flaw sizes, but only if additional stress analysis and inspections are performed. The method requires that the welding procedures are qualified for either of two minimum CTOD toughness levels: 0.005 in. or 0.01 in. The allowable defect depth is inferred from a given maximum applied strain. The actual defect depth may be determined from non-destructive testing or by considering the inherent size limitation imposed by the weld pass geometry.

The EPRG have also produced a set of guidelines for the assessment of girth weld defects that are determined at three levels. Tier 1 sets workmanship limits, which have been derived from existing welding standards. The Tier 2 limits are semiempirical, extending the workmanship limits based upon the results obtained from wide plate tests. A minimum Charpy (CVN) impact energy of 30 J and a minimum averaged CVN of 40 J are specified for both Tier 1 and Tier 2 assessments. The Tier 3 limits are also semi-empirical combining fitness for purpose failure criteria and the results of full-scale tests. Both a minimum Charpy impact energy and a minimum CTOD toughness are required for Tier 3. The minimum CTOD toughness value is 0.1 mm and the minimum averaged value 0.15 mm.

Appendix K of the Canadian pipeline standard CSA Z662 (2007) also permits a fitness for service approach in lieu of workmanship criteria. The methodology used is based on a modification of Miller's plastic collapse model and is in good agreement with test results. Appendix K also permits girth weld integrity to be assessed based on brittle fracture considerations.

Both the EPRG and the Canadian CSA approaches have their origins in the British Standard PD 6493 (1980), which has now been superseded by BS7910 (1999). The earlier British standard was developed for flat plates having through wall cracks and so needed to be modified for use with girth weld defects on pipelines. These modifications were largely semi-empirical in nature. The theoretical basis for the work is to be found in Miller's (1988) exhaustive review of limit load solutions for a wide range of weld geometries. Modern pipeline steels welded with modern techniques and consumables will generally produce welds of good toughness; hence, if a girth weld were to fail, it would be by plastic collapse rather than brittle fracture. Wang et al. (1996) recommended the use of the Miller solution based on extensive finite element analyses of pipes with large *D/t* ratios and girth weld defects up to 40% of the pipe circumference in length. The Miller solution has the form:

$$\frac{M_c}{4R^2 t\sigma_f} = \cos\left[\frac{\eta\beta\pi}{2}\right] - \frac{\eta\sin(\beta\pi)}{2} \; if\, \beta \leq \frac{1}{2-\eta} \qquad (6-77)$$

$$\frac{M_c}{4R^2 t\sigma_f} = (1-\eta)\sin\left[\frac{(1-\eta\beta)\pi}{2(1-\eta)}\right] + \frac{\eta\sin(\beta\pi)}{2} \; if\, \beta \succ \frac{1}{2-\eta} \qquad (6-78)$$

where M_c is the bending moment at plastic collapse = $4R^2 t\sigma_f$

The other parameters are,

R = pipe radius
t = pipe thickness
a = crack depth
c = half crack length
D = pipe diameter
$\beta = c/\pi R = 2c/\pi D$ the ratio of crack length to pipe circumference
$\eta = a/t$ = crack depth ratio
σ_f = flow stress

Equation (6-78) is not used in pipeline applications as it does not represent a practical service condition.

The plastic collapse stress for a girth weld is given by

$$\sigma_c = f\sigma_f\left[\cos\frac{(\eta\beta\pi)}{2} - \frac{\eta\sin(\beta\pi)}{2}\right] \qquad (6-79)$$

$$f = \frac{1}{1+\left(\frac{4}{\pi}-1\right)\frac{\eta\beta}{0.05}} \quad \text{if } \eta\beta < 0.05$$

$$f = \frac{\pi}{4} \text{ if } \eta\beta > 0.05$$

Failure Assessment Diagrams

The concept of a simple structural integrity assessment method for structures containing a defect was first proposed by the United Kingdom Central Electricity Generating Board in 1976 (Harrison et al. 1976). The procedure was referred to as the R6 method after the number of the report describing it, and it was based on the Dugdale strip yield model, which is applicable to elastic plastic materials. The failure of a flawed steel pipe is bounded by two modes, namely:

1. linear elastic fracture and
2. plastic collapse

As described previously, the actual failure mechanism is dependent on material properties, geometry, applied loads, and defect type and size. A failure assessment diagram (FAD) based on an engineering critical assessment, allows the simultaneous consideration of plastic collapse and brittle fracture and the interaction between them. There are three key components to the assessment.

1. The failure assessment curve (FAC) (a limit state)
2. The stress ratio S_r or L_r, which is defined as the ratio of the applied stress (load) to the plastic collapse stress (load) of the net section containing the defect
3. The toughness ratio K_r, which is defined as the ratio of the total applied stress intensity factor K to the material's plane strain fracture toughness value K_{Ic} as obtained from a small scale fracture test .If valid K_{Ic} data cannot be obtained directly because of specimen size limitations, equivalent toughness values can be derived from other elastic-plastic fracture toughness tests

The procedure is shown pictorially in Figure 6-64. The severity of a particular defect under any loading condition is judged in terms of the proximity of (L_r, K_r) in relation to a FAC. Fracture mechanics concepts are used to define the FAC. Below the FAC, the defect is safe; above the FAC failure is predicted.

In order to find the fracture assessment parameter (K_r), *a* linear elastic fracture assessment of the flawed pipe is performed, producing the K_r coordinate point.

The plastic collapse assessment parameter (L_r) is found following a stress analysis and its coordinate point plotted. The assessment determines the proximity of the flawed structure to failure in terms of plastic collapse.

The correct formulation of the FAC is vital because it

1. defines the elastic-plastic behavior of the structure and
2. defines a safety failure boundary as shown in Figure 6-65.

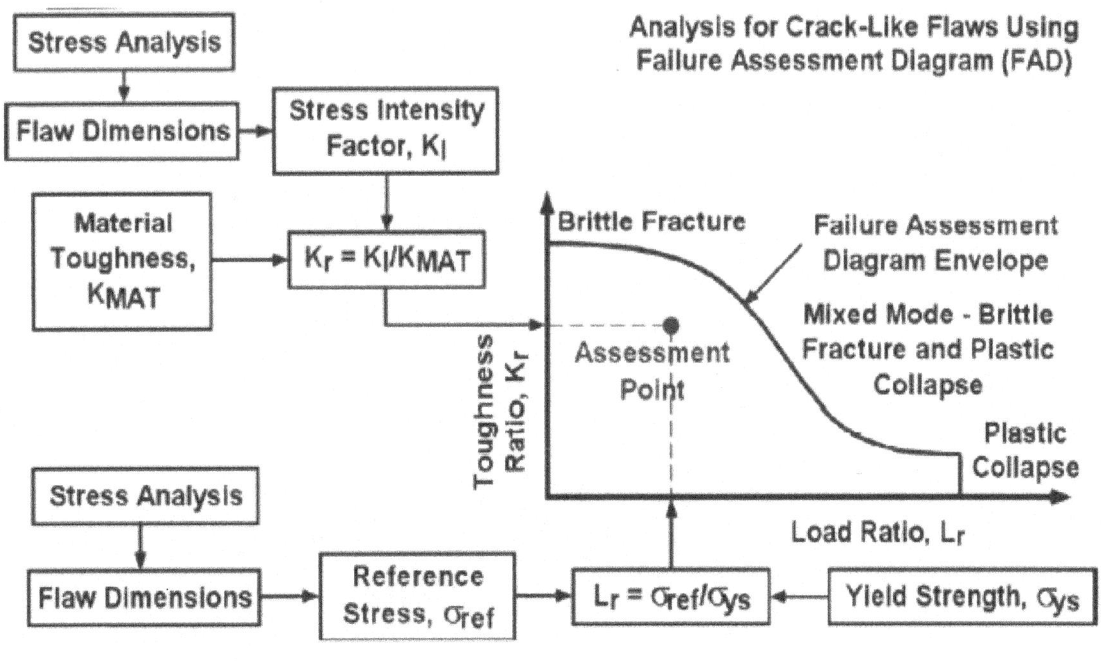

Figure 6-64. Schematic of the two-parameter failure assessment procedure

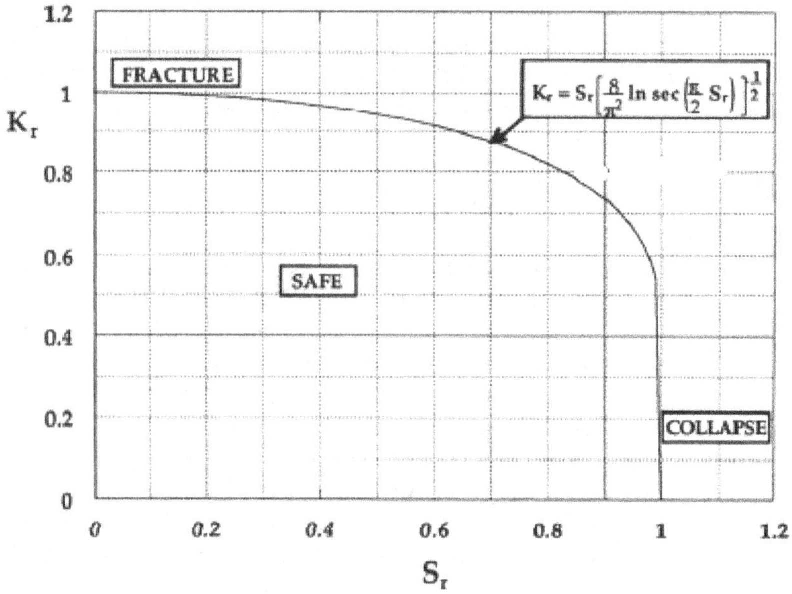

Figure 6-65. A strip yield FAD (Anderson 1995; reproduced courtesy of CRC Press)

A defect is normally assessed in a two-parameter failure assessment for a series of loads providing a load locus. An assessment provides information on two factors affecting the integrity of the structure:

1. Safety Margin
   Margins of safety are readily determined from the FAD
2. Mode of Failure.
   The location at which the load locus crosses the FAC determines the predicted failure mode of the defect.

Example:

To briefly illustrate the approach, consider a flat sheet of width 8 in. with a center crack of length 2 in. (Figure 6-66). Its fracture toughness K_{ic} is 70 ksi$\sqrt{\text{in}}$. Will this sheet fail if it is loaded to a nominal stress of 30 ksi? Assume the flow stress is 50 ksi.

$$\sigma_{collapse} = \sigma_{flow}(1 - a/W) = 50\ (1 - 0.25) = 37.5 \text{ ksi}$$

$$S_r = 30/37.5 = 0.8$$

$$K_i = \sigma\sqrt{\pi a}\left(\sec\left(\frac{\pi a}{2W}\right)\right)^{0.5} = 30\sqrt{\pi}\left(\sec\left(\frac{\pi}{8}\right)\right)^{0.5} = 55.3 \text{ ksi}\sqrt{\text{in}}$$

$$K_r = K_i/K_{ic} = 55.3/70 = 0.79$$

Plotting S_r and K_r on Figure 6-65 shows that this defect is almost touching the FAC indicating that it is barely safe.

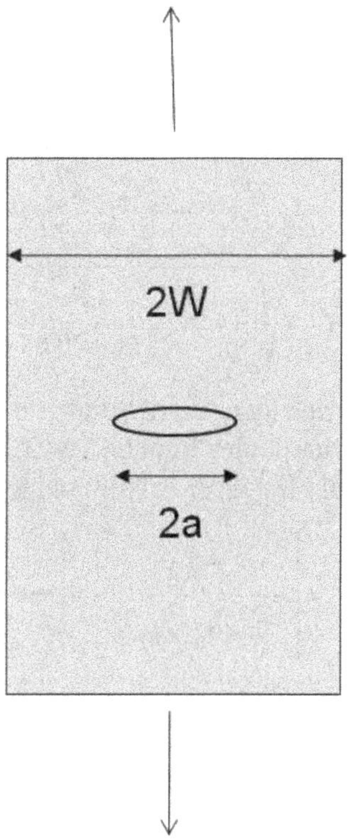

Figure 6-66. Crack example

Worked Examples

A. Using the log secant approach, determine the failure pressure of a pipeline 34 in. in diameter (863.9 mm) and wall thickness 9.53 mm, which has a part wall crack 4 in. long (100 mm) through 30% of its thickness. The steel is grade 359 and the full-size Charpy toughness is 25 J. Young's Modulus for the material is 200 GPa, and its ultimate tensile strength is 496.8 MPa. The flow stress may be assumed to be the mean of the specified minimum yield and the ultimate tensile stress.

Solution:

Input data

$E = 200{,}000$ MPa $\sigma_y = 359$ MPa $\sigma_{ult} = 496.8$ MPa $t = 9.53$ mm $C_v = 25$ J $A_d = 80$ mm^2

1. Separating the stress and toughness elements, which comprise Eq. (6-62), we can write it in SI units as

$$\frac{\left(\frac{C_v}{A_d}\right)E\pi}{8c\bar{\sigma}^2} = \ln \sec\frac{\pi}{2}\left[\frac{M_p\,\sigma_p}{\bar{\sigma}}\right]$$

2. Determine the appropriate value of Folias factor from defect length $L = 2c$

$$100 \leq L \leq \sqrt{50Dt} \leq \sqrt{50 \times 863.9 \times 9.53} \leq 641 \text{ mm}$$

$$\text{Hence } M_T = \sqrt{1 + 1.255\frac{c^2}{Rt} - 0.0135\left(\frac{c^4}{R^2t^2}\right)}$$

$$M_T = \sqrt{1 + 1.255\frac{50^2}{431.9 \times 9.63} - 0.0135\left(\frac{50^4}{431.9^2 \times 9.63^2}\right)} = 1.3257$$

From which $M_T^{-1} = 0.7543$

Now substituting from Eq. (6-63) into Eq. (6-74), it is possible to replace $\bar{\sigma}/\sigma_h$ by M_P and express M_P in terms of the damaged area A_0 as follows

$$M_P^{-1} = \left[\frac{1 - A/A_0}{1 - A/A_0 \cdot 1/M_T}\right] = \left[\frac{1 - d/t}{1 - d/t \cdot 1/M_T}\right] = \left[\frac{1 - 0.3}{1 - 0.3 \times 0.7543}\right] = 0.904$$

From which $M_P = 1.1053$

3. Determine the flow stress

$$\bar{\sigma} = \left(\sigma_y + \sigma_{ult}\right)/2 = (359 + 496.9)/2 = 427.9 \text{ MPa}$$

4. Evaluate the toughness component (note the need to correct for consistent units)

$$\frac{\left(\frac{C_v}{A_d}\right)E\pi}{8c\bar{\sigma}^2} = \frac{\left(\frac{25}{80}\right)200{,}000 \times \pi}{8 \times 50 \times 427.9^2} \times 10^3 = 2.6809$$

5. Solve for σ_p the stress at failure

$$2.6809 = \ln \sec\frac{\pi}{2}\left[\frac{1.1053\ \sigma_p}{427.9}\right]$$

From which $\sigma_p = 370$ MPa, the corresponding failure pressure is obtained by substituting into the Barlow equation

$$P_{fail} = \frac{2t\sigma}{D} = \frac{2 \times 9.63 \times 370}{863.9} = 8248 \text{ kPa}$$

We can evaluate the failure pressure for a wide combination of crack lengths and various crack depths. Figure 6-67 illustrates the results of such an evaluation for the parameters listed above. From Figure 6-67, we can see that at an operating pressure of 6070 kPa, a crack 40% of the through wall thickness reaches a critical length of around 190 mm.

B. Consider a 30-in. outside diameter natural gas pipeline having a nominal wall thickness of 0.375 in. and made from X52 (Grade 357) steel. The normal operating pipe of the pipe is 936 psig, and it has been found to contain a defect 5.5 in. long with a maximum depth 60% of the wall thickness. The design factor for the pipe may be taken to be 0.72, and there is no need to consider its operating temperature. Use the B31G method to determine the pipe's fitness for service.

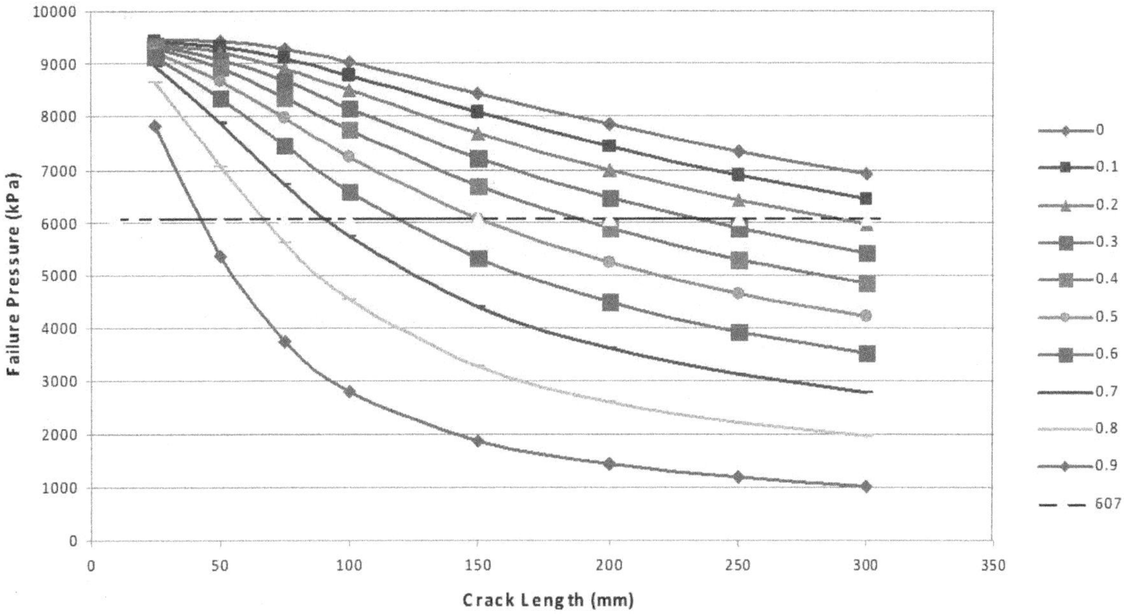

Figure 6-67. A typical graphical output of a part wall defect assessment; crack length vs. pressure for various crack depths

Solution:

Since the defect depth exceeds 10% but is less than 80% of the through wall in accordance with the B31G manual, an evaluation is required. The first check is on the validity of the approach since it is excessively conservative when the flaw length L is greater than $\sqrt{20Dt}$

$$L = \sqrt{20(30)(0.375)} = 15 \text{ in.}$$

Hence, we can use the maximum flaw length and depth in the evaluation.

$$L_{\text{allowable}} = 1.12\, B\, \sqrt{Dt}$$

Using Eq. (6-70), we first need to evaluate the term B

$$B = \left\{ \left[\frac{d/t}{1.1(d/t) - 0.15} \right]^2 - 1 \right\}^{0.5}$$

$$B = \left\{ \left[\frac{0.6}{1.1(0.6) - 0.15} \right]^2 - 1 \right\}^{0.5} = 0.6197$$

$$L_{\text{allowable}} = 1.12\,(0.6197)\,\sqrt{30\,(0.375)} = 2.328 \text{ in.}$$

Since the actual defect length is greater than 2.328 in., the metal loss is unacceptable, and hence, a safe operating pressure for the pipe must be found. This is given by Eq. (6-65)

$$P_{\text{safe}} = 1.1P\left[\frac{1-\frac{2d}{3t}}{1-M^{-1}\left(\frac{2d}{3t}\right)}\right]$$

P is the greater of the maximum allowable operating pressure or the calculated design pressure using the appropriate code conditions, $P = 2tSMYSF_dF_t/D$, where F_d is the design factor and F_t the temperature de-rating factor. In this instance, the pressures are the same, that is, 936 psig. The Folias factor is given by Eq. (6-51)

$$M = \sqrt{1 + 0.8\left(\frac{L}{\sqrt{Dt}}\right)^2} = 1+0.8(5.5/\sqrt{30.(0.375)})^2$$

$$M = 1.775$$

The safe operating pressure is therefore:

$$P_{\text{safe}} = 1.1.(936)\left[\frac{1-0.666.(0.6)}{1-1.775^{-1}\left(0.666.(0.6)\right)}\right]$$

$$= 798 \text{ psig or } 5500 \text{ kPa}$$

Example:

An oil pipeline 30 in. in diameter, connecting to a tank farm undergoes frequent and significant pressure cycles. The pertinent facts are the material is Grade 358 (X52), 0.375 in. thick with full size Charpy V notch upper shelf energy of 25 ft-lbs. The line cycles four times per day between operating pressures of 936 psig (6458 kPa) and 100 psig (690 kPa). The initial flaw size is 6 in. in length with a depth of 0.104 in. The flow stress can be taken as SMYS+ 10 ksi and Chapry specimen area = $(0.124'')^2$

Solution:

It is appropriate to use a Paris Law crack growth model and assume an elliptical flaw with a stress intensity factor given by $K_i = 1.12(\pi a/Q)^{0.5}$, where Q is a flaw shape parameter.

Crack growth is defined by: $da/dN = A(\Delta K)^n$, where N is the number of fatigue cycles and a reasonable value for A is 2.12×10^{-11} for ΔK in ksi units and $n = 4$.

The hoop stress at the highest operating pressure is found from the Barlow equation ($\sigma_p = pD/2t$) to be 37,440 psi and at the lower operating pressure = 4000 psi, making the stress excursion per cycle equivalent to 33.44 ksi.

The Folias factor M_T corresponding to the given flaw length is found in Eq. (6-50).

$$M_T = \sqrt{1+0.6275\left(\frac{L}{\sqrt{Dt}}\right)^2 - 0.003375\left(\frac{L}{\sqrt{Dt}}\right)^4}$$

$$M_T = \sqrt{1+0.6275\left(\frac{6}{\sqrt{30(0.375)}}\right)^2 - 0.003375\left(\frac{6}{\sqrt{30(0.375}}\right)^4} = 1.7244$$

and $M_T^{-1} = 0.58$.

To find the depth at which the flaw will fail at the highest operating pressure of 936 psig, we use the log secant method.

$$\frac{K_c^2\pi}{8c\bar{\sigma}^2} = \ln\sec\frac{\pi}{2}\left[\frac{M_P\,\sigma_P}{\bar{\sigma}}\right] \text{ where } K_c^{\,2} = \frac{12\,C_v\,E}{A_c}\text{, we can write,}$$

$$\frac{\frac{12\,C_v\,E}{A_c}\pi}{8c\bar{\sigma}^2} = \ln\sec\frac{\pi}{2}\left[\frac{M_P\,\sigma_P}{\bar{\sigma}}\right] = \frac{12\pi(25/0.124)(30)(10)^6}{8(3)(62{,}000)^2} = 2.4716$$

$$\ln\sec\frac{\pi}{2}\left[\frac{M_P\,\sigma_P}{\bar{\sigma}}\right] = 2.4716\text{ ; hence, } \frac{M_P\,\sigma_P}{\bar{\sigma}} = 0.94617$$

At the highest operating pressure, the Folias factor $M_P = 0.94617\ (62{,}000) = 1.56684$ and $M_P^{-1} = 0.63823$. Now M_P and M_T are related through:

$$M_P^{-1} = \left[\frac{1-A/A_0}{1-A/A_0\;1/M_T}\right]$$

Rearranging to solve for A/A_0, we have,

$$A/A_0 = \left[\frac{1-M_P^{-1}}{1-M_P^{-1}M_T^{-1}}\right] = \left[\frac{1-0.63823}{1-0.63823(0.58)}\right] = 0.57434$$

The cross-sectional area at failure A for an elliptically shaped flaw is $(\pi/4)La_{\text{final}}$, while the original defect area is given by Lt.

$$\text{Hence } a_{\text{final}} = \frac{0.57434\ (4)\ (0.375)}{\pi} = 0.274\text{ in.}$$

To find whether the final flaw will cause a rupture or simply leak, we need to determine the hoop stress at the failure pressure for a 6-in. through wall flaw.

$$\frac{M_T\,\sigma_{TW}}{\bar{\sigma}} = 0.94617\text{, from which } \sigma_{TW} = \frac{0.94617(62{,}000)}{1.7244} = 34{,}020\text{ psi}$$

Since the hoop stress of 34020 psi is less than the maximum hoop stress 37,400 psi, the final flaw will cause a rupture.

In order to determine the number of cycles and hence the time needed for the defect to grow to a critical size, the fatigue crack growth model is used

$$\int_{a_{\text{initial}}}^{a_{\text{final}}} \frac{da}{A(\Delta K)^n} = \int_{a_{\text{initial}}}^{a_{\text{final}}} \frac{da}{A\left(1.12\Delta\sigma\left(\frac{\pi a}{Q}\right)^{0.5}\right)^4}$$

$$N = \int_{a_{\text{initial}}}^{a_{\text{final}}} \frac{da}{A\left((1.12)(33.44)\left(\frac{\pi a}{Q}\right)^{0.5}\right)^4}$$

$$N = -\left[\frac{1}{A\left((1.12)^4(33.44)^4\left(\frac{\pi a}{Q}\right)^2\right)}\right]_{a_\partial}^{a_{\text{final}}}$$

An estimate for Q, can be obtained by substituting a mean value of crack depth/length, a/L, and the ratio of actual stress/yield stress (σ/σ_y) into the following equation:

$$Q = \left[0.8 + 0.77(a/L) + 4.25(a/L)^2\right] + \left[0.2 + 0.0.14(\sigma/\sigma_y) - 0.22(\sigma/\sigma_y)^2\right]$$

$$a_{\text{initial}}/L = 0.104/6 = 0.01733;\ a_{\text{final}}/L = 0.274/6 = 0.046, \text{ mean } 0.0317$$

$$\sigma_{\text{initial}}/\sigma_y = 37{,}000/62{,}000 = 0.597;\ \sigma_{\text{final}}/\sigma_y = 34{,}020/62{,}000 = 0.66, \text{ mean } 0.625$$

From this, a mean value of 0.9 for Q is obtained and substituting in the equation for N,

$$N = -\left[\frac{10^{11}}{(2.12)(1.5735)(1.2505)10^6(\pi/0.9)^2(0.274)}\right]$$

$$+$$

$$\left[\frac{10^{11}}{(2.12)(1.5735)(1.2505)10^6(12.185)(0.104)}\right]$$

$N = 18{,}917 - 7180 = 11{,}737$ cycles, which at 4 cycles per day would cause failure after 8.5 years.

REFERENCES

Anderson, T. L., 1995, *Fracture Mechanics, Fundamentals and Applications*, 2nd ed., CRC Press, Ann Arbor, MI.

API, 2005, Standard 1163, *In-line Inspection Systems Qualification Standard*, 1st ed., American Petroleum Institute, August 2005, 37 pp.

API, 2008, *Standard 1104 Welding of Pipelines and Related Facilities,* American Welding Society, January 1, 2006, 99 pp.

API, 2008, Specification 5L, *Specification for Line Pipe*, 44th ed., American Petroleum Institute, October 1, 2007, 176 pp.

ASME B31.4, 2002, *Pipeline Transportation Systems for Liquid Hydrocarbons and Other Liquids*, ASME, New York, NY.

ASME B31.8, 2003, *Gas Transmission and Distribution Systems*, ASME, New York, NY.

ASME B31.8G, 2009, *Manual for Assessing Remaining Strength of Corroded Pipelines*, A Supplement to ANSI/ASME Code B31.8 Code for Pressure Piping, ASME, New York, NY, October.

ASME B31.8S, 2004, *Managing System Integrity of Gas Pipelines*, ASME, New York, NY.

ASTM, 2001, *Standard Test Method for Measurement of Fracture Toughness*, E1820-01, American Society for Testing and Materials, West Conshohocken, Pennsylvania, PA.

Bjornoy, O. H., Cramer, E. H., and Sigurdsson, G., 1997, "Probabilistic Calibrated Design Equation for Burst Strength Assessment of Corroded Pipes," International Conference on Offshore and Polar engineering ISOPE'97, Honolulu, HI.

British Standards Institute BS 7910, 2005, *Guide to Methods for Assessing the Acceptability of Flaws in Metallic Structures*, British Standards Institution, July 27, 2005, 306 pp.

Burdekin, M. J., and Stone, D. E. W., 1966, "The Crack Opening Displacement Approach to Fracture Mechanics," Journal of Strain Analysis, Vol. 1, pp. 144–153.

Canadian Standards Association, 2007, *Z662-07 Oil and Gas Pipeline Systems*, Canadian Standards Association, Etobicoke, Ontario, Canada.

Chauhan, V., Ward, C., and Lethcoe, B., 2008, "Review of Methods for Assessing the Remaining Strength of Corroded Pipe," PHMSA Public Meeting on Anomaly Assessment and Repair Techniques, McLean, VA, October 22.

Cosham, A., and Kirkwood, M., 2000, "Best Practice in Pipeline Defect Assessment," Paper IPC00-0205 Proc., 3rd ASME International Pipeline Conference, Calgary Alberta Canada.

Darcis, P. G., Kohn, A., Bussiba, J. D., McColskey, C. N., McCowan, T. A., Siewert, R., Fields, R., Smith, R., and Merritt, J., 1998, "Crack Tip Opening Angle Measurement and Modeling Fracture Resistance in Low and High Strength Pipeline Steels," Paper No. IPC2006-10172, International Pipeline Conference, Calgary, September 25–29.

Det Norske Veritas, 1998, "Guidance for the Assessment of Corrosion in Line Pipe," DnV Unified Guidance Document, Version 4, May, Hovik, Norway.

Det Norske Veritas, 1999, "Recommended Practice Corroded Pipelines," DnV RP — F101, Hovik, Norway.

Dowling, S. D., and Sochie, D. F., 1982, "Simple Rain Flow Counting Algorithms," International Journal of Fatigue, Vol. 4, No. 1, January, pp. 31–40.

Driver, R. G., and Zimmerman, T. J. E., 1998, "A Limit States Approach to the Design of Pipelines for Mechanical Damage," Proc. 17th OMAE Conference, Paper OMAE98-1017.

Dugdale, D. S., 1960, "Yielding of Steel Sheets Containing Slits," Journal of Mechanics and Physics of Solids, Vol. 8. pp. 100–108.

Eiber, R. J., Maxey, W. A., Bert, C. W., and McClure, G. M., 1981, "The Effects of Dents on the Failure Characteristics of Linepipe," Battelle Columbus Laboratories, NG-18, Report No. 125, AGA Catalogue No. L51403, May.

Folias, E. S., 1965, "An Axial Crack in a Pressurised Cylindrical Shell," International Journal of Fracture Mechanics, Vol. 1, pp. 104–113.

Forman, R. G., Kearney,V. E. and Engle, R. M., 1967, "Numerical Analysis of Crack propagation in a Cyclic Loaded Structure," Trans ASME, Jnl. of Basic Engineering, Vol. D89, No 3, pp. 459–464.

Fowler, J. R., 1993, "Criteria for Dent Acceptability in Offshore Pipelines," PTC 7311 25th Offshore Technology Conference, Houston, TX.

Glover, A., and Rothwell, B., 2004, "Yield Strength and Plasticity of High Strength Pipelines," Proc. the International Pipeline Technology Conference, Ostend, Belgium, 2004, pp. 65–79. Published by Scientific Surveys Beaconsfield UK.

Griffith, A. A., 1921, "The Phenomenon of Rupture and Flow in Solids," Philosophical Transactions of the Royal Society, London A, Vol. 221, pp. 163–198.

Harrison, R. P., Loosemore, K., and Milne, I., 1976, "Assessment of the Integrity of Structures Containing Defects," CEGB Report No. R/H/6, Central Electricity Generating Board, United Kingdom.

Hopkins, P., Jones, D. G., and Clyne, A. C., 1989, "The Significance of Dents and Defects in Transmission Pipelines," Paper C376/049, Proc. International Conference on Pipework, Engineering and Operation, Institution of Mechanical Engineers, London, February.

Irwin, G., 1948, "Fracture Dynamics," Trans ASME, Vol. 40A, pp. 147–166.

Kanninen, M. F., and Popelar, C. H., 1985, *Advanced Fracture Mechanics,* Oxford University Press, New York, NY.

Kastner, W., Rohrich, E., Schmitt, W., and Steinbuch, R., 1981, "Critical Crack Size in Ductile Piping," Int. Jnl. Pressure Vessel and Piping, Vol. 9, pp. 197–219.

Kiefner, J. F., and Duffy, A. R., 1973, "Criteria for Determining the Remaining Strength of Corroded Areas of Gas Transmission Pipelines," AGA Operating Section on Transmission Conference American Gas Association.

Kiefner, J. F., and Vieth P. H., 1989, "A Modified Criterion for Evaluating the Remaining Strength of Corroded Pipe," Report to the Pipeline Research Committee of the American Gas Association, December 22.

Miller, A. G., 1988, "Review of Limit Codes of a Structure Containing Defects," Journal of Pressure Vessels and Piping, Vol. 32, pp.191–327.

Mohitpour, M., Golshan, H., and Murray, A., 2007, *Pipeline Design and Construction — A Practical Approach*, 3rd ed., ASME Press, New York.

O'Grady, T. J., Hisey, D. T., and Kiefner, J. F., 1992, "A Systematic Method for the Evaluation of Corroded Pipelines," PD Vol 46, Pipeline Engineering'92, ASME, New York, NY.

Orowan, E., 1945, *Notch Brittleness and the Strength of Metals*, Institute of Engineers and Shipbuilders, Scotland, pp. 165–215.

Paris, P. C., and Erdogan, F., 1960, "A Critical Analysis of Crack Propagation Laws," Journal of Basic Engineering Trans. ASME, Series D, Vol. 85, pp. 528–534.

Popelar, C. H., Grant, T., Bradley, W. and Stucker, G., 1996, "Development of a Procedure to Apply Leak Before Rupture Concepts to Gas and Hazardous Liquid Transmission Pipelines," Contract Report No. DTRS56-95-C-003, U.S. Department of Transportation Research and Special Programs Administration, Washington, DC.

Rolfe, S. T., and Barsom, J. M., 1977, *Fracture and Fatigue Control in Structures, Applications to Fracture Mechanics*, Prentice Hall Inc., Englewood Cliffs, NJ.

Roovers, P., Bood, R., Galli, M., Marewski, U., Steiner, M., and Zara, M., 2000,"EPRG Methods for Assessing Tolerance and Resistance of Pipelines to External Damage," Proceedings of the 3rd International Pipeline Technology Conference, Brugge, Belgium, May 21–24, ed. R. Denys, Elsevier Science, Belgium, Vol. II, pp. 405–425.

Rosenfeld, M. J., Pepper, J. W., and Leewis, K., 2002, "Basis for the New Criteria in B31.8 for Prioritization and Repair of Mechanical Damage," Paper IPC02-27122, Proc. 4th ASME International Pipeline Conference, Calgary, Alberta, October.

Sen, M., and Zhou, J., 2008, " Evaluation of Pipeline Stresses During Line Lowering," Proc. 7th International Pipeline Conference, Calgary, Alberta, October 2008.

Tada, H., Paris, P. C., and Irwin, G., 1985, *The Stress Analysis of Cracks Handbook*, Del Research Corporation, Hellertown, PA.

Wang, Y.Y., Sun, X., and Horsley, D. J., 1996, "Plastic Collapse Analysis of Girth Welded Pipes With Circumferential Cracks," Proc. 9th Symposium on Pipeline, Research Paper 41, Houston, TX, September 30–October 2.

Wang, Y. Y., Rudland, D., and Horsley, D., 2006, "Development of a FAD Based Girth Weld ECA Procedure — Part 1. Theoretical Framework, Paper IPC2002-27171, Part 2. Experimental Verification, Paper IPC2002-27173," Proc. 4th International Pipeline Conference Calgary, Alberta, Canada, September.

Wang, Y. Y., 2002, "Optimising Weld Integrity for High Strength Steels," Proc. Advanced Welding and Joining Technical Workshop, ed. T. A. Siewert, NIST, Boulder, CO.

Wells, A. A., 1961, "Unstable Crack Propagation in Metals: Cleavage And Fast Fracture," Proc. Crack Propagation Symposium, Vol. 1, Paper 84, Cranfield. United Kingdom.

Chapter 7

PIPELINE REPAIRS

When a pipeline operator becomes aware of a defect or anomaly in a pipeline and, as a result of performing an assessment such as was described in Chapter 6, has determined the defect poses a threat to the integrity of the pipeline, then a repair will become necessary. The defects that may require repairs include:

- General and corrosion pits
- Gouging
- Dents (smooth and kinked)
- Cracks (leaking or otherwise)
- Laps/slivers/laminations
- Arc strikes
- Spalling
- Welding defects (girth and longitudinal)
- SCC

The purpose of a repair is to restore the pipeline to full serviceability, although sometimes only a temporary repair can be made. In the event that the repair process does not reach the level of full serviceability, the operator will need to reduce the MOP of the pipeline.

It is of the utmost importance, when making a repair, to ensure the safety of the repair crew, since a defect could be about to fail.

Lowering the operating pressure is thus essential in circumstances where a defect is the result of actual, or assumed mechanical damage, or when it has been categorized as an "immediate repair condition" following an ILI. The pressure generally is reduced to 80% of the normal operating line pressure, unless the safe working pressure, calculated for the defect, is a lesser value. A pressure reduction is considered appropriate where *uncertainty* exists that an anomaly is safe at the operating pressure. This is particularly true at the time of excavation, as a matter of worker safety.

A common sequence to affect a repair to a pipeline would be as follows:

- First identify the location of the repair
- Obtain permits (depending on the location)
- Notify affected property owners
- Prepare a dig plan
- Arrange for outside Contract support, if required
- Ensure that all safety precautions are in place and personnel are fully trained
- Excavate the repair site
- Inspect the pipeline
- Check conditions of the pipeline coating
- Remove the coating
- Clean pipe if the defect is external
- Perform an ultrasonic test to check on wall thickness and/or defect location
- Perform a dye penetrant or magnetic particle test or take an etching of the anomaly
- Take dimensional data of the defect, i.e., depth, width, length, and the exact location in the pipe

- Compare this data with the information provided by the in-line assessment tool
- Repair the pipeline
- Return the site to its previous condition

It may be necessary to excavate a significant portion of the pipeline on either side of the pipe joint containing the defect.

Where SCC has been an issue in a pipeline, the industry practice has been to take advantage of the exposed section on any excavated pipe and check/investigate for indications of SCC.

Prior to any excavation work taking place, it is essential that the correct lock and tag out procedures are followed to isolate the section of line and that power to the cathodic protection rectifiers are turned off.

Before excavating in the vicinity of the line, the pipe joints must be located accurately and hand exposed, often to within 1 m on either side of the line segment, though this distance is generally set by company procedure and industry practice rather than a code requirement.

The work crew will wear appropriate personal protective equipment and if working near a gas line, or in a confined space, they should also carry gas detection equipment. Additionally, the excavated bell hole or trench must be suitably shored to prevent collapse and have the means for speedy egress at both ends of the excavation.

It is the industry's practice to undertake the repair after careful planning has occurred. This planning stage involves obtaining the following:

- Detailed knowledge of the pipe being repaired
- Pipe diameter, wall thickness, steel grade
- Pipe chemistry, toughness, and method of manufacture
- The terrain surrounding the defective pipe
- Site access
- Proximity of populated premises
- Necessary permits
- Full awareness of all safety issues

Permits may have to be obtained, and contact made with other utilities that may have facilities close by the repair site. These will need to be located. There may be a need to restrict public access to the area that will necessitate liaison with local authorities and landowners.

PIPELINE REPAIR CODES

Codes are there to safeguard life, property, and the environment. The nature and extent of the anomaly will be a major factor in determining the selection of the repair method(s) since these have to be made in compliance with both code requirements and the operator's own safe operating procedures. Table 7-1 provides a list of codes and standards, which make ref-

TABLE 7-1. Relevant codes and standards for making repairs

Gas Pipelines	Oil Pipelines
49 CFR 192	49 CFR 195
ASME B31.8	ASME B31.4
ASME B31.8S	API 1160
CSA Z 662	CSA Z 662
PRCI Pipeline repair manual	PRCI Pipeline repair manual
Company Operating procedures	Company Operating procedures

erence to repair requirements. The costs of repair and time to complete with the associated service impact inconvenience are other factors that affect the choice of repair scheme.

Additionally, the following codes are also applicable when repairing pipelines:

- ANSI/ASME B31G, 2009 Manual for Determining the Remaining Strength of Corroded Pipelines: Supplement to B31 Code for Pressure Piping
- ANSI/ASME PCC-2, 2008 Repair of Pressure Equipment and Piping
- API 570 2009 "Piping Inspection Code: In-service Inspection, Repair, and Alteration of Piping Systems"
- API RP 582, 2001 "Recommended Practice and Supplementary Welding Guidelines for the Chemical, Oil, and Gas Industries"
- API Standard 1104, 2008 "Welding Pipelines and Related Facilities"
- API RP 1107, 1991 "Pipeline Maintenance Welding Practices"
- API 1156, 1997 Effects of Smooth and Rock Dents on Liquid Petroleum Pipelines, Phase I
- API 1156-A, 1999 Effects of Smooth and Rock Dents on Liquid Petroleum Pipelines, Phase II
- API RP 2200, 1994 "Repairing Crude Oil, Liquefied Petroleum Gas, and Product Pipelines"
- API RP 2201, 2003 "Recommended Practice, Safe Hot Tapping Practices in the Petroleum & Petrochemical Industries"
- American Welding Society (AWS) D1.1/D1.1M, 2008 "Structural Welding Code-Steel"
- ISO 13847:2000 Petroleum and Natural Gas Industries — Pipeline Transportation Systems — Welding of Pipelines
- ISO 16708:2006 Petroleum and Natural Gas Industries — Pipeline Transportation Systems — Reliability-Based Limit State Methods

The repairs must be made by competent personnel working under experienced supervision. Maintenance welding procedures and welders are qualified in accordance with Appendix B in the 19th Edition of API 1104 (supersedes API RP-1107). The personnel involved should not only be trained but also briefed about issues specific to a particular repair. This includes in many jurisdictions a worker's right-to-know and to refuse working in what they may consider to be a dangerous situation.

REPAIR PROCEDURES

The type of defects that can, under appropriate circumstances, be repaired include internal and external corrosion, SCC, long seam defects, girth weld defects, dents, construction damage, and metallurgical defect features. The types of repair that can be made are:

- Cutout and replace
- Grinding or dressing
- Weld deposition
- Mechanical clamps (temporary sleeve)
- Metallic sleeves - snug fit and stand off sleeves (full encirclement)
- Nonmetallic (composite) sleeves
- Hot tapping, stopple, and bypass

The applications of the above repair techniques are summarized in Table 7-2 and are described therein:

TABLE 7-2. Summary of repair applications

Type	Superficial	Moderate	Severe	Extreme
Grinding/dressing	X	X		
Weld deposition	X	X	X	
Mechanical clamp (temporary)		X		
Full encirclement			X	X
Hot tap				X
Cutout				X

Pipe Cutout and Replacement

If the anomalies are extensive in nature, a cutout will be required necessitating replacement of the pipe with pre tested material taken from stock (Figure 7-1).

For many pipeline operators, cutout repair is the preferred method of repair. It removes the defective section, replacing it with new problem free pipe. It is, however, one of the most expensive options, especially when considering the environment, lost product, and outage downtime. Cutout is, however, the best solution if the circumstances are correct.

For cold cutting of the pipe to remove the defective section, sufficient pipe must be exposed to enable proper fit up to occur of the replacement section. Unless the stopple bypass technique illustrated in Figure 7-2 and described later is used, there is a need to interrupt product flow, depressurize the line and temporarily, de-activate it. If circumstance permits, this generally is the best possible solution, though it can be very expensive, and for this reason, performing cutouts is now becoming relatively less frequent.

Figure 7-1. Typical cold cutting and repairs

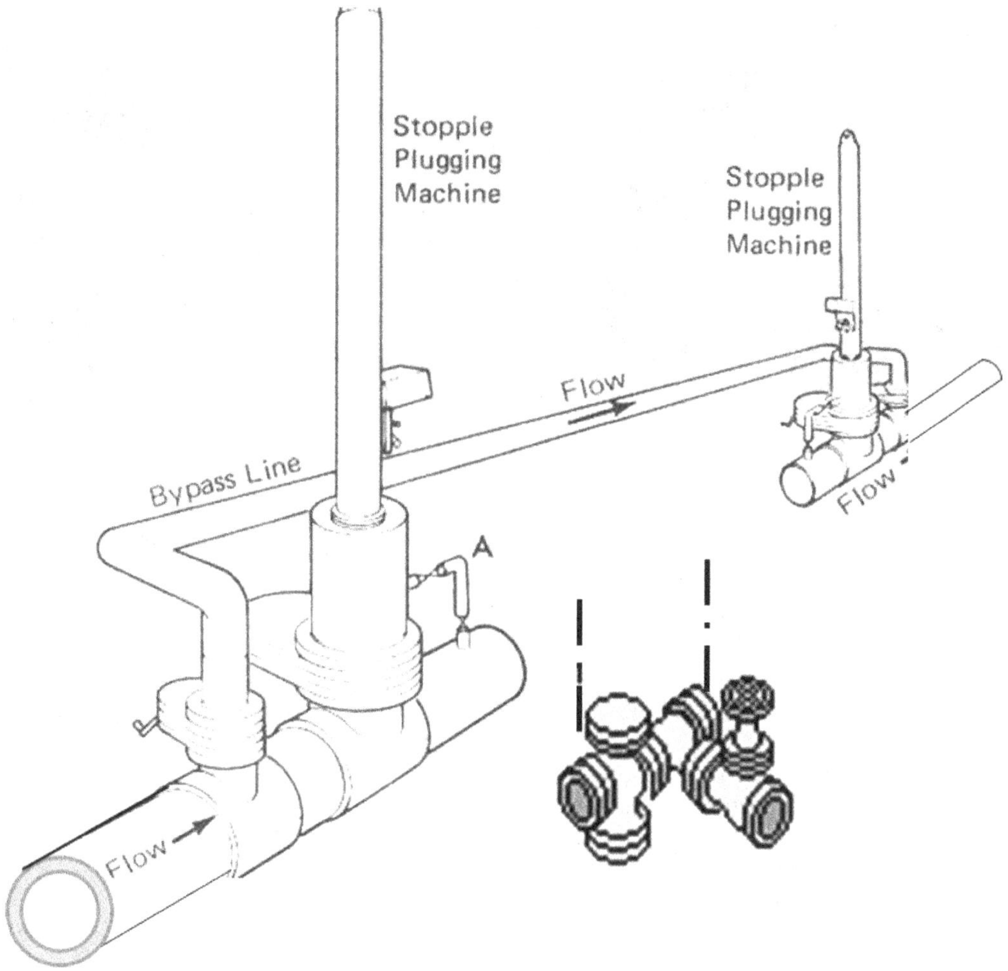

Figure 7-2. Stopple bypassing and cutting out (courtesy T.D. Williamson)

Grinding and Dressing

Grinding (Figure 7-3) is a very common form of repair that does not involve the application of any fitting to the pipeline. Superficial or low level defects such as gouges, superficial corrosion, abrasion, etc., are removed by grinding or dressing leaving a smooth profile free from stress raisers. Skill is required to perform this task, and it should only be done using a low powered grinder. The profile of the defect should be carefully measured and recorded. Grinding is also used to remove mechanical damage in dents provided the material removed does not amount to more than 40% of the wall thickness. Extreme care must be taken when evaluating the surface damage caused by denting to ensure all cracks have been removed.

When grinding is used to repair a weld defect, it is imperative then that the defect is completely removed down to sound metal and is free from cracks, shrinkage, and porosity. The use of magnetic-particle or dye-penetrant inspection of the weld cavity is strongly recommended.

Grinding Limitations and Procedure

Grinding generally can eliminate cracks and other concentrating anomalies as well eliminate excessive hard materials, weld deposits, etc. However, in circumstances, where

Figure 7-3. Dressing/grinding to a smooth/shape surface

grinding is the only option, the amount of material that can be removed must not reduce pressure carrying capacity of the pipeline below 100% SMYS.

Codes permit grinding to prescribed limits. For example, API 5 L (API 2009) allows grinding usually 8% to 12.5% of wall thickness for removal of imperfections.

Grinding should be effected when pressure has been reduced, and a safe trench and working environment exists. The following procedure is generally adopted by the industry:

- Ensure safe working environment
- Reduction of pressure to 80% of operating or "discovery" pressure level
- Removal of metal as required (to limit of ASME B31G, but not more than 40% of pipe wall thickness)
- Removal of all stress concentrations and cracks
- Verification of the extent of removal and ensuring smooth surface condition
- Removal of metallurgically altered material after any arc burn
- Undertake prior ECA on fatigue life of a dent (if applicable). For example, API RP 1156 (1999) may be consulted if rock dents are involved.

Engineering critical assessment involves gathering and assessment of the following information:

Pipeline design data
Nominal outside diameter, mm
Nominal wall thickness, mm
Minimum wall thickness, mm
Maximum depth of cover, m
Minimum depth of cover, m
Design life, years

Operating Conditions
Operating temperature
Operating pressure
Operating pressure at location of failure or repair (discovery pressure)

Installation Information
Pipe line installation and any strain/stress analyses report
Calculation of axial stress/strain due to local buckling/surface roughness

Girth welding procedure information
Maximum allowable pipe misalignment
Pipeline material certificates
Welding procedure qualification test data
Stress-strain curve for pipe material as available

In-service Welding Deposition

This repair technique is the most traditional method of repair under live conditions and is performed on all pressure containing sleeves, patches, and weld depositions.

Corrosion pits may be repaired using weld depositions, and the method is well suited to repairing small corroded areas that cannot be sleeved such as elbows and tees. In-service welding will require a high degree of control and therefore should be performed by a highly skilled welder to avoid risk of blow through.

It is inappropriate, and potentially dangerous, to use production welding procedures to make in service welds, since the fluid flowing condition in the pipeline can cause higher weld cooling rates leading to possible HIC. Weld cracking can be avoided by minimizing the presence of hydrogen in the welding atmosphere through the use of low-hydrogen electrodes such as E7018 or by using either flux core arc welding (FCAW) or gas metal arc welding (GMAW). Special in-service welding procedures have therefore to be developed and welders trained to use them. This often involves a staged learning process starting with welding on a specially developed rig flowing water through the test section at a prescribed rate.

The inherent problem within service welding is the requirement for preheating, which is not always possible especially for liquid lines. Repairing a corroded area with weld

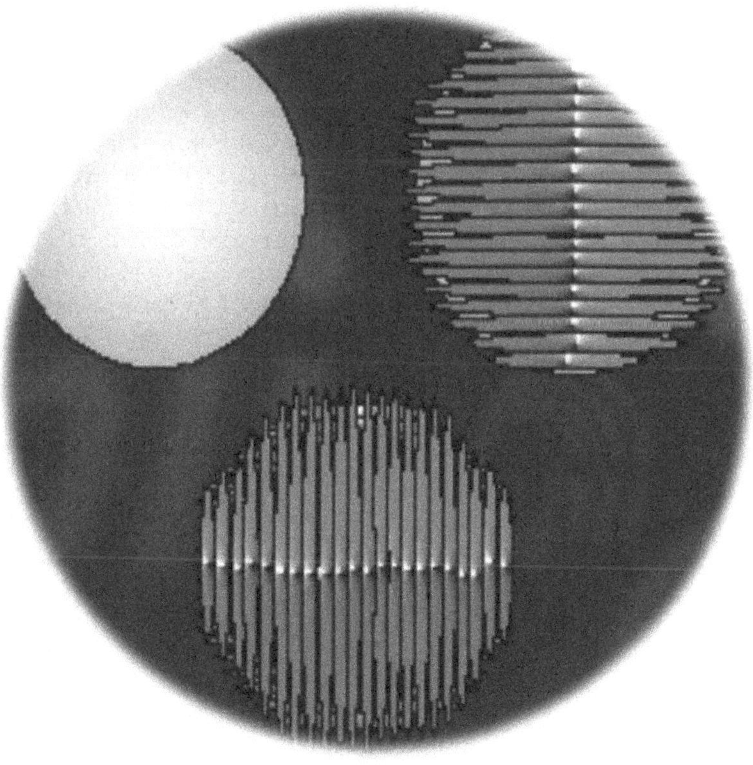

Figure 7-4. Weld deposition (after A. Clyne, DNV)

metal deposits may result in specific properties, which may be unlike the original surface properties because annealing may have taken place. Figure 7-4 illustrates the two-step process in depositing metal on a damaged area. Weldability of the parent pipe is critical, so it is important to ascertain the carbon equivalent (CE) of the pipe body before undertaking a repair procedure of this type. It is prudent to leave a period of time between the layering passes to ensure that cracking has not occurred. An assurance that cracking has been avoided is strengthened by using nondestructive testing (NDT) methods such as ultrasonics or magnetic particle testing of the weld, allowing 48 hours to 72 hours for delayed cracking effects. Clearly, if this was done on site, it could cause considerable operational delays, so whenever possible the welding procedures should be tested before hand and typical material coupons used in destructive tests to establish a crack-proof welding procedure.

It is equally important to be aware of the remaining wall thickness to avoid burn through. The process is labor intensive involving welders, assistants, and inspectors.

Mechanical Clamps

Mechanical, bolt on clamps (Figure 7-5) may be used to make mostly temporary repairs and are very versatile since they can accommodate pipe out of roundness. This type of mechanical clamp relies upon a good seal and may be installed on pipelines and laterals usually carrying petroleum liquids as a temporary measure to stop leaks. With this technique, it is still possible for movements in the pipeline to take place in the affected area without affecting the leak. However, it must be replaced with a permanent repair when practicable.

However, some of these types of clamps can be welded to make a permanent seal. The larger sizes of clamp are heavy and relatively expensive.

Permanent Sleeve Repairs

Three different generic types of sleeve repairs are currently available, two of which use a metal sleeve, the other a nonmetallic composite.

Sleeves are generally used to reinforce the damaged pipe by restraining the bulging, which might take place at the defect, rather than to share the hoop stress. The exception is the use of the pre heated compression sleeve, which, upon cooling, clamps in place around the pipe and relieves some of the hoop stresses. In these types of sleeves, the two halves of the heated compression sleeve are welded to a side bar so that there is no welding to the pipeline. Installation quality is verified by measuring the amount of shrinkage after cooling.

There are two type of sleeves, A and B. Type A sleeves are nonpressure retaining and the side seams are fillet welded joining both halves of the sleeve, as shown in Figure 7-6a.

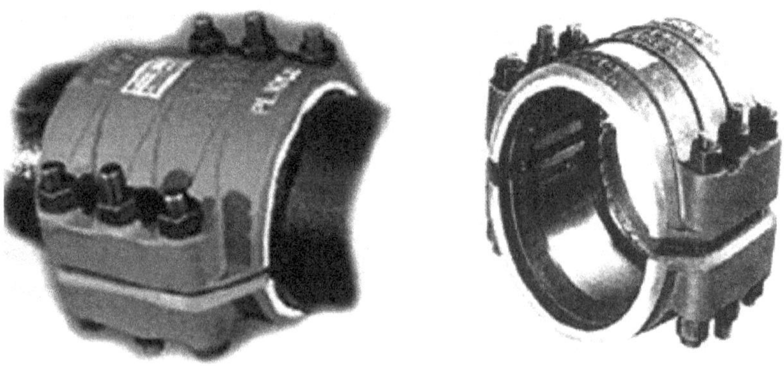

Figure 7-5. Temporary bolt on type sleeve/mechanical clamp

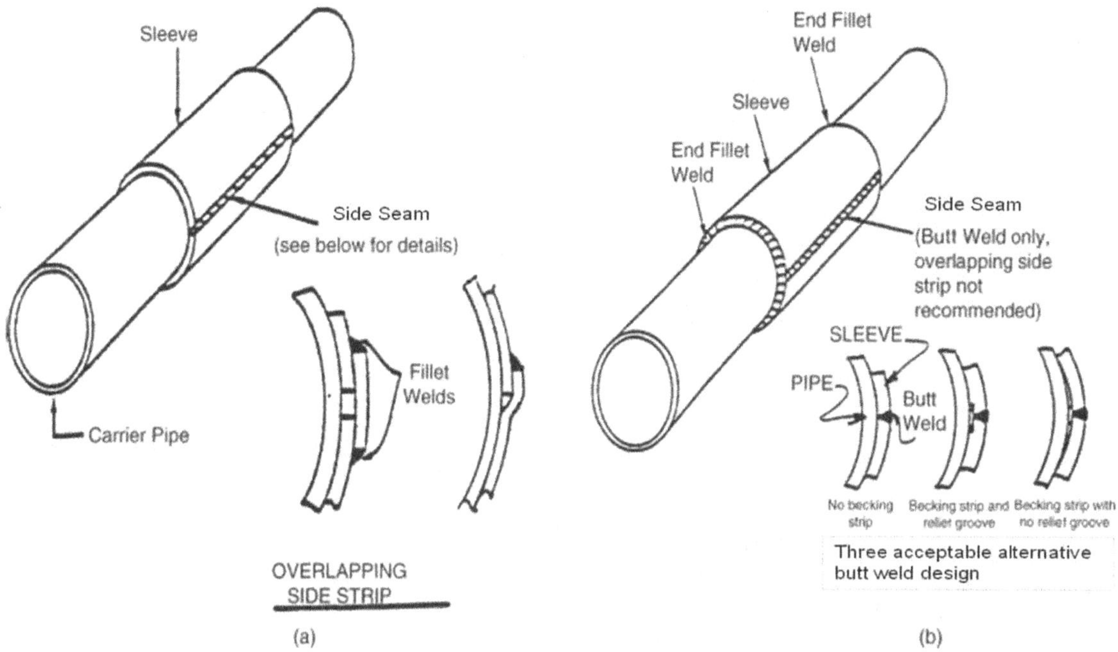

Figure 7-6. (a) Type A nonpressure retaining. (b) Type B pressure retaining sleeve

The end gaps are sealed to prevent the ingress of water. With Type B pressure retaining sleeves (Figure 7-6b), the side and end seams must use groove butt welds. The welding procedures for the end seams must be suited to the pipe metallurgy and the computed cooling rates to prevent cracking.

An example of compression sleeve is the PetroSleeve® shown in Figure 7-7. The following provides an outline of the application of PetroSleeve® repair techniques:

- Offsite engineering design including stress design
- Fabrication/assembly and infield installation
- Offsite assembly
- Grout application onto pipe
- Set PetroSleeve® onto pipe
- Apply clamping force on PetroSleeve®
- Heat assembly
- Complete the two fillet welds on top portion of the zipper
- Remove clamps
- Quality control assembly
- Generate report on assembly stresses
- Wrap sleeve and pipe (coating)
- Test if required
- Complete backfill

This type of sleeve (Smyth 1998, 2006) supports and reduces the stresses in the steel adjoining the defect. The installation of a PetroSleeve® can also eliminate flexing in the pipe material, additionally, it puts into compression the stress state in the material surrounding the defect. The result of these two actions works to eliminate fatigue, crack extension, and rupture. PetroSleeve® also is used for repair of pipeline internal defects (Smyth 2003).

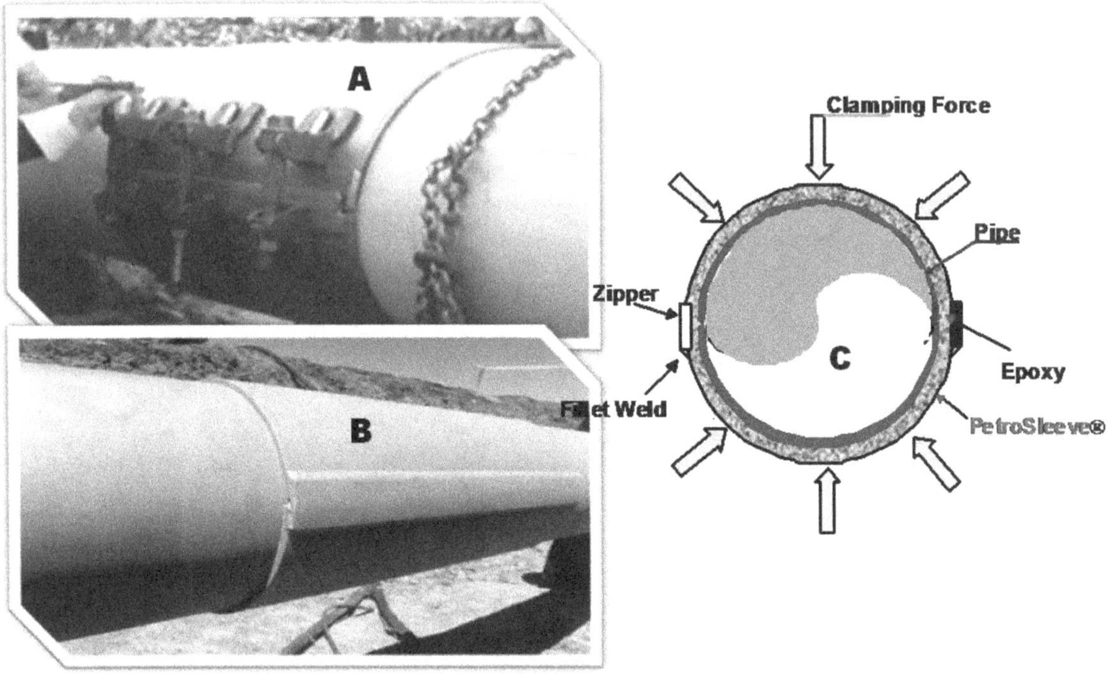

Figure 7-7. A PetroSleeve® compression sleeve. A. Clamping the sleeve halves. B. Completed sleeve. C. Principal of application (photos courtesy PetroSleeve®, http://www.petrosleeve.com)

Full-encirclement split sleeve repair includes "snug fit" and "stand off" designs. With the snug fit, both halves of the sleeve are in intimate contact with the pipe surface, and pretensioning is sometimes used to achieve fit up. The stand off sleeve is better suited to circumstances where the pipe is slightly out of round or has a very slight longitudinal curve. Hardenable polyester filler material is used to patch corroded areas under both types of sleeve and to fill the annular gap with the stand off fitting.

Full Encirclement Split Sleeve Repair. This is a snug fit design and is a common form of repair by the use of a full-encirclement split sleeve (Figure 7-8). By forming a snug fit

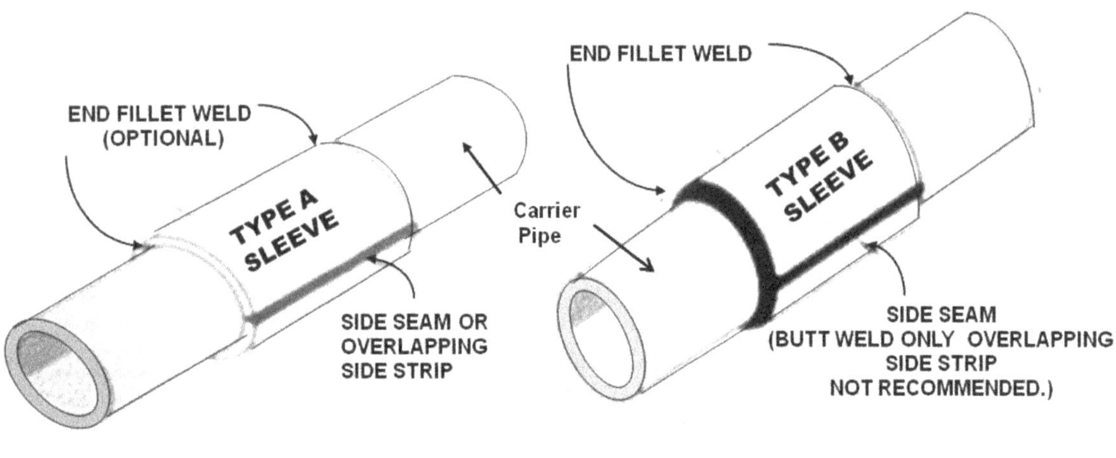

Figure 7-8. Full-encirclement split sleeve repair (Mohitpour et al. 2007)

around the carrier pipe, a properly installed split sleeve will provide support for the defective area such as a gouge or dent and gouge. The longitudinal seams are welded, and if total containment is required, fillet welds are directly performed at each end.

Type A sleeves are used for repairs of corrosion and dents and are as required to restrain the bulging of defective area. Such a sleeve does not carry much stress if the defect is relatively short with a defect length $L \leq (20Dt)^{1/2}$. The sleeve does not require high integrity side seams and is not welded to the carrier pipe, and therefore will not require any rigorous inspection.

Type B sleeves, on the other hand, are pressure containing sleeves and can be used for repair of (potential) leaks, gouges, and gouged dents. Type B sleeve must have the following characteristics:

- Must be able to have same pressure carrying capacity as the carrier pipe
- End must be welded to carrier pipe
- Side seams full penetration butt welds are required to achieve high integrity welds
- Low-hydrogen procedure must be implemented for end fillet weld
- Weld procedure and weld qualifications are required
- Rigorous weld inspections are necessary

A typical circumferential welded pressure containment sleeve is shown in Figure 7-9.

Split sleeve installation is highly specialized. Consideration must be given to application temperatures during welding to avoid potential hydrogen induced cracking. Consideration must also be given if the sleeve is subject to significant cyclic pressures to avoid potential fatigue cracks.

If gouges or dent with gouges are being repaired, the dent must be filled with epoxy to prevent bulging and reinforcement of the defect.

Epoxy Composition Filled Sleeve Repairs. This sleeve repair method is of the "standoff" type and developed to repair pipelines without the necessity for removal from service. This repair consists of two oversized half sections of semicircular steel shells that are joined together (by welding or bolting) to encircle the damaged area leaving a circular gap (Figure 7-10). This annular gap is sealed at each end of the sleeve using a simply applied fast setting material and then filled at a very low pressure with a high stiffness epoxy based grout. This type of sleeve can repair most nonleaking defects of all types.

Figure 7-9. Typical full-encirclement split sleeve: in welding (A) and completed (B)

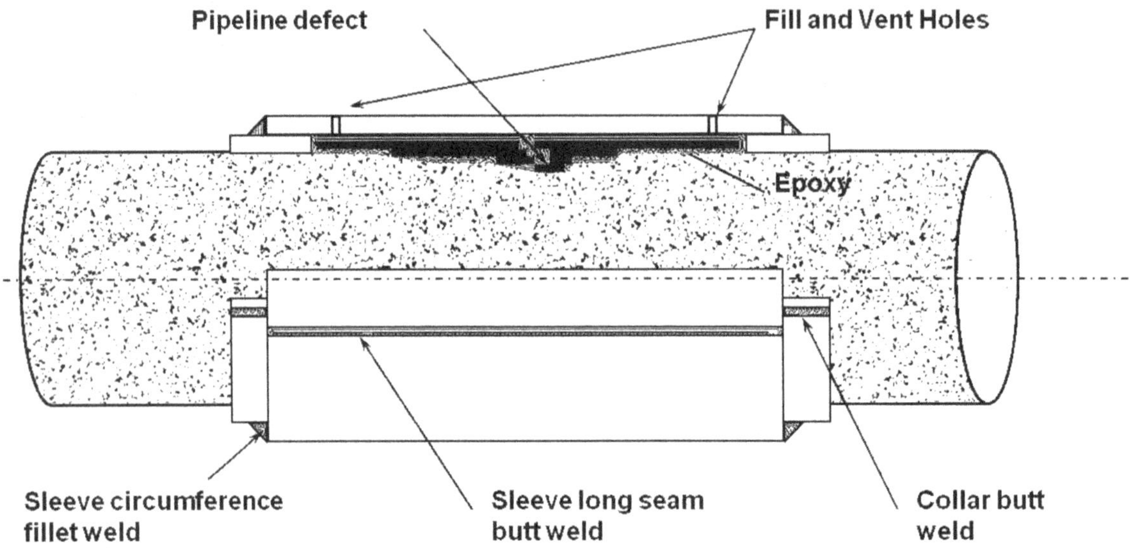

Figure 7-10. Schematic of epoxy composition filled sleeve repairs

There is no requirement for welding to the pipeline. Tests show that it provides good bulging restraint for the defective area and performs well under internal pressure loading and fatigue cycles.

The installation process for a typical epoxy sleeve involves the following (Figure 7-11):

1. Operating pressure reduction
2. Surface preparation
3. Fit-up
4. Welding of sleeve halves
5. Sealing of sleeve ends
6. Epoxy grout filling
7. Electrical continuity

The above sleeve methods can be used as permanent repair for:

- Corrosion to 80%
- Gouges, grooves, arc burns, cracks following grinding of dents, however, not on dents without grinding or on welds
- Weld slag or porosity
- Mill seam welds

Advantages of sleeve repairs are related to timing:

- work can be done by one welder
- sleeve welding can be done prior to shutdown
- shorter shutdowns or multiple sleeves can be undertaken
- final pouring is a two person job (flexible timing)
- there no steel sleeve to pipe contact

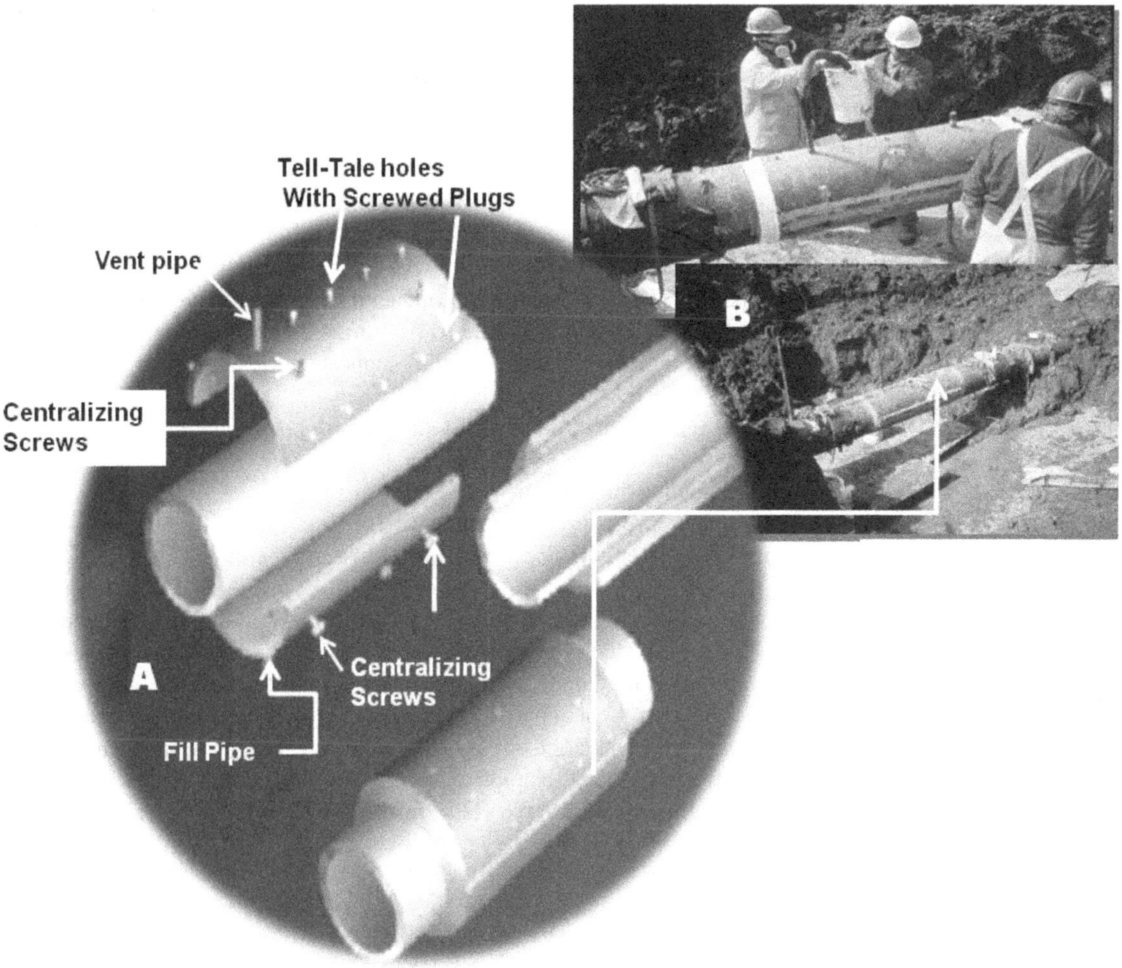

Figure 7-11. An epoxy filled sleeve repair: typical Schematic (A) and assembly (B)

Disadvantages are also related mostly to timing but also cost:

- requires two welders to install quickly
- most of the shutdown is for welding
- welding/installation will require shutdown
- multiple sleeves require long shutdowns
- cannot get updated information from MFL tool

Specially shaped coupling sleeves, known as Dresser couplings, or similar mechanical clamps may be used to repair girth welds. However, such clamps are temporary and the girth weld must be repaired permanently.

Nonmetallic (Composite) Sleeves

Composite wrap repairs are used for repair of relatively low level damage. Such tape wrap is impregnated with agents that are activated by contact with chemical accelerants; some variations of the tapes are activated by water. They are good reinforcements and relatively inexpensive.

Composite wrap repairs (see Figure 7-12) are mostly made using a polymer-based matrix reinforced with either a woven fabric or oriented fibers. Some wraps are preformed as coiled rolls, such as the Clock Spring™ (Figure 7-12), others are hand laid up as a wet bandage applied to the pipe surface. Both types are chemically bonded to the pipe and within each inter leaving layer as these are continuously wound around the pipe and are cured in place.

The Clock Spring® pipeline reinforcement system is a coil wrap of high-strength composite material (Hewing 1999), which consists of a fiberglass/polyester composite material coiled with adhesive in layers over a filler, reinforces steel pipe that has certain nonleaking defects. When properly installed, the system permanently restores the pressure-containing capability of the pipe. Adhesive filler is used to fill up the defect (corrosion, mechanical damage, or dents, etc.; Figure 7-13.) to assist with support and load transfer to the Clock Spring® installation.

Clock Spring® is typically a three-part system consisting of:

- High-strength composite structure
- High-performance two-part adhesive system
- High compressive strength, proprietary load transferring component

With this technology, repairs are made by wrapping a pipe defect with fiberglass composite-reinforced coil locked in place with a specially designed adhesive. Once these three elements are installed they bond together to form a very strong durable repair. The adhesive is an important part of the repair system and is strong enough to hold the repair in place and survive for 50 years in varied environments. A typical Clock Spring® installation consists of eight complete wraps of the pipe to form a monolithic, 0.50-in. thick structure. This composite wrap actually exceeds the yield strength of the original pipe. The strength

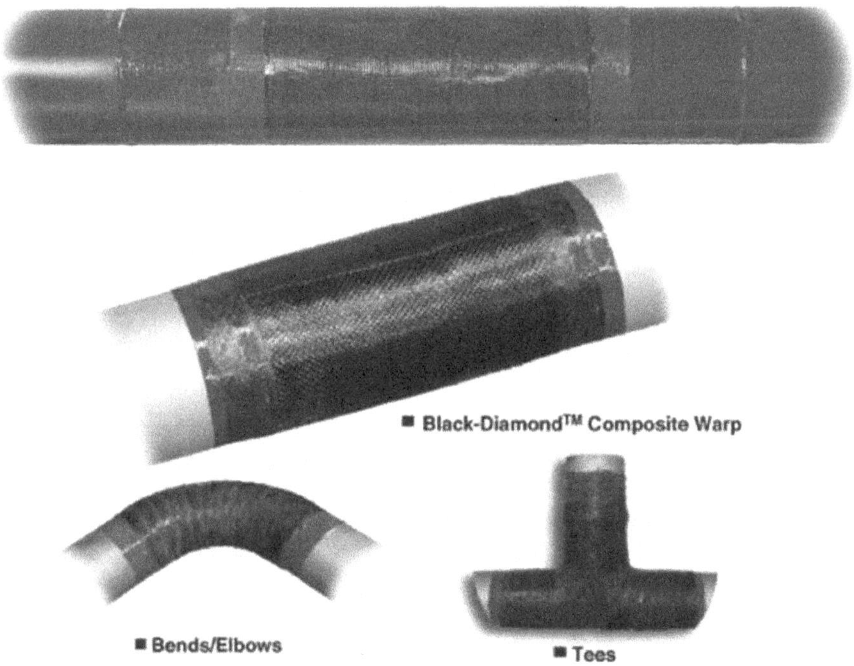

Figure 7-12. A composite wrap repair (Black Diamond is a trademark of TD Williamson Ltd; Morton, 2009)

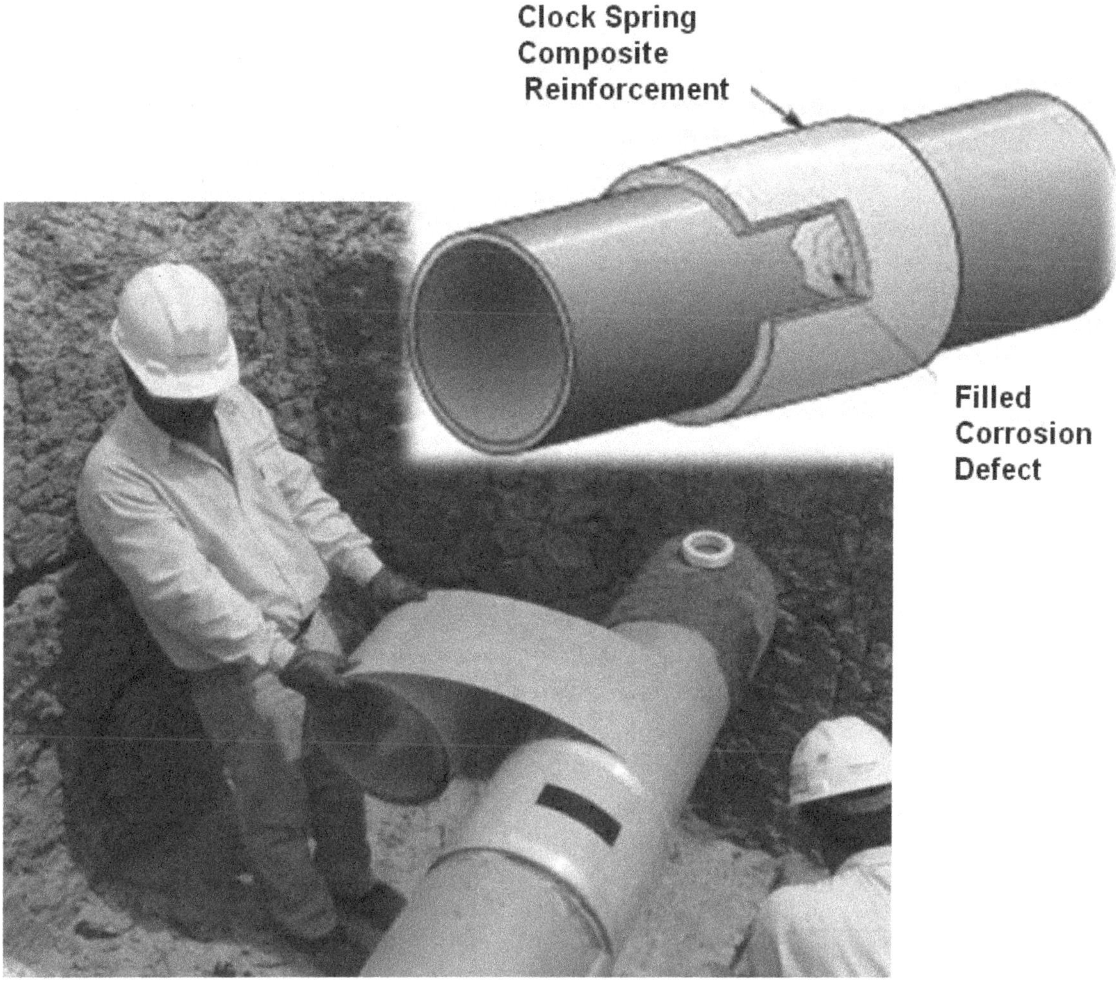

Figure 7-13. Clock Spring™ repair

and effectiveness of the Clock Spring® composite repair comes from the glass component, which is highly resistant to corrosion.

Armor Plate® Pipe Wrap (Figure 7-14) is another structural pipe repair system that uses multi-purpose fiberglass/epoxy wraps and curing agents. It consists of a woven fiberglass/epoxy resin material that results in a monolithic composite pipe reinforcement. The composite material Armor Plate Pipe Wrap is designed to repair straight pipe, elbows, tees, reducers, risers, high temperature, underwater, and offshore platform pipelines. Typical installation is shown in Figure 7-14A.

Armor Plate® Pipe Wrap is similar to StrongBack (Figure 7-14B) in that it is considered a good reinforcement system to repair relatively low levels of corrosion or mechanical damage; however, its hardening agent is chemical rather than water. StrongBack is a composite reinforcement sleeve water-activated, resin-impregnated, glass fiber remediation material. These products are based on a unique blend of liquid epoxy polymer and aliphatic polyamine curing agents, which are able to displace water from wet surfaces in order to make a permanent bond. Advantages and disadvantages of nonmetallic composite wraps are indicated in Table 7-3.

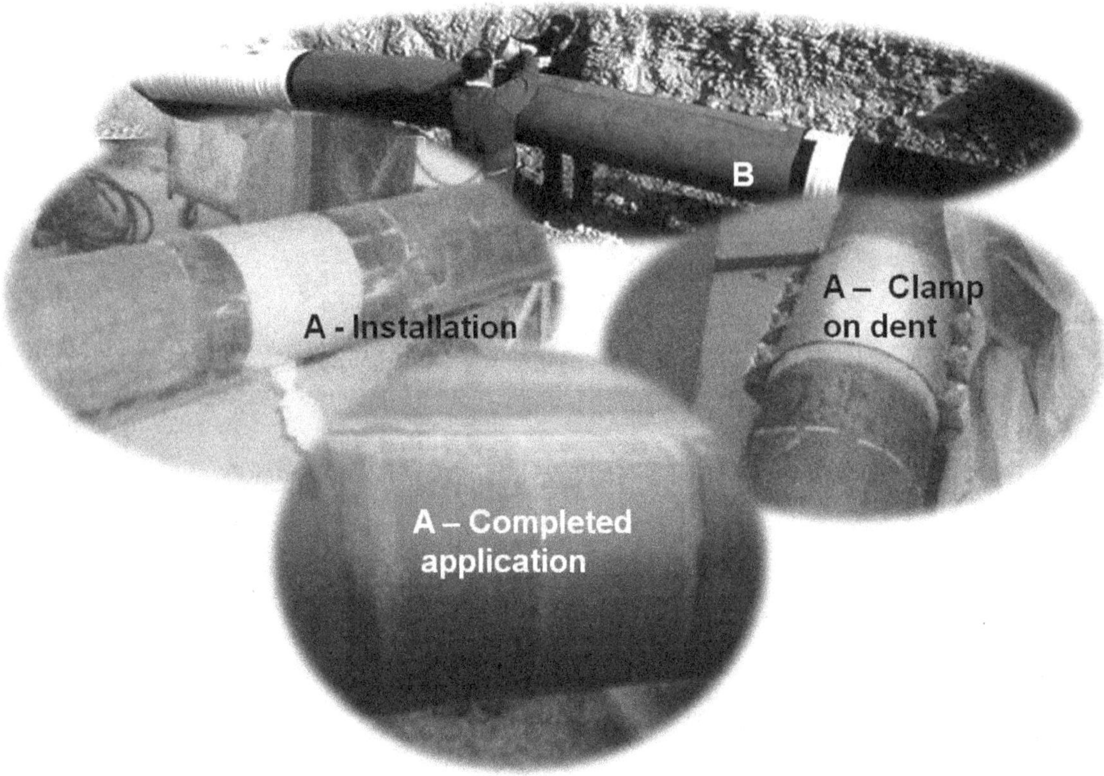

Figure 7-14. Repaired pipe line with Armor plated pipe wrap (A) and StrongBack (B)

The application process for a typical composite wrap sleeve involves the following:

1. Operating pressure reduction
2. Cleaning pipe surface to bare metal
3. Filling depressions with epoxy
4. Grinding seam weld crowns
5. Wrap material around the affected area. Use of wrap 50 mm beyond end of defects
6. Allow time to cure before restoring pressure, consult manufacturer for cure time and proper applications
7. Coat pipe under wrap with a non deforming coating
8. Coat the wrap just as you would do the steel sleeve

TABLE 7-3. Advantages and disadvantages of composite wrap repairs

Advantages	Disadvantages
• Ease of application o Light weight o No heavy and welding equipment required o Minimum shrink required for application • Non-magnetic, will not corrode • Can be applied on curved pipe	• Low stiffness and modulus require thicker sleeve • Potential susceptibility to long term degradation • Limited shelf life • Cannot be applied to leaky or crack-like defects • Each composite wraps are different • cannot be used to repair a crack-like defect or a defect in a brittle behaving material (apparent in older ERW pipes with bonding defects)

Integrity assurance/validation tests must include the following:

- Stress rupture test
- Creep and fatigue tests
- Chemical resistance test
- Cathodic protection resistance tests

Such testing will usually require the following:

- Creation of a long and narrow defect in the pipeline
- Pressurizing pipe to near failure condition
- Reducing pressure to 2/3 of near failure level
- Applying the reinforcing wrap repair at the "repair" pressure
- Testing

Composite wrap specification should generally refer to the following:

- Minimum tensile strength
- Minimum stiffness (modulus of elasticity)
- Minimum glass transition temperature (usually should be 28°C above maximum pipeline operating temperature)
- Minimum fiber content (usually 0.5 to 0.7 fiber/resin ratio by weight)

STOPPLE AND BYPASS

One way to repair a segment of a pipeline containing defects is to by pass the flow from the segment and cut the defective pipe out and replace it with a new one. The stopple and by pass (Figure 7-15) is an ideal technique to replace a section of pipeline while retaining continuity of production flow. However, it is extremely expensive and uses live welding, and fitting remain on the pipeline after repair and completion.

The process involves placing permanently four split tee fittings in pairs on either side of the section of pipe to be removed. Permanent sandwich type valves are then attached to the tees, through which a Hot Tapping tool is entered to cut through the pipe wall. Upon completion of the hot taps, a bypass section is created by attaching a length of pipe to the flanged ends of the two outermost valves. The process is completed by installing two stopple plug fittings through the inner pair of valves and purging/removing the fluid in the section of line between them before cutting away and replacing the unwanted pipe section. After the new section has been purged and/or filled with the fluid, the stopple plug fittings and the bypass line are removed and blind flanges fitted to all four valves.

Stopple and bypass involves the attachment of numerous Tees and Hot Tap fittings to the pipeline and is extremely expensive. For a simple two-position stopple operation, nine fittings are required, all of which have to be welded (Figure 7-16) to the live pipeline.

Two types of stopple fittings are used by the pipeline transmission industry. These are:

- Hot draw, full branch opening, snug fitting barrel, and a flange (Figure 7-15) and
- Fabricated, full sized nipple branch outlet welded to snug-fitting barrel and a flange.

The pipe where stopples will be fitted must be reasonably round to about 3.2 mm. Stopples should not be installed on bends and tees where there is strain on the pipe segment.

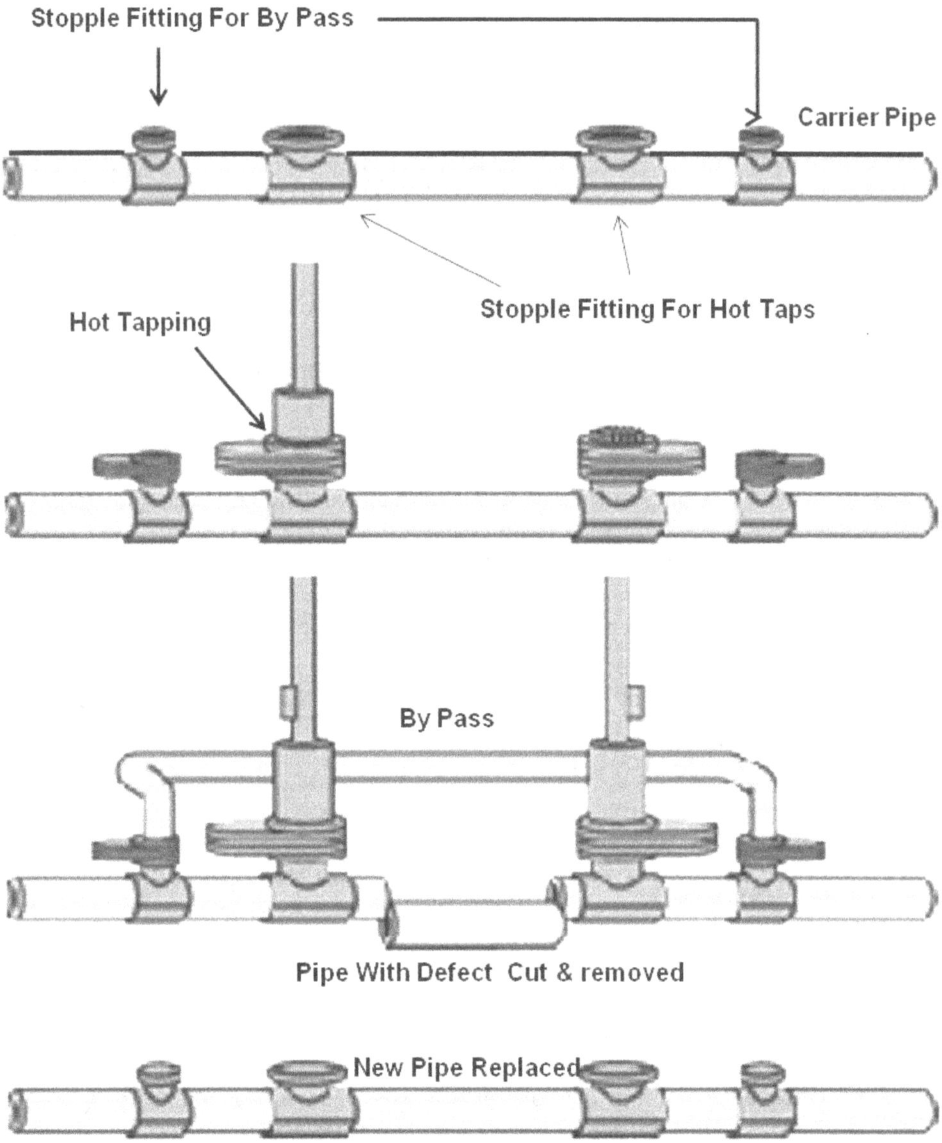

Figure 7-15. Double stopple and bypass repair technique (after T.D. Williamson)

When installing stopples, pipeline maintenance practices generally limit the line pressure to no more than 65% of the working pressure at the time of stopple installation. Industry practice does not allow this pressure to be exceeded even under transient conditions imposed by pipeline operation (such as the case of a liquid line).

Hot Tapping

The procedure for cutting into the pipeline and connecting such a branch piping while the pipeline has products under pressure and/or flowing condition is referred to as a "Hot Tap." A hot tapping machine (Figure 7-17) is used for undertaking the boring operation.

Welding that is required to attach a branch connection fitting to a pressurized pipeline is referred to as "Hot Tapping Welding."

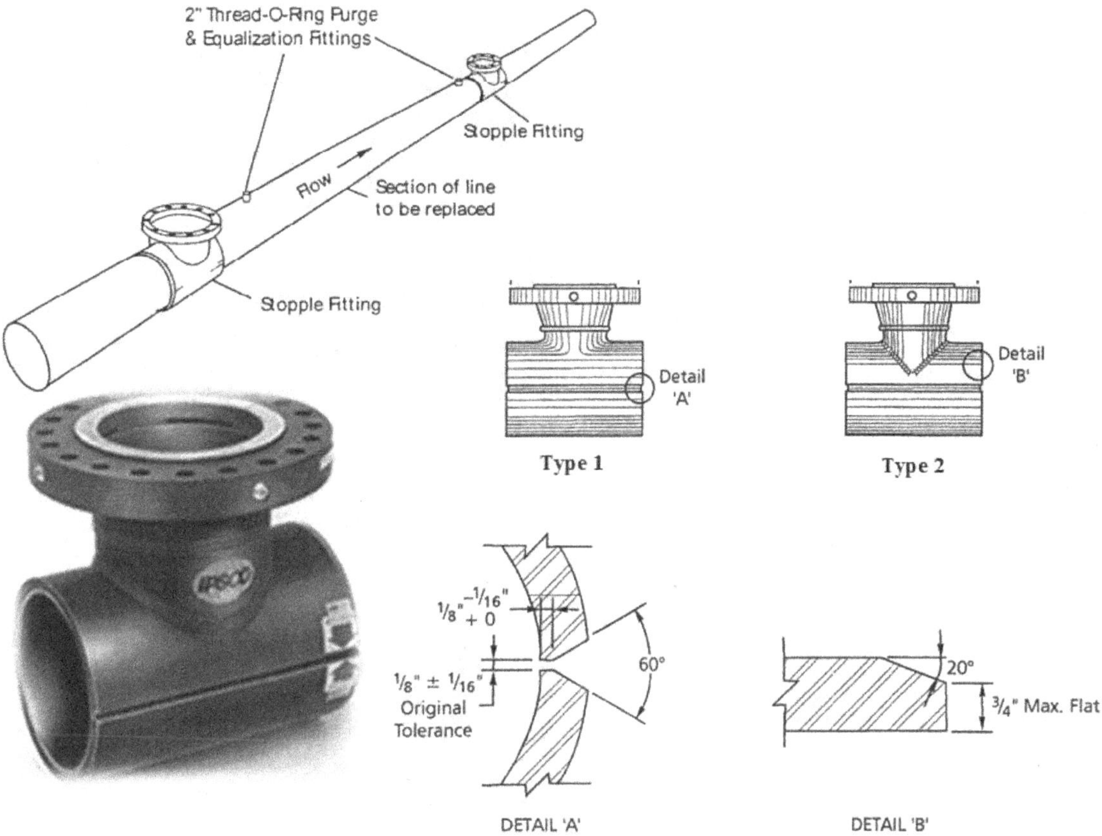

Figure 7-16. Stopple fitting and split tee weld configurations

Hot tapping involves placing a tee fitting over the defect area and welding it in place. It also may be used to remove a defect smaller than the tap size. The two halves of the split tee fitting are placed around the pipe and a valve assembly attached to the flange. The hot tap machine is affixed to the valve flange, and subsequently, the pipeline is drilled adjacent to the defect to equalize pressures between the pipe and the body of the tapping (drilling) machine. The tapping machine and valve are removed and a blind flange is used to plug the tee.

Hot tapping must always be undertaken as per manufacturers' procedure and specification to assure the safety and successful procedure. API 2201 (2003) fully describes safe hot tapping practices. Industry procedure for performing hot tapping typically includes:

- Preparation of
  - Tapping valve
  - Stubs/fitting and tapping unit
- Line pressure reduction (typically to 65% of operating pressure)
- Installing tapping unit, welding tees and units (as applicable)
- Undertaking tapping
- Tapping completion
- Line pressuring

Hot tapping attributes are shown in Figure 7-18. In general, the pipeline valve is first installed onto the stub. The valve must always be aligned, the bore free from dirt and

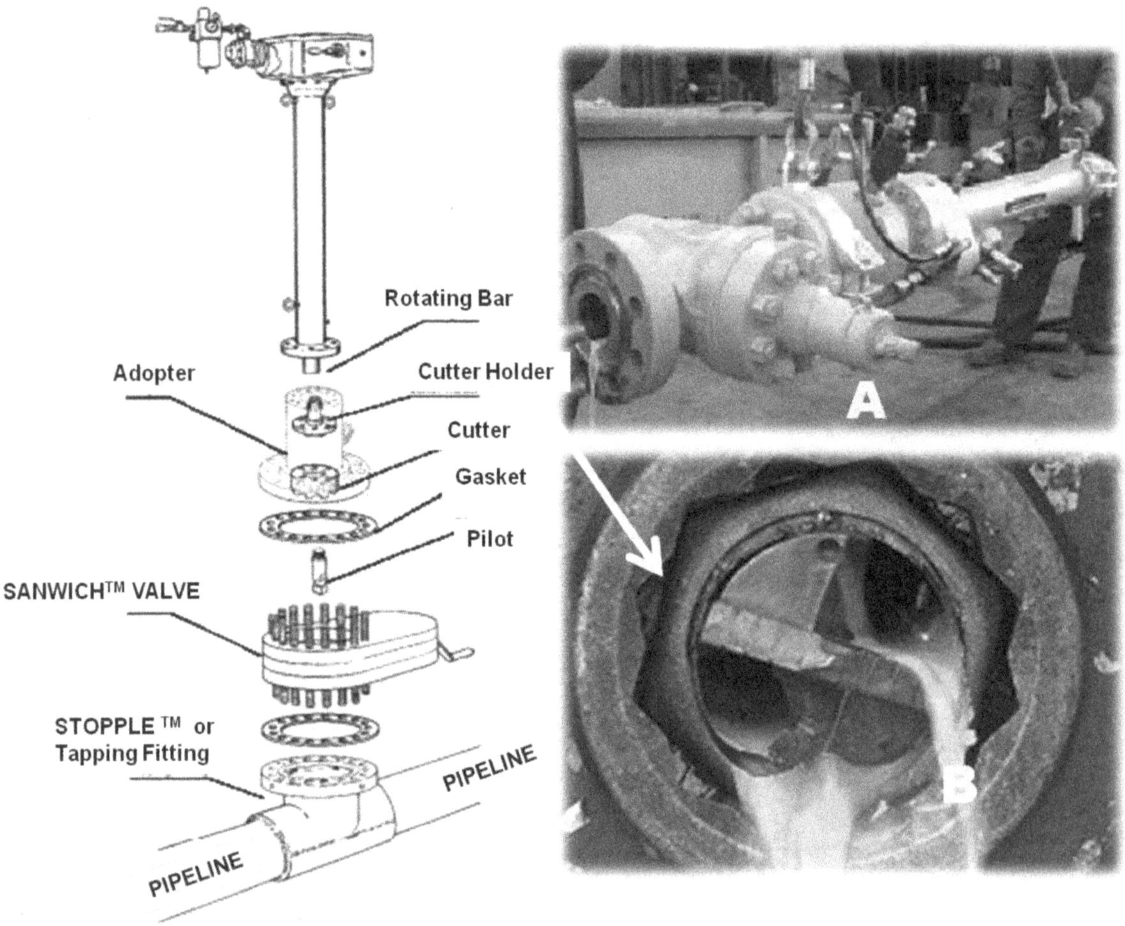

Figure 7-17. Exploded view of hot tapping machine (courtesy of T.D. Williamson) Typical hot tapping machine (A) and hot tapping through a companion flange (B) (courtesy of Weldcut Inc.)

obstructions and checked to ensure easy opening and closing. An x-ray (nondestructive test) must be taken of the weld.

Hot taps on a high-pressure facility must be performed by a crew qualified to make hot taps. Personnel performing hot taps should be:

- Familiar with the pressure limitations of the hot tapping equipment to be used
- Thoroughly trained in the mechanical procedures and safety precautions with the use of the equipment
- All hot tapping operations must be undertaken in accordance with the Owners Procedures and Manufacturers Recommendations

STOPPLE* AND HOT TAPPING PROCEDURE

This procedure is general in nature and only represents the major steps to be considered while doing a stopple and hot tapping application.

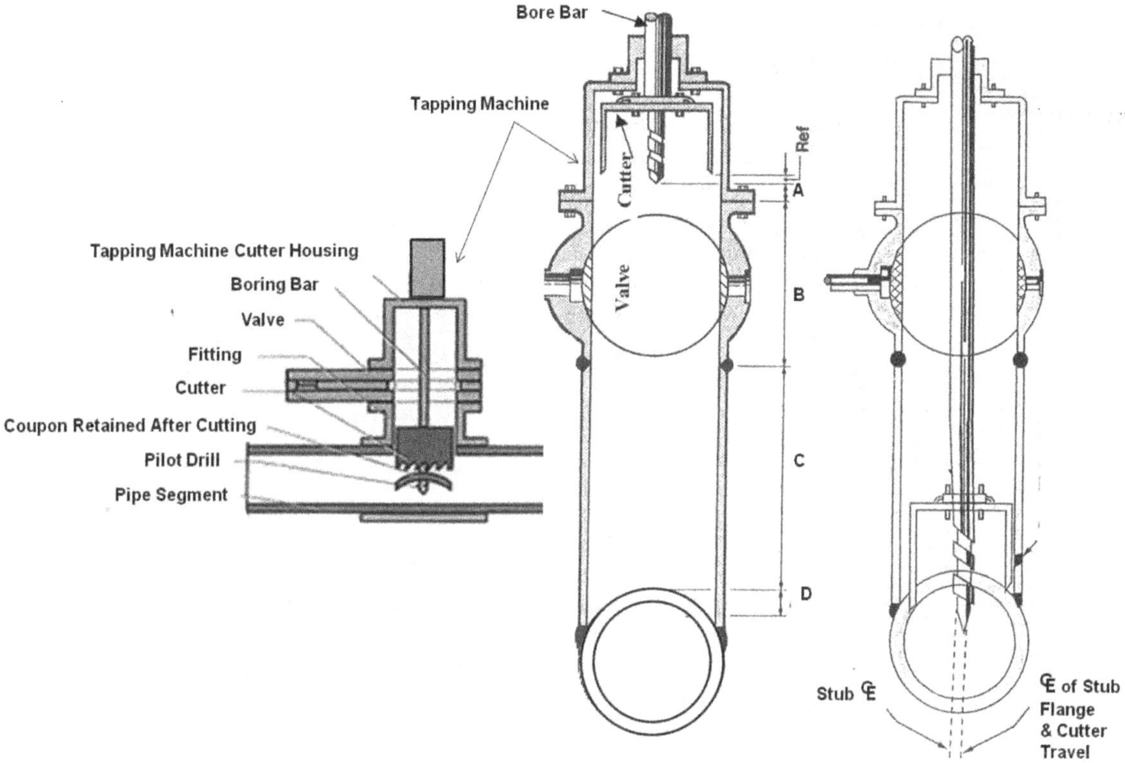

Figure 7-18. Hot tapping attributes and tapping operation

Stopple* Procedure

1. Attached stopple fittings
2. Check all safety issues including verifying line contents, pressure, temperature, pipe wall thickness, direction of flow, and flow rate
3. Check Lock-O-Ring flange on the stopple fittings
4. Install sandwich valve
5. Install equalization/purge valve on thread-o-ring fitting (Figure 7-16)
6. Make 2-in. equalization/purge hot-tap
7. Assemble hot-tapping machine
8. Make stopple plunger hot-tap
9. Install stopple sealing element
10. Install stopple plugging machine
11. Relieve pressure in the isolated section of line
12. Undertake tie-ins, repair, or installation of new bypass piping
13. Equalize pressure in the isolated pipe section
14. Retract stopple plugging head
15. Remove stopple plugging machine
16. Attach the lock-o-ring plug on the tapping machine
17. Install lock-o-ring plug in the lock-o-ring flanged fitting
18. Install 2-in. thread-o-ring completion plug
19. Remove all equipment

It may be noted that *Stopple is a trade mark of TD Williamson Ltd.

Hot Tapping Procedure

***Before Tapping*:**

- Check suitability for tapping:
    - o Inspect the pipe for external corrosion
    - o Verify internal defects with an ultrasonic meter or by radiography
    - o Verify proper tap/seam/joint relationships.
- Verify tapping equipment and that material are correct for the intended service
- Review of applicable maps and records by tapping crew
- Determination of the location of pipelines in the vicinity of the pipeline to be tapped
- Pipeline pressure reduction to a maximum pressure referred by the requirement of the tapping.

***During Tapping*:**

- Pipeline pressure verification
- Inspection of the recovered coupon
- Trouble shooting before and during tapping. Every hot tap is subject to certain possible tapping traps.
    1. *Valve not properly centered:* Cutter wedges in valve or will not enter stub
    2. *Pipe mis-alignment:* Cutter wedges on crotch or jams on cutting chips
    3. *Cutter travel overshoot or not travel enough* (Figure 7-17): Incorrect travel measurement may cause problems such as penetration of opposite side of the pipe, or too little travel preventing the completion of the tap
    4. *Pilot drill:* This retains the coupon through either mechanical retention devices such as latches, springs, wire clips, or threaded pilot drill. Should the coupon be lost, retrieval is only possible through pigging the line, blowing down (gas only) or depressurizing the pipe.

The following steps are usually followed to complete a hot tap:

1. Assembling the tapping machine according to the manufacturer's operating manual. Check must be made of the machine, unit control head, cutter, coupon (steel cut) retaining device and the power units, cables and hoses, while ensuring the coupon-retaining device is in working order
2. Calculating the distance of cutter travel to ensure the cut can be completed (without overshooting/cutting through the opposite side). The cutter can be retracted and the valve can be closed. This calculation must consider the following: A = tip of pilot drill to face of adaptor, B = valve length, C = stub length, D = distance to ensure coupon is severed and detached (see Figure 7-17).

   $$\text{Therefore, cutter travel} = A + B + C + D$$
3. Attaching the tapping machine to the valve adequately supporting and checking alignment
4. Opening the valve and extending and retracting the boring bar through the valve several times. This is to ensure that the cutter does not jam or drag.
5. Opening the bleed valve on the body of the tapping unit and on the body of the valve as applicable
6. Extending the boring bar until it meets the carrier pipe (so that it does not jam against carrier pipe)
7. Starting the cut

8. Stopping the tapping machine when the bit penetrates the carrier pipe and purge the valve and machine
9. When purging is complete, closing the body bleeds on the tapping unit and the valve and letting the pressure equalize.
10. Completing the cut
11. Retracting the boring bar
12. Closing the valve and depressurizing the tapping machine
13. Opening the valve body bleed to ensure the valve seals are functioning
14. Removing the tapping machine

PIPELINE PLUGGING

When repairing a pipeline, it is often necessary to isolate/bypass the affected segment of the pipeline to undertake the necessary repair work. Typically, for onshore cross country pipelines, repairs are undertaken while the line continues to be in service and this is the procedure that most pipeline companies follow. Also, plugging may be necessary for installing a new valve, a flange connection for a future lateral, or a connection to a temporary trap. The following describes some of the plugging alternatives. Method of plugging usually depends on the fluid service, gas, or high vapor pressure petroleum products (such as condensate, propane, etc.) or liquids including low vapor pressure fluids (such as crude oil, bitumen, water, etc.). However, some plugs can be used in either service. Types of plugs include the following:

- Mechanical plugs
- Freeze plugs
- Mud plugs
- Balloons or inflatable plugs

Mechanical Plugging

A mechanical plugging machine can temporarily stop the product in an operating pipeline, by acting as a block valve. When used in conjunction with a bypass line and another plugging machine, product can be re-routed around the isolated section of pipeline, while repairs, alterations, relocations, and instrument installations are made to the piping system. A typical mechanical plugging system and procedure is indicated in Figure 7-19.

Another type of mechanical plug is the ICE-PIG (patented) which were initially designed for use in the hydrostatic testing of pipelines. However these types of PIGs can also be used during pipeline cutout repairs of up to NPS 60 (60"). Its effectiveness and design address operator concerns of downtime and the effects of excessive low temperatures on the pipe, Figure 7.20.

Pig-locating transmitters are used to position the device within a pipe section or to either side of a repair area. As a small amount of residual fluid such as oil can compromise the integrity of an ice plug, thus the device incorporates a chamber purging system to remove residue and other contaminants from the freeze area.

Mechanical Vapor Plugs (V-PLUG) are another type of plug that are frequently used to isolate a segment of an offshore or onshore pipeline. These are used for a cut out to isolate explosive vapors in a pipeline.

Before use, there must be a full pipe opening for inserting the plugs and the pipeline and valving arrangement must allow the tool unobstructed travel to a receiving trap. A typical mechanical Vapor plug (V-PLUG) is shown in Figure 7.21. V- plugs use urethane sealing

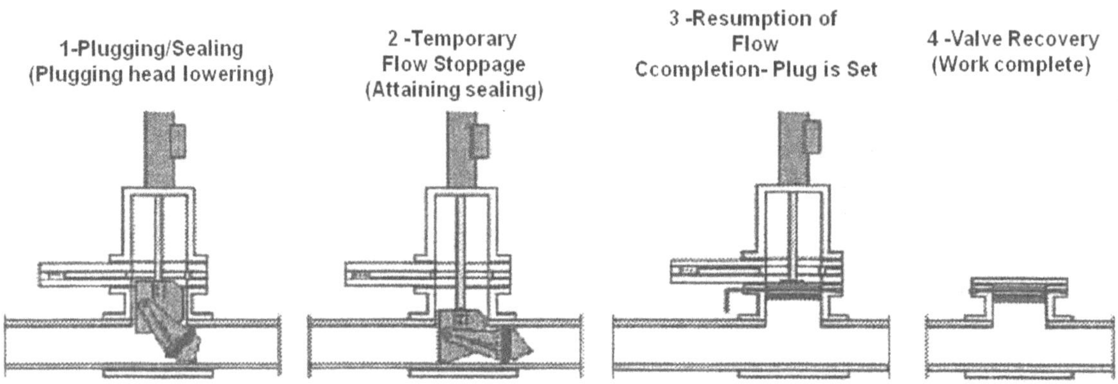

Figure 7-19. General mechanical plugging procedure machine (courtesy of T.D. Williamson)

element and a removable compressor mechanism. The operator compresses the sealing element, inserts the VPLUG into the pipe and then releases the compressor. The seal's rib design comes into direct contact with the pipe wall providing multiple seals. When required, a vent line may be attached and extended a safe distance away to release any gas build up.

Freeze Plugging

Freeze plugging (Figure 7-22) can be used to isolate pipe segments containing liquid product without the need to drain down the entire section of line. At the appropriate locations, a cooling jacket is placed around the pipe and liquid nitrogen circulated so as to freeze the contents. The frozen plug is held in place by virtue of the contraction of the pipe. Prior to

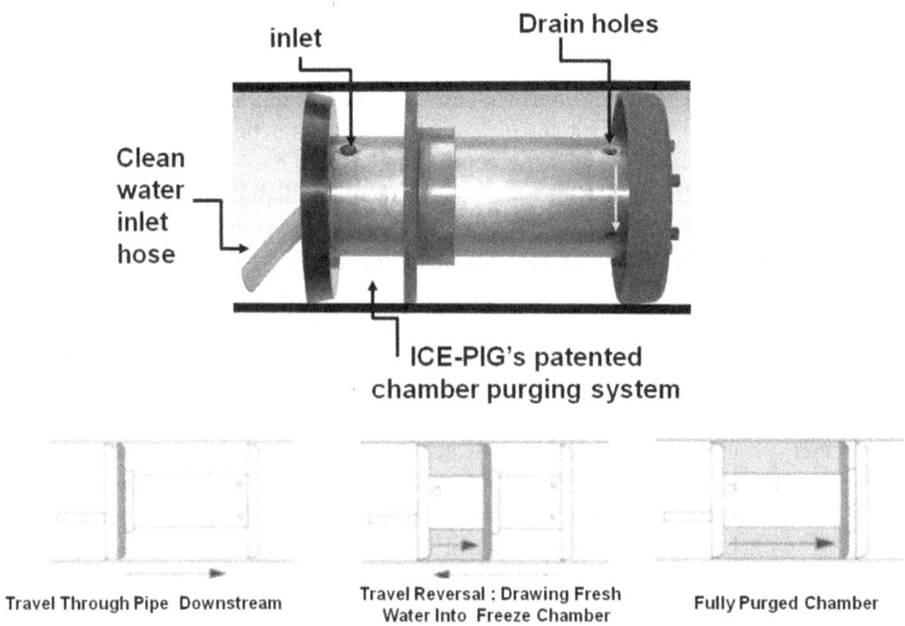

Figure 7-20. Mechanical ICE PIG Design and Plugging Process (courtesy PLP Pipeline Products & Services Inc., Edmonton, Alberta, Canada)

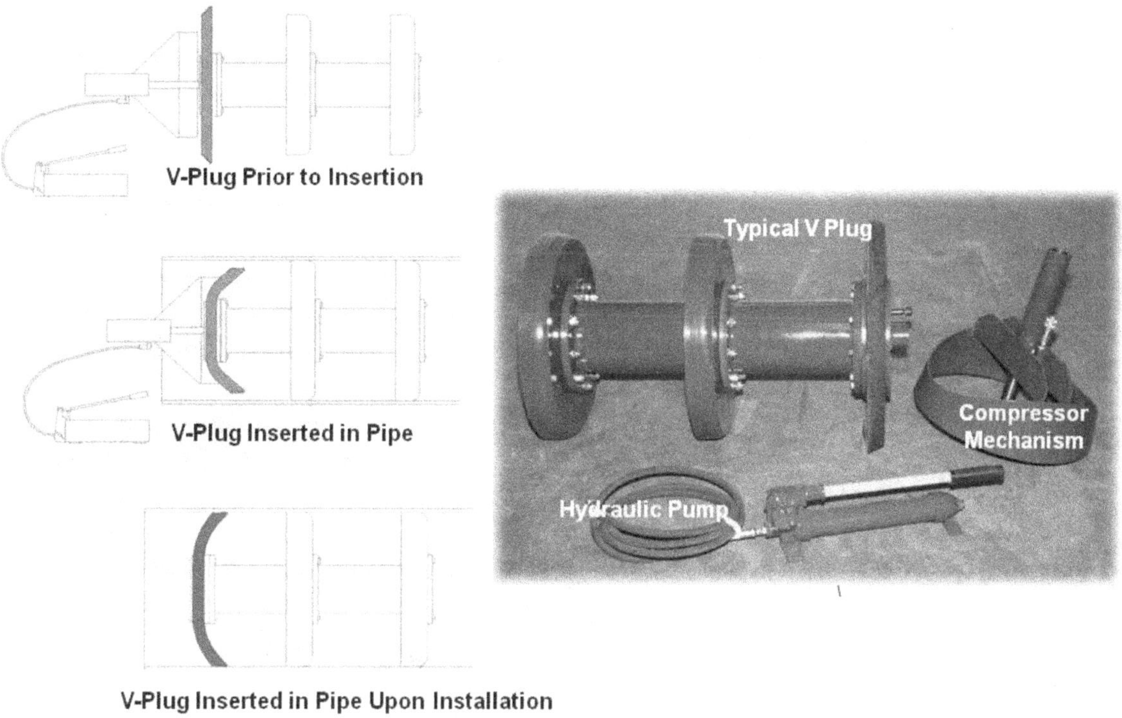

Figure 7-21. Typical Mechanical V-PLUG Operation (Courtesy PLP Pipeline Products & Services Inc. http://www.telusplanet.net/public/pipe/vplug.htm)

undertaking such a repair, the pipe should be subjected to a rigorous non destructive evaluation to ensure there is minimal cracking present.

Freeze plugging has the following challenges:

- Slow to form on large diameter pipelines or with controlled temperature
- Nonaqueous liquids can be extremely slow to freeze
- Possibility of a very small residual flow (or leak) can prevent freeze plug formation
- Pipe wall likely to be taken below transition/embrittlement temperature of pipe metal

If this technique is used, it is critical that the pipe wall temperature does not fall below a certain level/usually below the Charpy transition temperature of the pipeline material to avoid brittle fracture (Mohitpour et al. 2007). For details of application, refer to Mohitpour et al. (2010).

Mud Plugs

These plugs (Figure 7-23) are also used during pipeline and facility maintenance using bentonite or "mud-plugs" to seal the open end of pipe.

Disadvantages include the sloughing, drying, and improper installation of the mud plug. In some cases, this can and has led to the release of explosive or noxious gases (when

Figure 7-22. Freeze plugging a section of pipe

used in pipelines transporting flammable high vapor pressure fluids such as condensate) into the maintenance area. Other challenges include:

- movement due to internal pressure
- movement due to jarring of the line
- mud shrinkage due to job delays
- product contamination
- contamination of instruments
- time consuming
- costly (one time use)
- messy

Balloons or Inflatable Plugs

These are inflatable gasbags, with a polyester cover and nitrile bladder that can be inserted into the pipeline and inflated to create a plug in the pipeline segment for maintenance purposes. They have limited applications for hot work and high pressure situations.

These plugs are designed to provide a quick and simple means of performing a variety of functions and for use in a wide variety of pipe types and styles as well as job requirements. Some advantages would include:

- Will seal most slightly out of round pipes
- Negates the need to know exact pipe size

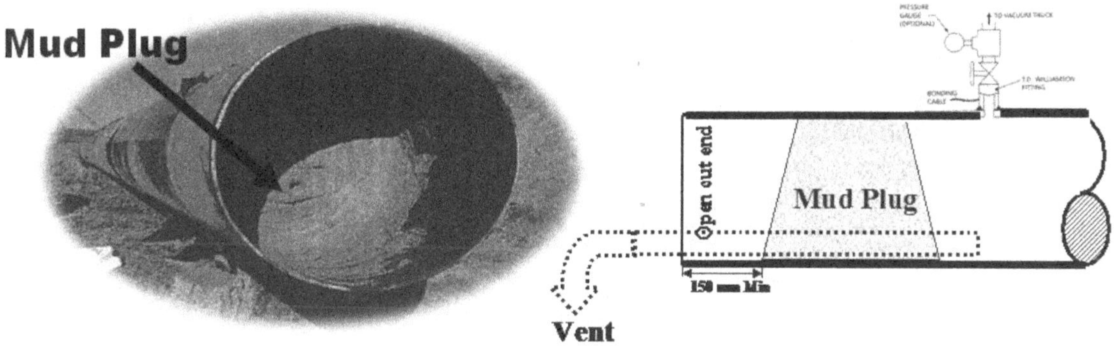

Figure 7-23. Schematic of a typical mud plug (Mohitpour et al. 2007)

- Will work in a variety of line sizes
- Excellent for standard low pressure air testing requirements

An example is the Hot Tap Pipe Plug Insertion Systems by Petersen®: a balloon plug for downstream of the hot tap and a somewhat larger balloon that inflates immediately below the hot tap (Figure 7-24).

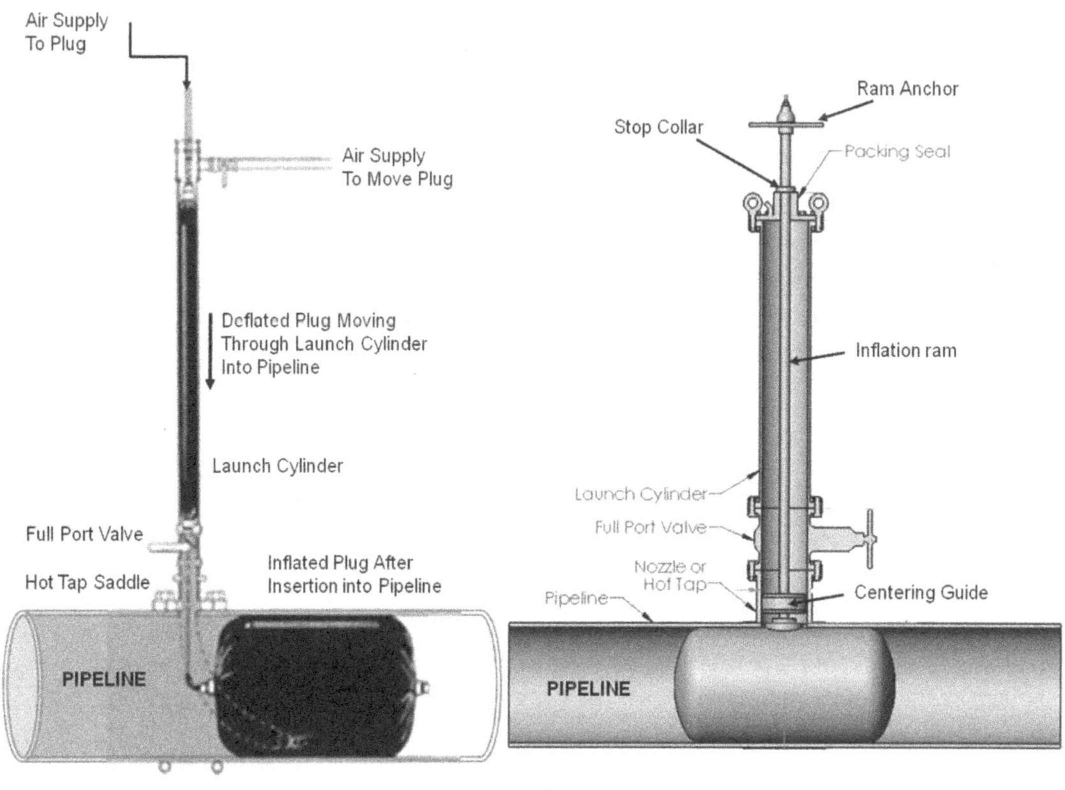

Figure 7-24. Typical Balloon/Inflatable Plug (courtesy of Petersen® , http://www.pipeplug.com)

PLANNING REPAIRS

An essential element of a successful repair is planning the repair work. This involves the following:

- Determining repair method
- Review of applicable laws, codes and regulations
- Obtaining permits
- Review of right of way and location of repairs
- Determining all adjacent utilities
- Determining the emergency and one-call system in place
- Determining the requirement for traffic control
- Protocol for limiting public access
- Review of procedure and the requirements:
  - Shut-down procedures
  - Safety-related condition and procedures
  - Drain-down / purging procedure
  - Procedures for maintain communication regarding pressure and flow
  - Safe trenching practices
  - Hot tapping procedures (if applicable)
- Tools and equipment requirement and types
- Personnel briefing requirements
- Material safety data sheet
- Leak/rupture spill/flammability hazard review
- Combustible gas indicator and testing
- Toxicity testing
- Checking confined space (trench) for oxygen
- Requirement for support and securing pipe
- Confirming wall thickness
- Requirement and procedures for turn off, lock, tag rectifiers and other appurtenances such as valves)
- Electrically bond separation points requirement
- Requirements for cold cut (unless area made safe)
- Requirement and tools for atmospheric monitoring
- Requirements for use of vapor seals and plugs (if applicable)

Other assessment includes:

1. A historical review of the pipeline (that existed) at the location of anomaly after the anomaly was present in its current state:

   - For a gas pipeline, no pressure level may be considered a historical level if it occurred more than 1 year (365 days) prior to the discovery of the anomaly.
   - For a liquid pipeline, no pressure level may be considered a historical level if it occurred more than 60 days prior to the discovery of the anomaly.
   - A previous hydrostatic test pressure can be used as a historical pressure level if it meets the time requirements.

2. Repair pressure: the pressure level at the location of the anomaly to be repaired at the time the repair is carried out

REPAIR METHOD SELECTION

The choice of which repair method to use is determined by careful consideration of all of the following,

- Codes, standards, regulations
- Owner strategies including
  - Company policies
  - Effectiveness in situation
  - Impact on service
  - Feasibility and availability
  - Cost and convenience
  - Training, quality, quality assurance
  - Safety

Compliance With Codes and Standards

The applicability of all of the above types of repair methods with respect to several pipeline codes are set out in Tables 7-4 to 7-6.

Repair Strategy

Inspections conducted per company's own integrity management plan (refer to Chapter 2) will identify anomalies/defects that must be evaluated. A number of these anomalies will require repair (API 1160 2001). However, any company policy and repair strategy must consider the long-term implication of service. The benefit of rehabilitating a pipeline through an integrity management program is to extend its useful life as is illustrated in Figure 7-25.

Pipeline rehabilitation strategies must be consistent with the requirement of the code and thus choice of repair. Figures 7-26 and 7-27 each indicate a simplified chart for strategizing the choice for repair of a leaking defect and for determining the appropriate choice for corrosion repairs.

API 1160 (2001) provides a ready reference for individuals determining the appropriate repair strategy for a certain type of defect in a certain location (seam, body, girth weld) of line pipe (Table 7-7). API 1160 stipulates that all repairs must be made with materials

TABLE 7-4. Permissibility of corrosion repair technique

<table>
<tr><th>Repair Method</th><th>ASME B31.4</th><th>ASME B31.8</th><th>CSA Z 662</th></tr>
<tr><td>Cutout</td><td colspan="3">Yes as complete cylinder</td></tr>
<tr><td>Grind out</td><td colspan="3">No</td></tr>
<tr><td>Reinforcement sleeve type A</td><td>Yes, if not leaking and not selective on an ERW seam</td><td>Yes, if not leaking</td><td>Yes, if not leaking</td></tr>
<tr><td>Pressure containing sleeve type B</td><td>Yes</td><td>Yes</td><td>Yes</td></tr>
<tr><td>Compression sleeve</td><td>Treat as type A or B</td><td>Treat as type A or B</td><td>Yes</td></tr>
<tr><td>Composite wrap</td><td>Yes, if not leaking and $d/t < 0.80$ and not selective on an ERW seam</td><td>No</td><td>Yes, if not leaking and $d/t < 0.80$</td></tr>
<tr><td>Mechanical clamp</td><td>Yes</td><td>Yes</td><td>No</td></tr>
<tr><td>Hot tap or fitting</td><td>Yes if fitting size $<$ NPS 3</td><td>Yes</td><td>Yes</td></tr>
<tr><td>Weld deposition</td><td>Yes</td><td>Yes</td><td>External corrosion only</td></tr>
</table>

TABLE 7-5. Permissibility of crack repair technique

Repair Method	ASME B31.4	ASME B31.8	CSA Z 662
Cutout		Yes as complete cylinder	
Grind out	Yes if grinding depth <40% wall thickness		Yes, per corrosion limit
Reinforcement sleeve type A	Yes, after grinding out and grinding depth <40% wall thickness	Yes	Yes, after grinding out
Pressure containing sleeve type B		Yes	
Compression sleeve	Treat as type A or B	Treat as type A or B	Yes, after grinding out
Composite wrap	Yes, after grinding out and grinding depth < 40% wall thickness	No	Yes, after grinding out
Mechanical clamp	Yes	No	Yes
Hot tap or fitting	Yes		Yes
Weld deposition		No	

that meet or exceed the MOP of the impacted portion of the pipeline and comply with applicable regulations.

The following notes are applicable to Table 7-7:

1. Pipe replacement is always an effective repair.
2. Use of weld deposition requires a minimum pipe wall thickness and control of welding parameters to prevent burn through. This generally prevents use of this technique in pipes with external metal loss >80% wall thickness except in heavy wall pipelines. API currently does not recommend use of this technique for wall thickness <0.181 in. (4.6 mm).
3. Metallic sleeves both bolted and weld-on are available for bends and fittings.
4. Special techniques utilizing multiple overlapping sleeves are required for bends.
5. A hardenable incompressible filler shall be used to fill the annular space between the dent and the sleeve.
6. Mechanical damage in a dent must be removed by grinding prior to installation of the sleeve.

TABLE 7-6. Permissibility of mechanical damage repair technique

Repair Method	ASME B31.4	ASME B31.8	CSA Z 662
Cutout		Yes as complete cylinder	
Grind out	Yes if grinding depth $d < 12.5\%$ wall thickness and d/diameter < 6%	Yes if grinding depth $d < 10\%$ wall thickness any length $d < 40\%$ wall thickness limited length and d/diameter <4%	Yes
Reinforcement sleeve type A	Yes, after removal by grinding and use of filler	Only with use of filler	Yes, after removal by grinding
Pressure containing sleeve type B		Yes	
Compression sleeve	Treat as type A or B	Treat as type A or B	Yes
Composite wrap	Yes, after removal by grinding and use of filler	Only if design is proven	Yes, after removal by grinding and not on a dent
Mechanical clamp	Yes	No	No
Hot tap or fitting		No	
Weld deposition		No	

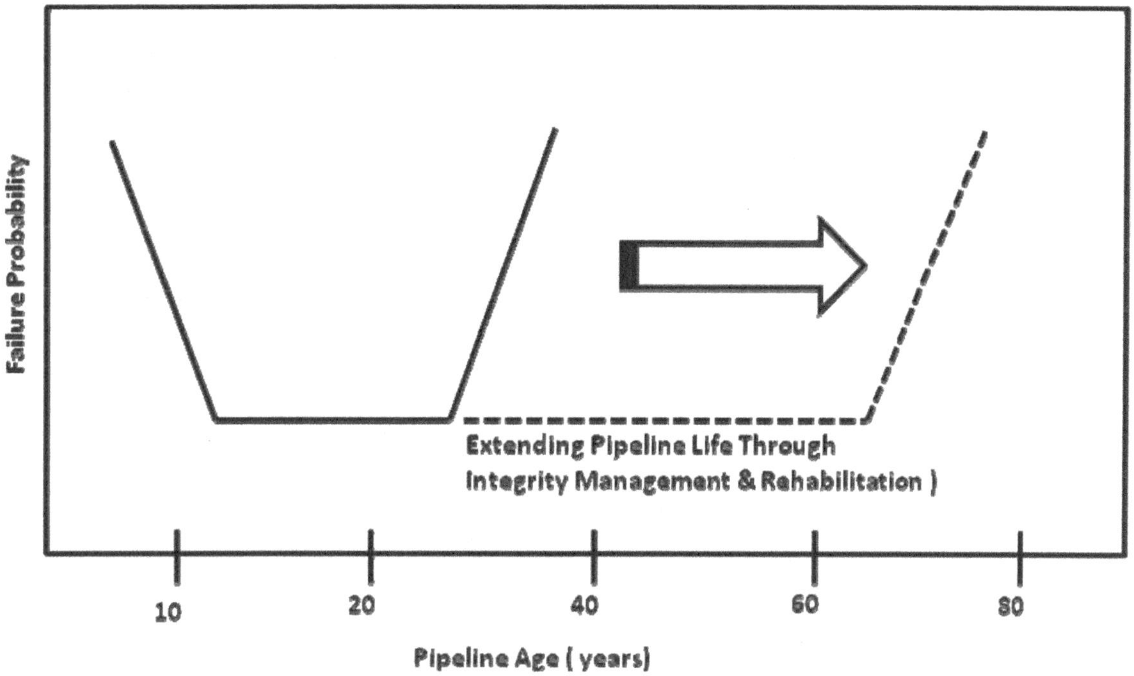

Figure 7-25. Extending Pipeline Useful Life Through Integrity Management and Rehabilitation

7. Only certain types of composite repairs when used with an incompressible filler are adequate for the repair of dents and such repairs must show by reliable engineering tests and analysis to permanently restore the serviceability of the line pipe.
8. Conservative industry practice is to limit the use of Type A and composite sleeves to external metal loss 80% of nominal wall. For the case of external metal loss

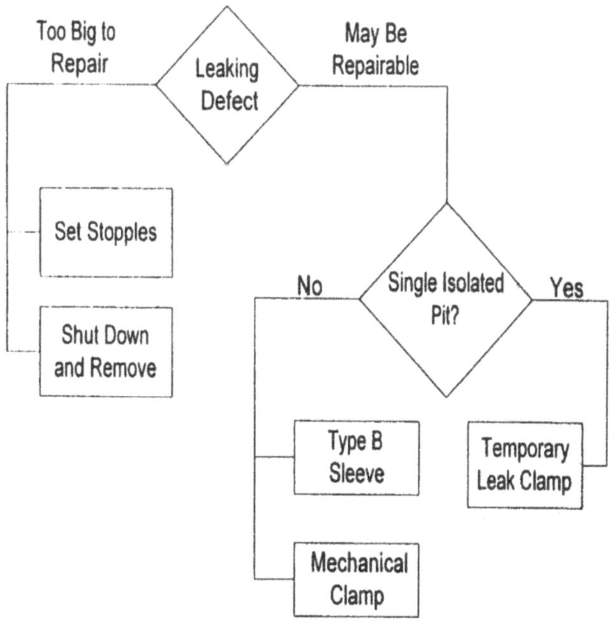

Figure 7-26. Choice of repair chart

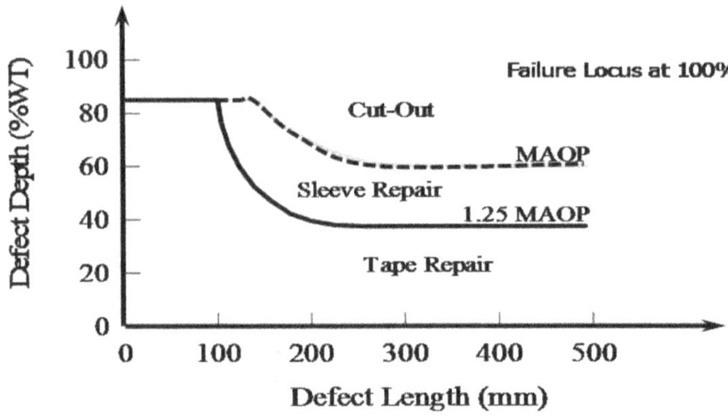

Figure 7-27. A typical intervention plot for repair of corrosion defects

>80%, a minimum wall must be present for Type A sleeves and composite reinforcement repair techniques. API currently recommends a minimum wall of 50 mils, precise NDT of pit depth, no internal corrosion, and sound engineering practice.

9. Other repair methods may be used provided they are based on sound engineering practice.

TABLE 7-7. API 1160 summary of commonly used permanent pipeline repairs (see Note 9; reproduced from API 1160 (2001) with permission)

Anomalies		Primary Repair Strategies (1)				
		Weld Metal Deposition (2)	Type A Sleeve	Type B Sleeve	Composite Reinforcement	Hot Tap
External metal loss ≤80% w.t.	Pipe seam	Yes	Yes	Yes	Yes	No
	Girth weld	Yes	Yes	Yes	Yes	No
	Pipe body	Yes	Yes	Yes	Yes	Yes
	Bend	Yes	Yes (3)	Yes (3)	Yes (4)	Yes
Internal metal loss 80% w.t.	Pipe seam	No	No	Yes	No	No
	Girth weld	No	No	Yes	No	No
	Pipe body	No	No	Yes	No	Yes
	Bend	No	No (3)	Yes (3)	No	Yes
External metal loss >80% w.t.	Pipe seam	Yes	No (8)	Yes	No (8)	No
	Girth weld	Yes	No (8)	Yes	No (8)	No
	Pipe body	Yes	No (8)	Yes	No (8)	Yes
	Bend	Yes	No (8)	Yes (3)	No (8)	Yes
Internal metal loss >80% w.t.	Pipe seam	No	No	Yes	No	No
	Girth weld	No	No	Yes	No	No
	Pipe body	No	No	Yes	No	Yes
	Bend	No	No (3)	Yes (3)	No	Yes
Leak, cracks, arc burns and girth weld flaws (12)	Pipe seam	No	No	Yes	No	No
	Girth weld	No	No	Yes	No	No
	Pipe body	No	No	Yes	No	No (10)
	Bend	No	No	Yes (3)	No	No (10)
	Thread collar	No	No	Not practical	No	No
Dents with stress concentrators	Pipe seam	No	Yes (5), (6)	Yes (6)	No	No
	Girth weld	No	Yes (5), (6)	Yes (6)	No	No
	Pipe body	No	Yes (5), (6)	Yes (6)	No	Yes (11)
	Bend	No	Yes (3), (5), (6)	Yes (3), (6)	No	Yes (11)
Plain dents	Pipe seam	No	Yes (5)	Yes	No (7)	No
	Girth weld	No	Yes (5)	Yes	No (7)	No
	Pipe body	No	Yes (5)	Yes	No (7)	Yes (11)
	Bend	No	Yes (3), (5)	Yes (3)	No	Yes (11)

10. Cracks that are not leaking can be hot tapped to remove the crack.
11. If entire dent can be removed, arc burns and girth-weld flaws can be repaired by grinding out the defect and/or Type A or B sleeves as long as the repairs are based on sound engineering practice.

Review of Table 7-7 and the above notes related to API 1160 will indicate the following

- It is important to review the defect type and size before deciding the choice of the appropriate repair methods.
- Repair methods can include repair or replacement.
- Defect types can include external and internal defects with greater than 80% metal loss or less than 80% metal loss.
- Compare the selected repair methods and the appropriate pressure testing technique with industry codes and standards for acceptable repairs and tests for particular types of defects, including
  - US DOT CFR 49 Parts 192 and195 Subparts C and D
  - ASME B31.4 Code, Section 451.6.2
  - ASME B31.8 Code
  - API RP 1156 (for engineering critical assessment, if dent involved)
  - API Standard 1160 and
  - API Standard 1107

ASME B31.4 and B31.8 both address testing of repairs to high-pressure pipelines covering the pressure testing of replacement pipe sections and examination of repair welds.

As indicated in Chapter 2, any integrity management program must include applicable mitigation activities to prevent, detect, and minimize the consequences of unintended releases. Mitigation activities do not necessarily require justification through additional ILI data. Mitigative actions can be identified during normal pipeline operation, during the initial risk assessment, during implementation of the baseline inspection plan, or during subsequent testing (API 1160 2001).

Steps involved for the defect assessment and the determination and conduct of the appropriate repair techniques for rehabilitating a pipeline segment are indicated as follows:

- External corrosion, refer to Figure 7-28
- Internal corrosion, refer to Figure 7-29
- Non-indented longitudinal cracks, refer to Figure 7-30
- Gouges and dents, refer to Figure 7-31
- Girth weld defects, refer to Figure 7-32

ISSUES RELATED TO PIPELINE REPAIR

Safety

Generally, to examine a defect, the pipeline must be excavated to make a proper assessment. Until the pipe segment affected with defects is excavated, exposed, and cleaned, one has no idea of the safety in the ditch and the excavation.

It is therefore very important that extreme precautions are taken while the excavation is taking place and thereafter during the examination and repair to assure personnel and public safety.

Pipeline codes, regulations, and industry publications do not include any guidance on what factors should be considered to determine how much, if any, the pressure should be

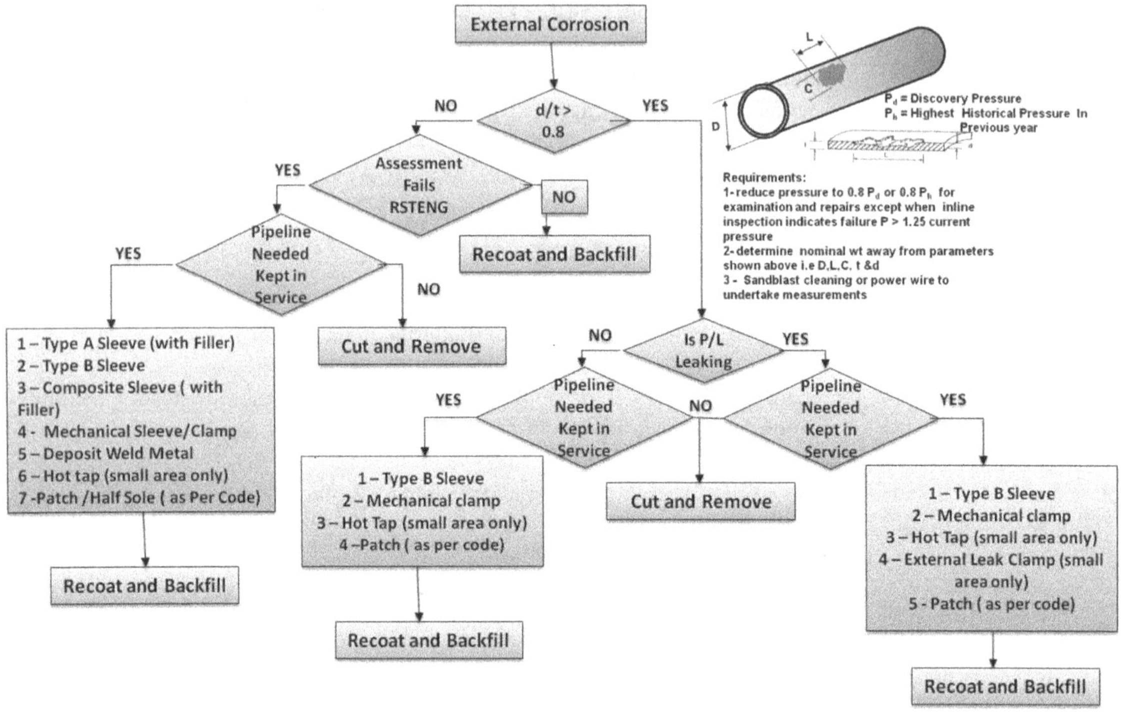

Figure 7-28. Assessment of Defect and Determination of Repair Techniques (for External Corrosion), After: Kiefner and Associates, undated

reduced from operating levels during excavation activities. Codes such as CSA Standard Z662 generally require that:

- "Pipeline pressures shall be at safe levels when work is being performed on in-service pipelines" and that
- "Excavation of piping suspected of containing defects and, if required, the subsequent permanent or temporary repair of such piping shall be performed after the piping is depressurized as necessary to an operating pressure that is considered to be safe for the proposed work."

While some companies reduce the operating pressure when excavating pipe for integrity evaluation of any pipeline IL) indications, leaks, or possible recent damage, other companies only reduce the pressure if the risk reduction achieved is considered to justify the resulting disruption in service and damage to customers (Coote and Keith 2004). However, the industry precaution is to reduce the line pressure to about 80% or less of the highest pressure seen over the previous month of operation.

When collecting field data from a defect, generally industry engages a qualified and experienced technician for the purpose and considers the following to assure safety issues as well requirement for proper repair assessment:

- Product in the line segment
- Pipeline segment mechanical information
  - o Grade, wall thickness, chemical composition
  - o Operating pressure

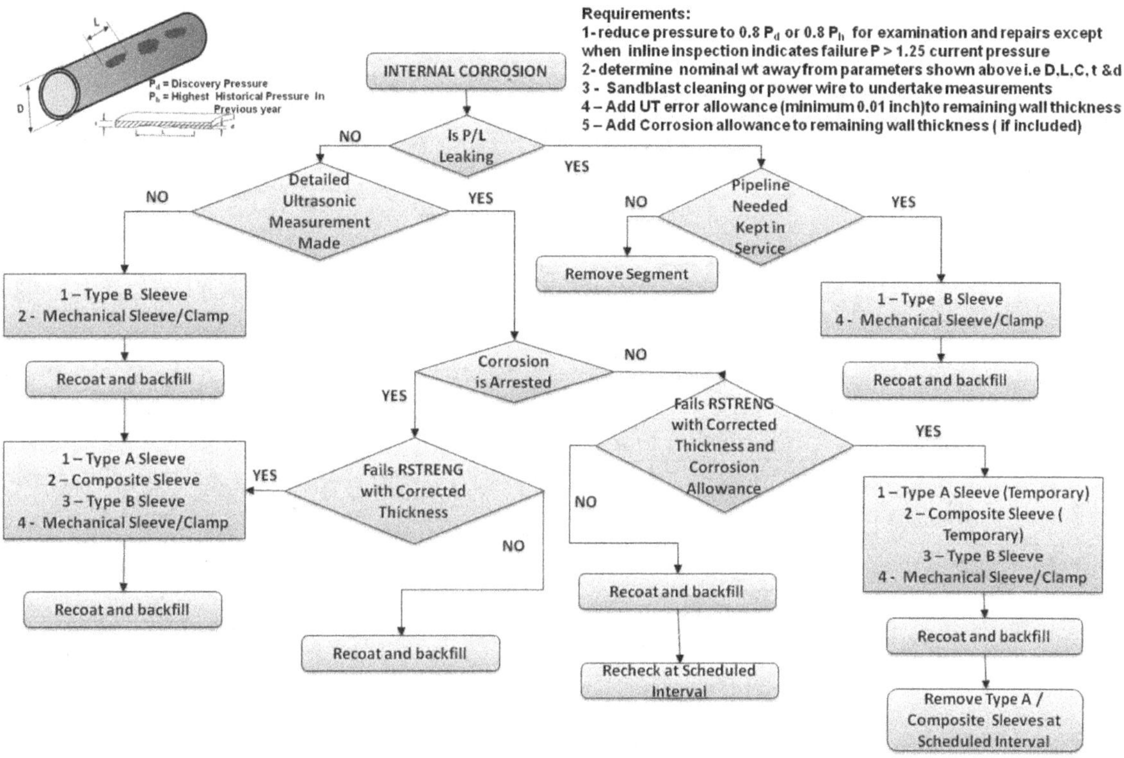

Figure 7-29. Assessment of Defect and Determination of Repair Techniques (for Internal Corrosion), After: Kiefner and Associates, undated

- Likely type of defect
- Assessment of line pressure causing rupture
- Extent of pressure reduction
- Type of repair and protocol for repair
- Repair schedule
- Purge and drain requirement, not always possible to repair a pipeline without taking it out of service
- Crew requirement and size
- Excavation method
- Site preparation work including any hand probing before use of mechanical excavation
- Extent of and condition for and after excavation
  - o Hand dug/dozer mechanical excavation
  - o Any mechanical excavator vibration effects (due to use of dozer or other devices)

  Note: It may be noted that the excavation procedures used by the industry include restrictions on the use of machine excavation near the pipe.
- Permit requirements
- Procedure requirement
- Ditch Safety
  - o Ingress and egress
  - o Signage and
  - o Potential for fire

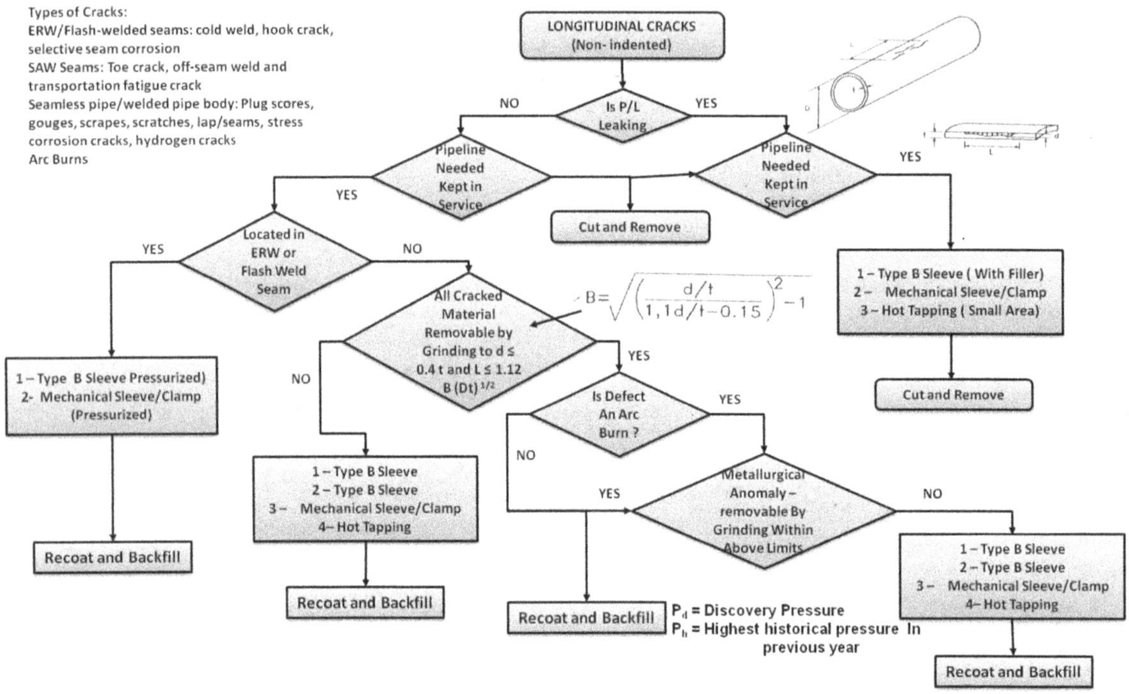

Figure 7-30. Assessment of Defect and Determination of Repair Techniques (For Non indented longitudinal cracks), After: Kiefner and Associates, undated

- First aide and medical emergency requirements
- Personal protection and self breathing apparatus
- Requirement for inspection and NDT
- Requirement for reporting and work completion

Figure 7-33 illustrates an example of a safe excavation site.

However, any repair assessment must be accompanied by a repair protocol (similar to Figures 7-28 to 7-32) to implement repairs that are commensurate with each defect as per code requirement and best engineering practice. Figure 7-34 indicates a simplified protocol (Smyth 2004).

Welding and Hot Tapping Issues (Line Under Pressure)

When welding sleeve repair or weld deposition buildup (Figure 7-35), there is a risk of

- burn through
- unstable decomposition of the flowing fluid and
- hydrogen cracking of metal (pipe and weld)

Burn through is a term commonly used by the pipeline industry to describe excessive melt through or a hole in the pipe caused by a root pass.

Factors affecting burn through are:

- Pipe wall thickness
- Welding parameters and process

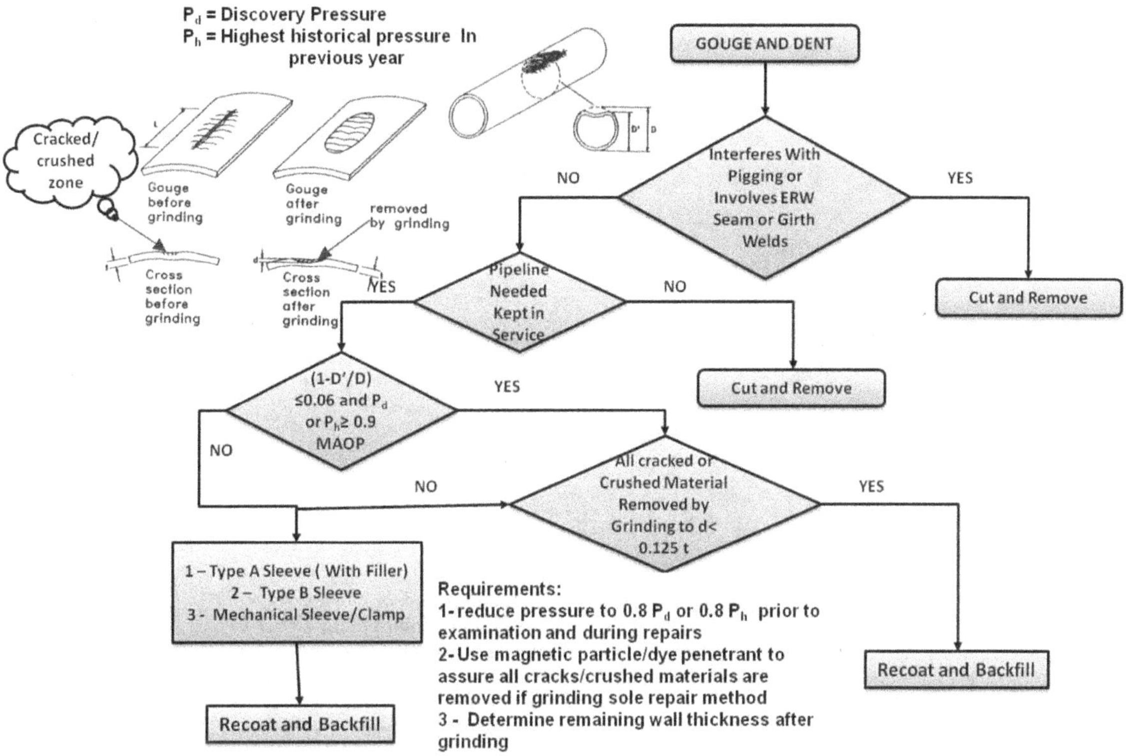

Figure 7-31. Assessment of Defect and Determination of Repair Techniques (Gouges and Dents), After: Kiefner and Associates, undated

- Heat input (HI), current level
- Pipeline operating parameters
- Thermal properties and temperature of contents
- Flow parameters
- Pressure (for gases), flow rate
- HI from weld versus heat removal by contents

Burn through may occur when welding onto a pressurized pipe if the un-melted area beneath the weld pool is not strong enough or does not have enough wall thickness to contain the flowing pressure. However, the greater the wall thickness of the pipe, the less the risk of burn through.

In practice, pipe wall thickness of as low as 4 mm (0.156 in.) have been successfully welded in service by careful selection of welding parameters and use of appropriate welding procedures. Normal accepted practical minimum is 4.8 mm (0.188 in.). The weld procedure and welding qualifications must include the following provisions for qualification and testing:

- highest rate of heat dissipation
- restraint level
- maximum Carbon Equivalent, CE of welded material

Unstable decomposition of the flowing hydrocarbon product can cause violent reactions when weld heat is transferred to the flowing product, which is under pressure. This

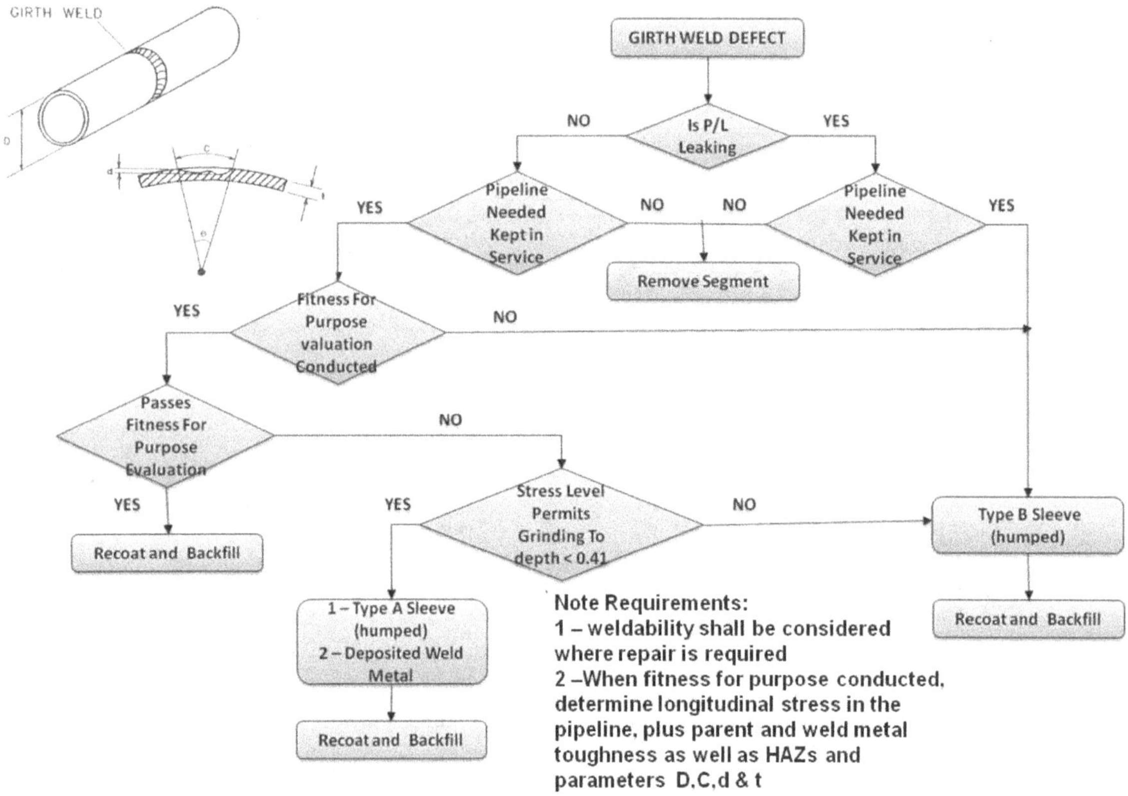

Figure 7-32. Assessment of Defect and Determination Repair of Techniques (For Girth Weld Defects), After: Kiefner and Associates, undated

can be avoided or risk lowered by limiting the internal wall temperature not to exceed the critical (self-ignition) temperature of the flowing products.

However, there is a risk of hydrogen cracking in weld situations if the flowing product increases the rate of heat dissipation from the weld, which tends to quickly remove heat from the pipe wall. The fast weld cooling rates combined with the high-CE of the pipeline (which tends to be higher for older pipelines) results in the development of hard, crack-susceptible weld microstructures.

The development of a hard microstructure makes repair welds on in-service pipelines particularly susceptible to hydrogen cracking. For hydrogen cracking to occur,

Figure 7-33. Typical safe excavation

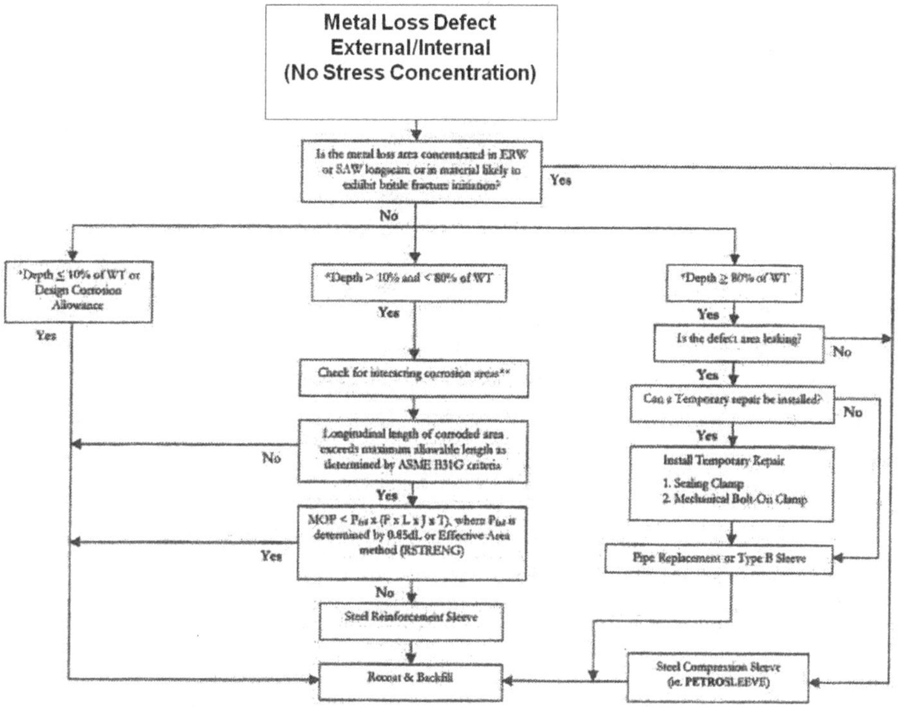

Figure 7-34. Typical defect assessment repair protocol (Smyth 2004)

three primary independent conditions must be satisfied simultaneously (LaMorte et al. 2007):

- Hydrogen in the weld
- A crack-susceptible weld microstructure
- Tensile stress acting on the weld

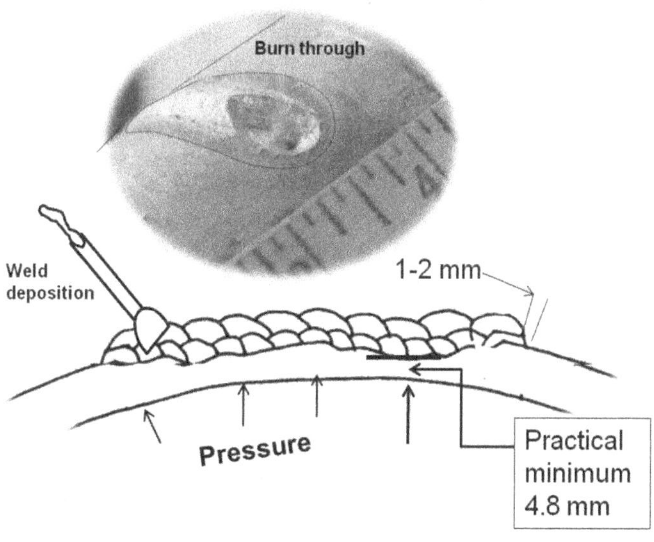

Figure 7-35. Burn through and typical weld deposition buildup

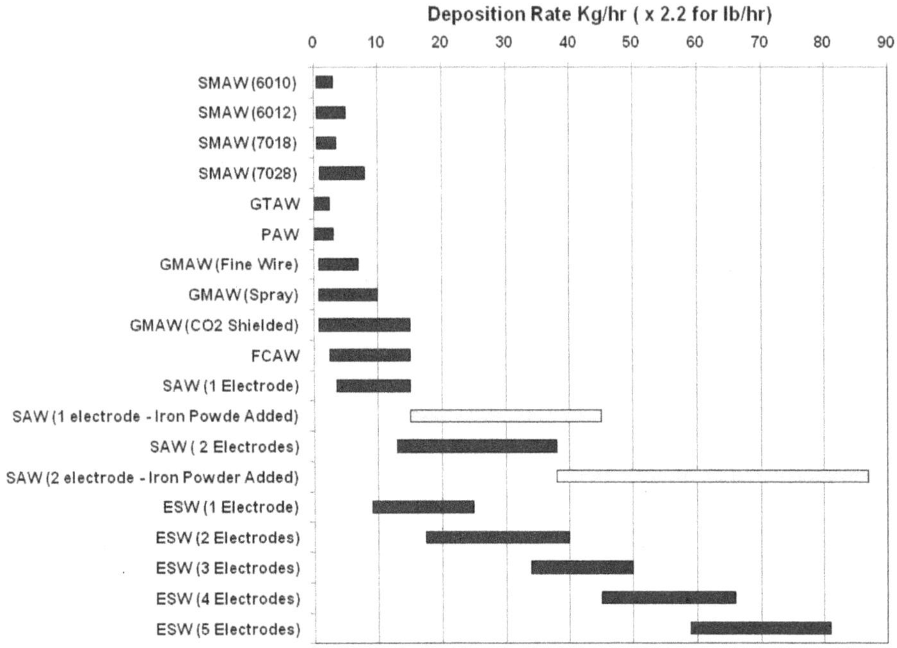

Figure 7-36. Deposition rates for typical pipeline welding processes — 100% duty cycle (LaMorte et al. 2007)

To prevent hydrogen cracking, at least one of these three conditions must be eliminated. Due to the nature of weld metal solidification, tensile stresses will always be present and must be assumed.

As a result, the first step taken toward avoiding hydrogen cracking of in-service welds is to minimize the hydrogen level by using low-hydrogen electrodes or a low-hydrogen welding process. Since low hydrogen levels cannot always be guaranteed, procedures that minimize the formation of crack-susceptible microstructures are also used.

The most commonly used options for preventing hydrogen cracking of in-service welds are: the specification of a minimum-required HI level and/or the use of a temper-bead deposition sequence or temper bead procedure. Weld deposition repair entails the use of a temper bead procedure because wall loss that necessitates a repair normally requires several weld layers to replace the wall loss (LaMorte et al. 2007).

Figure 7-36 contains a summary of deposition rates for several arc welding processes based on 100% duty cycle (without considering the effects of out-of-position welding or automation; LaMorte et al. 2007). The following welding processes may be noted:

- SMAW Shield Metal Arc Welding
- GTAW Gas Tungsten Arc Welding
- PAW Plasma Arc Welding
- GMAW Gas Metal Arc Welding
- FCAW Flux Cored Arc Welding
- SAW Submerged Arc Welding
- ESW Electro Slag Welding

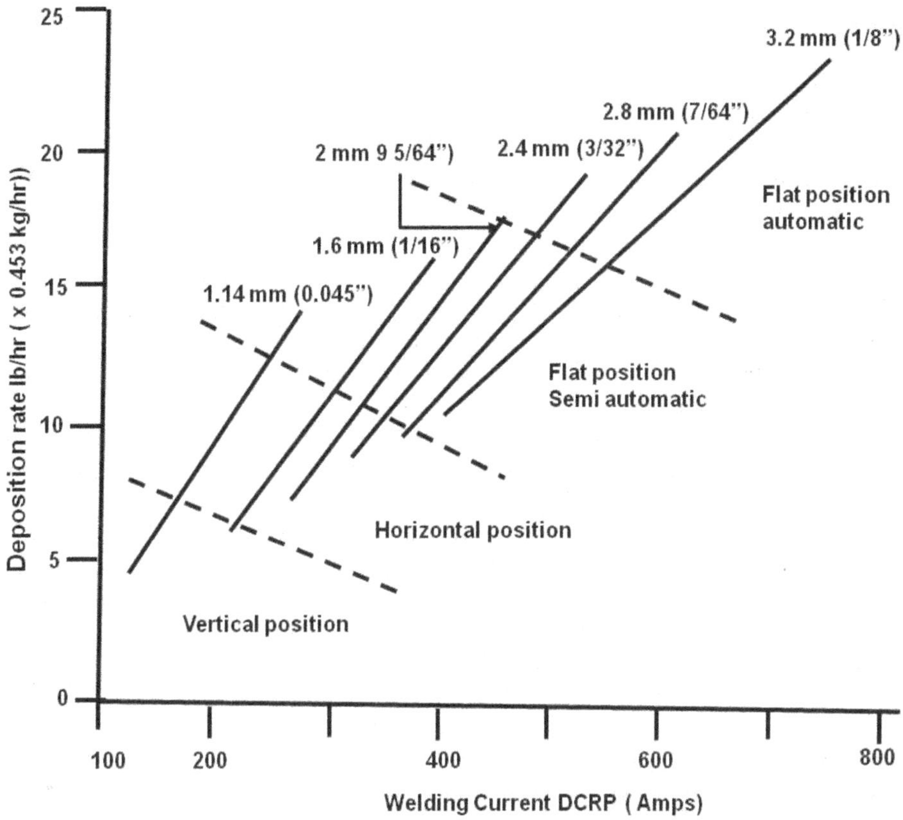

Figure 7-37. Deposition rates and current requirement typical pipeline welding processes (after LaMorte et al. 2007)

Figure 7-37 contains typical deposition rates for various GMAW filler metal diameters with the appropriate welding positions and amperage ranges. Current levels appropriate for use in semiautomatic and automatic processes are also noted in this figure.

HEAT INPUT (HI) EXAMPLE

HI ~ (current × voltage)/travel speed

HI = (Amps × Volts × 60)

(kJ/in.) (Travel Speed [in./min] × 1000)

At 110 A to 115 A ($^1/_8$-in.-diameter electrodes) and 22 V, the following HIs are required for various welding travel speed:

- 10 ipm => 15 kJ/in.
- 6 ipm => 25 kJ/in.
- 3.75 ipm => 40 kJ/in.

REFERENCES

American Petroleum Institute (API), 2009, *API 5L Specification for Line Pipe*, 44th ed., American Petroleum Institute, October 1, 2007, 176 pp.

American Petroleum Institute (API), 1999, *API RP 1156 — Effects of Smooth and Rock Dents on Liquid Petroleum Pipelines*, American Petroleum Institute, November 1, 1997, 242 pp.

American Petroleum Institute (API), 2001, *API Std 1160 — Managing System Integrity for Hazardous Liquid Pipelines*, American Petroleum Institute, November 1, 2001, 86 pp.

American Petroleum Institute (API), 2003, *API RP 2201 — Safe Hot Tapping Practices in the Petroleum & Petrochemical Industries*, 5th ed., American Petroleum Institute, July 1, 2003, 27 pp.

Coote, R. I., and Keith, J. K., 2004, "Pressure Reduction and Pipeline Excavation," Proceedings of ASME International Pipeline Conference(IPC), Calgary, Alberta, Canada, October 4–8.

Hewing M., 1999, "Differing Repair Methods — Options and Benefits" Proc. 2nd Latin American Pipeline Integrity Conf. Puerta La Cruz, Venezuela, Aug.

Keifner and Associates Undated "Workshop on Pipeline Reliability Assessment" Worthington, Ohio, USA.

La-Morte, C. R., Boring, M., and Porter, N., 2007, "Advanced Welding Repair and Remediation Methods," PRCI Report PR-185-04501, July 30.

Morton, A., 2009, "Wet-Applied Wrap Helps Restore Pipeline Systems," Pipeline and Gas Journal, July, http://www.tdwilliamson.com/en/MediaLibrary/Documents/resq_wrap_helps_restore_pipeline_systems.pdf.

Mohitpour, M., Golshan, H., and Murray, 2007, *Pipeline Design and Construction — A Practical Approach*, 3rd ed., ASME Press, New York.

Mohitpour, M., Van Hardeveld T., Szabo, J., and Pederson W., 2010, *Pipeline Operation and Maintenance — A Practical Approach*, 2nd ed., ASME Press, New York.

Smyth, R. J., 1998, "Repairing Pipe Defects (Cracking, Arc Burns, Corrosion, Dents) Without Operational Outages Using the Petrosleeve Compression Sleeve Repair Technique," ASME IPC International Pipeline Conference (IPC 1998) Calgary, Alberta, Canada, June 7–11.

Smyth, R. J., 2003, "Evaluation of the Ability of the Steel Compression Reinforcement Sleeve to Repair Active Internal Corrosion Defects," Paper No. IPC04-0044, Proceedings of ASME 5th International Pipeline Conference IPC2004 October, Calgary, Alberta, Canada.

Smyth, R. J., 2004, "Recommended Practice for Pipeline Repair," ASME International Pipeline Conference (IPC), Tutorial, Calgary, Alberta, Canada. Sept.

Smyth, R. J, and Cermak, L. E., 2006, "Testing to Determine Whether a PetroSleeve Can Be Used for an Electrical Connection to a Pipe for AC Mitigation," Proceedings of ASME 6th International Pipeline Conference IPC2006, Calgary, Alberta, Canada, September.

Chapter 8

INTEGRITY TECHNICAL SUPPORT

INTRODUCTION

The responsibility for providing technical support in managing the integrity of a pipeline system, including its facilities, is usually assigned to a group dedicated to that task. The size and structure of the group will vary depending on the amount and diverse nature of the company's pipeline facilities and the degree to which it contracts support from outside contractors and consultants. The principal activities and responsibilities of this functional group, however, will remain fairly consistent, irrespective of the previous considerations and will comprise the following broad support tasks:

1. Develop and manage the IMP (protecting existing/newly added facilities), in cooperation with the operations and maintenance functions
2. Provide ongoing planning and project execution technical support to the IMP
3. Perform engineering assessments with respect to the rehabilitation/re-use/change of service of existing facilities
4. Conduct incident Investigations
5. Miscellaneous activities

In this chapter, we will discuss each of the above listed tasks. Once the integrity program has been established and implemented within the company, ongoing tasks will fall into one of two categories, scheduled (planned) work or unscheduled activity. Scheduled work covers a broad range of routine activities from planning the frequency of right-of-way patrols, through ongoing monitoring of unstable ground conditions, to planning and coordinating the timing and conduct of ILI and in service hydrotesting. These aspects are addressed in the early sections of the chapter. Unscheduled work arises as a consequence of an integrity related issue such as a slope failure, or a loss of containment, which warrants some form of assessment or investigation, and this investigative process is described in Incident/Failure Investigation.

SUPPORTING THE IMP

In Chapter 2, a detailed description of the elements comprising an IMP and its connectivity with pipelines operations and maintenance was presented. A key part of such a program is the designation of roles and responsibilities for its various facets. The formal leader of the integrity technical support function will be designated as being responsible for the development and continuous improvement of the content and protocols within the IMP. He will review and approve assessment methods, integrity plans and budgets, data management needs, and changes to any technical procedures. His responsibilities will also include receiving and reviewing changes to operating procedures and ensuring that these are communicated through the Management of Change Process so that the requisite engineering assessment and planning procedures are updated. Depending on the organizational structure, he will be assisted in the above tasks by several team leaders who, in turn, are supported by Integrity specialists. Collectively, this group will be responsible for periodically performing a quality assurance review of the results and procedures of the IMP and report its findings to an officer of the company, generally the Vice President Engineering. The IMP review group meetings occur at least once a year or more frequently if the need arises. The group will also liaise with the preventive maintenance team and the regional maintenance managers to develop work plans and the associated resource planning and budget preparation.

Developing the Program

The technical support group is responsible for developing and administering the every day conduct of the integrity program designed to protect existing facilities. This involves all of the elements contained in Figure 8-1, i.e., identifying threats and then matching them with the appropriate response, balancing risk with the company's operating priorities. Chapter 2 contains a detailed description of an IMP, so it is unnecessary to repeat it here except to draw attention to the highly linked and closed loop, continuous improvement, nature of the various tasks shown in Figure 8-1. To date, very few companies have developed a good set of performance indicators, most being measures of activity rather than outcomes. As programs begin to mature, however, and enough relevant data is collected, trends will begin to emerge enabling meaningful trending to take place and leading indicators to be developed.

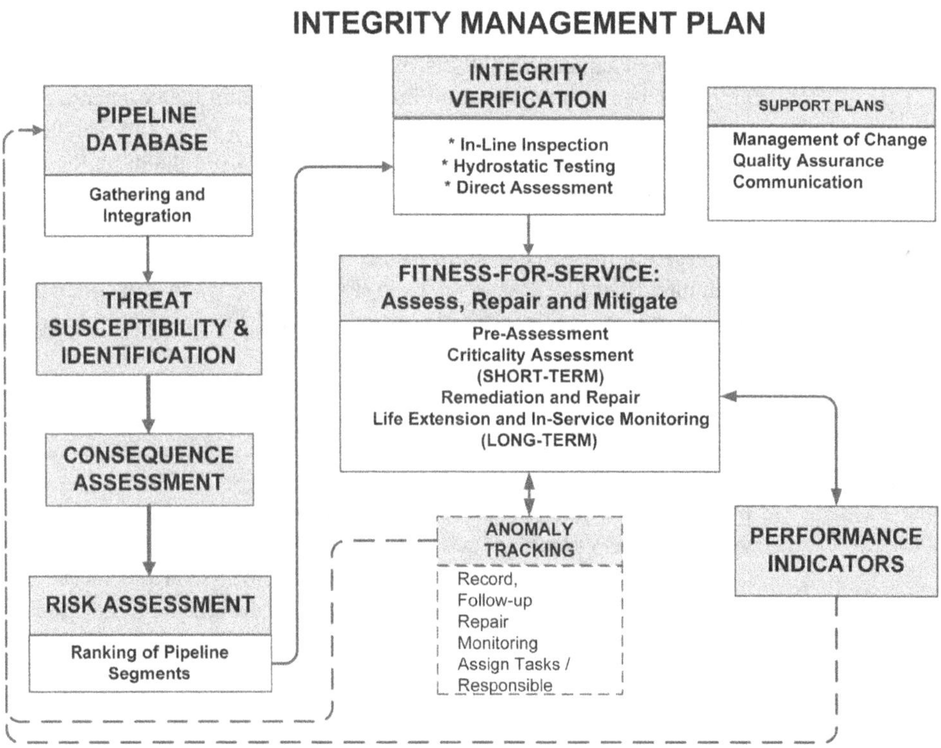

Figure 8-1. The tasks contained within an integrity management plan

Executing the Program

The technical support group, at a tactical level, is also responsible for the execution of the IMP, planning aerial patrol frequencies, leak detection activities, organizing and maintaining records of ILI inspections, excavations, defect assessments, and working with maintenance teams to plan repairs. In doing so, they will use a risk-based approach to inspection and maintenance planning such as is described in detail in Chapter 10.

Specifically, the group will develop and administer the pipeline inspection programs based on:

- the age of the pipeline system
- the elapsed time between inspections
- results from previous inspections
- excavation results
- results from CP surveys
- changes in population density
- external coating type
- soil corrosivity
- the nature of the flowing product

Where the product being transported has a propensity to corrode the internal surface of the pipe, the group will also be responsible for recommending prevention and repair measures for internal corrosion including the use of inhibitors and their concentrations and application methods.

Right-of-Way Surveillance

As part of the operation and maintenance of a pipeline system, maintenance staff will be assisted by the Integrity technical support group in conducting various activities along the pipeline. Among these activities will be right-of-way surveillance, airborne leak detection, planned integrity digs, line relocating, safety/emergency response exercises, and failure investigations. While the primary purpose of right-of-way surveillance is to check for unauthorized encroachments, which could lead to external damage, it also entails monitoring exposure of the pipeline at river crossings due to erosion, identifying issues related to forest fires, and monitoring potential slope movements. Right-of-way surveys are also known as line patrols and can be airborne or ground based. The frequency of such patrols is usually determined by the integrity technical support group and will form part of a larger pipeline health monitoring program. Seasonal variations in patrol frequency will be planned to account for predictable events such as the effect of swollen rivers due to heavy rains, or the occurrence of forest fires caused by lightning strikes. Often several pipelines will share a common right of way, and it is common in such cases for companies to pool their resources and share line patrol duties and costs.

Surficial erosion on portions of the right of way and washouts at creek and river crossings (see Figure 8-2) are further examined by ground patrols. Terrain monitoring should extend beyond the confines of the right of way since it is possible that deep seated soil movement could initiate some distance away from the pipe center line yet affect pipe soil interaction. Later in this section, we will examine one such circumstance as part of a more thorough discussion of ground monitoring.

The process of land restoration after pipeline ground disturbance will often involve reseeding and that activity, plus the ground warming conditions due to the transport of fluid in the pipeline, will change the color of vegetation, clearly demarcating the center line of the pipeline from its surroundings. This makes it easily visible from the air except of course

Figure 8-2. Erosion of a river bank exposing pipe (courtesy of Rafael Mora, NEB)

when there is snow cover or no vegetation. Many pipeline operators today have mapped the coordinates of their pipelines and recorded this information in a GIS system that can be linked to the navigational system of the airborne line patrol enabling the pilot to track the pipeline under any ground conditions. Pilots are also trained observers, and they can readily recognize unusual ground conditions such as staining caused by a mild escape of leaking oil or the discoloration of vegetation due to a small leak in a natural gas pipeline. Small gas leaks may sometimes cause local ground freezing due to cooling as the leaking gas quickly decompresses.

Aerial patrols looking for abnormal surface conditions are conducted of:

1. the entire ROW and
2. the land adjacent to the ROW (liquid pipelines) or
3. the entire ROW and
4. the land adjacent to and within 1000 ft. of the ROW (gas pipelines) usually at 14-day intervals but a minimum of 26 times per calendar year.

Abnormal surface conditions include:

- visible crude oil and/or vapor clouds
- indications of leaks, excavation or dredging
- encroachments
- unauthorized activity with the potential for ground disturbance (e.g., construction activity, seismic or forestry operations, mining, use of heavy equipment)
- debris (e.g., dumping of materials causing a load surcharge)
- exposed or floating pipe
- withering vegetation
- soil movement and vegetative cover (e.g., slides, erosion, sunken backfill, cave-ins)

- effects of ice
- water crossings and flooding damage to company property
- missing or poor condition of pipeline signs and markers

Any abnormal surface condition or activity observed by the pilot must be immediately reported by radio to the closest attended location as well to the Control center. These will be investigated by the maintenance personnel in the region as they are responsible for the remediation of abnormal conditions.

In densely populated areas and near industrial installations (e.g., a chemical plant, hazardous material storage area), regular ground patrol inspections of the ROW should be undertaken usually once per week. Any abnormal conditions observed must immediately reported, or if the pipeline is at risk, workers must take appropriate action along the ROW.

Ground marking of buried pipeline facilities (Figure 8-3) is an effective means of alerting landowners and the general public of the presence of a high pressure pipeline. These markers will provide a telephone contact number that should be called when a leak is discovered by the public. Note, from Figure 8-3, that there are sometimes climatic conditions such as deep snowfalls that may warrant markers to have extra height. The Technical Support will be involved, or at the very least be kept informed by the maintenance group, in any subsequent site follow-up to surface abnormalities.

Pipeline warning signs identifying the location of each pipeline on the right of way (i.e., one sign for each pipeline) should be installed:

- at road, railway, and water crossings on each side of the pipeline crossing
- in areas of population growth and/or construction activity at sufficient intervals to clearly identify the pipeline location
- in areas that are accessible to the public where the pipeline is above grade (e.g., riser, valves).

Figure 8-3. Pipeline ground markers

In Canada, ROW signs must identify high-pressure petroleum pipelines, providing the company name, the 24-hour emergency telephone number, and the commodity being transported on the sign. In the United States, the ROW signs must clearly state:

- "Warning Petroleum Pipeline"
- The company name
- A 24-hour emergency telephone number

For a gas pipeline, the ROW signs must clearly state: "Warning, Caution, or Danger," followed by, "Gas Pipeline," the company name and 24-hour emergency telephone number.

ROW signs may also identify (a) the meter post or kilometer post and (b) the pipe depth (on the back of the sign or on the signpost). Signs should be located vertical to the ROW, on one side of the pipeline, not directly over it.

Signs should also be posted at facility entrances or around facilities where the public has access. Minimum signage information includes:

- Identifying the operating company, facility name, and a 24-hour emergency telephone number
- A "No Smoking" sign
- If appropriate, a sign indicating "Danger, Hydrogen Sulfide Gas and Explosive Gases"
- A "No Trespassing," sign
- The legal land description and appropriate warning symbols

Remote valve sites should be identified by a sign facing the normal direction of public access to the site. Signs should identify (1) the company logo and (2) the milepost of the site. The sign should be attached to the fence surrounding the valve site, or in the absence of a fence, to a post installed in a conspicuous location.

The maintenance staff in a region is responsible for resolving any issues concerning reduced depth of cover on existing pipelines in their area. When reduced depth of cover on a section of pipeline is known or suspected, it should be investigated to determine the extent of the problem. This is done by analyzing previous depth of cover surveys, physical line probing or minor excavating. If the depth cover is too shallow, then remedial action, such as the line lowering discussed in Fitness for Service Assessments, may be warranted.

The condition of each pipeline crossing under a commercially navigable waterway, including any abandoned pipeline that crosses over, under, or through the waterway, should be inspected at intervals not exceeding 5 years. If any of the following adverse conditions are observed during underwater inspections, they should be reported as soon as possible:

- spanning pipelines (i.e., where support for the pipeline has been eroded)
- significant debris around or in contact with the pipeline
- pipe coating in poor conditions and abrasion
- dents and gouges
- any other condition that could affect the safety of the pipeline crossing (e.g., scouring of the river bottom, undercutting of river banks)

Navigable water crossings must be inspected by qualified underwater inspection contractors. Typical inspection methods include (1) visual checking using qualified divers or (2) remotely operated vehicle inspections operated by qualified technicians. Sonar imaging systems also may be used to supplement remotely operated vehicle inspections.

Leak Detection

The most widely used method of gas leak detection on a right of way is flame ionization, which can be deployed as a hand-held device or affixed to a vehicle. Either way, it is a tedious process requiring inspection teams to walk or drive slowly, along the right of way. Recently, some Integrity Support groups have been using more sophisticated leak detection technologies such as laser interferometry to detect the presence of methane in minute quantities in a dispersing plume above the pipeline center line or around compressor and metering facilities. The underlying operating principle of these devices is that methane, like any other gas, is unique in the way in which it will absorb the light energy passing through it. To detect the presence of methane, two light beams of differing, but specified wavelengths in the mid-infra red range, are created by pulsing a tunable laser, which is aimed at the above-ground center line of the pipe. An "online" beam is set to a wavelength that will be uniquely absorbed by methane, whereas the offline beam is at a wavelength that will not be absorbed. The difference in energy absorbed is directly related to the gas quantity and its relative concentration in air. The technology used is referred to as DIAL, an acronym for Differential Absorption LIDAR and the basic principle is shown in Figure 8-4.

Light Detection and Ranging (LIDAR) is an optical remote sensing technology that measures the properties of scattered light in order to determine the range and other information from a distant target. In this case, the reflected light beam signals are recorded by a receiver on board the aircraft and analyzed with respect to each other as well as taking into account the distance between the sensor and the ground. The amount of gas present in the path between the sensor and the ground is measured and expressed in parts per million per meter. This technology development, which was partly funded by PHMSA, has been successfully demonstrated in controlled blind testing, including over flights of residential areas (Brake et al. 2008). It is capable of detecting gas leaks of 2 specific cubic feet per minute (scfm) and greater, when the wind speed is less than 14 mph. A synchronized digital video recording is made of the pipeline corridor to ensure permanent documentation of the leak survey. Figure 8-5 illustrates the presence of the gas plume superimposed on an image of a sensor swath taken along the pipeline center line. This video image is also used as means of monitoring and recording unauthorized access to the right of way. Similar technology has also been deployed on vehicles and used to perform ground based leak detection surveys (Figure 8-6).

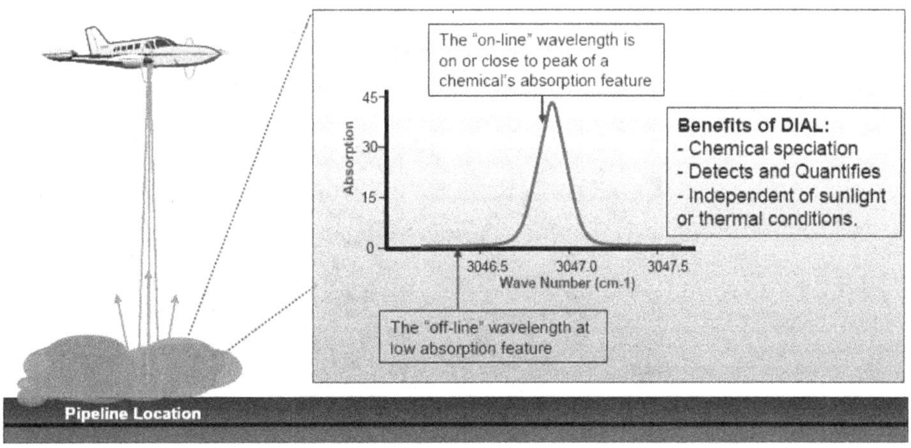

Figure 8-4. The operating principle of differential absorption LIDAR (DIAL) (Brake et al. 2008)

Figure 8-5. A pipeline leak detected using a DIAL sensor

An alternative means of providing real time leak detection capability is the use of buried distributed fiber optic sensors, though they have also much wider pipeline monitoring capabilities such as strain measurement and intrusion (Strong et al. 2008). The underlying principle of leak detection, using a fiber optic cable, is the detection of a temperature change

Figure 8-6. Street evaluating laser methane assessment (SELMA) (courtesy of Pergam Suisse)

in the surrounding soil, which effectively causes the fiber to strain and in turn affects the scattering of light within it. The change in temperature is confined to a small localized area near the leak and will be positive (warming) in the case of a liquids line where the flowing product is warmer than its surroundings, and negative (cooling) for a gas leak due to the pressure release. In the case of multi-phase pipelines, a combination of the two can occur, so detection in this circumstance is reliant upon a third effect, the change in thermal properties of the soil due to the absorption of the released fluid. These factors dictate the placement of the fiber cable relative to the pipe. In the case of an oil pipeline, the cable is placed below the pipe, whereas since cooling is greatest at the source of a gas leak, the cable is best positioned on the pipe surface.

Encroachment

Rizkalla et al. (2008) have developed a satellite-based ROW monitoring capability that has reached a commercial stage and is currently serving the pipeline industry (Mohitpour 2008). The service capability was developed within the PRCI technology development process in parallel with individual pipeline company funding of pilot studies and other advances. The encroachment management service (EMS) is shown pictorially in Figure 8-7. it processes satellite imagery of a given right of way communicating to operations or gas control staff "Alarm" notifications of potential mechanical damage before equipment reaches the ROW. An important technical component of the EMS is the automated software, which enables rapid detection and reporting of encroaching equipment.

Ground Movement

Detailed routing considerations will attempt to avoid areas where the ground conditions may be regarded as unstable. Sometimes, however, avoidance is not possible and the pipeline has to traverse through hilly terrain, in landslide prone areas, or even across seismically active fault zones. The rate of ground movement possible in each of these three categories is quite different. However, what is common is the possibility of causing distress to the pipeline to the degree that its integrity is compromised. Ground movement causes differential displacement of the buried pipe which, in turn, induces longitudinal and circumferential

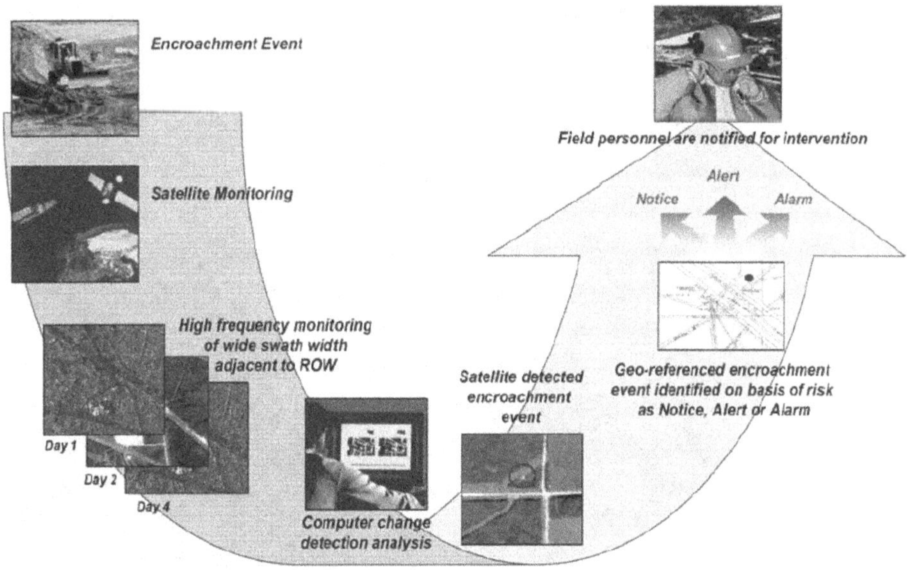

Figure 8-7. Pipeline encroaching monitoring process (courtesy of Via+; Mohitpour 2008)

straining in the pipe. These displacements are finite and from an integrity assessment view point are regarded as being displacement controlled rather than load controlled. It is possible therefore for the pipe to be strained to a critical level in one of three different ways. The tensile strains in the longitudinal direction may be sufficient to cause rupture, while compressive strains may result in wrinkling or buckling (Figure 8-8A). If the wrinkling is severe, localized tearing can occur such as is shown in Figure 8-8B.

Soil materials can move on a hillside through the development of slope instability. Under these conditions, failure of the slope material can occur on a range of time scales. Some types of mass movement involve rather rapid, spontaneous events. Sudden failures tend to occur when the stresses exerted on the slope materials greatly exceed their strength for short periods of time. Slow failures often occur when the applied stresses only just exceed the internal strength of the hill slope system. There are two types of slip movements: translational and rotational (Figure 8-9, A and B). With translational slips, movement occurs predominantly along more or less planar or gently undulatory surfaces and is frequently structurally controlled by discontinuities and variations in shear strength between layers of bedded deposits. The movement in a rotational slip occurs as a result of forces that cause a turning moment about a point above the center of gravity of the section. The surface of the rupture concaves upwards as can be seen from Figure 8-10.

Slope failure generally occurs under slow creep like conditions, although comparatively, rapid movements of the order of 6 cm/year have been observed (Rizkalla et al. 1999). This rate of movement is still sufficiently slow as to enable suspect areas to be continuously monitored. Slope monitoring forms part of the IMP and is one of the functions of the integrity technical support team. It was noted earlier that one of the tasks of an aerial patrol is to look out for ground movement. Figure 8-11 illustrates a fairly obvious indication of slope movement — the presence of tension cracks on the slope. The verticality of the trees immediately adjacent to the right of way can be a good indicator of whether the failure is surficial or deep seated. If the type of tree has a shallow root system and it is inclined from the vertical, we can make a reasonable assumption that the slip zone is shallow. There are two approaches for monitoring slope movements, (1) surficial measurement and (2) subsurface measurement.

Surficial measurement involves setting up survey markers on the surface of the slope. There are four common survey methods currently being used:

1. Total Station Surveying
2. GPS Surveying

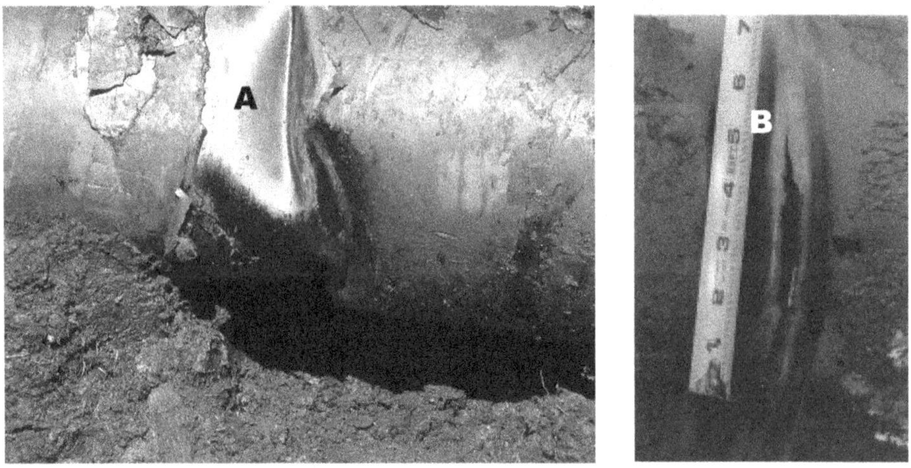

Figure 8-8. Buckling failure (A) and local tearing (B) (TMX Sumas)

Figure 8-9. A Translational slip failure (A) and a rotational slip failure (B)

3. InSAR Surveying
4. Slope Extensometers

Both Total Station Surveying and GPS surveying require installing monuments on the slope surface and a base station(s) on stable ground (Figure 8-12). The monuments are 1 in. square bars whose initial positions are carefully located with respect to the base station. Subsequent measurements are made at selected intervals of time, using either a conventional land survey (theodolite) or GPS triangulation.

Slope extensometers (Figure 8-13) are generally defined as an installation that runs along the surface of the slope (typically parallel to the direction of anticipated movement). They measure the overall deformation along the total length of the slope and are particularly useful when a slip or landslide suddenly starts to move rapidly, when it can provide an alarm. The operating principle of the system is shown in Figure 8-13. A zero expansion wire runs through a flexible PVC pipe staked to the surface of the slip or slide. It is fixed in position at the base of the slope. Its free end is attached through a pulley system to a pretensioning weight that moves along a vertical scale should the slope move (Figure 8-14).

Figure 8-10. The head of a rotational slip

Figure 8-11. Tension cracks on a steep slope

Slope inclinometers are a sub surface means of accurately measuring ground movement over time. The method of operation is shown schematically in Figure 8-15 and consists of drilling vertically in to the slope at a number of locations. A pipe containing four orthogonal longitudinal milled grooves is cemented into the drilled holes. The wheeled slope inclinometer probe, containing an accelerometer, runs to depth in two diametrically opposed grooves, and the resulting signal is integrated to provide a two-dimensional displacement plot. The procedure is repeated by running the probe in the other two grooves, which enables the full three-dimensional displacement of the vertical pipe to be obtained. In some probe arrangements, two accelerometers are fitted at right angles to each other, so that only one run is required. Three types of plots are produced from the probe runs:

Cumulative displacement: shows the cumulative displacements of the hole profile when compared to the initial reading and hence represents changes in the profile of the inclinometer pipe.
Incremental displacement: shows the incremental change at each measuring interval. If the slip surface if moving, a spike will be seen on this plot. This plot is useful for determining the depth of the slip surface.
Velocity plot: is a displacement versus time plot, measured over several increments.

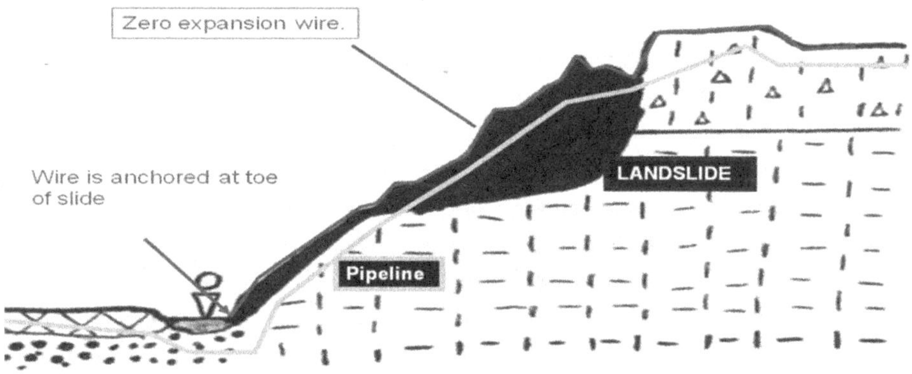

Figure 8-12. A surface slope extensometer

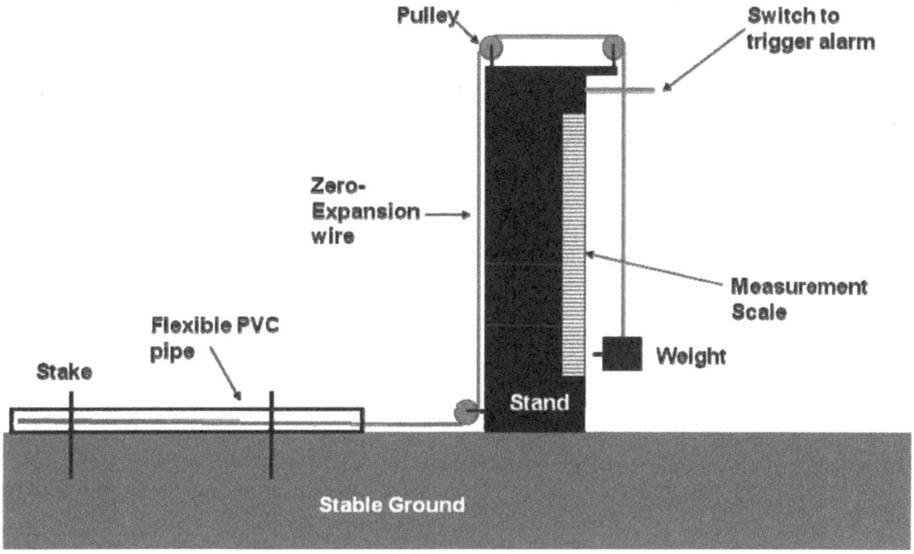

Figure 8-13. The measuring and alarm component of a slope extensometer

The soil strain meter (Figure 8-16) employs a potentiometer and a rod mounted between two anchors to monitor horizontal movements of the surrounding soil. These components are linked together so that movement of one anchor relative to the other causes a change in the output of the potentiometer. The initial reading of the strain meter is used as a datum. Subsequent readings are compared to the datum to calculate the magnitude, rate, and acceleration of movement. These devices are used to measure horizontal strain and monitor tension cracks on slopes.

Ground movement may also be measured using Synthetic Aperture Radar (SAR) images acquired from the RADARSAT-1 satellite. The radar signal can penetrate cloud cover and it is reflected off targets placed on the surface of the slope. The resulting SAR images are very consistent and comparatively inexpensive to obtain. The individual cost of each image has reduced considerably as more applications for their use has grown. A single SAR image in fine-beam mode measures 30 × 30 mi so it is possible to monitor several slopes on one image. The satellite's orbit permits a 24-day re-visit frequency, although higher revisit

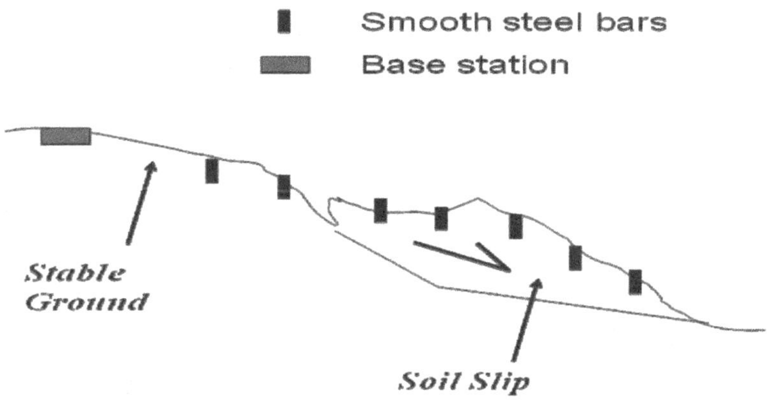

Figure 8-14. Total station survey

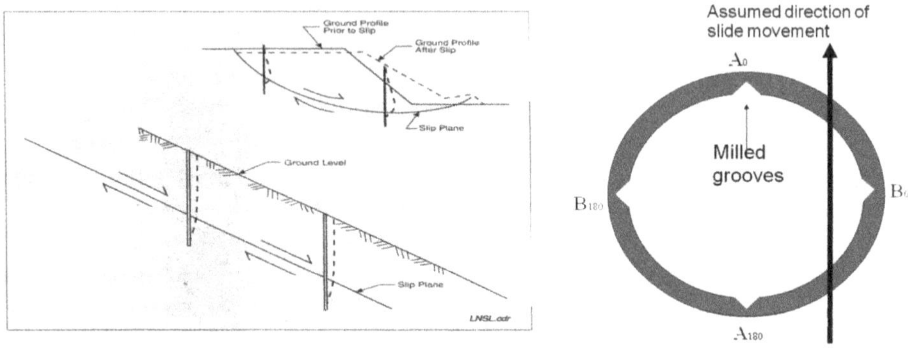

Figure 8-15. Principle of the slope inclinometer

frequencies are possible; however, the process is meant for use with relatively slow moving slopes. The aluminum tetrapod reflector shown in Figure 8-17 provides a phase stable target for the satellite radar signal. It has a perforated design making it less susceptible to the effects of wind and precipitation.

Accurate ground movement measurements are possible from satellites using "Interferometry" with SAR images (the InSAR process). Ground displacements are detected by comparing changes in the wavelength of back scattered signals between successive images, with subcentimeter accuracy regularly being achieved.

Mitigating slope movement usually involves the following methods either singly or in combination.

1. Unloading the slope crest to reduce the driving forces
2. Supporting the slope toe to increase the resisting forces
3. Controlling ground water by reducing pore water pressure through surface water control

Stress to the pipe itself may be relieved by digging a ditch alongside.

Figure 8-18 illustrates another type of geotechnical hazard — a rock fall (Figure 8-18) that requires a special type of mitigation. Often rock falls are triggered by excessive rain or snow melt, so a simple precaution is to attempt to provide surface water control through the use of catch basins or ditches placed higher up on the rock face. Diverting the rock away from the right of way may be achieved through the use of deflection berms, or constructing

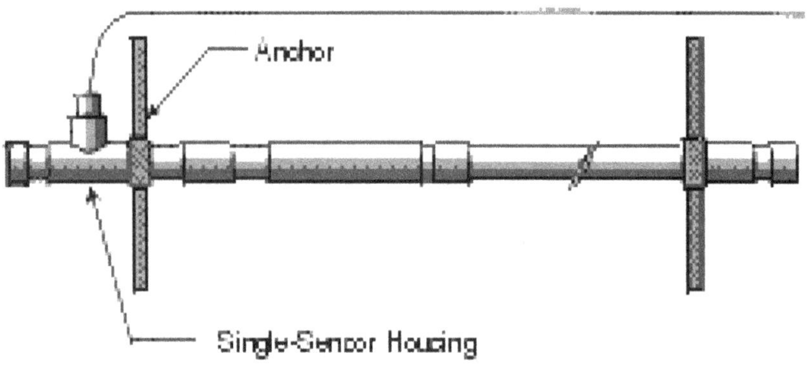

Figure 8-16. Soil strain meter

Figure 8-17. INSAR target reflector

rock fall nets (gabions) at the base of the slope. Depending on the nature of the rock surface, it may be possible to stabilize it using rock bolts or shotcrete.

DETERMINING RE-INSPECTION AND RE-TEST INTERVALS

The industry uses several means to establish defect growth rates and from these determine the associated re-inspection and hydrostatic re-testing intervals. The following section contains an overview of each approach.

Figure 8-18. A rock fall

Consider first establishing re-inspection intervals for a pipeline where the primary threat is metal loss due to corrosion. A number of factors must be considered to determine "fitness-for-purpose" revalidation intervals for the pipeline and the rationale for establishing the intervals should be reviewed and documented. Updates to the inspection program are triggered by completion of an ILI run and the results obtained from the associated excavation program. New "field metric" data is added and the re-inspection interval validated using this information. In the absence of a baseline, the industry standard data for growth rates given in Table 8-1 should be a useful starting point.

Historic Approach — Derivation of Interval Between Previous Inspections

Historically, qualitative processes can be employed in determining re-inspection intervals for pipelines. These processes involve consolidating and analyzing all available information about the line conditions (e.g., CP results), defect behavior, failure history, and prior ILI results, as available, to develop estimates regarding corrosion growth rates. In general, the maximum corrosion growth rate was established for any given pipe segment and used to determine an appropriate inspection interval. To do so requires a number of assumptions to be made. For example, the inspection interval was based on the time for a corrosion feature to grow from a depth of 50% to 80% wall thickness assuming a growth rate that was double the maximum estimated. The rationale for that assumption was that following an ILI, all external corrosion features with a depth of 50% or greater would be excavated. Based on depth alone, all features in the field with a depth of 80% wall thickness or greater would be repaired. In order to account for the effect of length and depth, not considered by this approach, the inspection interval could be conservatively defined as half of the calculated time period. To illustrate, suppose the inspection interval for a section of the pipeline can be calculated based on a 50% through-wall feature growing to 80% through-wall. Given a corrosion rate of 0.30 mm/yr (assuming say a wall thickness: 9.7 mm), it would take

TABLE 8-1. Industry Suggested Corrosion Growth Rates

NACE		ASME	Pipeline Research Committee — AGA		Materials Performance	
Recommended Practices[1]	1994–1997, Study Suggested[3]	1991–2000 Study, Suggested[4]			External Corrosion of Line Pipe[7]	
Pitting corrosion rate (mm/yr)	Maximum external corrosion growth rate (mm/yr)	Maximum corrosion growth rate for 10% deep external flaw (mm/yr)	Typical corrosion rates in soils[5] (mm/yr)	Typical corrosion rates in soils with CP[6] (mm/yr)	General corrosion rates (mm/yr)	Pitting corrosion rates (mm/yr)
0.3089 (min) 0.4064 (max)	0.2997 (max)	0.2743	0.5080 or less	0.2540 or less	0.2032	0.7112 (max)

Notes:

1. Pipeline External Corrosion Direct Assessment Methodology, NACE Standard Recommended Practice RP0502
2. Periodic Re-Verification Intervals for High-Consequence Areas, GRU Report No. GRI-00/0230
3. Pipeline Corrosion Management, Corrosion 2001 Paper No. 1627
4. Approach to Assessment of Corrosion Growth in Pipelines, Proceedings of IPC'02 4th International Pipeline Conference, Paper No. IPC2002-27243
5. An Assessment on the Criteria for Cathodic Protection of Buried Pipelines, PRCI Report No. PR-3-129
6. Time Dependence on the Minimum Requirements for Cathodic Protection Against Pitting Corrosion in Natural Soils, white paper report, c.2000
7. External Corrosion of Line Pipe — A Summary of Research Activities, Materials Performance, March 1996

approximately 10 years for such a feature to grow to that depth. To provide a significant degree of conservatism to the ILI interval, a 5-year interval would then be selected.

Appropriate inspection intervals can be established following each ILI inspection. First, the ILI data needs to be verified using field excavation results, then the following factors should be considered:

- Total number of corrosion defects
- Corrosion defect density
- Corrosion depth distribution
- Defect rupture pressure ratio (RPR) distribution (see Chapter 6)
- Corrosion configuration/length/ orientation
- Time since last ILI inspection
- Change in the above indicators since the last ILI inspection
- Existence of any corrosion failure since previous inspection

Each of the above metrics can be reviewed to determine an appropriate inspection interval.

Interval Validation Using Field Metrics

As sequential ILIs are completed (based on the revalidation intervals developed through the processes described above), additional data becomes available to validate the selected intervals. This process serves as the primary metric upon which the assessment interval can be increased, diminished, or maintained at prior levels. Each sequential ILI run and associated defect repair program provides field data used to help refine these revalidation intervals.

It is usual that the ILI program be such that it ensures that revalidation for any given line segment occurs prior to any defect within that segment exceeding predefined tolerances (Figure 8-19). These tolerances can be line specific, but are created to ensure consistent margins of safety below fitness-for-purpose criteria.

Generally, evaluation of the inspection interval is tied directly to results of the most recent ILI run and its associated excavation program. In the event that defect severity, as

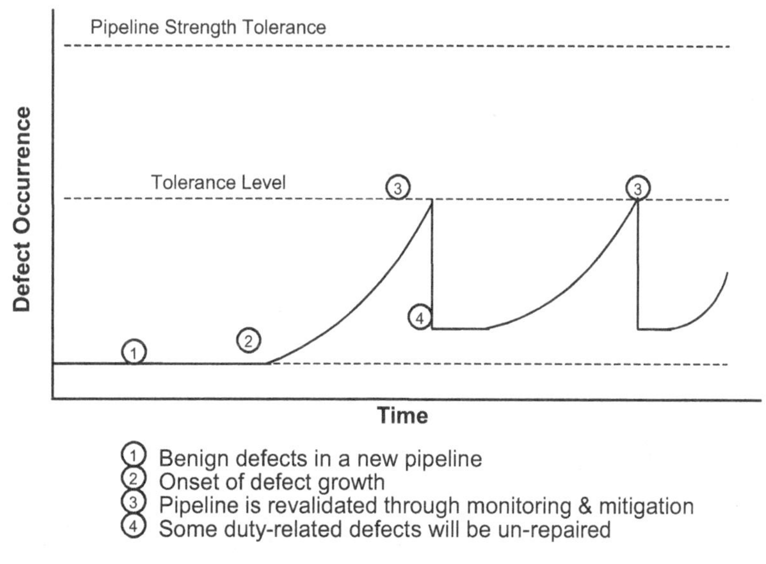

Figure 8-19. Validation assessment interval

indicated by the ILI tool and verified through field analysis, exceeds the defined tolerances, re-assessment intervals must be reviewed. Any revalidation interval that allows time for one or more defects to grow beyond these defined tolerances may be unacceptably long and therefore will have to be shortened.

We can readily determine if the actual corrosion growth rate is more aggressive than that predicted from the last inspection interval. This is the case when the excavation results show one or more defects with a RPR <0.8 and/or external corrosion greater than 80% of pipe wall thickness and/or internal corrosion greater than 50% of pipe wall thickness. This would exceed our defined tolerances and so the subsequent inspection interval for that segment must be shortened to accommodate this higher growth rate. In general, the subsequent revalidation interval is usually reduced by 1 year.

Conversely, if defect growth based on the prior inspection interval resulted in a manageable number of defects meeting the criteria for excavation, but few to none of these defects have an RPR <0.8 and few to none of the defects exhibit corrosion greater than 80% of the pipe wall thickness (or greater than 50% for internal corrosion), the interval as previously developed is validated by the field data and no changes are required.

In the event that the inspection interval used for the last revalidation of the line segment showed few or no features meeting excavation criteria and no features exceeding defined tolerances, the interval can be extended. The degree to which the interval can be extended is influenced by other factors including statistical methods for establishing corrosion growth rates.

Run Comparisons

Growth monitoring can be performed by comparing the results obtained from successive and or multiple ILIs. There is a four-step process involved in lining up the data from successive or multiple runs:

1. Alignment of the inspection runs in which tie points are identified and data is collected between these tie points.
2. Tolerance windows are developed (Figure 8-20). These involve creating clustering and boxing rules and defining tolerances for both axial and circumferential defects.
3. The type of match required is identified, that is, 1 to 1, 1 to many, or 1 to 0.
    - 1:1 This the easiest to deal with and allows one to one comparisons of depths, lengths, etc.

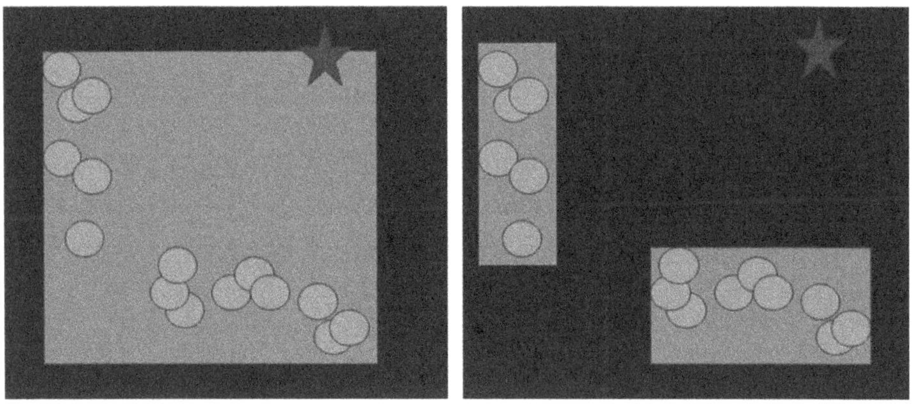

Figure 8-20. Tolerance windows showing clustering and boxing

- 1: many requires "rules" on what to compare (e.g., deepest indications)
- 1:0 this occurs both forward (that is a previously reported indication disappears) and backward (new indications show up)

There are different levels of possible comparisons that can be made:

- Run-to-run, which provides an understanding on relative "resolution"
- Joint-to-joint, which is often used to identify regions along a line where change may be taking place and allows identification and characterization of "orphan joints"
- Feature-to-feature, which is more difficult to do, but provides a basic QA/QC check and provides more detailed understanding
- Raw data (measured signals), which is the most difficult and prone to error but potentially the most valuable

Alignment is central to conducting any of the comparisons referred to above. Perhaps the simplest way of achieving alignment between separate ILI runs is to align girth welds. Matching joint lengths allows missed girth welds to be taken into account, while knowing the distance to upstream or downstream girth welds is useful for detailed pit-to-pit matching. Tool odometer inaccuracies over large distances can introduce significant errors and uncertainties. Circumferential orientation and drift is important and the long seam is used as a marker. Clustering rules for defects should be considered when defining tolerance windows such as are shown in Figure 8-20.

Making sense of comparisons is not an easy task. In most cases, indication-to-indication matching will show:

- Some indications grew, while others shrunk
- New indications appeared, while others disappeared
- Some indications "moved" around, or "moved" from the outside to the inside diameter
- Some indications changed type (e.g., metal loss became a mill feature and vice versa)

If the differences between successive results are generally within the tolerances of the tools, no change may have occurred. *Any* indication with a large change between two runs, however, should be a candidate for further investigation, regardless of whether the indication apparently grew or shrunk.

Statistical Analysis

GE-PII (2010) has developed a quantitative statistical analysis towards determination of the appropriate inspection interval based on a systematic analysis of all available data on the corroded condition of the pipeline system. First, the data is gathered, prepared, checked, and examined for trends along the pipeline. By observing variations in corrosion along the pipe and the effects of parameters such as coating type and installation dates, localized corrosion activity and critical sections of the pipeline are identified. A probabilistic methodology is then used to estimate the corrosion depth and length growth rate distributions. Conservative higher percentile values of these growth rates are used to grow these defects both in length and depth. The defects that remain after an excavation and repair are "grown," up to predefined tolerance criteria in terms of depth and RPR. The tolerance criteria defines a safe level at which the pipe should be re-examined. Therefore, the time required for remaining defects to reach the tolerance criteria is recommended as the safe re-inspection interval.

Pit Matching

Pit matching is the most accurate statistical means available to establish corrosion growth rate. To the extent possible, it matches each pit between two runs (limited by tool technology, defect interaction, etc., as discussed above in the run to run comparisons) and uses statistical means to evaluate the distribution of defects in question.

While providing excellent corrosion growth information, pit matching is both labor and capital intensive. Besides field data, corrosion growth rates (and associated re-inspection intervals) that have been derived through pit matching, provide the most quantitative analysis, and is the next most heavily weighted consideration in establishing future revalidation intervals.

Figure 8-21 illustrates a typical process for the assessment growth.

The above discussion concentrated on corrosion defects, though the run-to-run comparisons could be generalized for planar cracks. Running a crack detection tool in a gas line, however, has its logistical problems, and for this reason, many gas pipeline companies use hydrotesting as the means of managing crack defects. Fessler and Rapp (2006) published a method for determining hydrotest intervals based on a time versus pressure plot. Their methodology was validated by testing and extended by Sen and Kariyawasam (2008) to account for variable crack growth rates. The central premise is that the hydrotest will remove critical or near critically sized defects so that all cracks remaining in the line after the test are considerably smaller than the size of crack needed to initiate a failure at maximum operating pressure. It is possible using the plastic collapse approach described in Chapter 6 to determine from the pipe properties the critical length of a defect at the test pressure. Fessler's method does not require the actual dimensions of a crack be known, it is sufficient that the crack is large enough to fail at MOP. Further, he assumed that the crack size is directly proportion to pressure and that the rate of crack growth is constant. Hence, a change in crack size over time is analogous to the change in pressure over time. This is reasonable

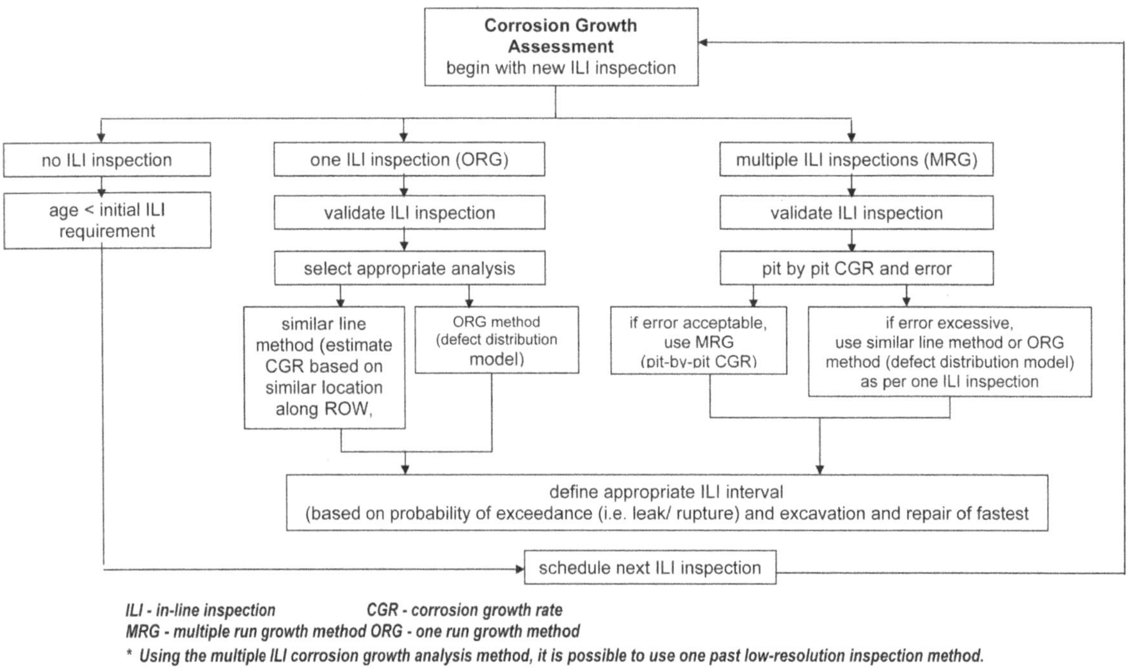

Figure 8-21. Corrosion growth assessment process

for circumstances where the average growth rate is extrapolated over short intervals and that the difference between the material flow stress and the operating stress is relatively small. Knowing the hydrotest history of a pipeline it is then possible to construct a plot of time versus pressure plot. In this manner, we are able to compute a crack growth rate by comparing the maximum flaw size at the time of the previous test, to the critical size at the current operating pressure and dividing their difference by the time interval.

FITNESS FOR SERVICE ASSESSMENTS

When determining if a pipeline section is structurally adequate, there are a number of limiting conditions, known as limit states, that should be examined. In general, they may be grouped into three categories:

1. A leak limit state
2. An ultimate limit state
3. A serviceability limit state

Clearly, among the many purposes a particular pipeline may serve, it is critical that containing the transported product is uppermost. Hence, our interest in the leak limits state condition. Similarly, one would have an expectation that the pipeline is not going to fail in a catastrophic manner — the ultimate limit state. These considerations lead directly to the "leak before rupture" operating philosophy that was addressed in Chapter 6. The ongoing safe operation of a pipeline requires the monitoring and assessment of its condition and the mitigation of any undesirable outcomes. This may involve the use of ILI tools or, where such pigging operations are not feasible, for whatever reason, an alternative form of assessment such as hydrotesting.

A very important factor is to assess the structural adequacy and remaining life of pipeline segments containing defects identified during ILIs, or following dedicated hydrotests. In addition, it is also required to assess a pipeline's fitness for service under varying circumstances such as a change of service, reversal of flow direction, reactivation after a suspension of service, and rehabilitation. Ongoing operational issues such as line relocations will require the group to provide engineering services to the maintenance function. All of these topics are addressed in the following section.

The term "fitness for service," has a specific meaning in pipeline and plant integrity assessments. It involves a multi-disciplinary engineering and risk analysis of portions of the pipeline or associated equipment. This is conducted in response to identified flaws to ensure safe, reliable, and continued service for a defined period of operation, e.g., until the next scheduled outage. The pipeline may contain flaws such as localized thinning due to corrosion or crack like features. It may not meet current engineering design standards, or it may be contemplated to run under more severe conditions than previously. The outputs from a fitness for service assessment are:

1. A decision to run, alter, repair, monitor, or replace a pipe section or piece of equipment and
2. guidance on suitable future inspection intervals (this is addressed in Determining Re-inspection and Re-test Intervals).

The fitness for service assessment applies analytical methods involving stress analysis and/or fracture mechanics such as those described in Chapter 6, to evaluate flaws, damage, and the effects of cyclical loads. The material properties, the results of nondestructive examinations (NDEs) used to detect size and characterize flaws, and details of the working environment are also needed. Lastly, an integrity assessment also requires knowledge of past operating conditions as well as a forecast of future conditions. API 579 (2007) Recommended Practice sets out a process to be followed when assessing plant integrity, which may be summarized in the following eight steps:

1. Identify the type of flaw and the material damage mechanism.
2. Determine the applicability and limitations of the assessment procedures.
3. Define the data requirements for each defect assessment type.
4. Apply assessment techniques and acceptance criteria.
5. Evaluate remaining life or limiting flaw size and establish an inspection interval.
6. Apply remediation methods to the extent possible and practical.
7. Employ in-service monitoring procedures when remaining life and inspection interval cannot be adequately established.
8. Document all information used and decisions made in Steps 1 through 7.

The eight steps outlined above for the conduct of Plant Integrity assessments can be repeated or may be customized for assessing pipelines in a continuous process to identify, assess, and mitigate risks to new or in-service pipeline sections or systems:

1. Preassessment: Severity determination and prioritization. Identify all threats and likely consequences. Risk Assessment/Management (High impact area prioritization)
2. Identifying susceptibility to flaws, characterization of ILI results
3. Data Integration
4. Integrity Assessment: analysis of ILI results (distribution, density, orientation, correlation to susceptibility, and estimation of remaining strength)
5. Development of short-term excavation and ILI validation programs
6. Remediation, repair, and assessment of ILI performance
7. Remaining life extension and in-service monitoring. Development of long-term excavation and re-inspection program
8. Documentation of all information used and decisions made in Steps 1 through 7

It is important to stress that integrity assessments should only be undertaken by individuals technically competent to perform the task, API 579 sets out as a minimum qualification that the engineer involved "shall have an Engineering degree and at least 2 Years relevant experience, while Inspectors are qualified per API Codes."

The key engineering disciplines required for fitness for service assessments include:

- Materials Science or Metallurgy
- Structural and/or Applied Mechanics
- Fracture Mechanics
- NDE including ILI
- Process Engineering

The above methods are deterministic in nature. A less common approach in engineering assessment is the use of structural reliability concepts, though these are now permitted in some pipelines codes. The Canadian pipeline standard Z662, in its 2007 edition, allows this approach, in specific circumstances, as described in its non-mandatory Annex O, "Reliability based design and assessment of onshore non-sour natural gas transmission pipelines." The principal purpose in the development of the Annex is to encourage its use in the design of those pipeline projects where the working stress approach would prove to be impractical and/or too costly (Nessim and Zhou 2005). The methodology involved has been widely used for many years in other structures, such as bridges and buildings, as well as in offshore pipelines. Its logical extension in each of these applications is to use the same approach for operational conditions, for example, in re-rating pipelines (Francis and Senior 1998; Senior et al. 1998), pipeline rehabilitation, and dealing with class location changes. In whatever circumstance, the methods are to be used, it is essential that the as built physical properties of the pipe are known as well as the operating and repair history of the pipeline and its current condition as determined, for example, by ILI. Caleyo et al. (2003) and Stephens and Nessim (2006) have shown how the method may be used to manage metal loss corrosion damage as identified through ILI. The procedure is set out in pictorial form in Figure 8-22. More generally, the various steps for assessing the response of the pipe to a broad range of threats are:

1. Identify the relevant failure modes (see Chapter 2).
2. For a specific-defect type, determine its dimensions accounting for both the POD and sizing accuracy (see Chapter 5).
3. Develop the limit state functions from the physical parameters that will result in pipe failure (see Chapter 6).
4. Account for the uncertainties in the mathematical model and in the physical and operational parameters used as input to it.
5. Calculate the probability of failure (P_f) and hence the reliability($1 - P_f$) for each of the identified failure modes.
6. Compare the calculated reliabilities against the target reliability set out in the requisite standard to ensure safe operation.
7. Determine sufficiency or the need to implement remedial action to ensure safe operation.

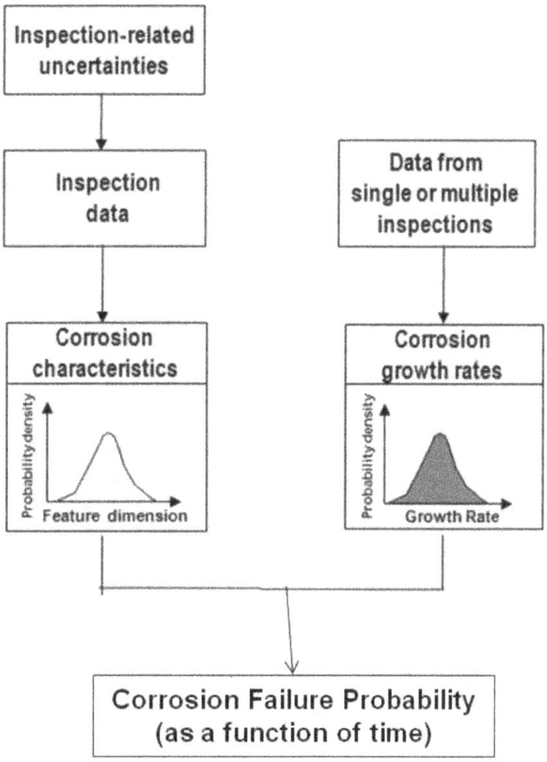

Figure 8-22. Determining the failure probability of a corrosion defect as a function of time (courtesy of C-FER Technologies)

As was noted earlier, to be effective in assessing the adequacy of operational natural gas transmission pipelines, the structural reliability approach requires a lot of detailed information and is data intensive, which limits its use to those companies with meticulous records keeping. It also requires experienced personnel to assemble and construct the various probability distributions and perform the calculations.

A considerable effort has gone into developing the target reliability figures and to establish the evaluation lengths that are appropriate for a risk-consistent reliability assessment. The code calibration was performed against a wide range of existing pipe sizes, operating pressures and class locations. The clear inference is that since these operating pipelines have withstood the various threats, their design must have been at least adequate and perhaps excessive. However, the prospective user must exercise caution as to whether their pipeline parameters falls within the calibrated range. It will be noted too, that the approach has currently been confined to sweet natural gas transmission pipelines, where the consequence portion of the risk assessment is more readily determined than for sour gas or liquid pipelines. The reader is directed to the paper by Nessim et al. (2006) for a detailed discussion of how the reliability targets are to be adapted and the evaluation lengths employed, when assessing location-specific damage (such as corrosion defects), to ensure that the public safety risk criteria underlying the targets are met.

An example of the use of the method in the assessment of a corroded pipeline and the effect of introducing staged repairs to remove defects can be seen in Figure 8-23, which has been taken from Stephens and Nessim (2006). The horizontal line on the graph represents the target reliability, which for a Class 1 location has been set at 1.0×10^{-3}/km-year. This value should not be exceeded; a condition which is violated by the curved line representing the calculated reliability of a corroding pipeline segment deteriorating progressively over time due to inattention. The effect on the calculated reliability of the segment due to making staged repairs is shown by the zigzag line. Removing or repairing corrosion defect results in an increase in reliability (a local downward trend in the line); however, as time passes, the remaining defects in the segment become more malevolent, decreasing the reliability and necessitating further intervention. The timing of each calculated intervention can be linked to the planned maintenance activities; so equipped with this information, the pipeline operator can focus on critical defects and maintain the prescribed level of reliability.

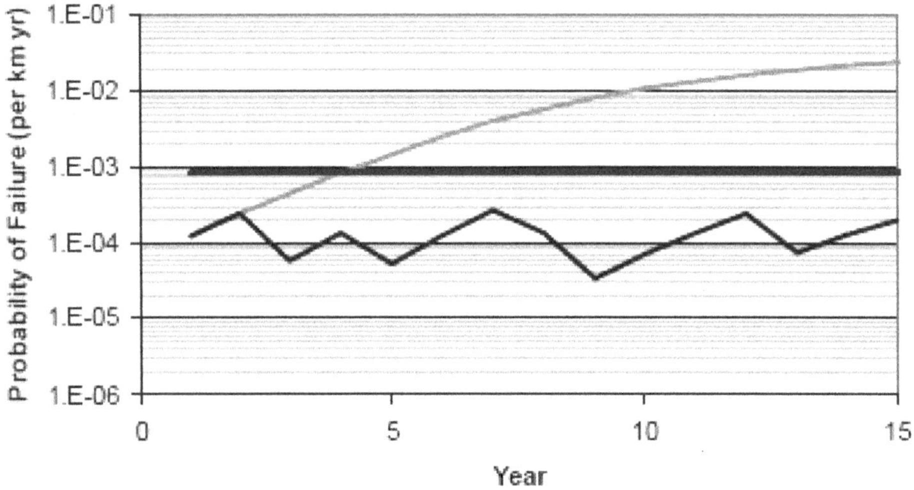

Figure 8-23. Segment reliability compared to general rupture reliability target — with staged defect repair (Stephens and Nessim, 2006)

Guidance on Change of Service and Rehabilitation

The useful operating life of a well-constructed and maintained pipeline is well beyond five decades; however, it is conceivable that, during its lifetime, business conditions may change, requiring a new set of operating conditions for the pipeline. These changes can range from re-activating a pipe section that has been temporarily taken out of service, through reversing the original flow direction or changing the transported product, to refurbishing and updating the pipeline to meet new demands.

Re-activation requires a careful consideration of how the pipe has been maintained since it became de-activated. Normal operating practice requires these sections to be purged of their content and to be filled with an inert gas such as nitrogen. The CP should remain active, so that if there has been an insignificant change in the inert gas pressure one may safely assume that the pipe section has remained leak tight. If, however, there has been a change in the gas pressure and/or a break in the application of the CP, a regulating authority may require a hydrotest or an ILI, to be conducted to re-qualify the pipe for its intended purpose.

Before discussing rehabilitation or a change in service, it may be useful to define what is meant by both terms.

Rehabilitation refers to the objective of restoring a pipeline and the associated above ground facilities, to the original design performance specification, or the most recently implemented design performance specification. It may also be undertaken as a result of deterioration in portions of the pipeline system, preventing the original performance specification from being achieved; or it could be to keep pace with amendments to the pipeline regulations, even though these are rarely applied retro-actively.

A change of service refers to any modification to the original or current design process specification of the pipeline. It may, or may not, require physical changes to be made to the pipeline facility. However, in all cases, a thorough system audit and design review must be conducted to determine the impact of the intended change and to identify the measures needed to mitigate such an impact. This same process is used with rehabilitation, but with the desired outcome of validating the new status against the original design intent.

As a minimum, a change to any of the following parameters would be classified as a change in service:

1. The transported fluid (e.g., oil to gas or vice versa) or the physical state of fluid conveyed (single to multi-phase or vice versa)
2. Design pressure of all or part of the pipeline system
3. Design temperature of all or part of the pipeline system
4. A change in flow direction

It should be noted that changes made to process parameters that fall within the original design limits may not necessarily be defined as a change of service — if the integrity level of the pipeline is known to comply with the original design intent.

Whenever a change in service is being sought, a common first step will be to establish the initial design basis for the pipeline and to compare this with the revised conditions. The purpose for doing so is to ensure that the inherent risk level is maintained, or reduced, and that there is sufficient structural adequacy to withstand the forces arising from its new operating condition.

Rehabilitation and change of service are both assessed using the protocols established for integrity management; hence, both will be required to follow the same management of change process to identify and mitigate any change in the risk profile or potential environmental impacts (see Chapter 2). For example, recoating portions of a pipeline using an over the ditch method, as part of a rehabilitation project, can have comparable environmental impacts as that found during the original construction.

In order to satisfactorily rehabilitate a pipeline or change its service, its current structural integrity status must be established, and then used, together with the impact of future operating parameters to estimate remaining life. This requires conducting a comprehensive baseline assessment of the pipeline segment, or facility, and its history and involves the following steps:

1. Data gathering the pipeline construction records; operating history and its current condition as determined from ILI.
2. Data review — determine the completeness of data and identify shortfalls
3. Design basis — the intent of the original use and design parameters
4. Technical analysis — determine the current level of compliance with the rehabilitation or change of service objectives, and the remedial work required to close any gaps
5. Risk analysis — based on inadequacy of available information or noncompliance with technical objectives
6. Feasibility — review of risk and economic criteria acceptability

The results of the baseline assessment may indicate that further inspection or testing is required in order to complete the record. The effect on the pipeline risk profile of other data gaps that have been filled through assumptions and inferences needs to be assessed to ensure that it remains acceptable. This is an area where the reliability based approach described previously might prove helpful.

There may have been regulatory changes enacted since the pipe was originally placed in service and the impact of these needs to be understood in order to determine their impact on the economic and technical feasibility of what is being proposed.

Addressing each of the six steps above in turn, the first of these, gathering the data from construction records and the repair and operating history can prove to be quite a challenge. Gaps in the data record should be highlighted so that an alternative provision can be made to address the shortcoming; for example. the pipe material properties may be found by sampling pieces, removed as part of a repair procedure, while the actual location of the pipe center line may be more accurately found from ILI runs rather than as built drawings.

The original design basis should be confirmed and it consists of two components: process design and engineering design.

The purpose of the original process design basis was to define the performance objectives of the pipeline, e.g., the flow rate, and any factors that may affect the quality of the transported product. In addition, those factors that may influence the environmental and social impact of the pipeline will have been described.

The process design basis ideally includes a description of the:

- Fluids to be transported
- Physical conditions under which the fluid(s) are to be transported (e.g., gas, liquid, or multiphase)
- Physical properties of each transported fluid (i.e., density, vapor pressure, flash point, etc.)
- Corrosivity of each transported fluid
- Soil conditions
- Details of internal corrosion monitoring and control (e.g., coatings, chemical injection, etc.)
- External corrosion monitoring and control (e.g., pipe coatings, field joint coating, CP, ac mitigation, etc.)
- Operating and control philosophy
- Construction plans
- Inspection and test plan

The engineering design basis describes the design codes used, the design pressure, and certificated maximum allowable operating pressure; the flowing temperature; design flow rates; flange ratings; material selection; buoyancy control measures; isolation and emergency shut down valves; pressure control; and means of providing overpressure protection etc.

The technical analysis to be performed will vary in accordance with the type of change of service. For example, when the direction of flow is to be reversed, it is immediately obvious that the valve and pumping arrangements along the length of the system will need to be reviewed and, where necessary, changed out. The pressure drop in a gas line between compressor stations is relatively insignificant since transmission lines usually work on an optimum compression ratio of about 1.3. Liquids lines, however, can experience a considerable pressure drop between pumping stations, so the reversed hydraulic gradient will subject the former downstream end sections to much higher pressures than previously. Accordingly, an integrity assessment should be carried for defects detected in these portions of the pipeline. For the same reason, a new risk assessment should be performed.

Changing the product in the line or seeking an increase in the certificated operating pressure both require an engineering assessment to be performed. The latter is the least complicated of the two and requires a complete defect assessment to be conducted, using the results of an ILI, or a recent hydrotest, to verify that all defects of a critical size at the new operating pressure have been removed. Generally, companies will attempt to demonstrate that an RPR of at least 1.25 has been achieved. Pressure control and over-pressure protection will have to be changed to be compatible with the new operating pressure.

A change in flowing product, say from gas to liquids, may necessitate the construction of storage tanks and a change in the type of main line block valve used as well as the valve actuator system. The pipeline will now be more sensitive to the presence of small leaks although larger leaks should be easier to determine using SCADA and a mass balance algorithm. Nevertheless, a leak detection test will be required and the regulator may also request a detailed line fill plan to ensure that the pipeline is not over pressured. To prevent the liquid filling the line from developing slack flow on downhill sections, back pressure is controlled by filling the line ahead with nitrogen. This also serves to purge the line of air.

An increase in population density near a gas pipeline will trigger a review of the class location, and this may result in a reclassification. If this is the case then, to remain compliant with the pipeline standard requires either a lowering of the operating pressure or an increase in the pipe wall thickness. Both of these measures will have an economic as well as an operational impact on the pipeline. It is for this reason that several companies in Canada have recently sought a waiver to the standard and instead have attempted to establish using the reliability-based design approach, alluded to previously, an equivalent or reduced level of risk, if the primary threat, that of external interference, is mitigated e.g. using slabs placed over the line.

Line Relocation

There are a number of pipeline maintenance activities that, in order to be carried out safely and with minimal risk to the pipeline, will require assistance from the Integrity Technical Support group. Among these is the relocating of the pipeline to remediate an inadequate depth of cover, or to accommodate a foreign crossing of the pipeline. The assistance will be in the form of the specialized stress analysis associated with line lowering. Specifically, an assessment will be made to determine the size of bell hole excavation required, the number and spacing between the pipe support points, and the lowering depth maximum intervals. Line lowering can be undertaken without taking the pipeline out of service. To do so, however, some detailed calculations to avoid overstressing the pipe are required; these can be found in Mohitpour et al. (2007). API 1117 (2008) provides guidance on lowering distances and on the lengths of the transition pieces required. Before undertaking any calculations, the following site-specific pieces of information and field data will be required to enable a working procedure to be developed:

1. the location of roads, power lines, drainage tiles, and any other pipelines in the immediate area
2. the soil conditions
3. the height of the water table
4. the pipe grade, diameter, and wall thickness
5. the repair history and current condition of the pipe
6. the location of any sleeves, valves, fittings, and bends in the vicinity
7. details of the axle weights of any equipment likely to be working over or near operating pipelines
8. the position of the pipeline center line obtained from an as built survey or a recent intelligent pig run

The information obtained in step (8) will be reconfirmed by exposing an entire section of the pipeline *before* permitting any pipeline movement to occur. The elevation of the top of the pipe is measured at approximately 3-m intervals, closer if any valves or appurtenances (attachments) are present. As the excavation proceeds, the elevations at these discrete points will continue to be surveyed, so that the incremental changes can be compared with the deflected profile calculated in the stress analysis.

The distances permitted between supports will clearly depend on the type of product being carried in the pipe and should not exceed those determined by calculation.

If there are attachments present in the section being lowered, then these will need to be adequately supported. Figure 8-24 illustrates a typical means of supporting the pipe in a cohesive soil as the excavation proceeds. As noted in the drawing, the distance between the wooden cribbing or support material should not exceed 20 m (65 ft.).

Figure 8-25 illustrates the pipe support condition for cohesive soils, where the distance between the support materials, whether soil plugs, cribs, or air bags should not exceed 9 m (30 ft).

Once the new trench profile is established and the pipeline is exposed, all the girth welds will be examined for defects using ultrasonic techniques. The pipe surface will be visually inspected for corrosion and mechanical damage and the coating checked both visually and with a holiday detector for impairments. If defects are found, these will be assessed by the Integrity Support personnel prior to continuing the lowering procedure. In order to keep the combined stresses in the pipe at an acceptable level, it will be necessary to lower the internal pressure. The 20% reduction from the MOP in a gas line is less than the 50% needed in an oil line during lowering since the effect of self weight of product is much

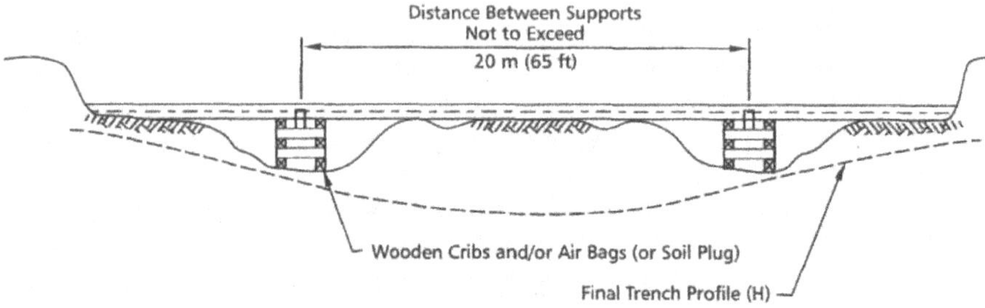

Figure 8-24. Pipe support in a noncohesive spoil (courtesy of Enbridge Inc.)

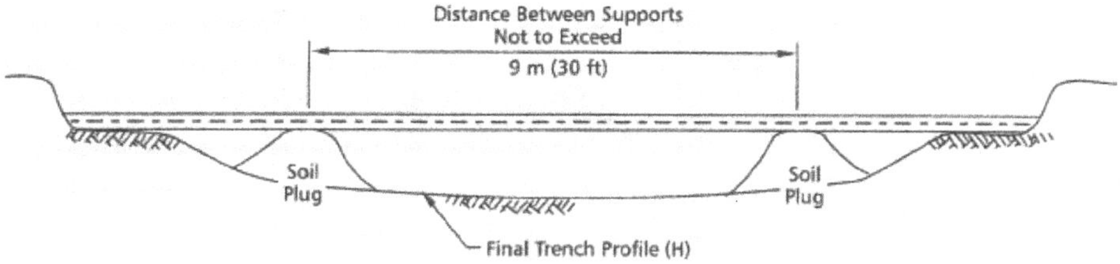

Figure 8-25. Pipe support in a cohesive spoil (courtesy Enbridge Inc.)

less. For safety reasons, only essential workers should be in the excavated bell hole during line lowering. When side booms or cranes are used for lowering in a filled pipe, it is good practice to use an appropriately sized spreader bar so that the correct sling spacing can be maintained. The maximum vertical lift in the pipe, as measured from the horizontal, will occur at the mid-point of the sling spacing and has a limiting value. For example, a 12–in. diameter pipe of 7-mm wall thickness will have a maximum vertical lift of 0.75 m (2.5 ft) and a sling spacing of 8.4 m (28 ft). As a rule of thumb for each 4-in. increase in pipe diameter, the allowable vertical lift limit will increase by 30 mm (1.2 in.) and the sling spacing by 1.2 m (4 ft).

Line lowering is done in stages so as to minimize bending stresses. The procedure will be to lift the pipeline near one support and remove one layer of cribbing or soil i.e., 150 mm (6 in.), before placing the pipe back on the support and proceeding to the next support. This is repeated continuously until the pipe is finally resting on the new trench bottom. An alternative to cribbing is to use air bags whose size and load carrying capacity must first be checked to ensure they can provide sufficient support for the pipe. For safety reasons, air valve controllers, which can be remotely operated from outside the bell hole, are used to inflate/deflate the bags. The air bags are placed in the excavation on top of a flat surface such as piece of plywood and they should have adequate side bracing to prevent the pipe section from rolling off. (When the air bag is inflated to its maximum pressure, the actual surface area of the bag in contact with the pipe will be quite small). In order to lower the pipe in an orderly fashion in accordance with the stress calculations, an amount of air is slowly released from each bag in sequence, until the entire length of pipe is resting on the deflated bags at the final grade. The bags are removed individually by lifting the pipe sufficiently to allow them to be withdrawn.

At the conclusion of the line relocation and before backfilling, the profile of the pipe section is surveyed and all welds are checked again for the appearance of cracking.

INCIDENT/FAILURE INVESTIGATION

A failure investigation is an extreme form of an engineering assessment and is the means by which the Integrity specialist hopes to determine the root cause and underlying factors leading to the loss of integrity of the pipe. Figure 8-26 illustrates the broad range of possible failure modes that can be investigated. Conducting an incident/failure investigation is obligatory in all pipeline standards. CSA Z 662-07, for example, contains the following clause,

Failure Investigations

"Leaks and breaks shall be investigated to determine their causes. Measures to prevent the occurrence of leaks or breaks due to similar causes shall be identified and implemented."

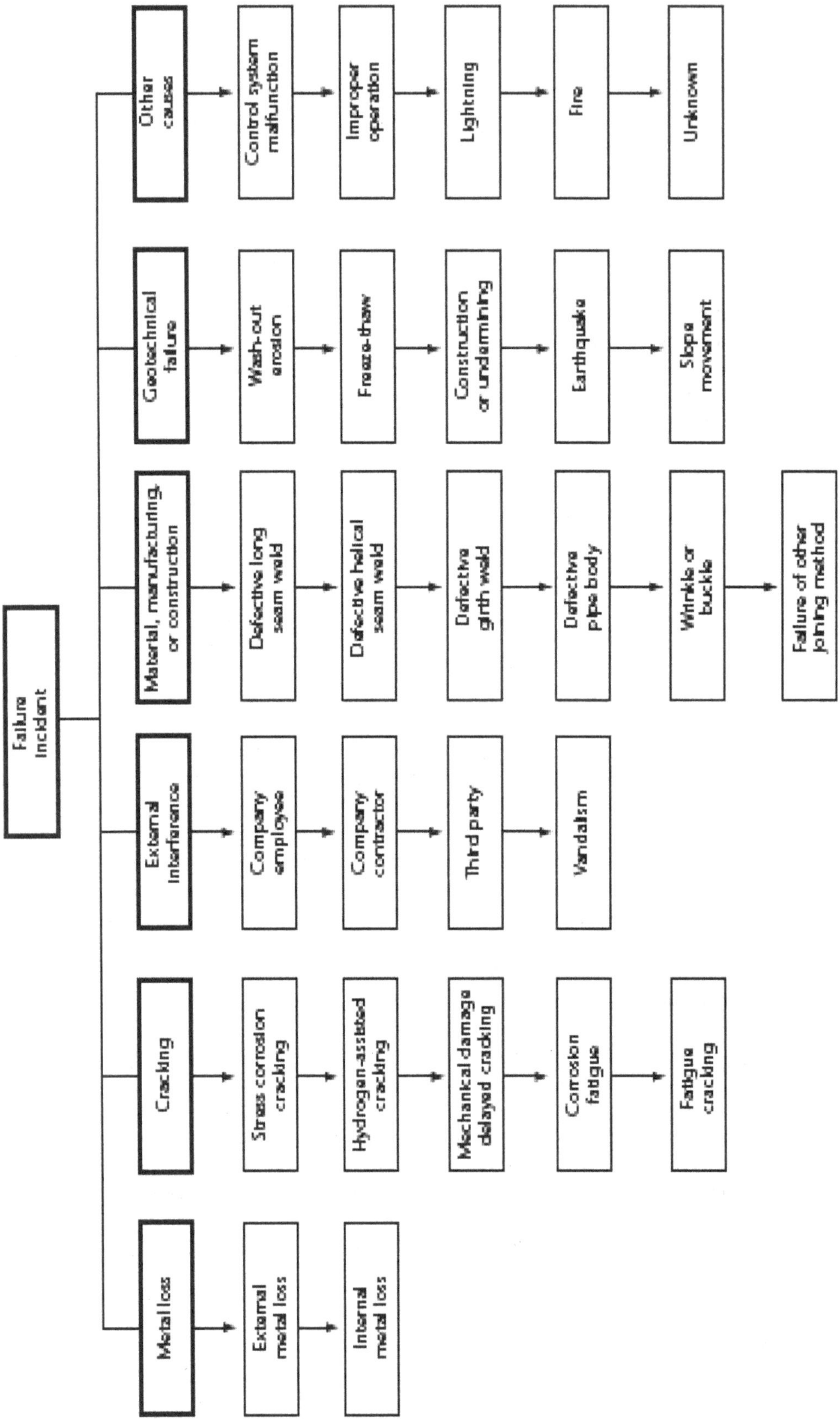

Figure 8-26. Potential causes of pipeline failure

Further, its Annex N states, "*Operating companies shall establish procedures for investigating and reporting failure and external interference incidents*" (see Clause 10.3.3). Such procedures should include:

a. the recording of incident information as specified in Clause H.5.2; and
b. an analysis to determine what changes are needed to improve the effectiveness of the pipeline IMP

Annex H, which is nonmandatory, contains a detailed list of elements relating to the reporting of an Incident which is defined as "*any unplanned release of product due to failure of a pipe or component.*" For example, H.5.2.2 provides details on how to record the physical location of an incident. H.5.2.3 describes how to specify the incident date and time, while paragraphs H5.2.4 and H5.2.5, respectively, set out requirements for recording the prevailing weather conditions and pipeline operating conditions at the time of the incident.

In keeping with other forms of forensic analysis, it is essential that the investigator obtains as much physical and anecdotal evidence as possible to assist in developing likely failure scenarios since, in many instances, the cause of failure is not immediately obvious. Chapter 9 describes the emergency response procedures that follow a reported incident and/or loss of containment. Curtailing flow to the affected pipeline section will be the first step and notifying first responders will follow shortly thereafter. Except in circumstances where a leak has been found as part of an investigative dig, it is unlikely that the integrity specialist will be the first on the scene. It is sensible, nevertheless, for this individual to check the hazards associated with the released product such as explosive levels (LELs), toxicity. e.g., H_2S or benzene, upon arrival at the site so as to ensure his own safety and that of others. The high level steps involved in performing an investigation are as follows:

1. Establish that it is safe for any personnel to enter the failure site.
2. Secure the failure scene to preserve evidence.
3. Determine if there are other pipelines in close proximity which may also have been damaged.
4. Start the collection of physical evidence from the site.
5. Protect and preserve all damaged surfaces.
6. Interview any witnesses.
7. Obtain the operational time history for the past 60 days.
8. Obtain a copy of the pipeline alignment sheet.
9. Obtain any relevant construction and operating records pertaining to the section of pipe.
10. Determine the results from any previous ILI results for the pipeline section and any operations and maintenance activities that have occurred.
11. Remove a sufficient length of pipe for the purpose of subsequent metallurgical examination.
12. Prepare test protocols for the metallurgical examination.
13. Establish a chronology of events.
14. Identify causes and contributory factors.
15. Conduct an engineering assessment based on the information received to determine if the pipeline is fit for a return to service.

Given that the above steps will take time and the operator will need to return the repaired pipeline to service as soon as possible, a preliminary determination of cause is warranted. This may conclude, for example, that the primary cause was generalized corrosion due to a coating failure, raising an immediate concern for the condition and structural adequacy of the remainder of the pipeline in the area. This might prompt the operator to voluntarily reduce the operating pressure upon its return to service, averting the possibility of being directed to do so by the pipeline Regulator. It must be emphasized, however, that no matter how dire the need for supply, the operating priority must always be safety and the protection of the environment through the avoidance of a repeat failure. The next priority will be to obtain uncontaminated site evidence before initiating the repair process.

A visit to the failure site preferably before any excavation has started will provide the investigator with a clearer understanding of circumstances that may have contributed to the failure. Observation of such features as soil instability, piping displacement, and failure orientation supported by sketches, photographs, measurements, and other pertinent data will assist in the subsequent metallurgical investigation and the development of sound conclusions. When conducting an investigation, the investigator, and others should take appropriate safety precautions when working around the exposed pipeline, ensuring that the upstream supply valve has been closed and continually monitoring toxicity and LELs. He should also ensure that there is adequate access and egress from the bell hole excavated around the failed pipe section. It is generally appropriate to expose the pipe to the nearest girth welds upstream and downstream of the damaged area as these serve as reference markers for both as built chainage and the ILI data. The resulting bell hole can be quite large and may on occasion fill with water and require continuous pumping to ensure its removal.

Field Evidence Gathering

Securing the site around the damaged zone is a priority and may require the assistance of the police service to cordon off the area depending on its size and location (Figure 8-27).

For some incidents, legislation requires that the site conditions remain undisturbed until government agencies complete their site investigation. Pipelines transporting liquids will decompress very quickly after rupturing, so the damaged pipe will remain in place. This is not the case with gas pipelines, where the stored energy in the compressed fluid may result in debris being scattered a considerable distance from the initiation point. Its complete recovery, particularly for high-pressure, large-diameter pipelines, can be very onerous, if not impossible, in some cases. It is important to survey the end-to-end distance between the arrests to assist in the reconciliation of the ejected fragments with the missing section of pipe. As many fragments as can be recovered will be removed under supervision from the site and "re-assembled" at a convenient location. If the pipe has a clearly defined longitudinal seam weld, this will greatly aid in the re-assembly process. This is not so easily achieved with spiral welded pipe. The resulting three-dimensional jigsaw puzzle is a very useful source of information in attempting to determine the initiation point and to recreate the event. For a gas pipeline rupture,

TABLE 8-2. Failure investigation equipment list

Recording	Measuring	Cleaning	Sampling
Camera, lenses	Dial depth gauge	Wire brush	Test kit for corrosion products
Film	Micrometer (25 mm/1 in.)	Nylon brush	Sample storage bags
Batteries	Vernier caliper	Flat file	Envelopes
Video camera	50 ft. (15 m) tape	Round file	Re-sealable liquid containers
Electronic flash	12 ft. (5 m) tape	Scraper	Clip board and grid paper
Flashlight	O.D. (5 m) tape	Emery paper	Dessicant
I.D. tags	Feeler gauges	Knives (2)	Soft, clean cloths
Flagging tape	Metal crayons		Packing materials
	felt tip pens		
	Dye penetrant test kit		
	Magnetic particle (MP) test kit		
	Ultrasonic thickness meter		
Personal Protective equipment			

Figure 8-27. The aftermath of a buried ruptured gas pipeline

a detailed survey will also be made of the crater, the location of each recovered fragment (recorded on a grid plot of the site), and the burn area to assist with future modeling of the blast process. Other surveyed features will include assessing distances to nearby dwellings and any damage caused to nearby property such as equipment/structures. A survey will also provide distances for new pipe requirements once tie-in locations are established. Before any materials are removed from site, environment sampling of soils, trapped water, and corrosion deposits should also take place. Ideally, these should be taken where no contamination/thermal alteration due to fire may have occurred. Samples of water in the ditch can be taken too, but this might be somewhat useless if there has been ingress of ground/surface water. The coating condition should also be examined where possible.

The remaining ends of the pipe in the case of a gas line are examined in the field and covered with a protective coating prior to their removal from the damage site. Often with a liquids pipeline rupture, the fracture surfaces can be covered with a split rubber hose to prevent damage and preserve features for the subsequent metallurgical examination. Under no circumstances should protective coatings be applied to fracture surfaces, as this may introduce chemical constituents and thus contaminate the fracture surface. A formal custody transfer process should be used when handing over physical evidence to a third party such as a metallurgical laboratory and resemble (Table 8-3).

Incident number		
Description		
Received from (Name)	(Position)	(Address)
Received at (Location)		
Received by (Name)	(Position)	

TABLE 8-3. Physical evidence and chain of custody

Date and Time	Reason for Transfer	Transferor	Recipient	Signature of Recipient
1/1/20xx	Preparation of test samples	Metallurgist X	Workshop supervisor	

Depending on the location and time of day, the event itself or circumstances leading up to it may have been witnessed, so it is important to determine if this is the case and if so, to record the witness observations. In addition, the investigator should identify and record any historical information provided by others concerning similar failures that may have occurred in the region. Field personnel will be an invaluable source of information because they are familiar with the day-to-day operation of the facilities. Since this group will be involved in the implementation of any subsequent mitigation efforts, it is important to include them at appropriate points in the investigation.

External damage such as a line strike or the possibility of vandalism or sabotage is an instance where a witness statement will be particularly important. These statements may not always be forthcoming as some initiating events and the cost of consequential damage may lead to protracted legal argument, prompting a witness to remain silent unless they have access to legal counsel. In such circumstances, the information disclosed to counsel remains privileged. A company, however, should experience little difficulty in obtaining statements from its own employees. The appropriate investigative safety agency, for example, the NTSB in the United States or the TSB in Canada, may choose to investigate the incident depending on its perceived importance or severity. (see Chapter 2). Incidents involving fatalities or serious injury are investigated by the appropriate health and safety authorities and the police, the latter to determine whether a criminal act was involved. Information obtained by these public bodies can often be made available to the investigator in due course.

Irrespective of the presence of witnesses, an extensive photographic and video record should be made of the damaged pipe and its immediate environs and a sketch made of the failure site. A noting should be made of the time, weather (including wind direction), ground conditions, (wet or dry), soil conditions (type, depth of cover, evidence of instability), and terrain (topography, vegetation).

Visual evidence of the "as found" failure conditions may prove critical in the subsequent analysis and compilation of the final report. Digital photography has largely supplanted the use of photographs taken with a conventional 35-mm camera. If, however, the latter type of camera is used, a higher speed 200 ASA film will minimize image distortion from movement as well as maximizing the depth of field. In addition, when an extensive number of photographs are planned, it is good practice to use multiple rolls of films with a minimum number of exposures such as 10 or 12. This is done to minimize the loss of photographic evidence in the event a problem occurs in the capturing of the images or in the development process of a roll of film. Notes should be made for each photograph as it is taken. This practice is helpful when reviewing the photographs at a later date or in the communication of observations linking each image to others. When multiple rolls of film are used, each roll should be identified in accordance with the prepared notes. This practice allows for an accurate correlation between the notes and the photographs upon film development.

The scale of the subject in a photograph or video may be easily determined by including an object within the image for reference. The practice of providing a scale is particularly helpful with close-up photography, and a pencil, coin, or finger will usually suffice as their relative size and dimensions are known. Including an image of a rule or tape measure allows direct measurements to be taken from the photographs (Figure 8-28).

Sketches and diagrams that provide overall and detailed information on spatial relationships, arrangements, measurements, and dimensions are helpful in the communication of observations specific to the failure incident to others and in subsequent analysis. Typically, sketches and diagrams may include an overall plan view, details of the location, and orientation of a failed component or observations specific to it. Cross-references are also made with the photographs. Once the "as found" conditions have been recorded, the selec-

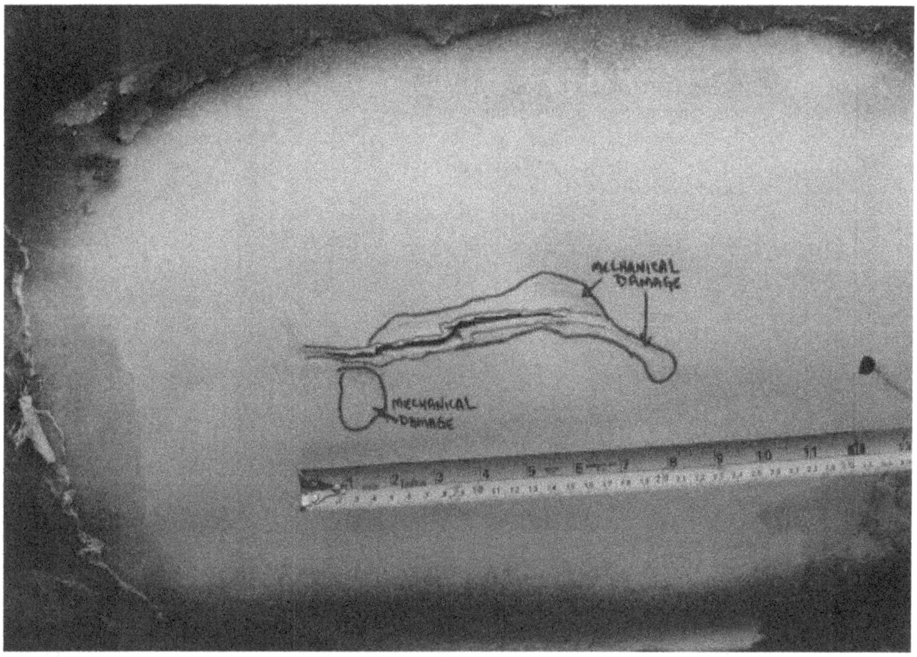

Figure 8-28. Measurement image

tion of physical evidence for analysis may then be performed. Physical evidence typically involves material items such as:

- failed components or equipment
- samples of liquids and lubricants
- remnant fragments, debris, and deposits

The selection of such physical evidence for analysis should be comprehensive, clearly identifying the original location of the evidence. Cross-references can be made to previous recordings such as photographs, sketches, vendor drawings, blueprints, or diagrams. Tables 8-4 to 8-8 provide typical details of the type of information collected for various types of failure.

In the event of a catastrophic failure where damage to components, equipment, or the pipeline is extensive, the site investigation may focus on the determination and the gathering of evidence associated with the primary failure event. A distinction between the primary failure event and resultant secondary damage is necessary in order to minimize subsequent analytical efforts. In order to conduct an investigation of a catastrophic failure, it may be

TABLE 8-4. Failed component data

Pipe:			
O.D.	w.t.	Grade	Spec
Other components			Material spec
Hydrotest pressure		MOP	
Year installed:		Years in service	
Manufacturer			
Comments			

TABLE 8-5. Leak or rupture

Date					Company Representative:		
Maintenance contact person				Witnesses			
Failed component service history							
Pressure at time of failure				Normal range of operating pressure			
Ambient temperature at time of failure							
Depth of cover							
Soil type	Sand	Clay	Rock		Peat	Loam	other
Soil conditions (around pipe)	Wet	Dry	Frozen		Stable	Unstable	
Soil color	Frost depth				Ph (optional)		
Was ice formed around leak site	Yes		No	If yes, depth of ice			
Coating type	Enamel		Polyethylene tape		Yellow jacket	FBE	Other
General coating							
Was there fluid under coating?			Test kit results				
Pipe surface conditions		No corrosion	Light general corrosion		heavy general corrosion		Pitting localized corrosion
Affected pipe length:			Repair date:				
Temporary or permanent?							
Replacement pipe:							
Specification							
Pipe length:		Outage time:			Radiographer		
Comments (description of corrosion products)							

necessary to complete a detailed inventory of all physical evidence. This practice ensures all evidence, such as small remnant fragments, is accounted for in the investigation.

Care should be exercised to prevent any post failure damage to physical evidence during the course of removal, handling, and transportation. Damage to failed components may

TABLE 8-6. Contact damage

Damage Caused By:	Equipment		Contractor		Third Party	Rock	
Does damaged area contain weld?					Yes	No	
If yes:			Girth		Seam	Fillet	
Circumferential location of damaged area							
Damage description			Plain dent		Gouge	Dent and gouge	
Method/equipment employed							
Damage assessment			Nondestructive examination MP inspection Liquid penetrant (LP) inspection				
Plain dent							
Maximum dent depth:			Was crack detected?			Yes	No
Stress concentrator in dent?	Yes	No	Prior to repair?			Yes	No
Is dent repairable?	Yes	No	After repair?			Yes	No
Applicable Code/Standard							
Gouge							
Maximum gouge depth:			Was crack detected?			Yes	No
Minimum w.t. in damaged area			Prior to repair?			Yes	No
Gouge repairable?	Yes	No	After repair?			Yes	No
Applicable code/standard							
Dent and gouge							
Maximum dent depth:					Was crack detected?	Yes	No
Minimum w.t. in damaged area					Prior to repair	Yes	No
Is damage repairable?			Yes	No	After repair	Yes	No
Applicable code							

TABLE 8-7. Leak details

Location	Pipe Body	Seam Weld	Girth Weld	Tap Connection
Other components				
Circumferential position:				
Pipe surface condition around leak				
Identification of failed pipe section and weld				
NDT used	MP inspection		LP inspection	
	Radiography			
Applicable repair specification				
Type of repair				
Result of NDT		Prior to repair		
		After repair		
Apparent cause of failure				
Comments				

render the evidence ineffectual or falsify results. Damage typically having the most detrimental impact on analysis takes the form of mechanical contact or corrosion of fracture surfaces. Another potential source of error involves the contamination of liquid and lubricant samples.

Metallurgical evaluations of fracture surfaces may provide significant information on the mode of failure and thus the root cause. For failed machinery components, such as pumps or compressors, an elemental analysis of liquids and lubricants may also prove helpful in that regard.

Minimizing the risk of damage during component/fracture surface removal is critical to a successful metallographic examination. The use of tools should be avoided in the removal of a pipe section or component where mechanical contact will occur with a fracture surface. Removal must therefore be made remote from any fracture surface. If removal involves cutting, such as employing an oxyacetylene torch, the fracture surface must be protected from the effect of heat and any molten metal spatter. Typically, hot cut distances should be a minimum of 100 mm from the fracture surface.

The site investigator should not attempt to physically fit fracture surfaces together, as this generally causes surface rubbing and may severely damage the fracture surfaces on a microscopic scale and make detailed analysis difficult or even impossible. This is probably the most common way that evidence is damaged.

TABLE 8-8. Rupture details

Fracture Length	Number of Fragments	
Fracture mode	Brittle	Ductile
Fracture surface features		
Fracture surface orientation w.r.t. pipe surface		
Minimum all thickness of fracture surface	Location of fracture origin(s)	
Location of fracture arrest points		
Fracture propagation direction(s)		
Characteristic of fracture paths(s)		
Did any mechanical damage occur		
Did ignition occur?		
Apparent cause of failure		
Comments		

No attempt should be made to clean fracture surfaces since this may inadvertently remove chemical constituents considered relevant to the analysis. Furthermore, any means of manual cleaning, such as brushing or the application of aggressive cleaning solvents, may result in microscopic damage to the fracture surface. As noted earlier, the use of protective coatings may also inhibit the effective cleaning of fracture surfaces during the course of analysis. Under ideal circumstances, fracture surfaces should be gently dried with a clean soft cloth, wrapped in either clean soft cloths or low lint tissues, packed in an appropriately sized re-sealable plastic bag, and clearly identified. Latex gloves should be worn so as to avoid physically touching failure surfaces with oil and dirt from the skin.

Sampling of liquids and lubricants should be conducted in such a way as to minimize the potential for contamination of the product. Utensils used in the collection of samples should be cleaned and cleaned upon reuse. Liquids and lubricants should be stored in clean re-sealable plastic containers and clearly identified.

It is extremely important to forward all physical evidence to an exclusive source such as a metallurgical laboratory for subsequent analysis. Often, evidence is inadvertently lost or distributed in order to satisfy a number of varied interests. This practice is not conducive to accurate analysis, as no benefit is derived from a restriction of essential evidence.

All the evidence removed from the site should be clearly tagged and contain supporting documentation such as its original location and the date of its removal from site. Additional useful information to accompany the evidence would be a contact name, a copy of the completed failure investigation checklist and duplicates of photographs and other recordings. Essentially, any readily available information that may be of benefit to the laboratory analysis should be included. All evidence needs to be handled with care and appropriately packaged to minimize damage in transit. It is very important to provide the name of a contact person for receipt of the evidence and to clearly identify the destination.

In order to make a repair, the damaged ends will be removed and these are sent for analysis. Any metallurgical investigation should follow an agreed-upon test protocol, established in conjunction with the safety board, and the appropriate pipeline regulator. A sample protocol would typically include examination of the damaged surfaces and determining the pipe material mechanical properties.

Conducting Metallurgical Investigations

In order to ensure that a logical and thorough metallurgical examination is undertaken as part of the failure investigation, it is necessary to follow a previously established test protocol. The investigating laboratory will set up a working file containing all of the relevant documentation and original copies of all test results and photographs. A storage area, identified by the incident number, is set aside to receive the failed component and all unsampled material should be stored in this assigned area. If the incident is being investigated by a regulator or safety authority such as the Transportation Safety Board, there may be conditions placed upon the disposal of materials. Usually, no material may be disposed of until clearance has been given by the designated investigator from the Transportation Safety Board or the appropriate Regulatory authority. Under no circumstances should any material be discarded before a failure report has been issued.

The first step in the metallurgical investigation is to conduct a thorough visual examination of the failed component, carefully noting and documenting the following items:

- The location of the failure (i.e., on a girth weld, fillet weld, or the pipe body)
- The direction of product flow
- The initiation point of the failure
- Evidence of mechanical damage
- Evidence of material defects

TABLE 8-9. Laboratory investigation visual and NDE examination

General Observation
Failure location
Failure orientation
Relevant dimensions
Type of NDE test
Purpose of test
Results
comments

- For fractured components, the fracture appearance (i.e., ductile versus brittle)
- The diameter and wall thickness of the pipe at the failure location
- The depths of any dents and/ or gouges
- Label and record the position of all components and fragments
- Record the positions where samples are to be taken

Sometimes it will be useful to create a negative mould of the damaged surface for future reference.

Very often, the exposed surfaces of failed components will need to be nondestructively examined to determine whether any material defect or cracks are present, and if this is the case, their identities and locations will be documented. The methods used for these examinations may include ultrasonic, radiographic, magnetic particle (MP) inspection, and these must be completed before any destructive examination is started. If significant indications are found during the MP inspection the crack should be carefully broken open and the cross section examined to determine the cause.

Fractographic examination is the next step with the exposed fracture surfaces visually examined by the naked eye and a stereo microscope (and in more complex cases, a scanning electron microscope) to determine the following:

- The fracture initiation region
- The identity and location of any material defects associated with the fracture
- The mechanism of cracking (i.e., ductile, brittle, fatigue)
- Deformation

For comparative purposes, it is useful to also examine a representative nearby area containing sound metal and a weld zone.

In order to minimize the risk of destroying evidence, this examination needs to be performed and photographs need to be taken prior to sectioning or otherwise altering the fracture surface. Cross-sections should be prepared from such regions as the fracture

TABLE 8-10. Laboratory investigation — fractography

Macroscopic Examination of Fracture Surfaces	
Fracture length	Fracture mode
Fracture surface features	
Fracture initiation site(s)	
Fracture arrest point(s)	
Fracture propagation direction(s)	
Minimum wall thickness of fracture surface	Fracture surface orientation
Microscopic examination of fracture surfaces	
Comments	

Location of samples taken for testing____________________

TABLE 8-11. Laboratory investigation—mechanical properties

	Strength		%Age	%Age	
	Tensile	**Yield**	**Elongation**	**Reduction in Area**	**Hardness**
Requirement					
Specimen no.					
Notch toughness					
	Test Temperature (°C)		Absorbed Energy Full-Size Equiv (J)		%Age Shear
Specimen					
Hardness profile					
Comments					

Note:

Carbon equivalent = $C + \left[\frac{Mn}{6} + \frac{Si}{24} + \frac{Cu}{15} + \frac{Ni}{20} + \frac{Cr+Mo+V+Cb}{5} + 5B\right]$

initiation site, or areas containing apparent material defects. These specimens will undergo metallographic examination to evaluate:

- Unacceptable microstructure or any harmful changes in the original microstructure
- Material imperfections
- Fracture mechanism
- Micro hardness measurements may also be required to identify locally hardened areas of metallographic specimens and if so, these are taken at a number of sampling points at the failure location and at a representative area of sound metal.

It is customary to complete a full mechanical testing and chemical analysis of the failed component for the dual purpose of determining the defining properties of the material and whether the applicable specification has been met. This task requires the following steps to be undertaken:

- Select the type, locations, and number of test specimens.
- Provide instructions regarding preparation of specimens and testing required.
- Direct machine shop in preparing test specimens.
- Perform or arrange for tests.
- Maintain records of identity and location of test specimens
- Evaluate the test results.

Tensile testing will follow the procedure set out in ASTM A370 and fracture toughness testing done in accordance with ASTM E23 to determine the brittle to ductile transition curve. The chemical composition of the pipe material is found using a mass spectrometer. All microspecimens will be identified using the assigned incident number and stored in a labeled microstorage cabinet.

Collecting Background Information

The background data collection phase is conducted away from the failure site with the type and amount of pertinent information sought being highly dependent on the nature of the failure and the extent of the subsequent analysis required. As may be seen from Table 8-12, the background information required will include the construction history of the pipeline.

TABLE 8-12. Background information

Construction	Service	Maintenance
Manufacturer	Normal/abnormal loading	Schedules
Manufacturing date	Accidental overloads	Inspections
Manufacturing process	Cyclic loads/vibrations	Repairs/modifications
Fabrication	Temperature gradients	Overhauls
Installation date	Changes in operations	Calibrations
Materials and grades	Corrosive environments	
Welding procedures		
Design drawing and specifications		

Knowledge of the service and maintenance history will help in the reconstruction of the events leading to the failure and involves obtaining the most recent pressure history of the pipeline. In the case of a liquids pipeline, the pressure spectrum for the past 60 days of operation (usually) can be as meaningful as any pressure extremes.

Many large companies nowadays use extensive geographical information systems to maintain a data base containing much relevant information to the investigator. For example, the age and method of manufacture of the line pipe, method of coating, a history of any repairs made to the pipe, and importantly the location and nature of previously observed defects on the pipe. Alignment sheets showing the as built location are useful to the investigator and may be available electronically or in hard copy. The result of the last ILI record should be examined focusing initially on the damaged area to determine if any anomalous situation was present. Often with the benefit of hindsight, such indications are present in the record but their significance has been misinterpreted. Should this be the case, then the inspection records will be re-analyzed to determine if other potential defects have also been misinterpreted.

It is extremely important to reconcile the location of ILI anomalies and tool behavior with the alignment sheet record. For example, in a recent failure investigation, the signal indicating a dent caused by an unreported line strike was misinterpreted as tool lift off at a bend. Later correlation of the damage location with the alignment sheet showed no bends were present.

Having obtained and processed the relevant information, the penultimate step in the failure investigation is to determine the failure mechanism and the circumstances that led to it. The methods described in Chapter 6 are used in this context. Finally, a report is prepared whose conclusions may initiate a change to the IMP.

Reporting

The final step in a failure investigation is report preparation. The first report, however, will be an incident summary that is prepared by the pipeline operating company within 24 hours of the incident occurring. It will briefly describe the incident type (i.e., leak, rupture, mechanical, or corrosion damage), whether a fire or injuries were involved, the incident date, the facility name and legal location, and chainage, if available. Details will be provided of the licensed maximum operating pressure, the nominal wall thickness, grade, diameter and length of the pipe, its in-service date, and coating type. If possible, the incident summary should include details of the repair plan including the estimated time required for repair and the type of repair (i.e., cutout, sleeve, tape repair, etc.). In the case of a cutout, the length of replacement pipe, replacement pipe grade, nominal wall thickness. and coating type and material should be given. The results of nonspeculative findings of the investigation to date will be supplied. A sample incident sheet can be found in Table 8-13.

This incident summary will be followed some few weeks later by the issuance of an interim report. In the case of a simple incident, the investigation may have been completed

TABLE 8-13. Incident background information

Incident No.	Repair No.	
Facility No.	Location	
Facility name		
Chainage	Proximity to highway	
Date and time of discovery	Date and time reported	
Circumstances of discovery		
Type of failure		
Rupture	Leak	Damage

Notes: More specific information would include:

1. details of the number and severity of any casualties
2. a description of the incident and the extent of any damage to the commodity pipeline, the environment, and other property
3. a description of the nature of the fluid contained in, or released from the commodity pipeline and any specific measures taken by the operator to protect the public
4. the anticipated repair method to be used and the expected time of arrival of repair equipment on site
5. the name, address, and title of the person making the report

by this point. For a more complex investigation, the interim report will present the results from the investigation to date and any preliminary conclusions and recommendations. It allows each recipient the opportunity to comment on the investigation preliminary findings and recommendations and to provide feedback on any errors and omissions. It is rare for an interim report to be more than a few pages in length, but in structural form, it will be similar to the final report and will contain a recommendations section with an estimated completion date and a list of personnel assigned for implementation. A typical structure of this type of report is as follows:

Summary Page
1.0 Background
2.0 Conclusions
3.0 Costs (of fluid lost, of repairs, to producers from lost production)
4.0 Failure Investigation
5.0 Discussion
6.0 Recommendations

After a further time lapse, the final report for a more complex incident will be completed containing the results of the onsite and metallurgical investigations, the estimated cost of the incident, conclusions, and recommendations.

The preliminary, the interim, and the final reports may be made available to the safety authority and the pipeline regulators through the defined regulatory process

Often the independent investigation conducted by the safety authority will take longer than the company investigation; however, once prepared, it will be submitted to the company and the relevant pipeline regulator for comment before becoming a public document.

The degree of detail required in reporting varies between jurisdictions, for example, in the United States companies, are required to report in the following cases:

- incidents involving third parties or third party pipelines
- incidents resulting in leaks (of any magnitude)
- incidents resulting in ruptures
- incidents where the cost of remediation exceeds \$250,000 (operating) or \$500,000 (capital)

In Canada, on federally regulated pipelines, there is no requirement to estimate the financial consequences of an incident, and reporting is limited to:

- incidents involving external damage to pipelines
- incidents resulting in leaks over 1.5 m^3 on liquid pipelines
- incidents resulting in any loss of containment on a gaseous medium pipeline
- incidents resulting in injuries

Record of Operating Incidents

A record of operating incidents should be compiled by the Integrity Technical support group and maintained in its files and preferably in a GIS database. This record lists all incidents that have occurred during the current year and is updated with each new incident by the person responsible for the investigation.

This record would include all the pertinent information on the incident.

Implementing Preventive measures

The team member responsible for conducting the investigation should undertake a follow-up to ensure that adopted recommendations measures for the investigation are being implemented and that the relevant procedural changes are documented in the Management of Change process within the company's Integrity Management plan.

Legal Issues

An important step in managing integrity is identifying all the potential hazards (threats) that could adversely impact the integrity of the pipeline and developing a rigorous IMP to deal with them. The legal consequences of not doing so should an incident occur is clearly set out in the following excerpt from an article written by Nidd et al. (2007) on the litigation consequences of pipeline integrity management choices.

"*In the event of a significant failure, however, the pipeline company will have to counter allegations of inadequate assessment and maintenance and defend its IMP. The only chance of defending the company is to show it did everything it reasonably could to identify and prevent failures. In order to convince a jury the company undertook all reasonable measures, those measures need to be objectively verifiable as sufficient to identify likely corrosive influences or other weaknesses on a pipeline.*

An operator's IMP for the inspection of its pipelines and identification of potential problem areas, therefore, should be based on a scientific and objective methodology. This is important because competent expert proof will be essential in defending the reasonableness of a pipeline company's integrity management decisions, and an expert's opinions must be based on objective, verifiable evidence.

Sound pipeline integrity management and the presentation of evidence at trial in defense of an operator's activities are inextricable. if the operator's primary method of corrosion inspection, for example, is too loosely based on the subjective analysis of projected growth rates or leak histories, rather than inspection methods such as ILI or hydrostatic testing which provide objective data, then any expert testimony in defense of the operator is less likely to convince a jury and even risks exclusion as evidence by the court for being scientifically unreliable."

Hence, great care should be taken in providing technical support to an IMP to ensure adequacy or completeness in processes and effectiveness in its implementation.

REFERENCES

API 579-1 ASME FFS-1, 2007, *Fitness for Service*, American Petroleum Institute, May, p. 1128.

API 1117, 2008, *Recommended Practice for Movement of In Service Pipelines*, American Petroleum Institute, July.

ASTM E23-07ae1, *Standard Test Methods for Notched Bar Impact Testing of Metallic Materials*, STM International, June 1, 2007, 28 pp.

ASTM A370-09ae1, *Standard Test Methods and Definitions for Mechanical Testing of Steel Products*, ASTM International, June 1, 2009, 47 pp.

Brake, D. E., Stearns, S. V., and Ronsky, N. D., 2008, "Enhanced Pipeline Leak Detection Using Airborne Differential Absorption LIDAR," Paper No. IPC2008- 63650, Proceedings of IPC2008 7th International Pipeline Conference, Calgary, Alberta, Canada, September 29–October 3.

Caleyo, F., González, J. L., and Hallen, J. M., A study on the reliability assessment methodology for pipelines with active corrosion defects, Int. J. Pres. Ves. & Piping 2002; 79: 77–86.

Fessler, R. R., and Rapp, S., 2006, "Method for Establishing Hydrostatic Re-test Intervals for Pipelines with Stress Corrosion Cracking," Paper IPC 20006-10163, Proceedings of ASME 6th International Pipeline Conference, Calgary, Alberta, Canada, September.

Francis, A., and Senior, G., 1998, "The Use of Reliability Based Limit State Methods in Uprating High Pressure Pipelines," Proceedings of IPC 1998 2nd International Pipeline Conference, Calgary, June.

GE-PII, 2010, "Post Inspection Evaluation" http://www.gepower.com/businesses/ge_oilandgas.

Kiefner, J. F., 1973, *Failure Stress Levels of Flaws in Pressurised Cylinders*, American Society for Testing and Materials, ASTM STP 536, pp. 461–481.

Kiefner, J. F., and Eiber, R. J., 1984, *Investigative Procedure for Examination of Failures in the Field*, American Gas Association.

Mohitpour, M., 2008, *Energy Supply and Pipeline Transportation — Challenges and Opportunities*, ASME Press, New York, NY.

Nessim, M. A., and Zhou, W., 2005, "Target Reliability Levels for Design and Assessment of Onshore Natural Gas Pipelines," Prepared for Gas Research Institute, GRI-04/0230, March.

Nessim, M. A., and Zhou, W., 2006, "Reliability Based Design and Assessment for Location-Specific Failure Threats," Paper No. IPC06-10097, ASME Proc. 6th International Pipeline Conference, Calgary, Alberta, Canada, September 25–29

Nidd, P., Sammons, A., and Towers, J., 2007, "Litigation Consequences of Pipeline Integrity Management Choices, Part 1," Pipeline and Gas Journal, Vol. 234, No. 6, June, pp. 40–46.

Rizkalla, M., 2008, *Pipeline Geo-Environmental Design and Geohazard Management*, ASME Press, New York, NY.

Rizkalla, M., Trigg, A., and Simmonds, G., 1999, "Recent Advances in the Modelling of Longitudinal Pipeline/Soil Interaction for Cohesive Soils," Proceedings OMAE Conference, Vol. 5, Pipeline Technology, ASME, New York, NY.

Sen, M., and Kariyawasam, S., 2008, "Analytical Approach to Determine Hydrotest Intervals," Paper IPC2008-64357. ASME Proceedings 7th International Pipeline Conference, Calgary, Alberta, Canada, September 29–October 3

Senior, G., Hopkins, P., and Francis, A., 1998, "Uprating the Design Pressure of In-Service Pipelines Using Limit State Design And Quantitative Risk Analysis," 2nd International Onshore Pipelines Conference, IBC, Istanbul, December 2–3.

Stephens, M., and Nessim, M., 2006, "A comprehensive Approach to Corrosion Management Based on Structural Reliability Methods," Paper IPC2006-10458, Proceedings of ASME 6th International Pipeline Conference, Calgary, Alberta, Canada, September 25–29

Strong, A. P., Sanderson, N., Lees, G., Hartog, A., Twohig, R., Kader, K., Hilton, G., and Khlybov, A., 2008, "A Comprehensive Distributed Pipeline Condition Monitoring System and Its Field Trial," Paper No. IPC2008-64549, Proceedings of IPC2008 7th International Pipeline Conference, Calgary, Alberta, Canada, September 29–October 3.

Chapter 9

EMERGENCY RESPONSE PLANNING AND INCIDENT MANAGEMENT

Pipeline emergency response planning (ERP) and execution comprise organization, processes, tools, and facilities required to respond to an emergency situation. An emergency situation can be minor or major. Major emergencies result from an incident causing rupture, release of containment/spill, fire/explosion, injury, fatality, loss of property, impact to the public and the environment, and finally cleaning out/decontamination, repairs, and restoration of service. While minor situation can be regarded an emergency from a response point of view (e.g., bandaging a wound to stop it bleeding and prevent infection, due to an industrial accident), it will not require a high level of alertness and mobilization of resources (including third-party agencies). This chapter is not all inclusive. It outlines process response to major pipeline-related emergencies. While a typical release of pipeline containment is outlined as an incident response design tool, physical tools/equipment and facilities, etc., required for responding to major emergencies are not covered.

BACKGROUND

Pipeline construction and operation impact the environment in which the facilities are located and traverse (Figure 9-1). Therefore, various factors have to be considered when designing, constructing, operating, and maintaining a pipeline facility including the long-term environmental issues, which must cover cumulative environmental impacts caused by a pipeline system expansion/growth and operation. To effectively mitigate impact of integrity issues as a consequence of pipeline operation on the public and the environment emergency, response and incident management plans must be in place.

Each pipeline system has unique features, and each fluid and contaminant has its own characteristics and properties. Because of the uniqueness of the system and the contaminant therein, environmental impact will be different and any response to an emergency situation will be specific to its operation.

Pipelines are, however, the safest and most reliable way to transport hydrocarbon energy products, including natural gas, crude oil, liquid petroleum products (LPG, condensate, refined batched products), chemical products, slurries, carbon dioxide, monoxide, etc. Pipelines are also use for renewable energy services such as the transportation of hydrogen, ammonia, etc. (Babu et al. 2009). Unlike road, rail, air, and marine transportation methods, pipelines are the only transportation system where the system transport the product continually through the enclosed system, i.e., the pipe and normally below ground or sea level.

However, no matter how well a pipeline facility is designed, maintained, and operated, it can still experience emergency situations: Pipeline emergencies may arise from the following (HSE 1996; Boatwright 2007; Figure 9-2):

- Intrusion/damage by third party (accidental or carelessness)
  - Third-party inflicted damage (instantaneous/immediate fail)
  - Previously damaged pipe (delayed failure mode)
  - Vandalism/terrorism

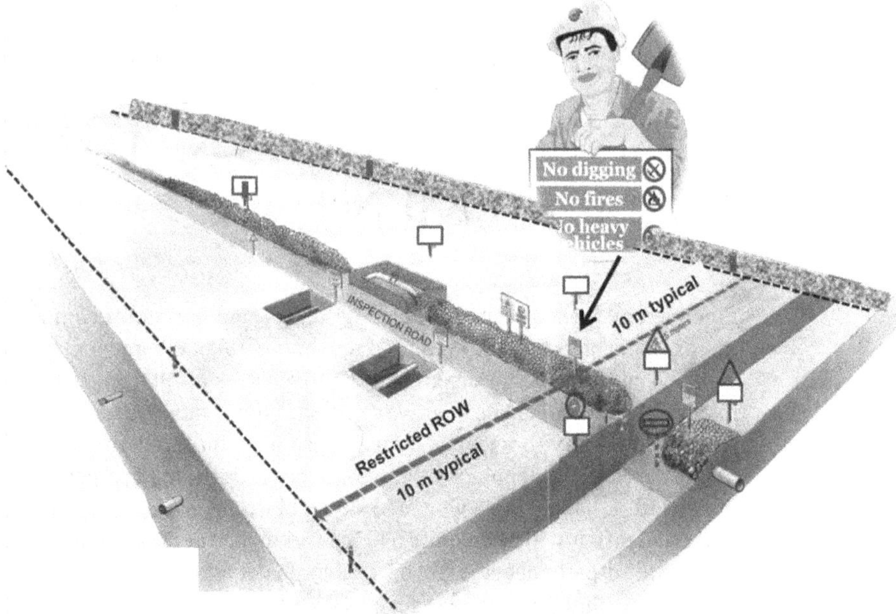

Figure 9-1. A typical pipeline right of way arrangement

- External/natural hazards, including:
  - Soil movement including settlements/heave
  - Earthquakes
  - Heavy rain or flooding
  - Cold weather
  - Lightning
- Other events such as failure of adjacent facilities, deliberate or malicious acts
- Corrosion (both internal and external)
- Environmental cracking and stress corrosion cracking
- Mechanical failures including:
  - Material defects, including wrinkle bend or buckle
  - Manufacturing-related defects
    - Defective pipe seam
    - Defective pipe
- Construction-induced defects, including weld defects
- Fatigue
- Operational errors

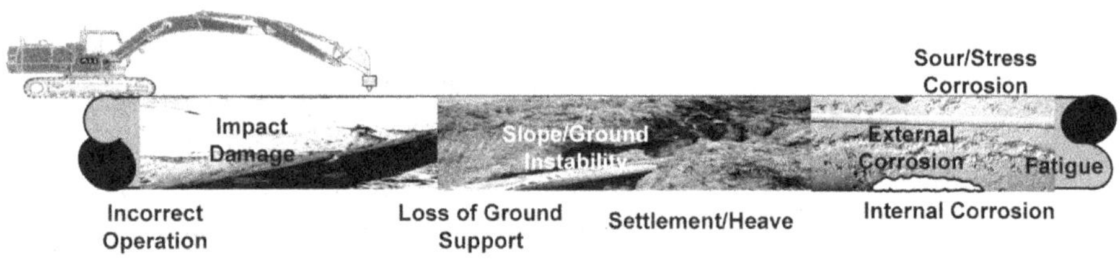

Figure 9-2. General causes of pipeline failure

- Maintenance problems
- Miscellaneous equipment and pipe
  - Gasket O-ring failure
  - Stripped threads/broken pipe/coupling fail
  - Control/relief equipment malfunction
  - Seal/pump packing failure
- Worker/operational error or events causing a fire, explosion, accidental release, or operational failure that disrupts normal operating conditions (Figure 9-3)

Therefore, risks to pipeline during construction, operations, and maintenance are real. If an incident does occur, both the operator and local emergency responders along the pipeline route must be ready to respond. As an example, pipelines carrying high-vapor pressure (HVP) fluids such as natural gas liquid, there are three potential risks associated with

Figure 9-3. Emergency response situations in pipeline operation

Figure 9-4. Potential risks associated with the failure of the pipeline and the release of (NGL) gas into the surrounding environment.

the failure of the pipeline and the release of (NGL) gas into the surrounding environment (Figure 9-4). This includes:

- vapor cloud dispersion close to ground
- the risk of fire
- a vapor cloud explosion
- the possibility of toxic effects of an un-ignited airborne vapor cloud where small quantities toxic of substances (such as mercury in the gas/product) are present

Therefore, well thought-out emergency plans must be in place. Such a plan must include measures to promptly and effectively shut down the pipeline, dispatch responders, and take measures designed to protect the safety of the public, employees, and the environment. Response activities include proper notification of local, state and federal agencies, and control, containment, and recovery of any released product.

Any environmental impact must consider the following:

- Environmental setting
- Threshold/acceptance limits
- Extent and nature of impact
- Mitigation measures
- Impact after mitigations
- Nearby inhabitant facilities (homes, schools, playgrounds, churches, etc.)

DEFINITION OF EMERGENCY

An emergency is any serious situation or occurrence that happens unexpectedly and demands immediate action or assistance.

Such situations, because of severity or size of impact, render normal response mechanisms inadequate. In the event of such a crisis, a more comprehensive and immediate response is required to manage the crisis and respond to the needs of the situation, which may include adjacent community. Activation of an emergency Incident Command System (ICS) and the appropriate emergency response (ER) team under these circumstances provides the community adjacent to the facilities with an appropriate response capacity in the event of such an emergency.

In order to protect the health, safety, and welfare of people or to limit damage to property, the environment, or company operations, the pipeline industry defines emergency as an unforeseen or imminent event that requires:

- prompt coordination of resources
- special communications
- and heightened authority for employees

To effectively assess ER and the required environmental protection in the event of a leak or pipeline rupture, the pipeline industry considers the following for ERP and ER:

- Effective policies
- Procedures for meeting regulatory needs including any Permitting Considerations
- Tools and equipment
- Identification of ER centers and need for additional center personnel and facilities as pipeline systems grow/expanded
- Class room and field training needs to respond to emergency situations covering:
  - Fluids handled
  - Establishing high-consequence areas
  - Pipeline safety issues and safe repair techniques
  - Fluid release (leak or rupture)
  - Flammable and safe work zone identification
  - Risk and HAZOP
  - Leak surveillance, illegal tapping
- Need for ER field exercises
- Third-party ER training needs and programs

ER COOPERATION — ESSENTIAL REQUIREMENTS

By combining efforts, pipeline companies and ER entities usually provide a more complete training program and response to an emergency situation.

To implement an effective ER pipeline companies, often undertake full-scale mock emergency exercises at their facilities (compressor/pump station) sites in conjunction with fire departments, police, and nearby medical centers as well as emergency call centers. Such a mock scenario could involve real situations such as a large tornado that could cause damage to the facility resulting in a minor/major release of natural gas, etc. A mock emergency is a good test of ER procedures, training programs, crisis management, and ICS.

In many countries, local state/federal/national laws dictate that pipeline companies must have detailed and comprehensive ER plans in place, which have been approved by their respective regulatory or transportation departments. Common to all such plans is a requirement to have a central unified command system (typically known as ICS) available at all times for handling ER. This system must identify the ER personnel who have the authority to make operational and financial decisions. The ICS also requires the company to make available an emergency control center. Additionally, some companies' ER teams have authority for enabling local contractors such as specialists to be hired as and when needed. The emergency command center must contain current copies of the company's approved ER procedure/manual including regional and site and pipeline facilities drawings and maps. Many companies make use of their GIS database information to physically display maps of the affected neighborhood so that the presence of potential hazards, the location of other utilities, and the proximity of hospitals, etc. can be quickly identified. In some cases, evacuation centers are identified in the event of major incidents.

As with any emergency, the primary goals are to secure the site, preserve evidence, protect the safety of the public and surrounding area, and where possible, contain the release.

It is important that the company maintains detailed records that are readily accessible of the contact information of local landowners who may need to be evacuated as well as a HCA that must be protected during an emergency. This is particularly the case in circumstances of a sour gas, natural gas, pipeline leak, or break, where the product, being lighter than air, will rise, bubble up, and dissipate.

In the case of HVP fluid release (such as propane, carbon dioxide [CO_2], etc.), the product is heavier than air and will collect in low-lying areas close to the ground. With sour gas, it may be more prudent to have a shelter in place policy rather than evacuation, but this needs to be determined beforehand. It is only a properly thought through ER procedure that can proactively foresee effective response needs for such emergency situations and share it with first responders.

Liquid (low or HVP) releases must be quickly contained to prevent extensive environmental damage, and the clean-up operation should begin as soon as possible, using a variety of means to recover the product. This includes other products such as fertilizers, acids, and petrochemicals that are also transported by pipelines. These will require Material Safety Data Sheet (MSDS) information to be on hand and neutralizing agents and protective clothing available. Subsequently, measures will be required to remediate and monitor the affected areas, but this is generally outside the scope of the ER plan. However, it is vital that pipeline company field personnel meet at frequent intervals with the local emergency service providers, throughout their system, to keep them fully informed of any changes made to ER policies and procedures. It is also useful to stage regular table-top (i.e., class room type) exercises or replicate full alerts to test the adequacy of emergency procedures and ICS.

In addition to protecting the general public, the well-being of company personnel employees must also and at all times be ensured in the event of an uncontrolled release or damage to a facility that could potentially cause accidents or fatalities. Hydrocarbons can potentially be explosive and can displace air in confined spaces, which can cause asphyxiation. All facilities within buildings, which have pipeline or equipment that handle/contain such a fluid must be equipped with detection devices and fire fighting and life-saving equipment.

These detection devices sound an alarm when a leak is detected and will automatically shut down the facilities *before* any leak can reach a concentration that supports combustion or an explosion in the presence of an ignition source or can cause serious asphyxiation. These detection devices can instantly communicate to a facility control office for immediate response.

ELEMENTS OF ER

The US Department of Transportation, CFR 49 (US DOT 2009), generally defines ER requirement for pipeline systems. The regulation states that each operator shall establish written procedures to minimize hazards resulting from pipeline emergency: It provides guidelines for having tools and for actions directed to protecting the public first and then the property. Additionally, it states that:

Each operator shall establish and maintain liaison with appropriate fire, police, and other public officials to:

- Learn the responsibility and resources of each government organization that may respond to a gas pipeline emergency

- Acquaint the officials with the operator's ability in responding to a gas pipeline emergency
- Identify the types of gas pipeline emergencies of which the operator notifies the officials
- Plan how the operator and officials can engage in mutual assistance to minimize hazards to life or property

Additionally, section 1CFR stipulates the requirements for "Emergencies" to be covered in the appropriate manual as follows:

- Receiving and identifying and classifying notices of events that need immediate response by the operator or notice to fire, police, or other appropriate public officials
- Prompt and effective response for each type of emergency
- Having personnel and equipment, instruments, tools, and material available as needed at the scene of emergency
- Taking necessary actions such as emergency shut down
- Control of released hazardous liquid at an accident scene to minimize the hazards, including a possible intentional ignition in the cases of flammable highly volatile liquid
- Minimization of public exposure to injury and probability of accidental ignition by assisting with evacuation
- Notifying fire, police, and other appropriate public officials of hazardous liquid emergencies and coordinating with them all responses
- Provide tools to assess the extent and coverage of vapor cloud and determine hazardous area
- Provide for a postaccident review of procedures and all activities to determine the effectiveness of emergency procedures and corrective action taken to establish deficiencies.

ROLE OF EMERGENCY MANAGEMENT

As previously defined, the role of Emergency Management is to provide a prompt and coordinated response to emergencies.

EMERGENCY SITUATIONS IN PIPELINE TRANSMISSION

Pipeline companies generally consider an emergency to be any unplanned occurrence that could or does result in:

- serious injury or threat to the safety of the public or company personnel
- major or significant damage to property and operations
- a potential impact to the environment

WHAT IS AN ERP?

An ERP is a document that helps direct and coordinate the actions to be taken in the event of an emergency. The plan must include information about the facility as well as detailed procedures and responsibilities for emergency actions as defined in Role of Emergency Management.

ER GUIDELINE MANUAL

The ERP manual will generally contain the following information:

- Critical infrastructure protection by defining ICS
- Preparedness, response, and recovery
- Communication and information
- Information for protection of the public, personnel, and infrastructure including the environment.

An ERP manual for a pipeline network must therefore and as a general rule cover the following:

Introduction

- Relationship between emergency preparedness and response
- Definitions
- Emergency management organizational structure
- Applicable standards
- Process and documentation maintenance
- Review and approval process flow chart
- Annual review meeting
- User's comments
- Distribution list

Pre-emergency Planning and Emergency Preparedness General

- Emergency preparedness process
- Accountability/responsibility
    - Senior emergency management specialist
    - Emergency management specialist
    - Support groups
    - Regional centers (as applicable)
    - Regional emergency preparedness coordination (if applicable)
    - Regional emergency preparedness group (if applicable)
    - Emergency management review group
    - Health, safety, and environment
- 24 Hour emergency answering service
- Activation criteria
    - Emergency activation flow chart
    - Emergency management escalation to crisis management
- Notification and reporting
- External emergency services, roles, and responsibilities
- ER contractors (if applicable)
- Training
    - Employee training
    - Emergency service awareness
    - Local authority awareness
- Conducting exercises
- Planned inspections and third-party audits

Emergency Notification and Reporting Procedures

- Notification and reporting procedures
- Notification procedures
- Reporting procedures
- Claims notifications

Safety Precautions

- Personnel responsibilities
- Safety equipment
- Spill site safety
- Water safety (crossings)
- Fluid characteristics (sweet versus sour, propane versus CO_2, etc.)

Public Relations

- Role of the media
- Handling media and public inquiries
- Documenting inquiries
- Controlling access
- Preparing media releases and statements
- Liaison procedures

General Emergency Procedures

- Incident logs
- Command posts
- On-site communications
- Site security
- Protecting the public
- Environmental procedures
- Financial authority
- Waste management (if applicable)
- Postemergency inspection
- Post mortem
- Sensitive areas
- Demobilization

Leak Response Guidelines

- Identifying leaks
- Safety precautions
- ER procedures
- Control and clean-up procedures

Other Emergencies

- Hazardous substance spills including HVP fluids
- Fire or explosion
- Medical emergencies
- Bomb or terrorist threat
- Natural disasters
- Aviation emergencies
- Computer disaster recovery plan

Appendices

- Definition of terms
- Abbreviations
- Forms

Tables and Illustrations

The following tables and illustrations are also necessary to be included in the text or as an appendix as appropriate:

Tables:

- Responsibilities of company personnel in an emergency
- Emergency notification responsibilities
- Dangerous goods quantities or levels for immediate reporting
- Guide to selecting containment, recovery and clean-up techniques for land
- Guide to selecting containment/abatement techniques for releases, spill or blowdowns (as applicable), noise
- Guide to selecting protection techniques for sensitive or HCAs
- Guide to ranking areas of environmental sensitivity
- Guide to selecting clean-up techniques for sensitive areas
- Call list, names, numbers, titles, back ups, etc.
- Others as necessary

Illustrations:

- System map
- ER organization chart
- Emergency notification chart
- Reporting structure — preliminary incident report
- Reporting structure — detailed report
- Access; control point maps
- Other schematics as necessary

INCIDENT COMMAND SYSTEM

The ICS is an emergency management structure originally developed for fighting forest fires. It is generally adapted by pipeline companies for pipeline emergency situations, including gas pipeline ruptures/leaks causing fire, HVP fluid, and oil spill response. Gas pipeline ruptures causing explosion and fires are similar to wild fires in that they involve the mobilization of labor and equipment in response to a catastrophic event spread by the affects of weather, and they harm the natural and manmade environment.

While every response during an emergency event cannot be fully anticipated, the ERP must provide guidance for responding to such events, which must include details on ICS for pipeline operator with respect to preparedness, internal and external communications, and prompt and effective responses with actions directed toward protecting people first, then the environment, and finally property.

Because pipeline incidents require coordination of information and resources among all responders, the ICS is one of the most important "best practices" in the ERP (PA 2007). The ICS provides common terminology, organizational structure and duties and operational procedures among operator personnel and various federal, state, and local regulatory and response agencies that may be involved in response operations.

The ICS is technically defined as, "The combination of facilities, equipment, personnel, procedures, and communications systems operating within a common organizational structure with responsibility for the management of assigned resources to effectively accomplish stated objectives pertaining to an incident." Definition is simply "An emergency response management system."

ICS is intended to rapidly transform the chaos of an emergency into a well-managed response. If correctly implemented, it enables the minimization of an incident's consequences through the rapid, effective, coordinated response of all available resources. For personnel involved in the response, it answers two key questions: "Who's in charge?" and "What's my job?"

ICS works well in ER situations because it organizes essential information and resources into a logical structure from which required actions can be planned and carried out. Further, it provides a preplanned ER structure that is flexible enough to handle any type of incident and adaptable to handle any size of incident.

Each element of an ERP therefore must be carefully thought through and written to correspond with the emergency situations that may arise in constructing, operating, and maintaining pipeline facilities. It must reflect the fluid/product it transports and must cover an organized system of roles, responsibilities, and procedures for commanding and controlling of emergency operations. Key considerations in incident action plans are thus:

- Type of pipeline
- Product transported
- Nature of the incident
- Exposures
- Environmental conditions
- Safety considerations
- Isolation of pipeline and facilities
- Incident potential
- Infrastructure impacts

An element that must be contained in the ERP is the ICS reflecting the requirement of US Department of Transportation, CFR 49 related to coordination/liaison/maintenance of activities between with appropriate fire, police, and other public officials with the pipeline company operating and maintaining the facilities.

A typical ICS applicable to pipeline operations must include the following:

- ICS Development and Management Team
- ICS Roles and Responsibilities (Key Personnel)
- Incident Commander

- Operation Chief/Office/Supervisor
- Safety Officer
- Public Relations/Liaison Officer
- Project Coordinator
- Environmental Officer
- Logistics
- Procurement Services
- Planning
- Documentation and Records
- Resources
- Technical Specialists
- ICS "Ground Rules"
- Span of Control (defining the extent of ER in supervision)

Reasons for ICS

- Pipeline emergencies require coordination of Information and resources among the various players in order to safely and efficiently resolve the emergency situation.
- US OSHA section 1910.120(q) (US DOL 2008) requires that public safety and industrial ER organizations use a "Nationally Recognized Incident Command System (ICS)."
- ICS is an organized system of roles, responsibilities, and procedures for the command and control of emergency operations.
- Requirement for incident management system documentation.
- ICS provides management structure and discipline related to emergency events.

The emergency organization can be divided into Command, Plans, Operations, Logistics, and Finance/Procurement Sections, each having prescribed functions, roles, and procedures (Figure 9-5). The Command Section sets objectives, monitors performance, and controls the response. In multijurisdictional events, Command is often referred to as "Unified Command" and includes representatives of affected agencies working in cooperation to provide clear, nonconflicting direction to the response. The Plans section provides the daily work plan to Operations in response to command objectives. Plans administer a disciplined planning function, defining the situation, keeping track of resources, and producing a daily Incident Action Plan. The Operations section carries out the action plan with labor and equipment in the field. Logistics orders and tracks requested resources and manages communications, transportation, and support functions. Finance/procurement tracks costs and administers claims. It also has the authority to hire local contractors and specialists, as and when needed.

ICS has common characteristics that permit different organizations to work together safely and effectively in order to bring about a favorable outcome to the emergency.

- Division of labor
- Lines of authority
- Unity of command
- Optimum span of control
- Establishment of line and staff functions within the organization

ICS Roles and Responsibilities (Key Personnel)

The following provides an outline of requirements/responsibilities from key groups or individuals in emergency management organization (Figure 9-5):

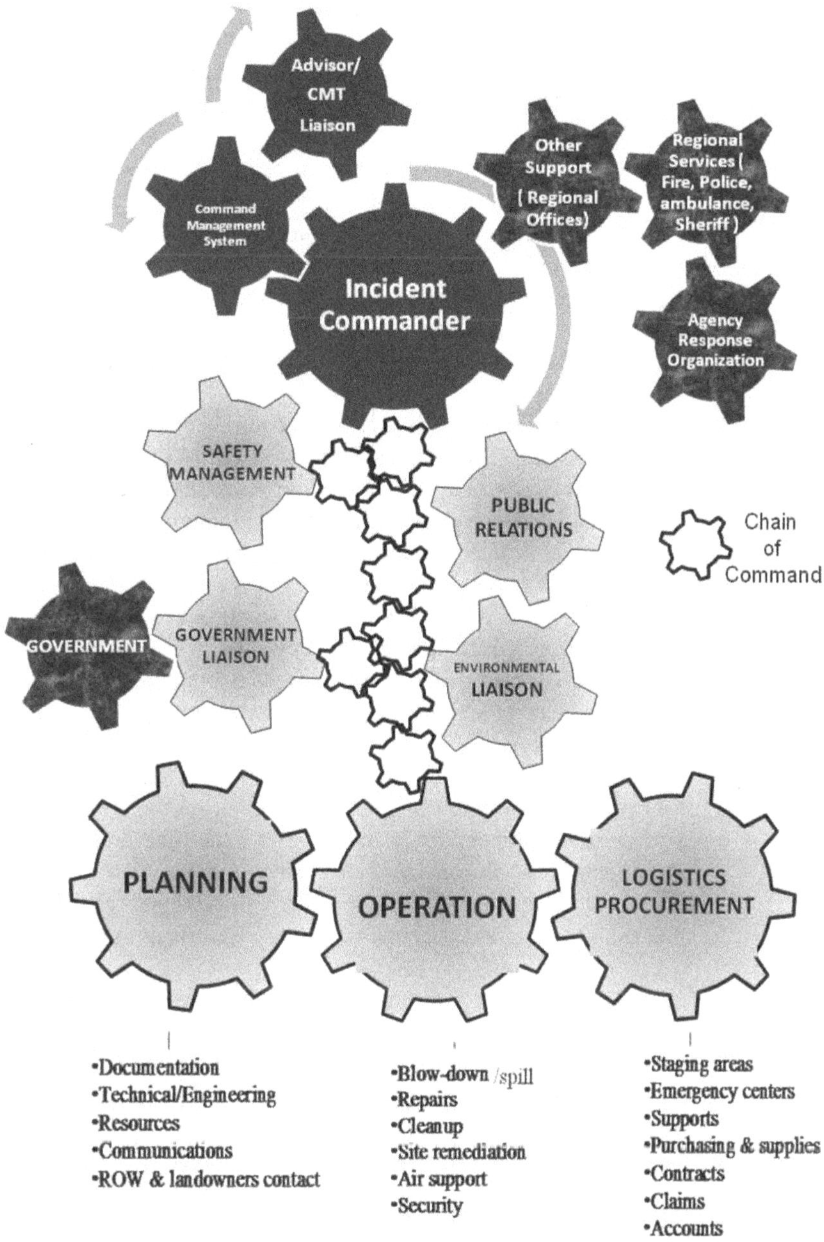

Figure 9-5. Typical emergency management organization — roles and responsibility (modify to suit prevailing conditions). CMT= Command/Crisis Management Team

Incident Commander

The Incident Commander is in charge of the incident, planning, and directing all aspects of company response. The Incident Commander is also the First Responder who initially investigates the emergency incident. The Incident Commander is the one ICS position that must be filled at all times during an incident. Key responsibilities include:

- Identifying type of emergency and providing additional details (e.g., severity, location and nature of emergency, weather conditions)

- Ensuring source of a spill/leak is shut down or isolated and the users of the facility are notified so operations can be adjusted accordingly
- Establishing ER organization, assigns tasks, and briefs key personnel
- Establishing the Command Post and maintaining communication regarding emergency status
- Assessing emergency and coordinating planning and conduction of response operations, including selection of initial and subsequent control points, as required
- Notifying essential contacts, following emergency notification procedures, and maintaining contact, as required
- Ensures response strategy corresponds with Area Contingency Plan
- Maintaining contact with Crisis Management Team (CMT) if activated. It is normal by the industry for the Pipeline Operations Management to establish initial contact with CMT
- Determining whether additional equipment and manpower are required
- Identifying and acquiring any special resources (e.g., sampling equipment, legal assistance, specialists, contractors, consultants, etc.)
- Anticipating potential changes in situation and identifying the need for any additional requirements (e.g., relief or back-up crews, special equipment for night operations, decontamination facilities)
- Ensuring records of on-site costs and product recovery (as applicable), and keeping a chronological log of emergency events
- Ensuring a post mortem of emergency operations
- Establishing and maintaining contacts with local ER agencies (police, fire, government, etc.)

Operations Supervisor

The Operations Supervisor, usually a Pipe Line Maintenance (PLM) Supervisor or alternate, directs all operational aspects at an emergency site. The ERP provides general guidance for these operations, but the ability of the Operations Supervisor to improvise and innovate is the key to a successful ER. Key Responsibilities may include:

- Obtaining briefing and maintains contact with Incident Commander
- In consultation with the Incident Commander, ensuring mobilization of sufficient manpower resources and equipment to the emergency site and control point locations
- Allocating sufficient on-site response personnel to adequately manage emergency
- Coordinating the ER operations
- Overseeing control point access, site preparation, deployment of equipment, clean-up, and repair
- Ensuring accurate information on changing conditions and events is conveyed to key ER personnel
- In consultation with the Incident Commander, requesting additional personnel, outside assistance and special resources (e.g., relief crews, special equipment for night operations, decontamination facilities)
- Relieving personnel and equipment not required from the site when safe to do so
- Arranging for storage or disposal of waste and recovered material
- In consultation with the Planning group, ensuring site remediation is undertaken once ER is complete
- Assuming responsibilities of other response team members as appropriate

Safety Management Supervisor

Generally, the Safety Supervisor identifies and evaluates on-site hazards and ensures the safety of ER operations. The Safety Supervisor must be familiar with safety procedures and emergency operations. The Safety Supervisor can alter, suspend or terminate those operations that he/she considers to pose an immediate or imminent threat to life or health and the environment. If the operations are not immediately dangerous, but are considered unsafe, the Safety Supervisor evaluates the situation and recommends corrective action to the Incident Commander. Responsibilities may include:

- Obtaining briefing from the Incident Commander
- Ensuring the emergency site is inspected and determined to be safe for workers (monitor for breathing hazards, take explosimeter readings, etc.)
- Advising the Incident Commander of any special safety issue
- Ensuring the Site Safety Plan is implemented and updated, as required
- Ensuring necessary safety equipment is acquired and/or used
- Ensuring external contractors are briefed on safety precautions
- In consultation with Logistics, ensures measures for public safety are in place
- Identifying nearest medical facilities and transport method
- Ensuring injured personnel are properly cared for; and that an analysis of safety aspects of the emergency is undertaken and reported

Public Relations Supervisor

Generally, Public Relations Supervisor handles on-site inquiries from the media and the public, allowing ER personnel to perform their duties without interruption. The Public Relations Supervisor ensures all distributed information is factual and is coordinated through appropriate company personnel. Key responsibilities are:

- Obtaining briefing from the Incident Commander
- Maintaining contact with the Incident Commander or Operations Supervisor regarding emergency status and updates and also maintaining contact with the Company Management, Safety, and Environment others as necessary
- Handling all media contact and general inquiries from the public
- Providing factual written or oral accounts of the emergency and response efforts, as well as additional comments as required
- Arranging for or preparing background information for use in briefings, interviews, etc.
- Ensuring information or written releases of a controversial or sensitive nature are cleared through company management
- Establishing Media Center, if required and arranging for necessary work space, materials, telephone, and staffing
- Arranging meetings between media representatives and ER personnel
- Arranging escorted tours of emergency site for media, regulatory, and others and provides safety equipment, as required
- Responding to special requests for information from the media and the public

Liaison

Liaison personnel are usually appointed to provide contacts between company ER personnel and other on-site response agencies. Other agencies may include fire departments, law enforcement agencies, and federal, provincial, state, county, and municipal governments as required. The functions may include:

- Obtaining briefing from the Incident Commander
- Identifying contacts from each on-site emergency agency and establishing communication links and maintaining contacts, as required, throughout the emergency
- Updating the Incident Commander regarding the presence and involvement of on-site response agencies

- Identifying and addressing agency concerns requiring resolution
- Arranging meetings between company ER personnel and other agency representatives, as required
- Monitoring ER operations to identify current or potential inter-organizational problems

Environment Supervisor

An Environment Supervisor is generally responsible for ensuring that all matters relating to the protection of the environment are addressed. The responsibilities may include:

- Reporting to and obtain briefing from Incident Commander
- Collecting pertinent environmental information regarding the site where emergency situation exists
- Advising Incident Commander of any specific environmental considerations
- Ensuring adequate wildlife and environmental protection measures are taken in accordance with ERP and contacting appropriate protection agency, as necessary
- Observing emergency site layout in accordance with ERP Depending on emergency leak/spill ensuring that a protective zone is established to minimize secondary contamination
- Identifying and discussing environmental issues and mitigative measures
- Developing containment (if applicable), cleanup, and protection strategies and provide recommendations to the Operations Supervisor and Incident Commander
- Evaluating waste management options (treatment and disposal)
- Obtaining all necessary approvals and permits
- Arranging for contractor or specialty environmental services as required
- Ensuring all environmental aspects of the emergency are analyzed and reported and appropriate action is undertaken
- Assisting Incident Commander in developing daily environmental protection action plan and maintaining a logbook of daily activities
- Coordinating environmental assessment

Logistics

The Logistics provides facilities, services, and material support, as required, in support of ER operations. Depending on the nature and complexity of the emergency, subordinate personnel may assume the responsibilities of this position. Requirements include:

- Obtaining briefing from the Incident Commander
- Organizing the Logistical requirements and assigning work locations and preliminary work tasks to section personnel
- Identifying service and support requirements for ER personnel (accommodation, food, etc.)
- Arranging for services and items required at the Command Post (secretarial services, office supplies, etc.)
- Controlling equipment and material distribution in consultation with the Incident Commander, ensuring the availability of adequate communications equipment (e.g., handsets, mobile radios, telephones)
- Arranging for acquisition and mobilization of special communication resources, air to ground communications
- Protecting personnel and property from theft
- Ensuring access to and from the site is controlled
- Protecting property from loss or damage by instituting security measures
- Advising on current service and support capabilities and estimates future service and support requirements

Procurement

The Procurement provides financial, accounting, and claims support during ER operations. Depending on the nature and complexity of the emergency, subordinate personnel may assume the responsibilities of this position. Responsibilities may include:

- Obtaining briefing from the Incident Commander
- Organizing financial needs and resources
- Identifying and ordering supply and support needs for Procurement
- Ensuring personnel time records are kept
- Ensuring maintenance of a daily record of equipment time
- Maintaining record of all charges or credits for fuel, parts, and services
- Preparing and signing contracts and land-use agreements
- Establishing contracts with supply vendors
- Determining need for company injury and claims specialists and contacts personnel as required
- Providing records security
- Ensuring all cost records are current and complete before demobilization

Planning Group

The Planning Group provides planning support to emergency operations, which allows ER personnel to:

- Better understand the current situation
- Maintain control of emergency resources
- Predict the course of emergency events
- Identify the need for additional or specialized resources
- Maintain a historical record of emergency events

Depending on the nature and complexity of the emergency, the Planning Group may also supervise specialized resources not assigned to other sections of the emergency organization. Key responsibilities may include:

- Obtaining briefing from the Incident Commander
- Organizing the Planning Section
- Establishing and supervises the documentation
- In consultation with the Incident Commander, identifying the need for specialized information (e.g., weather data, technical data)
- Advising key personnel of any significant changes in emergency status; ensures information collection and reporting requirements are met
- Ensuring the distribution of incident information to appropriate personnel
- Supervising any technical specialists under section control

Documentation Controls and Resources

The Documentation for ER situation rests with the Planning Group and may consist of either a single recorder or a group of individuals. The Documentation maintains accurate records of emergency operations, provides duplication services for ER personnel, and stores incident files for legal, analytical and historical purposes. Key responsibilities include:

- Obtaining briefings
- Establishing work area
- Establishing and organizes emergency files
- Establishes copying services

- Responding to requests for information
- Providing copies of forms and reports to authorized agencies
- Maintaining the incident log of emergency operations
- Retaining and filing reports, etc.
- Receiving and filing information provided by ER personnel
- Checking accuracy and completeness of records
- Noting errors or omissions in records and contacting appropriate ER personnel
- Maintaining, retaining, and storing incident files for postemergency use and forwards incident files to regulatory authorities as appropriate

In addition, the Planning Group is responsible for maintaining a master record of all on-site resources and personnel including:

- Establishing check-in system at the Command Post
- Preparing and maintaining command post displays, including organization charts and incident status information
- Maintaining communication with all emergency sites to obtain resource status updates
- Maintaining a master record of all on-site resources, personnel, and documentation

Technical Specialists

Technical specialists are advisors with special skills who may provide support to ER personnel. They may be the pipeline company's own personnel from various departments or contractors hired specifically for ER operations. Technical specialists may function as separate group, reporting to the Incident Commander, or may be assigned to other parts of the emergency organization, reporting to the Operations, Procurement/Logistics or Planning Groups. Technical specialists may include:

- Accountants
- Engineering specialists
- Industrial hygienists
- Lawyers
- Meteorologists
- Public affairs specialists
- Purchasing agents
- ROW agents
- System integrity specialists

The number and type of technical specialists needed depend on the nature and complexity of the emergency.

ICS "Ground Rules"

The following implementation rules provide additional considerations for the use of ICS.

Top–down Development

- The emergency organization structure starts with the top position (Incident Commander) and works down as required.
- The lead position (Incident Commander) must always be filled.
- The Incident Commander "opens" positions immediately below him/her when warranted based of the requirements of the situation.

- When any position is not "opened," the person occupying the position directly above assumes all responsibilities.
- Once the incident has reached its maximum level and is subsiding, demobilization occurs based on similar principles.

Modular Development

- The ICS system is intended to be developed and demobilized in groups or "modules."
- Modules are often required to specific tasks or areas (e.g., a new module may be set up under the Operations Chief if a second PLM crew arrives and is responsible for activities at another location).

Span of Control

- Any individuals directly supervising others can look after a set number of people before efficiencies and effectiveness are reduced; this is especially true under the high-stress, extremely active conditions of ER.
- As such, a "rule of thumb" has been established: a range of three to seven people is the target for anyone supervising others in an incident; five is considered optimum.

Common Terminology

- Terms being used by any individuals involved in the response must be defined in the same way so that all participants are "on the same wavelength" when communicating (e.g., incident "Level" designations, functional areas, etc.).

Unified Command

- When a number of stakeholders have a degree of jurisdiction in an incident (or believe that they have), incorporating a "Unified Command" system into the command functional area accommodates all input. This ensures that decisions are made based on several perspectives and all pertinent government regulations.
- Unified command can be used when an incident crosses geographic areas, political boundaries, or government department jurisdictions and/or involves multiple owners and operators.
- Under a unified command system, the Command functional area is expanded to encompass one individual representing each jurisdictional area. However, one person from this group must be selected as the Incident Commander.
- The unified command group meets regularly and frequently to ensure that input from all parties is considered before decisions are made. Again, however, the Incident Commander is still responsible for implementing actions (but may also delegate).

STEPS AND SUCCESS FACTORS IN RESPONDING TO AN EMERGENCY

Effective ER must contain the following steps:

1. Site Management Control and Containment Problem Identification
2. Hazard and Risk Evaluation

3. Selection and Use of Proper Personal Protective Clothing and Equipment
4. Information Management and Resource Coordination
5. Response Objectives
6. Decontamination and Clean-Up Operations
7. Incident Termination

Pipeline emergencies are handled safely and effectively when responders are dependent on a predetermined and approved system for managing the event

The above seven step process can ensure that the basic tactical functions to be evaluated and implemented are in place. It is a flexible guideline and the process can serve as a general framework for systematically resolving a pipeline emergency.

Critical success factors in responding to an emergency are:

- Avoiding commitment or positioning personnel and units in a hazardous position. Consider an escape route
- Establishing command and an Incident Command Post
- Establishing a Staging Area
- Establishing an isolation perimeter
- Establishing a hot zone
- Avoiding entry into the emergency affected area without necessary personal protective equipment (PPE) appropriate for the incident
- Isolating all personnel until emergency decontamination can be established
- Initiating public protective actions (PPAs) as necessary

CLASSIFICATION OF EMERGENCY INCIDENTS

Pipeline industry defines an incident as an event affecting company operations that may be an emergency or crisis.

DOT 49 CFR 191.3(1) defines emergency as:

(1) An event that involves a release of gas from a pipeline . . . and
 (i) a death, or personal injury necessitating in-patient hospitalization; or
 (ii) estimated property damage, including cost of gas lost, of the operator or others, or both, of $50,000 or more

However, pipeline industry defines responding to an emergency depending on the "Level of Alert" and generally it is pipeline company, either the control center or field offices, which decide whether or not the incident is an emergency.

If an emergency condition exists, it will be determined whether the "Level of Alert" is Level 1, Level 2, or Level 3, with Level 3 being the highest level of alert, Table 9-1 and notifications will begin accordingly.

HEALTH, SAFETY, AND ENVIRONMENTAL ASPECTS OF PIPELINE CONSTRUCTION

Aspects of Pipeline Construction As Related ER

Health and safety considerations as well as environmental evaluations of a pipeline system are an integral component of design and construction that requires special planning and implementations to ensure effective long-term safe pipeline facilities meeting stakeholders'

TABLE 9-1. ER levels of alert

	Criteria	Example of Responses
LEVEL 1	An emergency that has the potential to escalate, but does not meet any Level 2 or 3 Alert Criteria. Meets all of the following conditions: ✓ No immediate threat to the public ✓ No serious threat to health and safety of workers (PPE may be required) ✓ Minimal environmental impact ✓ Impact confined to company jobsite ✓ Creates little or no media interest ✓ Handled entirely by company personnel	• First responder will assess and confirm the situation. • The first responder confirms level of alert with the Crisis Manager on call. • Additional company personnel may be placed on standby. • External notifications are made as required, complying with appropriate regulatory reporting. • Emergency responders prepare for Level 2 or 3 Alert responses.
LEVEL 2	An emergency that does not meet any Level 3 Alert Criteria, but meets any of the following conditions: ✓ Definite risk to the public, workers, or the environment ✓ Confirmed incident. Expected to be under control quickly ✓ Requires involvement of external emergency services, state, and federal agencies ✓ Public safety actions are likely required ✓ Moderate environmental impact that extends or has the potential to extend beyond company ROW ✓ Creates local/regional media interest	• Level 1 alert duties are continued as appropriate. • Additional internal and external resources are activated. • Lead regulatory agencies are notified. • Additional notifications are made as required. • Sheltering or evacuation may be required. • Planned ignition or the product may be required. • The Crisis Management Plan is activated. • Emergency responders prepare for Level 3 Alert response.
LEVEL 3	An emergency the meets any of the following conditions: ✓ Causes serious threat to the public, workers, and/or the environment ✓ Requires extensive involvement of external emergency services, federal, and/or state agencies ✓ Significant and ongoing environmental impact with extent beyond company ROW ✓ Creates national media interest	• Level 1 and 2 Alert responses are continued as appropriate. • The company ER is at maximum • The Crises Management Plan is fully activated including executive involvement. • External government agencies are extensively involved. • The company continues to be a liaison with key government agencies and sends representative(s) to a government command center if established.

expectations. The latter includes the owners, the public, and as well the environment (Figure 9-6).

"In pipeline construction safety is paramount. Think safety at all times. Six-ton pipes are very unforgiving things when they are not treated with the respect they deserve."

The four corners of most successful construction project achievements are:

- Cost effective and timely installations
- Reliability of constructed facilities
- Safety with least environmental impact
- Harmony with community

Prior to construction, resources in the immediate area of the pipeline route are identified and assessed to determine potential impacts on pipeline construction as well as the long-term stability of the pipeline and sustainability of resources that affect the environment adjacent to and along the right of way (ROW). Site-specific resources identifications and evaluations include:

- wildlife (Figure 9-7)
- fisheries
- water crossings
- forest cover
- archaeological and

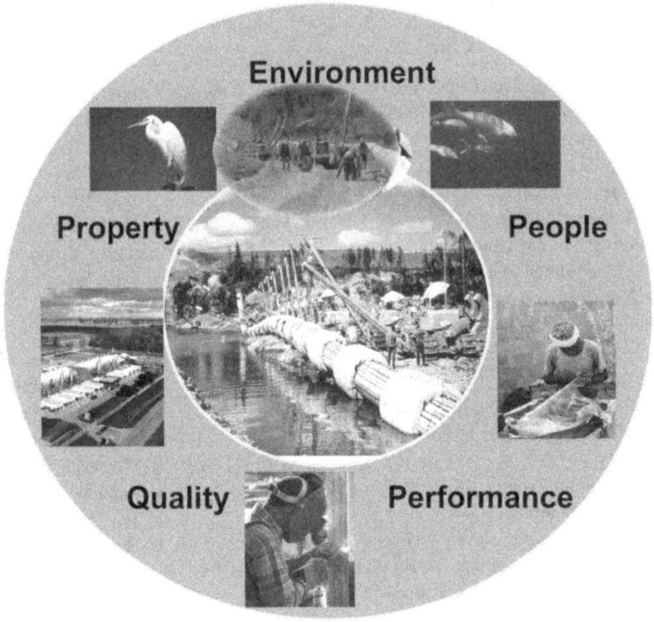

Figure 9-6. Health, safety, and environment (HSE) — relationship with pipeline construction

- paleontological resources
- soil and vegetation (for oil handling and reclamation)
- land use (avoiding conflicts with owners and other users)
- construction timing

Figure 9-7. A Stake holder in the pipeline ROW

Protection procedures are based upon these resource assessments and are then integrated into pipeline construction techniques.

Timing for pipeline construction is also taken into consideration in the evaluation of resources, as seasons can effect the selection of the most appropriate mitigative procedures. Construction techniques, which are effective during the summer, may not be appropriate or effective in the winter. Alternatively, dealing with site-specific resource impacts may create conflicting timing constraints with those of proposed construction schedules.

Specialized techniques to reclaim the construction ROW are implemented following mainline construction. Every attempt is made to match existing species for re-vegetation purposes to ensure successful reclamation of noncultivated lands and to return cultivated lands to their previous agricultural productivity.

During construction of the pipeline, environmental inspection is ongoing to ensure compliance with environmental design and protection procedures and to maintain consistency with various regulatory approvals. Problems that are identified during the design and construction phase of the project are reviewed extensively, and these may initiate further environmental research, which in turn may modify construction as it proceeds.

In selecting a pipeline ROW that is economical and complies with environmental regulations, an environmental impact assessment is usually undertaken for the purpose of determining and/or developing environmental quality management guidelines for pipeline construction and operation. These guidelines usually include the following:

A. Compliance
   - ✓ Legislation compliance
   - ✓ Environmental guideline compliance
   - ✓ Environmental coordination, audit and training
   - ✓ Recommendation by volcanologist/geotechnical/seismic consultants

B. Guidelines for soil erosion protection
   - ✓ Erosion process and types
   - ✓ Erosion risk assessment
   - ✓ Erosion protection

C. Guidelines for water quality protection
   - ✓ Baseline water quality
   - ✓ Site selection
   - ✓ Water analysis/quality index
   - ✓ Impact and mitigation measures

D. Guidelines for archeological heritage protection
   - ✓ Historical resources
   - ✓ Archeological studies
   - ✓ Regulations

E. Environmental protection resources and methods
   - ✓ ROW preparation and pipe installation
   - ✓ ROW width during construction
   - ✓ Grading
   - ✓ Rip Rap
   - ✓ Sand plugs
   - ✓ Temporary cover
   - ✓ Sediment/silt traps
   - ✓ Canals/disturbances
   - ✓ Volcanic areas (if applicable)
   - ✓ Permanent physical erosion control method

- ✓ Soft plugs
- ✓ Re-vegetation/re-grading
- ✓ Drainage system
- ✓ Berm/gabions
- ✓ Ditch foam plugs
- ✓ Agronomic erosion control
- ✓ Re-vegetation measures schedule
- ✓ Procedures and methods
- ✓ Re-vegetation plan

Codes

World Standards related to HSE issues affecting pipeline construction include:

- ISO 14001, ISO 9000
- BSI 8800
- OHSAS 180001
- American Petroleum Institute (API)
- Canadian Standard Association (CSA)
- NFPA
- OSHA, OHSAS
- Kyoto Protocol

Challanges of Pipeline Construction Safety

Work Site Safety

In preparing the work site necessary for construing pipeline facilities, there are a number of issues that must be reviewed and addressed. The review of these issues will ensure proper tools and safety gears are available for an accident-free environment and that the adverse effect of any safety related incidence is minimized: Some of these are:

- PPE Review
- Workplace Hazardous Material Information System (WHMIS) Review
- Hazard Recognition and Control
- Front End Operational Hazards and Controls
- Rigging and Hoisting Hazards and Controls
- Pipe and assembly Hazards and Controls
- Specialized Work Hazards and Controls
- Back End Operational Hazards and Controls
- Vehicle Operation
- Availability of Proper Equipment and Tools
- Correct Equipment and Tool Operation
- ER

Other Special Safety Consideration During Construction

The majority of accidents in the workplace including during pipeline construction are the result of human error. The number of accidents can be reduced if safety precautions and safe work habits are adopted. To ensure a hazard-free work environment, pipeline industry has adopted strict rules that must be carried out in accordance with the Owners/Operators Health, Safety and Environmental policies and procedures.

This must be carried out strictly in accordance with current industry regulations and the Operator's Health, Safety and Environmental policies and procedures.

Additionally, contractors undertaking pipeline construction are responsible for the provision and use of fire fighting appliances as part of their site fire precaution procedures.

Significant hazards, which may arise from the works during pipeline construction include, but are not limited to:

1. Tripping arising out of:
    - Materials and debris littering the site due to inadequate "housekeeping"
2. Manual handling arising out of:
    - Long pipe lengths
    - Specialist equipment
    - Plant machinery
3. Electricity arising out of:
    - Use of power tools
    - Contact with buried power cables
    - Contact with overhead power cables
4. Explosion arising out of:
    - Pressure testing operations
    - Working with gas cylinders
    - Working in the proximity of live gas pipelines
5. Fire resulting from:
    - Welding, burning operations in connection with pipework and equipment connections
6. Noise and vibration arising out of:
    - Disc cutting
    - Breaking out of hard surfaces
7. Collapse arising out of:
    - Improper temporary support of excavations
    - Improper support of building or fence structures during removal or replacement
    - Improper use of machinery on uneven /sloping ground
    - Improper stacking of materials on uneven/sloping ground
8. Airborne matter resulting from:
    - Grit blasting
    - Grinding/disc cutting
    - Painting
    - Sand/dust storm
9. Contact with moving machinery arising out of:
    - Restricted working width
    - Moving equipment and pipework
    - Earthworks/excavations
    - Working in areas with public vehicular access
10. Harmful substances arising out of:
    - Handling cement
    - Painting
    - Chemicals used on agricultural land
    - Solvent based bonding materials
    - Undeclared chemicals or elements, inherited from previous operator
    - H^2S emissions
11. Pollution and damage arising out of:
    - Interference with underground service
    - Incorrect identification of plant and equipment
    - Incorrect use of equipment
12. Confined spaces

Precautions for Material Use During Construction

Pipeline Construction Contractor must obtain all Health and Safety literature (e.g., WHIMS, TDG). Such information includes MSDS in relation to the specific materials.

The contractor must ensure that all recommendations regarding use, application, transport and storage are complied with in consideration of works, persons, and the general public.

Examples of such products include, but are not be limited to the following:

- Cement
- Solvent-based paints and solvents within confined areas
- Tar-based products

Precautions for Construction Operations

Additionally, in pipeline construction, the following precautions must be considered in integrated construction performance/operations:

- Working near overhead electricity and telephone lines
- Working and crossing buried services
- Working in and crossing public roads/railroads
- Traffic control on site
- Lifting Operations and Loading activities
- Working on slopes
- Testing
- Commissioning
- Thrust boring (if used)
- Working at high elevations
- Installation of equipment in areas of restricted space
- Working in or near airports

INDUSTRY SAFETY GUIDELINES FOR PIPELINE CONSTRUCTION

The following represents pipeline industry guidelines and practices for constructing pipeline facilities:

General Guideline

- Adopt the "buddy" system
- Protect yourself and your fellow employees at all times.
- Know how to administer first aid. First aid supplies must be available at all worksites. Minor abrasions and cuts, the most common incidents, are easily handled by basic first aid. Accident-related injuries of a more serious nature must also be anticipated and ER plans formulated. This is especially critical in remote locations.
- Ensure adequate fire protection equipment is available at all worksites and sufficient personnel are trained in the correct operation of the equipment.
- Always wear sensible clothing (e.g., protective coveralls). Weather and work conditions must be taken into account.
- Study the work you are expected to accomplish. Planning and pre-job safety meetings are the key to accident prevention.
- In compressor buildings, keep exits clear from tools and equipment in the event emergency evacuation is required.

- Erect barricades, flags, and warning devices around excavations or danger areas. Good communication with fellow workers or visitors is a must for a safe working environment.
- Erect screening around welding operations. Restrict access to essential personnel only or provide suitable eye protection.
- Maintain good housekeeping
- Do not clutter the work area. This is one good way to reduce fire risks.
- Be sure spills of oil, grease, and contaminants are cleaned away. Slips leading to back injuries are common. Wear suitable nonslip safety footwear to reduce this hazard.
- Stay away from equipment being hoisted. Wear head protection on site.
- Do not jump off equipment. This often leads to ankle and foot injuries.
- Follow safe excavation procedures including using a probe and/or hand shovel to establish locations of pipelines before excavating. Underground cables, conduits, or pipelines should be exposed by hand before excavating or trenching with any power-operated equipment. Failure to take this one simple step can result in severe injury to employees and damage to facilities.
- If in doubt about a work procedure, seek advice from a supervisor or more experienced employee.

Small Tool Safety

- Wear suitable approved safety glasses or goggles.
- Choose the right tool for the job. Avoid the temptation to take short cuts.
- Make sure the tools you select for the job are in good condition with handles that are sound and securely fitted. Cutting edges should be sharp and true. All tools should be free from dirt and grease. Repair or replace worn, broken or damaged tools.
- Use tools properly. An improperly used tool can be more dangerous than using a wrong one. Wrenches should be pulled, not pushed. Hammer handles should not be choked. A hammer is built and balanced to be used with a full handle. Unless special equipment is used, work such as cutting and chiseling should be done away from you, not toward you.
- Carry sharp tools facing away from you.
- Put tools away as soon as you are finished with them. Tools should be cleaned before they are put on tool boards or in boxes. Protect sharp cutting edges. Do not overcrowd tool boxes.

PPE

- Head Protection

  Hard hats are designed to absorb initial impact when an object strikes the shell. The inner headband and cap assembly is suspended inside the shell. Approved hard hats should be worn where employees are subject to injury from swinging, falling or moving objects. Many head injuries occur after objects are deflected upward from below head level (e.g., from machinery, lathes). Absence of obvious overhead hazards does not necessarily eliminate the need for head protection.

  All hard hat components including shells, headbands, and sweatbands should be inspected daily for signs of damage (e.g., dents, impact damage, penetration).

  Abrasive materials such as strong cleaning solutions and petroleum products must not be used to clean hard hats as they can deteriorate the strength of the hard hat material.

  Stickers should not be applied on hard hats because the glue can eat the plastic shell.

Hard hats must not be painted because paint will react with plastic and eventually cause the shell to become brittle and lose its structural integrity.

Holes must not be drilled in the shell of the hard hat for any reason. Provisions are usually already made on the hard hat for attachment of accessories (e.g., ear muffs, face shields).

Hard hats should not be stored loosely in vehicles since they can become deadly projectiles in a vehicle accident.

Hard hats that have exceeded their useful life or expiration date are to be discarded.

- Eye Protection
  A few seconds of carelessness can lead to serious eye injury or even the loss of an eye. Eye protection should be worn whenever an employee is engaged in or observing:
  - Power grinding, buffing, or wire brushing, whether there is a built-in eye shield or not on the grinder. Safety goggles and a face visor must be worn when grinding.
  - Drilling or chipping stone, concrete, paint, pipe, or metal.
  - Any other activity where there is danger of injury to the eye.

  There are many types of eye protection available for use in the workplace. Employees should select the appropriate eye protection for the job. Goggles, glasses, and face shields are the most common types used. Prescription lenses can be arranged as required. Lenses are usually made of glass or scratch-resistant plastics. Also, various tints are available depending on the type of activity. Damaged or heavily scratched goggles and shields are to be discarded.
- Hearing Protection
  Generally, industry takes measures (when designing facilities) to eliminate noise hazards in the workplace. However, it is not always possible to reduce noise to an acceptable level (e.g., during blowing down a gas pipeline). Therefore, hearing protection is required when it is not reasonably practical to decrease the noise or isolate an employee from a noise hazardous area.

  Hearing protection should be worn when working around noisy machinery and near turbines and blow-downs. Personnel in such areas should be kept to a minimum.

  Once noise-related hearing loss occurs, it is generally nonreversible. Difficulty hearing in a group conversation is often the first indication to an individual that hearing damage has occurred. At this point, the individual can only protect what hearing remains. Noise-exposed employees should have annual audiometric tests to check their hearing.

  Approved ear plugs, while offering minimum hearing protection, are not a substitute for ear muffs. In fact, in high-noise areas, a combination of plugs and muffs is recommended. Cotton, wool, or similar padding is not considered approved or adequate protection.
- Hand Protection
  Many types of safety gloves are available. Industry advises selecting suitable safety gloves, particularly when working with chemical, corrosive-type products. When working on valve sites and at pump/compressor stations, bruised and cut hands are common incidents when protective work gloves are not used.
- Foot Protection
  Industry advises wearing approved safety footwear to protect against injury to the feet and ankles. Footwear ranges from lightweight safety shoes to heavy duty safety boots. The soles are available in many types, including nonslip. If the correct type of footwear is used, the steel toe will deflect falling objects away from the foot and reduce the risk of serious injury. Steel toe running shoes are not recommended.

Hot Work Safety Precaution

Permit Requirement

- Hot work in the pipeline transmission industry is any work, which uses fire, naked flame, spark-producing tools or equipment or other sources of ignition.
- Hot work permit procedures ensure all safety precautions have been taken before any hot work is carried out.
- Hot work operations usually occur at stations. A hot work permit must be obtained from the local operating personnel. It must be clearly understood who is in charge of the operation at the site. The local operator and the employee completing the hot work discuss the job to be done and agree upon the correct procedures to be followed.
- The local/field operator conducts all safety checks outlined on the permit. Both individuals sign the permit prior to work commencing. The operator keeps a copy of the permit on site. It is valid for a maximum 24-hour period.
- The permit is only valid for the conditions at the time of issue. If the circumstances on site change and become hazardous, the permit is cancelled and a new request may be submitted. New safety procedures may be required or the job may be postponed.

Pre-job Safety Meeting

Industry advices for all personnel at the hot work site must be present for the pre-job safety meeting. The reason being that they must have a clear understanding of what safety procedures are to be followed. Emergency plans should be discussed and arrangements made so sufficient personnel are available at the site with the required equipment. Fire extinguishers with trained operators must be on standby at the site.

Pre-job Tests

Prior to any hot work, tests must be made to determine whether the local atmosphere contains any flammable substance in a quantity sufficient to create an explosive atmosphere. Proper calibrated gas and oxygen measuring equipment must be used. Hot work cannot proceed when test readings indicate levels above acceptable safety limits.

Confined Space Entry

A confined space is any enclosed or partially enclosed space with restricted access. Due to its design, construction, location, atmosphere, materials, or substances in it, or other conditions, a confined space may become hazardous to any employees required to enter it.

Industry advises conducting a pre-job safety meeting before entry into a confined space. Good planning is essential to ensure the work procedure is completed safely. Arrangements must be made so rescue workers and suitable safety equipment are available at the scene. Emergency rescue procedures must be defined and clearly understood for the "worst case scenario."

- Entry into a confined space in the pipeline transmission industry is a calculated risk if hydrocarbon vapor is present; if at all possible, alternative work methods should be used.
- The main risks in confined space entry include fire, explosions and oxygen-deficient atmospheres.

Examples of confined space entry include:

- Certain internal pipe inspections.
- Indoor installations where hydrocarbon vapor ventilation may be limited in the event of a leak (e.g., meter stations).
- Any facility where gas may displace available oxygen for breathing (e.g., drywells).

Key to Successful HSE Implementation in Pipeline Construction

"Prevention is better than one ounce of cure."

To successfully prevent incidents causing injury, fatality, and or adversely impacting the environment, the following is considered as a precursor for successful pipeline construction:

- Good information on pipeline locations and full familiarity with construction techniques
- Familiarity with government safety regulations and safe work practices prior to commencing work
- Having safe work permit in place
- One-call system (or an effective communication system) in place
- Adoption of industry's best practices and program
- Proper education campaign for the public
- Familiarizing all personnel with site and work conditions
- Prior research on condition of ROW and identification of all hot spots
- Favoring land-use solutions
- Proper health, safety and environmental protection program in place
- Proactive thinking for reducing natural hazard damage
- Foreseeing and managing human factors/errors, the cause and effects
- Rigorous implementation of HSE program during construction response and recovery plans in place
    - o Put in place effective response plans
    - o Recovery and clean-up are fundamental to incidence response

RISKS DURING PIPELINE CONSTRUCTION — ELEMENTS AFFECTING SAFETY

There are many physical risks involved when undertaking constructing pipeline facilities. Most personnel undertaking the construction works are aware of these inherent/voluntary risks associate with pipeline construction. However, the general public residing adjacent to pipeline ROW and in many instances those persons visiting construction ROW (including those affiliated with construction works such as pipe haulers, etc.) while construction is proceeding, are unaware of such risks. It is no coincidence that responsible operator and construction contractor put in place a comprehensive safety patrolling/inspection (Figure 9-8), program that would ensure general public and site personnel health and safety while systematic ensuring protection of the environment/ROW that can be caused from work and environmentally induced issues. These patrolling/inspection cover many items but also include the following:

- Transportation of dangerous goods
- Waste management
- Water quality
- Noise pollution
- Air emission/air quality
- Vegetation management
- Soil and land reclamation
- Erosion control
- Climate changes and weather conditions

Figure 9-8. Safety patrol during pipeline construction

- Contamination
- Vehicle and equipment operation
- Trenching and excavation

Some of the situations that could be encountered and mishaps that could be proactively anticipated or avoided during pipeline construction can be summed up as follows:

- Inadvertent damage to older facilities during construction of new installation
- Accidental discharge, environmental quality
- Water pollution, trench dewatering
- Environmental noise pollution
- Electrical shock caused by equipment operator hitting electrically lived power source
- Pipe transportation causing injury including fatality
- Unexpected thunder storm, lightening — lack of electrical grounding
- Tailpipe emission
- Fish and wild life habitat, un-seasonal construction
- Change in watercourses and flooding
- Unanticipated soil/land movement caused by unusual precipitation
- Unsafe drilling and blasting
- Trenching/excavation in soft/boggy/sandy soil leading to ditch wall caving
- Crossings
- Constructing pipelines in common ROW with power lines
- Unknown archeological/paleontological resources
- Constructing in heavy/deep slopes

Some of the above are discussed below:

Landslide During Construction

A landslide involves the down slope movement of a mass of soil or rock in response to gravity (Figure 9-9). There are many different types and sizes of landslides. They range in scale from small slumps involving only a few cubic yards of material (such as can be seen on many road cuts) to massive slides of several square miles and millions of cubic yards of earth. Similarly, the rate of down slope movement is highly variable; mudflows (relatively fluid masses of saturated soil) may move at velocities in excess of 30 miles (50 km) per hour, while large landslide masses may move intermittently at rates of only a few inches per year.

Although the characteristics of landslides are highly variable (Figure 9-10), the location and timing of landslide movements are not random. These features are at least partly predictable and are related to weather cycles, local geologic conditions, slope angle, and human-induced changes to slope conditions.

Landslide Risk Factors During Construction

During construction, the following factors and associated risks can affect the safety during pipeline construction:

Types of geologic terrain are particularly prone to slide activity (Miller and Koloski 1999).

- Sloping terrain underlain by deeply weathered rock, shale, or glacial deposits layered permeable and nonpermeable materials on a slope
- Steep slope. The steeper the slope in relatively weak materials, the greater the risk of sliding. Most landslides occur on relatively steep slopes (see above photograph)
- Shallow slopes (14% or 8 deg.) involving the presence of seams or layers of bentonite clay
- Size of a landslide mass and the direction of slide movement relative to a pipeline alignment; the greatest risks occur where the direction of slide movement is across (perpendicular to) the orientation of the pipeline
- Weather. Annual or seasonal precipitation that is 30% or more above normal often results in increased landslide activity. Initiation of slide movement typically occurs

Figure 9-9. Landslides in/along pipeline ROW

Figure 9-10. Landslide characteristics

during or immediately after a period of intense precipitation. Rain on wet snow (the "rain-on-snow" event) is particularly noted for landslide initiation.

Once started moving, large landslides (larger than a few acres) may move slowly for months before movement stops. Smaller slides typically move for shorter durations. Once landslide movement has occurred, the failure surface at the base of a landslide mass is weaker than the surrounding materials and is forever sensitive to renewed activity.

Options for hazard management include the following:

- Stabilization of landslides (this is often infeasible for large slide masses)
- Rerouting to avoid landslide areas
- Subsurface rerouting to carry pipelines beneath active landslides using directional drilling technology
- Surface and subsurface monitoring of ground conditions to detect slide movements
- Excavation to expose the pipeline and allow rebound of the steel (if existing pipeline)
- Installation and monitoring of strain gauges on the pipe (for existing pipeline)

Also, for ROW that is likely to be subject to land movement and instability, prior geotechnical evaluation will likely render information that would assist in the stabilization of ROW. Such geotechnical evaluation would include:

- Boreholes to identify
  - Lithology (rock or soil types)
  - Stratigraphy
  - Unconfined compressive strengths
  - Potential penetration rates
  - Identification of potential fluid losses zones

- Packer Tests to identify
  - Potential fluid loss zones
  - Maximum annular pressure for the drill path
- 3-D Seismic Data, Resistivity, and Ground Penetrating Radar (GPR) to identify
  - Consistency of stratigraphy
  - Information on the size and extent of fractures in three dimensions
  - Anomalies such as boulders, caverns and density pockets

Avoiding Steep Slopes. The best way to assure the integrity of the pipeline in a steep slope is firstly to reroute the pipeline in as far as economically possible. If not, select ROW along the slope in a hairpin fashion and install the pipeline in native soil and away from the slope drop side as far as possible. See Figure 9-11 below:

Archaeological Investigation. What do you do when a pipeline is in the middle of construction and the corridor route suddenly requires rerouting across a major waterway? (Figure 9-12) Cultural resource protection laws are still required, and river settings are known "hot spots" for archaeology.

A need to make a location adjustment during construction across a river valley often arises during the construction of a pipeline. Generally, historic preservation law has to be satisfied while construction is being rerouted.

Only proactive archeological and geotechnical assessments would ensure the avoidance of the reroute.

Excavation

Incidences occurring during pipeline excavation can be serious in nature if pipeline trenching particularly in unstable ground is not well planned and the work is not carried properly and with care (Figure 9-13).

Injuries sustained by a former Canadian wheelchair basketball team member and founding executive director of the Canadian Wheelchair Basketball Association, Reg McClellacan, can attest to an example of such an incidence in pipeline construction. He was injured in a pipeline construction accident near Red Deer, Alta, in 1972 when the side boom on a Caterpillar broke loose and fell on him (Coleman 2004).

Hazardous liquid pipeline accident and natural gas incident summaries published by National Transportation Safety Board (NTSB 2003; Figure 9-14) indicated that between 15% and 18% of accidents incidence are as a result of excavations. This had resulted between 7% and 12 % of total damages reported.

Figure 9-11. Steep pipeline ROW slopes

Figure 9-12. Re-routing 36-in. gas pipeline in North Dakota, USA

Figure 9-13. Excavation and pipeline rupture

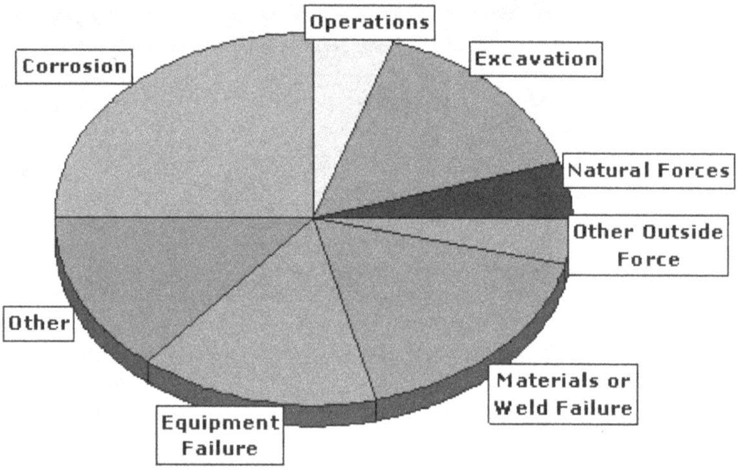

Figure 9-14. Contribution of pipeline excavation/backfill to overall pipeline incidence/accidence (NTSB 2003)

EMERGENCY INCIDENCES

Table 9-2 provides some examples of situations that can be regarded as hazard situations and for which ERP would potentially be required (Callan 2008; Pipeline Safety Foundation 2002; TSB 2007).

PIPELINE INCIDENT MANAGEMENT

Management of an incident that causes an adverse impact to the operation, operation and maintenance personnel, the public, and the environment must be met by a rigorous emergency planning control system. The system must consider the design, construction/installation conditions through to operational techniques, as well as impact of adjacent resources/activities on the facilities. This has to be coupled to the schedule of maintenance, inspection frequency and techniques, and any necessary repairs implemented as well as the design of emergency control systems and processes in place.

Pipeline emergencies are controlled through incident management tasks as long as incidents controls are well thought of in advance. The aim must be toward:

- Long-term safe and reliable operation
- Long-lasting system

TABLE 9-2. Examples of situations causing hazard leading to ER

Date and Location	Emergency Incidence	Details
1985 — Midland, TX	Burning bulldozer 	Bulldozer clearing mesquite from pasture land with a root plow, hit an 8-in. natural gas liquids pipeline, which was 22 in. deep. The impact cut the top off a section of the line. The bulldozer operator escaped before the gas ignited, which happened about 3 min after impact. The bulldozer engine, which had not been turned off, caused the ignition. The farmer had not called for a location before starting to clear the land to convert it to cropland.
1990 — Simms, MT	Gravel pit death scene 	The operator of this bulldozer died of burns a few hours after he hit a 12-in. natural gas pipeline near Simms, MT.
1991 — Bradenton, FL	Trench fire 	A "mis-located" gas line was hit by a contractor installing cable for traffic signals. The 4-i. line sustained a 6-in. gash. The escaping gas ignited, forcing the evacuation of a shopping mall. No one was injured.
1995 — Huntington Township, PA	Roadside hotspot 	A crew looking for a water leak struck a gas line and an ensuing explosion destroyed the backhoe and dump truck. The accident occurred when a large rock fell out of the backhoe bucket and struck the ground close enough to the plastic line to cause damage.
1999 — Bellingham, WA	Oil pipeline explosion (Olympic Pipeline)	Oil pipeline explosion causes two children and a young man to be killed. Explosion caused by incorrect pipeline operation.
2000 — Carlsbad, NM	Natural gas pipeline explosion	El Paso Natural Gas pipeline exploded, killing 12 members of an extended family that were camped nearby. The cited cause was corrosion-caused failure of the high-pressure transmission pipeline (more than half the wall thickness had rusted away), which had not been inspected since its construction a half-century ago.

TABLE 9-2. (Continued)

Date and Location	Emergency Incidence	Details
2007 — Burnaby, British Columbia	Crude oil pipeline, third-party damage  	Transmountain (24 in.) Westridge Dock Transfer Line was struck and punctured by an excavator. This caused approximately 234 m^3 of crude oil release (210 m^3 recovered). Crude oil flowed into Burrard Inlet Bay via the Burnaby storm sewer system. Eleven houses were sprayed with crude oil; many other residential properties required restoration (TSB 2007, Report Number P07H0040).
2002 — Brookdale, Manitoba Canada	NPS 36 gas pipeline failure	The rupture was caused at a zone of near-neutral (low) pH stress corrosion cracking. Following the rupture, the sweet natural gas ignited causing a crater line at the rupture site. There were no injuries. As a precautionary measure, approximately 100 people were evacuated from the occurrence area within a 4-km radius, including the village of Brookdale, for a period of 1 day (TSB 2002, Report Number P02H001)

- Least impact on the environment
- Efficient and economic system
- Well-thought-out control systems and action plans for responding to emergencies

Incident control can be defined as the process of identifying, recording, classifying, and progressing incidents until the affected service is restored, providing that the design of the method of controlling the incident is correct and has undergone a previous analysis. However, while it may take some times to effect incident resolution, it may be noted that the collection of data to identify causes of incidents is a secondary objective of Incident Control. The Incident Control is essentially a task of Incident Management.

Causes of an Incident

Causes of pipelines incidents have been listed previously as per Figure 9-2. However, Figure 9-15 (A and B) provides a comparison of pipeline failure causes in the USA between the 1980s and 2000–2003. It is interesting to note that while in the 1990s, internal/external corrosions were the least (15%) cause of pipeline failure, the 2000–2003 has seen corrosion to be the prevalent failure cause (44%). On the other hand, failures due mechanical damage have seen a decrease in magnitude.

Thus, the prevalent mode of pipeline failure is a moving target. However, such a change could reflect underlying issues; for example, aging pipelines leading to more corrosion or improvement in technology leading to better mechanical design of pipeline as well as incident management techniques.

While corrosion has been noted to be on the increase, research shows that safety management, legislative/regulation, and accidents have been the key concern of the pipeline

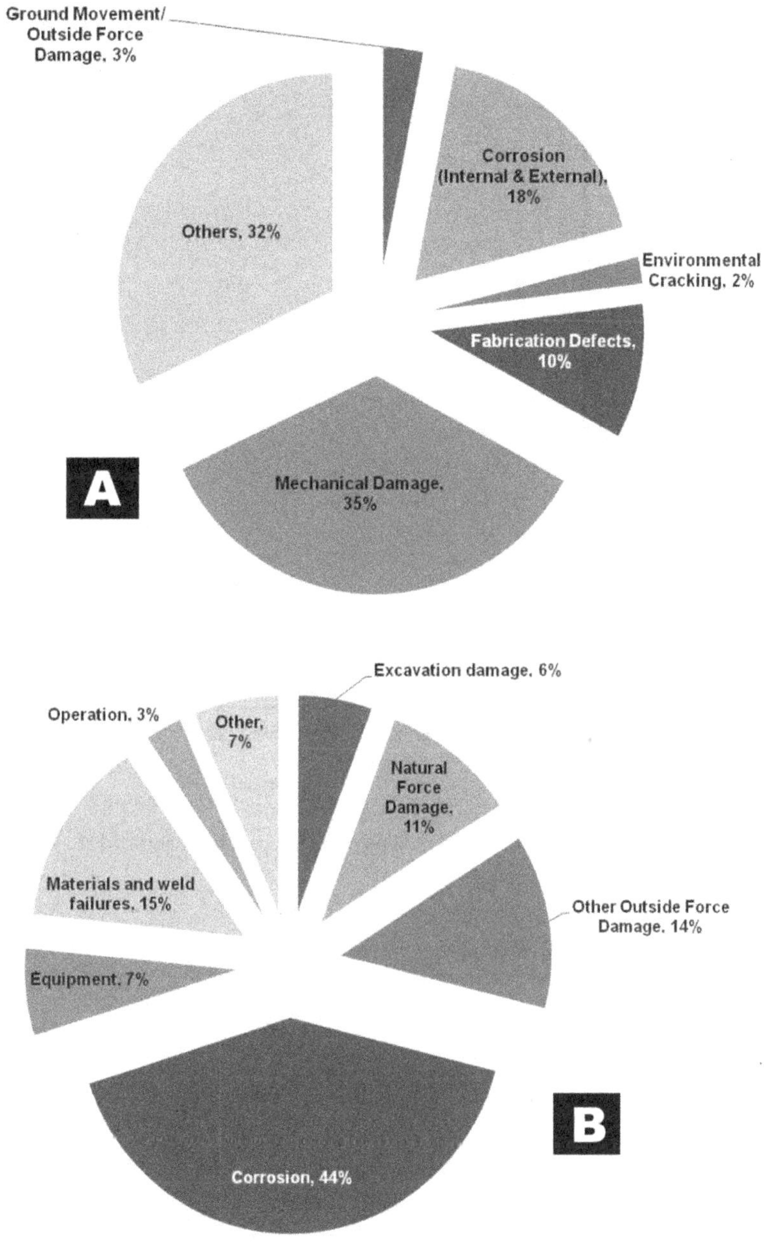

Figure 9-15. US natural gas pipeline failure causes: 1984–1990 (A) (PRCI 2003) and 2000–2003 (B), US Office of Pipeline Safety (Lee Chon Gee 2008)

industry in the past 15 years to 25 years (Figure 9-16; Sandoval et al. 2001), it is clear that safety management has taken a dramatic increase in the past 10 years and currently leads the issues facing the industry. While this statistic is generated for the United States, it is also relevant to the pipeline industry worldwide because the tendency has been toward more regulations and better incident management.

The implication of Safety Management in Pipeline Incidence Reduction is evident when incidence issues are analyzed by reviewing pipeline failures and the resulting injuries and damage over time (Zimmerman et al. 2008) as indicated in Figure 9-17.

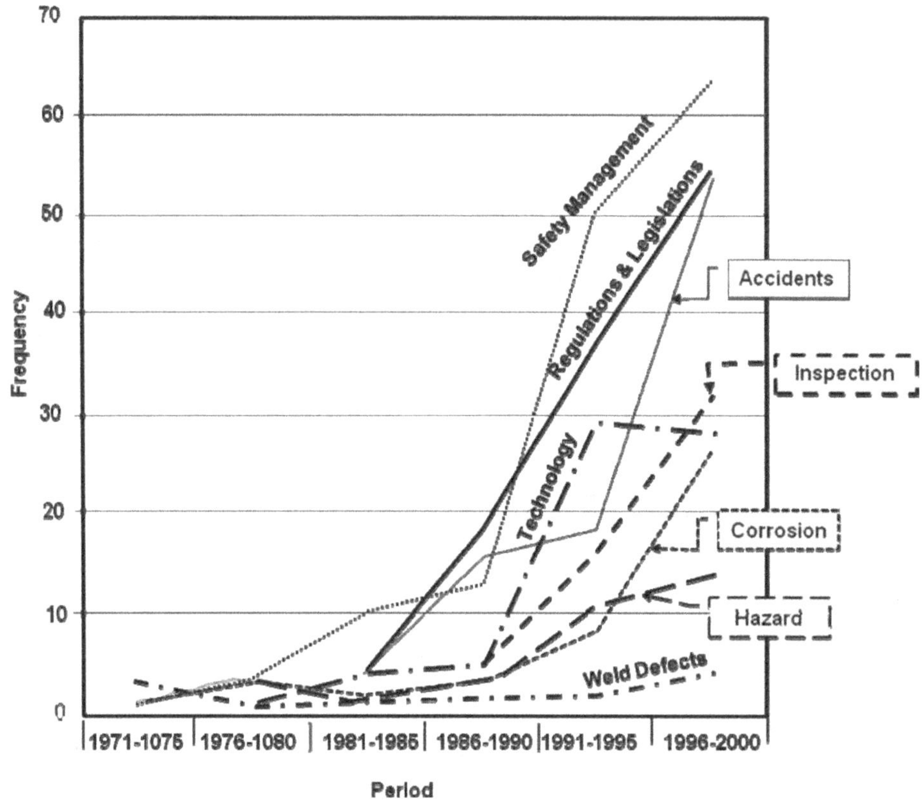

Figure 9-16. Key historical pipeline issues (after Sandoval et al. 2001)

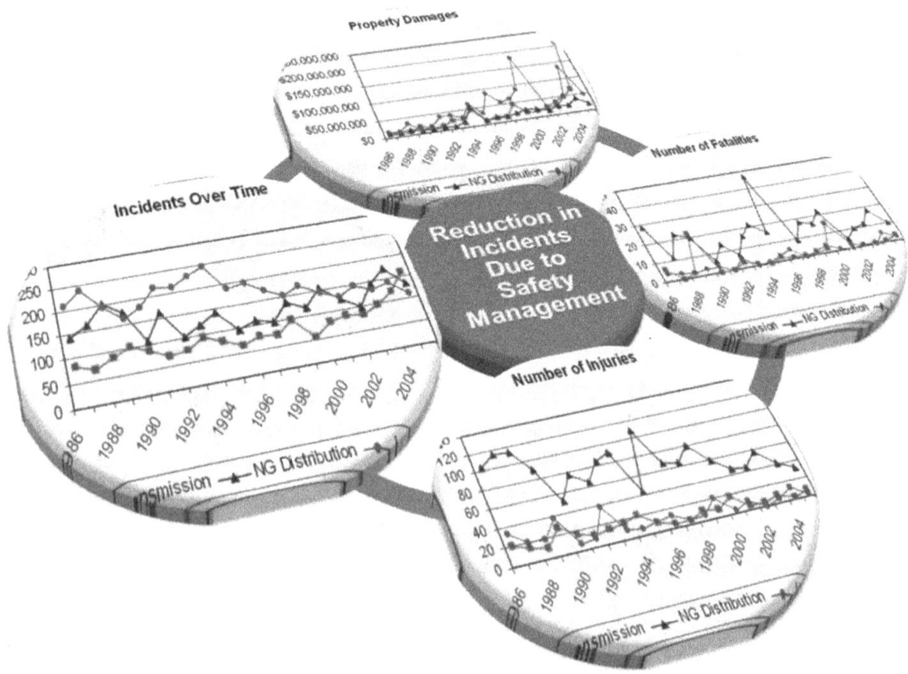

Figure 9-17. Implication of safety management in pipeline incidence reduction 1986–2008 (reproduced from Zimmerman et al. 2008)

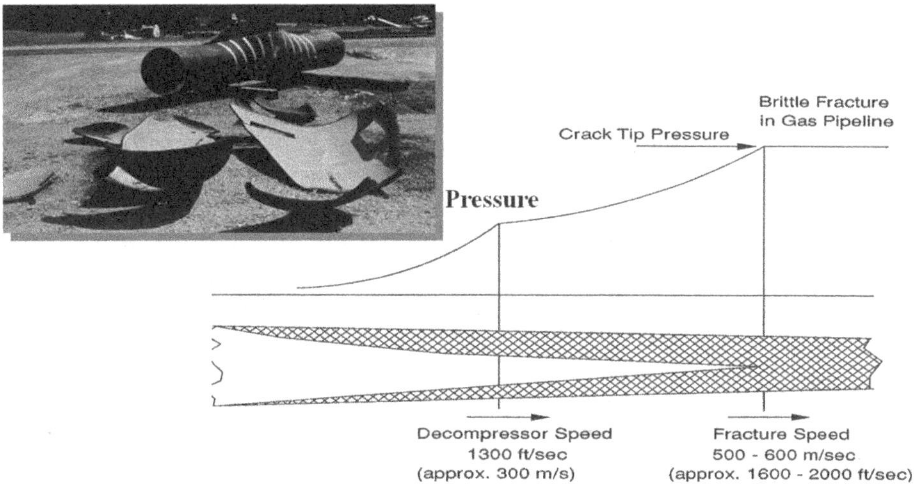

Figure 9-18. Brittle failure fracture in a gas pipeline mechanism and arrest

Failure and Consequences

An example of a failure in a gas pipeline is brittle fracture where lack of fracture control properties is evident (Mohitpour et al. 2008). Cracks in such a pipeline can be caused by a number of mechanisms including third-party damage, lack of proper fracture control capability (appropriate toughness and ductility), pipe defects, etc. Brittle fracture can spread rapidly; indeed the speed of brittle fracture in a gas pipeline is greater than the gas decompression velocity (Figure 9-18). Thus, any crack tip sees a constant pressure and the length of the failure can be very extensive; in fact, it stops only when it encounters something that slows crack velocity. The failure surface exhibits chevron-type markings pointing toward the origin of the crack.

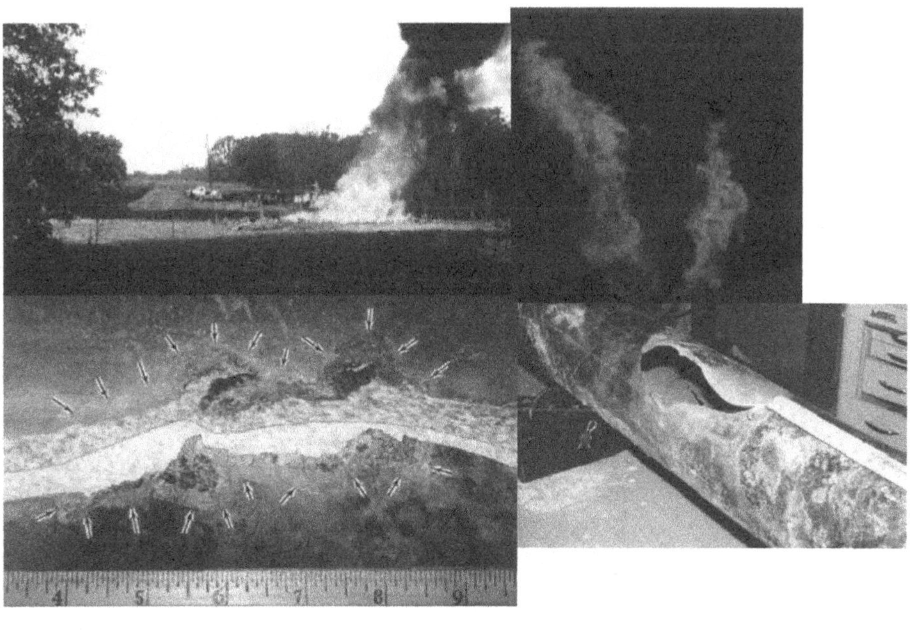

Figure 9-19. Incidents and causes in gas pipeline failures

Consequences of Gas Pipeline Failure

Some of consequent causes of an incident and resulting incidents are depicted in Figure 9-19.

The momentum of gas released by a pipeline failure is such that the backfill at the failure location will normally be ejected resulting in a crater and the gas will be vented to atmosphere in the form of a high-velocity jet (Figure 9-20). The extent of the crater and nature of the jet will depend on the failure mode and orientation of the release. The high-velocity gas escaping from the crater will normally be directed upward into the atmosphere. The momentum of the jet means that it will mix with the atmosphere and disperse without any drifting of flammable concentrations. It is therefore unlikely that ignition will occur unless there is an ignition source close to the dispersing jet.

Although an asphyxiant natural gas is nontoxic, there will be no hazard associated with toxicity as a result of the release. However, in all cases where there is a significant hazard to the surroundings, debris will be ejected significant distances from the crater, an over-pressure pulse will be produced in the atmosphere, and substantial sound energy will be generated by the failure.

Incident Management Design

Incident Site and Scene

The specific location of the actual incident is referred to as the "Site"; the immediate general location around the site is referred to as the "Scene." Two cordons would be set up, an inner around the Site and an outer around the Scene. A number of the functions and activities carried out actually happen at the Scene and are described in Figure 9-21.

Potential Hazard in Gas Pipelinez Failure

Significant damage to a natural gas pipeline that results in a pipeline puncture or pipeline rupture would lead to a pressurized release of natural gas. This may give rise to a thermal

Figure 9-20. Momentum of gas released resulting in a crater caused by natural gas pipeline failure (Mohitpour et al. 2005)

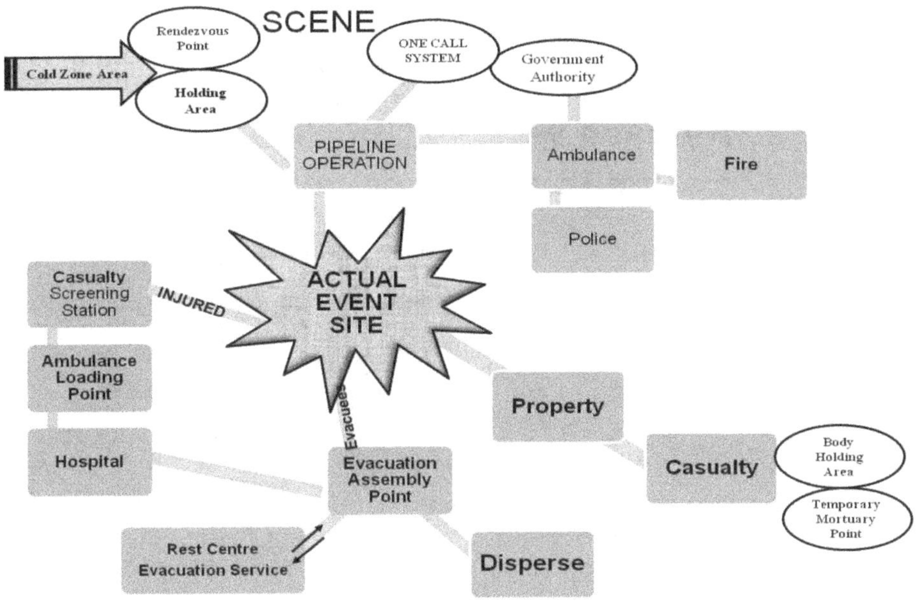

Figure 9-21. Site and scene of an event (after UK, Kent County 2005)

radiation hazard to individuals in the vicinity of the pipeline if the gas subsequently ignites.

The thermal radiation levels, depending on the magnitude of the release may be sufficiently large to cause injuries to individuals out of doors in the vicinity of the pipeline. People indoors are usually shielded from the effects of the thermal radiation, but the radiation levels, again depending on the magnitude of the release, may be sufficiently large to cause buildings to catch fire and collapse.

It is therefore important to design an evacuation location where personnel can be evacuated to and shielded from hot thermal radiations during and emergency situation where natural gas release cold likely get ignited. To undertake such a design, the following must be considered:

- Size of release
- Duration of release
- Extent of flammable cloud dispersion
- Likelihood of ignition
- Duration of thermal release in the event of an ignited release

Natural Gas Pipeline Fire Duration

The duration of the natural gas pipeline fire would vary depending on a number of factors including:

- the size of the pipeline failure
- pipeline diameter
- the distance from the failure to the nearest isolation valves (or emergency flow restrictive devices)
- control philosophy on isolation valve closures

It should be assumed that the gas release could potentially continue for some time (tens of minutes), and in some cases, it may be a number of hours before the gas leak is isolated and the hazard removed. However, the rate of gas release would reduce with time (and

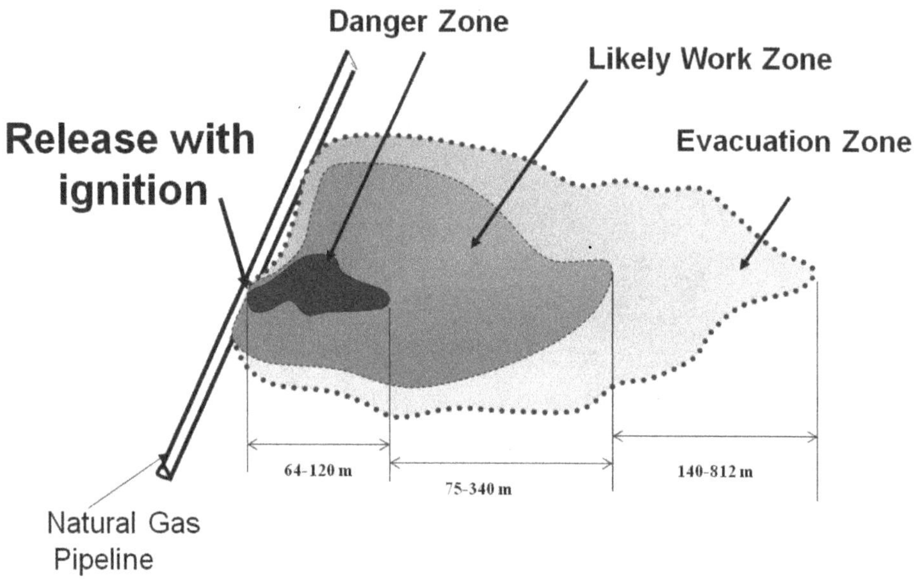

Figure 9-22. ERP — approximate affected zone in case of fire/explosion

hence the consequences of an ignited gas release) as the pipeline de-pressurizes. The latter intern affect the hot/danger zone, work zone, and evacuation/cold zone. These zones for a natural gas release, Figure 9-22 can be explained as follows:

Hot zone: This the zone that is highly explosive and extremely dangerous. Once ignited, it generates intense heat radiation. This zone should only be approached and entered by trained personnel who are appropriately equipped with PPE and self-contained breathing apparatus (SCBA). HSE 2006 define this one as "ZONE 0," a place in which an explosive atmosphere consisting of a mixture with air or flammable substances in the form of gas, vapor, or mist is present continuously or for long periods or frequently:

Warm zone: This zone can be generally considered as nonexplosive but could change to explosive depending on circumstances. Personnel can enter this zone without SCBA. The warm zone is also an area where personnel can be affected by the force of an explosion in and the heat from the hot zone. HSE 2006 defines this zone as "ZONE 1," a place in which an explosive atmosphere consisting of a mixture with air or flammable substances in the form of gas, vapor, or mist is likely to occur in normal operation occasionally.

Cold zone: This is the radial area that is generally safe for personnel and for performing work. Usually, this is where the emergency command post is stationed and rescue equipment is available. HSE 2006 refer to this zone as "ZONE 2," a place in which an explosive atmosphere consisting of a mixture with air or flammable substances in the form of gas, vapor, or mist is not likely to occur in normal operation but, if it does occur, will persist for a short period only.

Flammable Cloud. It should be recognized that if a release of natural gas occurs, there is the potential for ignition to occur at any time.

Steps thus must be taken in the event of a pipeline puncture or rupture that potential ignition sources are not introduced into the area around the release (i.e., the Scene; Figure 9-21) where gas could potentially be present in flammable concentrations. The extent of the flammable gas cloud is difficult to predict, as it is dependent on the:

- release pressure
- size
- orientation of the release
- atmospheric conditions

Flammability Limit and Extent

Flammability limits, also called flammable limits, or explosive limits give the proportion of combustible gases in a mixture, between, which limits this mixture is flammable.

The flammable range is expressed by lower explosive limit ("LEL") or lower flammability limit ("LFL") and upper explosive/flammability limit ("UEL" or "UFL"). In this range, generally, work does not take place. However, while ignition is still possible above UEL, work can be undertaken with appropriate safety gear (PPE and SCBA).

LEL is the concentration of the components (such as natural gas) in the air below which the propagation of a flame will not occur on contact with an ignition source. The UEL is the concentration of the component in the air above which the propagation of a flame will not occur on contact with an ignition source.

LEL and UEL for some products transported through pipelines are provided in Table 9-3 below.

It may be noted that upper explosive limit can change in the presence of other gases such as oxygen. Usually with oxygen present, UEL will increase to a high limit).

Flammability limits of mixtures (LFL_{mix}) of several combustible gases can be calculated using Le Chatelier's mixing rule for combustible volume fractions x_i:

$$LFL_{mix} = \frac{1}{\sum \frac{x_i}{LFL_i}} \text{ and similar for UEL/UFL.}$$

Temperature and pressure also influences flammability limits. Higher temperature results in lower LEL/LFL and higher UEL/UFL, while greater pressure increases both values.

Fire and Explosion

A hydrocarbon release occurs from a pipeline when an incidence such as rupture causes a release. In pressure of an ignition source, fire may result followed by an explosion. Type of fire that is likely to occur include (Babu et al. 2007):

- Jet fire
- Pool fire and
- Flash fire

TABLE 9-3. Flammability limit of some products transported through pipelines (L'Air Liquide, 1976) — % of gas in air

Product	LEL (%)	UEL (%)	Flame Temperature (°C)	Flame Speed (m/s)
Acetylene	1.5	100		
Methane (natural gas)	5	15	2,810	3.9
Benzene	1.2	8	1,690	
Ethane	3	12.4	1,960	0.4
Ethylene	2.7	28.5		
Hydrogen	5	14.5	2,830	14.36
Propane	2.3	10.5	2,820	3.31
Butane	1.8	8.4	1,970	0.37
Petrol	1.4	7		
Diesel	0.7	5		
Jet fuel	0.5–0.7	5–8		

Jet fire is typically associated with release of natural gas. Jet fire scenario depends on the diameter of pipeline and the leak size. If the natural gas pipeline has a leak at a certain point in its path and given that an ignition source is available, jet fire can occur. Jet fire is usually not associated with rupture of pipeline.

Pool fire occurs when a flammable liquid spills and forms a pool over a surface and given that an ignition source is present. Pool fire is usually not associated with natural gas.

Flash fire is a delayed ignition of a cloud of flammable vapor. It occurs due to a sudden intense fire caused by ignition of flammable materials and is characterized by high temperature, short duration, rapidly moving flame front, and considerable shock waves. Flash fire is usually more prominent with leak and accumulation of natural gas rather than rupture.

Natural gas is lighter than air and the majority of pipeline failures are expected to lead to gas jets that are orientated vertically upward. The extent of the flammable gas cloud at ground level for the majority of situations would therefore be expected to be negligible or only of the order of a few meters. However, heat radiation can be extensive and must be avoided while the gas is still flowing (Figure 9-23).

Explosion (Vapor Cloud Explosion/Late Explosion). Explosion is a sudden release of energy that causes a very rapid rise in pressure. Explosion causes pressure shock wave to travel outward radially. This is followed by negative or under pressure (Figure 9-24). Shockwave decreases as energy is absorbed in the atmosphere. However, most of the damage is done by the shock wave.

Late explosion usually occurs if hydrocarbon fluid is accumulated in a confined space or area. Late explosion occurs when a flammable material is dispersed in atmosphere, and after some delay, the ignition of the vapor cloud occurs in presence of an ignition source and when the appropriate gas/air ratio level is reached.

However, in case of underground cross country pipelines, the release of natural gas can find confinements in the form of dense vegetation and buildings. However, because of this and lighter nature of natural gas, this is relatively a rare occurrence.

It has been reported that the gas released from most natural gas pipeline failures does not ignite. Additionally, the likelihood of ignition depends on the nature and size of the gas release. Statistics show that there is a 50% likelihood of ignition for ruptures in large diameter pipelines (greater than 457 mm) and a 20% likelihood for smaller diameters. In the case of leaks, these figures reduce to 10% regardless of the pipeline diameter.

However, for natural gas pipelines explosion, in all cases where there is a significant hazard to the surroundings, debris will be ejected significant distances from the crater (that such incidences create), an over-pressure pulse will be produced in the atmosphere, and substantial sound energy will be generated by the failure.

Figure 9-23. Jet fire from a natural gas pipeline after rupture

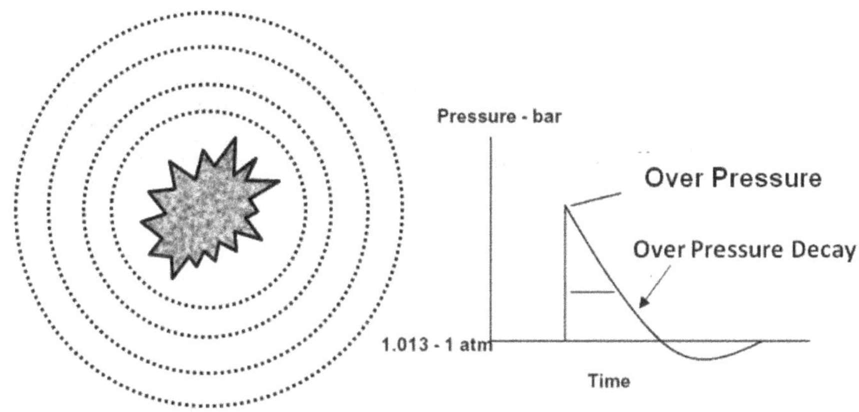

Figure 9-24. Shock waves created by explosion

Natural Gas Pipeline Emergency Planning Distances

Hazard range must take into account the likelihood of an event occurring. Such a likelihood must therefore be taken into account for determining the distances to be used for detailed emergency planning and hazard control.

The extent of the emergency planning distance, following an ignited natural gas pipeline failure, would vary from pipeline to pipeline and be dependent on a number of parameters, the principal ones being:

1. The individual risk contours below which risk of a specified level of harm is considered to be trivial and for which emergency plans need not be prepared
2. The leak size (if the failure results in a full bore pipeline rupture then this would be dependent on the pipeline diameter)
3. The orientation of the release
4. The pipeline pressure
5. The local wind speed and wind direction

A- Rural Area — Natural Gas Pipelines Operating Up to 75 Bar. Based on an individual risk of 1 chance per million (cpm) for rural pipelines and for vulnerable or large populations. Table 9-4 and Figure 9-25 below provide examples of worst-case emergency planning distances for rural and suburban pipelines (UK, Kent County Council 2005).

The following may be noted:

1. Distance to the threshold of noise pain determined in accordance with methods in IGE/SR/23 (Venting of Natural Gas, IGEM 2007) guidelines.
2. Distance equal to 1 building proximity distance (BPD) based on data relating to actual events (Most debris will be contained within this distance but pipe fragments may travel further.)
3. Estimates based on a high likelihood of window breakage but little structural damage.
4. Distance to an individual risk of 1 cpm in a pipeline designed to the minimum (thermal dose, or TD) TD/1 conditions.
5. Distance to an individual risk of 0.3 cpm in a pipeline designed to the minimum TD/1 conditions.
6. Thermal hazard range equal to 7 BPD for ruptures and escapes distance for leaks.

TABLE 9-4. Worst-case emergency planning distance from natural gas pipeline ruptures in rural area (after UK, Kent County 2005)

		Emergency Planning Distances					
					Thermal Radiation		
Type of Event	Pipeline Diameter (mm)	Noise (m) (1)	Debris Throw (m) (2)	Overpressure (m) (3)	Normal Localities (m) (4)	Sensitive Localities (m) (5)	Thermal Hazard Range (m) (6)
0 – 25 mm leak	All	3	5	5	4	5	5
25 – 75 mm leak	All	5	12	6	9	10	11
75 – 150 mm leak	All	18	20	10	72	81	90
Rupture	168	40	20	10	75	85	140
	324	64	33	17	75	105	230
	457	64	37	21	108	117	290
	610	64	53	26	150	170	365
	762	64	64	33	195	260	455
	914	64	80	40	220	265	560
	1067	64	95	48	230	275	665
	1219	64	117	58	250	340	812

ERP: Approximate Affected Zone in Case of Fire/Explosion. The emergency planning distances given below are based on casualty criteria of 1060 TD units. Casualty in this context corresponds to the definition given for a dangerous dose in the UK Health and Safety Executive land-use planning guidelines and can be expected to produce the following effects: "severe distress to almost everyone, a substantial fraction requires medical attention and some people are seriously injured and require prolonged treatment, any highly susceptible people might be killed."

The above criteria would be anticipated to result in effects ranging from skin blistering to death. These casualty criteria also correspond to that used by the Health and Safety Executive for determining the consultation distance for land-use planning.

B - Suburban Area — Natural Gas Pipelines Operating Up to 75 Bar. As per calculation for rural area but based on an individual risk of 0.3 cpm for suburban pipelines and

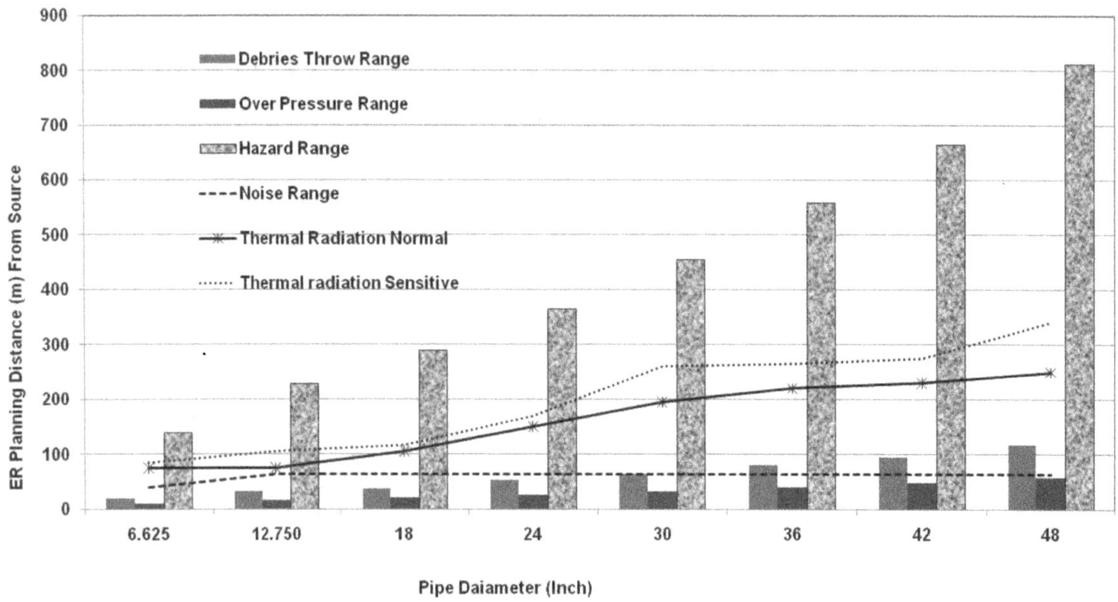

Figure 9-25. Emergency planning distance from pipeline ruptures (75-bar operating pressure)

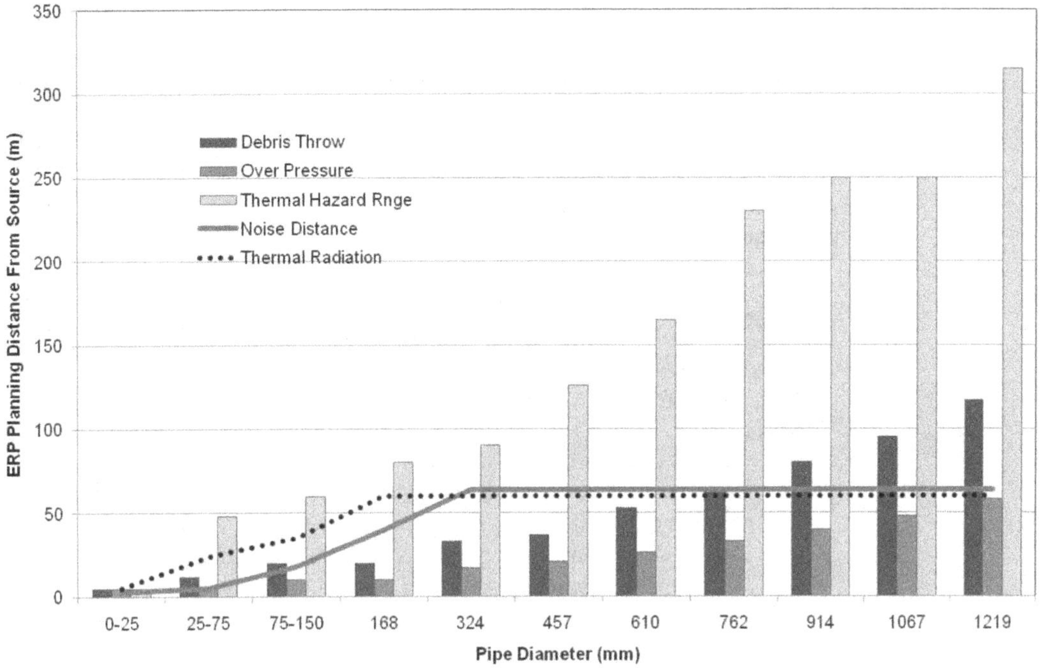

Figure 9-26. Emergency planning distance from pipeline ruptures — suburban area (75-bar operating pressure)

for vulnerable or large populations, Figure 9-26 below provides examples of worst-case emergency planning distances for rural and suburban pipelines.

Notes:

- Noise distance to the threshold of pain is based on IUK IGE/SR/23 (Venting of Natural Gas) guidelines (IGEM 2007).
- Debris throw distance (equal to 1 building proximity) is based on data relating to actual events as defined by UK, 2005. It is anticipated that most debris can be contained within this distance but pipe fragments may travel further.
- Over-pressure estimate is based on a high likelihood of window breakage but little structural damage.
- Thermal radiation distance is equal to an individual risk of 0.3 cpm in a pipeline designed to the minimum TD/1 conditions or the hazard range for a 75-mm to 150-mm leak if rupture does not reach this level.
- Hazard range distances are for zero casualty probability.

Important Note: The hazard range is less than that for rural areas due to the increased shelter provided by housing in the more densely populated suburban areas.

Consequences of Natural Gas Release

The release from a pipeline can be from a full rupture or a leak. Consequences that can be expected (UK, Kent County 2005) are:

- Thermal radiation
  - Rupture: Natural gas ruptures will normally result in an initial burst of high thermal radiation lasting for about 30 seconds followed by a decay phase in which the thermal radiation level will drop substantially over the next 15 minutes or

so. The thermal radiation level will then remain steady until the supply of gas is cut-off.

- o Leak: Unimpeded leaks of natural gas can result in steady thermal radiation levels. Impeded leaks generally behave as a smaller version of a pipeline rupture. As a result of the smaller opening involved in gas release, the area affected even by impeded leaks will be limited compared to a pipeline rupture.

- Noise: Natural gas will be released at sonic velocity until it is almost exhausted from the pipeline and will generate substantial noise levels, which may be heard at great distances. The approach illustrated in IGE/SR/23 (IGEM 2007) can be adopted to predict noise levels for venting of natural gas. A value of 130 dB is generally chosen to determine the hazard range since this represents the threshold of pain.
- Debris throw: Experience has shown that most of the debris ejected as a result of natural gas pipeline rupture would fall within one rural BPD of the pipeline (Figure 9-27).
- Over-pressure: Natural gas, released into the atmosphere as a result of the failure of a cross-country pipeline, generally does not form explosive mixtures. However, the decompression of the gas would produce an over-pressure in the atmosphere. There would also be an over-pressure generated as a result of flame propagation through the gas cloud if ignition occurs. In an open situation the over-pressure resulting

Figure 9-27. Debris throw in natural gas pipeline rupture

from either decompression or burning would not normally be expected to exceed about 40 mbar at ½ BPD. At this level of over-pressure a high degree of window breakage can be expected, but little structural damage would result.

Heat Affected Zone/Thermal Exposure

There are a number of ways by which the heat affected zone due to an ignited fluid from a pipeline rupture or leak can be calculated. This depends on whether it is jet or pool fire.

As stated previously, natural gas pipeline failures are expected to lead to gas jets that are orientated vertically upward. The heat radiation can, however, be extensive and must be avoided while the gas is still flowing out of the pipe. Emergency planning work zone must thus be located outside this radiation area such that effective and orderly control of the fire can be implemented while gas flow is continuing from the ruptured pipeline.

The Heat Affected Zone Radius. The zone radius can be calculated from the following empirical equation (ESA, 2005):

$$R = [(C \times Q_{CH4}/(k \times E\%)]^{0.55}$$

where

R = zone radius (assumed a sphere), m
C = 1840, a constant of proportionality
Q_{CH4} = volume flow rate of methane, m^3/s (actual)
k = safety factor applied to the LEL
= 0.5, for secondary grade releases or
= 0.25, or primary grade releases
$E\%$ = LEL, % vol/vol

The above equation takes into account obstructions caused by proximity to the ground, walls or other objects. It is only applicable to freely ventilated outdoor locations and assumes a wind-speed sufficient for turbulent diffusion. Also, the zone radius is measured from the point of release in all directions, and is thus independent of the density of the release. The density of the release can be incorporated by considering the actual flow "Q" over time from the rupture areas.

The above equation has been derived for the determination of heat-affected zone radius in landfill environment, and additionally, it is derived for a wind speed of 2 m/s or higher. So if wind speed is lower, then some "layering" will likely to occur. However, by selecting an appropriate safety factor (k), this equation can provide a reasonably conservative result for estimating the heat affected zone radius purposes.

Typically, in gas pipeline when a rupture occurs, the line pack condition changes over time and until the flow is curtailed in the affected pipeline segment (e.g., by activating/closing the adjacent up/down stream flow restrictive devices/valves). However, the flow from the rupture is in choked condition (i.e., discharging sonically at the speed of sound).

Thus, the following formulae can be utilized incrementally to (1) determine the pressure drop (i.e., pressure-time curve as the gas from the pipeline is depleted; Mohitpour et al. 2007) and then (2) calculating the flow over time considering line pack (Mohitpour et al. 2005) as the pressure in the pipeline is reduced from the operating pressure (i.e., pipeline pressure before the rupture) to the atmospheric condition.

A. Rupture Pressure–Time Curve. The time (t) required to blow down a segment of the pipeline can be calculated from the following formula (Mohitpour et al. 2008).

$$t = \frac{5.5V\ (\mathrm{ft}^3)}{A_v C_d\ (\mathrm{in.}^2)\ C_d} \sqrt{\frac{S_\partial}{ZT} 1n \frac{P_2\ (\mathrm{psi})}{P_1}}\ (\mathrm{seconds})$$

where:

A = cross-sectional rupture flow area, in.^{2}
A_v = smallest cross-sectional area of rupture (i.e., vena contract area), in.^{2}
C_d = coefficient of discharge to convert A to effective flow area, dimensionless
P = pressure of internal gas in system, psia
R = gas constant, 1545 mol. wt.
S_g = specific gravity of gas, dimensionless
T = temperature of gas in system, °C
t = Flow-down time, sec
V = system volume involved in blow-down, ft^3
Z = compressibility factor of gas, dimensionless
1 = initial system condition for time interval being considered
2 = final system condition for time interval being considered

B. Line Pack Volume. Line pack in a given pipeline can be calculated from a method described by Katz et al. 1959.

$$V_0 = \frac{A\,L z_0\,T_0 P_c}{P_0\,T}\,\frac{\int_0^{P_{r,1}} (P_r/z)^2 \cdot dP_r - \int_0^{P_{r,2}} (P_r/z)^2 dP_r}{\int_0^{P_{r,1}} (P_r/z) \cdot dP_r - \int_0^{P_{r,2}} (P_r/z)\, dP_r}$$

where

V_0 = volume of gas in pipeline measured at T_0 and P_0 (ft^3)
A = cross sectional area of pipeline, ft^2
L = pipeline segment length, ft
T = temperature, °C
P_c = pseudo-critical pressure of gas, psia
P_0 = standard pressure, psia
T_0 = standard temperature, °C
z_0 = compressibility factor at base conditions

Value of $\int_0^{Pr} (P_r/z)^2 \cdot dP_r$ and $\int_0^{Pr} (P_r/z)\, dP_r$ can be obtained from Tables A-7 of Katz et al. (1959) or obtained as follows for typical pipeline pressure ranges.

Alternatively, a simpler method described by Mohitpour et al. (2007) on the subject of "Determining Gas Leakage Using Pressure Drop Method" (p. 98) can be followed for line pack calculations.

Example:

For a 36-line, wall thickness t = 0.53, (Figure 9-28), (1)- calculate blow-down time and draw-down after pipeline rupture (at the 12" lateral junction) as indicated in diagram below; (2) calculate safe thermal distance for all ER activities assuming the following gas composition

Nitrogen	N_2	0.0055
Carbon Dioxide	CO_2	0.0001
Methane	C_1 (CH_4)	0.8004
Ethane	C_2	0.1744
Propane	C_3	0.0188
Pentane	nC_4	0.0005
Pentane	iC_4	0.0003

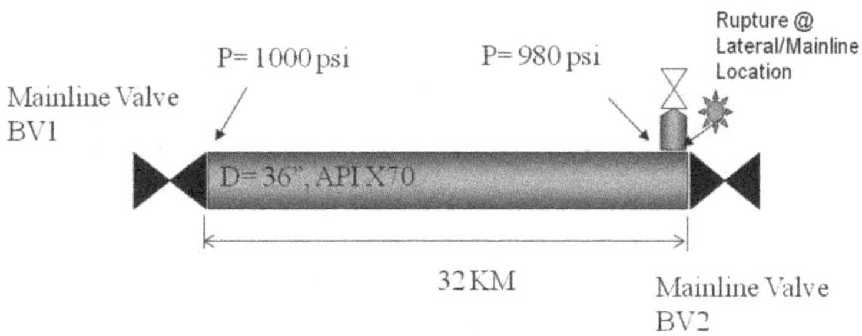

Figure 9-28. Sample problem, blow-down, heat-affected zone radius calculation

Results are shown in Table 9-5 and Figure 9-29.

The above calculation can be compared with the work performed by Taufik and Wisianto (2007) (Figure 9-30). Taufik and Wisianto assume a constant flow of ignited gas, while in real emergency situations, pipeline controls effect a shut-down of up/downstream valves adjacent to the location of the rupture, hence reducing the radiant heat exposure distance with time. It, however, may be noted that the maximum safe distance as calculated by the two techniques are comparable.

Taufik and Wisianto suggest the following expression for the calculation of radian heat radius (Figure 9-31):

$$R = [2348\ P \cdot D^2)/I_w]^{1/2}\ \text{ft}$$

where

I_w = threshold heat intensity, Btu/hr-ft^2
P = pipeline pressure (psia)
D = pipeline diameter, in.

For a threshold heat intensity of 5000 Btu/hour ft^2, the above expression will reduce to:

$$R = 0.685[P \cdot D^2]\ 0.5\ \text{ft}$$

TABLE 9-5. Tabulation of results for safe distance from heat radiation source

Time (sec)	*P* (psi)	Line Pack 10^6 m^3	Change (in) Line Pack 10^6 m^3	Change (in) Line Pack m^3	Draw-down in 600 sec	Distance From Source (m) LEL	UEL
0	990	1.7180	0.000	0	1,718		
600	682.1042	1.0820	0.636	636,000	1,060	951.2273608	649.7074
1,200	469.9658	0.6960	0.386	386,000	643	722.7796853	493.6731
1,800	323.8037	0.4530	0.243	243,000	405	560.3590245	382.7365
2,400	223.0989	0.2960	0.157	157,000	262	440.6844527	300.9964
3,000	153.7138	0.1920	0.104	104,000	173	351.359141	239.9854
3,600	105	0.1230	0.069	69,000	115	280.3818649	191.5065
4,200	72.96994	0.0770	0.046	46,000	77	224.3363903	153.2263
4,800	50.27586	0.0450	0.032	32,000	53	183.7449484	125.5015
5,400	34.63977	0.0250	0.020	20,000	33	141.8892254	96.9132
6,000	23.8666	0.0100	0.015	15,000	25	121.124811	82.73068
6,600	16.44395	0.0000	0.010	10,000	17	96.91319692	66.19366

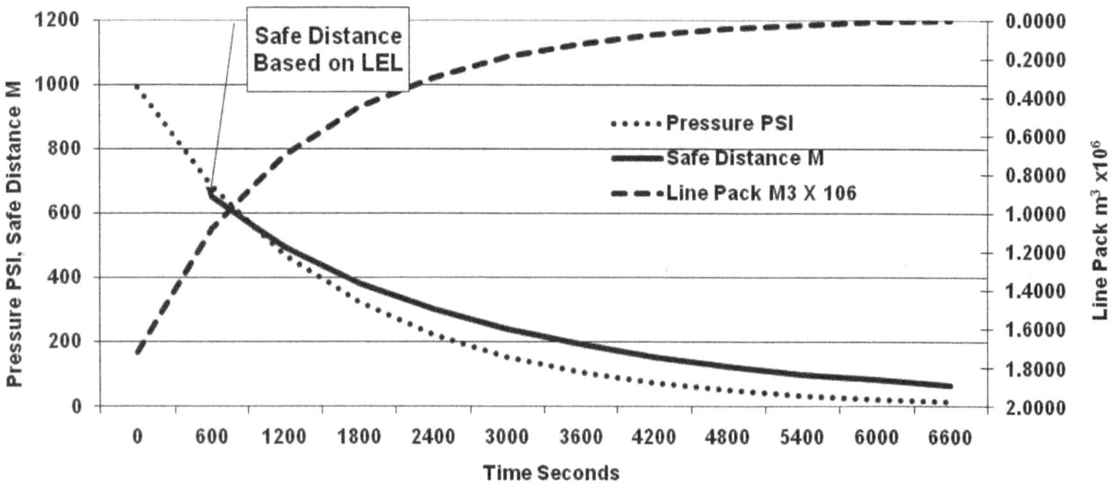

Figure 9-29. Sample problem result safe heat affected zone ($K = 0.5 * 15\%$)

Energy Vapor Cloud Explosion. The energy E in vapor cloud explosion can be calculated from the following equation:

$$E = \alpha H_c M$$

where

E = energy in vapor cloud explosion (J, TNT as applicable)

α = energy converted into blast wave usually between 0.03 and 0.1 based on previous explosions

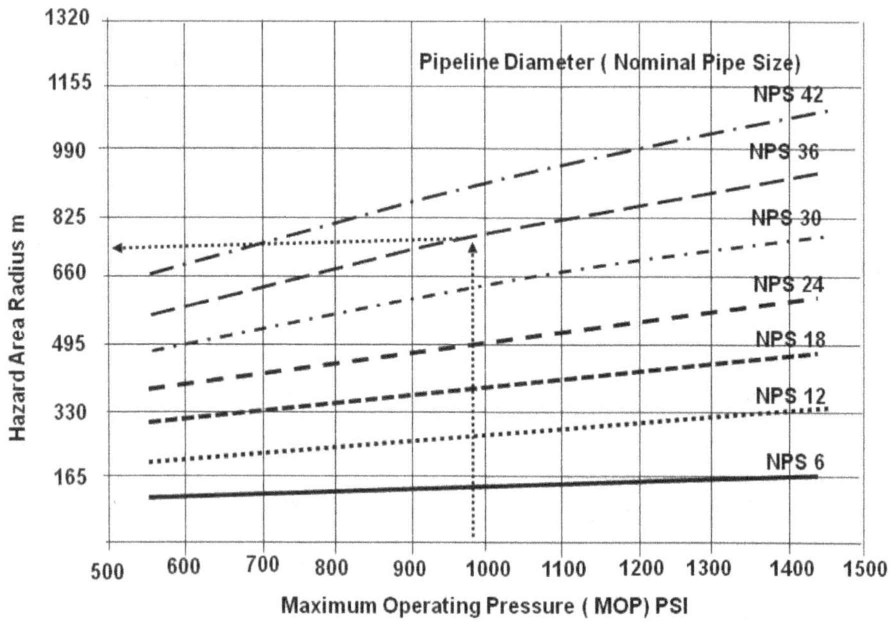

Figure 9-30. Hazard radius area as a function of diameter and pipeline pressure (after Taufik and Wisianto 2007)

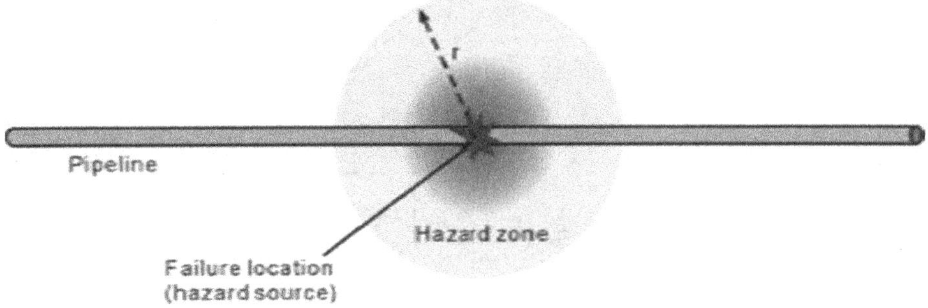

Figure 9-31. Hazard area related to heat intensity (after Taufik and Wisianto 2007)

H_c = heat of combustion of flammable gas; for natural gas, this will be about 1000 Btu/SCF (37.26 mJ/m^3)
M = mass of gas in flammable cloud (volume m^3 × density kg/m^3)

The mass M depends on the relative size of flammable cloud, which depends on wind velocity and terrain condition as well as release rate.

Table 9-5 provides total volumetric line pack as well as the incremental release until the pipeline segment is depleted. Assuming a calm day (no wind), mass flow can be considered as the total pipeline line pack, which upon an instant ignition can cause an explosion. However, the lower value of α (energy converted to blast) may be considered as released into an unconfined space. In this way it would be possible to ascertain whether the explosion wave can cause damage nearby such as window breakage.

A gram of TNT releases 980 to 1100 calories upon explosion. To define the ton of TNT, 1000 thermo-chemical calories = 1 g TNT = 4184 J.

Considering that 3% (i.e., α = 0.030) of the volume of the initial line pack given in Table 9-5, then initial TNT in gas will be

$$= 0.03 \times (1.718 \times 10^6) \times 37.27 \times 10^6/(4184 \times 10^9) = 0.46 \text{ ton TNT}$$

This value is considered as E_{TNT}.

Blast effects can also be related to the equivalent weight of TNT using by the relationship between the distance from the source, the charge weight, and the over-pressure caused by the blast wave, including the reflected shock wave. Figure 9-32 (Zalosh 1995) gives the relationship between over-pressure and "scaled distance" (Dsc). Scaled distance is the distance at which the over-pressure is calculated divided by the cube root of the TNT charge weight.

$$z = R/(E_{TNT})^{1/3}$$

where

z = scaled distance, m/(kg)$^{1/3}$
R = distance at which the over-pressure is calculated, m
E_{TNT} = weight of TNT, kg

Using the equation, $z = R/(E_{TNT})^{1/3}$, then scaled down TNT pressure z at various radius can be calculated from the scaled over-pressure equation given above.

From the above, the effects of pressure on humans and the environments can be assessed. Generally, human beings are capable of withstanding relatively high dynamic

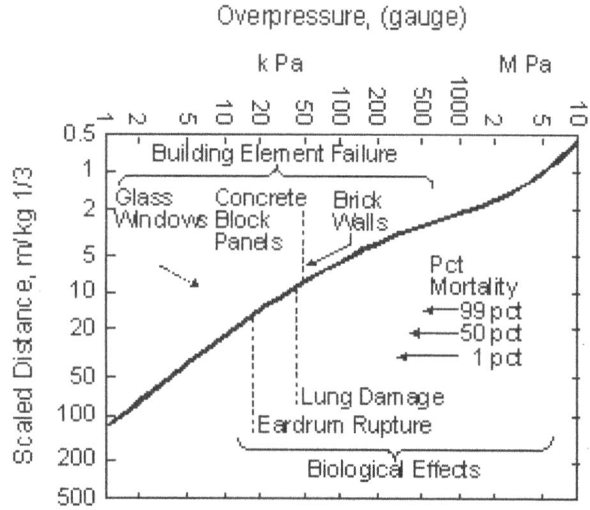

Figure 9-32. Ideal blast wave over-pressure vs. scaled distance (Zalosh 1995)

pressures and considerably higher static pressures. When people are fatally injured as a result of blast waves, it is usually because of falling objects, rather than the pressure associated with the blast wave. Table 9-6 summarizes the pressure effects of blast waves on humans (Fischer et al. 1995), which also depends on the impulse of the blast wave. With the exception of smoke, gas, and explosions, fires seldom reach pressures as high as those listed in Table 9-6.

Effects of Pressure on Components. Table 9-7 lists typical failure pressures of elements making up building as windows, walls, and doors.

Based on Table 9-5, assuming that the maximum TNT effect or 0.46 ton of TNT (from the rupture) remains constant until the pipeline is fully depleted of the compressed natural gas, the blast effect from the rupture was calculated (Table 9-8 and plotted as Figure 9-33).

Table 9-8 indicates that for the example given, the distance of about 100-m to 125-m radius from the pipeline can be considered as “Hot Zone” and not be approached. However, in reality soon after the occurrence of rupture, the adjacent upstream and downstream emergency flow restrictive devices (valves) will close. The gas within the isolated segment will then deplete from the pipeline through the rupture opening. The line pack thus depletes as pressure within the pipeline is reduced to atmospheric conditions (see previous Figure 9-29).

TABLE 9-6. Pressure effects on humans

Pressure (kPa)	Effect
35	Limit for eardrum rupture
70	Limit for lung damage
100	50% eardrum rupture
180	1% mortality
210	10% mortality
260	50% mortality
300	90% mortality
350	99% mortality

TABLE 9-7. Typical failure pressures of some building elements

Element	Typical Failure Pressure (kPa)
Glass windows	2–7
Room doors	2–3
Light partition walls	2–5
50-mm-thick breeze block walls	4–5
Unrestrained brick walls	7–15

Heavy Gases and Liquids Pipeline Safety and ER Issues

Unlike natural gas where a release in a ruptured pipeline is jetted upward, the release in a ruptured pipeline transporting heavy gases (such as propane, LPG, LNG, etc.) will result the emitting fluid to form a vapor cloud close at low spots extending beyond the release location and as well pooling some of the fluid in a liquid form close to the release area.

In a rupture of a low vapor pressure (LVP) fluid (such as oil, diesel, kerosene) pipeline, the spill flows from the pipeline. and (depending on topography) drains off until all fluid within the pipe accounting for the change in the elevation profile are evacuated. In a HVP heavy fluid pipeline (such as LPG, LNG, condensate, etc.), all the fluid within the pipeline will drain off irrespective of elevation profile. Unlike LVP, for which undulation in the elevation profile will help prevent total drainage of the pipeline upon rupture, the entire fluid contained in a section of a pipeline will boil and vaporize (over time) upon pipeline rupture and depressure to atmosphere. In these cases while similar calculations can be performed for determining safe heat-affected zone where related ERs preparations can be performed, the time calculation for fire abatement would be longer. In such cases (as LPG), typically appropriate "Emergency Flow Restricting Devices" (i.e., valves) are closed and the vapor cloud is ignited at safe distance.

In addition, the following are applicable to HVP liquids such as LPG and LNG when upon release to atmosphere generate immediate vapors and primarily stay close of to the ground and accumulate in low points:

- Liquid spills can be designed to be confined — so the very cold liquid can cool the substrate it covers over time.
- How much the substrate cools depends on its properties and the HVP liquid vapor rate.
- Typical confinement/low point may temporarily confine the LNG/LPG and similar vapors

TABLE 9-8. Calculated blast effect from point of rupture to safe radiation heat distance (refer to Figure 9-32 for scaled distance)

Distance From Rupture (m)	Scaled** Distance *z* (m)	Wave pressure (kPa)	Failure Effects
650	84.20	1.5	Components
493	63.86	3	Components
383	49.61	5	Components
300	38.86	7	Components
239	30.96	10	Components
191	24.74	13	Components
153	19.82	17	Components
126	16.32	25	Biological/Component
96	12.44	35	Biological
83	10.75	45	Biological
66	8.55	60	Biological
1	0.13	120	Biological

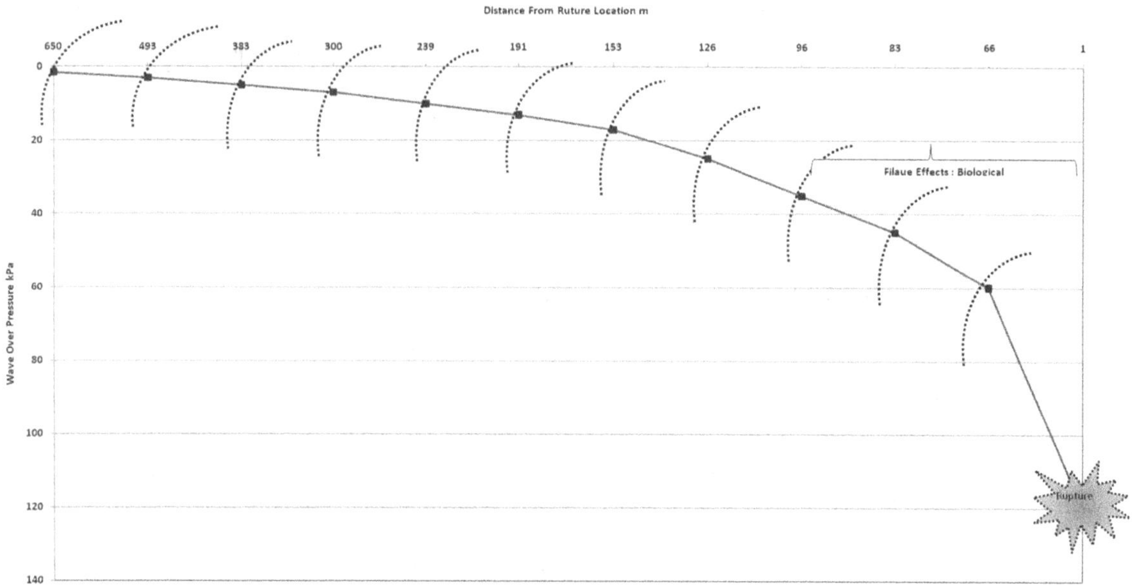

Figure 9-33. Blast front pressures (kPa) at various distances from pipeline rupture location

- As LNG/LPG vapor is formed, it mixes with air in the confined/low point space until gas/air outflows from the confinement/low point.

Appendix A to this book describes the significant safety issues of LPG pipelining including a discussion on ER and associated equipment needs and repair techniques. Similar approaches should be expected for rupture leading to release of other similar heavy HVP fluids.

However, where pool of flammable fluid (such as oil) is expected from a pipeline rupture, treatment of resulting fire in presence of ignition sources and ensuing heat radiation would be treated in a different way than natural gas. This is primarily due to generation of smoke, which may mask the heat radiation to an extent, but at the same time, it can cause reduction in oxygen and hence a suffocation issue.

Heat Generation in Pool Fire

The terminology "Thermal Radiation" sometimes can pose confusion and considered as synonymous to that from a "nuclear bomb." Thus, the wording "Radiant Heat" is expressed to be more appropriate (Raj 2006).

Pool fire occurs when a flammable liquid spills and forms a pool over a surface or water (Figure 9-34).

However, generally there are similarities in burning characteristics of an LNG pool fire and an oil fire (Figure 9-35; Raj 2005).

Radiant heat from pool fire can be calculated using a method described by UK Health and Safety Executives (Rewand Hulbert 1996). Other sources for reference and calculations are:

DOT 49 CFR 193 (US DOT 2009) and NFPA 59 A (NFPA 2004, 2007)

CFR 193 and NFPA refer to two exclusion zone requirements based solely on consequences for radiant heat from pool fire. These are:

1. Thermal radiation hazard from a pool fire (ignition at release); 5 kW/m (causing second-degree burns in 20 seconds)

Figure 9-34. Unconfined LNG fire on pool of water (10,000 gallons, diameter = 18 m, height = 80 m (Spicer and Martin 2005; Raj 2005, 2006)

2. Vapor dispersion hazard (ignition at the maximum extent of the flammable cloud); LFL/2 (flammable pockets of gas)

CFR 193 and NFPA also suggest worst-case meteorological conditions that are: observed for the site or default of 2 m/s (4.5 mph) wind speed.

Basic considerations for LNG pool fire suggested are:

- Meteorological conditions to be considered
- Spill size (total amount lost); spill rate (how quickly it is lost)
- Hazard endpoints (thermal hazards: 5 kW/m or 1.4 kW/m)
- Vapor dispersion hazards: LFL or LFL/2

The industry, however, uses the following thermal radiation levels for fire modeling:

- 38 KW/m^2 - tank walls distortion
- 12.5 KW/m^2 - wood, paper, paint ignition

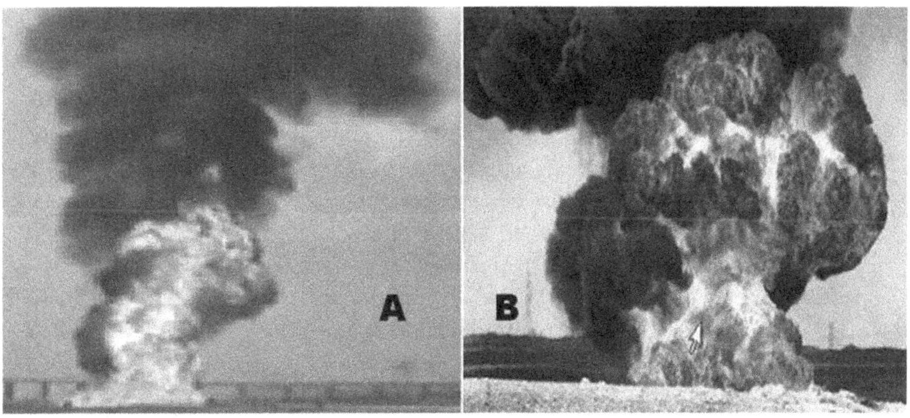

Figure 9-35. Similarities of an LNG pool fire (A) and an oil pool fire (B) (Raj 2005, 2006)

- 4-5 KW/m^2 - limit for people to escape, second degree burn
- 1.6-3 KW/m^2 - soften cables for sustained flaring
- 0.7 KW/m^2 - hot summer sunshine

Liquid Pool Fire Calculation. For determination of safe distance (due to radiant heat generation) from a pool fire of a hydrocarbon liquid, the method described by Wagner (1999) can be utilized. Other methods are due to Rew and Hulbert (1996) or Iqbal and Salley (2003).

While the method of Iqbal and Salley (2003) is the most comprehensive approach, the Wagner (1999) method will be illustrated below to provide an understanding of heat generation and radiation from a pool liquid hydrocarbon fire.

Spills on Land. A spill on land in a cross country pipeline can occur as a consequence of pipeline rupture. The rupture could be partial or full and the spill and the extent of its spread will depend on the following:

- Pipe diameter
- Operating pressure
- Products transported
- Product flow through the pipeline prior to rupture
- Volume of product released during the incident/accident
- Location of adjacent upstream and down stream flow restrictive devices (i.e., valves) and whether valves are automated for closure in rupture situations
- Pipeline elevation profile and location of the spill
- Rupture opening size
- Product burning rate (producing local fire rather than dispersion)
- Product heat content
- Size and likely volume of spill
- Size of likely burn area resulting from a fire
- Atmospheric condition

The flammability or explosive hazards associated with the transportation/storage of large quantities of liquid hydrocarbons after a large spill or a catastrophic failure reside in the energy stored in a given quantity of liquid. Upon evaporation and ignition, this energy can be released in the form of either a deflagration or a detonation wave with devastating consequences. It is the energy content that will provide the heat including radiant heat. A rough measure of heat generated from various sources is indicated below (Karlsson and Quintiere 1999):

Fuel	Heat Release Rate
A burning cigarette	5 W
A typical light bulb	60 W
A burning candle	80 W
A human being at normal exertion	100 W
A burning wastepaper basket	100 kW
A burning 1 m^2 pool of gasoline	2.5 MW
Burning wood pallets, stacked to a height of 3 m	7 MW
Burning polystyrene jars, in 2-m^2 cartons 4.9 m high	30–40 MW
Output from a typical reactor at an NPP	3,250—3,411 MW

The flame temperatures of selected fuels are listed below to indicate the extent of the flame temperature generated by a liquid hydrocarbon fire ball. See also Table 9-3.

Fuel Source (Flames)	Flame Temperature °C (°F)
Benzene	921 (1,690)
Gasoline	1,026 (1,879)
JP-4	927 (1,701)
Kerosene	990 (1,814)
Methanol	1,200 (2,192)
Wood	1,027 (1,881)

The energy content along with the liquid density of some common liquid hydrocarbons is illustrated in Figure 9-36.

Fire development is generally characterized in terms of heat release rate (HRR) versus time. Thus, determining the HRR (or burning rate) is an essential aspect of a fire hazard analysis (FHA).

The burning (or mass loss) rate along with constants for various fuels are depicted in Table 9-9.

Atmospheric Condition. Energy transmission from a radiant heat depends on atmospheric condition, (relative humidity, cloud coverage, etc.). Thus, calculation of radiative heat will require as estimate atmospheric transmissivity factor (i.e., fraction of energy transmitted). This depends on the content and water vapor pressure in atmosphere. Water vapor is almost always present in the surrounding air. For radiative heat transmission, water partial pressure calculations will be required These calculations are given below:

Saturation Pressure of Water Vapor. Water partial pressure (P_w) is the saturation pressure of the water vapor in moist air varies with the temperature of the air vapor mixture and can be calculated from the following expression:

$$P_w = [\, e^{(77.3450 + 0.0057\,T - 7{,}235/T)}]/T^{8.2} \qquad (9-1)$$

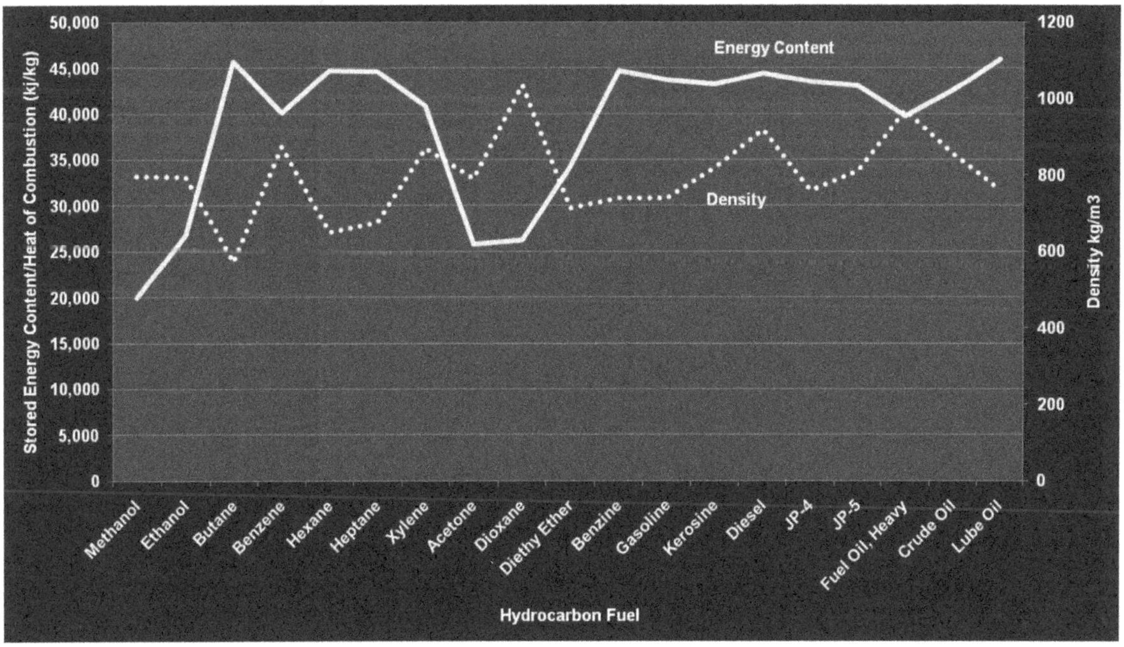

Figure 9-36. Typical energy content and density of some common liquid hydrocarbons (SFPE 2002)

TABLE 9-9. Large pool fire mass loss/burning rate of liquid hydrocarbons

Fuel	Mass Burning Rate m'' (kg/m² sec)	Empirical Constant $k\beta$ (m⁻¹)
• Methanol	0.017	100
• Ethanol	0.015	100
• Butane	0.078	2.7
• Benzene	0.085	2.7
• Hexane	0.074	1.9
• Heptane	0.101	1.1
• Xylene	0.090	1.4
• Acetone	0.041	1.9
• Dioxane	0.018	5.4
• Diethyl ether	0.085	0.7
• Benzene	0.048	3.6
• Gasoline	0.055	2.1
• Kerosene	0.039	3.5
• Diesel	0.045	2.1
• JP-4	0.051	3.6
• JP-5	0.054	1.6
• LPG	0.120	1.1
• Fuel oil, heavy	0.035	1.7
• Crude oil	0.0335	2.8
• Lube oil	0.039	0.7

where

P_w = water vapor saturation pressure, Pa
e = the constant 2.718
T = temperature of the moist air, K

Water vapor density: expressed as:

$$\rho_w = 0.0022 p_w / T \tag{9-2}$$

where

p_w = partial pressure water vapor, Pa, N/m²
ρ_w = density water vapor, kg/m³
T = absolute dry bulb temperature, K

Saturation pressure and density of water vapor for common temperatures are depicted in Table 9-10 and Figure 9-37 below.

TABLE 9-10. Saturation pressure and density of water vapor

Temperature		Saturation Pressure				Density	
(°C)	(°F)	(Pa)	(mm Hg)	(psia)	(in Hg)	(kg/m³)	10⁻³ (lb/ft³)
0	32	603	4.6	0.09	0.18	0.005	0.3
10	50	1212	9.2	0.18	0.36	0.009	0.59
20	68	2310	17.4	0.33	0.68	0.017	1.08
30	86	4195	31.7	0.61	1.24	0.03	1.9
40	104	7297	55.1	1.06	2.15	0.051	3.2
50	122	12210	92.2	1.8	3.6	0.083	5.19
60	140	19724	149	2.9	5.82	0.13	8.13
70	158	30866	233	4.5	9.11	0.2	12.3
80	176	46925	354	6.8	13.8	0.29	18.2
90	194	69485	525	10.1	20.5	0.42	26.3

Note: 1 lb/ft³ = 16.018 kg/m³, 1 kg/m³ = 0.0624 lb/ft³, 1 mm Hg= 133.3223684 Pa

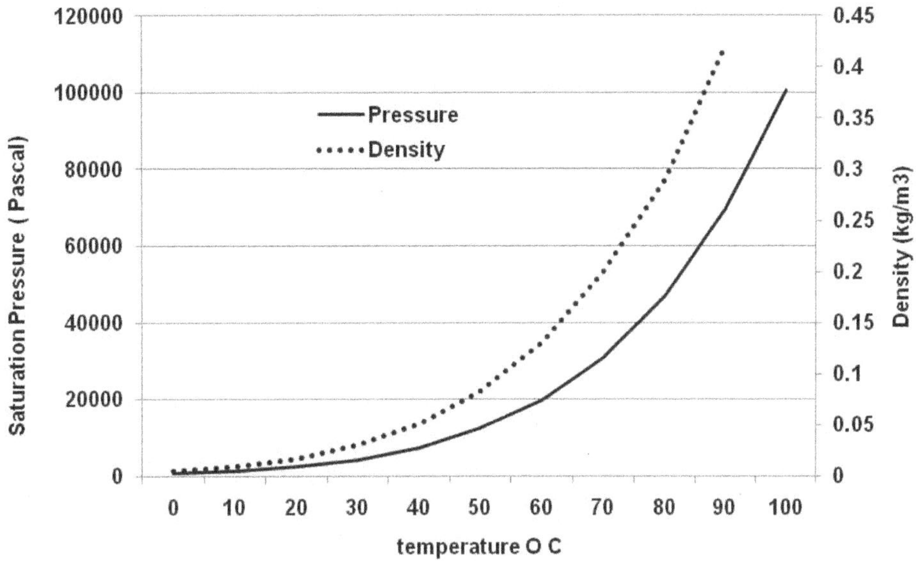

Figure 9-37. Saturation pressure and density of water vapor

Modeling Pool Fires. As indicated previously, pool fires are commonly associated with large releases of liquid hydrocarbons. Such releases can be from a pipeline due to a rupture. The liquid release is contained within an area, ground depressions, low points, etc.

To estimate the damage effects from pool fires, several types of calculations are required. In such calculations, the following must be considered:

To model a pool fire and determine the extent of fire radiation for ER purposes (i.e., safe distance determination) the following will be required:

- The pool fire diameter
- Total heat release
- Fire height based on its diameter
- Atmospheric condition (relative humidity, cloud coverage)
- Surface emissive power
- Point Source View Factor (i.e., the ratio of the thermal radiation on a target surface to that at the fire surface; it is a function of the distance from the emitting source of radiant heat
- Incident thermal radiation and resulting thermal effects

It is important to note that the pool fire diameter influences the fire height and therefore the view factor, which in turn influences the extent of radiant heat radius (Figure 9-38).

Pool Fire Volume/Area/Diameter

The pool fire diameter depends on the spill volume that is released from a pipeline segment. This volume as indicated previously depends on the following:

- Topography
- Drain off points
- Pipe Size
- Pipe length drained off
- Flowing product pressure and flow rate at the time of rupture/drain off
- Drain off volume while line remains under pressure

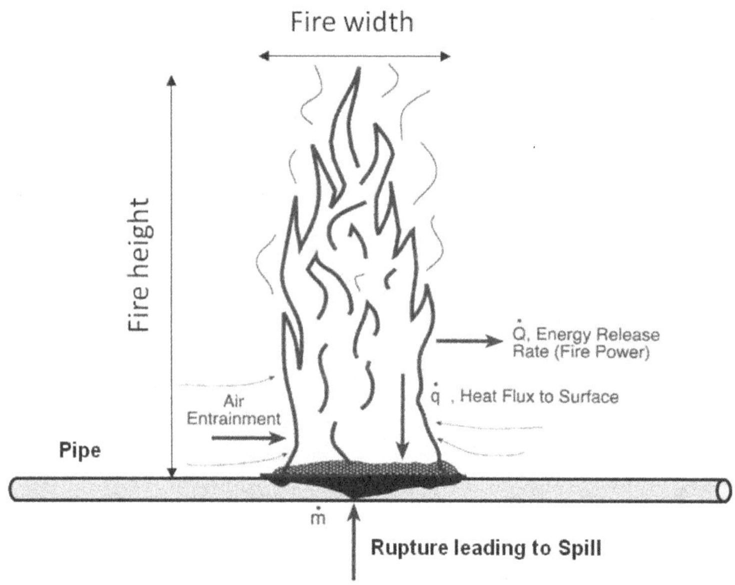

Figure 9-38. Pool fire model

The total volume drained off is thus the spill volume while line remains pressured (i.e., during the time that it takes for detecting the rupture, plus and the time that it takes for the adjacent valves/pumps to be closed and flow of product brought to a halt). To this amount, we add the drain off volume due to the elevation profile change (from high to low points). Generally, undulation in the elevation profile will help prevent total drainage of the pipeline upon rupture. However, in a HVP pipeline, the drainage will take place from the entire segment affected.

Volume drainage for HVP products from a pipeline rupture calculation can follow the method as described by Mohitpour et al. (2004) and (2005).

While the volume can be estimated, the most difficult part is the estimate of the spill spread. This depends on drain location where rupture or leak occurs, the geometry and geotechnical as well as topography and surface conditions and permeability of the surrounding terrain. Typically, a radius of 200 m, 25 mm thick, has been considered as the maximum target to limit a spill spread on water (US Coast Guard 33 CFR 155). This is equivalent to a volume of 3140 m^3 (19750 BBLS; Mohitpour et al. 2004).

On land (unless the spill is in or near a river or in swampy areas) because of the possibility of absorption properties of soil, it is likely that the radius of the spill spread will be much smaller. Actual offshore pipeline oil spills over the period 1991 to 2000 (caused by various incidences including anchor drops, corrosion, hurricane, etc.; US DOI-MMS 2002) indicates a range of oil spills between 1211 and 8210 BBLS. The Average is 3566 BBLS (566 m^3). With the 25-mm thickness, the spill radius would be about 84 m. Assuming that oil spill reach on land would be approximately half as much as that on water, then an assumption of 50 (calculated as 84/2 m plus 16% safety factor) meter maximum for onshore pipeline spillage would be reasonable.

Estimated Radiant Heat Release

This is given by;

$$Q = M_b E_c A$$

where

M_b = burning rate, kg/m^{2}s
E_c = heat of combustion, kJ/kg
A = pool area, m^2
Q = estimated radiant heat release, kW

Typical values for hydrocarbon heats of combustion and burning rates are provided previously in Figure 9-35 and Table 9-9.

Now assuming gasoline ($M_b = 0.055$ kg/m^2 and $E_c = 43{,}700$ kJ/kg) and a pool area with the 50-m radius (equivalent area 7850 m^2), then estimated radiant heat will be:

$$Q = 0.055\,(43{,}700)(7850) = 18{,}867{,}475 \text{ kW}$$

Assuming a radiant fraction for smoky pool fires of 0.35 (a conservative assumption since 65% of the radiant energy is assumed to be attenuated by the smoke; Figure 9-39), then the radiant heat, Q_r, is

$$Q_r = 6{,}603{,}616 \text{ kW}$$

Point Source View Factor

Point source view factor, F_p, can be estimated assuming a point source on a spherical surface a distance x away (Figure 9-39)

$$F_p = 1/(4\pi x^2)$$

where

F_p = point source view factor, length^{-2}
x = distance from the point emitting source of radian heat to target (length in m)

Assuming the distance x from the source is 200 m (roughly two football fields in length) and calculating the view factor for that distance assuming no wind and thus no flame tilt:

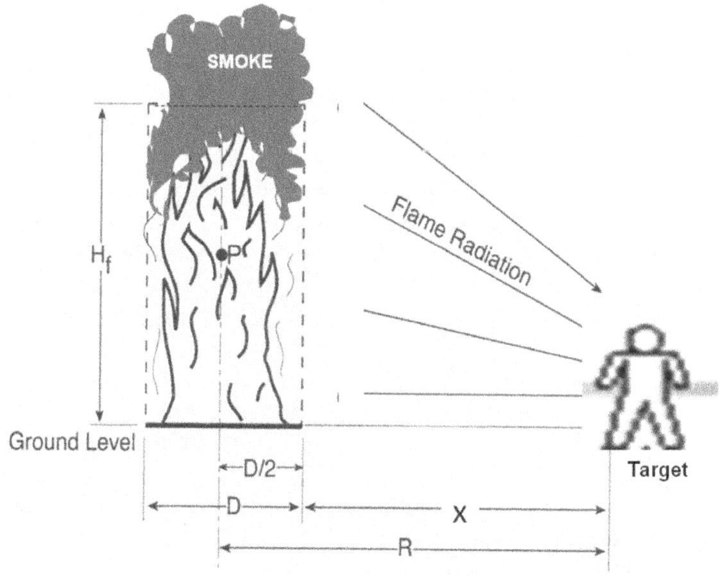

Figure 9-39. Hydrocarbon liquid pool fire/ configuration and heat radiation

$$F_p = 1/(4\pi(200)^2) = 1.99 \times 10^{-6}\ m^{-2},\ say = 2.0 \times 10^{-6}\ m^{-2}$$

Surface Emissive/Transmissivity Power

The following correlations accounts for atmospheric transmissivity (Pieterson and Huerta 1985)

$$T_r = 2.02(P_w x)^{-0.09}$$

where

T_r = atmospheric transmissivity (fraction of energy transmitted: 0 to 1)
P_w = water partial pressure, Pa
x = path length, distance from flame surface to target, m

Water partial pressure can be calculated from the Expression (1) given previously as follows:

The saturation pressure of water vapor in moist air at 20°C (273°K) can be calculated as:

$$P_w = e(^{77.3450 + 0.0057(273 + 20) - 7235/(273 + 20)})/(273 + 20)^{8.2}$$

$$= 2310\ Pa$$

T_a = the ambient temperature, °K

Substituting into the transmissivity equation

$$T_r = 2.02(2310[200])^{-0.09} = 0.624\ approx.$$

The thermal flux is therefore

$$Q_t = T_r F_p Q_r = (0.624)(2.0 \times 10^{-6}\ m^{-2})(6{,}603{,}616\ kW)$$

or $$Q_t = 8.24\ kW/m^2$$

Safe Distance Calculation

Safe distance is the distance that that one would not experience pain and second-degree burns are generally given at the threshold of 4 kW/m^2. However,

$$Q_t = T_r F_p Q_r = (0.624)[1/(4\pi x^2)\ m^{-2}](6{,}603{,}616\ kW) = 8.24$$

Solving for x^2 when $x^2 = (0.624)[1/(4\pi)](6{,}603{,}616\ kW)/8.24$

$$x^2 = (0.624/8.24)(1/4\pi)(6{,}603{,}616) = 39{,}815$$

Therefore, x = 200 m, which is inside the pain threshold of 4 kW/m^2 (=286 m). As previously indicated, the "Thermal radiation Level of between 4–5 kW/m^{2}" is the limit for people to escape, second-degree burn (DOT 49 CFR 193 (US DOT 2009) and NFPA 59 A (NFPA 2004, 2007).

Therefore, a safe distance of approximately 300 m must be observed to assure no lethal exposure (burns).

Flame Height Estimation

The flame height for circular pool fire (Figure 9-38) can be calculated using the following correlation of (Thomas 1963):

$$H_f/D = 42[M_b/(\rho_a(gD)^{0.5})]^{0.61}$$

where

H_f = visible flame height, m
D = equivalent pool diameter, m
M_b = burning rate, kg/m^{2}s
ρ_a = ambient air density (typically 1.2 kg/m^3 at 20°C and 1 atm)
g = acceleration of gravity 9.81 m/s^2

Substituting 100 m for the approximate pool diameter (50-m radius) and 0.055 for the burning rate (M_b) into the above formula, then a flame height of 78 m results.

Needless to say, this is a flame height of some significance in terms of the amount of radiant heat that it can transmit.

GLOSSARY OF TERMS AND ACRONYMS RELATED TO ER, SAFETY, AND RISK

Accident	An undesired and unplanned event that results in injury to people, damage to property, damage to the environment, loss of product quality, or loss of production.
Acute hazard	A hazard that can have either an immediate or delayed effect (with short-term or prolonged consequences) due to a single exposure to an accident, such as exposure to a gas explosion, fireball, or a release of an acutely toxic material.
Acutely hazardous material	In general, materials that pose an acute hazard, e.g., a substance possessing toxic content.
Assessments	An assessment is a less formal review of HSE procedures and practices directed at operational enhancements and at monitoring progress toward completion of audit recommendations.
Audit	An audit is an intensive review of policies, standards, procedures, and HSE management systems to ensure that they are adequate to meet the needs of the operation and are in compliance with all company and legislated standards.
Basic cause analysis (Root cause analysis)	A process to identify basic (personal and job/system factors) and contributing causes that can result in an undesired event.
Blow-down	A release of natural gas from pipeline or facilities that may occur intentionally or by accident due to rupture or leak.
Community	A group of people having common ties or interest who are subject to the same laws and live in the same locality.
Company	A legal entity, incorporated under the laws of the jurisdiction in which the entity operates, that conducts legal business for the purpose of earning revenues and profits for the entities' owners.

Confidentiality	Spoken to in confidence; entrusted with information.
Critical	A subject or condition that could result in a significant loss to a person, property, process, or the environment.
Critical equipment	Equipment whose failure cannot be tolerated for health, safety, environmental, or functional reasons and for which specific practices and procedures need to be specified to eliminate or reduce the probability of failure to acceptable levels.
Distribution	All activities involving the sale and transfer of raw materials, products or services in all geographic jurisdictions.
Emergency	An unforeseen combination of circumstances or the resulting state that calls for immediate action to prevent or resolve injury, illness, or damage to facilities, equipment, or the environment.
Emission	A release of any substance or product onto the air, land, water, or groundwater.
Equipment	A physical component or item installed at a facility that carries out a specific function (see also Critical Equipment).
Exposure	An instance where an employee or contractor is, or was, subject to some effect, influence, or safety hazard, or was in contact with a hazardous chemical or physical agent at a sufficient concentration, duration, and intensity to produce an injury/illness.
External force	Third-party intrusions into the pipeline corridor, earth movements (including scouring and earthquakes).
First aid	Any one-time treatment and subsequent observations of minor scratches, cuts, burns, splinters, etc., which do not ordinarily require medical care.
Hazard	The potential to cause harm.
Hazard control analysis	A method to determine the degree of hazard that results when various control measures fail.
Hazardous facilities	Facilities fixed in a single location that contain acutely hazardous materials with the potential to pose a significant, acute risk to public safety beyond the facility's boundaries.
Hazardous materials	Compounds, mixtures or products that (when stored in certain quantities or containers, spilled, or when burning) create specific hazards to people and/or the environment.
Hazard range	The distance from the pipeline within which the surrounding population could suffer a specified level of harm in the event of release of pipeline fluids following loss of containment. The hazard range may be a function of a variety of factors including the pressure in the pipeline, the nature of the fluid, local topography and weather conditions, and magnitude of the failure.

Hazardous waste	Any waste that may constitute a danger to the environment or human health.
Hazard/hot zone	The area around the source of a hazard that exposes the public to a significant risk.
Health hazard	Any chemical, biological, ergonomic, or physical agent that has the potential to adversely affect employee health.
Health/medical personnel	A physician or a registered nurse.
Health monitoring	Activities carried out on an essentially healthy employee population.
Health promotion	Health education and health services targeted at prevention of identified health conditions/illnesses through modification of risk factors and/or behaviors.
Heavier gas liquid	The heavier fractions of natural gas liquids, which are extracted as a marketable by product during oil and gas processing, and consist of pentanes, hexanes, and heavier components, including natural gasoline.
Highly sensitive land uses	Land uses that meet one or more of the following criteria: (a) Land uses whose onsite population cannot be readily evacuated or otherwise adequately protected from serious harm through methods such as sheltering in-place. These include schools, hospitals, clinics, nursing homes, multiple-family housing exclusively for the elderly or handicapped, stadiums, and arenas, but do not extend to Family Day Care if residential subdivisions *per se* are considered to be compatible with the hazardous facility in question. (b) Land uses that serve critical lifeline functions such as water supplies if exposed to a significant risk that will curtail its lifeline function for a critical period of time. (c) Other types of other land uses, such as high-density residential development, may also be determined to be incompatible in vicinity to an existing or proposed hazardous facility.
HSE	Health, Safety and Environment. Acronym used to describe people, departments and initiatives in force across the organization.
Health surveillance	A subset of monitoring, which is the strategy used to determine the health outcome of a group known to be at higher risk of an adverse effect because of a specific exposure. Blood pressure screening in the general population is an example of monitoring, while audiometric testing of workers exposed to noise is an example of surveillance.
HSE audit	Audits carried out by internal/external personnel using established protocols based on HSE standards and guidelines.
Hydrogeology	The study of the inter-relationships of geologic materials and processes with water especially ground water.

Incident	An unplanned event that may or may not result in undesirable consequences.
Incident management process	A formalized system for managing an HSE incident. The Incident Management Process includes: controlling the incident; reporting the incident; taking immediate corrective action; investigating the incident; documenting the investigation results; and making recommendations.
Incident rates	The number of injuries, illnesses, lost time cases, lost workdays, and recordable incidents related to a common exposure hours base of 100 full-time workers that enables inter-industry comparisons and trend analysis over time or comparisons among firms regardless of size.
Inspections	Inspections are undertaken at regular intervals by facility personnel to detect and remedy conditions and operating practices that may expose the company to unacceptable levels of risk. Facility and often divisional management personnel regularly participate in HSE inspections.
Industrial hygiene monitoring	Refers to a specialized form of workplace exposure monitoring. Industrial hygienists do not monitor biological systems, but rather environmental agents (i.e., physical, biological or chemical).
Liquefied petroleum gas (LPG)	The lighter fractions of natural gas liquids, which are extracted as a marketable by-product during gas production or processing, liquefied under pressure, and consist primarily of propane and butane.
Key transportation route	The route that the business/facility regularly uses to move goods or materials during their normal activities.
Learnings	Knowledge and information obtained as a result of incident investigations, about causes of incidents, and recommendations for actions to prevent similar incidents from recurring.
Loss	Loss is the result of an accident that results in unintended harm or damage (to people and/or equipment and/or the environment).
Lost time cases	Injury or industrial illness cases in which the employee is unable to return to work the next scheduled shift.
Lost work days	The number of workdays (consecutive or not), beyond the day of injury or illness that the employee was away. The total amount does not include the day of the injury or onset of the illness or any days on which the employee would regularly have been scheduled off.
Maintenance	The combination of all technical and administrative actions, including supervision actions, intended to retain an item, or restore it to a state in which it can perform a required function.

Management practice/	The method by which the components of a standard are introduced and implemented at the operational level.
Management system	Is that part of the overall management system, which includes the organizational structure, planning activities, responsibilities, practices, procedures, processes, and the resources for developing, implementing, achieving, reviewing, and maintaining a policy.
Medical aid	Any treatment of injuries administered by a physician or registered professional personnel. This type of treatment does not include first aid treatment (i.e., one-time treatment and subsequent observations).
Nonhazardous area	An area where there is zero or a negligible probability of a potentially explosive atmosphere being present.
Occupational hygienist	An individual having certification/registration by the American Board of Industrial Hygiene, the Canadian Registration Board of Occupational Hygienists Board, or other international organization, or possessing an equivalent combination of education, training or experience to enable them to perform work competently.
Occupational illness	Any abnormal condition or disorder, other than one resulting from an occupational injury, caused by exposure to factors associated with employment. It includes acute and chronic illnesses or disease that may be caused by inhalation, absorption, ingestion or direct contact (e.g., repetitive stress). The basic difference between an injury and an illness is the single instantaneous exposure concept.
Occupational injury	Any injury, such as a cut, fracture, sprain, and amputation, which results from a work-related event or from a single instantaneous exposure in the work environment. Note: Injuries resulting from animal, insect, or snake bites or from one-time exposures to chemicals (including chemical burns) are considered injuries.
Operable	The ability to start and stop equipment as required and the ability of the equipment to operate at the rate capacity.
Performance indicator	An established, measurable parameter that provides a complete indication of the performance level for the associated subject.
Personal protective equipment (PPE)	The general term used to describe specialized clothing (e.g., fire resistant clothing, steel toe boots, safety helmets, work gloves) and other apparatus worn by workers to protect themselves from occupational exposures. PPE reduces the exposure by creating a barrier between the person wearing the equipment and the hazard.
Policy	A statement of fundamental principles and objectives established by senior management.
Pollution	Any physical, biological, or chemical substance that, when released into the environment, results in adverse or harmful conditions.

Populated areas	An area characterized by land uses other than agriculture, open-space (including undeveloped, infrequently visited parkland), or hazardous facilities of similar nature (including onshore oil field operations and consolidated sites for oil/gas facilities), considering both present and reasonably anticipated future development. Populated areas also incorporate frequently traveled transit routes, including highways, roads, bike paths, and pedestrian paths unless transitory populations are sporadic or often absent.
Practice(s)	Generally accepted (by industry and/or government agencies) method for completing a task or job that is repeated numerous times.
Primary prevention	Measures that reduce or eliminate the likelihood of an accident in a manner that is inherently safe (without dependence upon safety systems), including redesign of production systems, reductions in chemical inventories, or substitution of hazardous materials with nonhazardous or significantly less hazardous materials.
Procedure(s)	Statements of who does what, how and in what sequence. Procedures may be mandatory or discretionary.
Process-related accidents	Accidents that occur as a result of process/equipment failures, including: ➢ Explosion/implosion ➢ Fire ➢ Exposure to hazardous materials ➢ Electrical shock ➢ Material spills ➢ Ionizing radiation
Process safety	A discipline that focuses on prevention of physical situations with the potential for human injury, damage to property or, damage to the environment through the release of chemical or mechanical energy.
Protocol	Provides guidance with respect to execution of identified topics and requirements.
Public safety	Protection of the general public from the acute, immediate effects caused by a single exposure to an accident, generally resulting in severe biological harm, including fatality. Public safety does not extend to issues of occupational safety or to chronic effects from prolonged exposure to a public health hazard.
Purpose	A statement of purpose or reason for existence of a standard.
Qualitative assessment	An evaluation based on opinion or experience without the benefit of full numerical methods.
Quantitative assessment	An evaluation based on the use of numerical methods.

Recordable cases	Cases that include fatalities, lost time, medical aid, or restricted work.
Regulatory requirements	All regulations, codes, and standards that have the authority of law.
Residual risk	Any risk remaining after decreasing frequency and consequence as low as reasonably practical.
Restricted work	An injured worker is able to return to the workplace but is unable to perform his/her "normal" work assignments as a result of work-related or off-the-job injury or an aggravation of a prior injury.
Risk	The combination of the expected frequency (events/year) and consequence (effects/event) of a single accident or group of accidents. It can be also defined as the probability (or likelihood) of harm actually occurring and the severity of its consequences.
Root cause	The job and personal factors (such as inadequate engineering, lack of knowledge or skill) from which substandard acts and conditions originate. Root causes are most frequently the result of inadequate safety systems, inadequate system standards, or inadequate compliance with standards.
Rupture blow-down	A release of natural gas from pipeline that may occur by accident due to rupture or leak.
Secondary prevention	Safety systems and programs such as leak detectors, double-walled vessels, emergency flares, and sprinklers that prevent accidents from occurring during a deviation in the normal operations of a facility.
Sensitive information	Patents, copyrights, trademarks, process descriptions, technical documents, licensed or proprietary software, external communications prior to release, and internal company confidential information (e.g., legal, financial, audit, strategic business plans, financial reports, employee records).
Significant risk	A compound measure of the probability and number of fatalities or serious injuries (generally referred to as societal risk) that is due to a deviation from normal operating conditions and that exceeds the threshold for environmental significance.
Societal risk	Risk to a group of people, expressed in terms of the distributed frequency of events that cause multiple casualties or, when appropriate, the frequency of casualties at a specific location.
Spill	A release, leak, discharge, or deposit of a liquid, solid substance or product from its intended containment onto the land, into the air or into waters or groundwater.
Stakeholder	All persons who are affected by any aspect of operations. Includes employees, customers, the general public, special interest groups, and government agencies.
Standard	A statement of measurable requirements that must be achieved.

Strategy	Long-term plan that forms the framework for actions to meet objectives and goals.
Toxicological review	A review of the toxicity (i.e., acute and long-term health effects and neurotoxicity) of the product and all its constituents in the specific form involved
Trend/group data analysis	Analysis and interpretation of health surveillance data to identify patterns or trends among groups of employees potentially at risk. This interpretation may suggest an adverse effect and the need for further action.
Vocational	Being trained in a skill or trade to be pursued as a career
Waste	A product, substance, class of substances, or mixture of substances that is intended for disposal or recycling.
Waste prevention	Reduction in the amount of waste generated through elimination, reduction, reuse, recycling, and recovery.
Wellness	A general description encompassing health programs and services, which are directed at employee's well-being in the workplace.
Work permit system	A system requiring a site permit used to review and authorize any potentially hazardous and/or nonroutine task. The work permit system facilitates: the identification of hazards that may be associated with the task; identification, specification, and documentation (in the work permit) of operational procedures and protective measures to reduce or eliminate identified hazards, and identification and documentation of procedures to ensure that the specified operational procedures and protective measures are implemented while the task is being undertaken.
World class	Any policy, standard, practice, procedure, or program that would be considered effective in addressing its subject matter by International companies.

ABBREVIATIONS

The following are acronyms are generally utilized in EMR by the pipeline industry

ANSI	American National Standards Institute
EAP	Employee Assistance Program
EPA	Environmental Protection Agency (USA)
FTW	Fitness to Work
HAZID	Hazard Identification
HAZOP	Hazard and Operability Study (One Method for Identifying Hazards)
ISO	International Standards Organization
LTD	Long-term Disability
MSDS	Material Safety Data Sheet(s)
NIOSH	National Institute of Occupational Safety and Health
OSHA	Occupational Safety and Health Administration
PPE	Personal Protective Equipment

SCBA	Self-Contained Breathing Apparatus
TLV©	Threshold Limit Values
TSCA	Toxic Substances Control Act (EPA)
WC	Workers' (or Workmen's) Compensation
WHMIS	Workplace Hazardous Material Information System – Canadian legislation dealing with the preparation and distribution of chemical hazard information

REFERENCES

Babu, R. J., Subramanian, R., and Birajdar, C., 2009, "Risk Associated with Cross-Country Natural Gas Pipeline," IORS 2009, The 2009 IEEE/RSJ International Conference on Intelligent Robots and Systems, St. Louis, MO, October. www.cholarisk.com/uploads/IORS-2009.pdf.

Boatwright, G., 2007, "Crisis Management: The Need for a Comprehensive Approach," Pipeline and Gas Journal, May, http://www.nixonassoc.com/crisis.pdf.

Callan, M., 2008, "Pipeline Emergency Response Operations," National Association of State Fire Marshals (NASFM), www.pipelineemergencies.com and http://www.pipelineemergencies.com/pdf/InstructorGuide6.pdf.

ESA (Environmental Service Association), 2005, "Area Classification for Landfill Gas Extraction, Utilisation and Combustion," Industry Code of Practice: ESA ICoP 2, http://www.esauk.org/publications/reports/ICoP2_-_Landfill_Gas.pdf.

HSE (Health & Safety Executives), 1996, "The Pipelines Safety Regulations — Further Guidance on Emergency Plans for Major Accident Hazard Pipelines," http://www.hse.gov.uk/pipelines/emergencyplanpipe.pdf.

HSE (Health & Safety Executives), 2006, "Drilling Into Landfill Waste," Industry Code of Practice, ESA ICoP 4.

IGEM (Institute of Gas Engineers and Managers), 2007, "Venting Natural Gas," http://www.igem.org.uk/download/document/SEPTEMBER_2007_PUBLICATION_LIST.pdf.

Iqbal, N., and Salley, M. H., 2003, "Fire Dynamics Tools (FDTs), Quantitative Fire Hazard Analysis Methods for the U.S. Nuclear Regulatory Commission Fire Protection Inspection Program," U.S. Nuclear Regulatory Commission Office of Nuclear Reactor Regulation Washington, DC, 20555-0001, Document NUREG-1805, Vol. 1, http://www.nfpa805.com/Fire_Modeling/Fire_Modeling_Information.htm.

Karlsson, B., and Quintiere, J., 1999, *Enclosure Fire Dynamics*, CRC Press, Boca Raton, FL.

Katz, D. L., Cornell, D., Vary, J. A., Kobayashi, R., Elenbass, J. R., Poettmann R. H., and Weinaug, C. F., 1959, *Handbook of Natural Gas Engineering*, McGraw-Hill Book Company, New York, NY.

L'Air Liquide, 1976, *Gas Encyclopedia*, Elsevier, The Netherlands.

Lee C. G., 2008, "Introduction to DNV," Indopipe Conference, Jakarta, Indonesia, June 10-12, http://www.iee- c.com/indopipe2008/PDF%20Indopipe2008/TS%203%20-%201%20Lee%20Chon%20Gee%20-%20DNV.pdf.

Miller, J. A., and Koloski, J. W., 1999, "Landslide Risks to Buried Steel Pipelines," Pipeline & Gas Journal, January, http://www.geoengineers.com/fullarticle.asp?articleid=143.

Mohitpour, M., Golshan, H., and Murray, A. 1., 2007, *Pipeline Design and Construction — A Practical Approach*, 3rd ed., ASME Press, New York.

Mohitpour, M., Szabo, J., and Van Hardeveld, T., 2005, *Pipeline Operation and Maintenance — A Practical Approach*, 1st ed., ASME Press, New York.

Mohitpour, M., Trefanenko, B., Tolmasquin, S., and Kossatz, H., 2004, "Valve Automation for Oil Pipeline Safety," ASME 5th International Pipeline Conference (IPC 2004), Calgary, Alberta, Canada, October.

Mohitpour, M., Von Basseheim, G., Braun A., 1998, "Route Selection for Project Success," ASME-OMAE International Pipeline Conference (IPC), Palliser Hotel, Calgary AB, Canada, June 9–13.

Mudan, K. S., 1984, "Thermal Radiation Hazards From Hydrocarbon Pool Fires," Proc. Energy Comb. Sci. Vol. 10, No. 1, pp. 59–80 (ISBN 0360-1285).

NFPA (National Fire Prevention Association), 2004, *NFPA 59: Utility LP-Gas Plant Code*, NFPA, Quincy, MA.

NFPA (National Fire Prevention Association), 2007, "Report on Proposals F2007 — NFPA 59 LP-Gases at Utility Gas Plants," http://www.nfpa.org/assets/files/PDF/ROP/59-F2007-ROP.pdf.

NTSB (National transportation Safety Board), 2003, "Pipeline Failure Causes," http://www.corrosion-doctors.org/Pipeline/Pipeline-failures.htm.

PA (Pipeline Association for Public Awareness (PA)), 2007, "Pipeline Emergency Response Guidelines," http://www.pipelineawareness.org/files/papa/ERguidelines-11.pdf.

Pieterson, C. M., and Huerta, S. C., 1985, "TNO Report B-4-0222," PO Box 342 7300 AH, Apeldoorn, The Netherlands.

Pipeline Safety Foundation, 2002, "Who's Watching Pipelines 1998," http://www.specialaudience.com/websites/pipelinesafetyfoundation_org/index.html.

PRCI (Pipeline Research Council International), 2003, "Next Generation Design and Assessment Methods for Onshore Oil and Gas Pipelines," R&D Forum, Tyson Corner, VA, April.

Raj, P. K., 2005, "Large LNG Fire Thermal Radiation — Modeling Issues and Hazard Criteria Revisited," Presented at the AIChE Spring Meeting, April 10–14, Session LNG VI — Risk & Safety.

Raj, P. K., 2006, "NFPA 59A Vapor Dispersion and Thermal Radiation Models in the light of New Data," AGA-NFPA 59A Standards Workshop, Charlotte, NC, October 10.

Rew, P. J., and Hulbert, W. J., 1996, *Development of Pool Fire Thermal Radiation Model*, Report #96, HSE Books, Sudbury, UK.

Sandoval, A., Beruvides, M., and Wiesner, T., 2001, "State of the Art Analysis of Current Research Trends in Pipeline Safety," Paper #T5a 5, Presented at AIChE Spring National Conference, New York, April 22–26.

SFPE (Society for Fire Protection Engineers), 2002, *Handbook of Fire Protection Engineering*, 3rd ed, pp. 3–26.

Spicer, III, T. O., and Martin, R. E., 2005, "Estimating the Consequences of a Liquefied Natural Gas Spill," California Coastal Commission Meeting Workshop, April 14, 2005, http://www.coastal.ca.gov/energy/lng/spicer-spill-lng.pdf.

Taufik, A., and Wisianto, A., 2007, "An Analysis of Gas Pipeline Explosion — Lessons Learned" 3rd Asian Pipeline Conference at Exhibition ASCOPE & MGA, Kuala Lumpur, Malaysia, November 21–22.

Thomas, P. H., 1963, " 'The Size of Flames From Natural Fires," Ninth Symposium (International) on Combustion, Academic Press, New York, pp. 844–859.

TSB (Transportation Safety Board of Canada), 2002 and 2007, "Pipeline Reports," http://www.tsb.gc.ca/eng/rapports-reports/pipeline/2007/index.asp.

UK, Kent County Council and Medway Council, 2005, "An Emergency Plan for Major Accident Hazard Pipelines," http://www.kent.gov.uk/NR/rdonlyres/03A12373-2A96-4D24-8A26-A0421FDBB768/0/PipelinesPlanApril2005.pdf.

US DOI-MMS (US Department of interior- Minerals Management Services), 2002, "OCS Oil Spill Facts," http://www.energy4me.org/use/environment/documents/oilspillfacts_mms.pdf.

US DOL (Department of Labor), 2008, "PART 1910 Occupational Safety and Health Standards," http://www.osha.gov/pls/oshaweb/owastand.display_standard_group?p_toc_level=1&p_part_number=1910.

US DOT, 2009, "Code of Federal Regulation (CFR 49) PART 193 — Liquefied Natural Gas Facilities: Federal Safety Standards." http://ecfr.gpoaccess.gov/cgi/t/text/text-idx?c=ecfr;sid=62e0ade542bd11162e41f385f87c6f31;rgn=div5;view=text;node=49%3A3.1.1.1.5;idno=49;cc=ecfr

US DOT, 2009, "Code of Federal Regulation (CFR 49) PART 194 — Response Plans For Onshore Oil Pipelines," http://ecfr.gpoaccess.gov/cgi/t/text/text-idx?c=ecfr&rgn=div5&view=text&node=49:3.1.1.1.6&idno=49.

Wagner, J. P., 1999, "Flammability Analysis of Hydrocarbon Vapors at Kittitas Terminal, Local Communities and Chicago, Milwaukee — St. Paul Railroad Tunnel," Report prepared for Cities of North Bend and Snoqualmie, Washington, Cascade Columbia Alliance, http://www.efsec.wa.gov/oplarchive/oplpft/Cca/jpw-2.pdf.

Zalosh, R. G., 1995, "Explosion Protection," *SFPE Handbook of Fire Protection Engineering*, 2nd ed., Section 3, Chapter 16, ed. P. J., DiNenno, National Fire Protection Association, Quincy, MA.

Zimmerman, R., Restrepo, C. E., and Simonoff, J. S., 2008, "Regression Modeling and Prediction of Causes and Consequences of Oil & Gas Pipeline Accidents and Electric Power Outages," Second Annual Homeland Defense and Security Education Summit, University of Maryland University College, March 19.

Chapter 10

PIPELINE RISK MANAGEMENT

INTRODUCTION

Managing assets equates to managing risk. To manage risk effectively, risk must first be assessed and then controlled as part of the integrity management plan for the pipeline system. The incorporation of risk management into pipeline integrity management planning is described in Role of Risk Assessment in Integrity Management Planning, and control and mitigation of risk is discussed in Types of Risk Assessment.

The methods described in these sections are generally found to effectively manage the vast majority of pipeline system assets by functioning principally as risk-based inspection (RBI) tools. However, in addition to its RBI function, risk management can be used as a general O&M tool for pipeline systems to identify areas where risk is above some predetermined threshold and should be reduced as a matter of prudence or of company policy. Risk management also provides a way of evaluating the cost and effectiveness of future maintenance programs and of rolling up pipeline system risk to corporate risk to be controlled as required at corporate level. At the department level, risk management techniques provide a rational basis for setting budgets — both totals and proportions of the O&M total allocated to various facility types (pipelines, compressor and pump station, meter stations, etc.), or simply to highlight segments of the pipeline needing particular O&M attention. In addition to addressing corporate risk, risk based planning can be used at the corporate level as powerful decision making tool — to project tariff changes and asset valuation for example.

This chapter considers pipeline risk management in a general sense that addresses all of the foregoing applications.

Figure 10-1 illustrates the application of general risk management principles to scenario evaluation (Wickenhauser and Playdon 2004). The figure shows the variation in risk of financial loss along the centerline of a 250-km gas pipeline. Such a plot identifies high-risk segments. A similar postmitigation plot can be used to demonstrate risk reduction.

Risk management in general comprises the following elements:

- Risk analysis in which threats are identified, probabilities and consequences are determined, and risk is estimated
- Risk assessment in which the estimated risk is evaluated against some predefined criteria to determine how the risk should be managed
- Risk control in which remediation, mitigation, and prevention measures are specified and applied

The above elements are illustrated as a process flow in Figure 10-2.

The process shown in Figure 10-2 is in fact quite generic and can be applied to a wide range of risk management scenarios. In principal, for example, the same basic methodology is used by police forces worldwide when dealing with such threats as domestic violence, and the US Federal Aviation Agency in analyzing comparative safety procedures. However, it should be noted that the purpose of the present chapter is limited to the application of risk management to pipeline integrity management planning.

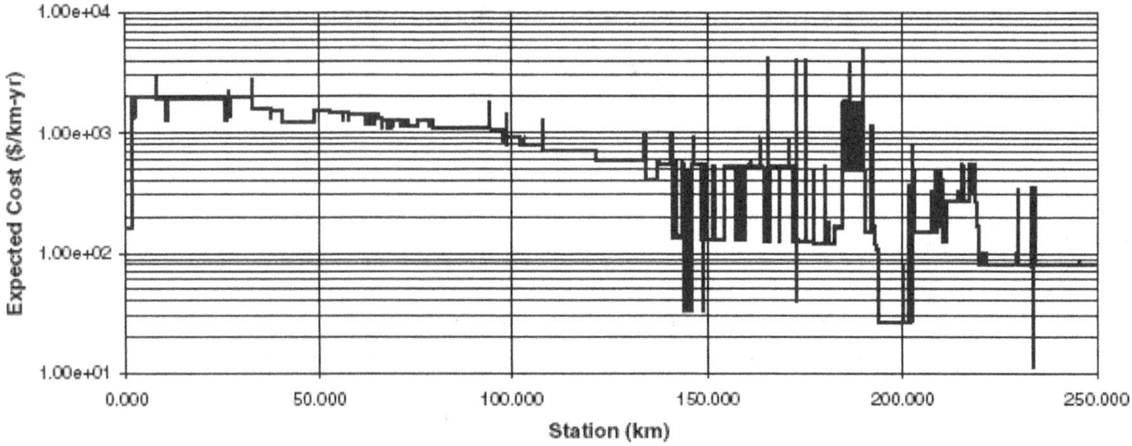

Figure 10-1. Typical risk profile for a gas transmission pipeline

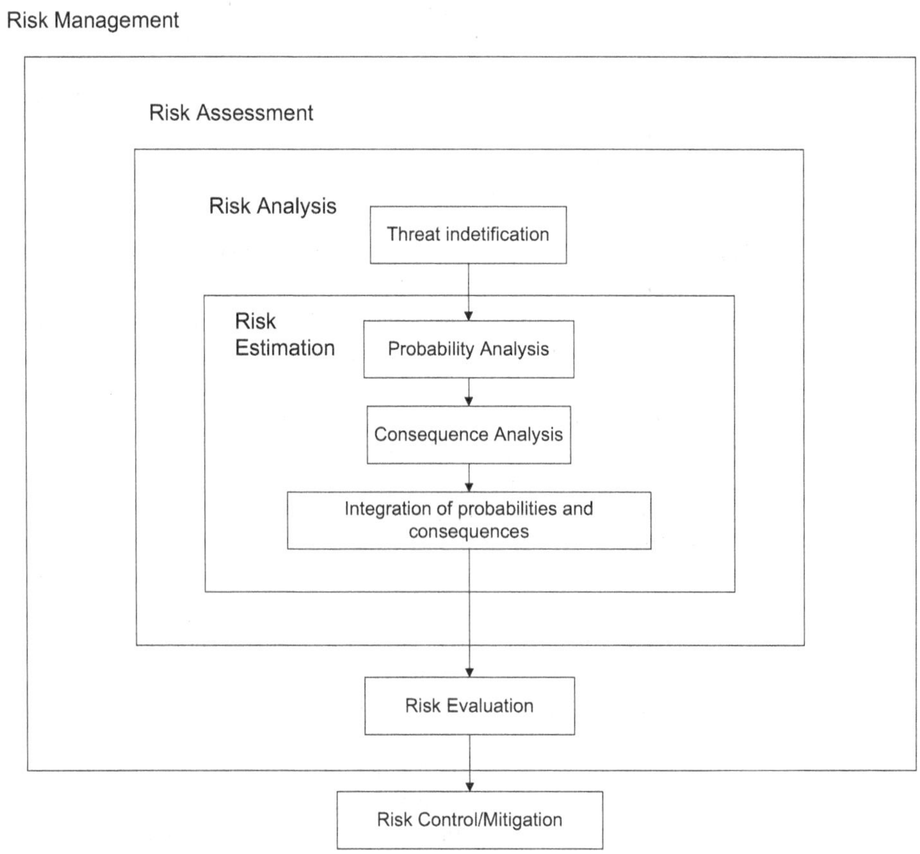

Figure 10-2. General risk management schematic

RISK ASSESSMENT

Risk assessment, as part of risk management, is one of several tools used to evaluate the integrity of a pipeline system with a view to taking actions to avoid undesirable consequences resulting from pipeline failures. These tools form the link between information assembled during data collection and integration and the physical actions to be taken — in general comprising measures to assess pipeline system integrity, modifications to O&M procedures, and/or physical risk mitigation and risk prevention measures. Other tools include system monitoring (via SCADA, for example), fitness for purpose studies, preventative maintenance, and continual review of material selection, repair welding, pipe handling, and other standard operating procedures.

The output from a practical risk assessment need not be complicated. Considering the wide variability and level of uncertainty associated with the factors that determine the integrity of a pipeline, the simpler and more robust the risk assessment can be made, the more useful it will be as pipeline integrity management tool. The context within which risk assessment is used in pipeline integrity management is illustrated by the following guidance given by ANSI/ASME B31.8S (2004):

> "Risk assessment methods should be used in conjunction with knowledgeable, experienced personnel (subject matter experts [SMEs] and people familiar with the facilities) that regularly review the data input, assumptions, and results of the risk assessments. Such experience based reviews should validate risk assessment output with other relevant factors not included in the process, the impact of assumptions, or the potential risk variability caused by missing or estimated data. These processes and their results shall be documented in the integrity management plan."

Risk is a fact or life for all engineering structures — and pipelines are no exception. Since risk ultimately can be expressed in monetary terms (setting aside the totally valid aversion we have to endangering life, damaging the environment, or not complying with delivery contracts), one might argue that the objective of risk management ought to be to minimize risk. The following example is intended to illustrate why this in general is not the case.

In this example, a segment of a 6.0-mm wall thickness pipe was found during integrity assessment to have a maximum defect depth of 20% of the wall thickness. Examination of the corrosion shape and dimensions of the defects in the segment indicated that the nature of the corrosion is pitting (see Figure 10-3) (Pipeline Operators Forum 2005). "*L*" is the corrosion length measured along the pipe axis, and "*W*" is the width measured in the circumferential direction.

The common ultimate limit state specified for maximum depth of pitting defects is a depth of 80% of the wall thickness. That is, when the defect depth reaches 80% of the wall thickness, the pipe is considered to have failed (DNV RP-F101 in fact allows pit depths up to 85% of the original wall thickness (DNV 1999)).

Let us suppose from previous work that it is known that the corrosion growth rate distribution for this type of corrosion has an average of 0.20 mm/yr and a lognormal distribution. We wish to estimate the optimum time to assess the pipeline segment using high-resolution ILI. Assume the following:

- The current cost of the ILI run is $100,000.
- The consequences of a leak failure at the location of this segment are calculated to be equivalent to $1,000,000.
- The rate of return on investments is taken as 10% per year.

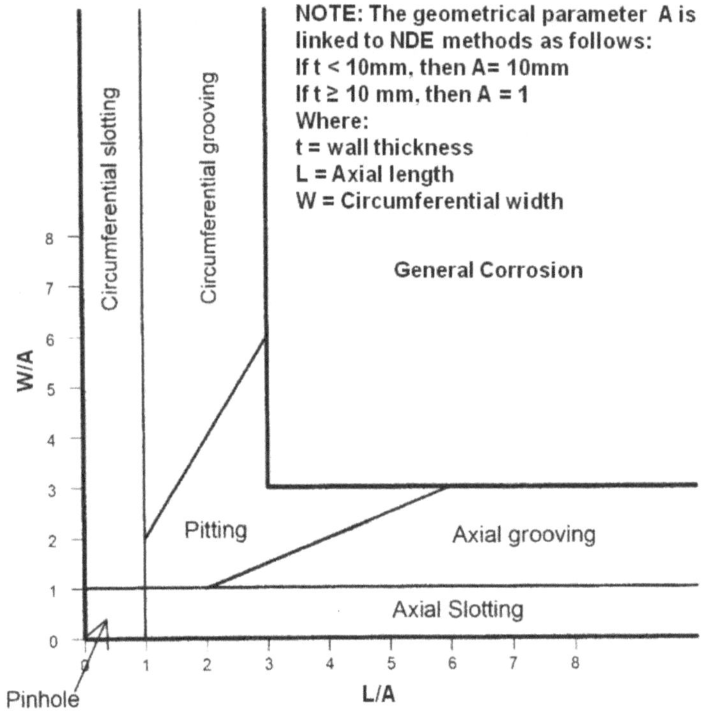

Figure 10-3. Graphical representation of metal loss defects classification

Clearly as the integrity assessment (ILI) is postponed further into the future, the lower is the net present cost of the inspection. However, at the same time, the probability of reaching or exceeding 80% of the wall thickness increases.

Since net present cost of the run decreases with time but risk (cost of failure) increases with time, we want to investigate the possibility of an inspection interval that would minimize the total cost (net present value of the future cost of the ILI run plus expectation value of the consequences). Note that the expectation value of the consequences is the probability of exceeding the failure limit state multiplied by the expected cost of reaching the consequence — this latter cost being set at a constant value of $1,000,000 in this example. The result of the calculation to investigate the optimum integrity assessment interval is shown in Figure 10-4.

Figure 10-4 shows that the optimum re-inspection interval minimizing total O&M costs is around 8 years where the probability of exceeding the stated criterion at the 8-year mark is about 4.10E-03. If company policy, for example, specified that the probability of failure should not exceed (say) 1.0E–04, then the inspection interval would have to be reduced to less than 5 years.

Figure 10-4 can be transposed to produce the familiar risk optimization plot. This is shown in Figure 10-5.

Figure 10-5 shows that for this example, the optimum risk level is around $4,100 — indicating that taking measures to lower the risk below this level in fact increases the expected expenditures. Therefore, it cannot be said that the best practice is to minimize risk. Rather as a general rule, the best plan is to seek the optimum level of risk consistent with prescribed safety, environmental, and operational constraints. It was noted above for, example, that company policy might dictate that POF in a stated time frame should not exceed a certain value. If such a constraint is applied to the optimum inspection interval calculation, the process is referred to as Constrained Optimization in which the constraint will always be found to have increased the minimized total expected expenditure.

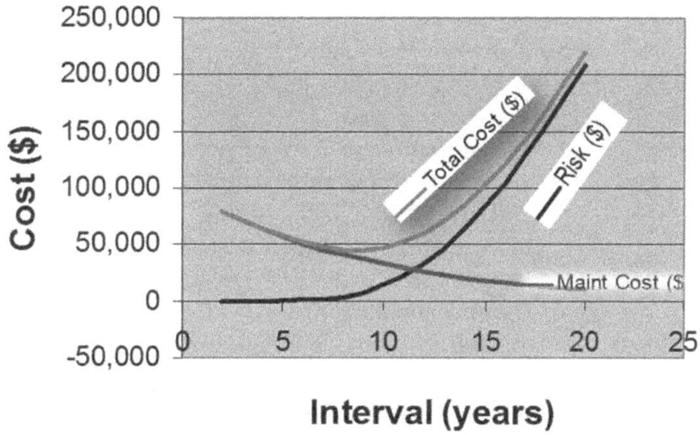

Figure 10-4. Estimate of optimum integrity assessment interval (ILI)

Comparing Figures 10-4 and 10-5, it can be seen that the practical approach to applying risk optimization is to treat the maintenance action (in this case, ILI) as the independent variable since control of this variable is the means by which the risk is controlled. Nonetheless, Figure 10-5 is commonly used (qualitatively) to illustrate that the underlying optimization relates to risk level.

Threat Identification

As noted in the above introductory example, before risk can be calculated threats to integrity have to be identified and the POF due to these threats and the expected consequences of failure estimated. It is now standard industry practice to address threats to pipeline integrity as in the following categories (ANSI/ASME B31.8S (2004)):

1. Time dependent threats:
   a. External Corrosion*
   b. Internal Corrosion*
   c. Stress Corrosion Cracking*

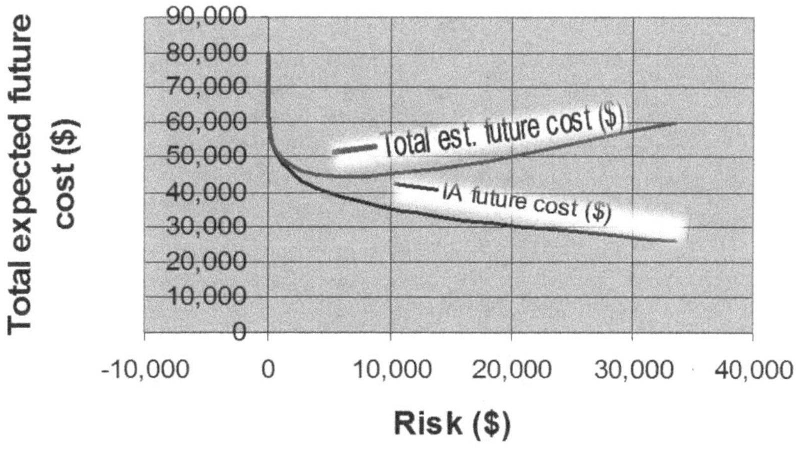

Figure 10-5. Risk optimization plot example

2. Stable threats:
   a. Manufacturing-related defects
      i. Defective pipe seam*
      ii. Defective pipe
   b. Welding and fabrication-related defects
      i. Defective pipe girth welds*
      ii. Defective fabrication welds
      iii. Wrinkle bends or buckles
      iv. Stripped threads or broken couplings
   c. Equipment failures*
      i. Gasket and o-ring failures
      ii. Control and relief equipment malfunction
      iii. Seal or pump packing failures
      iv. Other miscellaneous failure
3. Time-independent threats:
   a. Mechanical damage
      i. Damage inflicted by first, second, or third parties* (instantaneous/immediate failures)
      ii. Previously damaged pipe (delayed failures)
      iii. Vandalism and sabotage
   b. Incorrect operations
      i. Incorrect operational procedures
   c. Weather-related and outside force*
      i. Cold weather
      ii. Lightning
      iii. Heavy rains or floods
      iv. Earth movements

Those threats marked with an asterisk (*) are generally found to be the most commonly encountered globally on pipeline systems. Sources of pipeline failures statistics vary somewhat from one another and as a function of time. Figure 1-37 (Chapter 1), for example, shows a wide variation in cause percentages from year to year — there being a coefficient of variation (standard deviation divided by mean) ranging from 91% for Natural Force Damage to 15% for Material Failure. Figure 1-36 of Chapter 1 also illustrates that there can be differences due to geographical location and possibly reporting rules.

The relative rates of major threats identified by ANSI/ASME B31.8S (2004) are generally as shown in Figures 10-6, 10-7, 10-8, and 10-9.

Examining Figures 10-6, 10-7, 10-8, and 10-9, it is evident that product and geographical location and reporting have a significant effect on relative failure rates. However, the figures generally illustrate that the major causes are in general mechanical damage and external corrosion followed by (in no specific order):

- Internal corrosion
- Defective seam welds
- Defective girth welds
- Weather and outside force

SCC and equipment failure generally comprise a lower percentage of the total and are included in "Other."

The proportion of Other in the UKOPA distribution is significantly higher than in the OPS statistics. This is due to region-specific threats and reporting procedures. The Other category in UKOPA is broken down as follows:

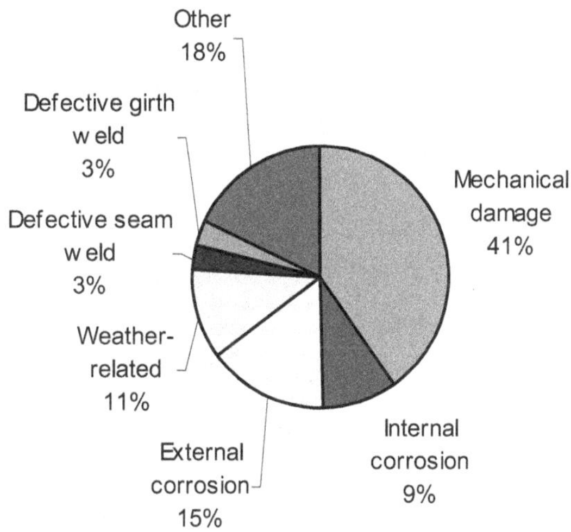

Figure 10-6. Pipeline failures by causes for Natural Gas transmission (Vieth et al. 1997)

Town gas internal SCC	50.0%
Pipe defect	22.4%
Unknown	8%
Pipe-fitting welds	4%
Leaking clamps	3.4%
Threaded joint	1%
Electric cable arc strike	1%

Failure due to incorrect operations is not included in the list of major threats above. However, it can be in some cases a significant cause as shown, for example, in Figure 10-9 (Table 14-4, Muhlbauer 2004).

The Incorrect Operations threat has to be addressed if it is required to comply with ASME B31.8S. But this does not imply that it has to be included in every risk assessment. There may be specific cases (such as in HAZOP studies) where the functional relationship between incorrect operations and likelihood of failure can be determined. But this is not

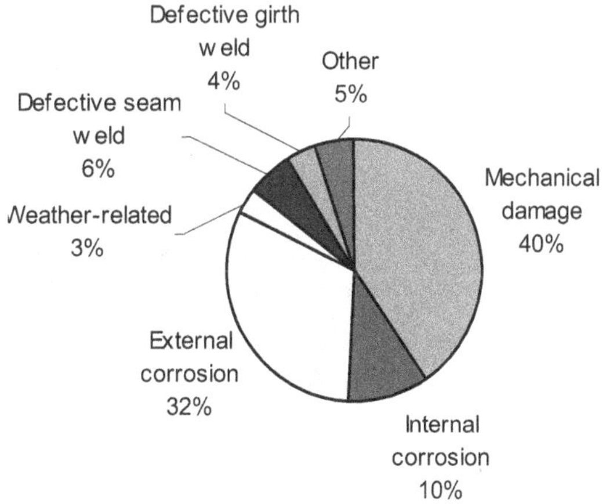

Figure 10-7. Pipeline failures by cause for Hazardous liquids transmission (Kiefner et al. 1999)

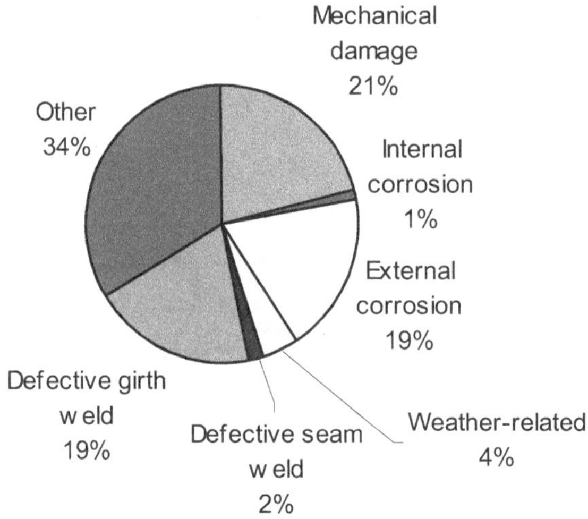

Figure 10-8. Gas and liquid pipelines failure by cause (UKOPA 2002)

generally the case in a system-wide risk assessment of the type used to develop pipeline integrity management plans. It is generally found to be more useful to address this threat by review and if necessary revision of general Operation and Maintenance procedures. This is essentially the guidance that is given in ANSI/ASME B31.8S (2004):

> "If the [risk] data shows the operation and maintenance are performed in accordance with operation and maintenance procedures, the procedures are correct and that the operating personnel are adequately qualified to fulfill the requirements of the procedure, no additional assessment is required. Deficiencies in these areas require mitigation as outlined below [in the ANSI/ASME B31.8S appendix]."

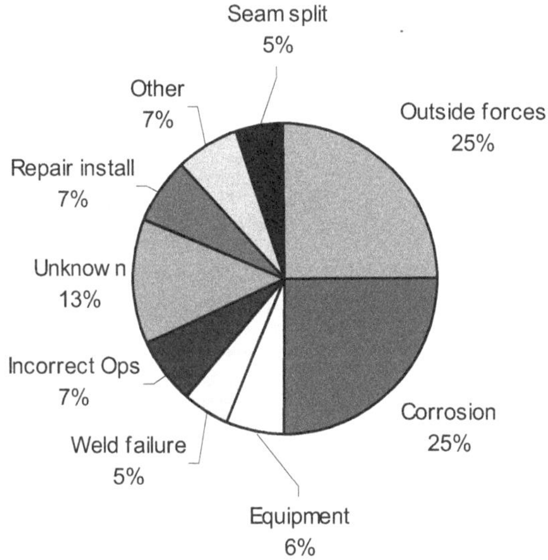

Figure 10-9. Common failure causes for US hazardous liquid pipelines (Muhlbauer 2004)

Although defective girth welds were listed above as a major threat, the threat more often presents itself in combination with the weather and outside force threat, since a girth weld that passes the construction/commissioning pressure test is unlikely to fail in service unless acted upon by an external force. (Girth welds can be a location of preferential corrosion, however — particularly if the girth weld coating is deficient.) Guidance on how to address the interaction with other outside force effects is again in ANSI/ASME B31.8S (2004):

> "The existence of these construction–related threats alone does not pose an integrity issue. The presence of these threats in conjunction with the potential for outside forces significantly increases the likelihood of an event. The data must be integrated and evaluated to determine where these construction characteristics co-exist with external or outside force potential."

The same is true in general for manufacturing defects involving seam welds. Some additional effect has to be present or develop as a rule to cause an in-service failure of seam weld defects. Since the seam weld defects can act as stress intensification factors, it is most often a fatigue action due to cyclic internal pressure fluctuations that leads to failure. For this reason, seam weld failures are sometimes classified as a fatigue threat. Experience shows that such fatigue failures are considerably more likely in liquid pipelines than in gas pipelines due presumably to the higher number of pressure cycles and higher strain rates found in liquid lines.

Comparing Figures 10-7 through 10-9, it is clear that using failure statistics to determine baseline failure rates (failures per threat per km-yr; see QRA methods), statistics must be taken from an appropriate peer system (i.e., is, one having the same product, same domain —onshore/offshore/above ground/buried, etc., and the same general geomorphic conditions) It is also clear that an appropriate timeframe has to be chosen and trends accounted for (as illustrated in Figures 10-10 and 10-13, see later). Figure 10-13 in particular provides an example of a useful strategy that can be used to smooth failure data to provide a more representative basis for prediction of failure likelihood — that is, the use of a rolling 5-year average rate rather than the more widely varying year to date statistic. The rolling 5-year approach also has the intrinsic benefit of putting more weight on recent failure histories — which we would expect to be more representative of current performance.

Consequence Analysis

Risk is present in all activities involving events the results of which are to some extent uncertain and which may have undesirable consequences. In stating a risk, it is common practice to express it in terms of the uncertainty of the events and the undesirable consequences of the events. On the basis of historical data, for example, we might decide that the risk of dying in a motor vehicle accident (the undesirable consequence) is one in ten thousand per year (the uncertain event). Such an estimate of the probability of dying might have arisen, for example, from statistics that showed that in a country with a population of 6 million, on average 6000 people are killed in motor vehicle accidents each year.

In pipeline systems, the undesirable consequence usually involves a breach of integrity of the pipe wall in the form of a leak or a rupture resulting in damage to persons or property including:

- Safety impact (fatalities or injuries)
- Environmental impact
- Property damage
- Impact to system reliability
- Financial impact (repair, remediation, lost product, etc.)
- Quality of life

- Public perception/company reputation
- Regulatory intervention

In practice, it is found that to manage pipeline system integrity effectively, the above list can most often be re-grouped as follows:

- Safety
- Environment
- Financial

An example of undesirable consequences associated with failures in liquid pipelines (oil spills) is shown in Figure 10-10. These spills for the most part lead to financial impact and environmental damage. Figure 10 also illustrates the improvement in pipeline integrity that has taken place over the last few decades — largely due to the progressive adoption of practical risk management strategies.

Although every pipeline failure results in an undesirable consequence to some extent, it has become common practice to identify areas adjacent to the pipeline that if affected by a pipeline failure would have particularly serious consequences. Following the usual US DOT terminology, these areas are referred to as HCAs. Regulations will generally require that particular attention be paid to risk mitigation in HCAs.

The determination of a HCA depends on specific regulatory requirements and on the product being transmitted. For example, the US DOT pipeline integrity rules specify the following 2 methods that can be used to define HCAs for gas pipelines (Pipeline Safety: Pipeline Integrity Management in High Consequence Areas (Gas Transmission Pipelines). Subpart 'O' — Pipeline Integrity Management §192.9):

Method 1 — any of the following:

- Class 3 location
- Class 4 location
- Any area outside of Class 3 or Class 4 where the potential impact radius (PIR) is greater than 200 m and encompasses 20 or more buildings intended for human occupancy
- The area within a potential impact radius that contains an identified site

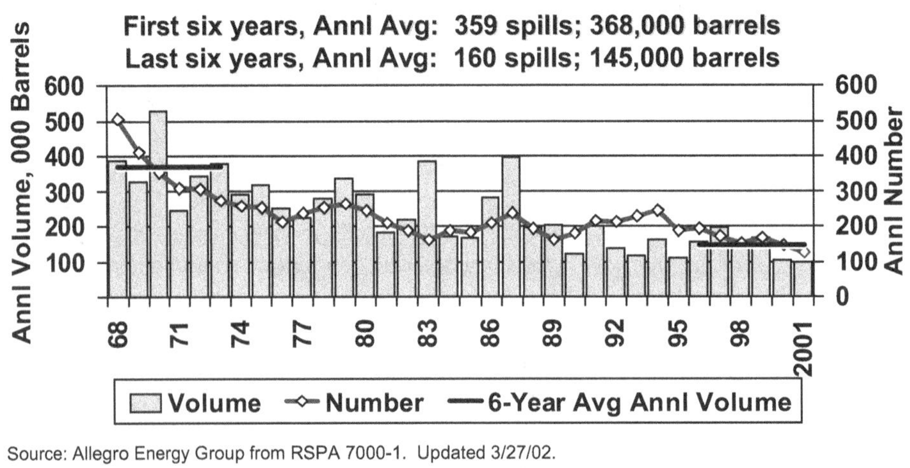

Figure 10-10. Plot of historical spills. Liquid pipelines 1968–2001

Method 2 — either of the following:

- An area within the PIR that contains 20 or more buildings intended for human occupancy
- The area within a potential impact radius that contains an identified site

The PIR is defined in Subpart "O" and ANSI/ASME B31.8S for natural gas in terms of the pipe outside diameter (OD, inches) and the MAOP (psi) of the pipeline as follows:

$$PIR = 0.69 * OD * \sqrt{MAOP} \quad ft \qquad (10-1)$$

Using an OD in mm and an MAOP in kPa, the PIR in meters is given by the following expression:

$$PIR = 0.003153 * OD(mm) * \sqrt{MAOP(kPa)} \quad m \qquad (10-2)$$

RISK ASSESSMENT AS PART OF AN INTEGRITY MANAGEMENT PLAN

The purpose of this chapter on pipeline risk management is not to present a comprehensive methodology that can be applied generally to the management of any pipeline system asset. Such methodology is presented in full in other sources — for example, Muhlbauer (2004), Nessim and Stephens (1995a), Nessim and Stephens (1995b), Sutherby et al. (2000), Fenyvesi et al. (2004a), and Dawson and Colquhoun (2009). The objective rather is to give guidance on practical steps that can be taken in the management of risk that results in an optimized program that assures safe and reliable operation of a pipeline system in a manner that does not damage the environment.

Having said that, there will be occasions in the operation of pipeline systems where pipeline failure risk must be quantified in the same way as any other type of financial or business risk that the operator has to deal with. Detailed application of these quantitative risk assessment (QRA) methods is outside the scope of this book. However, the general principals are presented in a practical manner so that the integrity management practitioner can make best use of specialist help when applying QRA methodologies.

The principal use of risk assessment in developing and executing integrity management plans for pipeline systems is to prioritize schedules of physical integrity assessment methods. These physical assessments (often simply referred to simply as Integrity Assessments) mainly comprise the following:

- In-service pressure testing (e.g. hydrotests)
- ILI with intelligent inspection pigs (ILI)
- Direct assessment (DA) comprising:
    - Preassessment (risk assessment)
    - Indirect examinations
    - Direct examination
    - Analysis and information integration
- ROV-based assessments for offshore pipelines
- Site investigation and monitoring programs for weather-related outside forces including geotechnical loadings

It is because of this interaction between risk assessment and integrity assessment that the use of risk assessment in integrity management planning can be considered to comprise risk-based integrity planning (or Risk-Based Inspection – usually referred to by its initials, RBI).

Assessing Risk — Deterministic QRA Example

Having introduced the need to control and optimize risk while operating a pipeline system, it is convenient now to review the concept and common understanding of risk in that context. As discussed further in Types of Risk Assessment, there are various types of risk assessment that can be used with respect to pipeline integrity management. The type that perhaps most succinctly illustrates the concept of risk is what can be referred to as Deterministic QRA (Dawson and Colquhoun 2009). Note that whereas QRA could in principal equally stand for qualitative risk assessment, by convention the term is used in the pipeline industry to mean Quantitative Risk Assessment. To avoid confusing terminology, qualitative/relative risk assessments are referred to as non-QRA in this chapter.

In pipeline integrity management (as in all common usage of the word), risk is considered to increase as POF increases. Risk is also considered to increase as severity of expected consequences of failure increase. Therefore, the effects of probability of an undesirable event occurring and the associated consequences have to be combined and taken into account in assessment of risk levels. For example, an individual being hit by a meteor would appear to have a probability of around 1 in 12.8 million per year. Whereas the consequences of such an accident are clearly grave, a chance of 1 in 12.8 million per year is so remote that very few people lose sleep over becoming a victim. The risk therefore is said to be low.[1] On the other hand, the probability of being involved in a minor "fender bender" during the first snowfall of winter can be considered to be high. The consequences of such events are generally relatively minor. We might say then (subjectively) that the risk associated with such accidents is moderate. This idea then of balancing probability with consequences leads to a general, quantitative definition of risk as follows:

$$\text{RISK} = (\text{POF}) \times (\text{Expected consequence of failure}) \qquad (10-3)$$

Note that since Probability does not have units (or at least is unit-less per length per time), risk is expressed in the units of the consequence. For example, suppose that from historical data, we estimate the POF for mechanical damage to pipelines on a given pipeline system to be 0.12E-03/km-yr, an estimate of the POF in the next 5 years for a 12-km segment of the pipeline then would then be:

$$\begin{aligned}\text{POF (12 km, 5 years)} &= 0.12\text{E-03 per km-yr} \times 12 \text{ km} \times 5 \text{ years} \\ &= 0.0072 \qquad (10-4)\end{aligned}$$

This is not an index. It is a quantitative estimate of POF within 5 years. It is single-valued and therefore is deterministic. Another form of QRA is discussed in Section 10.3.4 in which the probabilistic distribution of parameters is accounted for in terms of probability density functions. This latter approach is generally referred to as Probabilistic Risk Assessment Methodology (PRAM) (Kumamoto and Henley 2000; Bedford and Cooke 2001). Since in practical applications to pipeline problems PRAM is most often formulated in terms of reliability, it is referred to in this chapter as reliability-based QRA.

[1] Estimates vary by several orders of magnitude (Lapaz 1958; Jayjock 2004). For likelihood estimates of a meteorite strike while taking off in an airplane, see "Consider a Spherical Cow" by John Harte (1988).

With respect to the above POF calculation, suppose through knowledge of the physical layout of the pipeline system with respect to populations and property, it is possible to estimate that the conditional probability of a rupture given a failure $P(R|F)$ is 0.3 (say) and that the cost related to a rupture is $1,000,000 and that related to a leak is $500,000.

The consequence of failure (leak or rupture) can be calculated as follows:

$$Q = 0.3 \times \$1{,}000{,}000 + (1 - 0.3) \times \$500{,}000$$
$$= \$650{,}000 \qquad (10-5)$$

Then the risk (K) in 5 years is given by:

$$K = 0.0072 * \$650{,}000$$
$$= \$4{,}680 \qquad (10-6)$$

The foregoing example illustrates the deterministic QRA concept. Clearly, risk values as low as that given by Expression (10-6) only have a strict physical meaning when applied to a large pipeline system. In themselves, such numbers function as risk measures that allow comparisons of potential failure impacts between pipeline segments. Note too that expression (10-5) explicitly accounts for the uncertainty of failure mode (rupture versus leak). In the practical application of this methodology, however, it is often desirable to keep the risk calculation for rupture and for leak separate to reflect the specific remediation measures associated with each in the integrity management plan.

The above example of deterministic QRA was presented to illustrate the basic concept of risk. As will be explained later, other methods (non-QRA and reliability-based QRA for example) are also used with respect to pipeline integrity management. For reference, the advantages of deterministic QRA over these other methods include the following:

- Requires only deterministic values (whereas reliability-based methods require probability density function distribution of parameters)
- Follows common rules of statistics and thereby avoids arbitrary rules and assumptions generally required in non-QRA methods
- Risk results are compatible with other risk studies. That is, probabilities of failure, expected consequences, and risk all have the same meaning and physical significance as they would in any classical risk assessment.
- Does not need probability distributions or other specialized input. The pipeline and environment data are essentially the same as those used in non-QRA methods and required routinely to manage pipeline system integrity. Representative estimates of failure rates are generally available, and simple models can be used to adapt these rates to specific pipeline segments.
- Since fewer arbitrary assumptions are involved in the basic formulation, we should find that the method is generally more accurate than non-QRA — both in terms of order and magnitude of ranking. This advantage, however, might not greatly impact the execution of a pipeline integrity management plan as will be discussed later.

Also for reference, some of the disadvantages of the method might include the following:

- The method requires software to implement. This is perhaps may not be a great disadvantage since essentially all practical risk assessment applications now involve software solutions.
- The method relies on approximate cause-effect relationships to relate system attributes to POF. The value and accuracy of the results therefore are quite dependent on the accuracy and validity of these relationships for which comprehensive formulations are not easily available.

- Apparent precision may be false and/or misleading. That is to say, the implied underlying processes that govern failure might not exist uniformly. The damage considerations in some cases might therefore be more appropriately considered chaotic than probabilistic. Cases have been observed in determining the corrosion growth rate of internal corrosion defects for example where significant growth has occurred in one defect and not in another similar defect only some 75-mm distant. Both of these defects would clearly be subjected to the same operating environment, but due to some very local and unquantifiable variation in steel and/or surface conditions, growth occurred in one case by not the other. Such behavior cannot be rigorously addressed by statistical methods. Another such example of the foregoing is given in Limitations of the Probabilistic Inference of Operational Data.
- The apparent precision may be of little additional benefit to IM planning. As will be discussed later, the risk assessment methodology addressed by this section is predicated upon the eventual assessment of the integrity of pipeline segments by physical means. These assessments will generally be carried out in a 5-year to 7-year timeframe following risk assessment. Taken together with the previous bullet, damage prediction on real pipeline systems within this time frame most likely will only be able to indicate probable order of failure rather than provide precise estimates of failure dates. In this sense, deterministic QRA may offer little advantage over simpler non-QRA methods in routine application to pipeline integrity management planning.
- The method is generally more complex than non-QRA methods.

Limitations of the Probabilistic Inference of Operational Data

The information required to perform QRAs is often derived from historical data. This derivation must be done with care. For example, once repeat inspection results are available, it is in principle possible to estimate defect growth rate by comparing the results of the two inspections. This is one way that seemingly solid historical data can be made available for use in QRAs. However, the following corrosion growth rate calculation illustrates some of the pitfalls that have to be avoided in order to arrive at useable growth rate estimates.

In the example (taken from an actual ILI assessment), the depth of a feature (comprising a cluster according to the clustering rules provided to the ILI vendor) was compared between 2 successive runs. Details of the two runs are summarized in Table 10-1.

Both ILI reports clearly refer to the same feature (cluster). The question is: Is the feature growing? The depth in fact appears to be shrinking slightly. (This small variation might be expected between successive runs due to tool tolerance.) Although both the width and the length have increased slightly it could be concluded that the feature is not growing.

However, examination of the ILI signal as shown in Figure 10-11 tells a different story.

The upper portion of Figure 10-11 corresponds to the ILI signal for the 2001 run in Table 10-1, and the lower portion corresponds to the ILI signal from the 2006 run.[2] It can be seen that the 26% WT pit most likely is not growing. However, there is new corrosion that has grown from zero to 29% of the wall thickness in 5 years. The wall thickness in the example was 8.51 mm, so the indicated growth rate is 0.48 mm/year. A corrosion growth rate of 0.30 mm/yr would be considered significant. The new corrosion therefore ought to be investigated immediately, but might have been totally missed had only feature matching been done.

This example illustrates two important aspects of data evaluation. First (as noted above), it illustrates that the possible uncertainty is calculating corrosion growth rates.

[2] Example courtesy of GE Oil & Gas, PII Pipeline Solutions.

TABLE 10-1. Feature depth comparison between two sequential ILI runs

Item	2001 ILI Report	2006 ILIReport
Absolute distance (m)	2,749.08	2,749.17
Peak depth (%wt)	26%	29%
Length (mm)	246	277
Width (mm)	196	203

Two adjacent corroding areas, subjected to essentially identical environments, exhibit significantly different corrosion growth. In the risk assessment, these corrosion areas are so close they would certainly have been treated as identical and exposed to the same corrosive environment. What has to be recognized is that for the most part in performing pipeline risk assessments, we use deterministic calculations expressed in probabilistic terms to describe processes that may be more appropriately considered chaotic. The level of predictability of damage to pipelines is discussed further in Choice of Risk as an Indicator of Severity of a potential failure when addressing the reasons for and merits of using risk as a measure of potential failure seriousness. The potential for inaccuracies inherent in theoretical likelihood of failure predictability models leading to risk ranking errors is one of the many reasons why risk assessment results must be carefully reviewed and decision taken thereon by experienced and knowledgeable pipeline integrity management practitioners (see also Risk Assessment).

The second aspect illustrated by the example relates to the necessary rigor with which data are derived. In the example case and initial review of the data suggested, the feature reported in 2001 was not growing. However, when signal matching was performed by the ILI vendor, the case was found to be quite the opposite. As will be discussed later, the simpler risk assessment methods (which when performed carefully and correctly are quite adequate for integrity management planning) are generally less sensitive to data quality issues due to the largely qualitative nature of the methodology. However, when more rigorous methods are used (QRA and model-based semi-QRA for example), considerable care has to be taken when reviewing data sources to ensure that the benefits of these methods are not negated by poor data due to methods' strong dependence on data quality. One practical way to avoid the impact of rogue data is through data integration by using related data sets to validate individual data elements.

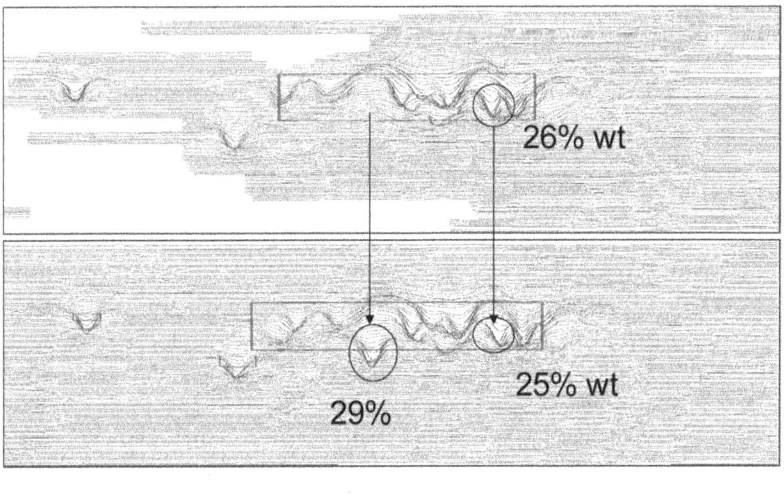

Figure 10-11. Signal matching comparison between 2 sequential ILI runs

The central role of data management in the integrity management process is illustrated graphically in Figure 10-12. Each element of the process requires and generates data. The hallmark of any semi-automated integrity management system is the functionality it provides to manage and analyze data. Integrating and comparing complementary data sources can often readily identify data anomalies of the type described above. The review and validation that must be carried out on the results of any theoretical risk assessment as described in Risk Assessment is greatly facilitated by the use of an advanced data management system linked to a modern GIS. Not to downplay the important role risk assessment plays in pipeline integrity management, the use of a high-quality GIS puts the control of the integrity management process firmly in the control of the integrity management practitioners ultimately responsible for the results of analyses and the implementation of these results into the integrity management plan. An essential feature of the GIS selected for the system is that it should display the risk profile graphically together with the pipeline system attributes and other risk management drivers such as HCA locations, ROW access, terrain details, and adjacent facilities.

Choice of Risk as an Indicator of Severity of a Potential Failure

We have shown that the common, quantitative understanding of risk is the product of POF and consequence of failure. Clearly, probability and consequence are in themselves valid indicators of the severity of a pipeline failure. We consider here the reasons why risk is chosen as the preferred indicator of severity.

ASME B31.8S (2004) defends the selection of risk as the criterion for identifying locations for integrity assessment and resulting mitigative actions as follows:

> "Examining both primary risk factors (likelihood and consequences) avoids focusing solely on the most visible or frequently occurring problems while ignoring potential events that could cause significantly greater damage. Conversely, the process also avoids focusing on the most catastrophic but unlikely events while missing scenarios that more likely to occur."

The defense offered by ANSI/ASME B31.8S (2004) can be illustrated by the following example using the simple models suggested in (Colquhoun et al. 2002). In this model:

L = likelihood of failure index
F = factors increasing likelihood
R = factors decreasing likelihood
Q = consequence index
K = risk index ($K = L * Q$)

Suppose an integrity assessment is to be scheduled for the upcoming year and we wish to decide between which of two lines we should assess first. These lines are:

- a natural gas line recently recoated with FBE in a Class 3 location, or
- a natural gas line coated with 8-year-old polyethylene tape (PE Tape) in a Class 1 location.

In the first case, we have a line with a good protective coat for which we would expect a low likelihood of failure. But it is in a Class 3 location where by original design (ASME B31.8 say) there could be a density of buildings intended for human occupancy of up to 45 in any mile length (1.60 km) of the line. In the second case, the PE tape coating would be expected to provide considerably less protection than the FBE — especially if there were any likelihood of soil stress on the coating due to geotechnical loading. The consequences for this stretch of line, however, would be lower than in Class 1 — but there still could be up to 10 building in a mile intended for human occupancy by original design.

Simplifying the risk assessment for the purposes of illustration (Colquhoun et al. 2002), we have:

$$L = F * R \qquad (10-7)$$

$$F = (\text{Soil\_Type} + \text{Soil\_Drain})/2 \qquad (10-8)$$

$$R = (R\,\text{coat} + R\,\text{CP})/2 \qquad (10-9)$$

$$R\text{\_coat} = (\text{Coat\_Type} + \text{Coat\_Cond})/2 \qquad (10-10)$$

$$\text{R\_CP} = (\text{CP\_Eff+CP\_Avail})/2 \qquad (10-11)$$

$$Q = \text{a function of Class Location only}$$
$$= 1.90 - 2.25 * (\text{Design Factor})\ (\text{say})$$

$$\text{Coat\_Type (FBE)} = 0.0\ (\text{say}) \qquad (10-12)$$

$$\text{Coat\_Type (PE tape)} = 1.0\ (\text{say}) \qquad (10-13)$$

Setting all other factors to the median value of 0.5, we can calculate the likelihood, consequence, and risk index values associated with the 2 scenarios as shown in Table 10-2.

On the basis of consequences, we would of course have chosen the Class 3 section to assess first. However, on the basis of likelihood of failure, we would probably have chosen the Class 1. In fact, the more serious scenario (the one potentially having the higher financial and social impact) would be the line in Class 3 since it has a risk index some 67% higher than the line in Class 1.

The risk index therefore provides a better and more balanced measure of impact severity than the likelihood or consequence indices of severity due to the potentially offsetting influences of likelihood and consequence. Having said that, it is also common practice to display and examine the components of risk also, since there may be policy or other reasons not captured in the risk assessment procedures that would over-ride the ranking based on risk. Should this be the case, the decision to over-ride the risk ranking must be recorded in the integrity management plan in the same way as SME decisions as described later in Non-QRA methods (a).

Risk values were used in the above example to prioritize maintenance actions. Risk assessment results when applied to pipeline systems should never be interpreted as justification for letting a pipeline segment fail. "Run to failure" is not a valid concept in pipeline integrity management. The goal we wish to achieve through good risk-based pipeline integrity management planning is that no one is ever harmed by a pipeline failure, nor is damage done to the environment, nor is system reliability impacted.[3] Risk management techniques provide a valuable tool to ensure that resources are allocated in the best manner to achieve this goal.

The foregoing argument is clearly valid if the damage mechanism is random and able to be described by statistical parameters and/or has a known probability density function. However, the example presented in Assessing risk-deterministic QRA suggests that certain damage mechanisms might not always behave in this way. The example given in Assessing risk-Deterministic QRA example related to corrosion. The same predictability challenges

[3] This aim of integrity management is stated succinctly for example in the BP Group Health, Safety, Security and Environmental goal: No accidents, no harm to people, and no damage to the environment.

TABLE 10-2. Sample scenario risk calculation

Class	Coating	L	Q	$K = L * Q$
1	PE tape	0.31	0.28	0.09
3	FBE	0.19	0.78	0.15

are noted, however, in the practical treatment of all threats — including mechanical damage, stress corrosion cracking, and outside force due to geotechnical loadings.

With respect to geotechnical loadings, the speaker in a keynote address in a Canberra, Australia, at a conference on probabilistic developments in Geotechnical Engineering in the mid-90s made the following observation (Rizkalla, M., private communication, 2008):

> "A geotechnical prediction is considered acceptable if it is within ±100% of the observed response. The prediction is good if it is within ±150% of the truth. If the prediction is within ±10% of the true response, it is likely that there is a mistake in the calculation."

With respect to corrosion (especially in stainless steels where passivation is more prevalent), Legat and Dolecek (Legat and Dolecek 1995) note that whereas general corrosion appears to be the result of a random (statistical) process, localized pitting corrosion is most likely generated by a deterministic chaotic process.

Do the potentially chaotic nature of damage due to threats to integrity and the low probabilities of failure negate the use of risk as a criterion? We believe not. The consequence component of risk can be estimated with acceptable engineering accuracy. The majority of the uncertainty lies in the estimation of failure likelihood, failure type (leak or rupture), and postfailure behavior of the product released (dispersion and ignition of gaseous or volatile flammable products for example). It is reasonable to assume a worst case for these last 2. In fact, this is the guidance that is provided in ANSI/ASME B31.8S (2004) for cases where the failure models do not accurately predict damage type or postfailure behavior of the product.

The uncertainty in the estimation of likelihood of failure is the more challenging issue. In practical pipeline risk assessment applications, however, this uncertainty does not prove to be problematic for the following reasons:

- The relatively coarse granularity that can be tolerated in RBI approaches.
- The risk assessment, which is supplemented in the majority of practical pipeline integrity applications of good engineering judgment
- The integrity of pipeline segments, which is ultimately confirmed by physical integrity assessments
- The generally low failure rates in pipeline systems

Integrity Management Planning

Before expanding on the role of risk assessment in integrity management planning, it is worthwhile reviewing briefly the main element of the integrity management planning process. These elements are illustrated in Figure 10-12. As noted previously, the risk assessment described in this chapter is an integral part of pipeline system integrity management planning in which the integrity of the system is determined through the use of physical integrity assessments.

Risk assessment's basic purpose is to guide other actions to be taken to assure the integrity of the pipeline system. As will be noted later, this guidance will principally be in determining the risk-prioritized order for physical integrity assessments. The basic integrity management process is summarized in Figure 10-12.

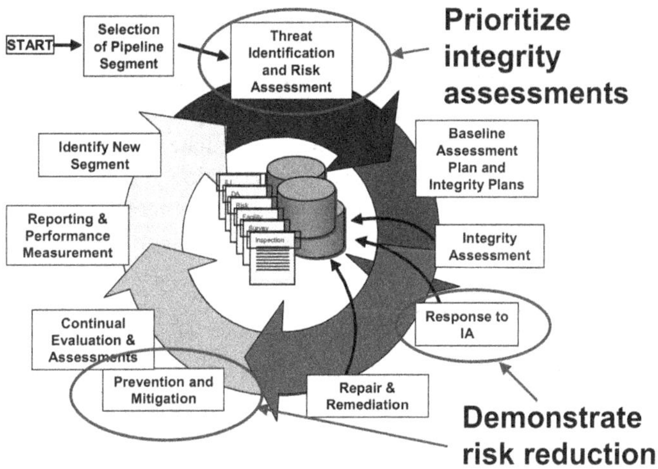

Figure 10-12. Pipeline integrity management process and relations to risk assessment (Colquhoun et al. 2006)

The entire integrity management process is described in detail in Chapter 2. Basically, the process has the following main steps:

- Pipeline segments that could affect HCAs or other risk receptors are identified. The methods to do this depend on the product being transported.
- Identify the applicable threats to integrity and perform a risk assessment to prioritize physical integrity assessments. The threats to integrity considered will generally include the following:
  - Mechanical damage (third party interference)
  - External corrosion
  - Internal corrosion
  - Stress corrosion cracking
  - Manufacturing defects (especially those associated with seam-welded pipe)
  - Construction defects (especially those associated with pipe joining methods)
  - Weather-related and outside force. This category of threat includes loads applied to buried pipelines due to earth movement. Note that such loads may be especially damaging in the presence of construction defects associated with pipe joining methods such as girth weld defects.
  - Incorrect operations (principally those that could lead to over-pressure of the pipeline)
  - Equipment failure
- Assemble a plan to perform physical integrity assessments on the pipeline system. This plan is often referred to as the Baseline Assessment Plan and typically will cover a rolling 10-year period.
- Perform the integrity assessments
- Respond to the results of the integrity assessments. The specific response will depend on the anomalies found during the assessments. In the case of metal loss defects due to corrosion for example, this step will usually include an estimate of the remaining strength of the defective area according to some predetermined criteria such as B31G (ANSI/ASME B31G 1991), RSTRENG (Kiefner and Vieth 1989), PCORR (Stephens et al. 1996, 1999; Stephens and Leis 1997; Leis and Stephens 1997), or DNV R- F101 (DNV 1999).

- Perform the necessary repairs and other remedial measures required to address current integrity concerns associated with the anomalies found. Such measures might include the use of repair sleeves in the case of metal loss defects and slope stabilization measures in the case of geotechnical, strain-induced loadings. This step is sometimes referred to as immediate or current integrity assurance.
- Application of other risk reduction and risk mitigation measures indicated by the risk assessment and the physical integrity assessments. These typically will be measures that affect the ongoing integrity protection of the pipeline such as the provision of concrete slabs over crossings or increased right-of-way surveillance in the case of the threat of mechanical damage due to third party interference. As shown in Figure 10-12, it most likely will be necessary at this stage to explicitly demonstrate risk reduction by performing either a further system-wide risk assessment or a site-specific risk assessment.
- Plan for continual integrity assessments. This step usually considers the expected rate of deterioration of the pipeline based on results of the physical integrity assessments to determine suitable intervals for future assessments or other interventions. Such deterioration might be associated with estimates of corrosion growth rates or slope movement for example. This step is sometimes referred to as future integrity assurance.
- Key performance indicators are identified and tracked to measure the efficiency of the integrity management plan and determine if the plan needs to be revised. Such key performance indicators might include:
  - Number of leaks/ruptures
  - Lengths inspected versus lengths planned for inspection
  - Delivery interruptions
  - Corrective and predictive maintenance spending trends
- New segments that could affect HCAs or other risk receptors identified during the (annual) application cycle of the integrity management plan are added to the integrity management database and included in the next system-wide risk assessment as the integrity management process cycle continues.

Again, the central position of the database in Figure 10-12 should be noted (and emphasized). Risk assessment in common with all other activities in the integrity management process relies heavily on high quality, currently updated data. The execution of each step of the process also continually produces data and integrity-related information. A well-structured data management system is essential to effective pipeline integrity management planning.

Role of Risk Assessment In Integrity Management Planning

As illustrated in Figure 10-12, risk assessment in integrity management planning has the following main objectives:

- Prioritization of pipelines and pipeline segments for scheduling of physical integrity assessment and other risk mitigation actions
- Demonstration of risk-reduction benefits
- Determination of the most effective mitigation measures for the applicable threats
- Assessment of the integrity impact of modified integrity assessment/inspection intervals
- Assessment of the use of or need for alternative integrity assessment methodologies

The application of the foregoing will minimize risk to HCAs and other risk receptors and will result in a more effective allocation of O&M resources.

The principal components of risk assessment methods are generally those given in ANSI/ASME B31.8S (2004) as follows:

- Identification of potential events or conditions that could threaten system integrity
- Evaluation of likelihood of failure and consequences
- Provide a risk ranking and identification of specific threats that primarily influence or drive risk
- Guide identification of integrity assessment and/or risk reduction or mitigation measures
- Provide data feedback loops to continually improve the integrity management process
- Provide a structure for and continuous updating of risk re-assessments

The output of the risk assessment is an Integrity Assessment Plan — also known as a Baseline Assessment Plan. This plan identifies the principal threats to be assessed, the basic timeline to assessment (generally to the nearest quarter of a year), and the tools to be use in the assessment. It has to be recognized that practically this IAP has to be modified to accommodate operational logistics such as seasonal access, tool availability, and available downtime windows.

Types of Risk Assessment

As noted in Accessing risk-Deterministic QRA example, risk assessment methods can be categorized broadly into:

- QRAs in which the probabilities of failure are expressed as classical probabilities (a number naturally occurring between 0 and 1.0) and the consequences are reducible to a common measure (such as a monitory value)
- Non-QRA methods in which likelihood (term used to distinguish the quantity from classical probability) and consequence are represented by arbitrary indices that only have meaning in relation to one another

The QRA methods can be further sub-categorized as:

- Deterministic QRA
- Rcliability based QRA

QRA Methods

Deterministic QRA methods have already been discussed in the example given in Assessing risk-Deterministic QRA and have wide applicability for both system-wide and site-specific risk assessment. In these methods, failure rates for each threat are initially derived from representative historical failure rates (baseline failure rates) from similar pipeline systems, and then are "dialed-up" or "dialed-down" for the subject pipeline segment depending on the attributes and parameters associated with that segment. For example, if the average failure rate due to external corrosion for a representative system is 0.96E-03 per km/year and the 90th percentile value is 2.00E-03 per km/year. We could assign a value of 2.00E–03 per km/year to old tape-coated pipe in corrosive soil in which the CP was known to be suspect, and zero to new FBE-coated pipe laid in dry, well-drained sandy soil with CP ON-potential always more negative than –850 mV. We would have to develop some representative function to scale/interpolate in between (say for coal-tar enamel coating 20 years old buried in moist clay soil with CP On-potentials occasionally more positive than –850 mV).

Strictly speaking in the foregoing, example, the POF (equal to the dialed-up/down failure rate) cannot be zero. However, within the operating realities of a real pipeline system, it is quite defensible to deem it so for the example given. Another approach is to decide on a "de minimis" POF level below which the POF is indeed taken as zero for the purposes of integrity management planning and operation of the integrity management process (derived from the legal term "de minimis non curat rex" — the law is not interested in trivial matters.) The value adopted by most US government agencies (e.g., RPA and FDA) is 1.0E–06/year. That is, if the POF falls on or below this value, it is considered to be zero. For example, applying a 0.9E-05 per km/yr failure rate to a 100-m segment, the POF would be 0.9E–06 per year, which we could consider to be zero using the de minimis argument (see also Muhlbauer 2004, Chapter 15 for the background and validity of the de minimis criterion) (Kelly and Cardon 1994).

An example of overall baseline failure rate statistics for Western Europe (UKOPA 2002) typical of those that might be used in a deterministic QRA is shown in Figure 10-13.

Reliability-based QRA generally is very data-intensive and therefore best suited to site-specific studies where the data to be collected are associated with one particular location. A common application of QRA-REL is to probability of exceedance calculations (POE). The following is an illustration of such a calculation.

With respect to Figure 10-14, suppose we are able to determine the probability density function of defect depths from examining many defect dimensions (say from an ILI run on the subject pipeline). We would expect the distribution to have the general shape of a lognormal plot as shown in Figure 10-14.

Suppose a defect depth is measured as 15% WT in 2008. Next we need to define a depth at which failure is deemed to have occurred (d^*). Knowing the distribution of depth measurements, we determine that the expected depth measurement for $P(d \le d^*)$ = (say) 95%. The mean of this second distribution is the expected depth measurement (shown as 35% WT in the example). If the WT were, say, 12.7 mm, the corrosion depth growth would be 0.20 * 12.7 mm = 2.54 mm. If we had determined that the mean corrosion growth rate was 0.30 mm/year, we would expect to have to intervene in 2.54/0.30 = 8.48 years. That is in 2016 in this case.

Note that the above numbers have been chosen for illustration only and are not intended to be representative of any ILI tool or any pipeline system or defect population.

The methodology needs to estimate a failure depth for the defect as noted (d^*). One such method of doing this might be to use the Modified B31G method or PCORR for calculating remaining strength of a defect that fails by yielding in remaining defect ligament.

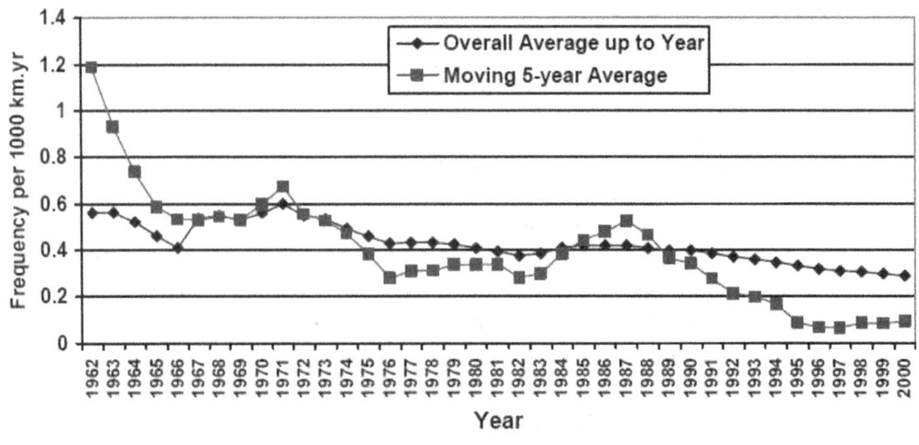

Figure 10-13. Overall baseline failure rates example (UKOPA 2002)

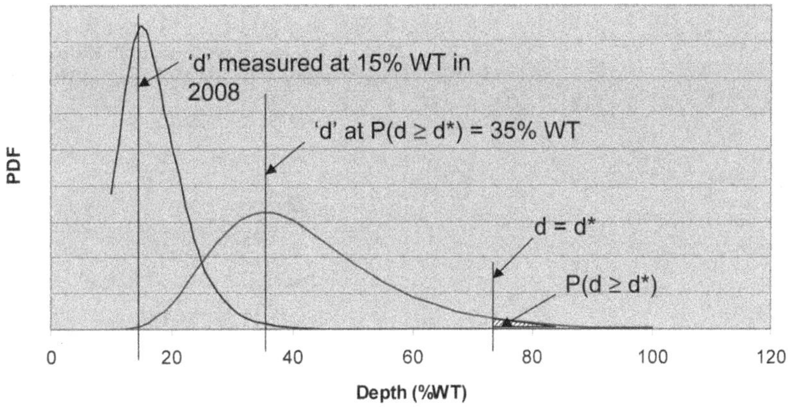

Figure 10-14. POE example for corrosion growth to specified failure depth

The process flow for the simplified POE example described above is summarized in Figure 10-15.

In fact, POE calculations tend to be a little more complex than that shown in Figure 10-15. Not only is there a PDF associated with peak depth measurement, but also with all other parameters involved in the calculation (e.g., defect length, SMYS, diameter, and wall thickness). The result of these distributions is that the "tolerable" corrosion depth, like the measured/actual depth, had a distribution associated with it as shown in Figure 10-16.

The general formulation of reliability-based QRA therefore can be represented as shown in Figure 10-17 as follows (courtesy of C-FER Technologies).

- Determine the required maximum operating pressure for the study period.
- Determine the probability density functions for pipe properties and dimensions.
- A probability of exceedance calculation is made by reducing the stated measurement uncertainties in the form of a probability density function for measured defect/flaw depths.
- The remaining strength of defects is modeled using some suitable criteria (such as B31G Modified or PCORR as mentioned earlier).
- Data from repeated inspections is used to calculate mean corrosion growth rates. This can be done on the basis of one inspection using "full life" or "half life" assumptions[4] for the growth pattern with time. However, ILI signal-to-signal matching using, for example, PII's RunCom™ or BJ Inspection Services GEMINIView™, by far gives the most accurate results (see also Fenyvesi et al. (2004b)).

Non-QRA Methods

These methods can be further subcategorized as follows:

- Qualitative methods
- Semi-quantitative methods

[4] Full-life corrosion growth rates assume that the growth took place linearly since commissioning to inspection. This assumption makes most sense for internal corrosion. For external corrosion, we usually use a half-life assumption — assuming (arbitrarily) that for the first half of its operational life the line was protected by good coating and/or effective cathodic protection.

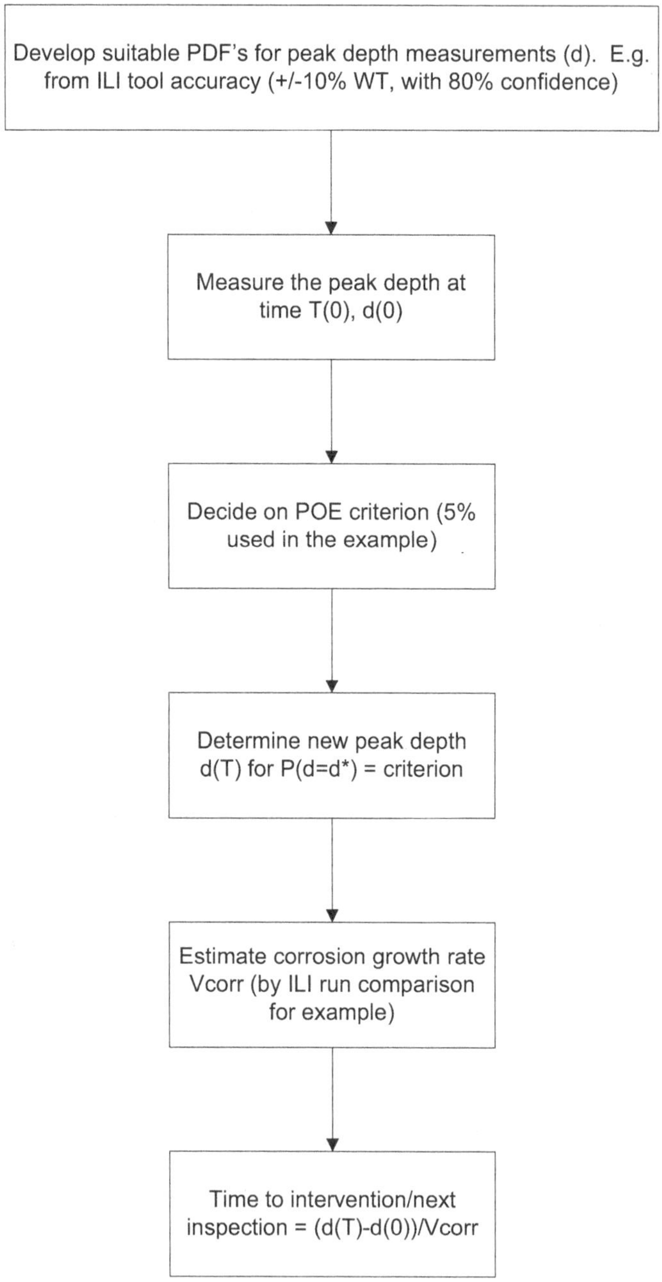

Figure 10-15. Flow diagram for simplified POE example for corrosion growth

The precise delineation between these two subcategories is not universally agreed upon. However, for the purposes of this chapter, we define semi-quantitative as referring to model-based methods in which probabilities and consequences are estimated using quasi-rigorous predictive models. Qualitative methods generally rely on the judgment of SMEs.

(a) Risk Matrix Methods

Perhaps the most common non-QRA method for general applications is still the risk matrix (Muhlbauer 2004). In this method, likelihood of failure and consequence are assigned

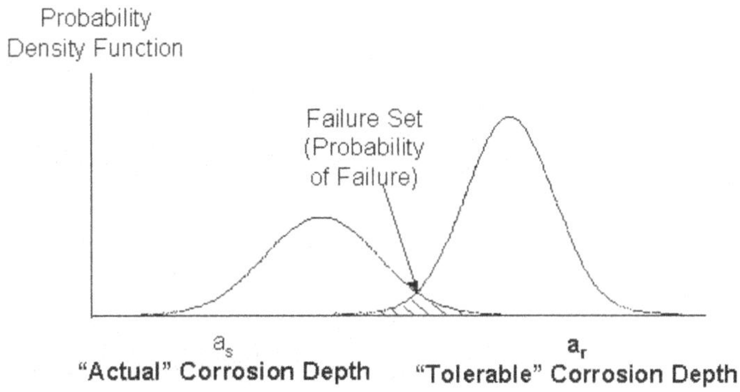

Figure 10-16. PDFs for measured/actual and tolerable corrosion depths

scores based on SME judgment and experience. (In non-QRA methods, the indices representing POF are usually referred to as likelihood indices to emphasize that they are not to be interpreted as classical probabilities.) An example illustrating the format of a risk matrix is shown in Figure 10-18. In this example, the assigned scores for likelihood and consequence are Low, Medium-Low, Medium-High, and High resulting in a 4 × 4 matrix. Other ranks of matrix can be used — 3 × 3 and 5 × 5 also being common. In practical applications involving SME teams, it is found that an even rank is preferred since this seems to avoid a tendency for team members to gravitate toward "medium." In the case of an even number, rank team members are forced to come down on one side or the other.

The SME team then has to decide which combinations of likelihood and consequence are to be rated High, Medium-High, Medium-Low, and Low. This decision is largely subjective and most likely will reflect the nature and level of risk aversion in the operator's company (which implies that the risk assessment matrix (RAM) need not be symmetrical as illustrated in Figure 10-18).

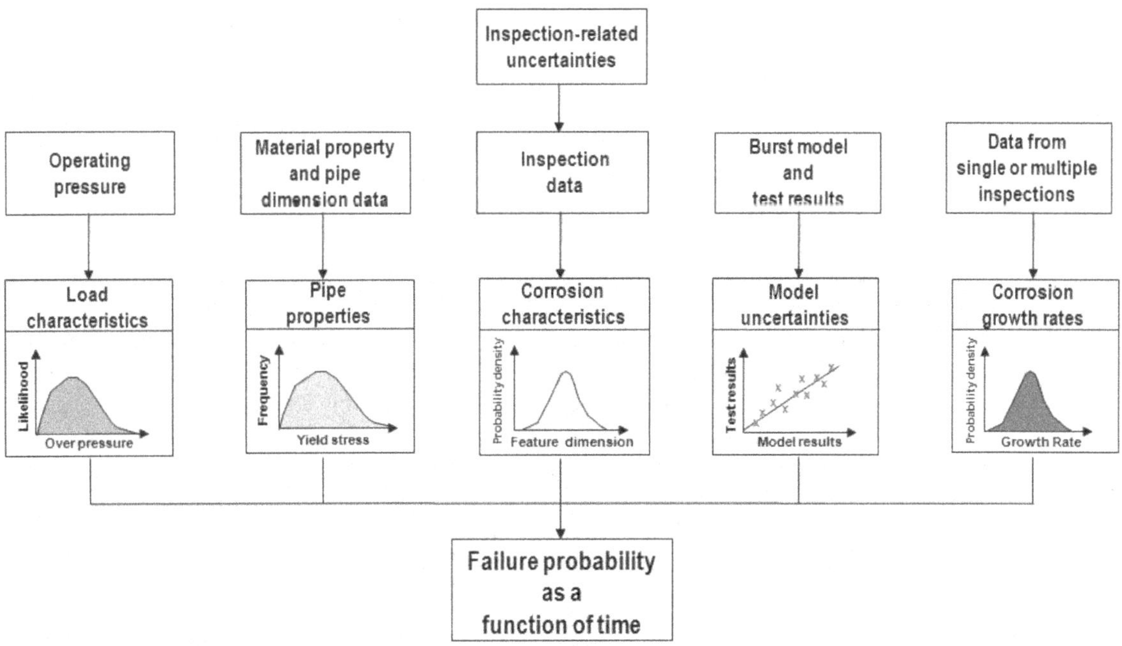

Figure 10-17. General application of reliability-based QRA (courtesy of C-FER Technologies)

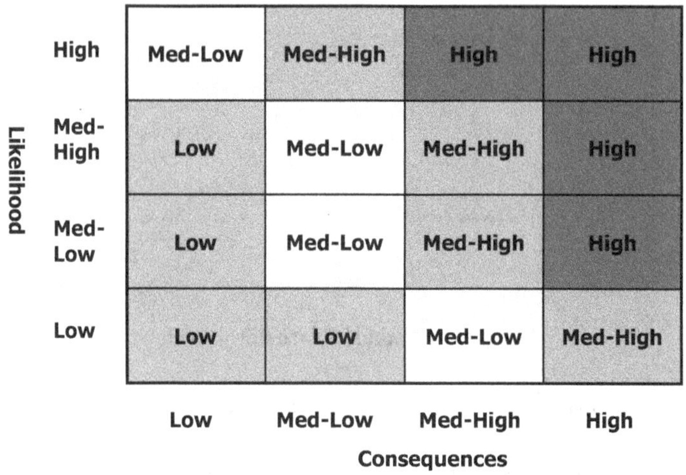

Figure 10-18. Illustration of 4 × 4 RAM

For example, the risk matrix in Figure 10-18 would most likely be one arrived at by SME's acting on behalf of an operator more averse[5] to adverse consequences than to high probabilities of failure.

The major advantages of the matrix methods include:

- Minimal reliance on software
- Leverages domain knowledge and experience of SME team
- Conceptually simple
- Flexible approach with wide application
- Simple graphical display of risk state

However, there are a number of disadvantages of the matrix method compared to other approaches, and these include the following:

- Highly subjective
- Susceptible to "group think"
- Inconsistent between assessments (different SME teams)
- Does not easily handle multiple consequences

Although matrix methods in principle provide adequate accuracy for development of pipeline integrity management plans, the disadvantages are generally found to outweigh the advantages. Since no automated or semi-automated framework is involved in the SME process, it is imperative to document the SME decisions clearly in the integrity management plan. The documentation should include the following elements:

- The names and functions of each member of the SME team
- The date(s) of the work session(s)
- A description of the scenario(s) being scored
- A summary of the basis for the scores developed
- The resulting risk matrix

[5] For a discussion on risk aversion, see Kumamoto and Henley (2000).

It should be noted that in addition to being a valid primary method of estimating risk level, the risk matrix is also a very useful tool for displaying risk results in general and will frequently be available in semi-QRA and QRA risk assessment software programs. The relatively course granularity of the classifications is in fact found to be beneficial in pipeline integrity management applications where frequently the decision to be made is whether to assess or intervene in segments immediately (high risk), or schedule assessment/intervention (medium-high, medium-low), or monitor and re-evaluated on the next application of risk assessment (low).

An example of the plotting of risk results on an RAM is shown in Figure 10-19. The risk assessment calculated the following risks for this scenario:

- Health and safety (HS):
  - Consequence = Level 2 (moderate)
  - Probability = C (rare)
  - Therefore, risk = Medium
- Environmental damage (ED):
  - Consequence = Level 3 (major)
  - Probability = D (probable)
  - Therefore, risk = Significant
- System interruption (SI):
  - Consequence = Level 4 (critical)
  - Probability = B (remote)
  - Therefore, risk = Medium

A decision model is required to make use of the risk rankings illustrated in Figure 10-19. This model in general comprises a set of predetermined actions to be taken to address each risk category. An example of such a decision model is given in Table 10-3.

Risk communication is addressed later in more detail, but it is worthwhile mentioning here that the use of the somewhat emotive labels for consequence shown in Figure 12

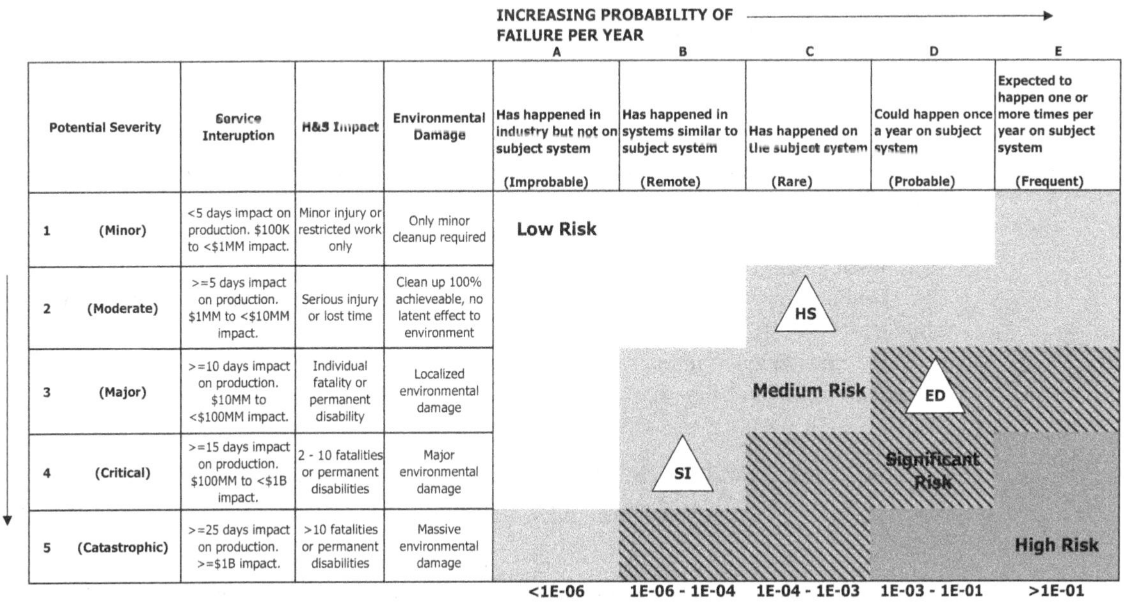

Figure 10-19. Practical example of RAM for major consequence categories

TABLE 10-3. Decision model example for assessed risk ranks

Risk Category	Definitions and Actions Required
Low	Low risk. Perform periodic monitoring of pipeline for applicable threats to verify status. Prepare and maintain planned preventative maintenance program.
Medium	Medium risk. Risk reduction measures to be developed and implemented within a stated time frame. Aim shall be to render subject pipelines low risk as far as reasonably practicable (ALARP)
Significant	Significant risk. Risk reduction measures to be identified and implemented with priority. Continued operation need to be reauthorized subject to preventative and/or remedial measures in place.
High	High Risk – Risk reduction measures to be evaluated and implemented with highest priority. Operations to be discontinued until preventative and/or remedial measures are in place. Authorization required to put asset back in service.

(Minor, Moderate, etc.) is contentious. Many people justifiably consider the possibility of even a single fatality to be "catastrophic" for example. The labels can be removed and the consequence severity levels referred to simply by number.

(b) Index Methods

Another important non-QRA approach is often referred to as the index method. In this approach, likelihood and consequences are usually expressed as weighted averages of variables that are in turn expressed as functions of component attributes. For example, the likelihood of failure due to external corrosion might be expressed as follows:

$$L\_EC = Sus\_EC * Sev\_EC \qquad (10-14)$$

where:

$$Sus\_EC \text{ (susceptibility to EC)} = 0.35 * ExtCoatType + 0.35 * GWCoatType + 0.15 * ExtCoatCond + 0.10 * FailEC + 0.05 * Age \qquad (10-15)$$

$$Sev\_EC \text{ (severity of EC)} = 0.40 * IA + 0.20 * IARes + 0.20 * PRL + 0.20 * FailEC \qquad (10-16)$$

Variables that might be used in the external susceptibility expression and the values that might be assigned to the attributes are given in Tables 10-4 and Table 10-5. These expressions, variables, and values are presented as illustrations of the methodology only. All such expressions, variables, and assigned values have to be developed specifically to suit individual pipeline configurations and operating environments.

Similar expressions can be derived for susceptibility and severity of all applicable threats.

The consequence index for societal impact for onshore gas lines might be given an expression similar to the following:

$$Q\_Soc = Prod\_Public * (25 * Radiation + 25 * ClassLoc + 25 * HighPop + 15 * NearPipeactivity + 10 * Cross) \qquad (10-17)$$

Variable that might be used in this expression and values that might be assigned to the attributes are presented in Table 10-4. Again, these expressions, variables, and values are given as examples of the methodology only.

TABLE 10-4. Variables and attribute values for external corrosion susceptibility

Variable Name	Variable Description	Attribute	Values
ExtCoatCond	External Coating Condition	Good condition	0.1
		Local disbonding / Damage	0.5
		Extensive disbonding / Damage	1
ExtCoatType	External Coating Type	FBE	0.1
		Three-Layer polyethylene	0.1
		Coal tar (mill applied)	0.2
		Extruded polyethylene	0.5
		Heat shrink	0.8
		Tape (Field applied)	1
CPAvail	Availability of CP System	High (>22 hrs)	0.1
		Medium (18 – 22 hrs)	0.5
		Low (<18 hrs)	1
CPEff	Efficiency of CP System	All sections protected to –950 mV	0.1
		All sections protected to –850 mV	0.2
		Some sections not protected to –850 mV	0.8
		Not protected	1
GWCond	GW Coating Conditon	Good condition	0.1
		Local dosbonding / Damage	0.5
		Extensive disbonding / Damage	1
GWCoatType	GW Coating Type	FBE	0.1
		Polyethylene	0.1
		Coal tar	0.2
		Tape / Heat shrink	1
FailEC	External Corrosion Failure History	No history of failures	0
		Evidences of EC not resulting in failure	0.2
		One failure (Leak)	0.6
		>One failure (Leak)	0.7
		One failure (Rupture)	0.8
		>One failure (Rupture)	1
Drainage	Drainage	Good	0.1
		Fair	0.5
		Poor	1
SoilType	Soil Type	Clay	1
		Rock	1
		Oher e.g. sand	0.2
Age	Age of Pipeline Section	(Current year – Commissioning year) / Max age	0 – 1

Once the likelihood indices for all applicable threats on the pipeline segment and consequence indices are calculated, the risk index for the segment for each threat can be calculated per Expression (1), and these risk indices added to give an overall risk index for the segment.

Many pipeline integrity management consultants have produced risk assessment methodologies and software products that can be incorporated in IM plans such as PII's PipeView^{TM} Integrity Risk (PVi), Dynamic Risk Assessment Systems Inc's RiskAnalyst^{TM}, and DNV's ORBIT^{TM} (all three of which form part of more comprehensive pipeline integrity management applications). One widely used non-QRA program representative of the index method is that developed by Muhlbauer (2004). Muhlbauer's approach assigns values to the following indices by evaluating the factors driving them and assigning an appropriate score:

- Third-party damage
- Corrosion
- Design
- Incorrect operations
- Leak impact factor

TABLE 10-5. Variables and attriute values for onshore gas societal impact expression

Variable Name	Variable Description	Options / Equation	Values
Prod_Public	Public Risk Product Factor	Natural Gas	0.5
		LPG	1
		NGL	0.7
		Propane	0.6
Radiation	Thermal Radiation following failure	OperatPress*(Diam/Max_Diam)*(Diam/Max_Diam)	0 – 1
		OperatPress – operating pressure	
		Diam – diameter	
		Max-Diam-maximum diameter	
ClassLoc	Class Location	Class 1 (<10 houses/mile)	0.1
		Class 2 (10–46 houses/mile)	0.3
		Class 3 (>46 houses/mile)	0.7
		Class 4 (multi-story buildings)	1
HighPop	Areas of High Population	Yes – Many	1
		Yes – One	0.8
		None	0
NearPipeactivity		None	0
		Low Risk of Activity	0.2
		Medium Risk of Activity	0.5
		High Risk of Activity	1
Cross	Number of Crossings	Normalized against maximum number of crossings	0 – 1

For example, the Third-party Damage index is evaluated on the basis of the following factors and assigned the following scores and weights:

- Minimum depth of cover 0 to 20 points 20%
- Activity level 0 to 20 points 20%
- Above-ground facilities 0-10 points 10%
- Line locating 0 to 15 points 15%
- Public education programs 0 to 15 points 15%
- Right–of-way programs 0 to 5 points 5%
- Patrol frequency 0 to 15 points 15%

These factors are evaluated qualitatively by considering the following driver descriptions:

Minimum depth of cover
- Soil cover
- Type of soil
- Pavement type
- Warning tape or mesh
- Water depth

Activity level
- Population density
- Stability of the area
- One-call system
- Other buried utilities
- Anchoring, dredging

Aboveground facilities
- Vulnerability
- Threats (traffic volume, etc.)

One-call system
- Mandated
- Response by owner
- Well-known and used

Public education	- Methods - Frequency
Right-of-way condition	- Signs - Markers - Overgrowth - Undergrowth
Patrol	- Ground patrol frequency - Ground patrol effectiveness - Air patrol frequency - Air patrol effectiveness

In addition to presenting a useful stand-alone risk management methodology specifically tailored for pipeline systems, Muhlbauer's book (2004) is a very useful risk management reference and information source (as the sub-title implies — Ideas, Techniques, and Resources). It should be noted that WKM Consultancy continues to develop its risk assessment methodologies (Muhlbauer 2006a, 2006b). The foregoing example, however, provides a good illustration of the practical operation of the index method.

(c) Model-Based Non-QRA

The non-QRA model-based models are similar to the non-QRA index methods described above except that the indices are calculated on the basis of semi-rigorous cause and effect relationships relating likelihood and consequence of failure to variables and attributes instead of the weighted average approach described above.

That is, Expressions (10-6), (10-7), and (10-8) are replaced by the following functions:

$$L = L(\text{ExtCoatType, GWCoatType, ExtCoatCond, FailEC, Age, IA, IARes, PRL, FailEC}) \quad (10-18)$$

$$\text{Q\_Soc} = \text{Q\_Soc(Prod\_Public, Radiation, ClassLoc, HighPop, NearPipeactivity, Cross)} \quad (10-19)$$

The precise formulations of the functions in Eqs. (10-18) and (10-19) tend to be considered intellectual property by the risk assessment software vendors. The functions generally take the form of algorithms making use of damage mechanism models derived from technical literature. For example, the likelihood of failure from internal corrosion in a sour gas line, might be derived from the NACE MR 0175 (2001) or the European Federation of Corrosion publication 16 (1995) (illustrated in Figure 10-20).

To construct an expression for an internal corrosion likelihood index based on EFC Publication 16, for example, one might relate the potential damage rate to the free water content in the product and the service domain shown in Figure 10-19. The inputs to such a model would be the partial pressure of H_2S in the product and the in-situ pH (possibly estimated from the CO_2 partial pressure interaction).

With respect to sweet service internal corrosion models, these might be based on the familiar de Waard and Milliams nomograph (1975) or the improved predictive model by de Waard et al. (1995).

The important point to stress with respect to non-QRA model-based methods is that, in common with non-QRA index methods, the quantities that represent likelihood of failure and consequence of failure are indices that only have meaning relative to one another. This is quite different from QRA methods in which the probabilities are true probabilities in the classical, statistical sense, and consequences are estimates (ultimately expressible in

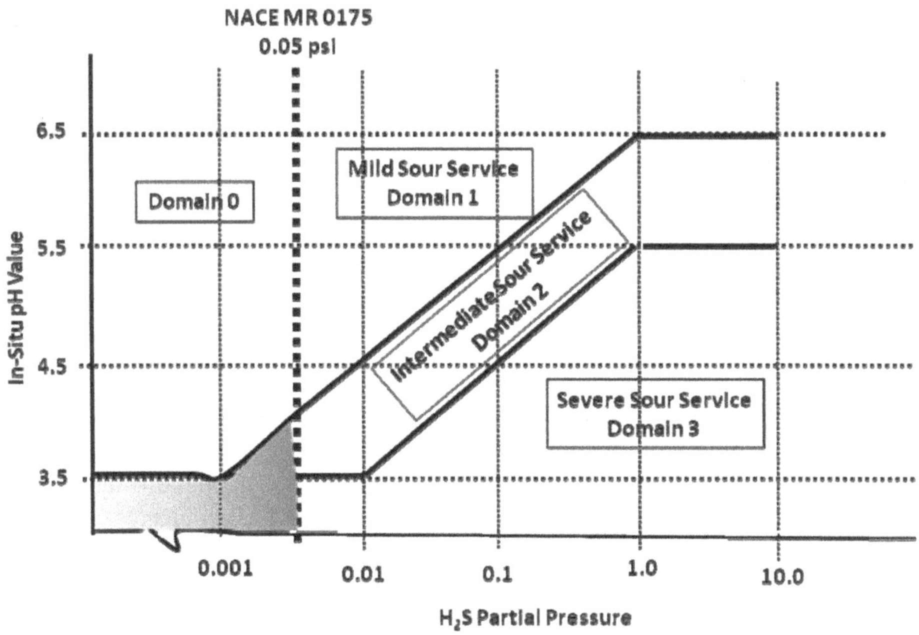

Figure 10-20. Use of sour domains per EFC publication 16 for estimation of sour gas internal corrosion susceptibility

monetary terms) of the expected impact of failure. The calculation flow for these indices is illustrated in Figure 10-21.

The major advantages of the non-QRA index and model-based methods include:

- Flexibility. They are able to address a wide range of application and risks.
- Repeatable/objective. They are largely insensitive to analysis team preconceptions.
- Readily programmed is designed for ease of application and integration with all other aspects of integrity management planning.
- Models need not be totally rigorous. They only need be as accurate as necessary to give a defensible risk ranking order.

Disadvantages associated with the methods include the following:

- Require software to implement
- Need/rely on approximate cause-effect relationships (model-based methods)
- Generally more complex than other non-QRA methods (such as SME-based matrix methods)

On balance, the advantages of the non-QRA index and model-based methods are found to significantly outweigh the disadvantages. These methods therefore tend to be the most common methods used in industry for system-wide risk assessment.

Selection of Risk Assessment Method

Role of risk assessment in integrity management planning describes the broad classification of risk assessment methods. Risk assessment as part of an integrity management plan describes how risk assessment is used to manage pipeline risk. The selection of which risk

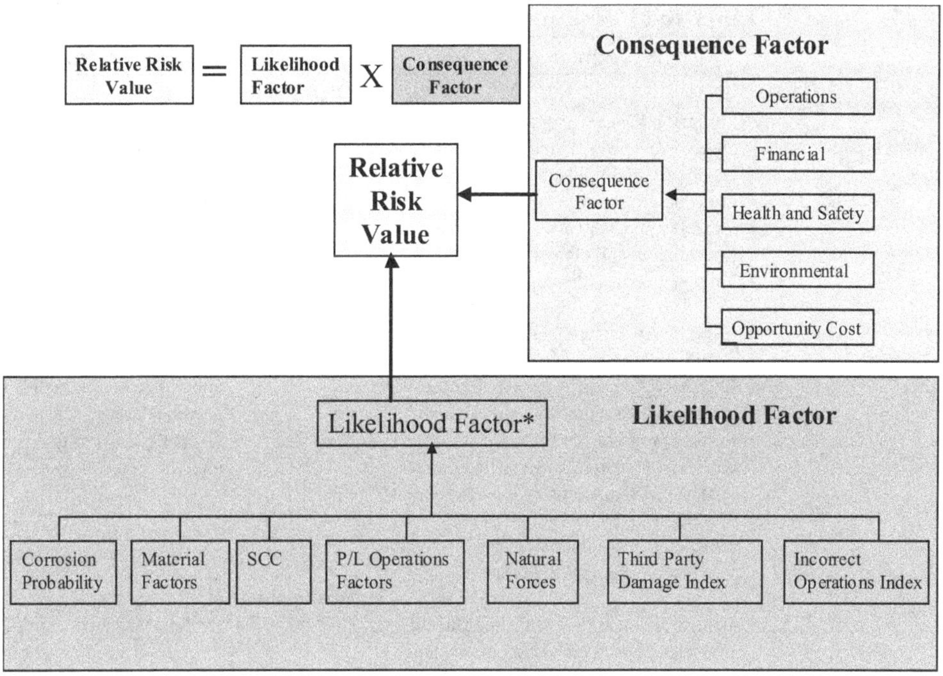

Figure 10-21. Calculation of relative risk indices

assessment method to use relates in general to the data available to perform the risk assessment and the precise use to which the results of the risk assessment are to put. Table 10-6 lists the principle advantages and disadvantages of each method and can be used therefore as a guide to selecting the most appropriate method.

The following suggested selection strategy is based on the advantage/disadvantage summary presented above in Table 10-6.

Non-QRA matrix methods:

- Although matrix methods in principle provide adequate accuracy for the development of practical integrity management plans, the disadvantages are generally found to outweigh the advantages
- Use only as an interim method until other less subjective methods are developed
- Use as a sanity check on results from other methods through review of results presented in RAM form

Non-QRA index and model methods:

- These are the most widely used methods at time of writing.
- They represent the minimum effort required to obtain objective, repeatable risk assessment results.
- They provide entirely adequate risk estimates for all pipeline integrity management purposes. However, these estimates are generally in the form of indices rather than true risk values in the classical sense.

Deterministic QRA methods:

- It is used if inconsistencies are found in the application of non-QRA methods
- It is used if risk results are required for comparison with other enterprise-wide risks (such as corporate financial risk and HSE quantitative risk requirements)

TABLE 10-6. Risk assessment methods comparison

Advantages	Disadvantages
(a) Non-QRA matrix methods:	
• Minimal reliance on software • Directly leverages domain knowledge and experience of SME team • Conceptually simple • Flexible approach with wide application (including nonpipeline applications) • Simple, graphical display of risk results	• Highly subjective • Susceptible to "group think" • Inconsistent between assessments (different SME teams) • Does not easily handle multiple consequences
(b) Non-QRA index and model-based methods	
• Flexible. Able to address a wide range of facilities and risks • Repeatable/objective. Largely insensitive to assessment team preconceptions • Readily programmed for ease of application and integration with all other aspects of IM planning • Models need not be totally rigorous, but rather only as accurate as necessary to give a defensible risk ranking order	• Requires software to implement • Requires the development of at least approximate cause-effect relationships between risk and damage considerations • Generally more complex than other non-QRA methods (such as SME-based matrix methods)
(c) Deterministic QRA	
• Describes risk in terms of deterministic estimates of probability and consequence of failure (no need for probability density distributions) • Follows common rules of statistics thereby minimizing the number of arbitrary assumptions associated with other methods • Uses information commonly available for pipeline integrity management • Generally more accurate that non-QRA methods	• Requires software to implement • Requires the development of cause-effect relationships between risk and damage considerations — generally more complex than those associated with non-QRA methods • The apparent precision my be misleading and not proportionately beneficial to integrity management planning • Generally more complex than non-QRA methods
(d) Reliability-based QRA	
• Engages all of the functionality of PRAM • Virtually eliminates the need for arbitrary assumptions • The most rigorous and accurate of all risk assessment methods • Provides the best estimate of residual risk (following the implementation of risk mitigation and prevention methods)	• Requires knowledge of the distribution of risk assessment parameters • Provides a level of precision in excess of that generally attainable in other elements of risk management (description of corrosion characteristics and soil types for example) • The apparent precision my be misleading and not proportionately beneficial to integrity management planning • Requires specific information from each assessment site making it highly data-intensive for system-wide risk assessment applications

- Since these methods use essentially the same inputs as those required for non-QRA index/model methods, they are suited to system-wide risk assessments that provide risk profiles for the subject pipeline system.
- There is evidence at time of writing that these methods may soon become industry standard.

Reliability-based QRA methods:

- These are best suited to site-specific assessments due to the need for extensive and higher quality site-specific information that will generally require site-specific investigations before the risk assessment can be undertaken.

- It is used on a site-specific basis if a defensible estimate is required of probability of exceedance of an established criterion is required.
- It is used on a reduced scale for system-wide risk assessment to calibrate deterministic-QRA methods.

RISK MITIGATION MEASURES

No matter what type of risk assessment methodology is used or what risk acceptance criteria are adopted, the last action in the cycle of risk management is risk control/mitigation. This step is represented in the integrity management process as risk mitigation and risk prevention measures. The precise measures of course depend heavily on the threats to be mitigated and the specifics of the pipeline systems.

The following list of possible risk control/mitigation measures (presented here as an illustration) relates to risk of mechanical damage resulting from form third-party interference (generally found to be the most frequent cause of failure in pipeline systems):

- Greater wall thickness (reduced chance of puncture and rupture)
- Higher material grade (reduced chance of rupture)
- Deeper burial (say to depths >1.5 m)
- Physical protection (such as the use of concrete slabs)
- Improved right of way marking
- Increased right of way surveillance (best practice is daily)
- Control of activities on and near right of way
- Safety management procedures

RISK ACCEPTANCE CRITERIA AND COMMUNICATION

One of the apparent strengths of QRA methodology is that it produces "real" numeric estimates of risk and POF. The use of these estimates, however, raises other issues that have to be addressed. Since these estimates can be compared with other daily risks it is logical that operators, members of the general public, and regulators will examine QRA pipeline risks in this light thereby requiring criteria upon which to accept or reject the risks. The acceptance criteria depend largely on the type of risk and the parties subjected to the risk versus the parties apparently benefiting from the risk.

For example, risk fatality or injury can be broadly classified into 2 types as follows:

- *Individual risk.* This is the risk at a given location to a hypothetical person located there unprotected 24-7 (24 hours a day, 7 days a week, 365 days a year, year-in, year-out). The concept of individual risk is somewhat academic, but it provides a very useful way of comparing and communicating risk. Individual risk can be defined as the frequency at which an individual may be expected to sustain a given level of harm from the realization of specified threats.
- *Societal risk.* This is the risk to groups of people who might from time to time be around the pipeline and therefore might be impacted by a pipeline failure. Societal risk criteria are predicated for the most part on increasing risk aversion in the case of multiple fatalities. Societal risk can be defined as the relationship between frequency and the number of people in a given population suffering from a specified level of harm from the realization of a given threat.

Typical criteria for individual risk are given in Figure 10-22.

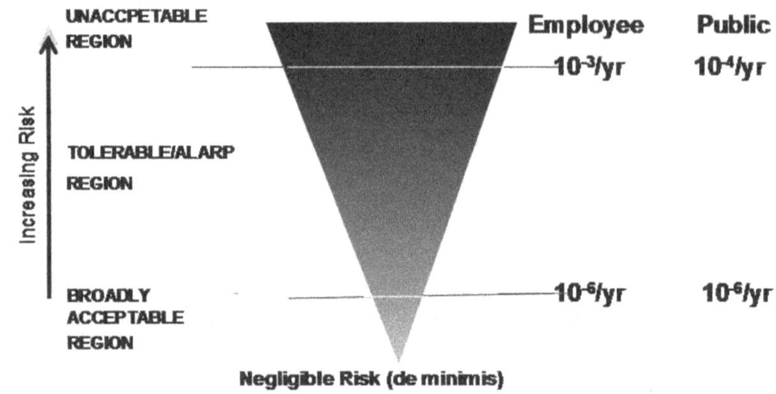

Figure 10-22. Typical criteria for individual risk

Recognizing risk aversion associated with multiple fatalities, societal risk acceptance criteria are usually expressed in terms of *F–N* relationships. That is, the tolerable fatality frequency (probability) *F*, for an expected number of fatalities, *N*. At typical *F–N* curve is given for illustration in Figure 10-23.

Figure 10-23 is taken from UK Health and Safety Executive (HSE), "Typical *F-N* curve. Hydrogen application. Risk acceptance criteria. Risk assessment methodology" (UK Health and Safety Executive (HSE)). It has to be emphasized that the criteria presented here are for illustration of the methodology only and in no way are to be interpreted as approval endorsement. Other *F-N* results are available which differ significantly from that shown in Figure 10-23 (see, for example, Krichsteiger (1998) and Muhlbauer (2004)).

In communicating risk, it is useful to keep in mind the following.

- Risk is regarded as more acceptable if (the cause or source) is:
  - Beneficial to the individual
  - Beneficial to the community
  - Adding economic benefit
  - Controlled by self, rather than others

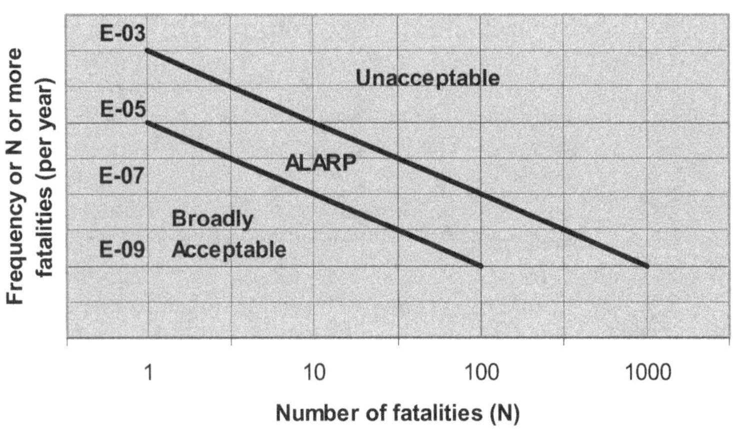

Figure 10-23. Typical *F–N* risk acceptance relationships for societal risk

- Risk is regarded as less acceptable if (the cause or source) is:
  - Involuntary rather than voluntary
  - Uncontrolled rather than controlled
  - Has potential for multiple injuries or fatalities in a single incident

In communicating risk levels and control measures generally, it is also worthwhile referring to comparative statistics such as the following (nominal annual individual risk of death in the UK):

Cause	Risk
Severe illness	3 in 1,000
Road accident	1 in 10,000
Lightning strike	1 in 10,000,000
Mining industry	1 in 10,000
Construction industry	0.9 in 10,000
Offices, shops	5 in 1,000,000
Rock climbing	8 in 1,000
Hang gliding	1.5 in 1,000
Gas pipeline failure	<1 in 1,000,000

It may be of limited comfort to members of the general public who might be affected by a pipeline failure to note that the risk is low, but it is worthwhile emphasizing and re-emphasizing that provided the procedures described in this section are performed properly and correct action is taken to implement them into the pipeline integrity management plan, the risk of mishap from a pipeline failure is effectively zero.

PIPELINE SECURITY RISK

Generally, pipeline networks are widespread, running alternately through remote and densely populated regions and areas with known security concerns. Consequently, these systems are inherently vulnerable to acts of vandalism and sabotage attack. Whereas pipeline operators have had security and emergency response programs in place for decades, there appears to be a growing awareness of the need to extend and formalize these to in response to increasing threats to security.

Taken as a whole, releases from pipelines cause relatively few annual fatalities. In the United States, for example, oil pipelines reported an average of 1.4 deaths per year from 1997 to 2001. Gas pipelines reported an average of 18.6 deaths per year during the same period (Trench 2009). It is difficult to make direct safety comparisons between pipelines and other transportation modes due to data limitations. Nonetheless, US DOT Office of Pipeline Safety (OPS) statistics suggest that pipelines have much lower fatalities per ton-mile of general freight moved than truck, rail, or waterborne transport. In general, the environmental safety record of oil pipelines is comparable to other transportation modes.

Accidental pipeline releases result from a variety of causes, including outside force (e.g., third-party excavation), corrosion, mechanical failure, control system failure and operator error. Natural forces, such as floods and earthquakes, can also damage pipelines. For United States, according to the DOT, of 183 gas pipeline accidents reported in 2002, outside forces were by far the leading cause, accounting for 46% of reported failures. Outside forces was also the leading cause of the 140 oil pipeline accidents in 2002, responsible for 32% of failures. (Bronzini and Goode 2008).

Pipeline security threats for the most part comprise:

- Acts of vandalism
- Illegal tapping off of product
- Terrorist attacks

It is not generally found feasible to address these threats in the same way as other threats to integrity since to a large extent they are random in nature and no well-established causal relationships appear to exist. Nonetheless, they are very real threats and have to be accounted for in the integrity management plan.

The physical plant associated with pipeline facilities may be damaged with explosives or by other mechanical means, disrupting flows and causing a release of pipeline contents. They are also vulnerable to "cyber-attack," or both physical and cyber attack. Some pipelines may also be indirectly disrupted by other types of terror strikes, such as attacks on regional electricity grids or telecommunications networks, which could in turn affect dependent pipeline control and safety systems.

Since pipelines supply fuel for vehicles, power plants, aircraft, heating, military bases, and other uses, serious disruption of a pipeline network poses additional "downstream" risks. Oil and gas pipelines have been a favored target of terrorists in the past several years. In Colombia, for example, Occidental Petroleum's Caño Limón pipeline has been bombed some 950 times since 1986, shutting it for months at a time and costing Colombia's government some $2.5 billion in lost revenues. One of these attacks in 1998 caused a fire and several fatalities (United States Government Accountability Office 2005).

There are other reported cases elsewhere:

- 1996, London, foiled plot to bomb gas pipelines and other utilities (36 explosive devices found).
- 1997, USA, Texas, foiled bombing of natural gas storage tanks at a processing plant
- 1999, Canada, Vancouver, foiled planning to blow up the trans-Alaska pipeline (with high explosives and timers for bombs)
- 2001, Alaska, high-powered rifle damage to the trans-Alaska pipeline causing shutdown and extensive economic and ecological damage.
- 2002, Saudi Arabia, foiled planned pipeline/oil terminal attack
- 2002-2004, Nigeria, Pakistan, Sudan, Myanmar, and Iraq, attacks on oil and gas pipelines.

Pipeline Safety and Security Regulation

Generally, various pipeline acts such as the US Natural Gas Pipeline Safety Act of 1968 (P.L. 90-481) (United States, Congress, House Committee on Interstate and Foreign Commerce 2006) and the Hazardous Liquid Pipeline Act of 1979 (P.L. 96-129) (US DOT PHMSA) specify government roles in pipeline safety. Under these statutes, appropriate government safety boards are given primary authority to regulate key aspects of pipeline safety: design, construction, operation and maintenance, and spill response and release planning. For example, in the US under Pipeline safety regulations, some with security implications are covered in Title 49 of the Code of Federal Regulations. This title:

- Authorizes DOT to inspect pipelines and enforce its regulations with fines, injunctions, and criminal penalties
- Requires operators to report incidents, safety-related conditions and annual summary data
- Prescribes minimum pipeline safety requirements, including operator qualifications for regulated functions
- Imposes oil spill response plan requirements to reduce environmental impact of accidental discharges
- Provides grants-in-aid to state pipeline safety compliance agencies that adopt damage prevention programs
- Requires operators to establish drug and alcohol programs.

In recent years, most jurisdictions worldwide have updated their regulations to include better/improved pipeline safety including terrorist threats. For example, US Pipeline Safety Improvement Act of 2002 (P.L. 107-355) (US DOL 2002) strengthens federal pipeline safety programs, state oversight of pipeline operators, and public education regarding pipeline safety. The integrity management rule as described in the rule for oil pipelines states that "identifying the need for additional preventive and mitigative measures, an operator must evaluate the likelihood of a pipeline release occurring and how a release could affect the HCA. This determination must consider all relevant risk factors..." Guidance for this rule also cites "security of throughput (effects on customers if there is failure requiring shutdown)" as a risk factor for establishing assessment frequency.

Likewise, system safety inspections requirement of the updated act are intended to provide data useful for related security inspections. For example:

- Development of integrity management analysis and program standards by DOT within 12 months of enactment, and gas pipeline operator implementation within 24 months of enactment
- Baseline integrity assessments of all HCA gas pipelines within 10 years, with re-inspections every 7 years thereafter
- US DOT to order safety actions for pipelines with potential safety problems and increases violation penalties.

The criminal provisions under 49USC60123 (US Department of Transportation 2002) as amended by P.L. 107-355 state that: "A person knowingly and willfully damaging or destroying, or attempting to damage or destroy, an interstate gas pipeline facility or interstate hazardous liquid pipeline facility shall be fined..., imprisoned for not more than 15 years, or both."

The provisions of P.L. 107-355 place significant new demands on the US DOT OPS. The act required gas pipeline operators with facilities in HCAs to adopt new integrity management program plans (by 2005), and to begin pipeline integrity assessments by mid-2006.

Pipeline Security Response

Pipeline operators have always sought to secure their systems. While their security programs traditionally tended to focus on personnel safety and preventing vandalism, some have been more comprehensive. For example, security at the trans-Alaska pipeline during the Gulf War included measures such as armed guards, controlled access, intrusion detection and dedicated communications at key facilities, as well as aerial and ground surveillance of the pipeline corridor. The surveillance program of one operator in Southeast Asia, for example, involves walking the 1200 km of line daily. A total of 120 people walk 10 km of the pipeline each day. This is a line particularly susceptible to illegal taps. The operator has also set up guard stations at strategic points and these are manned on a 24-7 basis.

The events of September 11, 2001 focused attention on the vulnerability of pipelines to different terrorist threats. In particular, the terrorist attacks raised the possibility of systematic attacks on pipelines by sophisticated terror groups in a manner that had not been widely anticipated before.

Since then various governmental jurisdictions and security departments have devised increased security measures and began identifying additional ways to deal with terrorist threats. For example, through the Interstate Natural Gas Association of America, formed a security task force to coordinate and oversee the industry's security efforts.

As a consequence of above security measure, a variety of industry security programs came under scrutiny by various jurisdictions. This scrutiny included developing common risk-based practices for incident deterrence, preparation, detection, and recovery. These assessments address issues such as:

- Spare parts exchange
- Critical parts inventory systems
- Security communications with emergency agencies
- Emergency response

The natural gas pipeline transportations companies reported significant commitments to bolster security at their critical facilities including:

- Strengthened emergency
- Contingency and business continuity plans
- Increased liaison with law enforcement
- Increased monitoring of visitors and vehicles on pipeline property
- Monitored pipeline flows and pressure on a continuous basis
- Increased employee awareness to security concerns
- Deployment of additional security personnel
- Encryption protocol standards to protect gas systems from cyber attack
- Incorporating redundancy in the delivery system to provide greater flexibility to redirect or shut down product flows
- Information sharing and analysis
- Better physical security
- Implementation of real-time threat alerts, cyber alerts and solutions
- Sharing of information between government and law enforcement agencies, technology providers, and other security associations

Pipeline Security Risk Assessment Plan (SRAP)

The practical method to treat security risk threats in integrity management is through a written and implemented pipeline SRAP. According to Borysiewicz (Borysiewicz et al. 2004), a Security Risk Assessment (SRA) provides a means of improving the security of pipeline systems and allocating operator resources to effectively:

- Identify and analyze actual and potential precursor events that can result in pipeline security related incidents

- Identify the likelihood and consequence of potential pipeline security related events
- Provide a comprehensive and integrated means for examining and comparing the spectrum of risks and risk reduction activities available
- Provide a structured, easily communicated means for selecting and implementing risk reduction activities
- Establish and track plan performance with the goal of improving that performance
- Establish standardized security conditions and protective measures for a broad range of security threats

In developing this guideline on pipeline security, certain guiding principles underlie the entire document. These principles are reflected in many of the sections and are provided here to give the reader the sense of the need to view pipeline security from a broad perspective.

- A pipeline security plan must be flexible.
- The integration of information is a key component in managing a pipeline security plan.
- Preparing for and conducting a SRA in managing pipeline system security.

SRA is an analytical process through which an operator determines the types of threats, events, or conditions which might impact pipeline security, the likelihood that these events or conditions will lead to a security related event, and the nature and severity of the consequences that might occur following an event. This analytical process involves the integration and analysis of the pipeline system and its facilities, the environment in which the pipeline operates, and risk mitigation methods available to the pipeline operator. SRA can have varying scopes, varying levels of detail, and utilize different methods. However, the ultimate goal of assessing security risks is to identify and prioritize the most significant risks so that an operator can make informed decisions about the security risks.

Pipeline SRAP Essential Elements

The SRAP has essentially the same structure as the basic pipeline integrity plan and therefore has a similar framework. Although all pipeline systems have design features and operating characteristics that are unique to each system, an effective pipeline security plan should have a solid base of common essential elements. This section describes a program that includes these essential elements and provides the basis for SRAP guidelines development. It is important to recognize that a pipeline security plan will be integrated with other operations processes (including the integrity management plan) and will change in an iterative manner as conditions change. Although the elements depicted in Figure 10-19 are shown sequentially for ease in illustration, in reality there is a significant information flow and interaction between the individual steps.

Initial Data Gathering. The first step in understanding the potential risks along the pipeline system is to assemble information about potential risks. In this element, the operator performs the initial collection, review, and integration of data that is needed to understand location specific risks to the pipeline security.

Initial SRA. In this element, the data assembled from the previous step is used to conduct a SRA of the pipeline system. The SRA begins with a systematic and comprehensive search to identify possible security risks to the pipeline system. The identification of potential risks should not be limited to a review of known risk categories, but should also include steps to look for new or unique manifestations of risks.

There is a significant variation in the detail and complexity associated with different SRA methods. Some Owner/Operators have found that an initial screening level SRA can be beneficial in terms of focusing resources on the most important areas. During a screening

SRA, an operator may limit the scope of the system to those portions of the system where a failure could have the most severe consequences.

The quantitative SRA methods are those where the characteristics of segments of the pipeline and the surrounding area are used to derive an actual estimate of the risk for that segment. Likelihood is estimated as the probability of a security related event along the segment over a given period of time. Actual expected levels of consequences in different categories (human, environmental, economic) are estimated and may be combined using some common metric (e.g., equivalent dollar cost).

After identifying the most significant risks on the system, the next step is to determine what mitigation actions or security measures might be desirable to reduce risk, and where assessment techniques such as facility or pipeline security assessments would be of the most value in identifying potential risk-threatening issues. The risk control and mitigation process involves:

- identification of risk control options that lower the likelihood of a pipeline system incident, reduce the consequences, or both;
- a systematic evaluation and comparison of those options; and
- selection and implementation of the optimum strategy for risk control.

Develop Security Plan. Using the output of the SRA, a plan is developed and maintained to address the most significant risks and assess the security of the pipeline system. This plan should include the mitigation risk control actions, as well as security assessment activities, e.g., inspections such as ROW patrol and traffic and personnel control. It provides a description of the various risk control options available, guidance to assist Owner/Operators in selecting a security assessment method, establishing a schedule for periodic security assessments, and employment of security mitigations.

Employ Security Mitigations. In this element, the security plan activities are implemented, the results are evaluated, and the necessary changes are made to assure risks that might lead to pipeline system failures are controlled.

Update, Integrate, and Review Data. After the initial security assessments have been performed, the operator has improved and updated information about the security of the pipeline system.

Pipeline Security Risk Reassessment. SRAs should be performed periodically to factor in recent operating data or threat intelligence; consider changes to the pipeline system design, e.g., new IT systems and new pipeline segments or facilities; and to analyze the impact of any external changes that may have occurred since the last SRA, e.g., adjacent facilities and changes in traffic flow.

Revise Plan. The security plan should be transformed into an on-going security assessment plan that is periodically updated to reflect new information and the current understanding of security risks.

Audit Plan. The operator should collect information and periodically evaluate the success of its security assessment techniques and other mitigation risk control activities.

Managing Change. Pipeline systems and the environment in which they operate are never static. A systematic process should be used to ensure that changes to the pipeline system are evaluated for their potential risk impacts prior to implementing the changes, and to ensure that changes in the environment in which the pipeline operates are evaluated.

As this final element indicates, managing pipeline security is not a one-time process. As implied by the loop in the lower portion of Figure 10-24 below, a security plan involves a continuous cycle of monitoring pipeline conditions, identifying and assessing risks, and taking action to minimize the most significant risks. SRAs must be periodically updated and revised to reflect current pipeline conditions so Owner/Operators can most effectively use their limited resources to achieve the goal of controlling risks and minimizing their impact.

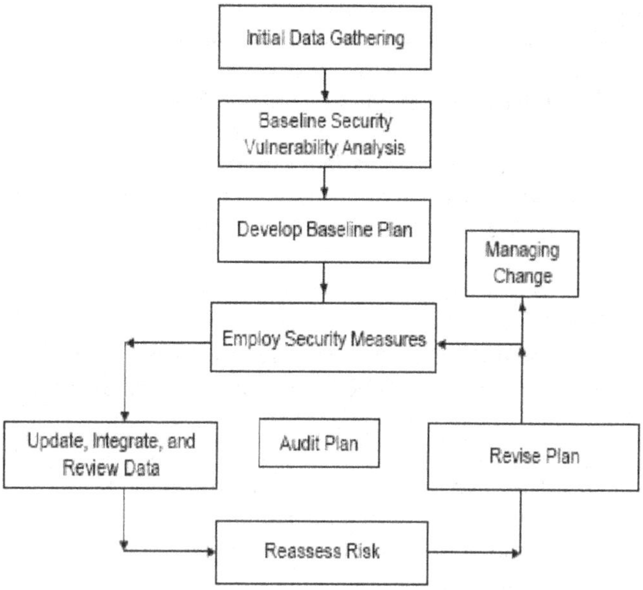

Figure 10-24. Framework for SRAP

When assessing the risks of a single large asset such as a cross-country pipeline, the system may be divided into geographical segments to compare the risks of respective pipeline segments to determine how to allocate resources across the pipeline system. The operator would decide how long the segments will be and the logical location of boundaries between segments. Factors that drive these decisions include:

- Scope of the SRA: that is, which assets are included/excluded from the assessment
- Equipment boundaries such as compressor/meter stations or block valves
- Geographical boundaries such as state lines or rivers
- Desired minimum/maximum length of any one segment, e.g., kilometer by kilometer
- How system databases are set up and organized; this is important since data will be transferred from one or several databases into the SRA method
- Operation changes, e.g., region, gas composition, pressures, and volumes
- Population density changes
- The presence of environmentally sensitive or population sensitive areas, e.g., HCA's, schools, water systems, waterways, third party facilities, and government installations

After completing the SRA for each pipeline segment, the results can be used to analyze risk factors in many different ways. First, the individual segments can be ranked: by total risk level, by individual likelihood category, or by consequence level. A varying risk profile along the pipeline system can be created, highlighting areas susceptible to particular risks. These rankings can be used by an operator to focus attention on potential high-risk areas. A number of comparative analyses can be performed, such as:

- Comparison of risks from different security risks along the pipeline
- Comparison of pipeline risks by geographic region
- Comparison of different pipeline system risks within a company
- Comparison of a pipeline risk profile with a predefined standard, such as compliance with regulatory directives or an operator defined standard

SUMMARY

Albert Einstein is reported as saying (Einstein 1934):

> "It can scarcely be denied that the supreme goal of all theory is to make the irreducible basic elements as simple and as few as possible without having to surrender the adequate representation of a single datum of experience."

This statement is often reduced to "Make everything as simple as possible, but no simpler." On that principle, the optimum choice of system-wide risk assessment method to be used as an integral part of pipeline integrity management planning is most likely index or model-based non-QRA or possibly deterministic QRA. Since these methods essentially comprises risk-based integrity assessment planning in which the integrity of the pipeline system is ultimately determined through physical integrity assessments, a fairly course assessment granularity is tolerable. Being formalized into computer software for the most part, the methods have the attraction of applying the operator's risk criteria on a consistent and impartial basis on every application.

If, however, site-specific data are available and site-specific risk assessments are require, or if quantitative risk results are needed for comparison or incorporation with other enterprise risk analyses, then a totally quantitative method will be more useful and defensible (either deterministic or reliability-based). Such a site-specific QRA assessment can be considered as a stand-alone study. Although generally not suited to routine application to the integrity management process, system wide QRA (either deterministic or reliability-based) can be incorporated into the integrity management plan if desired and appropriate data are available. The use of QRA methods is particularly applicable to situations where a quantitative estimate of risk reduction due to integrity management program actions is required.

Whether QRA or non-QRA methods are used, a measure of risk associated with potential pipeline failures is the optimum and most impartial criterion upon which to make operations and maintenance decisions affecting the integrity of pipeline systems. Adopting risk as the basic criterion avoids emotive attachment to actions designed purely to reduce the probability of pipeline failures or purely to reduce the consequences of pipeline failures.

REFERENCES

ANSI/ASME B31G, 1991, "Manual for Determining the Remaining Strength of Corroded Pipelines", American Society of Mechanical Engineers, January, 55 pp.

ANSI/ASME B31.8S (Supplement to ASME B31.8), 2004, "Managing System Integrity of Gas Pipelines", American Society of Mechanical Engineers.

Bedford T. and Cooke, R., 2001, "Probabilistic Risk Analysis. Foundations and Methods", Cambridge University Press, Cambridge.

Borysiewicz, M., Burgs, A., and Garanty, I., 2004, "Security Guidelines for Liquid Petroleum Pipelines," Centre of Excellence MANHAZ, Sect. 6.3, p. 35 of 40, http://manhaz.cyf.gov.pl/manhaz/warsztaty_11_2004/WP3/Wp3_ang/Security_guidelines_MB/material1_MB_security.pdf.

Bronzini, M. and Goode, M., 2008, "Hazmat and Pipelines, Transportation System Security and Safety," Fall George Mason University Department of Civil, Environmental & Infrastructure Engineering (CEIE), http://www.civil.gmu.edu/bronzini/080930%20Hazmat,%20Pipelines.pdf.

Colquhoun, I., Calvi, C., and MacPherson, H., 2006, "Management System Approach to Pipeline Integrity." Proceedings of IPC2006, International Pipeline Conference, Calgary, Alberta, September 25–29, Paper IPC 2006-10531.

Colquhoun, I. R., Choong, E., Kania, R., Gao, M., and Wickenhauser, P., 2002, "Rapid Analysis Tool for High-Level Risk Assessment Of Pipelines," International Pipeline Conference, Calgary, Alberta, Paper IPC02-27264.

Dawson, J., and Colquhoun, I., 2009, "Deterministic Quantitative Risk Assessment Development," Rio Pipeline Conference 2009, September 22–24, Paper No. IBP1070_09.

de Waard, C. and Milliams, D. E., 1975, "Carbonic Acid Corrosion of Steel," Corrosion, Vol. 31, pp. 131.

de Waard, C., Lotz, U., and Dugstad, A., 1995, "Influence of Liquid Flow Velocity on CO_2 Corrosion: a Semi empirical Model," CORROSION/95, Paper 128, Houston, TX.

DNV, 1999, "Recommended Practice RP-F101m," Corroded Pipelines.

Einstein, A., 1934, Journal of Philosophy of Science, Vol. 1, No. 2 (April), pp. 163–169.

Fenyvesi, L. L., Colquhoun, I., Kania, R., and Gu, B., 2004a, " A Risk Based Approach to Maintenance Planning Utilizing In-Line Inspection Data," International Pipeline Conference, Calgary, Alberta, October, 4–8, Paper No. IPC04-0178.

Fenyvesi, L., Lu, H., and Jack, T. R., 2004b, "Prediction of Corrosion Growth on Operating Pipelines," International Pipeline Conference, Calgary, Alberta, October 4–8, Paper No. IPC04-0268.

Harte, J., 1988, *Consider a Spherical Cow*, University Science Books, Sausalito, CA.

Jayjock, M. A., 2004, "Human Health Risk Assessment Process," PhACT, February 21.

Kelly, K. A. and Cardon, N. C., 1994, "The Myth of 10^{-6} as a Definition of Acceptance Criteria," EPA Watch Newsletter, Vol. 3, No. 17.

Kiefner, J. F., Kiefner, B. A., and Vieth, P. H., 1999, "Analysis of DOT — Reportable Incidents for Hazardous Liquid Pipelines — 1986 Through 1996," Final Report, The American Petroleum Institute, January 7.

Kiefner, J. F. and Vieth, P. H., 1989, "A modified Criterion for Evaluating the Remaining Strength of Corroded Pipe," AGA Catalog No L51609, PR3-805, December 22.

Krichsteiger, Ch. (ed.), 1998, *Risk Assessment and Management in the Context of the Seveso II Directive*, Elsevier, New York.

Kumamoto, H. and Henley, E. J., 2000, *Probabilistic Risk Assessment and Management for Engineers and Scientists*, 2nd ed., Wiley-IEEE Press, New York.

Lapaz, L., 1958, "Effects of Meteorites on the Earth," Advances in Geophysics, Vol. 4, pp 217–350.

Legat, A. and Dolecek, V., 1995, "Chaotic Analysis of Electrochemical Noise Measurement on Stainless Steels." Journal of the Electrochemical Society, Vol. 142, pp. 1851.

Leis, B. N. and Stephens, D. R., 1997. "An Alternative Approach to Assess the Integrity of Corroded Line Pipe — Part 1: Current Status, and Part 2: Alternative Criterion." Proceedings of the 7th International Offshore and Polar Engineering Conference, Honolulu, HI, May.

Muhlbauer, W. K., 2004, *Pipeline Risk Management Manual. Ideas, Techniques, and Resources*, 3rd ed., Elsevier Inc. Gulf Publishing, New York.

Muhlbauer, W. K., 2006a, "Enhanced Pipeline Risk Assessment. Part 1 — Probability of Failure Assessments Revision 2.1," KM Consultancy, WKM Consultancy Web site.

Muhlbauer, W. K., 2006b, "Enhanced Pipeline Risk Assessment. Part 2 — Pipeline Failure Consequences. Revision 1," WKM Consultancy, WKM Consultancy Web site.

NACE Standard MR175, 2001, "Metals for Sulfide Stress Cracking and Stress Corrosion Cracking Resistance in Sour Oilfield Environments."

Nessim, M. A. and Stephens, M. J., 1995a, "Optimization of Pipeline Integrity Maintenance," OMAE ASME. Paper No. OMAE-95-900.

Nessim, M. A. and Stephens, M. J., 1995b, "Optimization of Pipeline Integrity Maintenance Based on Quantitative Risk Analysis," Presented at the Reliability Conference, September 12–14, 1995.

Pipeline Operators Forum, 2005, "Specification and Requirements for Intelligent Pig Inspection Of Pipelines," Version 3.2, January.

Rizkalla, M., 2008, Calgary, Alberta. Private communication.

Stephens, D. R. and Leis, B. N., 1997, "Material and Geometry Factors Controlling the Failure of Corrosion Defects in Piping," 1997 Pressure Vessels and Piping Conference, Orlando, Florida, July.

Stephens, D. R., Leis, B. N., Kurre, J. D., and Rudland, D. L., 1999, "Development of and Alternative Failure Criterion for Residual Strength of Corrosion Defects in Moderate to High Toughness Pipe," Battelle Report to PRC International Report, A.G.A. Catalog Number L511794, January.

Stephens, D. R., Leis, B. N., and Rudland, D. L., 1996, "Influence of Mechanical Properties and Irregular Geometry on Pipeline Corrosion Defect Behavior," 9th Symposium on Pipeline Research, Paper 34, PRCI/AGA Catalog No. L51746, September.

Sutherby, R., Fenyvesi, L., Colquhoun, I., and Rizkalla, M., 2000, "System Wide Risk-Based Integrity Program at TransCanada," International Pipeline Conference, Calgary, Alberta, Canada.

The Institute of Materials, European Federation of Corrosion, 1995, "Guidelines on Materials Requirements for Carbon and Low Alloy Steels for H_2S-containing Environments in Oil and Gas Production," Publication No. 16.

Trench, C. J., 2009 "10 years of learning from PPTS" PPTS Workshop, January, Houston, TX, Sect. 6, p32 of 40.

UK Health and Safety Executive (HSE), "Typical F-N curve. Hydrogen Application. Risk Acceptance Criteria. Risk Assessment Methodology" as reported by Gerd P. Haugom, DNV Company.

UKOPA Pipeline Fault Database, 2002, "Pipeline Product Loss Incidents (1961–2000)," 2nd Report of the UKOPA Fault Database Management Group.

United States, Congress, House Committee on Interstate and Foreign Commerce, 2006, *US Natural Gas Pipeline Safety Act of 1968*, U.S. Govt. Print. Office, Washington, DC.

United States Government Accountability Office, 2005, "Security Assistance — Efforts To Secure Colombia's Caño Limón-Coveñas Oil Pipeline Have Reduced Attacks, But Challenges Remain," September, www.Gao.Gov/Cgi-Bin/Getrpt?Gao-05-971.

US Department of Transportation, 2002, "Rule 49USC60123," Pipeline and Hazardous Materials Safety Administration, 49 CFR Parts 190, 191, 192, 193, 194, 195, 198, and 199," http://www.phmsa.dot.gov/staticfiles/PHMSA/DownloadableFiles/Files/192-100%20(no%20number).doc.

US DOL (US Department of Labor), 2002, "Pipeline Safety Improvement — Section 6 — Protection of Employees Providing Pipeline Safety Information," http://www.osha.gov/dep/oia/whistleblower/acts/psia.html.

US DOT PHMSA (Pipeline Hazardous Material Safety Administration), 2006, *Pipeline Inspection, Protection, Enforcement, and Safety Act of 2006.*

Vieth, P. H., Morris, W. G., Rosenfeld, M. J., and Kiefner, J. F., 1997, "DOT-Reportable Incident Data Review — Natural Gas Transmission and Gathering Systems — 1985 Through 1995," Final Report, Contract No. PR-218-9603, PRC International, September 19.

Wickenhauser, P. L. and Playdon, D. K., 2004, "Quantitative Pipeline Risk Assessment and Maintenance Optimization," International Pipeline Conference, Calgary, Alberta, Paper No. IPC04-0451.

Appendix

Proceedings of ASME IPC 2006:
5th International Pipeline Conference
September 25th-29th 2006, Calgary, Alberta, CANADA

IPC2006-10032 PIPELINING LIQUEFIED PETROLEUM GAS (LPG)

Mo Mohitpour, Ph.D, P.E. FASME

Tempsys Pipeline Solutions Inc.

Andrew Jenkins, P.Eng.

TransCanada PipeLines Limited

Tom Babuk, P.Eng.

Empress International Inc.

ABSTRACT

Liquefied Petroleum Gas (LPG) is a mixture of light hydrocarbons, gaseous at normal temperature (15°C) and pressure (101.329 kPa) and maintained in the liquid state by increased pressure or lowered temperature. LPG is the generic name for "commercial butane" and "commercial propane". Because of its high heating values, high purity, cleanness of combustion and ease of handling, LPG finds very wide application in a large variety of industrial, commercial, domestic and leisure uses. The history of LPG goes back to the early 1900s. The first car powered by propane ran in 1913 and by 1915 propane was used in torches to cut through metal. Current global LPG consumption is over 200 million tonnes/annum.

Transportation of LPG by pipelines is environmentally friendly in that it entails less energy consumption and exhaust emissions than other modes of transportation. Worldwide, there are over 220,000 miles (350,000 kilometers) of petroleum, refined products and LPG pipelines. The majority are in the United States. Some refined products pipelines carry LPG in batch form. However, there are only about 8000 kilometers of single phase pipelines, of various diameters, that transport LPG (propane or butane) fluids (Mohitpour et al, 2006).

There are a number of codes that industry follows for the design, fabrication, construction and operation of LPG facilities. However, there are no regulations or legislation that specifically cite the pipeline transportation of the product.

From a safety point of view, although LPG is non-toxic, it can be very dangerous if not handled properly. A partial or complete rupture of an LPG pipeline, resulting in an accidental release, will cause issues related to evaporation, vapor cloud propagation and dispersion. Response to emergencies such as rupture and leak in LPG pipelining is thus critical and must ensure rapid action with respect to containment, control, elimination and effective maintenance/repair.

This paper provides an overview the code and regulatory requirements and summarizes the more significant aspects of the design, construction and safe operation pertaining to LPG pipeline systems. It covers the timeline and statistics of the global LPG business; the type of facilities that make up the industry; and the LPG properties pertinent to pipeline design. It also addresses the significant safety issues of LPG pipelining including a discussion on emergency response and associated equipment needs and repair techniques.

UNDERSTANDING LPG & MARKET

LPGs are hydrocarbon products in the C_3-C_4 range, propane and butane constituting the main components of them. These products, gaseous at normal temperature and pressure conditions, can be liquefied under high pressure and therefore can be handled very easily. The chemistry of commercial propane (C_3H_8) and butane (C_4H_{10}: n- and iso-butanes) is illustrated in Figure 1. Depending on the source and how it has been produced, small concentrations of other hydrocarbons may also be present in LPG.

LPG was not identified until the early 1900s. Prior to that time, gasoline would evaporate while in storage due to inadequate refining techniques. Dr Snelling discovered that the gases which evaporated could be condensed and stored as a liquid at moderate temperatures and pressures. By 1911, he had isolated and identified these gases as propane and butane, the two major components of LPGs. The timeline of major events leading to identification, use and the construction of the first LPG pipeline systems is provided in Table 1 (SJGS, 2002, Mohitpour, et al, 2000 & Govt of Canada).

Globally, more than 200 million tonnes of LPG are consumed annually and demand is growing. This is equivalent to approximately 10% of global natural gas consumption or approximately 6% of global oil consumption. North America is the largest producing region in the world, accounting for 30% of total output (with the US, as the largest single

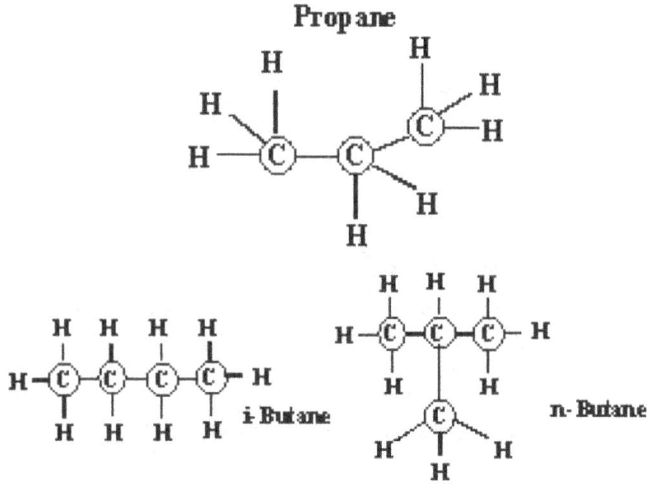

Figure 1. Chemical Make-up of Propane and Butane

TABLE 1. LPG Timeline

900 BC	Earliest development of pipeline by Chinese
347 AD	Oil wells drilled in China
1264	Mining of seep oil in Persia (Marco Polo)
1594	Oil wells hand dug in Baku, Persia
1735	Mining of oil sands in Alsace, France
1802	First commercial use of natural gas (J Watt steam engine factory)
1806	First gas mains to be laid in public street, London
1815	First production of oil in the USA
1846	Distillation of kerosene (Dr. Gesner, Canada)
1848	First modern oil well in Asia
1854	World's first oil company (Charles Tripp)
1854	First oil wells in Europe (Bóbrka, Poland: Ignacy Lukasiewicz)
1857	Invention of kerosene lamp (Michael Diezt)
1858	First oil well in North America (Ontario, Canada)
1859	First oil well in US (Titusville, Pennsylvania: Colonel E Drake)
1878	Invention of electric light bulb (Thomas Edison)
1879	First US long distance pipeline (Tidewater Pipeline, 174 km NPS 6, Pumped crude over the Allegheny)
1886	Gasoline-powdered automobiles (Karl Benz and Wilhelm Daimler)
1886	Louis V Aronson forms Art metal Works in New York (RONSON Metal Works)
1911	Identification of Profane & Butane (W Snelling, US Bureau of Mines)
1912	Development of first propane stove
1913	First car powered by propane
1913	First butane lighter "Wonderliter" by RONSON
1914	Patent granted for fuel to sustain a flame
1915	First use of propane in torches for metal cutting
1917	Formation of Phillips Petroleum Company (Oklahoma)
1918	Sale Propane Patent by Dr. Snelling to Frank Phillips
1920	First commercial sale LPG
1928	First automatic lighter patented in America
1940	First construction of LPG pipelines (Panhandle to Texas regions-James Harold Dunn)

producing country, representing 23%). The next largest producing regions are Asia/Pacific and the Middle East (Venn, 2004).

A breakdown of LPG consumption is provided in Figure 2.

Approximately 50% of the world's demand for LPG is in the domestic (residential and commercial) market. Other consuming sectors include refinery, chemical, industry, transport and agriculture (WLPGA, 2004).

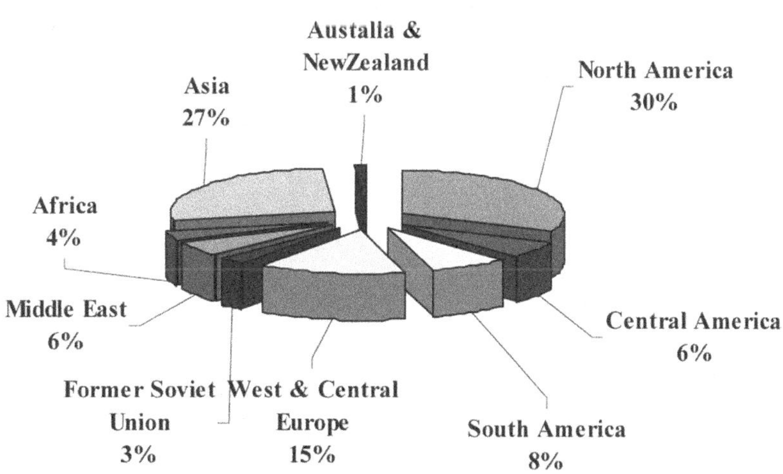

Figure 2. Breakdown of Global LPG Consumption

Propane is the most common liquefied petroleum gas. In the United States, the first commercial sales of propane were made in the 1920s. In 1922, when the US Bureau of Mines began tracking sales, 223,000 gallons were sold across the country. By 2004, US consumption of propane reached over 15 billion gallons for home, agriculture, industrial and commercial uses (PERC).

The majority of US supply is produced internally with imports making up about 10% of the country's supply. The primary mode of transporting propane within the United States is by approximately 70,000 miles (>110,000 km) of interstate pipelines in a batched form. This system is most developed along the Gulf Coast and in the Midwest. Imports come by pipeline and railcar from Canada as well as by sea from countries such as Algeria, Saudi Arabia, Nigeria, Venezuela and Norway (EIA, 2006).

The physical characteristics of typical commercial LPGs are summarized in Table 2 and Figure 3 (L'Air Liquide, 1976). The phase diagram for propane is shown in Figure 4 (L'Air Liquide, 1976).

Because of their characteristics, LPGs can be stored easily and transported efficiently. They are safe to be carried in tanks or cylinders as well as in pipelines. They have high heating values and burn relatively cleanly. In case of rupture of the container, LPGs will not ignite when combined with air unless the source of ignition reaches approximately 500°C. By contrast, gasoline will ignite when the source of ignition reaches only 220 to 260°C (a very narrow range).

Typically, propane is used for residential and commercial heating as well as automotive and industrial fuel gas applications (food production, ceramics and metallurgical applications, etc). Butane is used for seasonal gasoline blending, refrigeration (industrially known as R600a), cigarette lighters and portable stoves. At refineries, it is stored under pressure. Both propane and butane are also used as a raw material for petrochemical applications.

TABLE 2. Properties of Commercially Available Propane & Butane (L'Air Liquide, 1976, EIA, 1994)

Properties		Propane	Butane
Relative density (specific gravity) of liquid at 15C		0.50 to 0.51	0.57 to 0.58
Imperial gallons/ton at 15C		439 to 448	385 to 393
Litres/tonne at 15C		1965 to 2019	1723 to 1760
Relative density of gas at 15C and 101.325 kPa, Air = 1		1.40 to 1.55	1.90 to 2.10
Volume of gas (ft^3) per lb of liquid at 60F and 30 in Hg		8.5 to 8.7	6.5 to 6.9
Boiling point at atmospheric pressure °C (approx)		–42	–0.5
Vapour pressure (bar) at various temps		See Fig 3	See Fig 3
Latent heat of vapourisation (kJ/kg) at 15C		358.2	372.2
Latent heat of vapourization (Btu/lb) at 60F		154	160
Gas liberation/litre of liquid at 15C		311	239
Specific heat of liquid at 15C (kJ/kg °C)		2512	2386
Sulphur content per cent weight		Negligible to 0.02	Negligible to 0.02
Limits of flammability (% by vol of gas in a gas-air mixture to form a combustible mixture)		Upper 9.5 Lower 2.2	Upper 9 Lower 1.8
Ignition temp °C		470 to 600	420 min
Flame temp °C		1980	1970
Calorific values / higher	(MJ/m^3) dry	93.1	121.8
	(Btu/ft^3) dry	2500	3270
	(MJ/kg)	50.0	49.3
	(Btu/lb)	21500	21200
Calorific values / lower	(MJ/m^3) dry	86.1	112.9
	(Btu/ft^3) dry	2310	3030
	(MJ/kg)	46.3	45.8
	(Btu/lb)	19900	19700
Air required for combustion (m^3 to burn 1 m^3 of gas)		24	30

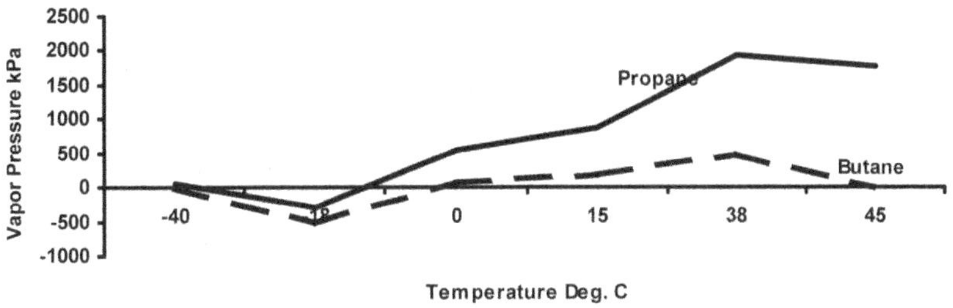

Figure 3. Typical Vapor Pressure of Commercially Available LPG

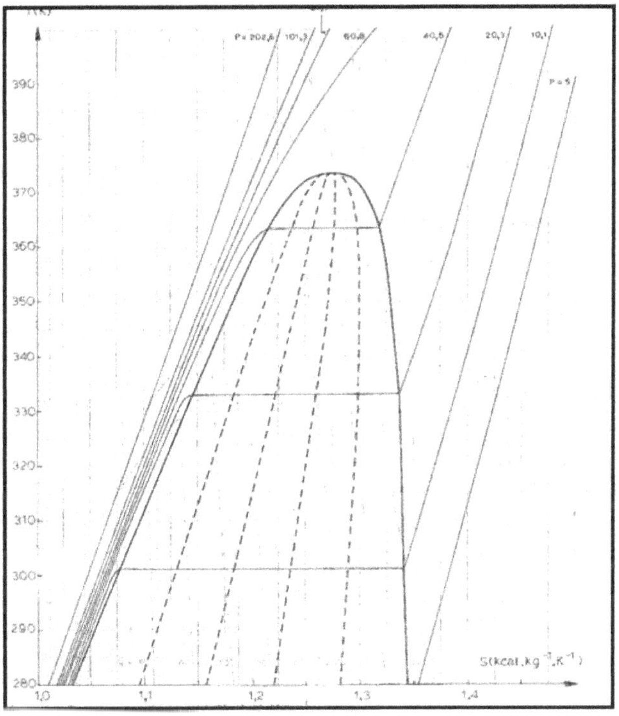

Figure 4. Propane Phase Diagram (L'Air Liquide, 1976)

LPG PIPELINES

The utilization of LPG as a fuel varies very widely within a country and from one country to another, depending on the cost (including transportation), accessibility and availability of the fuel in relation to alternative fuels, notably gasoline and diesel.

LPG production and distribution is schematically depicted in Figure 5. Normally, LPG is stored in liquid form under pressure in steel containers, cylinders or tanks. The pressure inside the container will depend on the type of LPG (commercial butane or commercial propane) and the external temperature. LPGs for domestic use are often transported from refineries by truck and distributed in 24 lb (10 kg) cylinders.

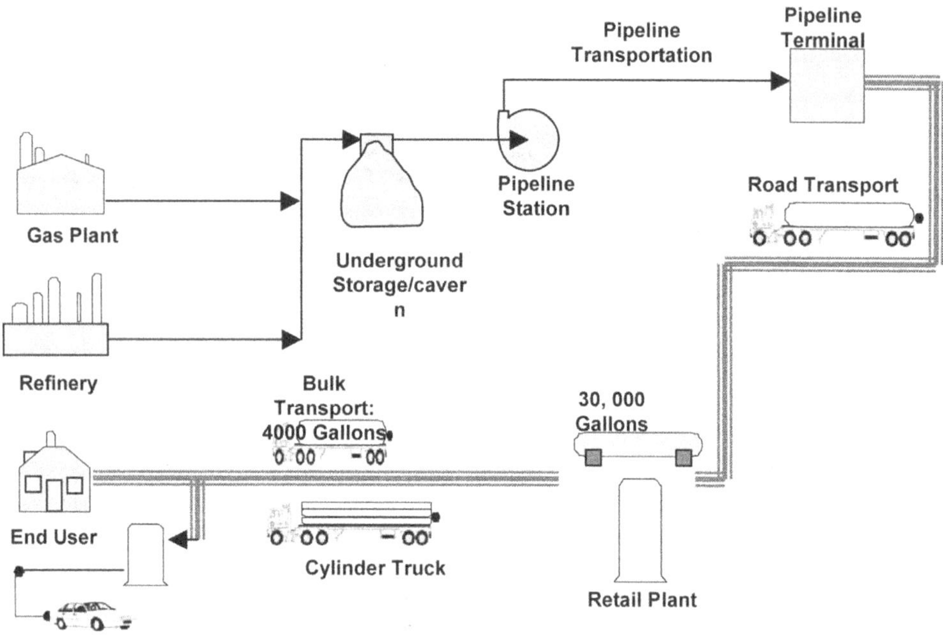

Figure 5. Typical LPG Transportation and Distribution

In 1940, James Harold Dunn was instrumental in constructing the first LPG pipelines from the Panhandle field to other regions of Texas and other regions of the USA in association with the Phillips Petroleum Company (TSHA, 2002).

Worldwide there are over 220,000 miles (>350,000 kilometers) of petroleum products, refined products and LPG pipelines. Some of the refined products pipelines carry LPG in a batch form. However there are only about 8,000 km of single-phase pipelines of varying diameter that transport LPG (propane or butane only). Percentage breakdown of these single-phase LPG pipelines by country is depicted in Figure 6 (those percentages appearing as zero are each less than one half of 1% of the total), (CIA, 2005).

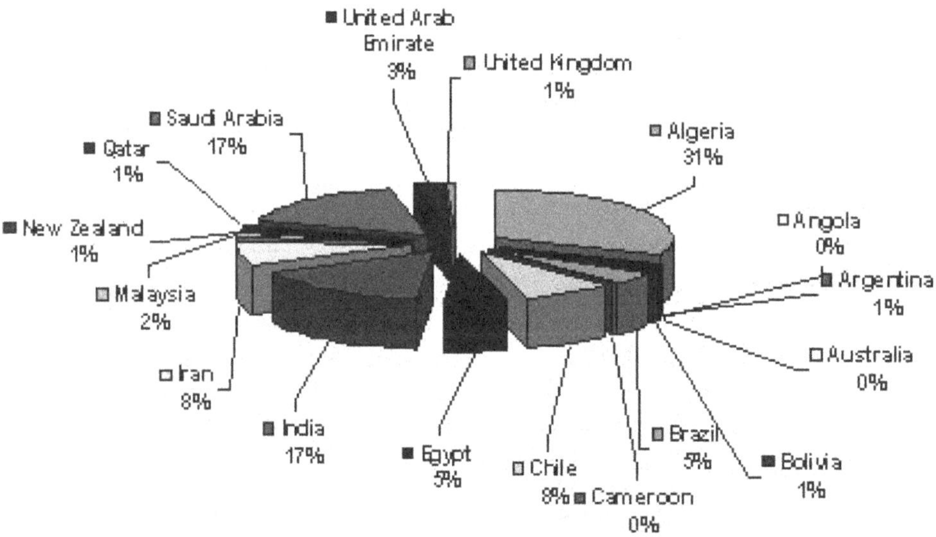

Figure 6. World Distribution of Single Phase Un-batched LPG Pipelines

TABLE 3. Typical Pipeline Properties of LPG Products

Commodity	Viscosity C/S-mm^2/s	Temperature °C	Density Kg/m^3
Propane	0.218	20	500.6
(@ 1000 kPa)	0.199	30	483.5
	0.166	52	446.9
Butane	0.237	25	560.8
(@470–520 kPa)	0.235	44	535.1
	0.212	52	529.0

Transportation of LPG by pipeline, both dedicated and batched, continues to grow as consumers and governments recognize the environmental advantages and other benefits of LPG.

The properties of typical LPG products, as relate to pipeline transportation, are summarized in Table 3 (Mohitpour et al, 2003).

CODES, STANDARDS & REGULATIONS

There are a number of codes that apply to the design, fabrication/construction and operation of LPG facilities. One specific code that is followed internationally is the National Fire Protection Agency "Liquefied Petroleum Gas Code (NFPA 58)" (NFPA 2004). However, the pipeline industry generally follows pipeline transmission codes such as ANSI/ASME B31.4 & B31.8; Canadian Standards Association (CSA) Z662-03; and BSI/European Codes BS-EN (European Committee for Standardization (CEN)) for the design of LPG transportation and storage facilities. Some of the latter codes are listed below:

BS-EN1442 Transportable refillable welded steel cylinders for liquefied petroleum gas (LPG) – Design and construction
BS-EN12252 Equipping of Liquefied Petroleum Gas (LPG) road tankers
EN12493 Welded steel tanks for liquefied petroleum gas (LPG) – Road tankers – Design and manufacture
EN14334 Inspection and testing of LPG road tankers

Additionally, the UK International Institution of Gas Engineers and Managers (IGEM) also provides a professional and policy forum and Recommendations on Transmission and Distribution Practices for engineers, technologists and managers working in the field of fuel gas technology. Some of the practices include the following:

IGE/TD/13 Pressure regulating installations for transmission and distribution systems
IGE/SR/18 Safe working in the vicinity of gas pipelines, mains and associated installations
IGE/SR/22 (2nd impression) Purging operations for fuel gases in transmission, distribution and storage
IGE/SR/23 (2nd impression) Venting of natural gas
IGE/UP/2 Gas installation pipework, boosters and compressors on industrial and commercial premises

In Canada, there are no regulations/legislation that specifically cite the pipeline transportation of LPG. There is reference, however, in the Onshore Pipeline Regulations of the

National Energy Board (NEB) to High Vapor Pressure (HVP) pipelines. For example, Part 1, Section 10 of the Regulations states:

(1) When an HVP pipeline is to be situated in a Class 1 location and within 500 m of the right-of-way of a railway or paved roadway, a company shall develop a documented risk assessment to determine the need for heavier wall design, taking into account such factors as pipeline diameter and operating pressure, HVP fluid characteristics, topography, and traffic type and density of the traffic on the railway or paved road.

(2) A company shall submit a documented risk assessment to the Board when required to do so under section 7.

The NEB further states that the goal of Section 10 is to ensure that an acceptable level of safety for HVP pipelines is provided for in the vicinity of railways and roadways.

The following standards apply to the testing of LPG products:

ASTM D1142	Dew Point of Water Vapor in Gaseous Fuels
ASTM D2713 - IP395	Dryness of Propane
ASTM D1267 – IP161 – ISO4256 – EN24256	Determination of Vapor Pressure of LP Gases
ASTM D1838 – IP411 – ISO6251 – EN26251 – NP M41	Copper Strip Corrosion by Liquefied Petroleum Gases
ASTM D1837	Volatility of LP Gases
IP59 Method C	Determination of Density or Relative Density of Gases by Schilling Effusiometer

DESIGN AND OPERATIONAL CONSIDERATIONS

Design and operational considerations for LPG pipelines are similar to those of liquid pipelines with specific consideration for HVP product transportation. Major issues include:

- hydraulics (setting pressure limits for phase control, or phase issues related to elevation changes, or water hammer effects, column separation, etc.);
- routing criteria such as High Consequence Areas (HCA);
- material fracture control properties (pipe, valves, fittings);
- sealing issues;
- automation and controls requirements (including leak detection, line break controls);
- measurement;
- pumping (maintaining the required vapour pressure);
- control of contaminants in the fluid stream;
- operations procedures such as pressuring or de-pressuring, flushing, isolations & lock out of parallel piping; and
- issues related to batch operation (if applicable).

Detailed considerations are provided elsewhere (Mohitpour et al, 2003 (new edition to be released 2006)).

Vapor pressures of LPG, as they relate to pipeline design and operations, are provided in Table 4 below:

TABLE 4. Propane and n-Butane Vapor Pressures

	Vapor Pressure kPa	
Temperature °C	Propane	Butane
−10	256	−4
0	388	40
10	552	95
20	757	172
30	1004	266
37.8	1218	362

SAFETY AND ENVIRONMENTAL CONSIDERATIONS

LPGs contain negligible toxic components. However, abuse (as with solvents) is highly dangerous. The central nervous system will be stressed if overexposure occurs. Headache, dizziness, narcotic effect, and unconsciousness can occur. Butane has an exposure limit of 800 ppm. In large concentrations, LPG is an asphyxiant (displaces air) and therefore confined space releases are a major concern.

In combustion, LPGs have lower particulate emissions and lower noise levels relative to diesel, making propane attractive as a transportation fuel in urban areas. Noise levels can be less than 50% of equivalent engines using diesel. Some of the safety and environmental issues related to LPG are highlighted in NPGA, 2003.

LPG RELEASE AND VAPOR CLOUD

When an LPG pipeline ruptures, two conditions occur simultaneously: 1) flow of LPG from and out of the pipeline and 2) formation of a vapor cloud from and extending beyond the release location. These are described below:

1) Flow Within Pipeline After Rupture

In a rupture of a low vapor pressure (LVP) fluid pipeline, the spill (such as oil) flows from the pipeline and (depending on topography) drains off until all fluid within the pipc accounting for the change in the elevation profile is evacuated. In a high vapor pressure fluid pipeline (such as LPG), all the fluid within the pipeline will drain off irrespective of elevation profile. Unlike LVP, for which undulation in the elevation profile will help prevent total drainage of the pipeline upon rupture, the entire LPG contained in a section of a pipeline will boil and vaporize upon pipeline rupture and depressure to atmosphere. Such an account of LPG vaporization has been given by Mohitpour, et al, 2003.

In a pipeline carrying HVP product, when a rupture occurs, the HVP fluid within the pipeline will experience an interface of liquid and gas. This interface will have the fluid liquefied under pipeline pressure on one side (higher than the vapor pressure of the fluid) and it will be at a lower pressure and gaseous on the other side (below the vapor pressure of the fluid). The location of the interface is time dependent and changes as the mass of fluid is removed from the pipeline. The mass of the fluid (liquid/gas) thus discharged at the rupture location is also time dependent and generally non linear with respect to time. Morrow et al, 1983, describe this phenomena and provide a technique for venting LPG through a segment of pipeline between isolation valves subject to a full rupture (Figure 7).

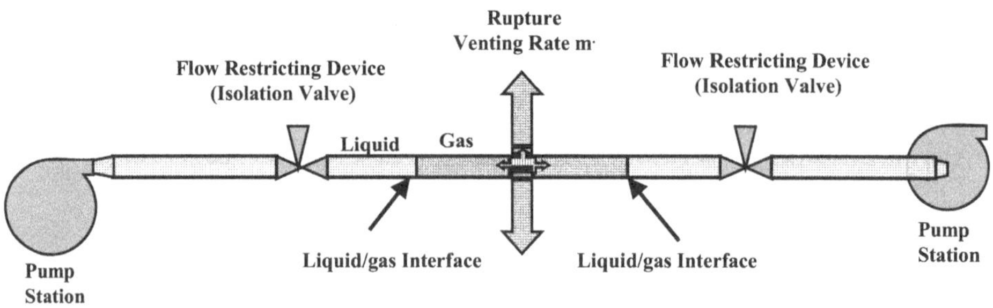

Figure 7. Pipeline Configuration and Rupture Schematic

2) LPG Movement From Rupture/Vapor Cloud Formation

The factors that affect the distance to which the flammable boundaries of vapor cloud travels include:

1. internal condition of the pipeline at the point of rupture: pressure, flow rate, fluid properties (the internal conditions change with time after rupture as described previously);
2. severity of rupture, full break versus partial rupture including orientation (top, side, bottom);
3. detection time versus pipeline shutdown time (pumps, isolation valves) and isolation conditions (valve spacing and valve closure timing);
4. surrounding terrain (affecting pooling & evaporation);
5. upwind environmental conditions (air velocity, terrain condition (roughness: buildings in semi urban environment versus cross country: farmland)) which affect the take up rate of vapor into the atmosphere; and
6. atmospheric conditions: wind velocity, direction, stability, pressure, temperature and humidity as well as precipitation (rain/snow).

Items 1, 2 & 3 affect the time-dependent mass-venting rate from the rupture. Items 2 & 4 affect the spreading of the liquid/vapor layers. Items 4, 5 & 6 affect the dispersion of the escaping vapor cloud. It is noted that item 3 (reaction time and isolation valve spacing) are the only controllable factors. Example of such factors in controlling liquid spill to increase oil pipeline safety is described by Platus et al 1974 & Mohitpour et al 2003 & 2004.

VAPOR AND PLUME DEVELPOMENT RISK

The LPG gas-to-liquid ratio for propane is very high. For example, propane under pressure can compress about 270 times its volume. By comparison, natural gas can be compressed about 600 times.

Risks associated with the pipeline transportation of LPGs include accidental release, evaporation, vapor cloud propagation and dispersion for a number of different accidents involving complete or partial rupture.

As is well known, the most dangerous accidents at LPG facilities are accompanied by the formation and growth of dense vapor clouds. Vapor clouds will form closest to the leak and will be visible, but the boundaries will be invisible and may not disperse readily. Also, the cloud can travel over a long distance producing a large hazardous zone, Figure 8.

If leaking LPG comes in contact with a person's skin, severe frost-bite can occur, resulting in a skin burn. Super-cooled LPG liquid and surface piping may also cause frost-

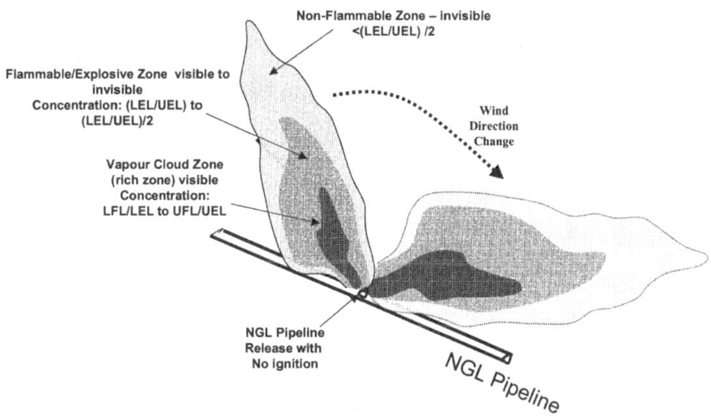

Figure 8. Release of Vapor Cloud From a Pipeline Rupture; Vapor Zones

bite. When a release is occurring, wind direction may change, causing the vapor cloud to be blown into an area previously considered safe. The radial distance of the leaking LPG must be considered.

If the flammable portion of the vapor cloud comes in contact with an ignition source, the characteristics of the resulting fire will vary depending on the release scenario and may result in a flash, jet or pool fire. In some cases, a fireball or even a vapor cloud explosion will occur (a popping noise may be heard as the LPG ignites and explodes).

Accident scenarios may include complete or partial rupture of the pipeline. As well, there may be some delay in the shut-down of pumps and the isolation of the affected pipeline section after leakage initiates.

During an LPG release, in most cases, wind conditions may change and this will influence the size, shape and direction of the vapor plume (wind speed, direction and product release rate). In general, the higher the wind speed, the smaller the flammable portion of the vapor plume.

Modeling of vapor cloud dispersion after LPG release to determine the extent of vapor travel requires specialized techniques. Since the molecular weight of LPG is greater than that of air, a dense gas dispersion model is commonly used to determine hazard ranges for ambient and low temperature releases. Such predictions by passive dispersion models may

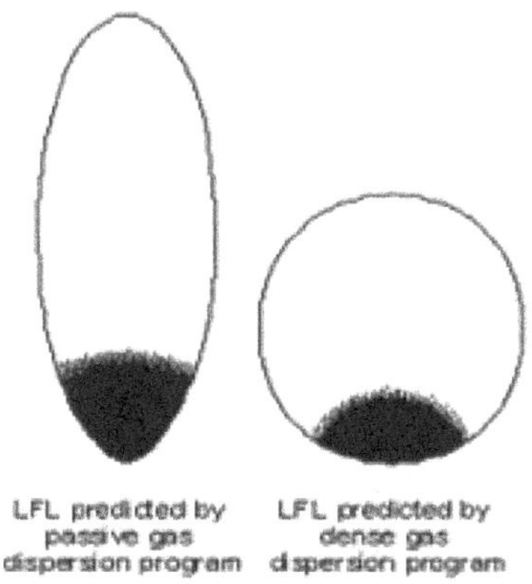

Figure 9. Passive and Dense LPG Vapor Dispersion Modeling (HSE, 2005)

be optimistic or pessimistic as indicated by Figure 9. A safety report that uses a passive dispersion model (based on Lower Flammability Limit, LFL) will overestimate the down wind extent of a flash fire but underestimate its width. Therefore, the use of a passive dispersion model must be properly justified, recognizing the inability of the model to correctly predict the width of the flammable cloud (HSE, 2005).

LPG PIPELINE MAINTENANCE

Maintenance requirements for LPG pipelines are similar to those of other liquid hydrocarbon pipelines. These generally include condition assessment; cathodic protection; leak and right of way surveys and maintenance (including geotechnical works); pipeline repair; failure analysis; locating and marking of facilities appurtenances; valve inspection and repair; regulator maintenance; and management of compliance records.

HVP and LVP liquid pipeline maintenance differ only as to how a response is initiated and how work is performed in an emergency situation that may involve a pipeline leak, rupture, or fluid release.

Liquid (low or high vapor pressure) releases must be quickly contained to prevent extensive environmental damage and the clean up operation should begin as soon as possible using a variety of means to recover the product. Subsequently, measures will be required to remediate and monitor the affected areas, but this is generally outside the scope of any contingency/emergency response planning.

EMERGENCY RESPONSE

Public safety is the top priority in any pipeline emergency response. "Emergency " is defined as any unforeseen combination of circumstances or disruption of normal operating conditions that poses a potential threat to human life, health, environment or property if not contained, controlled or eliminated. Types of emergencies include:

- vapor release inside or near a building;
- fire involving a pipeline facility;

Figure 10. An LPG Pipeline Fire, PVFD 2005. (Image Courtesy of KTAB-TV)

- explosion near or involving a pipeline facility;
- natural disasters;
- emergencies involving other pipelines; and
- arson/bomb threats.

There are three potential risks associated with the failure of an LPG pipeline and the release of LPG into the surrounding environment. These include: the risk of fire (Figure 10); a vapor cloud explosion; and the possibility of toxic effects of an un-ignited airborne vapor cloud (as a consequence, for example, of small quantities of mercury (if present) in the gas/product).

Regulations/Standards

Regulation, standards and safety rules for the transportation of LPG and the response to emergency situations include:

- ASME ANSI B31.4 "Pipeline Transportation Systems for Liquid Hydrocarbon and Other Liquids";
- API 1160 "Integrity Management of Liquid Hydrocarbon Pipelines";
- ASME B31.8S "Managing System Integrity of Gas Pipelines";
- US Department of Transportation (DOT), Code of Federal Regulations (CFR) - Regulation 195 (Pipeline Safety); and
- Occupational Health and Safety Act OHSA (Act 85 of 1993) and OHSA Process Safety Management.

Flaring LPG

Most pipeline operating plans permit the flaring and venting of associated gas under the following circumstances:

- for safety reasons;
- for unavoidable technical reasons (such as purge venting/emergency); and
- in emergencies (emergency transportation stops, compression, and others).

As most flaring situations in LPG pipeline transmission are related to emergencies, measurement and monitoring are not applicable. Flaring and venting under these circumstances does not normally require regulatory approval.

Responding to LPG Emergencies

Identifying LPG Leaks: Pressure drops related to LPG leaks are rarely identified at a pipeline control center, except in the event of a major rupture. This is due to the fact that LPGs have high vapor pressure and compressibility (e.g., propane can be compressed by as much as 5% if pressure is raised to 1,000 psi). Leaks are generally reported by someone in the vicinity of the leak site. Some indications of an LPG leak include:

- a cloud of steam or mist, caused by condensation and freezing moisture;
- ice build-up on exposed pipe and frozen ground around an underground pipe;
- brown vegetation, which is an indication of soil saturation;
- yellow-stained snow, which may be an indication of LPG accumulation under the snow; and
- odor of condensed LPG.

LPG Safety Precautions: These include:

- exploring for leaks (indoor, outdoor and night explorations);
- pipeline operation in an LPG leak situation;
- responding to small or large leaks;
- igniting an LPG cloud;
- pipeline repairs (no LPG ignition); and
- control of LPG release.

LPG Emergency Response Zones

When an LPG release incident occurs and whether or not an LPG fire is involved, it is normal practice to divide the area surrounding the release into hot, warm and cold zones (Figure 8). This allows for an organized response to the release including proper placement of emergency response personnel and equipment and safe and effective performance of necessary work. These zones are described below.

Hot Zone: This is the zone which is highly explosive and extremely dangerous and should only be approached and entered by trained personnel who are appropriately equipped with personal protection equipment (PPE) and self contained breathing apparatus (SCBA). Within this zone, LPG concentration levels will be between the lower flammability/ explosive limit (LFL/LEL) and upper flammability/explosive limit (UFL/UEL) of the vapor. *No equipment is installed in this zone.*

Warm Zone: Typically the zone has concentrations LFL to LFL/2. This means that the zone is generally non-explosive, but could change to explosive depending on circumstances. Personnel can enter this zone without self contained breathing apparatus. LPG concentration level is monitored at all times. Because there exists a small risk of explosion, no equipment is installed in this zone. The warm zone is also an area where personnel can be affected by the force of an explosion in the hot zone.

Cold Zone: Typically this zone will have vapor concentrations less than LFL/2 and therefore equipment and trailers can be set up without concern for explosion or fire. This is the radial area which is generally safe for personnel and for performing work. The emergency command post and rescue equipment are usually positioned within this area. Industry's experience indicates a range of distances to a safe work zone (LFL/2) to be 0.6 km for pipes up to NPS 8, 0.8 km for NPS 10-12, 1 km for NPS 14, 1.2 km for NPS 16 and 1.5 km for NPS 24.

CONCLUSION

Because of their high heating values, cleanness of combustion and ease of handling, LPGs find very wide application in a large variety of industrial, commercial, domestic and leisure uses. They are available as a by-product of petroleum refining and from natural gas production. Propane (C_3H_8) and butane (C_4H_{10}) are the main components of LPG.

Transportation of LPG by pipeline dates back to 1940. There are no regulations/legislation that specifically cite pipeline transportation of LPG. For the design, fabrication, construction and operation of LPG facilities, industry adheres to codes written for liquid hydrocarbons generally, for example ASME ANSI B31.4 "Pipeline Transportation Systems for Liquid Hydrocarbon". However, operation and maintenance of LPG pipelines, as with HVP pipelines, requires special attention in emergency situations involving fire, explosion or the possible toxic effects of an un-ignited airborne vapor cloud. Emergency procedures include the establishment of emergency zones. In the event of a major LPG release involving adverse atmospheric conditions and in which the vapor cloud cannot be safely and quickly ignited, nearby residences must be evacuated.

REFERENCES

CIA, The World Factbook, 2005, "*Pipelines*", http://www.odci.gov/cia/publications/factbook/fields/2117.html

EIA (Energy Information Administration), 1994 "*Alternatives to Traditional Transportation Fuels:An Overview*", Report DOE/EIA-0585/O - Distribution Category UC-98, June, http://tonto.eia.doe.gov/FTPROOT/alternativefuels/0585o.pdf

EIA (Energy Information Administration), 2006, "*Propane Prices: What Consumers Should Know Brochure*", http://www.eia.doe.gov/neic/brochure/propane04/Chapter1.htm

Government of Canada, Canada's Digital Collections; "Black Gold, Canada's Oil Heritage: Charles Tripp"; http://collections.ic.gc.ca/blackgold/people/trippstory.html

HSE (Health & Safety Executive), 2005 "*Safety Report Assessment Guide:LPG - Criteria*", http://www.hse.gov.uk/comah/index.htm

L'Air Liquide, 1976, "*Gas Encyclopedia*", Elsevier, The Netherlands

Mohitpour, M., Dawson. J., Babuk, T. & Jenkins, A., 2000 "*Concepts for Increased Natural Gas Supply – A pipeline Perspective*", presented at Forum 11, 16th World Petroleum Congress, Calgary, AB, Canada, June 11–15

Mohitpour, M., Golshan, H. & Murray, A., 2003, "*Pipeline Design & Construction – A Practical Approach*", 2nd Edition, ASME Press, New York.

Mohitpour, M., Golshan, H. & Murray, A., 2006, "*Pipeline Design & Construction – A Practical Approach*", 3rd Edition, (to be released), ASME Press, New York

Mohitpour, M., Trefanenko, Bill, Tolmasquim, S. T. & Kossatz, H, 2004, "*Valve Automation to Increase Oil Pipeline Safety*", AMSE 5th International Pipeline Conference, Hyatt Regency, Calgary, AB, Canada, Oct 4–8

Mohitpour, M., Trefanenko, Bill, Tiomno Tolmasquim, Sueli & Kossatz Helmut 2003, "*Oil Pipeline Valve Automation for Spill Reduction*", Rio Pipeline Conference & Exposition, Hotel Inter-Continental, Oct 21-24

Morrow, T. B., Bass, R. L. & Lock J. A, 1983, "*An LPG Pipeline Break Flow Model*", ASME Transaction, Jr. Energy Resources Tech. Vol. 105 pp 379-387, Sept

NPGA (National Propane Gas Association), 2003, http://www.npga.org/i4a/pages/index.cfm?pageid=466

PERC (Propane Education & Research Council), "*The History of Propane*", http://www.propanecouncil.org/files/The%20History%20of%20Propane.pdf

PVFD (Potosi Volunteer Fire Department), 2005, "*Pictures of 9/7/2000 LPG Pipeline Fire*", http://www.angelfire.com/tx/pvfd/pipeline.html

Platus, D. L., Mackenzie D. W. & Morse, C. P., 1974 "*Rapid Shutdown of Failed Pipeline Systems and Limiting Pressure to Prevent Pipeline Failure Due to Over Pressure*" Part 1, Report MRI-2628- TRI, Oct

SJGS (San Joaquin Geological Society), 2002, "*The History of the Oil Industry*", http://www.sjgs.com/history.html#ancient_to_present

TSHA (The Texas State Historical Association), 2002 "*The Handbook of Texas Online*", http://www.tsha.utexas.edu/handbook/online/

Venn, J., 2004, "*Rapid Access to Modern Energy Services Using LP Gas*", 19th World Energy Congress, Sydney, Australia, Sept 5-9

WLPGA (World LP Gas Association), 2004, "*Global LP Gas Statistics*", http://www.worldlpgas.com/v2/ressources.php?id=04

INDEX